Lactic Acid Bacteria: Genetics, Metabolism and Applications

Proceedings of the seventh Symposium on lactic acid bacteria: genetics, metabolism and applications,
1–5 September 2002, Egmond aan Zee, the Netherlands

Edited by:

R.J. Siezen,
Wageningen Center for Food Sciences, and NIZO food research, the Netherlands;

J. Kok,
University of Groningen, the Netherlands;

T. Abee,
Wageningen Center for Food Sciences, and Wageningen University, the Netherlands;

G. Schasfsma;
TNO Nutrition and Food Research, Zeist, the Netherlands

Reprinted from Antonie van Leeuwenhoek, 82(1–4) 2002

Springer-Science+Business Media, B.V.

Library of Congress Cataloging-in-Publication Data

A C.I.P. Catalogue record for this book is available from the Library of Congress.

ISBN 978-90-481-6141-6 ISBN 978-94-017-2029-8 (eBook)
DOI 10.1007/978-94-017-2029-8

Printed on acid-free paper

Originally published by Kluwer Academic Publishers in 2002
Softcover reprint of the hardcover 1st edition 2002

Table of Contents

Special Issue: **Lactic acid bacteria: genetics, metabolism and applications**
Proceedings of the seventh symposium on lactic acid bacteria: genetics, metabolism and applications, 1–5 September 2002, Egmond aan Zee, the Netherlands

Guest Editors: **R.J. Siezen[1], J. Kok[2], T. Abee[3] & G. Schaafsma[4]**
[1]Wageningen Center for Food Sciences, and NIZO food research, the Netherlands; [2]University of Groningen, the Netherlands; [3]Wageningen Center for Food Sciences, and Wageningen University, the Netherlands; [4]TNO Nutrition and Food Research, Zeist, the Netherlands

Antonie van Leeuwenhoek **82:** 1, 2002.

Lactic acid bacteria: genetics, metabolism and applications

Proceedings of the Seventh Symposium on Lactic Acid Bacteria: genetics, metabolism and applications, 1–5 September 2002, Egmond aan Zee, the Netherlands

R. J. Siezen[1], J. Kok[2], T. Abee[3] & G. Schaafsma[4]
[1]*Wageningen Center for Food Sciences, and NIZO food research, the Netherlands;* [2]*University of Groningen, the Netherlands;* [3]*Wageningen Center for Food Sciences, and Wageningen University, the Netherlands;* [4]*TNO Nutrition and Food Research, Zeist, the Netherlands*

Editorial

The Seventh Symposium on Lactic Acid Bacteria

Foods fermented with Lactic Acid Bacteria are an important part of the human diet. Lactic Acid Bacteria play an essential role in the preservation of foods and contribute to the nutritional, sensoric and health properties of food products and animal feed. The importance of Lactic Acid Bacteria in the production of foods throughout the world has resulted in a continued scientific interest over the last two decades in these micro-organisms by academic research groups as well as by industry. This research has resulted in a number of important scientific breakthroughs and has led to new applications. The most recent of these advances is the establishment of the complete genome sequences of a number of different Lactic Acid Bacterial species.

To communicate and stimulate the research on Lactic Acid Bacteria and their applications, a series of tri-annual symposia on Lactic Acid Bacteria was started in 1983 under the auspices of the Netherlands Society for Microbiology (NVVM), which was later also supported by the Federation of European Microbiological Societies (FEMS). The aim of these state-of-the-art symposia, is to offer a unique platform for universities, institutes and industry in this area of biotechnology, to present recent work, to obtain information on new developments and to exchange views with colleagues from all over the world on scientific progress and applications. The growing number of participants to these symposia has been a clear demonstration of the interest of the international industrial and scientific community in this area of research.

The 7th Symposium will be based on a number of plenary lectures that will review the scientific progress of the last years in the different areas of research on Lactic Acid Bacteria, and which are documented in this special issue of Antonie van Leeuwenhoek.

Antonie van Leeuwenhoek **82:** 3–27, 2002.

The cell membrane and the struggle for life of lactic acid bacteria

Wil. N. Konings
University of Groningen, Department of Microbiology, Groningen Biomolecular Sciences and Biotechnology Center, Kerklaan 30, 9751 NN Haren, The Netherlands (E-mail: W.N.Konings@biol.rug.nl)

Key words: lactic acid bacteria, metabolic energy generation, survival, membrane transport, multidrug resistance

Abstract

The major life-threatening event for lactic acid bacteria (LAB) in their natural environment is the depletion of their energy sources and LAB can survive such conditions only for a short period of time. During periods of starvation LAB can exploit optimally the potential energy sources in their environment usually by applying proton motive force generating membrane transport systems. These systems include in addition to the proton translocating F_oF_1-ATPase: a respiratory chain when hemin is present in the medium, electrogenic solute uptake and excretion systems, electrogenic lactate/proton symport and precursor/ product exchange systems. Most of these metabolic energy-generating systems offer as additional bonus the prevention of a lethal decrease of the internal and external pH. LAB have limited biosynthetic capacities and rely heavily on the presence of essential components such as sources of amino acids in their environment. The uptake of amino acids requires a major fraction of the available metabolic energy of LAB. The metabolic energy cost of amino acid uptake can be reduced drastically by accumulating oligopeptides instead of the individual amino acids and by proton motive force-generating efflux of excessively accumulated amino acids. Other life-threatening conditions that LAB encounter in their environment are rapid changes in the osmolality and the exposure to cytotoxic compounds, including antibiotics. LAB respond to osmotic upshock or downshock by accumulating or releasing rapidly osmolytes such as glycine-betaine. The life-threatening presence of cytotoxic compounds, including antibiotics, is effectively counteracted by powerful drug extruding multidrug resistance systems. The number and variety of defense mechanisms in LAB is surprisingly high. Most defense mechanisms operate in the cytoplasmic membrane to control the internal environment and the energetic status of LAB. Annotation of the functions of the genes in the genomes of LAB will undoubtely reveal additional defense mechanisms.

Introduction

Lactic acid bacteria (LAB) are facultative anaerobic bacteria with a fermentative metabolism (Kandler 1983). The primary source of their metabolic energy is supplied in the form of ATP by substrate level phosphorylation (Kandler 1983; Konings 1985). LAB have a relatively simple metabolism and make high demands on the nutritional composition of their growth media. Usually, in addition to carbon and energy sources, various amino acids, vitamins, nucleic acids and mineral components are required. The environments in which LAB can thrive are restricted by these requirements and are often associated with plants, meat and dairy products (Mundt 1982). The genome sizes of LAB are relatively small, which is consistent with their simple metabolic capabilities. The genome of *Lactococcus lactis* IL1403 contains 2 365 589 base pairs and encodes 2310 proteins (Bolotin et al. 2001). This is about half the size of the genome of *E.coli.* LAB are important bacteria for food fermentations. They play crucial roles in the fermentation of milk products, including cheese, vegetables such as sauerkraut and soy sauce and meat products such as sausages. LAB also contribute significantly to other fermentation processes such as the production of wine. The importance of LAB for food fermentation has strongly stimulated research of these bacteria and *L. lactis* belongs to the best-studied micro-organisms today.

L. lactis might serve as a model organism of physiological minimalism. It is adapted to a heterotrophic way of life while it has only rudimentary parts of many of the biosynthetic pathways. Yet, in spite of its simplicity, *L. lactis* has acquired an amazing variety of adaptation and defense mechanisms in order to cope with and to survive in continuously changing and often hostile environments. Many of these adaptation and defense mechanisms rely heavily on energy transducing systems in the cytoplasmic membrane. In particular, transport systems that translocate ions and substrates (*solutes*) from the external medium into the cytoplasm or excrete ions and end products of metabolism into the external medium, are crucial in the survival strategy (Konings et al. 1997). The importance of the cytoplasmic membrane for growth and survival of *L. lactis* is evident from the observation that about 17% of all proteins encoded by its genome are involved in biosynthesis and function of the cytoplasmic membrane (Bolotin et al. 2001).

The functional role of the cytoplasmic membrane

The cytoplasmic membrane is the only membrane present within the cell wall boundary in most bacteria and archaea (Lindsay et al. 2001). It forms a diffusion barrier and constitutes a physical barrier between the cytoplasm and the external medium. It defines a discrete entity, the cytoplasm that contains molecules belonging only to the organism. The diffusion barrier function of the cytoplasmic membrane restricts not only the diffusion of hydrophilic compounds across the membrane but also of lipophilic compounds (Bolhuis et al. 1996b). The barrier property allows the cytoplasmic membrane to serve two important functions in all living organisms. (i) Conservation of the integrity of the cytoplasm by preventing metabolites and ions from diffusing uncontrolled out of the cells and of external compounds and ions from diffusing in freely; (ii) Chemiosmotic energy transduction.

Inherent to the restricted ion-permeability of the cytoplasmic membrane is the potential to generate electrochemical ion gradients across the membrane by a variety of ion transport systems. The energetically most-important electrochemical ion gradient in most bacteria, including LAB, is that of protons (Mitchell 1966). Specific proton pumps in the membrane form these electrochemical proton gradients. A major proton pump in aerobic bacteria is the respiratory chain (Figure 1). Electron transfer in the respiratory chain can be coupled to the extrusion of protons from the cytoplasm to the external medium. As a result two gradients are formed:

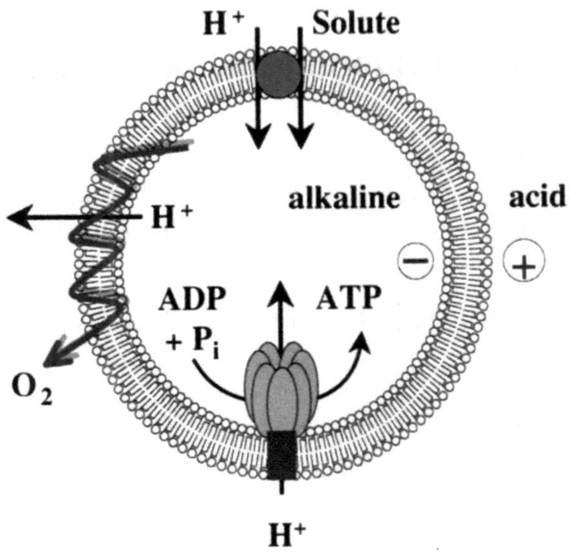

Figure 1. Energy transduction in aerobic bacteria. Protons are pumped by the respiratory chain from the cytoplasm across the membrane into the external medium resulting in the generation of a proton motive force, inside negative and alkaline. This PMF drives energy requiring membrane processes such as ATP-synthesis and solute transport.

(i) A pH gradient (ΔpH), inside alkaline versus outside, by the removal of protons from the cytoplasm and their accumulation in the external medium (Figure 1). The pH-gradient (expressed in mV) has a value of $-2.3RT/F\ \Delta pH$ (expressed in pH units) in which R is the gas constant, T, the absolute temperature and F, the constant of Faraday. The term $2.3\ RT/F$ is represented by Z and has a value of almost 60 mV. A pH difference of 1 pH unit therefore represents a force of almost –60 mV.

(ii) . A membrane or electrical potential, $\Delta\Psi$, by the translocation of the positive charge of the protons from inside to outside (Figure 1).

The sum of these gradients, the proton motive force or PMF equals: $\Delta\Psi - Z\,\Delta pH$ mV. This PMF exerts an inwardly directed force on the protons. Since cytoplasmic membranes have very low proton permeabilities, protons can only cross these membranes via specific transport proteins. The influx of protons, driven by the proton motive force, results in the release of energy, which can be used to drive metabolic energy-requiring membrane-bound processes (Figure 1). Well-known examples are ATP synthesis by the F_oF_1-ATPase, rotation of the flagellar motor and transport of solutes across the membrane (see below).

The function of transport proteins is to couple transport of charge/and or protons to an energy-requiring process, thereby transducing the metabolic energy of the PMF into chemical, mechanical or (electro-)chemical energy of other (ionic)solutes.

In view of the importance of these energy transducing processes for the function of a cell it is not surprising that a PMF of significant magnitude is needed for cell growth and viability. In most bacteria the PMF during logarithmic growth has a value of around −150 mV. When a carbon source is present a PMF of this magnitude can readily be maintained in aerobic organisms by the proton pumping respiratory chain. In fermentative bacteria the situation is drastically different. These organisms very often do not possess proton pumping electron transfer systems and PMF generation relies on the proton pumping F_0F_1-ATPase at the expensive of ATP, formed by substrate level phosphorylation. Even in the presence of carbon and energy sources these fermentative bacteria function under energy-limited conditions and additional supply of metabolic energy will directly be reflected by higher growth rates and growth yields.

This review presents the current knowledge of energy transduction and solute transport systems of LAB and in particular of *L. lactis* with special emphasis on the importance of the systems for energy metabolism and survival of LAB under transient conditions.

Energetic states of LAB

LAB are fermentative organisms and obtain their *metabolic energy* (energy which can be used for energy-requiring processes in the cell) mainly by substrate level phosphorylation. The metabolic energy supplied by fermentation is small. In the metabolism of glucose to lactate, two ATP are formed while an additional ATP can be obtained in further metabolism of pyruvate to acetate. This ATP is needed for the synthesis of cellular macromolecules and other energy requiring processes in the cytoplasm and in the cytoplasmic membrane and for the generation and maintenance of a PMF by the membrane bound F_0F_1-ATP (hydrol-/synthet-)ase (Maloney 1977). The supply of metabolic energy by fermentation is usually not sufficient to allow maximal rates of the biosynthetic machinery and this restricts the growth yield and the growth rate of LAB.

Moreover, in their natural habitats LAB often encounter drastic fluctuations in the composition of their environment. Consequently, LAB are often in a transient state with periods of abundant supply of energy followed by periods of starvation. In the presence of sufficient supply of energy the energetic conditions of the LAB will be optimal. For *L. lactis* these conditions have been found to be: intracellular ATP levels around 2.5 mM (the energy stored in the ATP pool is expressed as the phosphate potential $\Delta G'_p$; the $\Delta G'_p$ in *L. lactis* is ∼ −460 mV), PMF ∼ −150 mV and the internal pH between pH 7 and pH 7.5 (Konings et al. 1989). When the energy source runs out the rate of glycolysis rapidly decreases. Also the rate of PMF generation decreases rapidly, although the steady state level of the PMF is maintained for several hours (Figure 2) (Poolman et al. 1987g). This loss of activity results in a diminished rate of ATP synthesis and consequently of PMF generation. It also results in a decreased supply of metabolic energy for vital cellular processes and this directly affects viability. When the energy supply is restored within a short period of time glycolysis can start again and the internal ATP-pool and the PMF can be regenerated (Otto 1981; Konings & Otto 1983; Poolman 1987; Poolman et al. 1987 a;). This ability of *L. lactis* to survive periods of starvation is restricted from a few hours up to a few days, depending on the starvation media (Peterkofsky & Gadzar 1979) and the growth conditions preceding the starvation phase (Poolman et al. 1987a). Loss of viability is mainly caused by a loss of glycolytic activity due to the inactivation of glyceraldehyde 3-phosphate dehydrogenase (Poolman et al. 1987a).

From the considerations presented above, it is evident that both during fermentation and transient

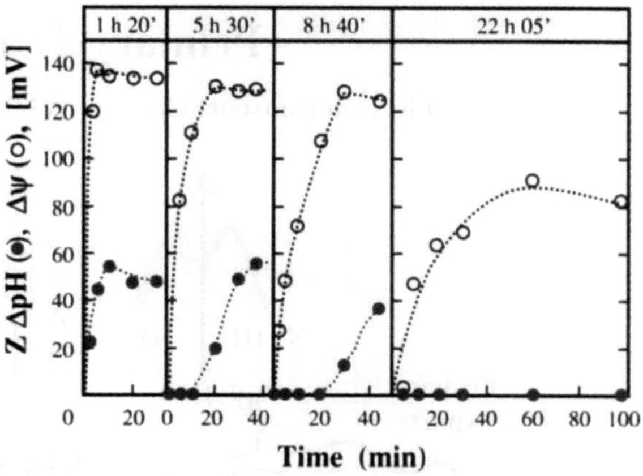

Figure 2. PMF generation in *Lactococcus lactis* Wg2 after various periods of lactose starvation (indicated in the top bar of the figure) (data from Poolman et al. 1987h).

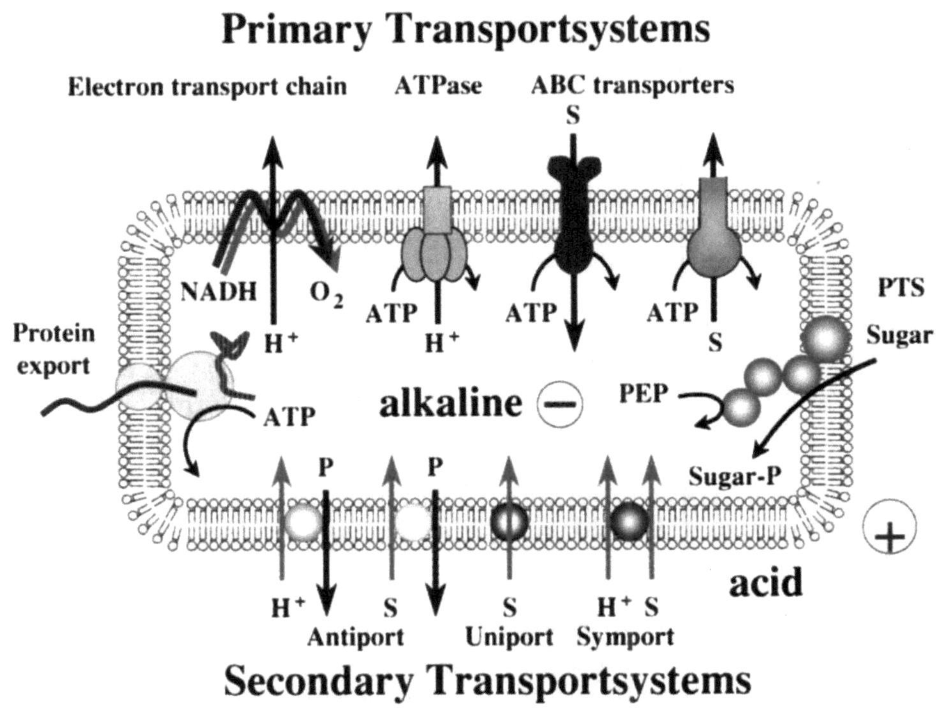

Figure 3. Different classes of transport systems in bacteria.

states the LAB could benefit considerably from a supply of additional usable sources of metabolic energy and such sources will directly contribute to growth and survival of the LAB. This review will describe a number of these additional usable energy sources of LAB.

Diversity of transport systems

Under glycolyzing and growing conditions solutes are taken up and end products of metabolism are excreted at high rates. The restricted permeability of the cytoplasmic membrane prevents high rates of passive fluxes even for solutes, which are significantly membrane permeable. Uptake of solutes and excretion of end products therefore need to be mediated by specific transport systems in the membrane.

The transport systems of LAB belong to four different classes (Figure 3)

(i) Primary transport systems, which use light, redox or chemical energy for the generation of a PMF or electrochemical gradients of other solutes. This class comprises the proton pumping electron transfer systems and F_0F_1-ATPase; the ion pumping P-type ATPases and the ATP-binding cassette (ABC)- transporters. Two categories of ABC-transporters exist: (a) the binding protein-dependent uptake systems and (b) the binding protein-independent extrusion systems.

(ii) Secondary transport systems which use the PMF for the uptake or efflux of solutes. This class comprises the uniport, the proton/solute symport and the solute/ H^+ or solute/solute(product) antiport systems (Figure 3).

(iii) The protein secretion system that uses ATP and PMF to drive proteins across the cytoplasmic membrane.

(iv) The phosphoenol-pyruvate phospho-transferase systems (PTS), which use phospho-enolpyruvate to concomitantly transport and phosphorylate sugars.

The genome of *L. lactis* IL 1403 encodes about 250 polypeptides that are directly involved in solute transport (Table 1). This means that well over 10% of all genes in its genome encode transport proteins, which reflects the important roles of transport processes in this bacterium. The products of these 'transport' genes of *L. lactis* are involved in a variety of transporters (Table 1). The largest class of transporters is secondary transport systems (73) followed by the ABC transporters (41). For comparison, *E.coli* has 194 secondary

Table 1. Transport systems in *Lactococcus lactis* IL1403 identified by genome annotation[a] and experimentally

Classes	Transporters[a]	Number of polypeptides[a]	Identified transporters
Primary transport systems	51	154	
F_0F_1-ATPase	1	7	1
Electron transport	1	12	1
ABC transporters	41	127	12
P-type ATPases	8	8	1
Secondary transport system	73	73	17
Protein secretion systems	1	8	1
PTS transporters	5	18	5
Total	130	253	38

[a] According to Bolotin et al. (2001).

transporters and 74 ABC transporters. The ratio of ABC transporters relative to the secondary transporters is therefore relatively high in *L. lactis.* (Binding protein-dependent) ABC transporters usually have very high affinities for their solutes and catalyze transport at high rates. The possession of such transporters gives *L. lactis* significant competitive advantage by allowing the organism to rapidly scavenge essential or growth-limiting solutes from dilute environments.

The substrates of many of these transporters are by no means known. A list of the transport systems, which have been studied in more or less detail, is given in Table 2. This represents about 30% of all transport systems of *L. lactis.* Clearly, many transporters still need to be characterized. The least studied transporters of *L. lactis* are the P-type ATPases. These transporters catalyze mainly the excretion of (cytotoxic) metals (Silver & Phung 1996).

The translocation of ions and other compounds (*solutes*) across the cytoplasmic membrane is energetically expensive, especially when the solutes have to be transported against a concentration gradient. According to the Nernst equation the energy of a solute (S) concentration gradient equals: $Z \log [S]_{in}/[S]_{out}$ mV, in which $[S]_{in}$ is the solute concentration in the cytoplasm and $[S]_{out}$, the solute concentration in the medium. A solute concentration gradient can only be generated and maintained at the expense of metabolic energy, usually ATP or a PMF. The driving forces for a variety of secondary and ATP-driven transporters are given in Figure 4. It should be realized that in all these transport processes the concentration gradients of all solutes or ions transported contribute to the driving force. Thus in all transport processes of solutes the ΔS ($=Z \log [S]_{in}/[S]_{out}$) contributes to the driving force. The $\Delta\Psi$ contributes to the driving force only when net charge is translocated in the transport process and the ΔpH is involved only when proton(s) are translocated. When solutes or charges are transported down their gradients their contribution to the driving force has a negative value; in the opposite direction their contribution has a positive value. When the total driving force is zero thermodynamic equilibrium is reached and no net transport will occur.

It has been stated above that ABC-transporters can usually accumulate solutes to high internal concentrations. The $\Delta G'_p$ is an important component of the driving force for solute transport in the F_0F_1-ATPase, the P-type ATPases and the ABC-transporters. The driving force supplied by $\Delta G'_p$ is usually significantly higher than the driving force supplied by the PMF, which explains the strong accumulative power of ABC transporters. The driving force for proton transport by the F_0F_1-ATPase is: $\Delta G'_p - n(\Delta\Psi - Z\Delta pH)$mV, in which n represents the number of protons transported per ATP hydrolyzed or synthesized. When the driving force supplied by $\Delta G'_p$ exceeds the driving force supplied by the PMF [$n(\Delta\Psi - Z\Delta pH)$ mV] ATP-hydrolysis will result in PMF generation; in the reversed situation the PMF will drive ATP-synthesis.

It should be realized that just like the F_0F_1-ATPase, transport processes are in principle reversible and that the direction of the transport processes depends on the direction of the driving force. This means that ATP or a PMF can be used to drive the uptake or excretion of a solute but that under conditions of high solute gradi-

Table 2. Transport systems in *Lactococcus lactis*

Class of transporter	Solute	Reference
PTS-transporters	Glucose and mannose	Thompson (1978)
	Lactose	Thompson (1979)
	Galactose	Thompson (1980)
	Sucrose	Thompson & Chassy (1981)
	Fructose	Thompson et al. (1985)
ABC transporters		
Binding protein-dependent uptake systems	Glutamate and glutamine	Poolman et al. (1987f)
	Asparagine	Truong & Poolman (1987)
	Aspartate	Poolman et al. (1987c)
	Phosphate	Poolman et al. (1987e)
	Glycine-betaine and proline	van der Heide & Poolman (2000a)
	Di- and tri-peptides (Dpp)	Sanz et al. (2001)
	Oligopeptides (Opp)	Kunji et al. (1993)
Extrusion systems		
	Anionic cytotoxic compounds	Molenaar et al. (1992)
	Anionic cytotoxic compounds	Glaasker et al. (1996b)
	Cationic cytotoxic compounds(LmrA)	Bolhuis et al. (1996b)
	Bacteriocins (LmrB)	Gajic et al. (2001)
P-type ATPases		
	Ca^{2+}	Ambudkar et al. (1986)
Secondary transporters		Driessen (1987); Poolman (1987)
Symport systems		
	Leucine, isoleucine, valine and methionine	Driessen et al. (1987a,b)
	Alanine and glycine	Driessen et al. (1987c)
	Serine and threonine	Driessen 1987
	Lysine	Driessen et al. (1989)
	Histidine	Otto et al. (1982)
	Proline	Driessen (1987)
	Cysteine	Driessen (1987)
	Tyrosine and phenylalanine	Poolman & Konings (1988)
	Lactate	Michels et al. (1979)
	α-Ketoglutarate	Marty-Teysset et al. (1996)
	di- and tri-Peptides (DtpT)	Hagting et al. (1994)
Antiport systems		
	Ca^{2+}	Driessen et al. (1985)
	Phosphate/hexose-6-phosphate	Maloney et al. (1984)
	Arginine/ornithine	Driessen et al. (1987d)
	Lysine	Driessen et al. (1989)
	Malate/lactate (MleP)	Poolman et al. (1991)
	Citrate/lactate (CitP)	Blandell et al. (1998)
	Cationic cytotoxic compounds (LmrP)	Bolhuis et al. (1994)

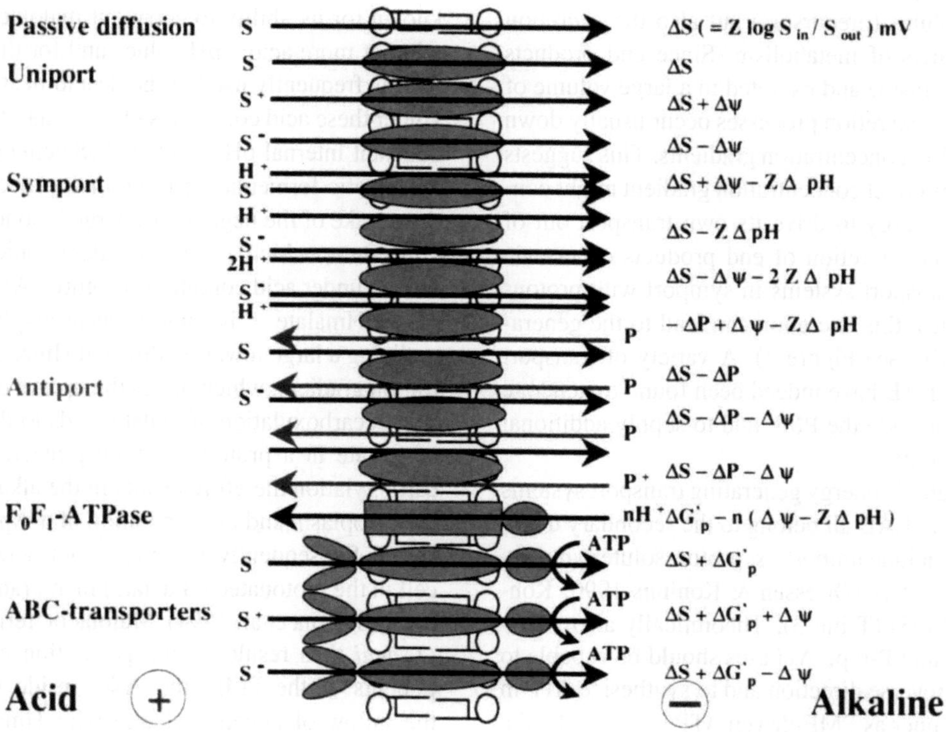

Figure 4. Driving forces for solute transport *via* different transport systems.

ents these gradients can drive the excretion of protons or charges and generate ATP or a PMF. When the total driving force has a negative value the transport process results in the accumulation of the solute at one side of the membrane at the expense of ATP or a PMF. On the other hand, when the driving force is positive ATP or a PMF can be generated at the expense of the solute and/or product gradient.

Metabolic energy supply by respiration

The fermentative nature of LAB has been brought into question by the observation that *L. lactis* can form a respiratory chain in the presence of hemin (Kaars Sijpesteijn 1970; Ritchey & Seeley 1976; Duwatt et al. 2001). These results are consistent with the presence of genes in the genome of *L. lactis* IL1403 encoding respiratory chain proteins (Bolotin et al. 2001). A similar effect of hemin was found for *Streptococcus faecalis* (Clarke & Knowles 1980). When oxygen is present electron transfer in this respiratory chain is coupled to the extrusion of protons and results in the generation of a PMF. Respiration can therefore contribute to the generation of metabolic energy in *L. lactis* (Blank et al. 2001; Duwatt et al. 2001) in a similar way as in obligate and facultative aerobic bacteria (Figure 1). As a result of this respiratory activity *L. lactis* has a greater growth yield and a remarkable improvement of survival, indicating that under fermentative conditions the metabolic energy supply limits growth. With a functional respiratory chain oxygen is beneficial instead of toxic for *L. lactis* (Duwatt et al. 2001).

Metabolic energy supply by secondary transport

In order to grow and to multiply bacteria have to take up and often to accumulate a variety of solutes. Transport processes are energetically expensive and consume a large fraction of the metabolic energy supply. An ABC transporter transports a solute at the expense of at least one ATP per substrate molecule (van Veen et al. 2000). Secondary transporters are energetically less expensive than ABC-transporters. Since the synthesis of one ATP by the F_0F_1-ATPase requires the influx of three to four protons (Jiang et al. 2001) the energy requirement of symport of a neutral solute with one proton represents one-third to one-quarter ATP-equivalents. Growth requires not only the uptake of solutes (precursors) but also the excretion of end products of metabolism. Since end products are produced inside and excreted in a large volume of medium these excretion processes occur usually down the end product concentration gradients. This suggests that the end product concentration gradient might contain enough energy to drive its own transport out of the cell. If the excretion of end products occurs via secondary transport systems in symport with protons and/or charges, this excretion can lead to the generation of a PMF (see Figure 4). A variety of transport processes in LAB have indeed been found to generate instead of consume the PMF and to supply additional metabolic energy.

The metabolic energy generating transport systems found so far in LAB all belong to the secondary transporters and include uniporters, proton-solute symporters and antiporters (Driessen & Konings 1990; Konings et al. 1995) (Table 3). Theoretically also ABC-transporters and P-type ATPases should be capable to work in the reverse direction and to synthesize ATP in a similar manner as PMF-driven ATP synthesis by the F_0F_1-ATPase. However, since the phosphate potential $\Delta G'_p$ is usually relatively high, energy conversion by the ABC-transporters in the reversed direction is energetically not possible.

Metabolic energy generating Uniport systems

The simplest mechanism by which solute transport can contribute to metabolic energy is via uniport systems in which either a negatively charged solute is taken up and rapidly metabolized internally or a positively charged end product of metabolism is excreted. The driving force for the translocation process is in both cases supplied by the solute gradient ΔS (*Z*. log $[S]_{in}/[S]_{out}$ mV) or product gradient $-\Delta P$ (*Z*. log $[P]_{in}/[P]_{out}$) mV and $-n\ \Delta\Psi$ (*n* represents the number of translocated charges). The charge translocation coupled to solute or product transport will lead to the generation of a $\Delta\Psi$, which at thermodynamic equilibrium will reach a value of: $\Delta\Psi = (\Delta S)/n$ mV. If in addition protons are consumed internally during solute metabolism or product formation the cytoplasm will alkalinize and the overall process will also lead to the generation of a ΔpH. In that case both components of the PMF, a $\Delta\Psi$ and a ΔpH, are generated.

Such PMF-generating uniport systems have been found in *Oenococcus oeni* (previously named *Leuconostoc oenos,* Dicks et al. 1995). This organism is well known for its ability to carry out malolactic fermentation at more acidic pH values and for this reason *O. oeni* is frequently used in the deacidification of wine. Under these acid conditions *O. oeni* maintains a rather constant internal pH of 5.8–6.3 (Salema et al. 1994). Malolactic fermentation in *O. oeni* proceeds through the uptake of the negatively charged mono-protonated $Hmalate^-$, which is the prevalent ionic species of malate under acid conditions (Figure 5A). In the cytoplasm $Hmalate^-$ is rapidly decarboxylated thereby realizing a large inwardly directed $Hmalate^-$ concentration gradient, which drives the uptake of $Hmalate^-$. The decarboxylation of malate leads to the formation of lactate in a proton consuming reaction. This decarboxylation therefore results in the alkalinization of the cytoplasm and the generation of a ΔpH, inside alkaline. Subsequently the end product lactate leaves the cell in the protonated (Hlactate) membrane-permeable form (Salema et al. 1994). Malolactic fermentation by *O. oeni* thus results in the generation of both components of the PMF: the $\Delta\Psi$, inside negative, by the inflow of negative charge with $Hmalate^-$ and a ΔpH , inside alkaline, by the internal consumption of protons in the decarboxylation reaction (Ramos et al. 1994). Another PMF-generating uniport system in *O. oeni* is coupled to the metabolism of citrate. This system involves a more complex metabolism and will be discussed below.

Metabolic energy generating proton-lactate symport systems

During glycolysis LAB continuously produce lactate at a high rate (up to 1 μmol/mg protein × min) and internal concentrations of lactate can be as high as 200 mM (Konings & Otto 1983). Lactate is a weak acid with a p*K* of 3.86. For most LAB the internal pH during glycolysis is above pH 7 (Poolman et al. 1987c) and essentially all lactate will be in the dissociated anionic form ($Lactate^-$). Unlike undissociated lactate (Hlactate) $Lactate^-$ is not membrane permeable and can only pass the membrane via a specific transport system (Figure 6). The lactate concentration gradient is directed from in to out and drives the lactate efflux process. When this $Lactate^-$ efflux occurs in symport with at least two protons the efflux process is electrogenic and can result in the generation of a $\Delta\Psi$, inside negative, and a ΔpH, inside alkaline. The energy of the lactate gradient is then converted to a PMF

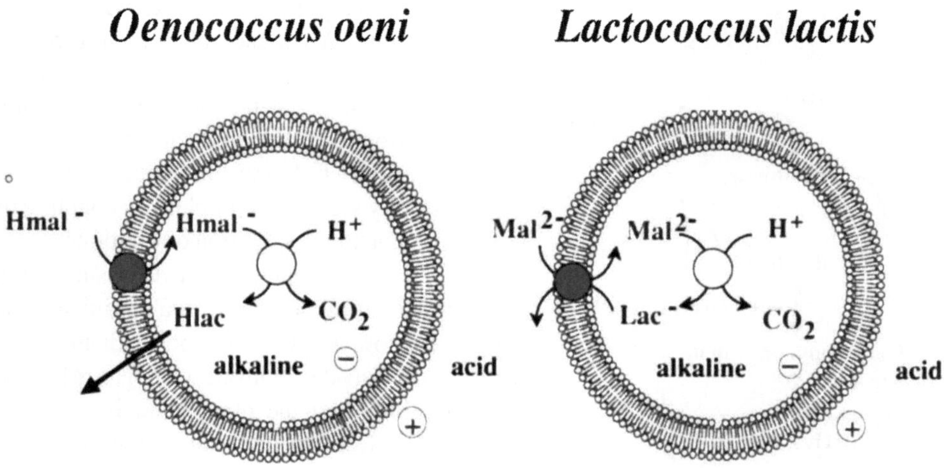

Figure 5. Metabolic energy conservation by malate transport and proton consuming decarboxylation. (A) *Oenococcus oeni*: uniport uptake of mono-anionic malate ($Hmal^-$), proton consumption by malate decarboxylation and passive efflux of undissociated lactate (Hlac). (B) *Lactococcus lactis*: exchange of di-anionic malate (Mal^{2-}) with mono-anionic lactate (Lac^-) and proton consumption by malate decarboxylation.

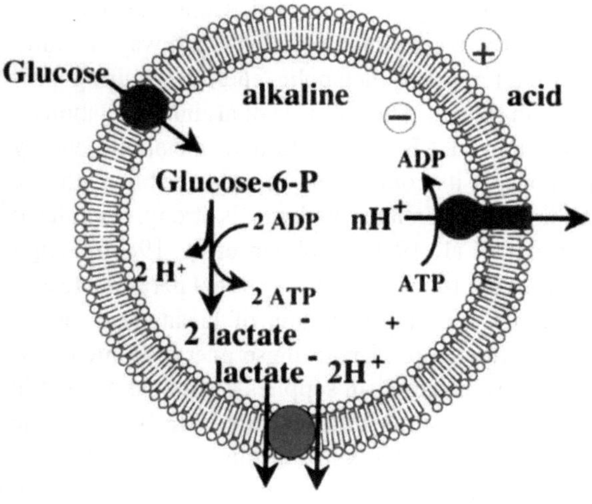

Figure 6. Metabolic energy conservation by electrogenic lactate efflux .

(Michels et al. 1979; Otto et al. 1980a,b, 1982; ten Brink & Konings 1982). This process has been called 'energy recycling' since the energy of the end product gradient is converted into metabolic energy in the form of a PMF. The number of protons excreted with lactate under optimal growth conditions of *L. lactis* has been found to be close to two (ten Brink & Konings 1982).

The metabolic energy gain from Lactate$^-$-proton excretion can be expressed in ATP-equivalents. With a proton/lactate stoichiometry of 2 and a proton/ATP stoichiometry in ATP-synthesis by the F_0F_1-ATPase between 3 and 4 (Jiang et al. 2001) the energy gain per lactate will be equivalent from 0.66 to 0.5 ATP. Glucose fermentation to lactate yields then 2.5 to 2.66 ATP equivalents instead of 2 and the additional energy yield by lactate efflux will be between 25 and 33%.

The beneficial effect of lactate excretion on the growth of *L. lactis* was demonstrated in mixed cultures of *L. lactis* and the lactate consumer *Pseudomonas stutzeri* (Otto et al. 1980a). Lactate consumption by *P. stutzeri* kept the lactate concentration in the medium low and as a result the molar growth yield of *L. lactis* increased significantly. Obviously, the gain of metabolic energy by lactate efflux depends strongly on the conditions inside and outside the cell. When the external pH decreases also the internal pH will decrease (see also Figure 11) (Poolman et al. 1987c) and more lactate will be present in the cytoplasm in the undissociated, membrane permeable Hlactate form. Consequently less Lactate$^-$ will efflux in symport with protons via the transport protein and the metabolic energy gain by lactate efflux will decrease. Furthermore, an increase of the lactate concentration externally as a result of the lactate production will decrease the driving force for the carrier-mediated efflux process and thermodynamic equilibrium will be reached at a lower proton motive force.

Metabolic energy by Precursor/Product exchange

During fermentation solutes (precursors) are taken up, metabolized intracellularly and the end products

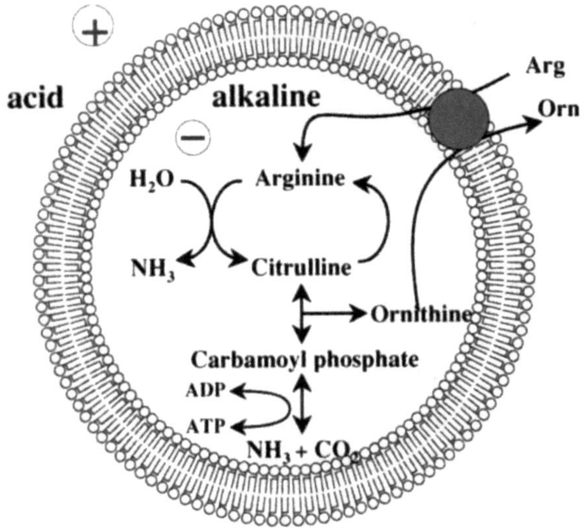

Figure 7. Arginine deiminase pathway in *Lactococcus lactis.*

of metabolism excreted. When the end products are structurally similar to the precursors these products can leave the cell via the same transport system by which the precursors are taken up. The transporters then catalyze precursor/product exchange. Energetically this is attractive for the cells since the transport process is driven by both the precursor gradient ΔS and the product gradient ΔP. Additional metabolic energy is not needed to drive this exchange process and the cells can use its metabolic energy for other purposes. Additional metabolic energy by the exchange process can be gained when the process is electrogenic, i.e., when net charge is translocated in the exchange process. As has been explained above, an exchange process that involves net translocation of positive charge from in to out or of negative charge from out to in will result in the formation of a $\Delta\Psi$, inside negative. Furthermore, when during end product formation protons are consumed internally the cytoplasm will be alkalinized and a ΔpH will be formed, inside alkaline. In that case precursor/product exchange coupled to metabolism results in the generation of both components of the PMF.

Electroneutral Precursor/product exchange

Metabolism of arginine to ornithine, ammonia and CO_2 (ratio 1:2:1) *via* the arginine deiminase pathway provides various LAB with an additional substrate level phosphorylation process (Abdelal 1979) (Figure 7). The arginine deiminase pathway requires two transport steps across the cytoplasmic membrane: the uptake of arginine and the excretion of the end product ornithine. When each of these transport steps would require energy in the form of a PMF or ATP the contribution to metabolic energy by arginine metabolism would not be significant. LAB and other bacteria (Driessen et al. 1987d; Poolman et al. 1987b) have cleverly solved this problem by coupling the uptake of arginine to the excretion of ornithine.

Since arginine in the cytoplasm is rapidly converted to citrulline, the internal arginine concentration is very low and the arginine gradient remains directed from outside to inside (Z log$[\text{Arg}]_{in}/[\text{Arg}]_{out}$ mV). Ornithine on the other hand is produced internally which leads to high concentrations inside while the concentration of ornithine in the medium will be low. Consequently the ornithine concentration gradient is directed from in to out ($-Z$ log $[\text{Orn}]$-$_{in}/[\text{Orn}]_{out}$ mV). Both the gradients of arginine and ornithine therefore contribute to the driving force for the electroneutral arginine/ornithine exchange process and additional metabolic energy is not needed. Consequently, ATP produced by the arginine-deiminase pathway can fully be used for other metabolic energy requiring processes in the cell. The sole role of arginine metabolism is to supply LAB with additional metabolic energy and none of its components is used for biosynthesis. The arginine deiminase pathway is strongly regulated (Konings et al. 1989; Poolman et al. 1987b). High activities are found when the external pH is decreased below pH 5 by the excretion of lactate and lactate fermentation stops. Under these energy-limited conditions Arg metabolism supplies additional metabolic energy and prevents a drop of the PMF below viable levels. An additional bonus of this process is that the production of ammonia prevents a lethal decrease of the internal and of the external pH (Marquis et al. 1987; Thomas & Batt 1968). The presence of the arginine deiminase pathway thus has significant survival value for these LAB.

Glucose or lactose represses the synthesis of the enzymes of the arginine deiminase pathway in *L. lactis*. This results in a sequential metabolism of these sugars and arginine (Crow & Thomas 1982). The glycolytic pathway and the arginine deiminiase pathway have different pH dependencies. At low pH the glucose PTS system and possibly other glycolytic enzymes are inhibited, whereas the arginine deiminase pathway is still active (Poolman et al. 1987b). Consequently, during growth of *L. lactis* the pH of the medium can oscillate, i.e., the pH decreases by glycolysis until sugar metabolism is inhibited where after the enzymes of the arginine deiminase pathway are derepressed leading to an increase of the pH by arginine metabolism. This ultimately stimulates again glycolysis resulting in repression of arginine metabolism and so on.

Metabolic energy by electrogenic precursor/product exchange

During fermentation a number of solutes (precursors) is taken up by LAB from the external medium via specific secondary transporters and internally metabolized to products, with a different charge than the precursors. When these products are excreted from the cells via the same transport proteins by which the precursors are taken up the exchange process will be electrogenic. Unlike the arginine/ornithine exchange system described above such an exchange process will result in the generation of a $\Delta\Psi$. Examples of such electrogenic exchange processes are found in a number of the decarboxylation processes that are catalyzed by LAB (Table 3; Poolman et al. 1991; Konings et al. 1995; Marty-Teysset et al. 1996 a,b).

As described above a precursor/product exchange process in which net positive charge is translocated from inside to outside or net negative charge from outside to inside can generate an electrical potential across the membrane. When thermodynamic equilibrium has been reached the driving force for translocation will be zero and ΔS - $\Delta P = n\Delta\Psi$ (*n* is the number of positive charges translocated outward or negative charge inward). The precursor/product exchange process therefore can result in the generation of a $\Delta\Psi$. The other component of the PMF, the transmembrane pH gradient, $-Z\Delta$pH, is formed during the decarboxylation of the precursor. In this enzymatic reaction protons are taken up from the cytoplasm resulting in an increase of the cytoplasmic pH and consequently the generation a transmembrane pH gradient (see above).

The process of precursor/product exchange coupled to decarboxylation does not involve the net translocation of protons across the membrane as is the case with true proton pumps, like the F_0F_1-ATPase. However, the overall results of the processes are the same, i.e. the generation of both components of the PMF. Thus, by compartimentalization of the decarboxylation pathway the free energy of the decarboxylation reaction can be converted into a PMF. An interesting feature of these precursor/product exchange processes coupled to decarboxylation is that the sole purpose of most of these processes (citrate fermentation is an exception) is to supply the cells with metabolic energy and that none of the components of the precursors serve other metabolic functions. In this respect the precursors have similar functions for the cells as gasoline has for a car. The best-studied decarboxylation coupled exchange systems of LAB are citrate fermentation, citrolactic fermentation, malolactic fermentation and histidine fermentation.

Malolactic fermentation

Some Lactococci, Lactobacilli, Leuconostocs, Oenococci and Pediococci carry out malolactic fermentation (Kunkee 1991). It has been shown above that malolactic fermentation in *O. oeni* involves an electrogenic uniport step coupled to a proton consuming decarboxylation step. The process in LAB that grow at higher pH values is different since malate is mainly present as Malate^{2-} above pH 6. For example during malolactic fermentation by *L. lactis* this Malate^{2-} is taken up and subsequently decarboxylated by malolactic enzyme to yield lactate (Figure 5B). Again due to the higher internal pH lactate is mainly present as Lactate^{-} and leaves the cell via the malate transporter. In *L. lactis* Malate^{2-}/Lactate^{-} exchange therefore is an electrogenic process in which negative charge is translocated from outside to inside and a proton is consumed in the decarboxylation process. As in *O. oeni* malolactic fermentation thus results in the generation of both components of the PMF.

In *L. lactis* this process can generate a PMF of sufficiently high value to drive ATP synthesis via the F_0F_1-ATPase (Poolman et al. 1991). An additional beneficial effect of malolactic fermentation is the external replacement of malate by lactate. Due to the pK difference of the carboxylic acid groups of malate and lactate, this conversion results in the alkalinization of the medium.

Histidine fermentation

Lactobacillus buchneri decarboxylates histidine to the biogenic amine histamine, which is excreted into the medium (Molenaar et al. 1993a). The p*K* of the imidazol-ring of histidine and histamine are 6.0 and 5.8, respectively, and the protonation state of the side chain of both molecules will have a similar pH dependency. Histidine enters the cells via the histidine transporter in the neutral form while the same transporter excretes histamine with one positive charge

Table 3. Metabolic energy supplying Solute transport systems in LAB

System	Organism	Reference
Uniport		
Malate influx	*O. oeni*	Salema et al. (1994)
Citrate influx	*O. oeni*	Ramos et al. (1994)
Proton-product symport		
Lactate efflux	*L. lactis*	Michels et al. (1979)
Electroneutral exchange		
Arginine/ornithine	*L. lactis*	Driessen et al. (1987d)
Lactose/galactose	*S. thermophilus*	Veenhoff et al. (2001)
Electrogenic exchange		
Malate/lactate	*L. lactis*	Poolman et al. (1991)
	Lb. plantarum	
Citrate/lactate	*Lc. mesenteroides*	Marty-Teysset et al. (1996)
	L. lactis	Bandell et al. (1998)
Aspartate/alanine	*Lb.* ssp. M3	Abe et al. (1996)
Glutamate/γ-aminobutyrate	*Lactobacillus* ssp.	Higuchi et al. (1998)
Histidine/histamine	*Lb. buchneri*	Molenaar et al. (1993)
Tyrosine/tyramine	*Lb. buchneri*	Konings et al. (1995)
Phenylalanine/phenylethylamine	*Lb. buchneri*	Konings et al. (1995)

due to the higher internal pH (Figure 8). This makes the histidine/histamine exchange process electrogenic, resulting in the net movement of a positive charge to the outside and the generation of a $\Delta\Psi$, inside negative. Additionally, a ΔpH is formed by the decarboxylation reaction as described above.

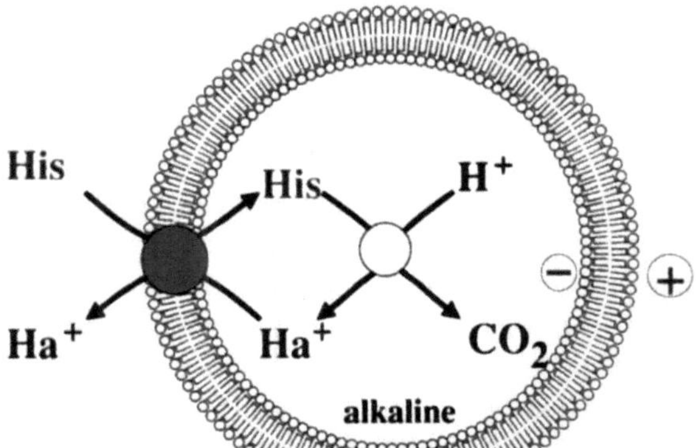

Figure 8. Metabolic energy conservation in *Lactobacillus buchneri* by exchange of mono-cationic Histidine (His) for di-cationic Histamine (Ha^+) and proton consuming decarboxylation.

Other decarboxylation processes

As early as 1924 it was suggested that decarboxylation of amino acids by micro-organimsms might fulfil a protective function against intracellular and extracellular acidification (Hanke & Koesler 1924; Lagerborg & Clapper 1952; Rodwell 1953). Several LAB perform amino acid decarboxylation at the end of food fermentation, when the carbon and energy sources have been consumed.

As a result of these decarboxylations biogenic amines are formed such as histamine, cadaverine, putrescine, tyramine and tryptamine (Gale 1946; Rice & Koehler 1976; ten Brink et al. 1990). Although these pathways have not all been studied with respect to their role in metabolic energy generation, in view of the results presented above, it is very likely that these processes result in the generation of a PMF in a similar way as has been described for malolactic and histidine fermentation (Table 3).

Direct evidence for PMF generation by the decarboxylations of aspartate to alanine and of glutamate to γ-isobutyrate has been presented for *Lactobacillus* ssp. In both exchange processes electrogenic exchange

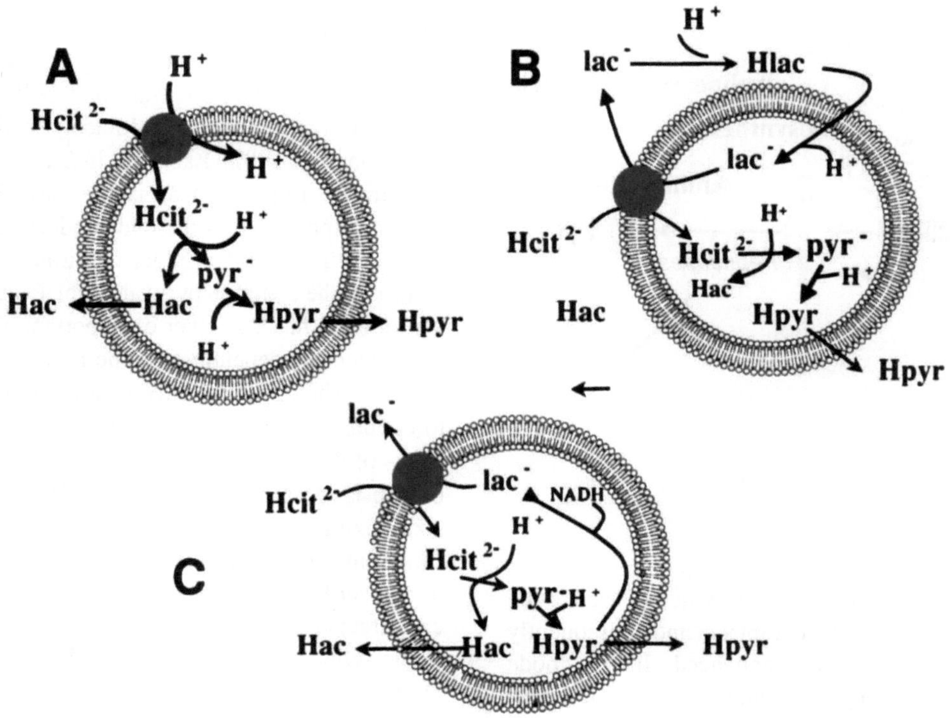

Figure 9. Metabolic energy conservation by citrate metabolism in *Leuconostoc mesenteroides* in the presence of: (A) citrate only; (B) citrate and lactate; (C) citrate and glucose. $Hcit^{2-}$, di-anionic citrate; Hpyr, pyruvate; Hac, acetate; Hlac, lactate; Lac^{-}, mono anionic lactate.

coupled to proton consuming decarboxylations result in $\Delta\Psi$ and ΔpH generation (Higuchi et al. 1993).

Citrate fermentation in Leuconostoc and Oenococcus spp.

Citrate fermentation-coupled PMF generation in LAB deserves some special attention due to the involvement of more complex metabolic processes. The ability to transport citrate is plasmid encoded in LAB (David et al. 1990; Kempler & McKay 1979). In *Leuconostoc mesenteroides* citrate is in the absence of another carbon source stoichiometrically converted to pyruvate according to the following reaction:

$$H_3citrate \rightarrow Hacetate + Hpyruvate + CO_2$$

Lc. mesenteroides grows around neutral pH values, at which $Hcitrate^{2-}$ is the most abundant citrate species (van der Rest et al. 1991). $Hcitrate^{2-}$ can be taken up in symport with one proton (see Figure 9A). During citrate metabolism protons are consumed leading to the generation of a ΔpH and ultimately to an increase of the medium pH. A similar PMF generation in the presence of citrate alone was also observed in *O. oeni* (Ramos et al. 1994). Citrate fermentation by *O. oeni* is in addition to malolactic fermentation important for wine production since citrate is a precursor of aroma compounds (e.g. acetoin and diacetyl) (Kunkee 1991). In the pH range of the natural habitat of *O. oeni*, i.e., pH 3.5–5.5, the predominant citrate species is $H_2citrate^{-}$. In this pH range the citrate transporter catalyses the uniport of monovalent $H_2citrate^{-}$, a process that is equivalent to symport of $Hcitrate^{2-}$ with one proton as catalyzed by *Lc. mesenteroides*. This implies that citrate metabolism is coupled to the generation of a $\Delta\Psi$, inside negative. Internally $H_2citrate^{-}$ is metabolized to the neutral end products acetic acid, acetoin and CO_2 and in this process one proton is consumed. Citrate metabolism in *O. oeni* therefore also leads to the generation of both a $\Delta\Psi$ and a ΔpH, independent of the F_0F_1-ATPase (Ramos et al. 1994).

Citrate metabolism by *Lc. mesenteroides* has been found to occur much faster in the presence of lactate without a change in the products of the pathway. The explanation for this increased rate of citrate metabolism was found to be the higher rate of citrate uptake (Marty-Teysset et al. 1996). In the presence of lactate the citrate transporter CitP catalyzes heterologous exchange between citrate and lactate, a process that is much faster than $Hcitrate^{2-}$/proton symport (Fig-

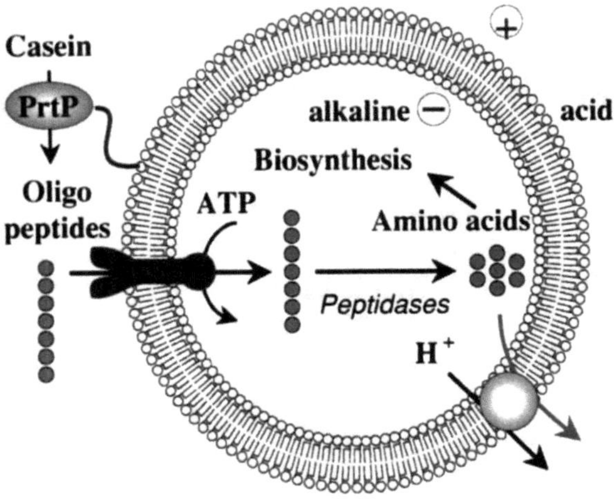

Figure 10. Proteolytic system of *Lactococcus lactis*.

ure 9B). Consequently, in the presence of lactate the rate of alkalization of the cytoplasm and subsequently of the external medium is enhanced. In this mode lactate functions as a proton shuttle, bringing in a proton when it enters the cell in its protonated state by passive diffusion and leaving the cell in its deprotonated state in exchange for $Hcitrate^{2-}$. The exchange of $Hcitrate^{2-}$ and $Lactate^{-}$ is an electrogenic antiport process in which a negative charge is taken up and a $\Delta\Psi$ is generated. This mode of citrate fermentation thus also leads to the generation of both components of the PMF.

The situation becomes more complex when glucose is present in the medium and co-metabolism of citrate and glucose occurs. The addition of citrate to glucose broth increases the specific growth rate and molar growth yield of *Lc. mesenteroides* (Lin et al. 1991). Under these conditions lactate is produced as an end product of citrate metabolism (citrolactic fermentation) and lactate from pyruvate instead of ethanol. The highest rates of citrate metabolism and the highest PMF, consisting of both a $\Delta\Psi$ and a ΔpH, are formed during this co-metabolism of citrate and glucose. When the $\Delta\Psi$, inside negative, becomes high, the flux through the pathway is inhibited, indicating that even under these conditions the transport step is controlling the rate of citrate metabolism to a large extent (Marty-Teysset et al. 1996). The generation of a PMF in the *Leuconostoc* and *Oenococcus* species consists of a number of metabolic steps and is not restricted to one step as in the fermentations involving a decarboxylation. A similar process of PMF generation by citrolactic fermentation occurs in *L. lactis* (Magni et al. 1999). In this bacterium a low level of constitutive expression of citrate transport usually exists but increased levels are induced by acid stress (Garci-Quintans et al. 1998). Citrate metabolism also contributes to flavor production (diacetyl and acetoin) in milk (Marshall 1987). Lactate is in *L. lactis* not formed by citrate fermentation as is the case for the *Leuconostoc* spp. but it is the end product of glycolysis. In *L. lactis* the end product of citrate fermentation feeds into the common pyruvate pool. Citrate metabolism in LAB has a number of important physiological consequences, depending on the organisms and their habitats. The different LAB have found various ways to couple citrate metabolism to the metabolic energy status of the cells by generating a PMF. Citrate metabolism is in addition coupled to pH homeostasis by alkalinizing the cytoplasm and eventually the medium. Citrate metabolism therefore may allow the organisms to recover from acid stress. Another important beneficial effect of citrate metabolism discovered in *L. lactis* is the relief of lactate toxicity (Magni et al. 1999). During fermentation of sugars by *L. lactis* lactate is accumulated and the medium is acidified. Both conditions inhibit growth and the inhibitory effects are synergistic. Because of the synergism the inhibitory effect of lactate and in general of weak organic acids (Houtsma et al. 1994) is usually ascribed to the undissociated acid that is membrane permeable and may compromise pH homeostasis of the cytoplasm. Other, specific inhibitory effects of the dissociated acids have also been suggested (Shelef 1994). Citrate metabolism in *L. lactis* specifically relieves the inhibition imposed by lactate and not the inhibition by acetic acid. This suggests that the inhibitory effects of both weak acids are not caused by negative effects on pH homeostasis, since in that case, the alkalinizing effect of citrate metabolism should also relieve the inhibition caused by any other weak acid. Mechanistically, the beneficial effects of citrate metabolism are due to the active removal of lactate from the cytoplasm. Since lactate is especially inhibitory at low pH, resistance to lactate toxicity provides a rationale for the induction of citrate fermentation pathway at low pH (Magni et al. 1999).

In *Lc. mesenteroides* citrate fermentation is directly involved in the synthesis of the essential amino acid aspartate. Aspartate can be converted to asparagine and participates in the biosynthesis of pyrimidines and purines (Marty-Theysset 1996).

The energetic consequences of proteolysis

Milk contains low concentrations of free amino acids and peptides and nitrogen is a growth-limiting factor for LAB in milk. The production of lactate is slow unless starter cells have proteolytic enzymes to degrade the milk proteins (α-, β- and κ-caseins) to peptides and free amino acids (Poolman et al. 1995). The process of casein degradation and utilization requires the concerted action of proteinase, peptidases, and amino acid and peptide transport systems. This whole complement of enzymes is termed the proteolytic system (Figure 10) (Konings et al. 1989; Smid et al. 1991).

Proteolysis of casein in *L. lactis* (Figure 10) starts with the hydrolytic action of the proteinase PrtP yielding a large number of oligopeptides, but no di- or tri-peptides or free amino acids. To obtain amino acids for biosynthesis polypeptides have to be transported across the cell membrane into the cells by the oligopeptide ABC transporter Opp. This transporter has extremely broad substrate specificity and can transport oligopeptides ranging from four up to 18 and possibly up to 35 amino acids (Detmers et al. 1998, 2000, 2001). Following breakdown by the extracellular proteinase PrtP and uptake via the oligopeptide transporters, the internally accumulated peptides are hydrolyzed further inside into amino acids by intracellular peptidases (Kunji et al. 1998).

The products resulting from the degradation of caseins by proteinases have been studied in great detail (Pritchard et al. 1993; Juillard et al. 1995a,b; Kunji et al. 1998; Flambard et al. 2000). The peptides obtained from β-casein by the digestion of proteinases from severall *L. lactis* strains have been identified (Juillard et al. 1995a; Kunji et al. 1998; Flambard et al. 2000). PrtP can degrade β-casein into more than 100 different oligopeptides ranging from four to 30 residues. Minimal amounts of di- and tri-peptides or free amino acids are produced. The degradation by various proteinases of β-, αsI- and αsII-caseins have also been studied, but only the major products have been identified so far (Visser et al. 1986; Bockelman et al. 1989; Monnet et al. 1992; Reid et al. 1994). These identified products are all oligopeptides and are derived mainly from the C-terminal regions of the caseins. Again no significant amounts of free amino acids and di- and tri-peptides could be detected in the hydrolysates.

The oligopeptides function as the main source of amino acids during growth in milk. Each amino acid taken up in this way saves the energy for its transport and biosynthesis. It was rather surprising to find that *L. lactis* accumulates via the oligopeptide transport system Opp only a limited number of all the different oligopeptides formed during proteolysis of β-casein. These peptides come only from two parts of β-casein, e.g., the region between amino acid 94 and 102 (GVSKVKEAM) and between 161 and 190 (SVLSLSQSKVLPVPQKQKAVPYPQRD-MPIQAF) (Juillard et al. 1995a; Kunji et al. 1998). These peptide fragments contain all amino acids, except for histidine, that are considered essential since they cannot be synthesized by *L. lactis* (Otto, 1981): glutamate, glutamine, methionine, valine and leucine/isoleucine. Histidine is most likely supplied from peptides derived from κ-casein (Flambard et al. 1998). Also the growth stimulating amino acids aspargine, proline, phenylalanine and alanine are present. Some non-essential amino acids are not present in these peptides: arginine, aspartate, cysteine, threonine, tryptophan and tyrosine (Kunji et al. 1993). These amino acids have to be synthesized from the available amino acids or from α-ketocarboxylates (Marty-Teysset et al. 1996).

It is quite remarkable given the competition between the oligopeptides for the single oligo peptide transport system that the whole complement of amino acids important for growth of *L. lactis* is accumulated via these oligopeptides. However, the amino acids are not supplied in the ratio required for protein synthesis. A number of amino acids are present in excess while the supply of essential amino acids such as methionine and valine can be growth-limiting (Juillard et al. 1995b; Flambard et al. 1998). The excess amino acids can serve as a source for the synthesis of other amino acids which are in short supply, can be converted to other metabolic intermediates and/or can be excreted across the membrane via specific amino acid transport systems. The amino acid transport systems of *L. lactis* ML3 have been studied extensively (Table 2). Additional amino acid transport systems are almost certainly present in *L. lactis*. Kunji et al. (1993) and Juillard et al. (1995b) showed that during growth of *L. lactis* in milk the available free amino acids are not used and that preference is given to the uptake of oligopeptides as sources of amino acids. This rather surprising observation strongly indicates that the activity and/or the expression of the amino acid transport systems are strongly controlled. The transcription of a number of components of the proteolytic system has shown to be down regulated by the presence of free di- and tri-peptides in the cytoplasm while free amino

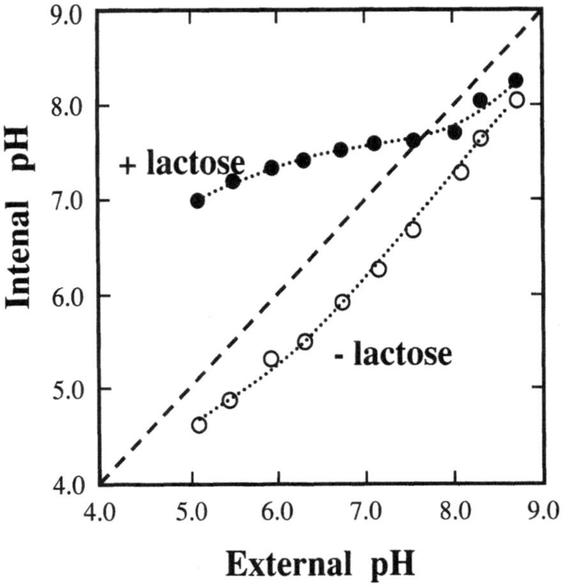

Figure 11. Internal pH of *Lactococcus lactis* ML3 in the presence (●) and absence (○) of 10 mM Lactose. Data from Poolman et al. (1987e) and E.J. Smid and Poolman, unpublished.

acids appear to exert the opposite effects (Sanz et al. 2001).

The metabolic energy that is needed for the uptake of the individual amino acids during growth in a chemically defined medium amounts to more than 10% of the metabolic energy generated during glycolysis (Poolman et al. 1987c). Energetically the uptake of oligopeptides via the Opp transport system is significantly less expensive than the uptake of the individual amino acids via their individual transport systems. Moreover, a number of amino acids, released in excess internally upon hydrolysis of the oligopeptides, will leave the cells *via* specific amino acid transport systems. Since this efflux occurs by proton symport transporters it can contribute to the generation of a PMF and supply metabolic energy. This additional supply of metabolic energy will be especially important when during fermentation the milk sugars have been depleted. In line with such metabolic energy saving process are the observations that growth of *L. lactis* in peptide containing media results in significantly higher cell yields and growth rates than in media containing the same quantities of amino acids in free form (Otto 1981; Juillard et al. 1995b).

The importance of controlling the internal pH

Homeostasis of the internal pH is essential for growth and survival of all biological cells. The main reason for the strong sensitivity to the internal pH is that many metabolic pathway enzymes function optimally around neutral pH and that their activities decrease rapidly at lower and higher pH values. Regulation of the internal pH in LAB is required since the external pH decreases continuously during growth as a result of acid production. Fermenting cells of *L. lactis* maintain an intracellular pH relatively constant around pH 7.5 (Figure 11). When the external pH increases above pH 7.5 the intracellular pH tends to follow the external pH (ten Brink et al. 1985; Poolman et al. 1987b).

A variety of processes in LAB affect the internal pH such as proton consumption or release during metabolism, PMF-generation by ATP-hydrolysis by F_0F_1-ATPase and PMF-driven solute transport processes. The mechanism of pH homeostasis has been well studied in *Streptococcus faecalis* (Kobayashi et al. 1986). The internal pH is mainly raised by the extrusion of protons by the F_0F_1-ATPase in combination with electrogenic uptake of K^+. The activity and the synthesis of the F_0F_1-ATPase decrease with increasing pH. Fine-tuning of the pH_{in} can be achieved by compensatory proton fluxes usually catalyzed by proton/sodium or potassium ion exchange systems. The mechanism of pH-homeostasis in *L .lactis* appears to be very similar to that of *S. faecalis.* At acid pH-values these regulatory mechanisms are, however, not sufficient to maintain the internal pH within the desired limits and additional proton removing processes are needed. It has been shown above that a number of those processes are present in LAB. The most potent systems are the arginine deiminase pathways and malate and citrate fermentations. The activities of these systems ultimately lead also to an increase of the external pH. An additional function of these systems thus lies in preventing acid-killing of the cells by the opposing effect of the acidification of the external pH.

The internal pH strongly affects the activities of solute transport systems. This has been studied extensively for the amino acid transport systems of *L. lactis* (Poolman 1987; Poolman et al. 1987b,c,d). With the exception of the transport systems for alanine and serine the activities of most amino acid transport systems increase with increasing pH and maximal activities are reached between pH 7.5 and 8.0. Since the metabolic energy cost for amino acid transport can amount up to 10% of the metabolic energy generated by glycolysis (Poolman at al. 1987a), a decrease of transport under low energetic conditions prevents the organism from wasting metabolic energy by 'unbalanced' transport processes.

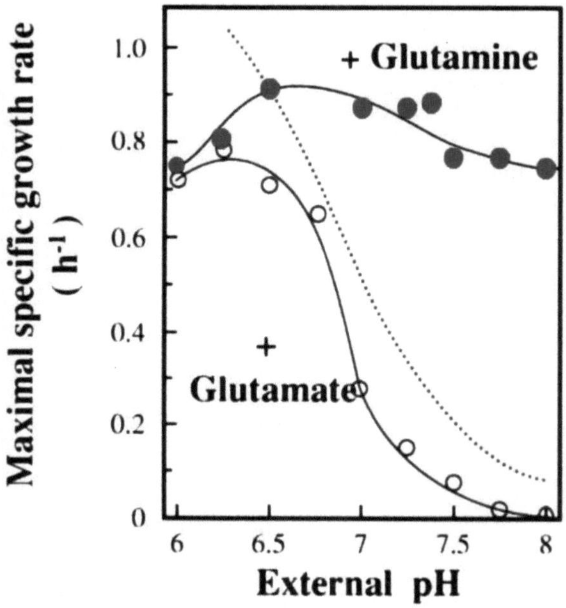

Figure 12. pH dependence of growth of *Lactococcus lactis* ML_3 in chemically defined medium at 30 °C with L-glutamate (○) or L-glutamine (●) as sources of L-glutamate for biosynthesis. The dashed line indicates the pH dependence of the predicted growth rate with L-glutamate assuming that uptake of undissociated L-glutamate is growth rate limiting (from Poolman & Konings 1988).

Effects of the medium pH on amino acid transport and growth

The effects of the external pH on the uptake of individual amino acids have also been studied (Poolman et al. 1987b,c, 1988; Konings et al. 1989). The optimum external pH for growth of most LAB varies between pH 6.0 and 6.5 and decreases rapidly at higher and lower pH values (Otto 1981). Since *L. lactis* can regulate its internal pH between pH 7.0 and 7.5 in the presence of an energy source (ten Brink & Konings 1982; Poolman et al. 1987c), the failure to grow at alkaline or acid pH values can therefore not be caused by a decreased internal pH but must be an effect of the external pH. The external pH can directly influence the affinities and activities of membrane bound processes such as the solute transport systems or it can affect the availability of essential nutrients (Poolman & Konings 1988). The affinity and rate of L-leucine uptake increases markedly with external pH with an apparent p*K* of about 7. This pattern of external pH-effects was not found for all amino acid transporters. The affinities of the alanine and serine transporters were essentially pH-independent in the pH range 5.5–8.0, while the rates of efflux of these amino acids decrease with increasing pH. The effect of the external pH on the different transport systems therefore varies considerably.

Solute transport systems usually have affinity for a particular ionic state of the solutes. Since the external pH affects directly the ionic state of the solutes the availability of the transportable solute will be strongly pH-dependent. An excellent example of this effect is found in the availability of the proper ionic state of glutamate for uptake by the glutamate/glutamine transporter of *L. lactis* (Poolman et al. 1987d,f,g).

Glutamate is an essential amino acid for *L. lactis*. At pH 6–7, glutamate is taken up by an ABC transporter at high rates [~15 nmol/min × (mg protein)] and accumulated in the cells to high levels. The substrate specificity of the glutamate transporter is restricted to protonated glutamate and the glutamate anion is not transported. Since above pH 7 the concentration of protonated glutamate decreases logarithmically with increasing pH the rate of glutamate uptake also decreases drastically. In addition to protonated glutamate the glutamate transporter can also transport glutamine with a high affinity (K_m about 2 μM). Between pH 6 and 8 the ionic condition of glutamine is not affected by the medium pH and high growth rates of *L. lactis* are observed up to pH 8.0 with glutamine as source of glutamate. However, when only glutamate is present in the medium the growth decreases drastically above pH 7 (Figure 12).

Osmoregulation

LAB often encounter changes in the osmolarity of their environment. A sudden increase of the medium osmolarity leads to loss of water from the cells. As a result turgor pressure will be lost, the concentration of internal solutes will increase and the cell volume will decrease. These hyperosmotic conditions lead to cell death unless the cells are able to counteract these effects by accumulating so called compatible solutes. Accumulation can be done by synthesizing or by taking up these compounds. When the osmolarity of the medium decreases these compounds have to be released rapidly in order to prevent the cell from bursting. The most important compatible solutes of LAB are glycine-betaine, carnitine and proline (Molenaar et al. 1993b; Glaasker et al. 1996a, 1998).

LAB have a limited capacity to synthesize compatible solutes. In *Lactobacillus plantarum* glycine-

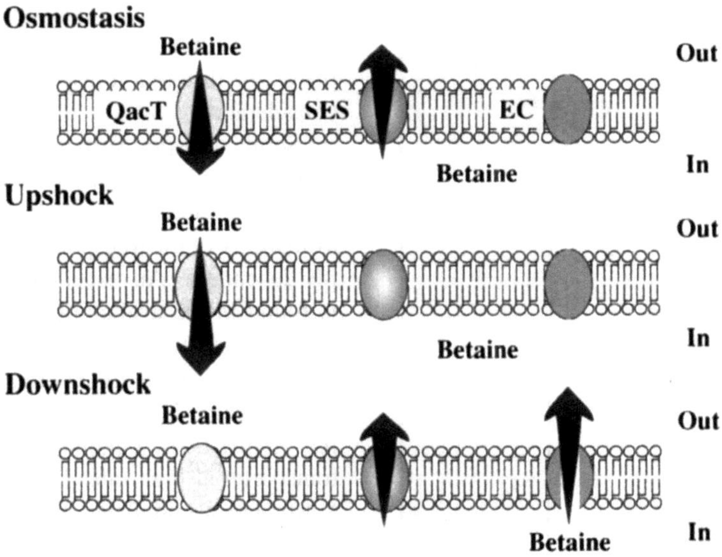

Figure 13. Osmoregulation in *Lactobacillus plantarum.* Under osmostatic conditions there is a basal level of glycine-betaine uptake and efflux. Upon osmotic upshock, the uptake system for glycine-betaine (QacT) is activated and the efflux inhibited. Upon osmotic downshock the QacT system is inhibited whereas a specific efflux system (SES) is activated and a putative channel (EC) is opened. The sizes of the arrows reflect the relative contribution to the total fluxes. (from Poolman & Glaasker 1998).

betaine and proline are taken up by the transporter QacT, which has high affinity for glycine-betaine (K_m = 18 μM) and low affinity for proline (K_m= 950 μM). This transport system is semiconstitutive and activated by osmotic upshock. QacT is only involved in the uptake of the osmolytes. When the osmolality of the medium decreases rapidly (osmotic downshock), these compounds have to be released from the cells rapidly. This is achieved by the opening of channels in the cytoplasmic membrane through which these osmolytes can efflux rapidly (Figure 13) (Glaasker et al. 1996a, 1998). *L. lactis* also has a high affinity glycine-betaine ABC-uptake system OpuA. This system is osmotically regulated at the levels of both expression and transport activity. The interesting observation was made that this regulation of the activity is not mediated via the turgor pressure but via changes in membrane properties (van der Heide & Poolman 2000a,b). This osmotic sensing mechanism appears to be strongly affected by electrostatic interactions between lipids and the transporter OpuA. Also the ionic strength of the cytoplasm serves as an osmotic signal, presumably by affecting lipid–protein interaction (van der Heide et al. 2002).

Resistance to cytotoxic compounds

LAB are often found on plants and in soil, environments that contain a wide variety of cytotoxic compounds, such as cationic alkaloids, which are excreted by plants (Stermitz et al. 2000). Growth of LAB in these environments will only be possible if the organisms have developed suitable protection mechanisms (Putman et al, 2000b). Resistance against cytotoxic compounds can be achieved by:

(i) synthesizing enzymes that degrade or chemically modify and inactivate cytotoxins (Davies 1994),
(ii) eliminating entry ports for hydrophilic drugs and
(iii) expressing drug efflux systems that export lipophilic drugs before these compounds have a chance to find their cellular targets (Lewis 1994; van Veen & Konings 1997).

In bacteria these resistance systems have been found to confer resistance to antibiotics (Putman et al. 2000b). Such antibiotic resistant organisms are specifically selected by the presence of antibiotics in their environment. Especially the excessive nutritive and therapeutic treatment of farm animals with antibiotics, amounting to one-half of the world's antibiotic production, has led to the selection of antibiotic-resistant micro-organisms that contaminate the food produced (Perreten et al. 1997). In many of these antibiotic

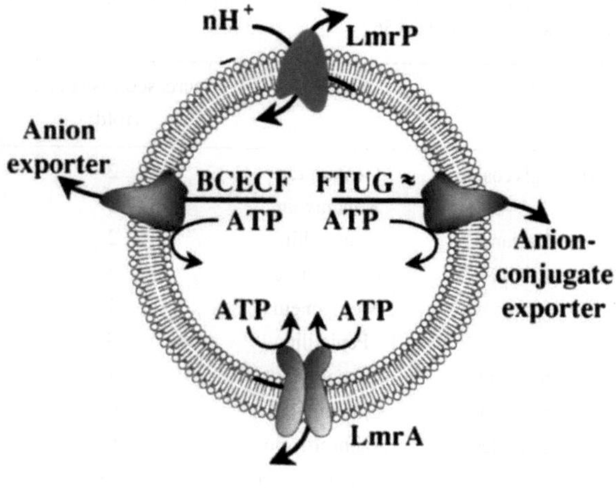

Figure 14. Multidrug resistant systems in *Lactococcus lactis*.

resistant organisms the resistance is conferred mainly by drug efflux systems. These drug-efflux systems are often very specific for a given drug or class of drugs, while others, the so-called multidrug resistance systems (MDRs), can handle a wide variety of drugs with different chemical structures and chemical targets. Also LAB have developed such fascinating MDRs for cationic and anionic lipophilic drugs (Figure 14). In *L. lactis* there are at least three MDRs for cationic drugs (van Veen et al. 1999). One of these MDRs is a PMF-driven electrogenic proton/drug exchange system LmrP (Bolhuis et al. 1996a) . A second MDR in *L. lactis* is LmrA, which belongs to the class of ABC transporters (van Veen et al. 1996), while the third MDR is homologous to LmrA and is called LmrB (Gajic et al. 2002). Two anionic multidrug transporters have been described so far in *L. lactis,* both of which are ABC transporters. One system excretes the fluorescent pH indicator 2′′,7′-bis-(2-carboxyethyl)-5[and 6]-carboxyfluorescein (BCECF) (Molenaar et al. 1992) while the second system has affinity for conjugated anionic drugs , such as conjugated *N*-(fluorescein thio-ureanyl)-glutamate (FTUG) (Glaasker et al. 1996b).

The cationic MDRs LmrA and LmrP have been studied most extensively. Both MDRs are inducible and higher expression levels are obtained when *L. lactis* is grown in the presence of increasing concentrations of toxic compounds (Bolhuis et al. 1994). LmrA and LmrP have overlapping specificity for lipophilic toxic compounds and deletion of one of the two systems leads to higher expression of the other system. Strains of *L. lactis* in which both MDRs have been deleted have so far appeared to be not viable, emphasizing the important physiological roles of these MDRs. It suggests that at least one MDR is essential for viability of *L. lactis,* even during growth in the absence of added toxic compounds.

The PMF-driven LmrP is a member of the major facilitator superfamily (Bolhuis et al. 1995) and is homologous to various multidrug transporters in pathogenic bacteria (van Veen & Konings 1997). The ABC transporter LmrA has significant sequence homology with ABC transporters of a variety of bacteria including several pathogenic ones (van Veen et al. 1998b; Poelarends et al. 2002 in press). HorA, a close homologe of the ABC-type multidrug resistance pump. LmrA was discovered in *Lactobacillus brevis*, a major contaminant of spoiled beer (Sami et al. 1997, 1998). This LAB can grow in beer in spite of the presence of antibacterial compounds (iso-α-acids) derived from the flowers of the hop plant *Humulus lupulus* L. Hop resistance of *Lb. brevis* is, at least in part, dependent on the expression of the *horA* gene, which is located on a 15-kb plasmid (Sami et al. 1997). The role of HorA in hop resistance was first suggested by a spontaneous mutant lacking this plasmid, which was sensitive to hop compounds. Reintroduction of the plasmid into this segregation mutant restored hop resistance (Sami et al. 1998). These complementation studies as well as the heterologous expression of the *horA* gene in *L. lactis*, demonstrated that HorA is involved in resistance to hop compounds. Moreover, almost all lactobacilli isolated as beer-spoilage strains possess *horA* homologues (Sami et al. 1997). In addition to conferring hop resistance, HorA confers resistance to the structurally unrelated drugs novobiocin and ethidium bromide (Sami et al. 1997). Drug transport studies in *L. lactis* and membrane vesicles and in proteoliposomes in which purified HorA was reconstituted identified this protein as a new member of the ABC family of multidrug transporters (Sakamoto et al. 2001).

A most surprising observation is the significant homology of LmrA with eukaryotic ABC MDRs and in particular with the human multidrug resistance P-glycoprotein. P-glycoprotein confers resistance to anti-cancer drugs and is one of the major causes of multidrug resistance of tumor cells during chemotherapy (Gottesman 1995). LmrA is equivalent to a half molecular version of the human multidrug resistance P-glycoprotein (van Veen et al. 1996). Recent studies with covalently linked dimers of LmrA suggest that LmrA, in analogy with P-glycoprotein, is active as a homo-dimer (van Veen et al. 2000, 2001).

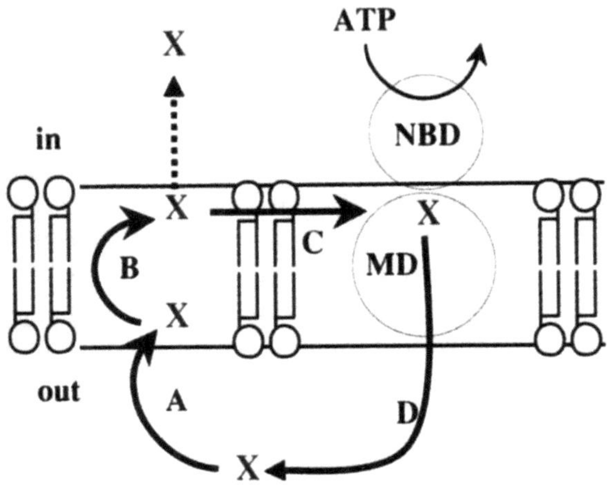

Figure 15. Mechanism of drug transport across the cytoplasmic membrane according to the vacuum-cleaner mechanism. (A) Drug molecules reaching the cell rapidly insert into the outer leaflet of the plasma membrane. (B) Flipping of the drug to the inner leaflet of the membrane is relatively slow and the rate limiting step in entry. (C) The drug binds to the MDR and (D), is excreted from the inner leaflet directly into the external medium. NBD, nucleotide binding domain; MD, membrane domain.

LmrA and LmrP have very broad substrate specificity and confer resistance to a variety of natural drugs and synthetic chemotherapeutic drugs. Many of the substrates are also substrates of the human P-glycoprotein and include antracyclines (daunomycin and doxorubricin), vinca alkaloids (vinblastin and vincristine) and cytotoxic agents (ethidium bromide, rhodamine 6 G and colchicin). A wide variety of compounds influence the activity of MDRs. These compounds are called modulators. The function of LmrA can be inhibited by a wide variety of structurally unrelated modulators, which also function as modulators for P-glycoprotein. Thus, the drug and modulator specificity of LmrA is very similar to that of P-glycoprotein. LmrA is thus both a structural and functional homologue of P-glycoprotein and an excellent model system for the studies of the molecular mechanism of medically important MDRs (van Veen et al. 1998a).

Among the cytotoxic compounds that can be excreted by LmrA and LmrP are members of different classes of antibiotics (Table 4) (van Veen et al. 1999; Putman et al. 2001). In fact, of all MDRs known to date, LmrA has the broadest substrate specificity. Seventeen of the 21 clinically most used antibiotics are excreted by LmrA. These observations strongly emphasize the important function of LmrA and LmrP in the resistance of *L. lactis* to cytotoxic compounds

Table 4. Effect of LmrA expression on the relative resistance to antibiotics of *E. coli* CS1562

Class	Antibiotic	Increased resistance[a] (fold)
Aminoglycosides	Gentamicin	2
	Kanamycin	3
β-Lactams	Ampicillin	2
	Ceftazidime	3
	Meropenem	1
	Penicillin	4
Glycopeptides	Vancomycin	1
Lincosamides	Clindamycin	14
Macrolides	Azithromycin	33
	Clarithromycin	23
	Dirithromycin	264
	Erythromycin	53
	Roxithromycin	35
	Spiramycin	35
Quinolones	Ciprofloxacin	2
	Ofloxacin	4
Steptogramins	Dalfopristin	163
	Quinupristin	31
	RP59500	55
Tetracyclines	Chlorotetracycline	28
	Demeclocycline	12
	minocycline	138
	Oxytetracycline	8
	Tetracycline	14
Others	Chloramphenicol	11
	Trimethoprim	1

[a]Data obtained from Putman et al. (2002a).

and antibiotics (Putman et al. 2000a). Evidently, the expression of LmrA and LmrP must have significant survival value for *L. lactis*. The mechanism by which LmrA and LmrP confer resistance to a wide variety of lipophilic cytotoxic compounds has been analyzed (Bolhuis et al. 1996). A common feature of the MDR substrates is their ability to intercalate between the lipids of biological membranes. Studies with lipophilic substrates (TMA-DPH and Hoechst 33342), which are fluorescent only in the lipid bilayer and not in the water phase, revealed the following sequence of events (Figure 15):

(i) a lipophilic cytotoxic compound intercalates rapidly in the outer leaflet of the cytoplasmic membrane;

(ii) subsequently it flips spontaneously from the outer leaflet to the inner leaflet of the membrane, a process that is relatively slow;

(iii) the MDRs LmrA and LmrP pick up the compound from the inner leaflet and pump it in a metabolic energy-dependent process directly into the external water phase.

The slowest step in this excretion process is the flipping of the lipophilic compound from the outer leaflet to the inner leaflet of the cytoplasmic membrane. This is a crucial step in the resistance mechanism since it allows LmrA and LmrP to keep the inner leaflet effectively free of the toxic compound thus preventing leakage of the toxic compound from the membrane into the cytoplasm (Bolhuis et al. 1996).

A mechanism by which the molecular machine LmrA works has been proposed on the basis of equilibrium binding experiments, photoaffinity labelling and drug transport assays. According to this proposal (van Veen et al. 2001) the dimeric LmrA mediates drug transport by an alternating two-site (two cylinder engine) mechanism. The transporter possesses two drug-binding sites: a transport competent site on the inner membrane surface and a drug-release site on the outer surface. The interconversion of these two sites, driven by the hydrolysis of ATP, occurs via a catalytic transition state intermediate in which the drug transport site is occluded.

Concluding remarks

Despite the relatively small size of their genomes a remarkably high number and variety of systems help LAB to survive in transient and often hostile environments. Most defense systems are located in the membrane. These systems either control the composition of the internal environment or are involved in the generation of metabolic energy. The functions of most membrane proteins are even for the best studied LAB *L. lactis* still unknown. The genomes of several LAB have or are currently sequenced. Functional genomics will undoubtedly reveal several other and often unexpected defense mechanisms in LAB.

Acknowledgements

This manuscript describes a major fraction of the research on LAB conducted over the last 25 years in the author's laboratory. The results of this research formed the basis for more than 180 scientific publications and for 31 Ph.D. theses. In this research about 75 scientists (Staff members, Postdoctoral fellows, Ph.D. and graduate students and technicians) have put in more than 275 man-year research time. Furthermore, with more than 90 scientists from national and internal research institutes was fruitfully collaborated on a variety of aspects of LAB. The author is very grateful for all the support, stimulating discussions and friendship received from all the individuals who shared with him the excitement of LAB research. He likes to thank in particular his colleagues Arnold Driessen, Juke Lolkema and Bert Poolman for their leadership, their creativity, their positive criticism and their loyal support during all these lucky years that we could work together. Research on LAB has been done in close interaction with the Molecular Genetics Research Group of Prof. Gerard Venema and his successor Prof. Oscar Kuipers and Jan Kok. Their generosity in sharing their research results, their willingness to assist with genetic problems and to supply strains and constructs has been crucial for our research progress. This review was written during a sabbatical leave of the author at the Dept. of Microbiology, University of Stellenbosch, South Africa. The suggestions and critical comments received from Prof. S. Silver, Chicago and Prof. Lee Lynd, Hanover, USA, have considerably improved the manuscript. Dr. Le Phung kindly prepared the figures for this review. The Netherlands Science Organization NWO and the European community have supported research on Lactic Acid bacteria extensively over the years.

References

Abe K, Hayashi H. & Maloney P.C. (1996) Exchange of aspartate and alanine. J. Biol. Chem. 271: 3079–3084.

Abdedal AT (1979) Arginine catabolism by microorganisms. Annu. Rev. Microbiol. 33: 139–168.

Ambudkar SV, Lynn AR, Maloney PC & Rosen B (1986) Reconstition of ATP-dependent Calcium transport from Streptococci. J. Biol. Chem. 261: 15596–15600.

Blank LM, Koebman BJ, Michelsen D, Nielsen LK & Jansen PR (2001) Hemin reconstitutes proton extrusion in an H^+-ATPase-negative mutant of *Lactococcus lactis*. J. Bacteriol. 183: 6707–6709.

Bockelman W, Monnet V, Geis A, Teuber M & Gripon J-C (1989) Comparison of cell wall proteineases from *Lactococcus lactis* subsp. *cremoris* AC1 and *Lactococcus lactis* subsp. *lactis* NCDO 763. Appl. Microbiol. Biotechnol. 31: 278–282.

Bolhuis H, Molenaar D, Poelarends GJ, van Veen HW, Poolman B, Driessen AJM & Konings WN (1994) Proton motive force-driven and ATP-dependent drug extrusion systems in multidrug-resistant *Lactococcus lactis*. J. Bacteriol. 176: 6957–6964.

Bolhuis H, Poelarends GJ, van Veen HW, Poolman B, Driessen AJM & Konings WN (1995) The lactococcal *lmrP* gene encodes a proton motive force-dependent drug transporter. J. Biol. Chem. 270: 26092–26098.

Bolhuis H, van Veen HW, Brands JR, Putman M, Poolman B, Driessen AJM & Konings WN (1996a) Energetics and mechanism of drug transport by the lactococcal multidrug transporter LmrP. J. Biol. Chem. 271: 24123–24128.

Bolhuis H, Molenaar D, van Veen HW, Poolman B, Driessen AJM & Konings WN (1996b) Multidrug resitance in *Lactococcus lactis*: Evidence for ATP-dependent extrusion from the inner leaflet of the cytoplasmic membrane. EMBO J. 15: 4239–4245.

Bolotin A, Wincker P, Mauger S, Jaillon O, Malarme K, Weissenbach J, Ehrlich SD & Sorokin A (2001) The complete genome sequence of the lactic acid bacterium *Lactococcus lactis* ssp. *lactis* IL1403. Genome Res. 11: 731–753.

Clarke DJ & Knowles CJ (1980) The Effect of Haematin and Catalase on *Streptococcus faecalis* var. *zymogens* growing in glycerol. J. Gen. Microbiol. 121: 339–347.

Crow VL & Thomas TD (1982) Arginine metabolism in Lactic Streptococci. J. Bacteriol. 150: 1024–1032.

David S, van der Rest M, Driessen AJM, Simons G & de Vos WM (1990) Nucleotide sequence and expression in *Escherichia coli* of *Lactococcus lactis* citrate permease gene. J. Bacteriol. 172: 5789–5794.

Davies J (1994) Inactivation of antibiotics and the dessimination of resistance genes. Science 264: 375–382.

Detmers FJM, Lanfermeijer FC & Konings WN (2001) Peptides and ATP binding cassette peptide transporters. Res. Microbiol. 152: 245–258.

Detmers FJM, Lanfermeijer FC, Abele R, Jack RW, Tampe R, Konings WN & Poolman WN (2000) Combinatorial peptide libraries reveal the ligand binding mechanism of the oligopeptide receptor OppA of *Lactococcus lactis*. Proc. Natl Acad. Sci. USA 97: 12487–12492.

Detmers FJM, Kunji ERS, Lanfermeijer FC, Poolman B & Konings WN (1998) Kinetics and specificity of peptide uptake by the oligopeptide receptor Opp A of *Lactococcus lactis*. Biochemistry 37: 16671–16679.

Dicks LMT, Dellaglio F & Collins MD (1995) Proposal to reclassify *Leuconostoc oenos* as *Oenococcus oeni* [corrig.] gen.nov., comb.nov. Int. J. Syst. Bacteriol 45: 395–397.

Driessen AJM (1987) Amino acid transport in lactic streptococi, Ph.D.thesis, University of Groningen.

Driessen AJM & Konings WN (1990) Energetic problems of Bacterial fermentations: Extrusion of metabolic endproducts. In: Krulwich TA (Ed) The Bacteria Vol XII (pp 449–478) Chapter 15, Academic Press, San Diego.

Driessen AJM, Hellingwerf KJ & Konings WN (1985) Light-induced generation of a proton motive force and Ca^{2+} transport in membrane vesicles of *Streptococcus cremoris* fused with bacteriorhodopsin proteoliposomes. In: Packer L (Ed) Recent Advances in Biological Membrane Studies (pp 439–462). Plenum Press, New York.

Driessen AJM, Hellingwerf KJ & Konings WN (1987a) Mechanism of energy coupling to entry and exit of neutral and branched chain amino acids in membrane vesicles of *Streptococcus cremoris*. J. Biol. Chem. 262: 12438–12443.

Driessen AJM, de Jong S & Konings WN (1987b) Transport of branched chain amino acids in membrane vesicles of *Streptococcus cremoris*. J. Bacteriol. 169: 5193–5200.

Driessen AJM, Kodde J, de Jong S & Konings WN (1987c) Neutral amino acid transport by membrane vesicles of *Streptococcus cremoris* is subjected to regulation by the internal pH. J. Bacteriol. 169: 2748–2754.

Driessen AJM, Poolman B, Kiewiet R & Konings WN (1987d) Arginine transport in *Streptococcus lactis* is catalyzed by a cationic exchanger. Proc. Natl. Acad. Sci U.S.A. 84: 6093–6097.

Driessen AJM, van Leeuwen C & Konings WN (1989) Transport of basic amino acids by membrane vesicles of *Lactococcus lactis*. J. Bacteriol. 171: 1453–1458. Duwat P, Sourice SS, Cesselin B,

Lamberet G, Vido K, Gaudo P, Leloir Y, Violet F, Loubiere P & Gruss A (2001) Respiratory capacity of the fermenting bacterium *Lactococcus lactis* and its positive effects on growth and survival. J. Bacteriol. 183: 4509–4516.

Flambard & Juillard V (2000) The autoproteolysis of *Lactococcus lactis* lactocepin III affects its specificity towards β-casein. Appl. Environ. Microbiol. 66: 5134–5140.

Flambard, Helnick, Richard & Juillard V. (1998) The contribution to the amino acid supply for *Lactococcus lactis* depends on the type of proteinase. Appl. Environ. Microbiol. 64: 1991–996.

Gajic O, G Buist, W Konings, L Topisirovic, J Kok and OP Kuipers (2001) MDR transporter protein involved in the secretion of and immunity against two lactococcal bacteriocins. Functional Genomics of Gram-Positive Microorganisms. Book of Abstracts (p 92). San Diego, CA.

Gale EF (1946) The bacterial amino acid decarboxylases. Adv. Enzymol. 6: 1–32.

Garci-Quintans NC, Magni C, de Mendoza D & Lopez P (1998) The citrate transport system of *Lactococcus lactis* subsp.*lactis* biovar *diacetylactis* is induced by acid stress. Appl. Environ. Microbiol. 64: 850–857.

Glaasker E, Konings WN & Poolman B (1996a) Osmotic regulation of intracellular pools in *Lactobacillus plantarum*. J. Bacteriol. 178: 575–582.

Glaasker E, Konings WN & Poolman B (1996b) The application of pH-sensitive fluorescent dyes in lactic acid bacteria reveals distinct extrusion systems for unmodified and conjugated dyes. Mol. Membr. Biol. 13: 173–181.

Glaasker E, Heuberger EHML, Konings WN & Poolman B (1998) Mechanism of osmotic activation of the quaternary ammonium compound transporter (QacT) of *Lactobacillus plantarum*. J. Bacteriol. 180: 5540–5546.

Gottesman MM, Hrycyna CA, Schoenlein PV, German UA & Pastan I (1995) Genetic analysis of the multidrug transporter. Annu. Rev. Genet. 29: 606–649.

Hagting A, Kunji ERS, Leenhouts KJ, Poolman B & Konings WN (1994) The di- and tripeptide transport protein of *Lactococcus lactis*: A new type of bacterial peptide transporters. J. Biol. Chem. 269: 11391–11399.

Hanke ME & Koesler KK (1924) Studies on proteinogenous amines. J. Biol. Chem. 59: 835–855.

Higuchi T, Hayashi H & Abe K (1993) Exchange of glutamate and α-aminobutyrate in a *Lactobacillus* strain. J. Bacteriol. 179: 3362–3364.

Houtsma PC, Kusters BJM, de Wit JC, Rombouts FM & Zwietering (1994) Modelling growth of *Listeria innocua* as a function of lactate concentration. J. Food Microbiol. 24: 113–123.

Jiang W, Hermolin J & Fillingame (2001) The preferred stoichiometry of c subunits in the rotary sector of *Eschericha coli* ATPase synthase is 10. Proc. Natl. Acad. Sci. U.S.A. 98: 237–241.

Juillard V, Laan H, Kunji ERS, Jeronimus-Stratingh CM, Bruins AP & Konings WN (1995a) The extracellular P2-type proteinase of *Lactococcus lactis* hydrolyzes β-casein into more than one hundred different oligopeptides. J. Bacteriol. 177: 3472–3478.

Juillard V, Le Bars D, Kunji ERS, Konings WN, Gripon JC & Richard J (1995b) Oligopeptides are the main source of nitrogen for *Lactococcus lactis* during growth in milk. Appl. Environ. Microbiol. 61: 3024–3030.

Kaars Sypersteyn A (1970) Induction of cytochrome formation and stimulation of oxidative dissimination by hemin in *Streptococcus lactis* and *Leuconostoc mesenteroides*. Antonie van Leeuwenhoek 36: 335–348.

Kandler O (1983) Carbohydrate metabolism in lactic acid bacteria. Antonie van Leeuwenhoek 49: 209–224.

Kempler GM & McKay LL (1979) Characterisation of plasmid deoxyribonucleic acid in *Streptococcus lactis* subsp.*diacetylactis*: evidence for a plasmid-linked citrate utilization. Appl. Environ. Microbiol. 37: 316–323.

Kobayashi H, Suzuki T & Unemoto T (1986) Streptococcal cytoplasmic pH is regulated by changes in amount and activity of a proton-translocating ATPase. J. Biol. Chem. 261: 627–630.

Konings WN (1985) Generation of metabolic energy by end-product efflux. Trends Biochem. Sci. 10: 317–319.

Konings WN, Poolman B & Driessen AJM (1989) Bioenergetics of solute transport in lactococci. CRC Crit. Rev. Microbiol. 16: 419–476.

Konings WN, Poolman B & Driessen AJM (1992) Can the excretion of metabolites by bacteria be manipulated? FEMS Microbiol. Rev. 88: 93–108.

Konings WN, Lolkema JS, Bolhuis H, van Veen WH, Poolman B & Driessen AJM (1997) Antonie van Leeuwenhoek 71: 117–128.

Konings WN & Otto R (1983) Energy transduction and solute transport in streptococci. Antonie van Leeuwenhoek. 49: 247–257.

Konings WN, Lolkema JS & Poolman B (1995) The generation of metabolic energy by solute transport. Arch. Microbiol. 164: 235–242.

Kunji ERS, Smid EJ, Plapp R, Poolman B & Konings WN (1993) Di-tripeptides and oligo peptides are taken up via distinct transport mechanisms in *Lactococcus lactis*. J. Bacteriol. 175: 2052–2059.

Kunji ERS, Hagting A, de Vries CJ, Juillard V, Haandrikman A, Poolman B & Konings WN (1993) Transport of β-casein derived peptides by the oligopeptide transport system is a crucial step in the proteolytic pathway of *Lactococcus lactis*. J. Biol. Chem. 270: 1569–1574.

Kunji ERS, Mierau I, Hagting A, Poolman B & Konings WN (1996) The proteolytic systems of lactic acid bacteria. Antonie van Leeuwenhoek 70: 187–221.

Kunji E.R.S., Gang G., Jeronimus-Strating C.M., Bruins A.P. & Konings W.N. (1998) Reconstruction of the proteolytic pathway of β-casein by *Lactococcus lactis* Mol. Microbiol. 27: 1107–1118.

Kunkee RE (1991) Some roles of malic acid in the malolactic fermentation in wine making. FEMS Microbiol. Rev. 88: 55–72.

Lagerborg VA & Clapper WE (1952) Amino acid decarboxylases of lactic acid bacteria. J. Bacteriol. 63: 393–397

Lewis K (1994) Multidrug resistance pumps in bacteria: variations on a theme. Trends Biochem. Sci. 19: 119–123.

Lin JP, Schmitt P & Divies C (1991) Characterization of a citrate-negative mutant of *Leuconostoc mesenteroides*: metabolic and plasmidic properties. Appl. Microbiol. Biotech. 34: 628–631.

Lindsay MR, Webb RI, Strous M, Jetten MSM, Butler MK, Forde RJ & Fuerst JA Cell compartmentalisation in planctomycetes: novel types of structural organization for the bacterial cell. Arch. Microbiol. 175: 413–429.

Magni C, de Mendoza D, Konings WN & Lolkema JS (1999) Mechanism of citrate metabolism in *Lactococcus lactis*: resistance against lactate toxicity at low pH. J. Bacteriol. 181: 1451–1457.

Maloney PC (1977) Obligatory coupling between proton entry and the synthesis of ATP by *Streptococcus lactis*. J. Bacteriol. 132: 564–575.

Maloney PC, Ambudkar SV, Thomas J & Schiller L (1984) Phosphate/hexose-6-phosphate antiport in *Streptococcus lactis*. J. Bacteriol. 158: 238–245.

Marty-Teysset C, Postma C, Lolkema JS, Schmitt P, Divies C & Konings WN (1996) Proton motive force generation by citrolactic fermentation in *Leuconostoc mesenteroides*. J.Bacteriol. 178: 2178–2185.

Marquis RE, Bender GR, Murray DR & Wong A (1987) Arginine deiminase system and bacterial adaptation to acid environment. Appl. Environ. Microbiol. 53: 198–200.

Marshall VM (1987) Lactic acid bacteria: starters for flavour. FEMS Microbiol. Rev. 46: 327–336

Michels PAM, Michels JPJ, Boonstra J & Konings WN (1979) Generation of electrochemical proton gradient in bacteria by the excretion of metabolic end products. FEMS Microbiol. Lett. 5: 357–364.

Mitchell P (1966) Chemiosmotic Coupling and Energy Transduction, Glynn Research, Bodmin, Cornwall, England.

Molenaar D, Bolhuis H, Abee T, Poolman B & Konings WN (1992) The efflux of a fluorescent probe is catalyzed by an ATP-driven extrusion system in *Lactococcus lactis*. J. Bacteriol. 174: 3118–3124.

Molenaar D, Bosscher JS ten Brink B, Driessen AJM & Konings WN (1993) Generation of a proton motive force by histidine decarboxylation and electrogenic histidine/histamine antiport in *Lactobacillus buchneri*. J. Bacteriol. 175: 2864–2870.

Molenaar D, Hagting A, Alkema H, Driessen AJM & Konings WN (1993) Characterization and osmoregulatory role of uptake systems for proline and glycine-betaine in *Lactococcus lactis*. J. Bacteriol. 175: 5438–5444.

Mundt JO (1982) The ecology of Streptococci. Microbial. Ecol. 8: 355–369.

Monnet V, Le Bars D & Gripon J-C (1986) Specificity of a cell wall proteinase from *Streptococcus lactis* NCDO763 towards bovine β-casein. FEMS Microbiol. Lett. 36: 127–131.

Otto R (1981) An ecophysiological study of starter streptococci. Ph.D. Thesis, University of Groningen

Otto R, Hugenholtz J, Konings WN & Veldkamp H (1980a) Increase of molar growth yield of *Streptococcus cremoris* for lactose as a consequence of lactate consumption by *Pseudomonas stutzeri* in mixed cultures. FEMS Microbiol. Lett. 9: 85–88.

Otto R, Sonnenberg ASM, Veldkamp H & Konings WN (1980b) Generation of an electrochemical proton gradient in *Streptococcus cremoris* by lactate efflux. Proc. Natl. Acad. Sci. U.S.A. 77: 5502–5506.

Otto R, Lageveen RG, Veldkamp H & Konings WN (1982) Lactate efflux-induced electrical potential in membrane vesicles of *Streptococcus cremoris*. J. Bacteriol. 149: 733–738.

Patchett RA, Kelly AF & Kroll RG (1994) Transport of glycine-betaine by *Listeria monocytogenes*. Arch.Microbiol. 162: 205–210.

Perreten V, Schwarz F, Cresta L, Boeglin M, Dasen G & Teuber M (1997) Antibiotic resistance spread in food. Nature 389: 801–802.

Peterkofsky A & Gadzar C (1979) *Escherichia coli* adenylate cyclase complex: Regulation by the proton electyrochemical gradient. Proc. Natl. Acad. Sci. U.S.A. 76: 1099–1103.

Poolman B (1987) Energy transducing processes in growing and starving *Lactococcus lactis*. Ph.D. Thesis, University of Groningen.

Poolman B & Glaasker E (1989) Regulation of compatible solute accumulation in bacteria. Mol.Microbiol. 29: 397–407.

Poolman B & Konings WN (1988) Relation of growth of *Streptococcus lactis* and *Streptococcus cremoris* to amino acid transport. J. Bacteriol. 170: 700–707.

Poolman B, Bosman B, Kiers J & Konings WN (1987a) Control of glycolysis by glyceraldehyde-3-phosphate dehydrogenase in *Streptococcus cremoris* and *Streptococcus lactis*. J. Bacteriol.169: 5887–5890.

Poolman B, Driessen AJM & Konings WN (1987b) Regulation of arginine/ornithine exchange and the arginine deiminase pathway in streptococci. J. Bacteriol. 169: 5597–5604.

Poolman B, Driessen AJM & Konings WN (1987c) Regulation of solute transport in Streptococci by external and internal pH values. Microbiol. Rev. 51: 498–508.

Poolman B, Hellingwerf KJ & Konings WN (1987d) Regulation of glutamate-glutamine transport by intracellular pH in *Streptococcus lactis*. J. Bacteriol. 169: 2272–2276.

Poolman B, Nijssen RMJ & Konings WN (1987e) Dependence of *Streptococcus lactis* phosphate transport on internal phosphate concentration and internal pH. J. Bacteriol. 169: 5373–5378.

Poolman B, Smid EJ & Konings WN (1987f) Kinetic properties of a phosphate-bond driven glutamate/glutamine transport in *Streptococcus lactis* and *Streptococcus cremoris*. J. Bacteriol. 169: 2755–2761.

Poolman B, Smid EJ, Veldkamp H & Konings WN (1987g) Bioenergetic consequences of lactose starvation for continuously cultured *Streptocococcus cremoris*. J. Bacteriol. 169: 1460–1468.

Poolman B, Driessen AJM & Konings WN (1988) Effects of the cytoplasmic pH on solute transport in bacteria. In: Whittenbury R, Gould GW, Banks JG & Board RG (Eds) Homeostatic Mechanisms in Micro-organisms (pp 117–129). Bath University Press.

Poolman B, Molenaar D, Smid EJ, Ubbink T, Abee T, Renault PP & Konings WN (1991) Malolactic fermentation: Electrogenic malate uptake and malate/lactate antiport generate metabolic energy. J. Bacteriol. 173: 6030–6037.

Poolman B, Kunji ERS, Hagting A, Juillard V & Konings WN (1995) The proteolytic pathway of *Lactococcus lactis*. J. Appl. Bacteriol. Symp. Suppl. 79: 65–75.

Pritchard GG & Coolbear T (1993) The physiology and biochemistry of the proteolytic system in lactic acid bacteria. FEMS Microbiol. Rev. 12: 179–206.

Putman M, van Veen HW, Degener JE & Konings WN (2000a) Antibiotic resistance: era of the multidrug pump. Mol. Microbiol. 36: 772–774.

Putman M, van Veen HW & Konings WN (2000b) Molecular properties of bacterial multidrug transporters. Microbiol. Mol. Biol. Rev. 64: 672–693.

Putman M, van Veen HW, Degener JE & Konings WN (2001) The lactococcal secondary multidrug transporter LmrP confers resistance to lincosamides, macrolides, streptogramins and tetracyclines. Microbiology 147: 2873–2880.

Ramos A, Poolman B, Santos H, Lolkema JS & Konings WN (1994) Uniport of anionic citrate and proton consumption in citrate metabolism generates a proton motive force in *Leuconostoc oenos*. J. Bacteriol. 176: 4899–4905.

Reid JR, Coolbear T, Pillidge CJ & Pritchard GG (1994) Specificity of hydrolysis of bovine κ-casein by cell envelope-associated proteinases from *Lactococcus lactis* strains. Appl. Environ. Microbiol. 60: 801–806.

Rice SL & Koehler PE (1976) Tyrosine and histidine decarboxylase activities of *Pediococcus cerevisiae* and *Lactobacillus* sp. the production of tyramine in fermented sausages. J. Milk Food Technol. 39: 166–169.

Ritchey TW & Seeley H (1976) Distribution of cytochrome-like respiration in streptococci. J.Gen. Microbiol. 93: 195–203.

Rodwell AW (1953) The occurrence and distribution of amino acid decarboxylases within the genus *Lactobacillus*. J. Gen. Microbiol. 8: 224–232

Salema M, Poolman B, Lolkema JS, Louriero Dias MC & Konings WN (1994) Uniport of monoanionic L-malate in membrane vesicles from *Leuconostoc oenos*. Eur. J. Biochem. 225: 289–295.

Sakamoto K, Margolles A, van Veen HW & Konings WN (2001) Hop resistance in the beer spoilage bacterium *Lactobacillus brevis* is mediated by the ATP-binding cassette multidrug transporter HorA. J.Bacteriol. 183: 5371–5375.

Sami M, Yamahita H, Hirono T, Kadokura H, Kitamoto K, Yoda K & Yamasaki M (1997) Hop-resistant *Lactobacillus brevis* contains a novel plamid harboring a multidrug resistance-like gene. J. Ferment. Bioeng. 84: 1–6.

Sami M, Suzuki K, Sakamoto K, Kadokura H, Kitamoto K, Yoda K (1998) A plasmid pRH45 of *Lactobacillus brevis* confers hop resistance. J. Gen. Appl. Microbiol. 44: 361–363.

Sanz Y, Lanfermeijer FC, Renault P, Bolotin A, Konings WN & Poolman B (2001) Genetic and functional characterization of *dpp* genes encoding a dipeptide transport system in *Lactococcus lactis*. Arch. Microbiol. 175: 334–343.

Shelef LA (1994) Antimicrobial effect of lactate: a review. J.Food Microbiol. 56: 2099–2103.

Silver S & Phung LT (1996) Bacterial heavy metal resistance: new surprises. Annu. Rev. Microbiol. 50: 753–789.

Smid E. J. (1991) Physiology and Implications of peptide transport in Lactococci. Ph.D. Thesis, University of Groningen.

Smid EJ, Poolman B & Konings WN (1991) Casein utilization by lactococci. Appl. Environ. Microbiol. 57: 2447–2452.

Spratt BG (1994) Resistance to antibiotics mediated by target alterations. Science 264: 388–393.

Stermitz FR, Lorenz P, Tawara JN, Zenewicz LA & Lewis K (2000) Synergy in a medical plant: antimicrobial action of berberine potentiated by 5"-methoxyhydnocarpin, a multidrug pump inhibitor. Proc. Natl. Acad. Sci. U.S.A. 15: 1433–1437.

Ten Brink B & Konings WN (1982) Electrochemical proton gradient and lactate concentration in *Streptococcus cremoris* cells grown in batch culture. J. Bacteriol. 152: 682–686.

Ten Brink B., Otto R., Hansen U.P. & Konings W.N. (1985) Energy recycling by lactate efflux in growing and non-growing cells of *Streptococcus cremoris*. J. Bacteriol. 162: 383–390.

Ten Brink B, Damink C, Joosten HMLJ & Huis in 't Veld J (1990) Occurrence and formation of biologically active amines in foods. Int. J. Food Microbiol. 11: 3–84.

Thomas TD & Batt RD (1968) Survival of *Streptococcus lactis* under starvation conditions. J. Gen. Microbiol. 50: 367–382.

Thompson J (1978) In vivo regulation of glycolysis and characterization of sugar:phosphotransferase systems in *Streptococcus lactis*. J. Bacteriol. 136: 465–476.

Thompson J (1979) Lactose metabolism in *Streptoccus lactis*: phosphorylation of galactose and glucose moieties *in vivo*. J. Bacteriol. 140: 774–785.

Thompson J (1980) Galactose transport systems in *Streptococcus lactis*. J. Bacteriol. 144: 683–691.

Thompson J & Chassy BM (1981) Uptake and metabolism of sucrose by *Streptococcus lactis*. J. Bacteriol. 147: 543–551.

Thompson J, Chassy BM & Egan W (1985) Lactose metabolism in *Streptococcus lactis*: studies with a mutant lacking glucokinase and mannose-phosphotransferase activities. J. Bacteriol. 162: 217–223.

Truong L & Poolman B (1987) unpublished

Van der Heide T & Poolman B (2000a) Glycine-betaine transport in *Lactococcus lactis* is osmotically regulated at the level of expression and translocation activity. J. Bacteriol. 182: 203–206.

Van der Heide T & Poolman B (2000b) Osmoregulated ABC-transport system of *Lactococcus lactis* senses stress via changes in the physical state of the membrane Proc. Natl. Acad. Sci. U.S.A. 97: 7102–7106.

Van der Heide T, Stuart MCA & Poolman B (2001) On the osmotic signal and osmosensing of an ABC transport system for glycine-betaine. EMBO J. 20: 7022–7032.

Van der Rest M.E., Abee T., Molenaar D. & Konings W.N. (1991) Mechanism and Energetics of a citrate transport system of *Klebsiella pneumoniae*. Eur. J. Biochem. 195: 71–77.

Van Veen HW & Konings WN (1997) Multidrug transporters from bacteria to man:similarities in structure and function. Semin. Cancer Biol. 8: 183–191.

Van Veen HW & Konings WN (1998b) The ABC family of multidrug transporters in microorganisms. Biochim. Biophys. Acta 1365: 31–36.

Van Veen HW, Venema K, Bolhuis H, Oussenko I, Kok J, Poolman B, Driessen AJM & Konings WN (1996) Multidrug resistance mediated by a bacterial homolog of the human drug transporter MDR1. Proc. Natl. Acad. Sci. U.S.A. 93: 10668–10672.

Van Veen HW, Callaghan R, Soceneantu L, Sardini A, Konings WN & Higgins CF (1998) A bacterial antibiotic-resistance gene that complements the human multidrug-resistance P-glycoprotein gene. Nature 391: 291–295.

Van Veen HW, Margoles A, Muller M, Higgins CF & Konings WN (2000) The homodimeric ATP-binding cassette transporter LmrA mediates multidrug transport by an alternating two-site (two-cylinder engine) mechanism. EMBO J. 19: 2503–2514.

Van Veen HW, Putman M, Margolles A, Sakamoto K & Konings WN (1999) Structure function analysis of multidrug transporters in *Lactococcus lactis*. Biochim. Biophys. Acta 1461: 201–206.

Van Veen HW, Higgins CF & Konings WN (2001) Multidrug transport by ATP binding cassette transporters: a proposed two-cylinder engine mechanism. Res.Microbiol. 152: 365–374.

Veenhoff LM, Heuberger EHML & Poolman B (2001). The lactose transport protein in a cooperative dimer with two sugar translocation pathways. EMBO J. 20: 3056–3062.

Visser S, Exterkate FA, Slangen CJ & de Veer GJCM (1986) Comparitive study of action of cell wall preoteinases from various strains of *Streptococcus cremoris* and αs1-, β- and κ-casein. Appl. Environ. Microbiol. 52: 1162–1166.

Visser S, Slangen CJ, Robben AJPM, Heerma WD & Haverkamp J (1994) Action of cell envelope proteinase (CEPIII-type) from *Lactococcus lactis* subsp. *cremoris* AM1 on bovine κ-casein. Appl. Microbiol. Biotechnol. 41: 644–651.

Antonie van Leeuwenhoek **82:** 29–58, 2002.

Discovering lactic acid bacteria by genomics

Todd Klaenhammer[1,*], Eric Altermann[1], Fabrizio Arigoni[3], Alexander Bolotin[4], Fred Breidt[5], Jeffrey Broadbent[6], Raul Cano[7], Stephane Chaillou[24], Josef Deutscher[8], Mike Gasson[9], Maarten van de Guchte[4], Jean Guzzo[10], Axel Hartke[25], Trevor Hawkins[11], Pascal Hols[12], Robert Hutkins[13], Michiel Kleerebezem[2], Jan Kok[14], Oscar Kuipers[14], Mark Lubbers[15], Emmanuelle Maguin[4], Larry McKay[16], David Mills[17], Arjen Nauta[18], Ross Overbeek[19], Herman Pel[20], David Pridmore[3], Milton Saier[21], Douwe van Sinderen[22], Alexei Sorokin[4], James Steele[23], Daniel O'Sullivan[16], Willem de Vos[2], Bart Weimer[6], Monique Zagorec[24] & Roland Siezen[2,26,*]

[1]*Department of Food Science, Southeast Dairy Foods Research Center, North Carolina State University, Raleigh, NC 27695-7624, USA;* [2] *Wageningen Center for Food Sciences, 6700AN Wageningen, The Netherlands;* [3] *Nestlé Research Center, Vers-chez-les-Blanc, 1000 Lausanne 26, Switzerland;* [4]*Génétique Microbienne, CRJ INRA, Domaine de Vilvert, 78352 Jouy en Josas cedex, France;* [5]*USDA Agricultural Research Service, Department of Food Science, North Carolina State University, Raleigh, NC 27695-7624, USA;* [6] *Department of Nutrition and Food Sciences, Utah State University, Logan, UT 84322-8700, USA;* [7] *Environmental Biotechnology Institute, California Polytechnic State University, San Luis Obispo, CA 93407, USA;* [8] *Laboratoire de Génétique des Microorganismes, INRA-CNRS URA1925, F-78850 Thiverval-Grignon, France;* [9] *Institute of Food Research, Norwich, UK;* [10] *Laboratoire de Microbiologie, UMR INRA-Université de Bourgogne, Equipe PG2MA ENSBANA 21000 Dijon, France;* [11] *Joint Genome Institute Production Genomic Facility, 2800 Mitchell Drive, Walnut Creek, CA 94598, USA;* [12] *Unité de Génétique, Batiment Camoy, Place Croix du Sud, UCL, B-1348, Louvain-la Neuve, Belgium;* [13] *University of Nebraska, Department of Food Science and Technology, Lincoln, NE 68583-0919, USA;* [14] *Department of Molecular Genetics, University of Groningen, 9751NN Haren, The Netherlands;* [15] *Fonterra Research Centre [Formerly the New Zealand Dairy Research Institute], Private Bag 11 029, Palmerston North, New Zealand;* [16] *Department of Food Science and Nutrition,University of Minnesota, St. Paul, MN 55108, USA;* [17] *Department of Viticulture and Enology, University of California, Davis, CA 95616-8749, USA;* [18] *Corporate Research FCDF, PO Box 87, 7400AB Deventer, The Netherlands;* [19] *Interegated Genomics Inc, 2201W Campbell Park Dr, Chicago, IL 600612, USA;* [20] *DSM Food Specialties, PO Box 1, 2600MA Delft, The Netherlands;* [21] *University California, San Diego, La Jolla, CA 92093-0116, USA;* [22] *Department of Microbiology, University College Cork, Cork, Republic of Ireland;* [23] *Department of Food Science, University of Wisconsin-Madison, Madison, WI 53706-1565, USA;* [24] *Flore Lactique et Environnement Carné, INRA-CRJ, Domaine de Vilvert, 78350 Jouy-en-Josas, France;* [25] *Laboratoire de Microbiologie de l'Environnement, EA 956 USC INRA, IRBA Université de Caen, F-14032 Caen Cedex, France;* [26] *Center for Molecular and Biomolecular Informatics, University of Nijmegen, PO Box 9010, 6500GL Nijmegen, The Netherlands (*Authors for correspondence; E-mail: klaenhammer@ncsu.edu, siezen@cmbi.kun.nl)*

Key words: lactic acid bacteria, genomics, Gram-positive bacteria, food, health, *Lactococcus, Lactobacillus, Streptococcus, Pediococcus, Leuconostoc, Oenococcus, Propionibacterium, Bifidobacterium, Brevibacterium*

Abstract

This review summarizes a collection of lactic acid bacteria that are now undergoing genomic sequencing and analysis. Summaries are presented on twenty different species, with each overview discussing the organisms fundamental and practical significance, environmental habitat, and its role in fermentation, bioprocessing, or probiotics. For those projects where genome sequence data were available by March 2002, summaries include a listing of key statistics and interesting genomic features. These efforts will revolutionize our molecular view of Gram-positive bacteria, as up to 15 genomes from the low GC content lactic acid bacteria are expected to be available in the public domain by the end of 2003. Our collective view of the lactic acid bacteria will be fundamentally changed as we rediscover the relationships and capabilities of these organisms through genomics.

Introduction (contributed by Todd Klaenhammer)

The lactic acid bacteria (LAB) represent a group of bacteria that are functionally related by their ability to produce lactic acid during homo- or heterofermentative metabolism. The acidification and enzymatic processes accompanying the growth of LAB impart the key flavor, texture, and preservative qualities to a variety of fermented foods. Industrial applications of the LAB rely on six key beneficial and non-pathogenic species; *Lactococcus* (milk), *Lactobacillus* (milk, meat, vegetables, cereal), *Leuconostoc* (vegetables, milk), *Pediococcus* (vegetables, meat), *Oenococcus oeni* (wine) and *Streptococcus thermophilus* (milk). Other members of the LAB, notably lactobacilli, occupy important niches in the gastrointestinal tracts of humans and animals and are considered to offer a number of probiotic benefits to general health and well being. These benefits include a positive influence on the normal microflora, competitive exclusion of pathogens, and stimulation/modulation of mucosal immunity. More recently, LAB are being used in the production of industrial chemical and biological products including biopolymers (*Leuconostoc* spp.), bulk enzymes (*Lactobacillus brevis*), ethanol, and lactic acid (*Lactobacillus casei, lactis, delbrueckii, brevis*) (Gold et al. 1992; Hofvendahl & Hahn-Hagerdal 2000). The LAB are also strong candidates for development as oral delivery vehicles for digestive enzymes and vaccine antigens (Wells et al. 1996; Pouwels et al. 1998; Steidler et al. 2000). Their innate acid tolerance, ability to survive gastric passage, and safety record during human consumption, are key features that can be exploited to effectively deliver biologics to targeted locations and tissues.

Realizing their practical significance in fermentation, bioprocessing, agriculture, food, and more recently, medicine, the LAB have been the subject of considerable research and commercial development over the past decade. Contributing to this explosion have been the recent efforts to determine the genome sequences of a representative collection of LAB species and strains. The first complete genome of the LAB group was published on *Lactococcus lactis* subsp. *lactis* IL1403 by Bolotin et al. (2001). The genome was 2.4 Mb in size and revealed a number of unexpected findings: biosynthetic pathways for all 20 amino acids, albeit not all are functional, a complete set of late competence genes, five complete prophages, and partial components for aerobic metabolism. Noting that some of these systems are not functional or complete, the genomic analysis of *Lactococcus* suggests an evolutionary trend toward minimization of the chromosome and elimination of unnecessary systems during adaptation to nutritionally complex environments, such as milk. At this writing, four other LAB genomes have been completed, (*L. plantarum, L. johnsonii, L. acidophilus, S. thermophilus*), and >20 more are in progress (Table 1) with expected completions for sequencing by the end of 2002. Additional genome sequencing is underway for microbes that are not considered as members of the LAB, but contribute important LAB-like properties either as probiotics (*Bifidobacterium longum, B. breve, and Brevibacterium linens*) or flavor adjuncts (*B. linens, Propionibacterium freudenreichii*). Among the total genome projects ongoing, there are several cases where genome sequences will become available for multiple strains of the same species, notably *L. lactis* (three strains), *L. casei* (two strains), *L. delbrueckii* (three strains), *S. thermophilus* (three strains), *Oenococcus oeni* (two strains) and *B. longum* (two strains).

Of the 29 genomes listed in Table 1 on LAB-type microbes, 11 genomes are being sequenced by the Department of Energy–Joint Genome Institute (JGI) in collaboration with the Lactic Acid Bacteria Genome Consortium (LABGC), composed of 10 US scientists representing seven universities in the U.S. As part of their microbial genomes program (see http://www.jgi.doe.gov/JGI_microbial/html/index.html) JGI will carry out genome sequencing of LAB species representing considerable diversity in ecological habitat (milk, meat, plants, GI tract) and roles (probiotic versus fermentation). All of the microbial genomes will undergo the same processing; first to generate a 10× coverage of each genome with a random shotgun small-insert library (3 kb), and to supplement this with ∼5× coverage of a large insert (40 kb) cosmid library. These data will be incorporated into an assembly and the assembled scaffolds

Table 1. Genome sequencing projects of lactic acid bacteria

Species	Strain	genome size (Mb)	Contact person	Institution
Sequencing completed				
Lactococcus lactis	IL1403	2.3	A. Sorokin	INRA and Genoscope, FR
Lactobacillus plantarum	WCFS1	3.3	M. Kleerebezem	WCFS, NL
Lactobacillus johnsonii	NCC533	2.0	D.Pridmore	Nestlé, CH
Lactobacillus acidophilus	ATCC700396 NCFM	2.0	T. Klaenhammer, R. Cano	North Carolina State University and Cal-Poly Technical University, USA
Bifidobacterium longum	NCC2705	2.3	F. Arigoni	Nestlé, CH
Sequencing ongoing				
Lactococcus lactis ssp. *cremoris*	MG1363	2.6	O.Kuipers, M.Gasson, D. van Sinderen	University of Groningen, NL; IFR, UK; UCC, IRL
Lactococcus lactis ssp. *cremoris*	MG1363	2.4	A. Sorokin, A. Bolotin	INRA, FR
Lactobacillus delbrueckii ssp. *bulgaricus*	ATCC11842	2.3	E. Maguin, M.vd Guchte	Genoscope and INRA, FR.
Lactobacillus delbrueckii ssp. *bulgaricus*	DN-100107	2.1	T. Smokvina	Danone Vitapole, FR
Lactobacillus sakei	23K	1.9	M. Zagorec, S. Chaillou	INRA, FR.
Lactobacillus casei	BL23	2.6	J. Deutscher, A. Hartke	INRA/CNRS, Caen University, FR
Lactobacillus helveticus	CNRZ32	2.4	J. Steele	University of Wisconsin, USA
Lactobacillus rhamnosus	HN001	2.4	M. Lubbers	Fonterra Research Center (formerly NZDRI) and ViaLactia BioSciences, NZ
Streptococcus thermophilus.	LMG18311	1.9	P. Hols	UCL, Belgium
Streptococcus thermophilus	CNRZ1066	1.8	A.Bolotin	INRA, FR, Integrated Genomics Inc, USA
Oenococcus oeni	IOEB84.13	1.8	J. Guzzo	Universities Dijon and Bordeaux-INRA-GENOME Express, FR
Bifidobacterium breve.	NCIMB8807	2.4	D. van Sinderen	University College Cork (UCC), IRL
Propionibacterium freudenreichii	ATCC6207	2.6	H. Pel, J. Sikkema	DSM Food Specialties, Friesland Coberco Dairy Foods, NL
Joint Genome Institute/Lactic Acid Bacteria Genome Consortium (JGI/LABGC)				
Lactobacillus gasseri	ATCC33323	1.8	T. Klaenhammer	North Carolina State University, USA
Lactobacillus casei	ATCC334	>2.2	J. Broadbent	Utah State University, USA
Lactobacillus delbrueckii ssp. *bulgaricus*	ATCCBAA-365	2.3	J. Steele	University of Wisconsin, Madison, USA
Lactobacillus brevis	ATCC367	2.0	M. Saier	University of California, San Diego, USA
Lactococcus lactis ssp. *cremoris*	SK11	2.3	B. Weimer, L. McKay	Utah State University and University of Minnesota, USA
Leuconostoc mesenteroides	ATCC 8293		F. Breidt, H.P. Fleming	USDA, North Carolina State University, USA
Oenococcus oeni	PSU1 ATCC BAA-331	1.8	D. Mills	University of California, Davis, USA
Pediococcus pentosaceus	ATCC25745	2.0	J. Steele, J. Broadbent	University of Wisconsin, Madison, Utah State University, USA
Streptococcus thermophilus	ATCC BAA-491	1.8	R. Hutkins	University of Nebraska, USA
Brevibacterium linens	BL2/ ATCC 9174	3.0	B. Weimer	Utah State University, USA
Bifidobacterium longum	DJ010A	2.1	D. O'Sullivan	University of Minnesota, USA

ordered by PCR primer walking to fill gaps. The scaffolds will undergo an automated annotation which will be performed by Oak Ridge National Laboratories.

It is noteworthy that as these sequences are generated, they will be placed in the public domain on the JGI website for common use. The timely and public availability of genome information for various LAB species will catapult our collective efforts to carry on with comparative and functional genomic analyses of the LAB group.

The explosion of available genome sequences for LAB will accelerate their exploitation in both traditional and non-traditional arenas. While phylogenetically closely related by their small genomes (~2–4 Mb), the LAB occupy a diverse set of ecological niches suggesting that considerable genetic adaptation has occurred during their evolution. Comparison of the genome sequences of multiple LAB species and strains is expected to provide a critical view of microbial adaptation and genetic events leading to their adaptation to specialized environments. Comparative genomics among the microbes sequenced thus far has already illustrated that essential housekeeping gene functions are widely conserved among microbes and horizontal gene transfer commonly occurs. An expected outcome of comparative genomics of LAB will be the definition of conserved and unique genetic functions in LAB that enable core functions, e.g., production of lactic acid, proteolytic and peptidase activities, survival at low pH , stress tolerance, production of antimicrobials, transport systems, cell signaling, and attachment/retention in dynamically mobile environments.

It is anticipated that IS-elements, bacteriophages, and mobile genetic elements provide the major routes through which horizontal gene transfer occurs, and the roadmap to the most interesting and practically significant genetic regions that underscore the unique and beneficial properties of the LAB. It is well documented that the LAB undergo conjugation, exist in phage-contaminated environments where gene transfer may occur by transduction, and harbor sets or remnants of competence genes for transformation (Bolotin et al. 2001). Understanding gene transfer, particularly in environments where LAB coexist, or compete, will provide one important view of their evolution, adaptation, and potential for unique applications. Conjugation has played a key role in the evolution and adaptation of *L. lactis* to a milk environment noting that many attributes for growth in milk, including lactose and casein utilization and bacteriophage resistance, are encoded by conjugative plasmid DNA in this species (Broadbent 2001).

With the availability of genomic maps of more than 25 LAB, comparative genomic analysis will identify critical similarities and differences and is expected to identify 'islands of adaptability', defined as key genetic regions that may be instrumental in the evolution of the various species to their specialized habitats, and their functions within those environments. Comparisons of food-grade LAB with other related Gram-positive pathogens (*Enterococcus faecalis*, *Streptococcus agalactiae*, *Streptococcus equi*, *Streptococcus pneumoniae*, *Streptococcus mutans*, *Streptococcus pyogenes*, *Listeria monocytogenes*) has already revealed many common features and will most certainly define the essential genetic differences between pathogens, non-pathogens, and commensals. Over the next year, our scientific community will be fundamentally empowered by the availability of numerous related, yet distinct genomes, to make these comparative analyses and define the similarities and differences that characterize the genomes of LAB.

This chapter presents a collection of summaries on the LAB organisms that are now in genomic sequencing or analysis. Each summary describes the organism and its roles in the environment or bioprocessing, the status of the sequencing effort as of March 2002 and, for some, selected interesting features that have been uncovered (Tables 2 and 3). Our view of the LAB will be fundamentally changed as we discover the relationships and capabilities of these organisms through genomics.

SUMMARIES OF SEQUENCING PROJECTS

Lactococcus lactis subsp. *cremoris* SK11 (contributed by Larry L. McKay and Bart Weimer)

Lactococci are mesophilic LAB that were first isolated from green plants. However, today they are used extensively in food fermentations, which represent about 20% of the total economic value of fermented foods produced throughout the world. This group of bacteria, previously designated the lactic streptococci (*Streptococcus lactis* subsp. *lactis* or *S. lactis* subsp. *cremoris*) was placed in this new taxon in 1987 by Schleifer. Lactococci gained notable interest because many of their functions important for successful fermentations are linked to plasmid DNA

Table 2. Genome statistics (March 2002)

Species	Strain	Genome			ORFs	Prophages		IS elements		RNA			Plasmids
		Size (Mb)	Contigs	%GC		Intact	Remnant	Total intact	Different types	rRNA operons	tRNA loci	tRNA number	Size kb (%GC)
L. lactis	IL1403	2.366	1	35.4	2310	5	1	43	6 (IS981, IS982, IS983, IS904, IS905, IS1077)	6		62	None
L. lactis	MG1363	~2.5	458 (>1 kb)	37.1	~2500	0	6	62	8 (IS981, IS982, IS905, IS904, IS946, IS1077G,ISS1, IS1675)	6	?	?	None
L. plantarum	WCFS1	3.308	1	44.5	~3050	2	2	12	2 (IS1165, IS1182)	5	5	62	1.9 (39.5%) 2.4 (34.3%) 36.1 (40.8%)
L.johnsonii	NCC533	2.022	1	34.6	1857	3	0	15	6 (ISL3, IS30)	6	4	87 ?	None
L. acidophilus	ATCC700396	1.99	1	34.7	1979	0	3	6	1 (IS3)	6			None
L. gasseri	ATCC33323	1.84	67	35.1	1803	1	1	3	3(ISL3, IS1397, IS1167)	>2		20	None
L. helveticus	CNRZ32	~2.4	~1000	37.1	> 2,300	?	≥ 1	?	> 5 (IS1201, ISL3 ISLh1, IS3, IS30)	≥3	?	?	None
L. delbrueckii ssp. *bulgaricus*	ATCC11842	1.8	3	50	?	?	?	Several		?	6 or 7	?	None
L. sakei	23K	1.772 (96%)	<150	41.4	>1792	?	?	12	3 (IS1520-types, new IS30-like, new IS150-like)	7	6	>58	None
L. casei	BL23	~ 2.6	563	46	?	?	Several	18	5 (IS1, IS3, IS30)	≥5	?	?	None
O. oeni	IOEB84.13	1.754	63	37.9	1784	4		3		2	2	43	3.9
P.freudenreichii	ATCC6207	2.64	12	67.4	2552	?		Many	?	2 (?)	2	44	None
B. breve	NCIMB8807 ATCC9174	~ 2.43	140	~ 53	?	0		~ 35	(IS2606,IS1383, IS1088,IS10,IS21)	~4	?	?	None
B. longum	NCC2705	2.256	1	60.1	~1730		Some?	16	5(IS3,IS21,IS30, IS256,IS200/IS605)	4	3	57	3.6 (64.8%)

Table 3. Encoded features

		L.lactis	*L.lactis*	*L. plantarum*	*L. rhamnosus*	*L. johnsonii*	*L. acidophilus*	*L. gasseri*	*L. helveticus*	*L. casei*	*B. longum*
strain		IL1403	MG1363	WCFS1	HN001	NCC533	ATCC700396	ATCC33323	CNRZ32	BL23	NCC2705
Biosynthesis pathways (de novo)											
Amino acids		all	all except Gln/Glu, Leu, Val, Ile, Met, His	all except Ile, Leu, Val	all except Ala, Leu, Lle, Val, Phe, Tyr	None	?	?	?	all except Ile, Leu, Val	all except Met and maybe Trp
Vitamins, cofactors		folate, menaquinone, riboflavin, thioredoxin	?	folate, molybdenum cofactor	folate, thiamin, heme		?	?	?	?	folate, nicotinate, thiamin
Nucleotides	Purines	present	present	present	present	absent	?	?	?	present	present
	Pyrimidines	present	present	present	present	present				present	present
Polysaccharides	gene clusters	1 ?	1 (?)	4	?	1	?	>2	1	?	1?
Bacteriocins	type	none	none	plantaricin	none ?	lactacin F	helveticin J	helveticin J, lactacin B	helveticin J-like	?	none
Bile salt hydrolases	number	2 ?	1 ?	4	none	3	3	2	1	1	1
Proteolytic system											
Cell-envelope proteinase		0	0	0	1	1	1	0	> 2	?	0
Peptidases		13	~14	19	29	25	>20	19	> 18	21	>20
Sugar transport											
PTS complete (EIIA+B+C)		6	?	25	?	?	?	9	> 6	31	1
Secreted proteins											
Lipoproteins		39	~30	~50	?	~40	?	?	?	~45	59
LP×TG-anchor proteins		6	~4	~25	~8	~15	?	?	?	?	8

(McKay 1985). Plasmids are commonly exchanged between strains via conjugation (McKay 1985; Dunny & McKay 1999) and with the chromosome by IS elements (Hughes 2000). Presumably, these exchanges and rearrangements mediate rapid strain adaptation and evolution but also add to the instability of important metabolic functions.

These bacteria are selected for use in fermentations based on their metabolic stability, their resistance to bacteriophage, and their ability to produce unique compounds – often from amino acid catabolism. The study of their physiology in adverse conditions such as low pH and high NaCl indicates that they adapt to these environments quickly and change their metabolism as a result of carbohydrate starvation (Stuart & Weimer 1998). Recent genome studies and physical maps indicate that bacterial genomes are very dynamic (Hughes 2000). These rearrangements are mediated by IS elements and result in gene duplication, translocation, inversion, deletion and horizontal transfer events. For example, an inversion encompassing approximately one-half of the chromosome of *L. lactis* ML3 occurred by homologous recombination between two copies of IS*905* (Daveran-Mingot et al. 1998). The response to these stresses, particularly to exposure to bacteriophage (Forde & Fitzgerald 1999), highlights the plasticity of the genome (Le Bourgeois et al. 1995; Delorme et al. 1994; Davidson et al. 1996). Establishing the links between environmental conditions, genome organization, and cellular physiology in lactococci will provide new and exciting information about the molecular mechanisms of these important bacteria. Advances that define the fundamental knowledge of the genetics, molecular biology, physiology, and biochemistry of lactococci will provide new insights and applications for these bacteria.

The importance of lactococci, specifically *L. lactis* subsp. *cremoris,* is demonstrated by its continual use in food fermentations (Garvie et al. 1981; Beimfohr et al. 1997). *L. lactis* subsp. *cremoris* strains are preferred over *L. lactis* subsp. *lactis* strains because of their superior contribution to product flavor via unique metabolic mechanisms (Sandine, 1988; Salama et al. 1991). The DNA sequence divergence between the subspecies is estimated to be between 20 and 30% (Godon et al. 1992). Of the many lactococcal strains used, *L. lactis* subsp. *cremoris* SK11 is recognized for the beneficial flavor compounds it produces (Lawrence et al. 1976). Although some progress in unlocking this strain's genetic secrets has been made, much more can be accomplished by using a genomics/proteomics approach. With this genome sequence, it will be possible to confirm the metabolic and evolutionary differences between subspecies of lactococci in order to identify the important characteristics that define this genus.

Acknowledgments

The Joint Genome Institute of the US Department of Energy is supporting the sequencing effort for this LAB genome. The efforts of K. Baldwin, L.-S. Chou, Y. Xie, T. Hawkins, S. Stilwagen, P. Richardson, and K. Kadner are gratefully acknowledged.

Lactococcus lactis ssp. *cremoris* MG1363 (contributed by Douwe van Sinderen, Mike Gasson, Jan Kok and Oscar Kuipers)

Lactococcus lactis ssp. *cremoris* MG1363 is a cured isolate of strain NCDO 712 (Gasson, 1983), and is the most widely used model *Lactococcus* strain in genetic, physiological and applied research all over the world. *L. lactis* is a prominent bacterium in cheese production and can also be used for various other applications such as a vehicle for oral vaccines or delivery of health-promoting factors via the GI tract. Even before the genome sequencing projects started already over 10% of its gene content had been identified and characterized by conventional methods. Recently the complete genome sequence of *L. lactis* ssp. *lactis* IL1403 was published (Bolotin et al. 2001), providing a great opportunity to compare the two genomes, with respect to their similarities and differences. Many important features, such as pathways leading to the formation of flavour compounds, carbon and nitrogen metabolic routes, gene regulation mechanisms, prophage or sex-factor occurrence, and stress responses can be compared when the genome sequence of MG1363 is also known. Functional studies using transcriptome and proteome analyses can also be applied in order to speed up characterisation of important novel targets.

Genome sequencing status and results

To obtain the complete nucleotide sequence of the genome of *Lactococcus lactis* subsp. *cremoris* MG1363 a shotgun sequencing approach was used. The sequence of 15 000 ends of inserts of pUC clones, ranging in size from 1 to 3 kb, have been determined, yielding 9 million bp of sequence. The obtained sequences were aligned with known sequences of MG1363 present in the NCBI nucleotide database (± 270 kb) and sequences present within the consortium (± 150 kb) using the Staden and Sequencer software packages. As the construction of a random library with inserts larger than 20 kb failed, the genome sequence of *L. lactis* subsp. *lactis* IL1403 was used as a template to start completing the genome sequence. After positioning of all the contigs on IL1403, forward and reverse primers were designed using an in-house developed software tool, in the ends of the contigs. The nucleotide sequences of the PCR fragments are currently being used to close the gaps. Annotation of the nucleotide sequence is being done with Glimmer, Blast, PFAM, Blocks.

Acknowledgments

We thank Sinead Leahy, Gerald Fitzgerald, Udo Wegmann, Claire Shearman, Girbe Buist, Richard Baerends, Aldert Zomer, Sacha van Hijum, and Anne de Jong for their contributions and support.

Lactococcus lactis ssp. *cremoris* MG1363 (contributed by Alexei Sorokin and Alexander Bolotin)

Two strains, IL1403 and MG1363, isolated from industrial sources and cured from plasmids, are commonly used to study genetics, physiology and molecular biology of *Lactococcus lactis*. These strains belong to two *L. lactis* strain clusters, called *lactis* and *cremoris*, distinct by the spectra of dairy fermentation products. We sequenced the entire genome of the strain IL1403 (Bolotin et al. 1999, 2001). To compare genomes of the two strains in the simplest way we decided to perform a co-linear scaffolding of the chromosome of strain MG1363. Co-linear scaffolding consists in random sequencing of a limited number of clones and Long Range PCR mapping of the co-linear regions of two genomes, one of which was completely sequenced earlier. In the co-linear regions, established by PCR, gene content and order should be approximately the same. The established co-linear scaffold of the MG1363 genome, which shares 85% identity with that of IL1403, allows easy access to 50% of the genes from this strain.

Genome sequencing status and results

Total DNA of *L. lactis* MG1363 was randomized by *Alu*I in conditions of non-complete digestion or by sonication and extracted after separation in an 0.8% agarose gel to give an average fragment size of 1.5 kb. It was then cloned into vector pSGMU2 cut by SmaI and the inserts from approx. 1500 random clones were sequenced by forward and reverse primers. Sequences of 513 of these clones, having a total non-redundant length of 0.3 Mb, were used to construct a co-linear scaffold of the strain MG1363 over IL1403. The total size of the co-linear genome parts was estimated to be around 1100 kb, that is approximately 45% of the entire genome (Bolotin et al. 2002). Oligonucleotide-directed sequencing over gaps amplified by LR PCR fragments of MG1363 and the entire sequence of a few regions of interest resulted in the accumulation of 1038 kb non-redundant sequence distributed in 733 contigs. The contigs were compared for homology, using BLASTx, with the complete protein set of *L. lactis* IL1403. Sequences of 523 contigs encode proteins that are more than 80% identical to proteins from IL1403. The sequences of the 733 contigs are available from GenBank as dbGSS (acc. Nos. BH770319-BH771051).

Acknowledgments

We thank Saulius Kulakauskas for giving us the *L. lactis* ssp *cremoris* MG1363 strain, Nathalie Galleron and Benois Quinquis for assisting in bank constructions and sequencing. Discussions with members of *Lactococcus* communities of Génetique Microbienne and UPLGA labs (CRJ INRA) and their interest to this work are very appreciated.

Lactobacillus plantarum WCFS1 (contributed by Michiel Kleerebezem and Roland Siezen)

The genus *Lactobacillus* encompasses a large number of different species that display a relatively large degree of diversity. Among these, *Lactobacillus plantarum* is a flexible and versatile species that is encountered in a large variety of environmental niches, including some dairy, and many vegetable or plant fermentations. Some strains of *Lactobacillus* are found as natural commensals of the gastrointestinal tract (GI tract), the oral cavity and the female uro-genital tract of animals and humans. The ability to adhere to specific regions of the GI tract is a property which has stimulated research aimed at the use of lactobacilli as delivery-vehicles for therapeutic compounds such as immunomodulators, antibodies, enzymes and vaccines (Marteau & Rambaud 1993 Hols et al. 1997; Slos et al. 1998). *Lactobacillus plantarum* arrives in the small intestine in an active state (Vesa et al. 2000) and is frequently encountered as a natural inhabitant of the human gastro-intestinal tract (Ahrne et al. 1998). Some strains are being marketed as probiotic. Previous studies using gel electrophoresis have indicated that *L. plantarum* has one of the largest genomes known among lactic acid bacteria (Chevalier et al. 1994; Daniel 1995). The Wageningen Centre for Food Sciences (WCFS) in collaboration with Greenomics in Wageningen, has determined the complete genome sequence of *L. plantarum* WCFS1, a single colony isolate of strain NCIMB8826, a strain of human origin.

Genome sequencing status and results

Sequencing was accomplished using a shotgun approach, while λ-clones and multiplex PCR were used to fill the gaps. The circular chromosome of *L. plantarum* WCFS1 consists of 3 308 274 bp with an average G+C content of 44.5%, and is among the largest of lactic acid bacteria. In addition, the strain harbours three plasmids of 36 069 bp (G+C content: 40.8%), 2365 bp (G+C content: 34.3%), and 1917 bp (G+C content: 39.5%), respectively. Automated prediction software (Glimmer, Genemark) was used to obtain a primary prediction of protein encoding genes, which was manually improved to result in a current list containing about 3050 genes.

The orientation of the majority of the genes is highly organized from the origin to the terminus of replication for both halves of the chromosome. Five rRNA-encoding operons, with very few sequence polymorphisms, could be identified scattered around the chromosome. Automated ORF analysis and functional annotation was performed using BioScout (LION) and Pedant-Pro (BioMax) software packages, followed by extensive manual curation. Functions were predicted for over 65% of the gene products. Major categories are proteins involved in energy metabolism (8%), cell envelope (8%), transport (13%) and regulation (9%). Over 2300 encoded proteins have expect scores better than 1e-10 to database entries, and nearly 90% of these are most similar to proteins of other Gram-positive bacteria, predominantly those encoded in the genomes of *Listeria, Lactococcus, Bacillus* and *Streptococcus*.

L. plantarum is a versatile and flexible organism and is able to grow on a wide variety of sugar sources. This phenotypic trait is reflected by the high number of predicted PEP-dependent sugar phosphotransferase systems (PTS; 25 complete) and other sugar transporters encoded by WCFS1. Strikingly, over 60% of these sugar transporters are located within 250 kb in both directions from the origin of replication.

Acknowledgments

We thank Jos Boekhorst, Bjorn Ursing, Richard van Kranenburg, Douwe Molenaar and Willem de Vos for their contribution to this work. Sequencing and assembly were performed by Greenomics (www.greenomics.com).

Lactobacillus rhamnosus HN001 (contributed by James Dekker and Mark Lubbers)

Lactobacillus rhamnosus is one of the few species of *Lactobacillus* that have been used as probiotic organisms in functional foods. In addition, this species is one of the two most common non-starter lactic acid bacteria found in New Zealand cheddar cheese; the other is *L. paracasei*. We have identified a strain of *L. rhamnosus*, designated HN001, that has both flavour-enhancing and probiotic attributes. It can be used as an adjunct during cheese manufacture to reduce adventitious microflora, accelerate cheese ripening, and improve cheese flavour. We have also demonstrated that HN001 has the 'prerequisite' properties of a probiotic, including confirmation of taxonomic classification, acid and bile resistance, adherance to intestinal cells, transient colonisation, and lack of any toxicity or detrimental effects (Prasad et al. 1998; Zhou et al. 2000a,b, 2001; Gopal et al. 2001; Sheih et al. 2001). A number of *in vitro* and *in vivo* tests, including and human clinical trials, have demonstrated the ability of HN001 to influence specific cytokines, NK cell activity, T cell stimulatory capacity and phagocytic ability, and influence the balance of the intestinal microflora in humans (Gill et al. 2000, 2001a; Tannock et al. 2000). HN001 also has antimicrobial activity against important gut pathogens (Gill et al. 2001b). *L. rhamnosus* HN001 is genetically accessible and we have developed tools and techniques for efficient gene disruption and over-expression. Therefore, HN001 is an ideal candidate for an in-depth study of factors that contribute to probiotic and flavour-enhancing attributes of lactobacilli.

Genome sequencing status and results

A draft genome sequence of HN001 was obtained by shotgun sequencing. Analysis of the data indicates a genome size of about 2.4 Mb, and an average G+C content of approximately 46.4%, which is within the range reported for other *Lb. rhamnosus* species. In addition, HN001 harbours two plasmids of 8754 bp (G+C content: 41.9%) and approximately 40 000 bp (G+C content: 43.7%). Automated ORF prediction and annotation were not used. Rather, the whole genome sequence was compared against publicly available databases and the output manually examined for genes of specific interest. In addition, metabolic pathways of particular interest were also identified. To date, over 900 candidate genes have been examined and almost 300 genes with potential involvement in flavour, survival and probiotic activity (e.g. gut adhesion, immune and anti-microbial activity) have been identified. Of these, 39 HN001 genes have been functionally characterised using a number of approaches, including gene knockout by disruption, overexpression, gene complementation and biochemical assay of purified recombinant protein.

Acknowledgments

We acknowledge the valuable contributes of Michael Collett, Christopher Pillidge, and Marie-Laure Delabre, and the financial support of our sponsors ViaLactia BioSciences NZ Ltd, the New Zealand Foundation for Research Science and Technology, and the Marsden Fund of the Royal Society of New Zealand.

Lactobacillus johnsonii La1 (NCC2761) (contributed by David Pridmore)

Lactobacillus johnsonii strains have been mainly isolated from the feces of humans and animals (Johnson et al. 1980; Fujisawa et al. 1992), suggesting that these bacteria constitute part of the natural intestinal flora. *L. johnsonii* La1 (formerly *Lactobacillus acidophilus* La1) is a Nestlé proprietary strain that has been extensively studied for its probiotic properties and is commercialised in the LC[1] fermented milk products. La1 shows immunomodulatory properties (Link-Amster et al. 1994; Schiffrin et al. 1995; Haller et al. 2000a,b), antipathogenic properties (Bernet et al. 1994; Bernet-Camard et al. 1997; Felley et al. 2001; Pérez et al. 2001) and the ability to interact with the host (Granato et al. 1999; Neeser et al. 2000). The determination and analysis of the complete genome sequence of *L. johnsonii* La1 has provided valuable targets for the investigation of the above interactions. *L. johnsonii* La1 is amenable to genetic manipulation, an important consideration in view of the wealth of information revealed in the genome sequence and the interest to investigate the probiotic features of this bacterium at the genetic level. Efficient transformation by plasmids, the ability to use pG+host9 and its temperature sensitive replication to produce genetic disruption of selected genes and the application of pG+host9:ISS*1* (Maguin et al. 1996) to produce random mutations are some of the tools available, making *L. johnsonii* La1 an important host for advanced genetic analysis.

Genome sequencing status and results

Random shotgun cloning and sequencing was supplemented with sequencing of BAC clones and long-range PCR products to finally produce a circular chromosome of approximately 2.022 Mbp. The computer-predicted *Sma*I digest closely matches the physical digest pattern. Bioinformatic analysis has identified the *L. johnsonii* La1 counterparts of the best-studied genes and pathways. The initial analysis of these results has not revealed any major surprises, except for the complete lack of amino acid biosynthetic pathways which appears to be compensated for by an increased number of amino acid transporters. Given the limited literature describing the natural habitat of this bacterium, a more profound analysis of the genetic potential of *L. johnsonii* La1 could provide important clues as to the nutritional composition of its environment. The structure of the *L. johnsonii* La1 chromosome is unusual in that both the plot of GC skew and the direction of the transcription of the majority of genes do not place the proposed terminus opposite the origin (approximately 1000 kb), but close to 1300 kb. This chromosome structure may be explained by the structure, orientation and position of two genes encoding large cell-wall anchored proteins in *L. johnsonii* La1. Both genes predict proteins with more than 100 copies of a conserved 10 amino acid repeat. This repeat is reflected in the gene sequences and may be the target for recombination. Inversion of the approximately 600 kb of sequence between the repeat would position the terminus opposite the origin. A second example of *L. johnsonii* La1 genome rearrangement has been observed by defining the direct repeats produced by IS element transposition. One pair of elements showed a perfect exchange of the direct repeat sequences. Confirmed by PCR, this event has produced an inversion of approximately 800 kb of the *L. johnsonii* La1 genome. Extension of this analysis to other *L. johnsonii* isolates has allowed the identification of some of the intermediates in the events leading to the *L. johnsonii* La1 genome configuration.

Acknowledgments

Anne-Cecile Pittet, Marie-Camille Zwahlen, Bernard Berger, Mark Schell, David Vilanova, Maria Karmirantzou, Frank Desiere and Josef Hermanns are sincerely thanked for their contribution to this work.

Lactobacillus acidophilus ATCC700396 / NCFM (contributed by Todd R. Klaenhammer and Raul Cano)

In 1900, Moro first isolated *Lactobacillus acidophilus* from infant feces. Meaning 'acid loving', *acidophilus* can be found in the intestinal tract of humans and animals, as well as infants consuming high milk, lactose, or dextran diets. Metchnikoff's 1906 work on *The prolongation of life: optimistic studies*, implicated a lactic acid bacillus, found in Bulgarian yoghurts, as the agent responsible for deterring intestinal putrefaction and aging. Later, *Lactobacillus acidophilus* was considered to be the most likely species to fulfill the base criteria expected of a probiotic cultures; survival through the GI tract, bile tolerance, acid tolerance, and antimicrobial production. Over the last century a considerable amount of research has been carried out on *L. acidophilus*, and a group of closely-related species clustered in a group known as the 'acidophilus complex' (Klaenhammer & Russell, 1999). Of the six species in the group, *L. acidophilus* continues to be the bacterium most often implicated in providing probiotic benefits and remains to be the species most commonly found in foods or supplements that contain probiotic cultures. In a recent survey, Clark et al. (2001) found that when *L. acidophilus* was recovered from probiotics samples, the majority of confirmed strains showed a genetic fingerprint resembling that of NCFM. The *L. acidophilus* NCFM culture has been used commercially since 1972 in a variety of products and probiotic dairy foods, like Sweet Acidophilus^{TM} milk. Its long history of safe use in commercial products for and human consumption, and its close genetic relationship to the neotype strain, ATCC4356 (Walker et al. 1996), were two major reasons that the *L. acidophilus* NCFM strain (deposited as ATCC700396) was selected for genome sequencing.

L. acidophilus is an obligate homofermenter and metabolizes hexoses primarily to lactic acid. Both the D- and L-isomers of lactate are produced. *L. acidophilus* NCFM metabolizes fructo-oligosaccharides (Kaplin & Hutkins 2000), which are prebiotic compounds that support the growth of beneficial gut bacterial. Some of the functional properties of this organism have been reviewed (Sanders & Klaenhammer 2001) and include: survival through the GI tract (Conway et al. 1982), reduction of fecal mutagenic enzymes (Goldin & Gorbach 1980; Goldin et al. 1980), adherence to intestinal tissues (Conway et al. 1987; Greene & Klaenhammer 1994), stimulation of IgA in mice by a culture cocktail that includes *L. acidophilus* NCFM (Tejada-Simon & Pestka 1999), and production of the bacteriocin, lactacin B (Barefoot & Klaenhammer 1983). In contrast to the other members of the acidophilus complex, there is considerable information on *L. acidophilus* NCFM related to its transformability by electroporation (Luchansky et al. 1988; Walker et al. 1996) and the development of genetic tools, that will be used for the functional genomic analysis of this species (Kullen & Klaenhammer 1999; Russell & Klaenhammer 2001a,b).

Genome sequencing status and results

Genome sequencing was accomplished using a shotgun random library approach and pUC18. PCR and direct genome sequencing (Fidelity Systems, Inc.) were used to fill gaps and polish selected regions. The genome has been assembled into a single contig and consists of 1.99 Mbp with an average GC content of 34.6%. The genome was automatically annotated by the prediction, database, and annotation software GAMOLA (Global annotation of multiplexed on-site blasted DNA sequences) to identify 1979 ORFs in a coding density of 89.9%. Predicted ORFs were manually verified against the on-site Blast and Pfam databases. Some features encoded in the NCFM genome are six copies of an IS3-type element, two bacteriocins, three distinct regions encoding competence genes, two *ldh* genes, two bile salt hydrolase genes, two operons for fructo-oligosaccharide (FOS) utilization, and no intact prophages. There is also a collection of genome sequence information available in public databases for various strains of the *L. acidophilus* species, including heat shock operons *groELS* and *dnaK*, a sigma factor *ropD*, bacteriocins acidocins 8912 and B, bile salt hydrolase, and β- and phospho-β-galactosidases. Comparison of the NCFM genome sequence with these available database sequences has revealed a number of interesting similarities and marked differences.

Acknowledgments

This project was supported by Dairy Management, Inc, the Southeast Dairy Foods Research Center, the California Dairy Foods Research Center, the Environmental Biotechnology Institute and Rhodia, Inc. The efforts of Eric Altermann, Mike Russell, A. Hamrick, T. Cox, and D. Doherty are gratefully acknowledged.

Lactobacillus gasseri ATCC33323, neotype (contributed by Todd R. Klaenhammer, Eric Altermann and Rodolphe Barrangou)

Recent developments in molecular taxonomy have revealed six different *Lactobacillus* species that comprise the 'acidophilus' group, which are widely considered as model probiotic organisms (Klaenhammer & Russell 2000). Among these, *Lactobacillus gasseri* appears to represent the major homofermentative *Lactobacillus* species that occupies the human GI tract (Mitsuoka, 1992; Kullen et al. 2000; Heilig et al. 2002). *L. gasseri* demonstrates good survival in the GI tract (Conway et al. 1982; Pedrosa et al. 1995) and has been associated with a variety of probiotic activities and roles including reduction of fecal mutagenic enzymes (Pedrosa et al. 1995), adherence to intestinal tissues (Conway et al. 1987; Greene & Klaenhammer 1994), stimulation of macrophages (Kitazawa et al. 1994; Kirjavainen et al. 1999; Kitazawa et al. 1994), and production of bacteriocins (Itoh et al. 1995).

Data are available on the transformability by electroporation (Russell & Klaenhammer 2001a,b), and expression of heterologous genes in *L. gasseri* (Cho et al. 2000), as well as on the genome sequence of one of its temperate bacteriophage (Altermann et al. 1999). *L. gasseri* is more amenable to DNA introduction and manipulation than other members of the acidophilus complex, leading to the development of genetic tools that will be essential to functional genomic analysis of this species (Kullen & Klaenhammer 1999; Russell & Klaenhammer 2001a,b).

Genome sequencing status and results

Genome sequencing of the *L. gasseri* neotype strain, ATCC33323, was carried out by the Department of Energy–Joint Genome Institute in collaboration with the Lactic Acid Bacterial Genomics Consortium. The strategy for genome sequencing was based on a shotgun random sequencing of a small genomic DNA insert (average 2.5 kb) library cloned into pUC18 at approximately 8-fold coverage, complimented by sequencing a large insert (35–45 kb) Fosmid library at a lower coverage. Gap closing is ongoing using a multiplex PCR approach. The genome assembly of *L. gasseri* currently consists of 67 contigs, representing a total of 1.84 Mbp, with an average G+C content of 35.1%. The sequence information is posted on the JGI Microbial Genomes web site at: (http://www.jgi.doe.gov/JGI_microbial/html/ lactobacillus_gas/lactob_gas_homepage.html) GAMOLA (*G*lobal *A*nnotation of *O*n-site *B*asted DN*A*-sequences) software was developed by E. Altermann to automatically predict coding open reading frames (ORFs) and annotate the draft sequence (http://www.cals.ncsu.edu/food_science/trk/main.html). GAMOLA relies on the available software Glimmer2 (Delcher et al. 1999), NCBI toolkit (National Center for Biotechnology Information Bldg 38A, NIH 8600 Rockville Pike Bethesda, MD 20894), Primer3 (Rozen & Skaletsky, 1997) and HMMER2.2g (Eddy 1998) and combines the results into functionally annotated DNA sequences in Genbank format.

For the *L. gasseri* sequence, the automated annotation predicted 1803 ORFs larger than 100 bp. The average gene length is 930 bp with a coding percentage of 89.1%. After the initial automated annotation, the genome was predicted to encode: 31 ABC transporters, 14 ion-ATPase transporters and two proton-antiporters; 21 PTS-systems (some partial), 16 stress responsive genes (sigma factor, chaperones, heat shock operons and chaperones), one intact prophage, 14 mobile elements (transposases, integrases and IS-elements), 32 genes involved in cell-cycle or cell-shape, eight genes implied in aggregation/adhesion, seven genes related to late competence, 40 peptidases (proteolytic system plus general peptidases), 20 tRNA genes, and at least two rRNA operons. Manual curation of the automated annotation will be carried out upon completion of genome sequencing and closure of the major gaps.

This project has revealed a number of gene systems that are likely to be important in the gastrointestinal survival and activity of this human probiotic species. The public availability of the *L. gasseri* genome will promote comparative genomic analysis among LAB.

Acknowledgment

The Joint Genome Institute of the US Department of Energy and the Lactic Acid Bacteria Genome Consortium is supporting this sequencing project. The efforts of R. Barrangou, E. Altermann, W. M. Russell, T. Hawkins, S. Stilwagen, P. Richardson, and K. Kadner are gratefully acknowledged. Additional support was provided by Dairy Management, Inc, the Southeast Dairy Foods Research Center and Rhodia, Inc.

Lactobacillus helveticus CNRZ32
(contributed by: Jim Steele and Jeff Broadbent)

Within the genus *Lactobacillus, L. helveticus* is part of the obligately homofermentative ('Group I') cluster, which can produce D- and L-lactic acid from hexose sugars via the Embden-Meyerhof pathway and is incapable of fermenting pentoses (Axelsson 1998). *L. helveticus* grows on a relatively restricted number of carbohydrates that includes lactose and galactose and typically requires riboflavin, pantothenic acid and pyridoxal for growth (Hammes & Vogel 1995).

Phylogenetically, *L. helveticus* is quite closely related (<10% sequence divergence) to *L. amylovorus*, *L. acidophilus*, *L. delbrueckii*, *L. acetotolerans*, *L. gasseri*, and *L. amylophilus*. Among these species, *L. helveticus* is most closely related (<2% sequence divergence) to *L. amylovorus* and *L. acidophilus* (Schleifer & Ludwig 1995). The genome size of *L. helveticus* has been determined to be 2.4 Mb by pulse-field gel electrophoresis (unpublished data). Approximately 40 chromosomal genes and four plasmids have been sequenced from *L. helveticus*. Methods for electroporation and gene replacement have been developed for *L. helveticus* CNRZ32 (Bhowmik & Steele 1993; Bhowmik et al. 1993); however, attempts to utilize these methods with other strains of *L. helveticus* have been unsuccessful.

L. helveticus is a component of 'thermophilic' starter cultures used in the manufacture of a number of fermented dairy products (Hassan & Frank, 2001). These cultures have an optimum growth temperature of approximately 42 °C and contain *Streptococcus thermophilus* with *L. delbrueckii* subsp. *bulgaricus* and/or *L. helveticus*. These cultures are utilized in the manufacture of Swiss-type and Italian-type cheese varieties. There has been a 325% increase in the past 20 years in the production of Mozzarella cheese in the United States to more than 2244 million pounds, with an economic value of nearly $1 billion (National Cheese Institute 1998). Additionally, strains of *L. helveticus* are commonly used as flavor adjunct cultures, where they are added to bacterial ripened cheese varieties to reduce bitterness and accelerate the development of beneficial flavors.

To have utility in the manufacture of fermented dairy products strains must be resistant to bacteriophage, have stable fermentation properties, and consistently produce products with acceptable flavor and texture attributes. A significant body of research is available concerning the proteolytic system of these organisms, with at least 15 components of the proteolytic system characterized at the nucleotide level and numerous single and double mutants lacking specific components studied in detail (Christensen et al. 1999). Limited information is available concerning a number of other industrially relevant metabolic pathways (i.e., carbohydrate fermentation) and enzymes (i.e., esterases). Genomic sequence analysis of *L. helveticus* will allow researchers to fill the significant gaps present in our understanding of the physiology of this organism by providing a comprehensive view of the enzymes and metabolic pathways potentially involved in industrially relevant phenotypes. This knowledge will allow researchers to develop more effective strategies to enhance the utility of these organisms in the manufacture Swiss-type and Italian-type cheese varieties as well as their utility as flavor adjunct cultures.

Acknowledgment

This project was supported by Dairy Management, Inc. through the Wisconsin Center for Dairy Research and the Western Dairy Center, College of Agricultural and Life Sciences and Chr. Hansen, Inc. Fred Blattner and George F. Mayhew from the Genome Center of Wisconsin are thanked for their assistance with genome assembly and annotation.

Lactobacillus delbrueckii subsp. *bulgaricus* ATCC BAA-365 (contributed by James Steele)

Within the genus *Lactobacillus, L. delbrueckii* is part of the obligately homofermentative ('Group I') cluster, which can produce D-lactic acid from hexose sugars via the Embden–Meyerhof pathway and is incapable of fermenting pentoses (Axelsson 1998). The *L. delbrueckii* species contains three subspecies, *L. delbrueckii* subsp. *delbrueckii*, *L. delbrueckii* subsp. *lactis*, and *L. delbrueckii* subsp. *bulgaricus*. *L. delbrueckii* subsp. *bulgaricus* grows on a relatively restricted number of carbohydrates and typically requires pantothenic acid and niacin (Hammes & Vogel 1995).

Phylogenetically, *L. delbrueckii* subsp. *bulgaricus* is closely related (<10% sequence divergence) to *L. amylovorus*, *L. acidophilus*, *L. helveticus*, *L. acetotolerans*, *L. gasseri*, and *L. amylophilus* (Schleifer & Ludwig 1995). The GC ratio of *L. delbrueckii* subsp. b*ulgaricus* (49–51%) is somewhat higher than that found among other species (34–46%) within this phylogenetic tree (Hammes & Vogel 1995). The genome size of *L. delbrueckii* subsp. *bulgaricus* has been determined to be 2.3 Mb by pulse-field gel electrophoresis (Leong-Morgenthaler 1990). Very few chromosomal genes (<15) have been sequenced from *L. delbrueckii* subsp. *bulgaricus*, however the complete sequence of a small cryptic plasmid and the partial sequence of a bacteriophage are known. Gene transfer systems for *L. delbrueckii* subsp. *bulgaricus* include two conjugation-based gene transfer systems (Rantsiou et al. 1999; Thompson et al. 1999) and electroporation (Serror et al. 2002).

L. delbrueckii subsp. *bulgaricus* is a component of 'thermophillic' starter cultures used in the manufacture of a number of fermented dairy products (Hassan & Frank 2001). These cultures have an optimum growth temperature of approximately 42 °C and contain *Streptococcus thermophilus* with *L. delbrueckii* subsp. *bulgaricus* and/or *L. helveticus*. These cultures are utilized in the yogurt, Swiss-type and Italian-type cheese varieties. There has been a 240% increase in the past 20 years in the production of yogurt in the United States to 1371 million pounds in 1997, with a wholesale value of over $1.1 billion (Milk Industry Foundation 1998).

Acknowledgments

The Joint Genome Institute of the US Department of Energy is supporting the sequencing effort for this LAB genome. The efforts of T. Hawkins, S. Stilwagen, P. Richardson, and K. Kadner are gratefully acknowledged.

Lactobacillus delbrueckii ssp. *bulgaricus* ATCC11842 (contributed by Maarten van de Guchte and Emanuelle Maguin)

Lactobacillus delbrueckii ssp. *bulgaricus* is a thermophilic lactic acid bacterium that is principally known for its use in yogurt production, where it assures milk fermentation in conjunction with *Streptococcus thermophilus*. As such, *L. bulgaricus* is one of the economically most important LAB.

For the type strain ATCC11842 a reproducible transformation procedure is available (Serror et al. 2002) as well as gene inactivation systems (unpublished) that will allow the exploitation of genome data. A Proteome reference map has been established (unpublished) which will facilitate and benefit from a genome analysis. In this context, the determination of the complete genome sequence of this strain was undertaken in a joint effort of the Centre National de Séquençage (CNS, Genoscope, Evry Fr) and the Institut National de la Recherche Agronomique (INRA).

Genome sequencing

Although the genome size of *L. bulgaricus* was previously estimated to be 2.3 Mbp, recent pulse field electrophoresis analysis revealed that the genome size of the type strain ATCC11842 was about 1.8 Mbp. The genome sequence has been determined using a shotgun sequencing approach. Contigs were assembled after sequencing of two plasmid libraries with differently sized inserts. Sequences from a mini-BAC library were subsequently used for scaffolding. The scaffold structure was validated by long range PCR, and links between the initial scaffolds were established by multiplex long accurate PCR. The average GC content of the *L. bulgaricus* genome is 50%, while regions that presumably result from horizontal transfer can be detected with GC contents as low as 31%.

Acknowledgments

We thank V. Barbe, S. Oztas, S. Mangenot, C. Robert, R. Eckenberg, R. Dervyn, A-M. Dudez, M. Zagorec, J. Weissenbach, and S.D. Ehrlich for their contribution and support. We thank Alexei Sorokin and Alexander Bolotin for helpful discussions.

Lactobacillus sakei 23K (contributed by Stéphane Chaillou and Monique Zagorec)

Lactobacillus sakei is the predominant LAB found on fresh meat and is widely used as starter in fermented meat products in Western Europe. It is also occasionally found on silage, sourdough and smoked fish. The physiology of *L. sakei* is still poorly understood when compared to dairy LAB. Recent studies seem to indicate that *L. sakei* has developed original metabolic and physiological traits to adapt to a meat environment (for a review see Champomier-Verges et al. 2002). To better understand the ability and fitness of *L. sakei* to grow on meat we have chosen the genomic approach. The *L. sakei* genome sequencing project was launched in year 2000, using strain 23K, isolated from sausage, as a model.

Genome sequencing status and results

A physical and genetic mapping of the chromosome was first established by pulse-field gel electrophoresis after digestion with rare-cutting enzymes (Dudez et al. 2002). This preliminary map allowed to estimate the chromosome size to be about 1.85 Mb. Seven *rrn* operons including one doublet were identified and their flanking regions characterised. The sequencing of 90% of the genome was accomplished by a classical shotgun strategy, where 8000 clones were first sequenced on their forward side, followed by reads on the reverse side for 2000 clones flanking contigs end, and 500 internal reads. The gap closure phase was carried out using ligation-mediated PCR and several customised BLAST strategies against related genomes such as *Bacillus subtilis*, *Bacillus halodurans* and *Listeria monocytogenes*. Remaining gaps are currently analysed with a set of other strategies such as multiplex PCR, sequencing on chromosomal DNA, and contigs ordering based on the physical map. Currently, 96% of the genome has been sequenced with an average base redundancy of 4×. A preliminary analysis of the genome content was carried out with the CDS prediction software SHOW (Nicolas et al. 2002) revealing 1792 putative protein-encoding genes for 1772 kp sequenced. Automated annotation followed by manual curation will be done with a improved version of Artemis software from the Sanger Center, UK.

Acknowledgments

This project was financed by the Institut National de la Recherche Agronomique (INRA).

Lactobacillus casei ATCC 334 (contributed by Jeffrey Broadbent)

Lactobacillus casei are rod-shaped LAB with a G + C content of 45–47% (Kandler & Weiss, 1986). Within the genus *Lactobacillus*, *L. casei* is clustered with facultatively heterofermentative ('Group II') species which produce lactic acid from hexose sugars via the Embden–Meyerhof pathway and from pentoses by the 6-phosphogluconate/phosphoketolase pathway (Axelsson, 1998). Growth occurs at 15 °C but not 45 °C, and requires riboflavin, folic acid, calcium pantothenate, and niacin (Kandler & Weiss 1986). Previously, four subspecies of *L. casei* were recognized: *L. casei* subsp. *casei*, *L. casei* subsp. *pseudoplantarum*, *L. casei* subsp. *rhamnosus*, and *L. casei* subsp. *tolerans* (Kandler & Weiss 1986). However, recent phylogenetic studies have led to proposals that members of the *L. casei* group be divided into three species: *L. rhamnosus*, *L. zeae* and *L. casei*, with *L. casei* ATCC 334 as the neotype strain for the latter species (Collins et al. 1989; Dellaglio et al. 1991; Dicks et al. 1996; Mori et al. 1997; Chen et al. 2000).

Members of the *L. casei* cluster are remarkably adaptive, and may be isolated from raw and fermented dairy products, fresh and fermented plant products, and the reproductive and intestinal tracts of humans and other animals (Kandler & Weiss 1986). Industrially, *L. casei* have application as human probiotics (health-promoting live cultures), as acid-producing starter cultures for milk fermentation, and as specialty cultures for the intensification and acceleration of flavor development in certain bacterial-ripened cheese varieties (Kosikowski 1982; Fox et al. 1998; Fonden et al. 2000).

L. casei ATCC 334 was originally isolated as an adventitious contaminant from Emmental cheese. This bacterium, which has an estimated genome size of

2.2 Mb (Ferrero et al. 1996; Tynkkynen et al. 1999), was selected for genome sequencing because: (a) it was isolated from ripening cheese; (b) it is amenable to transformation and genetic manipulation; and (c) it is the proposed type strain for *L. casei*. Sequence analysis of the *L. casei* ATCC 334 genome is expected to provide a comprehensive view of the enzymes and metabolic pathways that are potential contributors to cheese flavor development. In addition, improved knowledge of global gene regulation and integrative metabolism in *L. casei* will help answer long-standing questions regarding mechanisms for the health-promoting benefits of LAB, identify means by which LAB species grow in harsh environments , highlight the most rational strategies for metabolic and genetic improvements to industrial strains, and improve molecular biology resources for genetic manipulation of dairy lactobacilli.

Acknowledgment

The Joint Genome Institute of the US Department of Energy is supporting the sequencing effort for this LAB genome. The efforts of B. McManus, T. Hawkins, S. Stilwagen, P. Richardson, and K. Kadner are gratefully acknowledged.

Lactobacillus casei BL23 (contributed by Josef Deutscher and Axel Hartke)

Lactobacillus casei is a heterofermentative organism widely used in milk fermentation and is therefore of great biotechnological importance. In addition, it belongs to the few bacteria for which probiotic effects have been unequivocally established (for a critical review on probiotics see Marchand & Vandenplas 2000). 'Healthy' effects of this organism have been observed in numerous studies with men and animals. For example, *Lb. casei*, which survives transit through the gastrointestinal tract (Yuki et al. 1999), has been shown to positively affect the immune response (Matsuzaki & Chin 2000; Yasui et al. 1999), to enhance the resistance towards certain forms of diarrhea and to prevent infections by several pathogenic organisms (Wagner et al. 2000; Alvarez et al. 2001). Genetic tools, which allow the transformation of many *Lb. casei* strains and the construction of predetermined chromosomal mutants carrying deletions or point mutations (Dossonnet et al. 2000; Viana et al. 2000), are available, which together with its probiotic features makes this organism a very interesting subject for further genetic and functional genomic studies.

Genome sequencing status and results

The genome size of *Lb. casei* BL23 has previously been estimated to be 2.3 Mb, but recent pulsed-field gel electrophoresis analysis (in collaboration with the laboratory of M. Zagorec) revealed a genome size between 2.5 and 2.7 Mb. In order to sequence the genome of *Lb. casei* BL23 we used a shotgun approach (collaboration between the Laboratoire de Génétique des Microorganismes, INRA-CNRS, Thiverval-Grignon and the Laboratoire Microbiologie de l'Environnement, Université de CAEN). Sequencing *Lb. casei* DNA fragments with a mean size of 1.3 kb from both ends, present in about 8000 isolated clones, allowed us to determine about 90% of the complete genome. We presently attempt to close the about 560 remaining gaps. *Lb. casei* BL23 has been cured of plasmid pLZ15 and therefore contains only a single circular chromosome.

Lb. casei seems to be a versatile organism capable of adapting to variable growth conditions. Most intriguingly it seems to be able to utilize a large variety of different carbon sources. This is reflected by the relatively large number of predicted phosphoenolpyruvate:sugar phosphotransferase systems (PTS) (see Table 3). In *Lb. casei*, the coordinate action of the various carbon utilization systems was found to be controlled by a protein kinase/phosphatase (HprK/P) implicated in carbon catabolite repression and inducer exclusion (Dossonnet et al. 2000). Inactivation of only the phosphatase activity of this bifunctional enzyme had a drastic effect on the growth behavior by preventing the utilization of almost any carbohydrate (Monedero et al. 2001).

Acknowledgments

We acknowledge the participation of Yanick Auffray, Alain Mazé, Gregory Boël, Ivan Mijakovic, Jean-Christophe Giard, Jean Marie Laplace, Abdellah Benachour, Alain Rincé, Vianney Pichereau and the financial aide from the INRA, the Ministère de la Recherche and the Conseil Régionale de Basse-Normandie.

Lactobacillus brevis ATCC 367 (contributed by Milton Saier)

Lactobacillus brevis is a heterofermentative bacterium that can be isolated from many different environments. It is frequently used as a starter culture in silage fermentation, sourdough production and LAB-type beer fermentation. In beverages obtained by alcoholic fermentation, lactobacilli may contribute to the quality of the product but may also cause spoilage. Certain *L. brevis* strains are resistant to hop bittering substances such as isohumulone and are able to grow in beer. Their growth changes the turbidity, flavor and aroma of the beer (Richards & Macrae 1964). *L. brevis* strains involved in wine fermentation may produce biogenic amines by decarboxylation of precursor amino acids through the action of substrate-specific enzymes (Moreno-Arribas & Lonvaud-Funel 2001). The ingestion of foods containing high levels of such amines, particularly histamine and tyramine, can lead to several toxicological disturbances (ten Brink et al. 1990; Mariné-Font et al. 1995).

We have studied the involvement of HPr(ser-P) in the regulation of non-PTS permeases in *L. brevis* (Ye et al. 1994, 1995). When provided with an exogenous energy source such as arginine, galactose grown cells of *L. brevis* transport galactose and the non-metabolizable galactose analogue, thiomethyl-β-galactoside (TMG), by a sugar:H^+ symport mechanism (Romano et al. 1987; Djordjevic et al. 2001). *L. brevis* shows very low transformation efficiency by electroporation rendering this bacteria difficult to manipulate. The cryptic plasmid of 14 kb from *L. brevis* has recently been sequenced and will be used in the development of novel vectors and essential genetic tools.

Acknowledgment

The Joint Genome Institute of the US Department of Energy is supporting the sequencing effort for this LAB genome. The efforts of G. Lorca, T. Hawkins, S. Stilwagen, P. Richardson, and K. Kadner are gratefully acknowledged.

Pediococcus pentosaceus ATCC 25745 (contributed by Jeffery Broadbent and James Steele)

Phylogenetically *Pediococcus* and *Lactobacillus* form a super-cluster that can be divided in to two sub-clusters, all species of *Pediococcus* fall within the *Lactobacillus casei–Pediococcus* sub-cluster. Morphologically, pediococci and lactobacilli (rods) are distinct. The formation of tetrads via cell division in two perpendicular directions in a single plane is a distinctive characteristic of pediococci. *Pediococcus* can be described as 'the only acidophilic, homofermentative, LAB that divide alternatively in two perpendicular directions to form tetrads' (Simpson & Taguchi 1995). Lactic acid is produced from hexose sugars via the Embden–Meyerhof pathway and from pentoses by the 6-phosphogluconate/phosphoketolase pathway (Axelsson, 1998).

Strains of *P. pentosaceus* have been reported to contain between three and five resident plasmids (Graham & McKay 1985). Plasmid-linked traits include the ability to ferment raffinose, melibiose, and sucrose, as well as, the production of bacteriocins (Daeschel & Klaenhammer 1985; Gonzalez & Kunka 1986). Plasmids can be conjugally transferred between *Pediococcus* and *Enterococcus*, *Streptococcus*, or *Lactococcus* (Gonzalez & Kunka 1983). Electroporation has been utilized to introduce plasmids into pediococci, including *P. pentosaceus* (Kim et al. 1992; Caldwell 1996).

P. pentosaceus can be isolated from a variety of plant materials and bacterial-ripened cheeses. This organism is used as an acid-producing starter culture in sausage fermentations, cucumber and green bean fermentations, soya milk fermentations, and silage (Simpson & Taguchi 1995). *P. pentosaceus* are also a typical component of the adventitious or non-starter microflora of most cheese varieties during ripening (Beresford et al. 2001). In addition, it has been suggested that this organism may have value as an acid-producing starter culture in the dairy fermentations (Caldwell et al. 1996, 1998).

Genetic studies of *P. pentosaceus* have generated a limited quantity of information on plasmid and chromosomal encoded genes. With only one plasmid and eight unique chromosomal regions sequenced, the vast majority of genes encoding industrially important attributes have yet to be described. Genomic sequence analysis of *P. pentosaceus* genome will help fill key knowledge gaps by providing a comprehensive view of the enzymes and metabolic pathways related to: (1) acid and flavor production in fermented meat and vegetable foods; (2) mechanisms by which by *P. pentosaceus* and other non-starter LAB grow and direct flavor development in ripening cheese; and (3) mechanisms by which *P. pentosaceus* and related LAB spoil wine and other alcoholic beverages.

Acknowledgments

The Joint Genome Institute of the US Department of Energy is supporting the sequencing effort for this LAB genome. The efforts of T. Hawkins, S. Stilwagen, P. Richardson, and K. Kadner are gratefully acknowledged.

***Streptococcus thermophilus* LMD-9 (contributed by Robert Hutkins)**

Research during the past two decades has revealed that *Streptococcus thermophilus* has properties that make it one of the most commercially important of all LAB. *S. thermophilus* is used, along with *Lactobacillus* spp., as a starter culture for the manufacture of several important fermented dairy foods, including yogurt and Mozzarella cheese. It's use has increased significantly during the past two decades as a result of the tremendous increase in consumption of these products. According to USDA statistics, in 1998, more than 2.24 billion pounds Mozzarella cheese and 1.37 billion pounds of yogurt were produced, respectively, with a combined economic value of nearly $5 billion. This increase has led to new demands on the performance and production requirements of such starter cultures. Industrial strains, for example, should be insensitive to bacteriophage, have stable fermentation characteristics, and produce products having consistent flavor and texture properties. Although research on the physiology of *S. thermophilus* has revealed important information on some of these properties, including sugar and protein metabolism, polysaccharide production, and flavor generation, only recently has the genetic basis for many of these traits been determined.

Currently, several traits in *S. thermophilus* have been targeted for strain improvement programs (Delcour et al. 2000). Since bacteriophage are responsible for considerable economic losses during cheese manufacture, efforts are underway to engineer restriction and other phage-resistance systems into commercial strains. Enhancing stability and expression of exopolysaccharides that act as natural thickening agents has also attracted significant attention. Finally, *S. thermophilus* has an important role as a probiotic, alleviating symptoms of lactose intolerance and other gastrointestinal disorders.

The genome of *S. thermophilus* is 1.8 Mb, making it among the smallest genomes of all LAB. Although a moderate thermophile, it is phylogenetically related to the more mesophilic lactococci and has a comparable low G+C ratio (40%). Genes coding for metabolic pathways involved in sugar catabolism (Poolman et al. 1989; Vaughan et al. 2001), protein and peptide utilization (Fernandez-Espla et al. 2000; Garault et al. 2002), polysaccharide production (Almirón-Roig et al. 2000), the stress response system (Perrin et al. 1999), and phage resistance mechanisms (Solow & Somkuti 2000; Burrus 2001) have been sequenced and characterized. Although most strains do not harbor plasmids, other mobile elements have been reported (Guedon et. al. 1995), and techniques for gene transfer and mutagenesis have been developed (Coderre & Somkuti 1999; Baccigalupi et al. 2000).

Acknowledgments

The Joint Genome Institute of the US Department of Energy is supporting the sequencing effort for this LAB genome. The efforts of L. Durso, J. Goh, T. Hawkins, S. Stilwagen, P. Richardson, and K. Kadner are gratefully acknowledged.

***Streptococcus thermophilus* LMG18311 and CNRZ1066 (contributed by Alexander Bolotin, Ross Overbeek and Pascal Hols)**

Streptococcus thermophilus is one of the most economically important LAB used in the manufacture of yogurt and some Swiss- or Italian-type hard cooked cheeses. This Gram-positive, non-sporulating, catalase-negative, facultative-anaerobe coccus produces remarkable quantities of polysaccharides contributing to the rheological properties of fermented products. The comprehensive knowledge of *S. thermophilus* biology is important for the dairy industry but is rather limited (Delcour et al. 2000). The availability of the whole genome sequence of this species would strongly improve the knowledge of its metabolism and pave the way for engineering of new starter cultures and better control of existing fermentation processes.

The Life Science Institute at the Catholic University of Louvain (Belgium), the Laboratory of Genetic Microbiology, Jouy en Josas, INRA (France) and Integrated Genomics Inc, Chicago (USA) joined their efforts in determining the genome sequences of two strains of *S. thermophilus*.

Genome sequencing status and results

The complete sequences were determined by the random shotgun sequencing strategy followed by multiplex PCR described earlier (Bolotin et al. 2001). Sequences of the two strains were determined by construction of two independent sequence datasets containing 20 000 random and 1500 primer directed reads for LMG18311 strain and 28 000 random and 2000 primer directed reads for CNRZ1066 strain. The assembled genomes were analysed using the ERGO bioinformatics suite. Both genomes contain ~1.8 Mb encoding six rRNA, a compete set of tRNAs, several types of IS, and about 1800 open reading frames. They are organized as a single circular chromosome with 39% GC content, and show 95% nucleotide identity. Although the *S. thermophilus* species is characterized as GRAS (Generally Recognized As Safe), several features of its metabolism are similar to that of pathogenic Streptococci. Analysis of the genome has revealed several key aspects of the pathways of carbohydrates, amino acids, nucleotides, polysacharides and lipids metabolism. More than 100 genes involved in the transport of a variety of substrates such as peptides, sugars, metal ions, and cofactors have been identified. The presence of several phage and plasmid related genes indicates the importance of horizontal gene transfer for evolution of *S.thermophilus*. More information can be found at web sites: http://www.biol.ucl.ac.be/gene/genome/, http://spooky.jouy.inra.fr (restricted access), http://www.integratedgenomics.com (restricted access)

Acknowledgments

We thank Benoît Quinquis, Pierre Renault, Alexei Sorokin, and S. Dusko Ehrlich from Génetique Microbienne, INRA; Alla Lapidus, Eugene Goltsman, Mikhail Mazur, Gordon D. Pusch, and Michael Fonstein from Interegated Genomics Inc.; Bénédicte Purnelle, Deborah Prozzi, Katrina Ngui, David Masuy, Frédéric Hancy, Anne Bernard, Sophie Burteau, Marc Boutry, André Goffeau, and Jean Delcour from Unité de Génétique, UCL who participated in the sequencing projects of the two *S. thermophilus* genomes.

Leuconostoc mesenteroides subsp. *mesenteroides* LA81 (contributed by Fred Breidt)

Leuconostoc species are epiphytic bacteria that are wide spread in the natural environment and play an important role in several industrial and food fermentations. *Leuconostoc mesenteroides* is a facultative anaerobe requiring complex growth factors and amino acids (Reiter & Oram 1982; Garvie 1986). Most strains in liquid culture appear as cocci, occuring singly or in pairs and short chains; however, morphology can vary with growth conditions; cells grown in glucose or on solid media may have an elongated or rod-shaped morphology. Cells are Gram-positive, asporogenous and non-motile. A variety of LAB, including *Leuconostoc* species are commonly found on crop plants (Mundt et al. 1967; Mundt 1970). *L. mesenteroides* is perhaps the most predominant LAB species found on fruits and vegetables and is responsible for initiating the sauerkraut and other vegetable fermentations (Pederson & Albury 1969). *L. mesenteroides* starter cultures also used in some dairy and bread dough fermentations (Server-Busson et al. 1999). Under microaerophilic conditions, a heterolactic fermentation is carried out. Glucose and other hexose sugars are converted to equimolar amount of D-lactate, ethanol and CO_2 via a combination of the hexose monophosphate and pentose phosphate pathways (Demoss et al. 1951; Garvie 1986; Gottschalk 1986). Other metabolic pathways include conversion of citrate to diacetyl and acetoin (Cogan et al 1981) and production of dextrans and levan from sucrose (Alsop 1983; Broker 1977). Viscous polysaccharides produced by *L. mesenteroides* are widely recognized as causing product losses and processing problems in the production of sucrose from sugar cane and sugar beets (Tallgren et al. 1999). The first observation of the production of polysaccharide 'slime' from sugar, dates to the earliest days of the science of microbiology; Pasteur (1861) attributed this activity to small cocci, presumably *Leuconostoc* species. Commercial production dextrans and levans by *L. mesenteroides*, for use in the biochemical and pharmaceutical industry, has been carried out for more than 50 years (Alsop 1983; Sutherland 1996). Dextrans are used in the manufacture of blood plasma extenders, heparin substitutes for anticoagulant therapy, cosmetics, and other products (A;sp[1983; Kim & Day 1994; Leathers et al 1995; Sutherland 1996). Another use of dextrans is the manufacture of Sephadex gels or beads, which are widely used for industrial and laboratory protein separations (Sutherland 1996). We have selected *L. mesenteroides* subsp. *mesenteroides* LA81 for sequencing. The strain is widely available from a variety of culture collections listed as: ATCC 8293, DSM 20343, NCIB 8023, CCM 1803, NCDO 523, and NRRL B-1118. The strain is generally accepted as a neotype strain for this species; it produces dextran, and was originally isolated from fermenting olives.

Acknowledgments

The Joint Genome Institute of the US Department of Energy is supporting the sequencing effort for this LAB genome. The efforts of V. Plengvidhya, T. Hawkins, S. Stilwagen, P. Richardson, and K. Kadner are gratefully acknowledged.

Oenococcus oeni PSU-1 (ATCC BAA-331) (contributed by David Mills)

Oenococcus oeni (formerly called *Leuconostoc oenos*) is a lactic acid bacterium that occurs naturally in fruit mashes and related habitats (VanVuuren & Dicks 1993). *O. oeni* is employed commercially to carry out the malolactic conversion, an important secondary fermentation in the production of wine (Kunkee 1991). *O. oeni* is a facultative anaerobe and one of the most acid- and alcohol-tolerant LAB. *O. oeni* shares relatively little DNA homology with the other genera in the *Leuconostoc* branch of the LAB (Dellaglio et al. 1995).

Perhaps the most studied aspect of *O. oeni* is its ability to carry out the malolactic conversion. This involves uptake of malate, decarboxylation to L-lactic acid and CO_2, and subsequent export of end products. The malolactic conversion generates energy for the cell in the form of a proton motive force (Salema et al. 1996). Recently the genes encoding the malate decarboxylase (*mleA*) and malate permease (*mleP*) have been cloned and characterized (Labarre et al. 1996a,b). Other chromosomal genes from *O. oeni* that have been characterized include: histidine decarboxylase (Coton et al. 1998), α-acetolactic acid decarboxylase (Garmyn et al. 1996), stress-related genes (Jobin et al. 1997, 1999), genes involved in arginine metabolism (Tonon et al. 2001) and rDNA genes (Martinez-Murcia & Collins 1990). Several genetic elements from *O. oeni* have been characterized including bacteriophages (Gindreau et al. 1997; Gindreau & Lonvaud-Funel 1999), and plasmids (Fremaux et al. 1993). In addition, the conjugative transposon Tn916 has been mobilized into the *O. oeni* genome (Zuniga et al. 1996b).

Many researchers have examined the diversity of *O. oeni* strains within and around wineries. Studies have employed various molecular typing methods (protein profiling, plasmid profiling, RAPD, PFGE, rDNA RFLP, etc.) to discern regional differences in strains. An outcome of this analysis is the general view that *O. oeni* is a genetically homogenous species (Zapparoli et al. 2000).

The strain sequenced in this project, *O. oeni* PSU-1 (ATCC BAA-331), was originally isolated at Penn State University from red wine undergoing a spontaneous malolactic fermentation (Beelman et al. 1977). PSU-1 was shown to reliably induce the malolactic fermentation in wines and is currently employed commercially (Beelman et al. 1980). While the *O. oeni* is considered homogenous species, the strain PSU-1 was previously shown to be representative of the larger of two divergent groupings (Tenreiro et al. 1994). An extensive physical map of PSU-1 generated by Ze-Ze et al. 1998, 2000) has tremendously aided sequence compilation and genome scaffolding operations.

Acknowledgments

This project was supported by the US Department of Energy–Joint Genome Institute, the American Vineyard Foundation, and the California Competitive Grants Program for Research in Viticulture and Enology. The efforts of C. M. Joseph, K. Ranstiou, T. Hawkins, S. Stilwagen, P. Richardson, and K. Kadner are gratefully acknowledged.

Oenococcus oeni IOEB 8413 (contributed by Jean Guzzo)

Oenococcus oeni is a LAB most often responsible for an important step in the winemaking process, the malolactic fermentation (MLF). The main transformation is decarboxylation of malic acid to lactic acid. MLF leads to a natural decrease of acidity, together with an enhancement of stability and quality of wine. Normally, it occurs spontaneously after alcoholic fermentation, but in some wines the environment is so hostile to bacterial growth that MLF is delayed or even totally prevented. To solve this problem, winemakers use malolactic starters prepared by the industrial culture of selected strains. The laboratory of microbiology, UMR INRA, from Dijon University and the laboratory of biotechnology and applied microbiology, UA INRA, from Bordeaux University, in collaboration with GENOME Express, have determined the complete nucleotide sequence of the chromosome of *O.oeni* IOEB8413, a strain isolated from Bordeaux wine. The knowledge of the *O. oeni* genome represents a considerable value for developing new tools for the control of malolactic fermentation in wine.

Genome sequencing status and results

We first sequenced two random small insert and large insert libraries to favour the scaffold formation between sequence contigs during the assembly stage. Subsequently primers were designed for gap closure sequencing by gene walking and multiplex PCR approaches. The sequenced contigs were assembled using the Phred and Phrap software packages. The current assembly represents 1 753 879 nucleotides and a G+C content of 37.9%. At this stage, early annotation has been undertaken in order to improve the finishing step. The annotation was performed using the GenoAnnot^{tm} software (Geno*^{tm} platform), an integrated computer environment specialised in large-scale sequence annotation. A first pass in an automatic mode with well-dedicated annotation strategies revealed a total of 1784 putative coding sequences, 43 tRNA and 588 predicted rho-independant terminators. Functional annotation was then based on similarity searches with the Blast programs family. Preliminary Blast analysis of the current assembly revealed two ribosomal RNA operons, the presence of three different IS elements in several copies, and protein-coding genes belonging to prophages. Some functional categories of proteins are being studied in detail: stress proteins, transcriptional regulators, central carbon metabolism enzymes and transporters, proteins involved in nitrogen assimilation and the membrane anabolic pathway.

Acknowledgments

We thank Aline Lonvaud-Funel (University of Bordeaux) and Yves Vandenbrouck (GENOME Express) for their contribution to this work.

Brevibacterium linens BL2 Bio (contributed by Bart Weimer)

The *Brevibacterium* genus is a heterogeneous mixture of coryneform organisms that have particular application to industrial production of vitamins, amino acids for fine chemical production, and are commonly used in cheese production (Amador et al. 1999; Rattray & Fox 1999). This genus contains nine species from diverse habitats, such as soil, poultry, fish, human skin, and food. While *Brevibacterium linens* is phenotypically similar to *Arthrobacter globiformis,* cellular pigmentation, cell wall composition, DNA/DNA hybridization and 5s RNA analysis show that *Brevibacterium* is distinctly different (Park et al. 1987). PFGE analysis indicates that diversity within the species is related to polymorphisms in the 16S rRNA genes with genome sizes that range from 3.2 and 3.9 Mbp (Lima & Correia 2000).

B. linens is a non-motile, non-spore forming, non-acid fast, Gram-positive coryneform that tolerates high salt concentrations (8–20%) and is capable of growing in a broad pH range (5.5–9.5), with an optimum of pH 7.0. They also survive carbohydrate starvation and drying for extended periods (Boyaval et al. 1985). *B. linens* is unusual as they produce base as they grow, raising the pH to ~9.5 within 24–36 h. Recent interest in *B. linens* has focused around their ability to produce a self-processing extracellular protease (Rattray et al. 1997; Weimer et al. 2000), their ability to produce high levels of volatile sulfur compounds (Ferchichi et al. 1985; Dias & Weimer 1998a,b), bacteriocin production (Valdes-Stauber & Scherer 1996), cell-membrane-associated carotenoid pigment production (Arrach et al. 2001), and aromatic amino acid metabolism (Leuschner & Hammes 1998; Ummadi & Weimer 2001). These organisms are also noted for their metabolize heterocyclic and polycyclic ring structures–a trait that is not associated with other bacteria but is common in fungi. Of particular note is the degradation of insecticides (including DTT and DDE). These organisms also metabolize amino acids, particularly aromatic amino acids, to produce plant growth hormones.

Acknowlegments

The Joint Genome Institute of the US Department of Energy is supporting the sequencing effort for this LAB genome. The efforts of T. Hawkins, S. Stilwagen, P. Richardson, and K. Kadner are gratefully acknowledged.

Propionibacterium freudenreichii ATCC 6207 (contributed by Herman Pel)

Propionibacteria are high G+C Gram-positive bacteria belonging to the class of Actinobacteria that prefer anaerobic growth conditions and have a peculiar physiology. They produce propionate as their major fermentation product. Propionate fermentation yields more energy and, consequently, biomass than any other anaerobic microbial fermentation. Furthermore, propionibacteria utilize polyphosphate and pyrophosphate instead of ATP for several energy-dependent reactions and their metabolism is tuned to synthesize high levels of porphyrins, in particular B12.

DSM Food Specialties uses a classical dairy microorganism *Propionibacterium freudenreichii* to produce vitamin B12 for feed applications. Vitamin B12 is a highly complex but essential vitamin that can only be produced economically through the fermentation of microorganisms. Friesland Coberco Dairy Foods (FCDF) is a world player in cheese production. Propionibacteria have long been employed in the production process of Swiss-type cheeses for which they are indispensable for the typical eye-formation and production of characteristic taste components.

Genome sequencing status and results

The lack of fundamental physiological knowledge of propionibacteria hampers their efficient industrial exploitation. DSM Food Specialties and Friesland Coberco Dairy Foods have therefore in a combined effort sequenced and annotated the genome of the *P. freudenreichii* type strain ATCC 6207. A total of 2 641 522 base pairs were sequenced using a shotgun approach resulting in 12 contigs. The genome sequence revealed a GC content of 67% and was found to contain at least 2552 open reading frames. The annotated genome, including the resulting blueprint of the metabolic capabilities of the cell, forms a solid basis for the use of powerful tools such as transcriptomics and metabolomics to analyze the metabolic response of *P. freudenreichii* to genetic and environmental changes.

Acknowledgments

Jan Sikkema (FCDF), Jan Hunik (DSM Food Speciaties) and Albert van Ooyen (DSM Food Specialties) are greatfully acknowledged for their contribution and support to this work. The *Propionibacterium freudenreichii* genome was sequenced by Qiagen (www.qiagen.com) The genome was annotated by Biomax Informatics (www.biomax.de).

Bifidobacterium breve NCIMB 8807 (contributed by: Douwe van Sinderen)

Species of the genus *Bifidobacterium* are Gram-positive bacteria, strictly anaerobic, fermentative rods, often Y-shaped or clubbed at the end and contain DNA with a relatively high G+C content. They represent a major element in the microflora of the human gastrointestinal tract. They are considered to have a significant role in maintaining the good health of the human host, while there is mounting evidence pointing to the activity of these bacteria in mediating other positive health attributes such as the alleviation of lactose tolerance, stimulation/potentiation of the immune system, and production of vitamins and antimicrobials. Bifidobacteria have been shown to be the predominant species in the gastrointestinal tract of infants, and represent the third most numerous species encountered in the colon of adult humans, considerably outnumbering other groups such as *Lactobacillus* species. The role of these bacteria in human health has stimulated significant interest in the health care and food industries, and has highlighted the position of these bacteria in the development of functional and pharma foods, which would contain these bacteria as active ingredients.

Despite growing consumer interest, key aspects regarding *Bifidobacterium* species, such as metabolic activities (particularly relating to catabolism of prebiotics) and physiology are still poorly understood. The determination of the complete genome of the *B. breve* strain NCIM 8807 (National Collection of Industrial and Marine bacteria, Aberdeen, Scotland), an isolate from nursling stools, was undertaken as a first step towards the molecular analysis of a probiotic *Bifidobacterium* species. This plasmid-free strain was selected because it is reasonably easy to transform ($\pm 10^4$ transformants per μg of plasmid DNA), shows good adherence properties to epithelial cells, and exhibits reproducible growth properties and moderate tolerance to oxidative, osmotic and acid stresses.

Genome sequencing status and results

Random sequences were obtained using a small-insert (2–4 kb) plasmid bank and a large insert (20–35 kb) cosmid library generating a total of roughly 10.5 million of raw sequence data (approximately 4.4-fold redundancy). These sequences have each been checked for quality and contaminating cloning vector sequences, and subsequently been used for an overall assembly by means of a combination of the Staden software package, and PHRED/PHRAP. This resulted in an initial assembly into 376 contigs, which were then examined individually for internal joints to reduce the total number of contigs to 250. Gap closure by primer walking is now ongoing and has reduced the number of contigs to 140, representing a total DNA sequence of 2.43 Mb. Preliminary analysis of the obtained sequence has revealed quite a high number of IS elements, and a large number of genes predicted to be involved in the degradation of poly- and oligosaccharides, e.g., arabinogalactan endo-1,4-β-galactosidase and α-mannosidase. Four rRNA-encoding operons have sofar been identified, while the calculated G+C content of the genome is roughly 53%.

Acknowledgments

Funding of this work was obtained through the Higher Education Authority Programme for Research in Third Level Institutions, Cycle 1 (HEA PRTLI1), and BioResearch Ireland. Sinead Leahy, Sinead Ryan, Jose Antonio Moreno Muñoz, Des Higgins and Gerald Fitzgerald are gratefully acknowledged for their contributions to this work.

***Bifidobacterium longum* DJO10A (contributed by Daniel O'Sullivan)**

Bifidobacteria are anaerobic, Gram-positive, irregular or branched rod-shaped bacteria that are commonly found in the intestines of humans and most animals and insects. They were first isolated and described over 100 years ago from human feces and were quickly associated with a healthy GI tract due to their numerical dominance in breast-fed infants compared to bottle-fed infants (Tissier 1899, 1906). While they were first grouped in the genus *Bacillus,* the genus *Bifidobacterium* was proposed in the 1920s (Orla-Jensen 1924). However, there was not a taxonomic consensus for this new genus and for much of the 20th century, they were classified in the genus *Lactobacillus*, due to their rod-like shapes and obligate fermentative characteristics. However, the accumulation of studies detailing DNA hybridizations, G+C content and unique metabolic capabilities resulted in the resurrection of the *Bifidobacterium* genus. They are characterized by a unique hexose metabolism that occurs via a phosphoketolase pathway often termed the 'bifid shunt'. Fructose-6-phosphate phosphoketolase (F6PPK) is a key enzyme of the 'bifid shunt' and its presence is the most common diagnostic test for this genus, as it is not present in other Gram-positive intestinal bacteria.

The genus is comprised of 31 characterized species, 11 of which have been detected in human feces (Tannock 1999). *B. longum* is often the dominant species detected in humans and is the only species to regularly harbor plasmids. It is a leading member of the probiotic bacteria due to numerous studies that have provided a growing body of evidence for its role in a myriad of potential health benefits. These include diarrhea prevention in antibiotic treated patients (Black et al. 1991), cholesterol reduction (Dambekodi & Gilliland 1998), alleviation of lactose intolerance symptoms (Jiang et al. 1996), immune stimulation (Takahashi et al. 1998), and cancer prevention (Reddy & Rivenson 1993). This myriad of potential health benefits attributed to the *B. longum* species clearly illustrates that this species possesses many very interesting characteristics. It is anticipated that identification and functional analysis of the genetic determinants involved in these activities will strengthen the evidence for the involvement of *B. longum* in these significant health benefits. Selection of suitable strains for probiotic purposes is very difficult as inherent characteristics of strains of *B. longum* that are necessary for its survival and competition in the human large intestine are currently very poorly understood (O'Sullivan 2001). The use of the sequenced genome in microarray analysis should reveal the pertinent traits that are important for these bacteria to attain dominance in these complex ecosystems.

Acknowledgments

The Joint Genome Institute of the US Department of Energy is supporting the sequencing effort for this LAB genome. The efforts of JU-H. Lee, J. Halgersen, T. Hawkins, S. Stilwagen, P. Richardson, and K. Kadner are gratefully acknowledged.

Bifidobacterium longum NCC2705 (contributed by Fabrizio Arigoni)

Bifidobacteria are obligate anaerobes in the *Actinomycetales* branch of the high G+C Gram-positive bacteria. There are at least 32 species of bifidobacteria, largely isolated from the GITs of many mammals as well as chickens and honeybees (Biavati & Mattarelli 2001). They are among the first colonizers of the sterile GITs of newborns and predominate until weaning when they are surpassed by other groups (Harmsen et al. 2000; Favier et al. 2002). Although the bifidobacteria represent <6% of the adult fecal flora, their presence has been associated with beneficial health effects (Biavati & Mattarelli 2001). For example, some studies showed that infant formula containing bifidobacteria reduces incidence of diarrhea (Saavedra et al. 1994). These types of studies have led to widespread use of bifidobacteria as components of health-promoting foods (probiotics). Although foods containing bifidobacteria are widely consumed probiotics, there is only fragmentary information about the physiology, ecology, and genetics of any one species. To rapidly increase knowledge and understanding of bifidobacterial biology and their complex interactions with their human hosts and GIT microflora, we determined and extensively analyzed the genome sequence of a *B. longum* strain isolated from infant feces.

Genome sequencing status and results

The genome sequence of *B. longum* NCC2705 was determined by shotgun-sequencing of randomly-cloned, 1.5- and 3-kb fragments. ORFs were identified with ORPHEUS (Pedant-Pro; Biomax Informatics). Total genome sequence analysis of *B. longum* revealed a large 2 256 646-bp chromosomal replicon and a small 3626-bp plasmid replicon that is 89% identical to pKJ36 (Genbank AF139129). The average G+C content was 60.1%, although six regions had much lower G+C content. The chromosome contains four nearly identical *rrn* operons and 57 tRNAs. We identified 1730 probable coding regions, representing 86% of the genome. The average gene length was 1100 bp, and a specific or general function was assigned to approx. 70% of them. We were able to attribute 78% of the predicted proteins to a COG family. The percentages of predicted proteins assigned to COGs in each of the 18 defined functional categories (Tatusov et al. 2001) were similar to those for other prokaryotes, except for an unusual number (>8.5%) assigned to the carbohydrate transport-metabolism category. This suggests that *B. longum* is well adapted to take advantage of the wide diversity and fluctuations in the nutrient composition in the colon. In addition, we observed that gene duplication and horizontal gene transfer have played an important part in this physiological adaptation. Using SignalP (Nielsen et al. 1997), we identified approximately 200 proteins with probable Sec-type signal peptides. Of these 59 were predicted as surface-associated lipoproteins (PROSITE accession PS00013) and 26 as solute-binding proteins of ABC transporter systems.

Acknowledgments

M.A. Schell, M. Karmirantzou, B. Snel, B. Berger, D. Vilanova, G. Pessi, P. Bork, T. Pohl, G. Bothe, M.C Zwahlen, M. Delley are sincerely thanked for their contribution to this work.

References

Ahrne S, Nobaek S, Jeppsson B, Adlerberth I, Wold AE & Molin G (1998) The normal *Lactobacillus* flora of healthy human rectal and oral mucosa. J. Appl. Microbiol. 85: 88–94.

Almirón-Roig E, Mulholland F, Gasson MJ & Griffin AM (2000) The complete *cps* gene cluster from *Streptococcus thermophilus* NCFB 2393 involved in the biosynthesis of a new exopolysaccharide. Microbiology. 146: 2793–2802.

Alsop RM (1983) Industrial production of dextrans. In: Bushell ME (Ed) Progress in Industrial Microbiology (pp 1–42). Elsevier, New York.

Altermann E, Klein JR & Henrich B (1999) Primary structure and features of the genome of the *Lactobacillus gasseri* temperate bacteriophage (phi) adh. Gene 236: 333–346.

Alvarez S, Herrero C, Bru E & Perdigon G (2001) Effect of *Lactobacillus casei* and yogurt administration on prevention of *Pseudomonas aeruginosa* infection in young mice. J. Food Prot. 64: 1768–1774.

Amador E, Castro JM, Correia A & Martin JF (1999) Structure and organization of the rrnD operon of *Brevibacterium lactofermentum*: analysis of the 16S rRNA gene. Microbiology 145: 915–924.

Arrach N, Fernandez-Martin R, Cerda-Olmedo E & Avalo J (2001) A single gene for lycopene cyclase, phytoene synthase, and regulation of carotene biosynthesis in *Phycomyces*. Proc. Natl. Acad. Sci. 98: 1687–1692.

Axelsson L (1998) Lactic acid bacteria: classification and physiology. In: Salminen S & Von Wright A (Eds) Lactic Acid Bacteria: Microbiology and Functional Aspects, 2nd edition (pp 1–72). Marcel Dekker, New York.

Baccigalupi L, Naclerio G, de Felice M & Ricca E (2000) Efficient insertional mutagenesis in *Streptococcus thermophilus*. Gene 258: 9–14.

Barefoot SF & Klaenhammer TR (1983) Detection and activity of lactacin B, a bacteriocin produced by *Lactobacillus acidophilus*. Appl. Environ. Microbiol. 45: 1808–1815.

Beelman RB, Gavin A III & Keen RM (1977) A new strain of *Leuconostoc oenos* for induced malo-lactic fermentation in eastern wines. Am. J. Enol. Vitic. 28: 159–165.

Beelman RB, McArdle FJ & Duke GR (1980) Comparison of *Leuconostoc oenos* strains ML-34 and PSU-1 to induce malolactic fermentation in Pennsylvania red table wines. Am. J. Enol. Viticult. 31: 269–276.

Beimfohr C, Ludwig W & Schleifer K-H (1997) Mosaic structure of large regions of the *Lactococcus lactis* subsp. *cremoris* chromosome. System. Appl. Microbiol. 20: 216–221.

Beresford TP, Fitzsimons NA, Brennan NL & Cogan TM (2001) Recent advances in cheese microbiology. Int. Dairy J. 11: 259–274.

Bernet M-F, Brassart D, Neeser J-R & Servin AL (1994) *Lactobacillus acidophilus* La1 binds to cultured human intestinal cell lines and inhibits cell attachment and cell invasion by enterovirulent bacteria. Gut 35: 483–489.

Bernet-Camard M-F, Liévin V, Brassart D, Neeser J-R, Servin AL & Hudault S (1997) The human *Lactobacillus acidophilus* strain La1 secretes a non bacteriocin antibacterial substance active *in vitro* and *in vivo*. Appl. Environ. Microbiol. 63: 2747–2753.

Bhowmik T & Steele JL (1993) Development of an electroporation procedure for gene disruption in *Lactobacillus helveticus* CNRZ32. J. Gen. Microbiol. 139: 1433–1439.

Bhowmik T, Fernández L & Steele JL (1993) Gene replacement in *Lactobacillus helveticus* CNRZ 32. J. Bacteriol. 175: 6341–6344.

Biavati B & Mattarelli P (2001) The family Bifidobacteriaceae. In: Dworkin M, Falkow S, Rosenberg E, Schleifer KH & Stackebrandt E (Eds) The Prokaryotes (pp 1–70). Springer, New York.

Black F, Einarsson K, Lidbeck A, Orrhage K & Nord CE (1991) Effect of lactic acid producing bacteria on the human intestinal microflora during ampicillin treatment. Scand. J. Infect. Dis. 23: 247–254.

Bolotin A, Mauger S, Malarme K, Ehrlich SD & Sorokin A (1999) Low-redundancy sequencing of the entire *Lactococcus lactis* IL1403 genome. Antonie van Leeuwenhoek 76: 27–76.

Bolotin A, Wincker P, Mauger S, Jaillon O, Malarme K, Weissenbach J, Ehrlich SD & Sorokin A (2001) The complete genome sequence of the lactic acid bacterium *Lactococcus lactis* ssp. *lactis* IL1403. Genome Res. 11: 731–753.

Bolotin A, Ehrlich SD & Sorokin A (2002) Studies of genomes of dairy bacteria *Lactococcus lactis*. Sci. Aliments (in press)

Boyaval P, Boyaval E & Desmazeaud MJ (1985) Survival of *Brevibacterium linens* during nutrient starvation and intracellular changes. Arch. Microbiol. 141: 128–132.

ten Brink B, Damink C, Joosten HMLJ & Huis inþt Veld JHJ (1990) Ocurrence and formation of biologically active amines in foods. Int. J. Food Microbiol. 11: 73–84.

Broadbent JR (2001) Genetics of lactic acid bacteria. In: Steele JL & Marth EH (Eds) Applied Dairy Microbiology, 2nd ed. Marcel Dekker, New York.

Broker BE (1977) Ultrastructural surface changes associated with dextran synthesis by *Leuconostoc mesenteroides*. J. Bacteriol. 131: 288–92.

Burrus V, Bontemps C, Decaris B & Guédon G (2001) Characterization of a novel type II restriction-modification system, *Sth*368I, encoded by the integrative element ICE*St1* of *Streptococcus thermophilus* CNRZ368. Appl. Environ. Microbiol. 67: 1522–1528.

Caldwell S, McMahon DJ, Oberg CJ & Broadbent JR (1996) Development and characterization of lactose-positive *Pediococcus* species for milk fermentation. Appl. Environ. Microbiol. 62: 936–941.

Caldwell S, Hutkins RW, McMahon DJ, Oberg CJ & Broadbent JR (1998) Lactose and galactose uptake by genetically engineered *Pediococcus* species. Appl. Microbiol. Biotechnol. 49: 315–320.

Champomier-Verges M-C, Chaillou S, Cornet M & Zagorec M (2002) *Lactobacillis sakei:* recent developments and future prospects. Res. Microbiol. 153: 115–123.

Chen H, Lim CK, Lee YK & Chan YN (2000) Comparative analysis of the genes encoding 23S-5S rRNA intergenic spacer regions of *Lactobacillus casei*-related strains. Int. J. Syst. Evol. Microbiol. 50: 471–478.

Chevallier B, Hubert JC & Kammerer B (1994) Determination of chromosome size and number of rrn loci in *Lactobacillus plantarum* by pulsed-field gel electrophoresis. FEMS Microbiol. Lett. 120: 51–56.

Cho JS, Choi YJ & Chung DK (2000) Expression of *Clostridium thermocellum* endoglucanase gene in *Lactobacillus gasseri* and *Lactobacillus johnsonii* and characterization of the genetically modified probiotic lactobacilli. Curr. Microbiol 40: 257–63.

Christensen JE, Dudley EG, Pederson JR & Steele JL (1999) Peptidases and amino acid catabolism in lactic acid bacteria. Antonie van Leeuwenhoek 76: 217–246.

Clark RH, Russell WM & Klaenhammer TR (2000) Distribution of *Lactobacillus acidophilus* among a variety of cultured foods and probiotics. Abstracts, Annual IFT Meeting of the Institute of Food Technologists, Dallas, TX, 10 June, 2000

Coderre PE & Somkuti GA (1999) Cloning and expression of the pediocin operon in *Streptococcus thermophilus* and other lactic fermentation bacteria. Curr. Microbiol. 39: 295–301.

Cogan TM (1987) Co-metabolism of citrate and glucose by *Leuconostoc* spp.: effects on growth, substrates and products. J. Appl. Bacteriol. 63: 551–58.

Cogan TM, O'Dowd M & Mellerick D (1981) Effects of sugar on acetoin production from citrate by *Leuconostoc lactis*. Appl. Environ. Microbiol. 41: 1–8.

Collins MD, Phyllips BA & Zanoni P (1989) Deoxyribonucleic acid homology studies of *Lactobacillus casei*, *Lactobacillus paracasei* sp. nov., subsp. *paracasei* and subsp. *tolerans*, and *Lactobacillus rhamnosus* sp. nov., comb. nov. Int. J. Syst. Bacteriol. 39: 105–108.

Conway PL, Gorbach SL & Goldin BR (1987) Survival of lactic acid bacteria in the human stomach and adhesion to intestinal cells. J. Dairy Sci. 70: 1–12.

Coton E, Rollan GC & Lonvaud-Funel A (1998) Histidine carboxylase of *Leuconostoc oenos* 9204: Purification, kinetic properties, cloning and nucleotide sequence of the *hdc* gene. J. Appl. Microbiol. 84: 143–151.

Daeschel MA & Klaenhammer TR (1985) Association of a 13.6-megadalton plasmid in *Pediococcus pentosaceus* with bacteriocin activity. Appl. Environ. Microbiol. 50: 1528–1541.

Dambekodi PC & Gilliland SE (1998) Incorporation of cholesterol into the cellular membrane of *Bifidobacterium longum*. J. Dairy Sci. 81: 1818–1824.

Daniel P (1995) Sizing the *Lactobacillus plantarum* genome and other lactic bacteria species by transverse alternating field electrophoresis. Curr. Microbiol. 30: 243–246.

Daveran-Mingot ML, Campo N, Ritzenthaler P & Le Bourgeois P (1998) A natural large chromosomal inversion of *Lactococcus lactis* is mediated by homologous recombination between two insertion sequences. J. Bacteriol. 180: 4834.

Davidson B, Kordis N, Dobos M & Hillier A (1996) Genomic organization of lactic acid bacteria. Antonie van Leeuwenhoek 70: 161–183.

Delcher AL, Harmon D, Kasif S, White O & Salzberg SL (1999) Improved microbial gene identification with GLIMMER. Nuclaic. Acids Res. 27: 4636–4641.

Delcour J, Ferain T & Hols P (2000) Advances in the genetics of thermophilic lactic acid bacteria. Curr. Opin. Biotechnol. 11: 497–504.

Dellaglio F, Dicks LMT, du Toit M & Torriani S (1991) Designation of ATCC334 in place of ATCC393 (NCDO 161) as the neotype strain of *Lactobacillus casei* subsp. *casei* and rejection of the name *Lactobacillus paracasei*. Int. J. Syst. Bacteriol. 41: 340–342.

Dellaglio F, Dicks LMT & Torriani S (1995) The genus *Leuconostoc*. In: Wood BJB & Holzapfel WH (Eds) The Genera of Lactic Acid Bacteria, vol. 2, (pp 235–278). Blackie Academic & Professional, London

Delorme C, Godon J-J, Ehrlich SD & Renault P (1994) Mosaic structure of large regions of the *Lactococcus lactis* subsp. *cremoris* chromosome. Microbiology. 140: 3053–3060.

Demoss RD, Bard RC & Gunsalus IC 1951. The mechanism of heterolactic fermentation: a new route of ethanol formation. J. Bacteriol. 62: 499–511.

Dias B & Weimer B (1998a) Conversion of methionine to thiols by lactococci, lactobacilli, and brevibacteria. Appl. Environ. Microbiol. 64: 3320–3326.

Dias B & Weimer B (1998b) Purification and characterization of methionine β-lyase from *Brevibacterium linens* BL2. Appl. Environ. Microbiol. 64: 3327–3331.

Dicks LMT, du Plessis EM, Dellaglio F & Lauer E (1996) Reclassification of *Lactobacillus casei* subsp. *casei* ATCC 393 and *Lactobacillus rhamnosus* ATCC15820 as *Lactobacillus zeae* nom. rev., designation of ATCC 334 as the neotype of *L. casei* subsp. *casei*, and rejection of the name *Lactobacillus paracasei*. Int. J. Syst. Bacteriol. 46: 337–340.

Djordjevic GM, Tchieu & Saier MH (2001) Genes involved in control of galactose uptake in *Lactobacillus brevis* and reconstitution of the regulatory system in *Bacillus subtilis*. J. Bacteriol. 183: 3224–3236.

Dossonnet V, Monedero V, Zagorec M, Galinier A, Perez-Martinez G & Deutscher J (2000) Phosphorylation of HPr by the bifunctional HPr Kinase/P-Ser-HPr phosphatase from *Lactobacillus casei* controls catabolite repression and inducer exclusion but not inducer expulsion. J. Bacteriol. 182: 2582–2590.

Dudez A-M, Chaillou S, Hissler L, Stentz R, Champomier-Verges M-C, Alpert C-A & Zagorec M 2002. Physiscal and genetic map of the *Lactobacillus sakei* 23K chromosome. Microbiology, 148: 421–431.

Dunny G & McKay LL (1999) Group II introns and expression of conjugative transfer functions in lactic acid bacteria. Antonie van Leeuwenhoek 76: 77–88.

Eddy SR (1998) Profile hidden Markov models. Bioinformatics 14: 755–763.

Favier CF, Vaughan EE, De Vos WM, & Akkermans AD (2002) Molecular monitoring of succession of bacterial communities in human neonates. Appl. Environ. Microbiol. 68: 219–226.

Felley CP, Corthésy-Theulaz I, Blanco Rivero J-L, Sipponen P, Kaufmann M, Bauerfeind P, Wiesel PH, Brassart D, Pfeifer A, Blum AL & Michetti P (2001) Favourable effect of an acidified milk (LC-1) on *Heliocobacter pylori* gastritis in man. Eur. J. Gastroenterol. Hepatol. 13: 25–29.

Ferchichi M, Hemme D, Nardi M & Pamboukdjian N (1985) Production of methanethiol from methionine by *Brevibacterium linens* CNRZ 918. J. Gen. Microbiol. 131: 715.

Fernandez-Espla MD, Garault P, Monnet V & Rul E (2000) *Streptococcus thermophilus* cell wall-anchored proteinase: release, purification, and biochemical and genetic characterization. Appl. Environ. Microbiol. 66: 4772–4778.

Ferrero M, Cesena C, Morelli L, Scolari G & Vescovo M (1996) Molecular characterization of *Lactobacillus casei* strains. FEMS Microbiol. Lett. 140: 215–219.

Fonden R, Mogensen G, Tanaka R & Salminen S (2000) Effect of culture-containing dairy products on intestinal microflora, human nutrition and health – current knowledge and future perspectives. International Dairy Federation Bulletin number 352, IDF, Brussels.

Forde A & Fitzgerald D (1999) Bacteriophage defense systems in lactic acid bacteria. Antonie van Leeuwenhoek 76: 89–113.

Fox PF, McSweeney PLH & Lynch CM (1998) Significance of non-starter lactic acid bacteria in cheddar cheese. Aust. J. Dairy Technol. 53: 83–89.

Fremaux C, Aigle M & Lonvaud FA (1993) Sequence analysis of *Leuconostoc oenos* DNA: organization of pLo13, a cryptic plasmid. Plasmid 30: 212–23.

Fujisawa T, Benno Y, Yaeshima T & Mitsuoka T (1992) Taxonomic study of the *Lactobacillus acidophilus* group, with recognition of *Lactobacillus gallinarum* sp. nov. and *Lactobacillus johnsonii* sp. nov. and synonymy of *Lactobacillus acidophilus* group A3 with the type strain of *Lactobacillus amylovorus*. Int. J. System. Bacteriol. 42: 487–491.

Garault P, Le Bars D, Besset C & Monnet V (2002) Three oligopeptide-binding proteins are involved in the oligopeptide transport of *Streptococcus thermophilus*. J. Biol. Chem. 277: 32–39.

Garmyn D, Monnet C, Martineau B, Guzzo J, Cavin J-F & Divies C (1996) Cloning and sequencing of the gene encoding alpha-acetolactate decarboxylase from *Leuconostoc oenos*. FEMS Microbiol. Lett. 145: 445–450.

Garvie EI, Farrow JAE & Phillips BA (1981) A taxonomic study of some strains of streptococci which grow at 10 °C but not at 45 °C including *Streptococcus lactis* and *Streptococcus cremoris*. Zbl. Bakteriol. Hyg. I Abt. Orig. C 2: 151–165.

Garvie EI (1986) Genus *Leuconostoc*. In: Sneath PHA, Mair NS, Sharpe ME & Holt JG (Eds), Bergey's Manual of Systematic Bacteriology, vol 2, 9th ed. (pp 1071-1075). Williams and Wilkins, Baltimore, MD

Gasson MJ (1983) Plasmid complements of *Streptococcus lactis* NCDO 712 and other lactic streptococci after protoplast-induced curing. J. Bacteriol. 154: 1–9.

Germond JE, Delley M, D'Amico N & Vincent SL (2001) Heterologous expression and characterization of the exopolysaccharide from *Streptococcus thermophilus* Sfi39. Eur. J. Biochem. 268: 5149–5156.

Gill HS, Rutherfurd KJ, Prasad J & Gopal PK (2000) Enhancement of natural and acquired immunity by *Lactobacillus rhamnosus* (HN001), *Lactobacillus acidophilus* (HN017) and *Bifidobacterium lactis* (HN019). Br. J. Nutr. 83: 167–176.

Gill HS, Rutherfurd KJ & Cross ML (2001a) Dietary probiotic supplementation enhances natural killer cell activity in the elderly: an investigation of age-related immunological changes. J. Clin. Immunol. 21: 264–271.

Gill HS, Shu Q, Lin H, Rutherfurd KJ & Cross ML (2001b) Protection against translocating *Salmonella typhimurium* infection in mice by feeding the immuno-enhancing probiotic *Lactobacillus rhamnosus* strain HN001. Med. Microbiol. Immunol. (Berlin) 190: 97–104.

Gindreau E & Lonvaud-Funel A (1999) Molecular analysis of the region encoding the lytic system from *Oenococcus oeni* temperate bacteriophage variant phi10MC. FEMS Microbiol. Lett. 171: 231–238.

Gindreau E, Torlois S & Lonvaud-Funel A (1997) Identification and sequence analysis of the region encoding the site-specific integration system from *Leuconostoc oenos (Oenococcus oeni)* temperate bacteriophage phi-10MC. FEMS Microbiol. Lett. 147: 279–285.

Godon J, Delorme C, Ehrlich SD & Renault P (1992) Divergence of genomic sequences between *Lactococcus lactis* subsp. *lactis* and *Lactococcus lactis* subsp.*cremoris*. Appl. Environ. Microbiol. 58: 4045–4047.

Gold RS, Meagher Mm, Tong S, Hutkins RW, & Conway T (1996) Cloning and expression of the *Zymomonas mobilis* 'Production of ethanol' genes in *Lactobacillus casei*. Curr. Microbiol. 33: 256–260.

Goldin BR & Gorbach SL (1980) Effect of milk and *Lactobacillus* feeding on human intestinal bacterial enzyme activity. Am. J. Clin. Nutr. 39: 756–761.

Goldin BR, Swenson L, Dwyer J, Sexton M & Gorbach S (1980) Effect of diet and *Lactobacillus acidophilus* supplements on human fecal bacterial enzymes. J. Natl. Cancer Inst. 64: 255–261.

Gonzalez CF & Kunka BS (1983) Plasmid transfer in *Pediococcus* spp.: Intergeneric and intrageneric transfer of pIP501. Appl. Environ. Microbiol. 46: 81–89.

Gonzalez CF & Kunka BS (1986) Evidence for plasmid linkage of raffinose utilization and associated α-galactosidase and sucrose hydrolase activity in *Pediococcus pentosaceus*. Appl. Environ. Microbiol. 51: 105–109.

Gopal PK, Prasad J, Smart J & Gill HS (2001) In vitro adherence properties of *Lactobacillus rhamnosus* DR20 and *Bifidobacterium lactis* DR10 strains and their antagonistic activity against an enterotoxigenic *Escherichia coli*. Int. J. Food Microbiol. 67: 207–216.

Gottschalk G (1986) Bacterial Metabolism, 2nd ed. Springer, New York.

Graham DC & McKay LL (1985) Plasmid DNA in strains of *Pediococcus cerevisiae* and *Pediococcus pentosaceus*. Appl. Environ. Microbiol. 50: 532–534.

Granato D, Perotti F, Masserey I, Rouvet M, Golliard M, Servin AL & Brassart D(1999) Cell surface-associated lipoteichoic acid acts as an adhesion factor for attachment of *Lactobacillus johnsonii* La1 to human enterocyte-like Caco-2 cells. Appl. Environ. Microbiol. 65: 1071–1077.

Greene JD & Klaenhammer TR (1994) Factors involved in adherence of lactobacilli to human Caco-2 cells. Appl. Microbiol. 60: 4487–4494.

Guedon G, Bourgoin F, Pebay M, Roussel Y, Colmin C, Simonet JM & Decaris B (1995) Characterization and distribution of two insertion sequences, IS *1191* and iso-IS *981*, in *Streptococcus thermophilus*: does intergeneric transfer of insertion sequences occur in lactic acid bacteria co-cultures? Mol. Microbiol. 16: 69–78.

Haller D, Blum S, Bode C, Hammes WP & and Schiffrin EJ (2000a) Activation of human PBMC by non-pathogenic bacteria *in vitro*: evidence of NK cells as primary targets. Infect. Immun. 68: 752–759.

Haller D, Bode C, Hammes WP, Pfeifer AMA, Schiffrin EJ & Blum S (2000b) Non-pathogenic bacteria elicit a differential cytokine response by intestinal epithelial cell/leucocyte co-cultures. Gut 47: 79–87.

Hammes WP & Vogel RF (1995) The genus *Lactobacillus*. In: Wood BJB & Holzapfel WH (Eds) The Genera of Lactic Acid Bacteria (pp 19–54). Chapman & Hall, London.

Harmsen HJ, Wildeboer-Veloo AC, Raangs GC, Wagendorp AA, Klijn N, Bindels JG & Welling GW (2000) Analysis of intestinal flora development in breast-fed and formula-fed infants by using molecular identification and detection methods. J. Pediatr. Gastroenterol. Nutr. 30: 61–67.

Hassan AN & Frank JF (2001) Starter cultures and their use. In: Marth EH & Steele JL (Eds) Applied Dairy Microbiology, 2nd edition (pp 151–206). Marcel Dekker, Inc, New York.

Heilig HGH, Zoetendal EG, Vaughan EE, Marteau P, Akkermans ADL & deVos WM (2002) Molecular diversity of *Lactobacillus* spp. and other lactic acid bacteria in the human intestine as determined by specific amplification of 16S ribosomal DNA. Appl. Environ. Microbiol. 68: 14–123.

Hols P, Slos P, Dutot P, Reymund J, Chabot P, Delplace B, Delcour J & Mercenier A (1997) Efficient secretion of the model antigen M6-gp41E in *Lactobacillus plantarum* NCIMB 8826. Microbiology 143: 2733–41.

Hughes D (2000) Evaluating genome dynamics: The constraints on rearrangements within bacterial genomes. Genome Biol. 1: Reviews 0006.1–0006.8.

Hofvendahl K & Hahn-Hagerdal B (2000) Factors affecting the fermentative lactic acid production from renewable resources. Enzyme Microb. Technol. 26: 87–107.

Itoh T, Fujimoto Y, Kawai Y, Toba T & Saito T (1995) Inhibition of food-borne pathogenic bacteria by bacteriocins from *Lactobacillus gasseri*. Lett. Appl. Microbiol. 21: 137–141.

Jiang TA, Mustapha A & Savaiano DA (1996) Improvement of lactose digestion in humans by ingestion of unfermented milk containing *Bifidobacterium longum*. J. Dairy Sci. 79: 750–757.

Jobin M-P, Delmas F, Garmyn D, Divies C & Guzzo J. (1997) Molecular characterization of the gene encoding an 18-kilodalton small heat shock protein associated with the membrane of *Leuconostoc oenos*. Appl. Environ. Microbiol. 63: 609–614.

Jobin M-P, Garmyn D, Divies C & Guzzo J (1999) The *Oenococcus oeni* clpX homologue is a heat shock gene preferentially expressed in exponential growth phase. J. Bacteriol. 181: 6634–6641.

Johnson JL, Phelps CF, Cummins CS, London J & Gasser F (1980) Taxonomy of the *Lactobacillus acidophilus* group. Int. J. System. Bacteriol. 30: 53–68.

Kandler O (1983) Carbohydrate metabolism in lactic acid bacteria. Antonie van Leeuwenhoek 49: 209–224.

Kandler O & Weiss N (1986) Genus *Lactobacillus*. In: Sneath PHA, Mair NS, Sharpe ME & Holt JG (Eds.) Bergey's Manual of Systematic Bacteriology, vol 2, 9th ed. (pp 1063–1065). Williams and Wilkins, Baltimore, MD.

Kaplan H & Hutkins RW (2000) Fermentation of fructooligosaccharides by lactic acid bacteria and bifidobacteria. Appl. Environ. Microbiol. 66: 2682–2684.

Kim D & Day DF (1994) A new process for the production of clinical dextran by mixed-culture fermentation of *Lipomyces starkeyi* and *Leuconostoc mesenteroides*. Enzyme Microb. Technol. 16: 844–48.

Kim WJ, Ray B & Johnson MC (1992) Plasmid transfers by conjugation and electroporation in *Pediococcus acidilactici*. J. Appl. Bacteriol. 72: 201–207.

Kirjavainen PV, El-Nezami HS, Salminen SJ, Ahokas JT & Wright PF (1999) The effect of orally administered viable probiotic and dairy lactobacilli on mouse lymphocyte proliferation. FEMS Immunol. Med. Microbiol. 26: 131–135.

Kitazawa H, Tomioka Y, Matsumura K, Aso H, Mizugaki M, Itoh T & Yamaguchi T (1994) Expression of mRNA encoding IFN alpha in macrophages stimulated with *Lactobacillus gasseri*. FEMS Microbiol. Lett. 120: 315–321.

Klaenhammer TR & Russell WM (2000) Species of the *Lactobacillus acidophilus* complex. In: Robinson RK, Batt C & Patel

PD (Eds) Encyclopedia of Food Microbiology, Vol. 2, (pp 1151–1157). Academic Press, San Diego, CA.

Kosikowski FV (1982) Cheese and Fermented Milk Foods, 2nd edn. Kosikowski and Assoc., Brooktondale, NY.

Kullen MJ, Sanozky-Dawes RB, Crowell DC & Klaenhammer TR (2000) Use of DNA sequence of variable regions of the 16Sr-RNA gene for rapid and accurate identification of bacteria in the *Lactobacillus acidophilus* complex. J. Appl. Microbiol. 89: 511–518.

Kunkee RE (1991) Some roles of malic acid in the malolactic fermentation in wine making. FEMS Microbiol. Rev. 88: 55–72.

Labarre C, Diviès C & Guzzo J (1996a) Genetic organization of the *mle* locus and identification of a *mleR*-like gene from *Leuconostoc oenos*. Appl. Environ. Microbiol. 62: 4493–4498.

Labarre C, Guzzo J, Cavin JF & Diviès C (1996b) Cloning and characterization of the genes encoding the malolactic enzyme and the malate permease of *Leuconostoc oenos*. Appl. Environ. Microbiol. 62: 1274–1282.

Lawrence RC, Thomas TD & Terzaghi BE (1976) Reviews of the progress of dairy science: cheese starters. J. Dairy Res. 43: 141–193.

Leathers TD, Hayman GT & Cote GL 1995. Rapid screening of *Leuconostoc mesenteroides* mutants for elevated proportions of alternan to dextran. Curr. Microbiol. 31: 19–22.

Le Bourgeois P, Lautier M, van den Berghe L, Gasson MJ & Ritzenthaler P (1995) Physical and genetic map of the *Lactococcus lactis* subsp. *cremoris* MG1363 chromosome: comparison with that of *Lactococcus lactis* subsp. *lactis* IL1403 reveals a large genome inversion. J. Bacteriol. 177: 2840–2850.

Le Bourgeois P, Daveran-Mingot ML & Ritzenthaler P (2000) Genome plasticity among related *Lactococcus* strains: identification of genetic events associated with macrorestriction polymorphisms. J. Bacteriol. 182: 2481–2491.

Leong-Morgenthaler P, Ruettener C, Mollet B & Hottinger H (1990) Construction of a physical map of *Lactobacillus bulgaricus*. Proc. Third Symp. Lactic Acid Bact. A28.

Leuschner RG & Hammes WP (1998) Degradation of histamine and tyramine by *Brevibacterium linens* during surface ripening of Munster cheese. J. Food Prot. 61: 874–878.

Lima PT & Correia AM (2000) Genetic fingerprinting of *Brevibacterium linens* by pulsed-field gel electrophoresis and ribotyping. Curr. Microbiol. 41: 50–55.

Link-Amster H, Rochat F, Saudan KY, Mignot O & Aeschlimann J-M (1994) Modulation of a specific humoral immune response and changes in intestinal flora mediated through fermented milk intake. FEMS Immunol. Med. Microbiol. 10: 55–64.

Luchansky JB, Muriana PM & Klaenhammer TR (1988) Application of electroporation for transfer of plasmid DNA to *Lactobacillus, Lactococcus, Leuconostoc, Listeria, Pediococcus, Bacillus, Staphylococcus, Enterococcus,* and *Propionibacterium*. Mol. Microbiol. 2: 637–647.

Maguin E, Prevost H, Ehrlich SD & Gruss A (1996) Efficient insertional mutagenesis in lactococci and other gram-positive bacteria. J. Bacteriol. 178: 931–5.

Marchand J & Vandenplas Y (2000) Microorganisms administered in the benefit of the host: myths and facts. Eur. J. Gastroenterol. Hepatol. 12: 1077–1088.

Mariné-Font A, Vidal-Carou MC, Izquierdo-Pulido M, Venciana-Nogués MT & Hernández-Jover T (1995) Les amines biogénes dans les aliments: leur signification, leur analyse. Ann. Fals. Exp. Chim. 88: 11–140.

Marteau P & Rambaud JC (1993) Potential of using lactic acid bacteria for therapy and immunomodulation in man. FEMS Microbiol. Rev. 12: 207–222.

Martinez-Murcia AJ & Collins MD (1990) A phylogenetic analysis of the genus *Leuconostoc* based on reverse transcriptase sequencing of 16S ribosomal RNA. FEMS Microbiol. Lett. 70: 73–84.

Matsuzaki T & Chin J (2000) Modulating immune responses with probiotic bacteria. Immunol. Cell. Biol. 78: 67–73.

McKay LL (1985) Roles of plasmids in starter cultures. In: Gilliland SE (Ed) Bacterial Starter Cultures for Food (pp 159–174). CRC Press, Boca Raton, FL.

Milk Industry Foundation (1998) Milk Facts. USA.

Mitsuoka T (1992) The human gastrointestinal tract. In: Wood BJB (Ed) The Lactic Acid Bacteria, Vol. 1: The Lactic Acid Bacteria in Health and Disease (pp 69-114). Elsevier Science Publishers, Essex.

Monedero V, Poncet S, Mijakovic I, Fieulaine S, Dossonnet V, Martin-Verstraete I, Nessler S & Deutscher J (2001) Mutations lowering the phosphatase activity of HPr kinase/phosphatase switch off carbon metabolism. EMBO J. 20: 3928–3937.

Moreno-Arribas V & Lonvaud-Funel A (2001) Purification and characterization of tyrosine decarboxylase of *Lactobacillus brevis* IOEB 9809 isolated from wine. FEMS Microbiol. Lett. 195: 103–107.

Mori K, Yamazaki K, Ishiyama T, Katsumata M, Kobayashi K, Kawai Y, Inoue N & Shinano H (1997) Comparative sequence analyses of the genes coding for 16S rRNA of *Lactobacillus casei*-related taxa. Int. J. Syst. Bacteriol. 47: 54–57.

Mundt JO (1970) Lactic acid bacteria associated with raw plant food material. J. Milk Food Technol. 33: 550–553.

Mundt JO, Graham WF & McCarty IE (1967) Spherical lactic acid producing bacteria of southern-grown raw and processed vegetables. Appl. Microbiol. 15: 1303–1308.

National Cheese Institute (1998) Cheese Facts. USA.

Neeser J-R, Granato D, Rouvet M, Servin AL, Teneberg S & Karlsson K-A. (2000) *Lactobacillus johnsonii* La1 shares carbohydrate-binding specificities with several enteropathogenic bacteria. Glycobiology. 10: 1193–1199.

Nicolas P, Bize L, Muri F, Hoebeke M, Rodolphe F, Ehrlich SD, Prum B & Bessières P (2002) Mining *Bacillus subtilis* chromosome heterogeneities using hidden Markov models. Nucleic Acids Res. 30: 1418–1426.

Nielsen H, Engelbrecht J, Brunak S & von Heijne G (1997) Identification of prokaryotic and eukaryotic signal peptides and prediction of their cleavage sites. Protein Eng. 10: 1–6.

Orla-Jensen S (1924) La classificationdes des bactéries lactiques. Lait 4: 468–474.

O'Sullivan DJ (2001) Screening of intestinal microflora for effective probiotic bacteria. J. Agric. Food Chem. 49: 1751–1760.

Park YH, Hori H, Suzuki K, Osawa S & Komagata K (1987) Phylogenetic analysis of the coryneform bacteria by 5S rRNA sequences. J. Bacteriol. 169: 1801–1806.

Pasteur L (1861) Sur la fermentation visquese et la fermentation butyrique. Bull. Soc. Chim. Paris 11: 30–31.

Pederson CS & Albury MN (1969) The sauerkraut fermentation. NY State Agric. Expt. Sta. (Geneva, NY) Tech. Bull. Bulletin 824.

Pedrosa MC, Golner BB, Goldin BR, Barakat S, Dallal GE & Russell RM (1995) Survival of yogurt-containing organisms and *Lactobacillus gasseriA* (ADH) and their effect on bacterial enzyme activity in the gastrointestinal tract of healthy and hypocholorhydric elderly subjects. Am. J. Clin. Nutr. 61: 353–359.

Pérez PF, Minnaard J, Rouvet M, Knabenhans C, Brassart D, De Antoni GL & Schiffrin EJ (2001) Inhibition of *Giardia intestinalis* by extracellular factors from Lactobacilli: an in vitro study. Appl. Environ. Microbiol. 67: 5037–5042.

Perrin C, Guimont C, Bracquart P & Gaillard JL (1999) Expression of a new cold shock protein of 21.5 kDa and of the major cold shock protein by *Streptococcus thermophilus* after cold shock. Curr. Microbiol. 39: 342–347.

Poolman B, Royer TJ, Mainzer SE & Schmidt BF (1989) Lactose transport system of *Streptococcus thermophilus*: a hybrid protein with homology to the melibiose carrier and enzyme III of phosphoenolpyruvate-dependent phosphotransferase systems. J. Bacteriol. 171: 244–253.

Pouwels PH, Leer RJ, Shaw M, Heijne den Bak-Glashouwer MJ, Tielen FD, Smit E, Martinez B, Jore J & Conway PL (1998) Lactic acid bacteria as antigen delivery vehicles for oral immunization purposes. Int. J. Food Microbiol. 41: 155–67.

Prasad J, Gill HS, Smart JB & Gopal PK (1998) Selection and characterisation of *Lactobacillus* and *Bifidobacterium* strains for use as probiotics. Int. Dairy J. 8: 993–1002.

Rantsiou K, Phister T, McKay LL, Dunny G & Mills D (1999) Broad host range mobilization of plasmid derivatives by the lactococcal conjugal element pRS01. Proc. Sixth Symp. Lactic Acid Bact. E13.

Rattray FP & Fox PF (1999) Aspects of enzymology and biochemical properties of *Brevibacterium linens* relevant to cheese ripening: a review. J. Dairy Sci. 82: 891–909.

Rattray FP, Fox PF & Healy A (1997) Specificity of an extracellular proteinase from *Brevibacterium linens* ATCC 9174 on bovine beta-casein. Appl. Environ. Microbiol. 63: 2468–2471.

Reiter B & Oram JD (1982) Nutritional studies on cheese starter. 1. Vitamin and amino acid requirements of single strain starters. J. Dairy Res. 29: 63–68.

Richards M & Macrae RM (1964) The significance of the use of hops in regard to the biological stability of beer . II. The development of resistance to hop resins by strains of lactobacilli. J. Inst. Brewing 70: 484–488.

de Roos NM & Katan MB (2000) Effects of probiotic bacteria on diarrhea, lipid metabolism, and carcinogenesis: a review of papers published between 1988 and 1998. Am. J. Clin. Nutr. 71: 405–411.

Rozen S & Skaletsky HJ (1998) Primer3 http://www-genome.wi.mit.edu/genome_softwar e/other/primer3.html

Russell WM & Klaenhammer TR (2001a) Identification and cloning of *gusA*, encoding a new β-glucuronidase from *Lactobacillus gasseri* ADH. Appl. Environ. Microbiol. 67: 1253–1267.

Russell WM & Klaenhammer TR (2001b) An efficient system for directed integration into the *Lactobacillus acidophilus* and *Lactobacillus gasseri* chromosome via homologous recombination. Appl. Environ. Microbiol. 67: 4361–4364.

Saavedra JM, Bauman NA, Oung I, Perman JA & Yolken RH (1994) Feeding of *Bifidobacterium bifidum* and *Streptococcus thermophilus* to infants in hospital for prevention of diarrhoea and shedding of rotavirus. Lancet 344: 1046–1049.

Salama M, Sandine WE & Giovannoni S (1991) Development and application of oligonucleotide probes for identification of *Lactococcus lactis* subsp. *cremoris*. Appl. Environ. Microbiol. 57: 1313–1318.

Salema M, Capucho I, Poolman B, San Romão MV & Dias MC (1996) In vitro reassembly of the malolactic fermentation pathway of *Leuconostoc oenos* (*Oenococcus oeni*). J. Bacteriol. 178: 5537–5539.

Sanders ME & Klaenhammer TR (2001) Invited review: the scientific basis of *Lactobacillus acidophilus* NCFM functionality as a probiotic. J. Dairy Sci. 84: 319–331.

Sandine WE (1988) New nomenclature of the non-rod-shaped lactic acid bacteria. Biochemie 70: 519–522.

Schiffrin EJ, Rochat F, Link-Amster H, Aeschlimann J-M & Donnet-Hughes A (1995) Immunomodulation of human blood cells following the ingestion of lactic acid bacteria. J. Dairy Sci. 78: 491–497.

Schleifer K-H (1987) Recent changes in the taxonomy of lactic acid bacteria. FEMS Microbiol. Rev. 46: 201–203.

Schleifer KH & Ludwig W (1995) Phylogenetic relationships of lactic acid bacteria. In: Wood BJB & Holzapfel WH (Eds) The Genera of Lactic Acid Bacteria (pp 7-18). Chapman & Hall, London.

Serror P, Sasaki T, Ehrlich SD & Maguin E (2002) Electrotransformation of *Lactobacillus delbrueckii* subsp. *bulgaricus* and *L. delbrueckii* subsp . *lactis* with various plasmids. Appl. Environ. Microbiol. 68: 46–52.

Server-Busson C, Foucaud C & Leveau J-Y (1999) Selection of dairy *Leuconostoc* isolates for improtant technological properties. J. Dairy Res. 66: 245–56.

Sheih YH, Chiang BL, Wang LH, Liao CK & Gill HS (2001) Systemic immunity-enhancing effects in healthy subjects following dietary consumption of the lactic acid bacterium *Lactobacillus rhamnosus* HN001. J. Am. Coll. Nutr. 20(2 Suppl): 149–56.

Simpson WJ & Taguchi H (1995) The genus *Pediococcus*, with notes on the genera *Tetratogenococcus* and *Aerococcus*. In: Wood BJB & Holzapfel WH (Eds) The Genera of Lactic Acid Bacteria (pp 125-172). Chapman & Hall, London.

Slos P, Dutot P, Reymund J, Kleinpeter P, Prozzi D, Kieny MP, Delcour J, Mercenier A & Hols P (1998) Production of cholera toxin B subunit in *Lactobacillus*. FEMS Microbiol. Lett. 169: 29–36.

Solow BT & Somkuti GA (2000) Molecular properties of *Streptococcus thermophilus* plasmid pER35 encoding a restriction modification system. Curr. Microbiol. 42: 122–128.

Stuart M, Chou L-S & Weimer BC (1998) Influence of carbohydrate starvation and arginine on culturability and amino acid utilization of *Lactococcus lactis* subsp. *lactis*. Appl. Environ. Microbiol. 65: 665–673.

Steidler L, Robinson K, Chamberlain L, Schofield KM, Remaut E, Le Page RW & Wells JM (1998) Mucosal delivery of murine interleukin-2 (IL-2) and IL-6 by recombinant strains of *Lactococcus lactis* coexpressing antigen and cytokine. Infect. Immun. 66: 3183–3189.

Sutherland IW (1996) Extracellular polysaccharides. In: Rehm H-J, Reed G, Puhler A & Stadler P (Eds) Biotechnology, 2nd ed., Vol 6: Products of Primary Metabolism (pp 613–657). VCH, New York.

Tallgren AH, Airaksinen U, von Weissenberg R, Ojamo H, Kuusisto J & Leisola M (1999) Exopolysaccharide-producing bacteria from sugar beets. Appl. Environm. Microbiol. 65: 862–64.

Takahashi T, Nakagawa E, Nara T, Yajima T & Kuwata T (1998) Effects of orally ingested *Bifidobacterium longum* on the mucosal IgA response of mice to dietary antigens. Biosci. Biotechnol. Biochem. 62: 10–15.

Tannock GW (2000) Identification of lactobacilli and bifidobacteria. Curr. Issues Intest. Microbiol. 1: 39–50.

Tannock GW, Munro K, Harmsen HJ, Welling GW, Smart J & Gopal PK (2000) Analysis of the fecal microflora of human subjects consuming a probiotic product containing *Lactobacillus rhamnosus* DR20. Appl. Environ. Microbiol. 66: 2578–2588.

Tatusov RL, DNatale DA, Garkavtsev IV, Tatusova TA, Shankavaram UT, Rao BS, Kiryutin B, Galperin MY, Fedorova ND & Koonin EV (2001) The COG database: new developments in phylogenetic classification of proteins from complete genomes. Nucleic Acids Res. 29: 22–28.

Tejada-Simon MV & Pestka JJ (1999) Proinflammatory cytokine and nitric oxide induction in murine macrophages by cell wall and cytoplasmic extracts of lactic acid bacteria. J. Food Prot. 62: 1435–44.

Tenreiro R, Santos MA, Paveia H & Vieira G (1994) Inter-strain relationships among wine leuconostocs and their divergence from other *Leuconostoc* species, as revealed by low frequency restriction fragment analysis of genomic DNA. J. Appl. Bacteriol. 77: 271–280.

Thompson JK, McConville KJ, McReynolds C & Collins MA (1999) Potential of conjugal transfer as a strategy for the introduction of recombinant genetic material into strains of *Lactobacillus helveticus*. Appl. Environ. Microbiol. 65: 1910–1914.

Tissier H (1900) Recherches sur la flore intestinale des nourrissons (etat normal et pathologique) Paris Thèses: 1–253.

Tissier H (1906) Traitement des infections intestinales par la méthode de la flore bactérienne de l'intestin. Crit. Rev. Soc. Biol. 60: 359–361.

Tonon T, Bourdineaud J-P & Lonvaud-Funel A (2001) The arcABC gene cluster encoding the arginine deiminase pathway of *Oenococcus oeni*, and arginine induction of a CRP-like gene. Res. Microbiol. 152: 653–661.

Tynkkynen S, Satokari R, Saarela M, Mattila-Sandholm T & Saxelin M (1999) Comparison of ribotyping, randomly amplified polymorphic DNA analysis, and pulsed-field gel electrophoresis in typing of *Lactobacillus rhamnosus* and *L. casei* strains. Appl. Environ. Microbiol. 65: 3908–3914.

Ummadi M & Weimer BC (2001) Tryptophan metabolism in *Brevibacterium linens* BL2. J. Dairy Sci. 84: 1773–1782.

Valdes-Stauber N & Scherer S (1996) Nucleotide sequence and taxonomical distribution of the bacteriocin gene lin cloned from *Brevibacterium linens* M18. Appl. Environ. Microbiol. 62: 1283–1286.

Vaughan EE, van den Bogaard PTC, Catzeddu P, Kuipers OP & de Vos WM (2001) Activation of silent *gal* genes in the *lac-gal* regulon of *Streptococcus thermophilus*. J. Bacteriol. 183: 1184–1194.

Vesa T, Pochart P and Marteau P (2000) Pharmacokinetics of *Lactobacillus plantarum* NCIMB 8826, *Lactobacillus fermentum* KLD, and *Lactococcus lactis* Mg 1363 in the human gastrointestinal tract. Aliment. Pharmacol. Ther. 14: 823–828.

Viana R, Monedero V, Dossonnet V, Vadeboncoeur C, Perez-Martinez G & Deutscher J (2000) Enzyme I and HPr from *Lactobacillus casei*: their role in sugar transport, carbon catabolite repression and inducer exclusion. Mol. Microbiol. 36: 570–584.

van Vuuren HJJ & Dicks LMT (1993) *Leuconostoc oenos*: A review. Am. J. Enol. Viticult. 44: 99–112.

Wagner RD, Pierson C, Warner T, Dohnalek M, Hilty M & Balish E (2000) Probiotic effects of feeding heat-killed *Lactobacillus acidophilus* and *Lactobacillus casei to Candida albicans*-colonized immunodeficient mice. J. Food Prot. 63: 638–644.

Walker DC, Aoyama K & Klaenhammer TR (1996) Electrotransformation of *Lactobacillus acidophilus* group A1. FEMS Microbiol. Lett. 138: 233–237.

Weimer BC, Yi X & Brown R (2000) Autocatalytic processing of the protease from *Brevibacterium linens* BL2: a kinetic analysis for the degradation of casein. International Dairy Federation Biennial Cheese Flavor Conference, Banff, Alberta.

Wells JM, Robinson K, Chamberlain LM, Schofield km & Le Page RW (1996) Lactic acid bacteria as vaccine delivery vehicles. Antonie van Leeuwenhoek 70: 317–330.

Yasui H, Shida K, Matsuzaki T & Yokokura T (1999) Immunomodulatory function of lactic acid bacteria. Antonie van Leeuwenhoek 76: 383–389.

Ye JJ & Saier MH (1995) Cooperative binding of lactose and HPr(Ser-P) to the lactose:H^+ permease of *Lactobacillus brevis*. Proc. Natl. Acad. Sci. U.S.A. 92: 417–421.

Ye JJ, Reizer J, Cui X & Saier MH (1994) ATP-dependent phosphorylation of serine in HPr regulates lactose:H^+ symport in *Lactobacillus brevis*. Proc. Natl. Acad. Sci. U.S.A. 91: 3102–3106.

Yuki N, Watanabe K, Mike A, Tagami Y, Tanaka R, Ohwaki M & Morotomi M (1999) Survival of a probiotic, *Lactobacillus casei strain Shirota*, in the gastrointestinal tract: selective isolation from faeces and identification using monoclonal antibodies. Int. J. Food Microbiol. 48: 51–57.

Zapparoli G, Reguant C, Bordons A, Torriani S & Dellaglio F (2000) Genomic DNA fingerprinting of *Oenococcus oeni* strains by pulsed-field gel electrophoresis and randomly amplified polymorphic DNA-PCR. Curr. Microbiol. 40: 351–355.

Ze-Ze L, Tenreiro R, Brito L, Santos MA & Paveia H (1998) Physical map of the genome of *Oenococcus oeni* PSU-1 and localization of genetic markers. Microbiology 144: 1145–1156.

Ze-Ze L, Tenreiro R & Paveia H (2000) The *Oenococcus oeni* genome: Physical and genetic mapping of strain GM and comparison with the genome of a 'divergent' strain, PSU-1. Microbiology 146: 3195–3204.

Zhou JS, Shu Q, Rutherfurd KJ, Prasad J, Gopal PK & Gill HS (2000a) Acute oral toxicity and bacterial translocation studies on potentially probiotic strains of lactic acid bacteria. Food Chem. Toxicol. 38: 153–61.

Zhou JS, Shu Q, Rutherfurd KJ, Prasad J, Birtles MJ, Gopal PK & Gill HS (2000b) Safety assessment of potential probiotic lactic acid bacterial strains *Lactobacillus rhamnosus* HN001, *Lb. acidophilus* HN017, and *Bifidobacterium lactis* HN019 in BALB/c mice. Int. J. Food Microbiol. 56: 87–96.

Zhou JS, Gopal PK & Gill HS (2001) Potential probiotic lactic acid bacteria *Lactobacillus rhamnosus* (HN001), *Lactobacillus acidophilus* (HN017) and *Bifidobacterium lactis* (HN019) do not degrade gastric mucin in vitro. Int. J. Food Microbiol. 63: 81–90.

Zuniga M, Pardo I & Ferrer S (1996) Transposons Tn916 and Tn925 can transfer from *Enterococcus faecalis* to *Leuconostoc oenos*. FEMS Microbiol. Lett. 135: 179–185 .

Antonie van Leeuwenhoek **82:** 59–71, 2002.

Global control of sugar metabolism: a Gram-positive solution

Fritz Titgemeyer* & Wolfgang Hillen
*Lehrstuhl für Mikrobiologie, Friedrich-Alexander-Universität Erlangen-Nürnberg, Staudtstr. 5, 91058 Erlangen, Germany (*Author for correspondence; Tel.: +49-(0)9131-8528084; Fax: + 49-(0)9131-8528082; E-mail: ftitgem@biologie.uni-erlangen.de)*

Key words: carbon catabolite repression, carbon regulation, inducer exclusion, *Lactobacillus*, *Lactococcus*

Abstract

Bacteria utilise carbon sources in a strictly controlled hierarchical manner for which they have developed global control mechanisms that govern and coordinate carbon source-specific regulation. This is achieved via carbon catabolite repression (CCR), which is the result of global transcriptional control and inducer exclusion. A common mechanism for transcriptional control has evolved within the group of low-GC Gram-positive bacteria, including lactic acid bacteria. The seryl-phosphorylated form of the phosphotransferase HPr (HPr-ser-P) mediates CCR in concert with the pleiotropic regulator CcpA (catabolite control protein) by repressing or activating catabolite-controlled genes. HPr-ser-P can concomitantly trigger inducer exclusion by inhibition of carbohydrate-specific permeases. Histidyl-phosphorylated HPr (HPr-his P) is required for the transport of many carbon sources by the phosphotransferase system (PTS). In addition, HPr-his P controls carbohydrate-specific regulators and catabolic enzymes by phosphorylation. Thus, the ratio of HPr-his P/HPr-ser-P determines utilisation of a particular carbon source. This ratio is mainly adjusted by the bifunctional HPr kinase/phosphatase (HPrK/P), which itself is controlled by the metabolic state of the cell. As a result, the information about the metabolic state of the cell is combined with signals scoring the availability of carbon sources to fine-tune the expression of catabolic genes with the goal to optimise growth rate in any given mixture of nutrients. This review summarises the current understanding of carbon catabolite regulation in low-GC Gram-positive bacteria with special emphasis on lactic acid bacteria.

Introduction

When bacteria are exposed to a mixture of carbon sources they choose the substrate that yields a maximum profit for growth (Gunnewijk et al. 2001; Brückner & Titgemeyer 2002). Therefore, they have developed sophisticated mechanisms that enable them to sense the nutritional situation and adjust their catabolic capacities by regulatory responses that fall under the term carbon catabolite repression (CCR). Among those, global transcriptional control mechanisms and the protein control mechanism of inducer exclusion are probably the most important ones. Two principal solutions for global carbon control are currently recognised at molecular detail in the bacterial world. These have been described for species belonging to the enteric Gram-negative bacteria and the low-GC Gram-positive bacteria. Both mechanisms take advantage of the components of the phosphotransferase system (PTS) to integrate C-regulatory signals (Postma et al. 1993; Saier et al. 1996). They differ at the molecular level as transmission of CCR signals in enteric bacteria involves the PTS protein IIA^{Glc}. IIA^{Glc} P triggers the activity of the catabolite activator protein CAP that is required for gene activation of numerous catabolite-controlled genes (Figure 1; Saier et al. 1989). Enzyme IIA^{Glc} functions also in catabolic protein regulation via inducer exclusion. Low-GC Gram-positive bacteria, on the other hand, use the PTS protein HPr to exert inducer exclusion and to trigger the global regulator CcpA to confer CCR by global gene repression/activation (Figure 2). The general outcome of C-regulation is similar in both bacterial groups: preferential use of the carbon source allowing a maximal growth rate. In this review, we will first compare the principles of global carbon control in enteric and low-

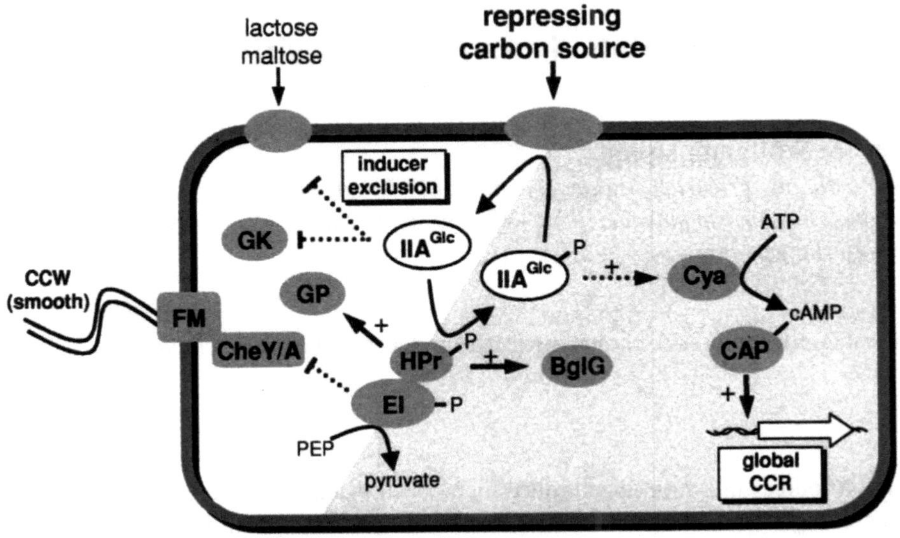

Figure 1. Carbon catabolite regulation in enteric Gram-negative bacteria. The model shows a bacterial cell under two conditions: (i) in the presence of a repressing carbon source (white background), and (ii) in the absence of a repressing carbon source (grey background). Regulation by activation (+) and repression (T) is indicated, whereby phosphorylation events are depicted by solid and other interactions by dashed arrows. The phosphorylation state of IIA^{Glc} is highlighted. Under repressing conditions, IIA^{Glc} is mainly dephosphorylated and mediates inducer exclusion. Under non-repressing conditions IIA^{Glc}-P stimulates cAMP synthesis and triggers, indirectly, activation of catabolite-repressed genes by the cAMP-CAP complex (global CCR). Abbreviations are as follows: BglG, antiterminator protein of the β-glucoside operon; CAP, catabolite activator protein; CCW, counter clockwise; CheY/A, chemotaxis proteins; Cya, adenylate cyclase; EI, PTS phosphotransferase enzyme I; FM, flagellar motor; GK, glycerol kinase; GP, glycogen phosphorylase; HPr, histidine-containing PTS phosphotransferase. For further explanations see text.

GC Gram-positive bacteria. We then will emphasise recent advancements in understanding the particular mechanisms in lactic acid bacteria (LAB) and discuss possible applications.

The enteric bacterial solution

The glucose-specific phosphotransferase enzyme IIA^{Glc} plays the central role in carbon catabolite regulation in *Escherichia coli* by transmitting C-regulatory signals via its phosphorylation state to various cellular targets (Figure 1). During PTS-mediated transport a phosphate group is transferred from phosphoenolpyruvate to the carbohydrate via the general phosphotransferases enzyme I (EI) and HPr, and one of more than 20 carbohydrate-specific enzyme II-ABC transport complexes (Postma et al. 1993). In the case of glucose transport, IIA^{Glc} transfers the phosphate group to the IIB domain of the permease $IIBC^{Glc}$ which catalyses transport and concomitant phosphorylation of the sugar. When other PTS substrates are imported, the phosphate group is drained to these systems with the consequence that IIA^{Glc} is mainly present in its unphosphorylated form. Recent work on carbon regulation exerted by non-PTS sugars showed that the phosphorylation state of IIA^{Glc} is also modulated (Hogema et al. 1998). Growth on lactose or glucose-6-phosphate changes the phosphorylation of IIA^{Glc}, presumably by influencing the PEP/pyruvate ratio. Thus, under all described conditions the amounts of $IIA^{Glc}/IIA^{Glc}{\sim}P$ reflect the physiological state of the cell and lead to the proper transmission of signals for global carbon control.

Global transcriptional control

The link between IIA^{Glc} and CCR is the stimulation of adenylate cyclase by $IIA^{Glc}{\sim}P$ to produce cAMP, which is required for the catabolite activator protein CAP to activate catabolite-controlled genes (Figure 1; Busby & Ebright 1999; Brückner & Titgemeyer 2002). This occurs usually in a coordinated fashion with substrate specific repressor or activator proteins, which respond to the presence of the specific carbon source. CAP is further involved in regulation of other cellular processes such as sigma factor selectivity (Colland et al. 2000; Ueguchi et al. 2001) and

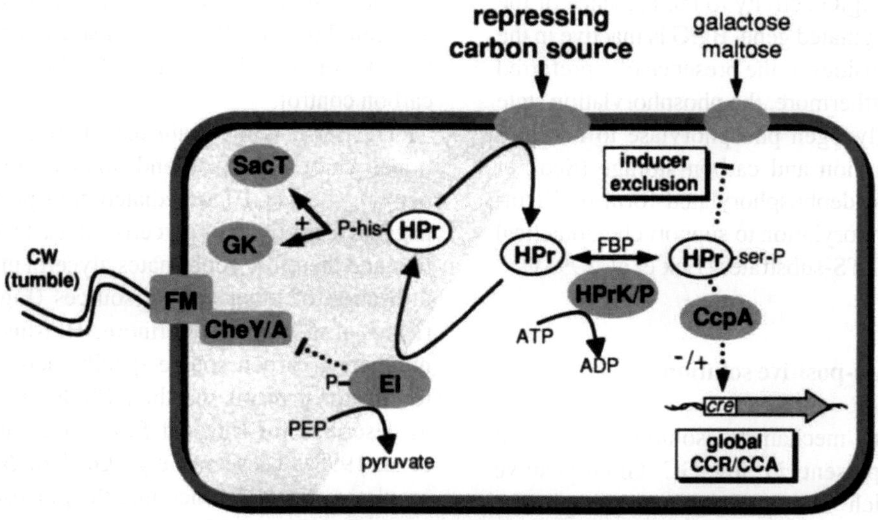

Figure 2. Carbon catabolite regulation in low-GC Gram-positive bacteria. The model shows a bacterial cell under two conditions: (i) in the presence of a repressing carbon source (white background), and (ii) in the absence of a repressing carbon source (grey background). Regulation by activation (+) and repression (T) is indicated, whereby phosphorylation events are depicted by solid arrows and other interactions by dashed arrows. The phosphorylation state of HPr is highlighted. Under repressing conditions, HPr is mainly serine-phosphorylated (HPr-ser-P), mediating inducer exclusion and CcpA-dependent carbon catabolite repression/activation (global CCR/CCA). Under non-repressing conditions HPr-his-P predominates and activates PTS-dependent transport, glycerol kinase (GK), and substrate-specific regulators like SacT. Abbreviations are as follows: CcpA, catabolite control protein; CheY/A, chemotaxis proteins; *cre*, catabolite responsive element; CW, clockwise; EI, PTS phosphotransferase enzyme I; FM, flagellar motor; FBP, fructose-1,6-bisphosphate; GK, glycerol kinase; HPr, histidine-containing PTS phosphotransferase; HPrK/P, HPr kinase/phosphatase; SacT, antiterminator protein of the sucrose operon. For further explanations see text.

flagellum biosynthesis (Soutourina et al. 1999). In total, CAP has been estimated to control more than 100 genes in *E. coli*, including its own.

Other pleiotropically acting factors of carbon regulation are Cra (formerly called FruR) and Mlc (Saier et al. 1996; Plumbridge et al. 2001). Cra is the repressor of the fructose-PTS operon and an activator of expression of the key genes for gluconeogenesis, the Krebs cycle, and the glyoxylate shunt. Its activity depends on the metabolites fructose-1-phosphate and fructose-1,6-bisphosphate. Mlc contributes to regulation of *malT*, encoding the activator of the maltose regulon, and to the genes encoding the PTS permeases specific for glucose and mannose. Mlc regulates expression of the *ptsHIcrr* operon (encoding HPr, EI and IIA^{Glc}) in conjunction with Cra and cAMP-CAP, indicating the interplay of these globally acting elements in *E. coli*.

Inducer exclusion

Unphosphorylated IIA^{Glc} exerts inducer exclusion by allosteric inhibition of cytoplasmic targets such as glycerol kinase, preventing the formation of the inducer glycerol-3-phosphate, or by blocking permeases specific for lactose, maltose, melibiose, or raffinose (Figure 1; Titgemeyer et al. 1994; Brückner & Titgemeyer 2002). The latter mechanism appears to be enhanced when the cognate substrate is present, suggesting an economic use of IIA^{Glc} molecules. The action of IIA^{Glc} on inducer exclusion targets is understood in molecular detail, e. g., by the X-ray structure of the complex of IIA^{Glc} with glycerol kinase, or by mutational analysis of the interaction of IIA^{Glc} with lactose permease LacY (Hurley et al. 1993; Seok et al. 1997). Analysis of the *E. coli* genome revealed several ORFs possessing a consensus sequence for IIA^{Glc} recognition. Thus, more regulatory interactions may exist (Sondej et al. 1999). A possible link between C-regulation and stress is indicated by the involvement of IIA^{Glc} in transcriptional and translational control of RpoS, the major sigma factor of the stress regulon (Ueguchi et al. 2001).

Other PTS components, beside IIA^{Glc}, are also involved in carbon regulation (Figure 1). The antiterminator protein BglG regulates expression of the genes for β-glucoside utilisation. It contains so-called PTS regulation domains (PRD) which are phosphorylated by HPr and the respective enzyme II in a complex manner (Stülke et al. 1998; Görke & Rak

1999). This links BglG activity to the presence of the substrate for the regulated gene. BglG is inactive in the absence of β-glucosides or the presence of a preferred PTS substrate. Furthermore, the phosphorylation state of HPr triggers glycogen phosphorylase to coordinate sugar consumption and carbon storage (Seok et al. 2001), and the dephosphorylated form of EI inhibits CheA phosphorylation to support chemotactical response towards PTS-substrates (Lux et al. 1995).

The low-GC Gram-positive solution

A totally different mechanistic solution for global C-regulation is present in low-GC Gram-positive bacteria, in which HPr, HPr kinase/phosphatase (HPrK/P), and the catabolite control protein CcpA are involved in sensing the physiological state of the cell and regulating carbon consumption (Figure 2; Stülke & Hillen 2000). The PTS protein HPr reflects by its phosphorylation state the nutritional situation of the cell and therefore can be considered as the functional equivalent of IIA^{Glc} of enteric bacteria. But other than IIA^{Glc}, HPr is phosphorylated at a second exclusively regulatory site, serine 46, by a unique HPrK/P, which is not present in enteric bacteria (Reizer et al. 1998; Kravanja et al. 1999). The balance between HPr-his-P and HPr-ser-P is adjusted as follows (Figure 2): PEP-dependent phosphorylation of HPr by EI yields HPr-his-P, which is required for PTS-mediated transport of carbon sources. As in enteric bacteria the PTS serves as a major uptake system of carbohydrates in low-GC Gram-positive bacteria. This is indicated by the presence of fifteen sugar-specific PTS permeases in *B. subtilis*, nine in *Lactococcus lactis* and eight in *Streptococcus pneumoniae* (Reizer et al. 1999; Bolotin et al. 2001; Hoskins et al. 2001). When PTS-substrates are metabolised, the level of glycolytic intermediates such as fructose-1,6-bisphosphate rises and stimulates HPrK/P to generate HPr-ser-P. This form is required for CCR (see below). Concomitantly the rate of sugar consumption by the PTS is reduced, because HPr-ser-P is not a substrate for EI and does not participate in uptake of carbon sources (Deutscher et al. 1994). As soon as the level of glycolytic intermediates drops and the level of inorganic phosphate rises, the bifunctional HPrK/P catalyses the reverse reaction by dephosphorylation of HPr-ser-P yielding HPr, which then leads to again increased PTS-dependent transport (Kravanja et al. 1999). Since the metabolic state of the cell is the signal determining the kind of HPr phosphorylation it is immediately apparent how non-PTS carbon sources also influence the ratio of HPr-his-P/HPr-ser-P which, in turn, exerts carbon control.

HPr-ser-P is the main player in global transcriptional carbon control and inducer exclusion (see below). Beside PTS-mediated transport HPr-his P-triggers the activity of glycerol kinase by phosphorylation and therefore coordinates glycerol utilisation with utilisation of other carbon sources (Figure 2; Charrier et al. 1997). Furthermore, HPr-his-P controls a number of carbon source-specific activator and antiterminator proteins via their PRDs in a similar way as described for BglG of *E. coli* (see above) (Stülke et al. 1998). These are, e.g., GlcT of *B. subtilis*, the regulator of *ptsGHI* encoding the glucose-specific enzyme II, HPr and EI; and SacT, the regulator of the sucrose-PTS (Stülke & Hillen 2000). As in *E. coli*, EI of *B. subtilis* serves also in signalling of chemotactical responses by interaction of EI P with the CheA signal protein (Garrity et al. 1998).

Transcriptional control exerted by CcpA

HPr-ser-P is a co-repressor for the regulator CcpA, which exerts global transcriptional control (Deutscher et al. 1995). CcpA was discovered as the factor that triggers glucose repression of the *B. subtilis* amylase gene *amyE* by binding to a *cis*-acting sequence termed catabolite responsive element (*cre*) (Henkin et al. 1991; Hueck et al. 1995; Miwa et al. 2000). Subsequent studies revealed that CcpA is generally involved in CCR of catabolite-controlled genes such as those required for the utilisation of gluconate, xylose or histidine. Interestingly, the position of the *cre* box with respect to the promoter varies suggesting different modes of transcriptional control by CcpA. The *cre* of *amyE*, e.g., is located within the promoter and, hence, binding of CcpA may prevent initiation of transcription. The *cre* sites of the *hut* (histidine utilisation) operon are within the coding region, where they cause a roadblock of transcription (Wray et al. 1994). Some genes like those of the gluconate operon (*gnt*) of *B. subtilis* or those of the xylose operon (*xyl*) of *B. megaterium* contain multiple *cre* boxes, which are thought to contribute synergistically to CCR depending on special environmental conditions (Gösseringer et al. 1997; Miwa et al. 1997). CcpA is also an activator of transcription. Examples are genes for glycolysis like phosphofructokinase, genes involved in assimilation of ammonium like glutamate synthase, and genes contributing to overflow metabolism such as acetate kinase (*ackA*) (Henkin et al. 1996; Faires et al. 1999; Tobisch et al. 1999). The mechanism of CcpA-mediated gene activation is less well defined due to the fact that in some cases a *cre* box is absent. However, *cre* sites are crucial for the activation of the genes *ackA* and *pta* (phosphotransacetylase) for acetate production (Turinsky et al. 1998; Presecan-Siedel et al. 1999). Activation of *pta* required in addition to the CcpA/HPr-ser-P complex the cofactor fructose-1,6-bisphosphate. Since these *cre* sites are located upstream of the promoter, it appears that CcpA-binding stabilises the activity of the RNA polymerase–promoter complex to enhance transcription. Another striking phenotype of *ccpA* mutants observed in *bacilli*, *staphylococci* and several LAB is a reduced growth rate. This is attributed in *B. subtilis* to the lack of *gltAB* activation encoding glutamate synthase, which leads to inefficient assimilation of a nitrogen source (Faires et al. 1999). In addition, the CcpA-dependent activation of glycolytic genes, which was demonstrated for the *gap* operon of *B. subtilis* and the *las* operon of *L. lactis*, may further contribute to this phenotype (Luesink et al. 1998; Ludwig et al. 2001).

The pleiotropic role of CcpA has been investigated by proteome analysis (Küster-Schöck et al. 1999; Tobisch et al. 1999). Out of 1500 visible proteins on two-dimensional gels, 39 were subject to CcpA-mediated repression and four were subject to CcpA-mediated activation. A similar analysis of the *Enterococcus faecalis* proteome revealed 16 overexpressed and six repressed polypeptides in a *ccpA* mutant (Leboeuf et al. 2000a,b). The majority of the 16 proteins were also identified as glucose starvation proteins, indicating that CcpA may regulate carbon source starvation. The role of CcpA in glucose-dependent repression and activation has been further addressed by transcriptome analysis (Moreno et al. 2001). About 300 of the *B. subtilis* genes were more than threefold regulated by CcpA in a direct or indirect manner with 75% of those genes being repressed by CcpA. In a similar study a much smaller number of CcpA-regulated genes has been reported (Yoshida et al. 2001). This can be due to differences in growth conditions and a more stringent regulation factor of fivefold applied to identify candidate genes. Another genome-wide study revealed 126 *cre* sites in the genome of *B. subtilis* (Miwa et al. 2000). *In vivo* testing of a representative fraction showed that at least two third of those are physiologically relevant. Taken together, the role of CcpA as the key regulator for CCR is emphasised by proteome and transcriptome analysis.

Until 1995, it remained unclear how CcpA receives C-regulatory signals. *In vivo* analysis of mutations in *ptsH*, in which the codon for the regulatory residue serine 46 was replaced by an alanine codon, were impaired in CCR of several catabolic systems, indicating a link between HPr and CCR (Deutscher et al. 1994). Subsequent *in vitro* studies demonstrated an interaction of HPr-ser-P with CcpA (Deutscher et al. 1995) and showed that the complex of CcpA with HPr-ser-P is required for efficient binding to *cre* sites (Küster-Schöck et al. 1999). An NMR study revealed that two HPr monomers bind to the CcpA dimer and led to the identification of amino acid residues of HPr that are involved in CcpA interaction (Jones et al. 1997). These are near the surface areas around the phosphorylation sites histidine 15 and serine 46. It may be noteworthy that other effectors of CcpA beside HPr-ser-P may be of physiological relevance. In *in vitro* experiments it was shown that addition of NADP and glucose-6-phosphate, respectively, increased the binding of CcpA to different *cre* target sites (Gösseringer et al. 1997; Miwa et al. 1997; Kim et al. 1998).

An HPr paralogue termed Crh (catabolite repression HPr-like protein) is present in *B. subtilis* (Galinier et al. 1997). Crh only has the serine 46 phosphorylation site but lacks histidine 15. It contributes in addition to HPr to the serine-phosphorylation dependent pathway. It appears that Crh is restricted to *bacilli*, because it was not found in the genomes of *Staphylococcus aureus*, *E. faecalis*, *S. pyogenes* and *L. lactis*.

In addition to CcpA, the regulators CcpB, CcpC, AbrK and GerE are involved in carbon regulation. However, their contribution to global carbon control appears to be restricted to specific growth conditions and remains to be elucidated further (Fisher et al. 1994; Chauvaux et al. 1998; Jourlin-Castelli et al. 2000; Moreno et al. 2001).

Carbon regulation in LAB

While most molecular data of carbon regulation originate from investigation of *bacilli*, there is increasing data from LAB. They underline the importance of understanding carbon regulation in these bacteria to elucidate the best nutritional conditions in dairy and other fermentation processes.

Table 1. CcpA of lactic acid bacteria

Organism	CcpA (aa)	Target systems	PID[a] / Reference
Enterococcus faecalis[b]	333	↓ galactose operon ↑ lactate dehydrogenase ↓ glc-starvation genes	g3676165 / Leboeuf et al. (2000)
Lactobacillus casei	333	↓ N-acetyl-glucosaminidase ↓ β-galactosidase ↓ sorbose operon ↓ lactose operon	g2668503 / Monedero et al. (1997) Gosalbes et al. (1999 Yebra et al. (2000
Lactobacillus delbrueckii subsp. lactis	333	↑ peptidase genes	g3024388 / Stucky et al. (1996) Schick et al. (1999
Lactobacillus delbrueckii subsp. bulgaricus	333	↑ peptidase Q	g7209172 / Morel et al. (1999) Morel et al. (2001
Lactobacillus pentosus	336	↓ xylose operon ↓ β-glucosidase ↓ β-galactosidase ↓ phosphoketolase	g5823363 / Lokman et al. (1997) Mahr et al. (2000 Chaillou et al. (2001 Posthuma et al. (2002
Lactobacillus plantarum	336	↓ β-glucosidase ↓ β-galactosidase	g14331034 / Muscariello et al. (2001)
Lactococcus lactis	332	↓ galactose operon ↓ β-glucoside operon ↑ glycolytic *las* operon	g4165285 / Luesink et al. (1998) Monedero et al. (2001b
Streptococcus gordonii	334	↓ amylase-binding protein A	g12240237 / Rogers & Scannapieco (2001)
Streptococcus mutans[c]	333	↑ α-galactosidase ↑ mannitol-1-P dh ↑ P-β-galactosidase	g2323342 / Simpson & Russell (1998)
Streptococcus thermophilus	333	↓ lactose operon ↑ lactate dehydrogenase	g7341263 / van den Bogaard et al. (2000)

The table shows all CcpA and CcpA-like systems of LAB that have been characterised in genetic and biochemical detail. The arrows indicate CcpA-dependent repression (↓) or activation (↑) of investigated target systems.
[a]PID, protein identification number.
[b]*E. faecalis* has been more thoroughly studied by proteome analysis, which revealed in addition a number of genes that are regulated in a CcpA-dependent way (Leboeuf et al. 2000b).
[c]Analysis of the CcpA-like protein RegM in *S. mutans* showed activation of the three listed catabolic enzymes, although it was expected that they are subject to CCR. This observation is opposite to the generally described repressor function of CcpA and suggests that RegM of *S. mutans* operates differently in carbon regulation.

Since 1996, genes encoding CcpA and CcpA-like proteins have been described from a number of LAB (Table 1). Mutation of *ccpA* showed phenotypes that were equivalent to those known from other low-GC Gram-positive bacteria. It was, e.g., shown that xylose and galactose utilisation are subject to CcpA-mediated glucose repression in *Lactobacillus pentosus* and *L. lactis*, respectively (Lokman et al. 1997; Luesink et al. 1998). CcpA-dependent gene activation was demonstrated for the glycolytic *las* operon of *L. lactis* (Luesink et al. 1998) and, as in *bacilli*, a reduced growth phenotype of *ccpA* mutants was also observed (Monedero et al. 1997; Luesink et al. 1998; Leboeuf et al. 2000). The genetic organization of *ccpA* genes in LAB is without exception *pepQ-ccpA*, where *pepQ*, which encodes a prolidase, and *ccpA* are always divergently transcribed (Mahr et al. 2000). Other low-GC Gram-positive bacteria do not have *pepQ* linked to *ccpA*. Northern blot analyses showed that *ccpA* is transcribed as a monocistronic unit and appears to be autoregulated in most LAB, presumably via *cre* located in the *ccpA* promoter region (Monedero et al.

1997; Luesink et al. 1998; Simpson & Russell, 1998; Schick et al. 1999; Mahr et al. 2000; Le boeuf et al. 2000; Morel et al. 2001).

The discovery of the *ptsK* gene (also termed *hprK*) encoding HPr kinase/phosphatase in *B. subtilis* in 1998 (Galinier et al. 1998; Reizer et al. 1998) was shortly followed by gene cloning reports of *hprK* from several LAB, namely *E. faecalis*, *Streptococcus salivarius*, *Lb. casei* and *Lactobacillus brevis* (Kravanja et al. 1999; Brochu & Vadeboncoeur 1999; Dossonnet et al. 2000; Djordjevic et al. 2001). In the course of genome sequencing projects, *hprK* was further found in *S. mutans*, *Streptococcus bovis*, *S. pneumoniae*, *S. pyogenes*, *Lb. delbrueckii* and *L. lactis*, indicating its general occurrence in LAB. Characterisation of *E. faecalis* HPrK/P revealed the bifunctional nature of the enzyme (Kravanja et al. 1999). The kinase activity of *E. faecalis* HPrK/P is stimulated by high ATP levels and is inhibited by high levels of inorganic phosphate, whereas low cellular ATP levels favour the phosphatase activity. In contrast, low ATP levels alone are not sufficient to switch on the phosphatase activity of HPrK/P of *B. subtilis*. Stimulation of the kinase reaction by the glycolytic intermediate fructose-1,6-bisphosphate was reported for *B. subtilis* and *Lb. casei*, while HPrK/P of *E. faecalis* and *S. salivarius* showed no stimulatory effect (Reizer et al. 1998; Kravanja et al. 1999; Brochu & Vadeboncoeur 1999; Dossonnet et al. 2000). Thus, the overall function of HPrK/P as a bifunctional kinase/phosphatase is conserved. However, stimulation and inhibition by cellular metabolites appears to be species-specific.

Carbon regulation of Lactobacillus casei

Lb. casei, one of the most studied species among LAB, was mutated in the genes encoding CcpA, EI and HPr (Monedero et al. 1997; Viana et al. 2000). *In vivo* analyses of such mutants revealed that the enzymes *N*-acetylglucosaminidase and phospho-β-galactosidase are subject to CcpA-dependent CCR exerted by several sugars. A *cre* site was found in the promoter region of the *lacTEGF* operon, which encodes the antiterminator protein LacT, the lactose-PTS permeases LacE ($IICB^{Lac}$) and LacF (IIA^{Lac}), and phospho-β-galactosidase LacG. Deletion of the *cre* site or a mutation of *ccpA* abolished CCR of a *lac* promoter fusion to the reporter gene *gusA* (Monedero et al. 1997). In the presence of glucose, the *ccpA* mutant showed a transcript that did not proceed through a terminator structure present upstream of the operon. Further mutational analysis revealed that LacT is required for induction of the *lac* operon by antitermination in the presence of lactose (Gosalbes et al. 1999). LacT, which belongs to the class of PRD-containing regulators, itself is apparently controlled by the common PTS elements and enzyme $IICB^{Lac}/IIA^{Lac}$ permease complex via phosphorylation at histidine residues located in the two PRD domains (Gosalbes et al. 2002). The *sor* operon required for utilisation of sorbose is also regulated by CcpA (Yebra et al. 2000). Glucose repression of this system was dependent on CcpA but also on PTS components including the mannose/glucose-specific enzyme II permease, which indicated that PTS-dependent transport of glucose is required for the CcpA signal transduction pathway. Strains carrying chromosomal *ptsH* alleles encoding mutations of serine 46 show a lack of diauxic growth in media containing glucose plus either lactose, maltose, or ribose (Viana et al. 2000). Furthermore, the activity of *N*-acetylglucosaminidase was completely deregulated from glucose repression. Thus, global CCR in LAB seems to operate via the CcpA/HPr-ser-P complex, which was confirmed *in vitro* with the purified components (Mahr et al. 2002). Surface plasmon resonance experiments demonstrated that only HPr-ser-P but no other HPr species bind to CcpA. Binding of this complex to the *cre* site of the *lac* operon was demonstrated by electrophoretic mobility shift assays. The role of *cre* in another LAB, *Lactobacillus plantarum*, was directly demonstrated. An *in vivo* footprint revealed that two G residues of *cre* in *bglH*, encoding a phospho-β-glucosidase, were protected in cells grown in the presence of glucose (Marasco et al. 1998).

HPrK/P of *Lb. casei* has been studied in detail by mutational and structural analysis (Dossonnet et al. 2000; Fieulaine et al. 2001). An *hprK* mutant showed a deficiency in CCR as determined by assaying *N*-acetylglucosaminidase activity in the absence and presence of glucose. Recently, the structure of a truncated, active *Lb. casei* HPrK/P protein was determined at 2.8 Å resolution, representing the first structure of a bacterial Ser/Thr kinase and lacking any resemblance to eucaryotic kinases (Fieulaine et al. 2001). The ATP-binding site (Walker motif) forms a typical P-loop and binds inorganic phosphate. In a parallel study, mutants with a reduced phosphatase activity were isolated (Monedero et al. 2001a). The mutations were located in the Walker motif or within a conserved C-terminal region that contacts the ATP-binding site of an adjacent subunit of the HPrK/P hexamer. The *Lb. casei* alleles were placed into the *B. subtilis* genome and it was shown that the strain carrying *hprKV267F* had a growth deficiency on PTS and many non-PTS carbon sources. This was attributed to the increased amounts of HPr-ser-P shutting off PTS transport activity due to the lack of free HPr. At the same time, CCR was fully switched on resulting in CcpA/HPr-ser-P-dependent repression of catabolic genes encoding utilisation of non-PTS carbon sources. This observation was supported by combination with mutations in *ccpA* and *ptsH*. An *hprKV267F ccpA* double mutant could grow on non-PTS carbon sources, whereas mutation of serine 46 in HPr restored growth on PTS and non-PTS carbohydrates.

CcpA-control of catabolic and central metabolic genes of Lb. pentosus

Lb. pentosus is a heterolactic lactic acid bacterium associated with the fermentation of cucumber, olives and Sauerkraut. Cloning and analysis of the *ccpA* gene revealed that its expression is autoregulated by a tandem promoter that comprises a *cre* site (Mahr et al. 2000). Catabolic gene regulation has been documented by demonstration of CcpA-mediated repression of the *xylAB* operon, encoding xylose isomerase and xylulose kinase (Lokman et al. 1997). The enzymes β-glucosidase and β-galactosidase, which are inducible by cellobiose and lactose, respectively, were also found to be under control of CcpA (Chaillou et al. 2001). These enzymes were repressed upon addition of the sugars glucose, mannose, or fructose, which are transported by a glucose/mannose/*N*-acetylglucosamine-specifi c enzyme II^{Man} and by a fructose-specific enzyme II. In a recent study, it was reported that CcpA is further involved in regulation of the central metabolic phosphoketolase pathway (Posthuma et al. 2002). The key enzyme of this pathway is xylulose-5-phosphate phosphoketolase (XpkA), that connects the upper part of the phosphoketolase pathway with the lower part of the glycolytic pathway by conversion of xylulose-5-phosphate into glyceraldehyde-3-phosphate. The sequencing of *xpkA* revealed a *cre* located within the promoter region. XpkA activity was high when *Lb. pentosus* was grown in the presence of gluconate or ribose, which are metabolised via the phosphoketolase route. It was low when glucose or fructose, which are metabolised via glycolysis, was additionally present. This effect was abolished in a *ccpA* mutant strain. On this basis, it was suggested that carbon source signals lead to CcpA-mediated repression of *xpkA* to adjust the activity of the phosphoketolase pathway (Posthuma et al. 2002).

CcpA-mediated repression/activation in L. lactis

The CcpA/HPr-ser-P-dependent regulation of CCR is also present in *L. lactis*, where a *ccpA* mutant was insensitive to glucose repression of the galactose operon (Luesink et al. 1998). A putative *cre* was found in the *gal* promoter exactly at the beginning of the *gal*-mRNA. The involvement of HPr-ser-P was indicated by phenotypes obtained with *ptsH* alleles (Luesink et al. 1999). Growth on galactose was strongly impaired when HPrS46D was present, while HPrS46A showed no effect on galactose fermentation. Since HPrS46D but not HPrS46A mimics a serine-phosphorylated HPr, this result is in agreement with CcpA/HPr-ser-P-mediated regulation. This was further substantiated by regulation of genes encoding a β-glucoside-specific PTS, which involves for proper regulation the regulator BglR of the antiterminator family (Monedero et al. 2001b).

CcpA-dependent catabolite activation was also found in *L. lactis* (Luesink et al. 1998, 1999). Growth on glucose resulted in higher activities of the glycolytic key enzymes phosphofructokinase (Pfk), pyruvate kinase (Pyk), and L-lactate dehydrogenase (Ldh), the genes of which form the tricistronic *las* operon. A *ccpA* mutation resulted in reduction of expression of this operon. This indicated that CcpA may be a transcriptional activator, most likely via *cre* located immediately upstream of the −35 region of the *las* promoter. The involvement of HPr-ser-P was indicated by overexpression of HPrS46D in the wild-type strain, which led to higher activities of pyruvate kinase and L-lactate dehydrogenase in galactose-grown cells. This was not the case when HPrS46A was overproduced. Since overexpression of HPrS46D in a *ccpA* deletion mutant did not result in enhanced enzyme activity, it was concluded that HPr-ser-P is required for CcpA-mediated activation of the *las* operon. However, deletion of *ptsH* led only to a 30% reduction of the glycolytic enzyme activities, indicating that regulation of the *las* operon is not exclusively dependent on an intact *ptsH* gene. Thus, it is possible that other effectors of CcpA are involved in *las* operon activation.

Taken together, the data demonstrate that carbon regulation in *L. lactis* includes CcpA-dependent gene repression and activation. The physiological function of glucose-dependent activation of glycolysis genes can be explained by an adjustment to increase the metabolic rate to better accommodate growth on the incoming sugar. Furthermore, activation of the *las* operon could be a mechanism for the prevention of end product accumulation, since the Las enzymes are required to generate glycolytic end products that can easily leave the cell (Luesink et al. 1998). The latter process has also been observed in *B. subtilis* for secretion of acetate and acetoin (Turinsky et al. 2000).

Special features of CCR of Streptococcus thermophilus

S. thermophilus, a bacterium that is used in starter cultures for yoghurt and mozzarella cheese, possesses a few unique features with respect to carbon regulation. In contrast to most other LAB, *S. thermophilus* prefers lactose over glucose and uses non-PTS transporters for the uptake of these carbon sources. Lactose is taken up by the permease LacS either by lactose/H^+ symport or by lactose/galactose exchange driven by the sugar concentration gradients (Gunnewijk et al. 2001). The latter mode shows that *S. thermophilus* uses the glucose moiety of lactose for growth and expels galactose into the medium. Growth on lactose causes carbon regulation mediated by CcpA as demonstrated by activation of a glycolytic gene encoding lactate dehydrogenase that, unlike in *L. lactis*, does not form an operon with the genes encoding pyruvate kinase and phosphofructokinase (van den Bogaard et al. 2000). This mechanism highlights the fact that a non-PTS carbon source can also trigger CcpA for carbon regulation.

CcpA is further involved in transcriptional regulation of LacS. When cells were grown in the presence of lactose, LacS levels were twofold higher, while a mutation in *ccpA* caused a sevenfold increase of LacS and led to massive expulsion of glucose, resulting in a slower growth rate and less cell mass. Thus, the down-regulation of *lacS*, presumably by CcpA/HPr-ser-P, reduces an overcapacity of lactose import in order to adjust a rate-limiting glycolytic flux that is optimal for cell growth (Gunnewijk & Poolman 2000; van den Bogaard et al. 2000). To make it more complex, the transport activity of LacS is additionally controlled at the protein level by the PTS via a IIA^{Glc}-like domain in the LacS C-terminus that is phosphorylated by HPr-his-P (Gunnewijk et al. 2001). This enhances the transport activity by an increase in V_{max} of the lactose/galactose exchange reaction. Based on the finding that the ratio of HPr-ser-P/HPr-his-P is shifted in the stationary growth phase towards more HPr-his-P, it was suggested that under these conditions the higher LacS activity can compensate for lower lactose concentrations. The multifaceted regulation of lactose utilisation by LacS in *S. thermophilus*, which includes autoregulation at the gene and protein level, illustrates how a cell achieves finely tuned control of carbon source utilisation (Gunnewijk et al. 2001).

Inducer exclusion/expulsion in LAB

Two mechanisms of carbon regulation operating at the protein level have been reported in members of LAB. These are inducer exclusion and inducer expulsion which have not been detected in other low-GC Gram-positive bacteria so far (Saier et al. 1996; Viana et al. 2000). Based on *in vitro* analyses, a direct role of HPr-ser-P has been proposed to drive both processes (Ye et al. 1994a,b).

Inducer exclusion describes the allosteric inhibition of a catabolic protein, e.g. a permease, which prevents the entry or formation of the inducer for the respective catabolic genes (Saier 1989). *In vivo* investigation of HPr mutants provided strong physiological evidence that HPr-ser-P confers inducer exclusion in *Lb. casei* (Viana et al. 2000). This was demonstrated by maltose uptake studies of strains carrying different *ptsH* alleles. The wild-type strain and a *ccpA* mutant showed a marked decrease in maltose uptake when glucose was added, which indicated a CcpA-independent regulatory mechanism. It was abolished when the strain carried mutations at HPr serine 46 or at the adjacent position isoleucine 47. The same observation was made in an *hprK* mutant, which is unable to generate HPr-ser-P (Dossonnet et al. 2000). Similar results were reported for *L. lactis*, where maltose and ribose uptake were relieved from glucose inhibition in an HPrS46A mutant (Monedero et al. 2001b). A third example comes from *Lb. brevis*, where it was indirectly shown that the galactose/H^+ symporter is under carbon control involving HPr-ser-P (Djordjevic et al. 2001). However, a direct demonstration of allosteric inhibition of a permease by HPr-ser-P remains to be given.

The mechanism of inducer expulsion describes the rapid expulsion of a previously internalised non-preferred carbon source, e.g., the lactose-derivative methyl-β-D-thiogalactoside (TMG), when exposed to a preferred carbon source afterwards (Reizer et al.

1983; Saier et al. 1996). This is realised by a two-step process, which includes dephosphorylation of the internalised sugar-phosphate and subsequent expulsion. At present, no *in vivo* evidence supports a role of HPr-ser-P in inducer expulsion, as has been suggested from *in vitro* studies (Ye et al. 1994b). Comparison between isogenic wild type and HPrS46A mutant strains of *Lb. casei*, *L. lactis* and *Lb. brevis* revealed no significant changes regarding inducer expulsion (Dossonnet et al. 2000; Djordjevic et al. 2001; Monedero et al. 2001a). Therefore, it remains to be seen what molecular basis underlies the phenomenon of inducer expulsion observed in many LAB.

Carbon regulation and proteolysis

Many LAB possess a complex proteolytic system to efficiently break down caseins, the major milk proteins. The proteolytic system of *L. lactis*, e.g., involves a cell wall bound protease, several peptide transporters, and more than a dozen cytoplasmic peptidases (Kunji et al. 1996; Guédon et al. 2001). Casein breakdown and sugar fermentation contribute to the flavour and texture of dairy products. The fact that *ccpA* in LAB is always linked to a divergently transcribed *pepQ* gene encoding a proline-specific peptidase led to the suggestion that expression of this and perhaps other *pep* genes may be regulated by CcpA and may therefore be coordinated with carbon regulation (Stucky et al. 1996; Morel et al. 1999; Mahr et al. 2000). This hypothesis has been addressed in the thermophilic *Lactobacillus delbrueckii ssp. lactis*, which contains a 152-bp intergenic region between *pepQ* and the *ccpA* homolog *pepR1* (Schick et al. 1999). PepQ activity and mRNA levels were twofold higher when *Lb. delbrueckii* was grown on lactose compared to glucose. The *cre* site partly overlaps the −35 region of the *pepR1* promoter, has a distance of 16 nt to the *pepQ* promoter and was required to mediate the regulatory effect of PepR1. In addition, PepR1 interacted with the promoter regions of several other proteolysis genes, namely *pepX* and *pepI*, and *brnQ* encoding a branched chain amino acid transporter. Inspection of the promoter sequences revealed the presence of *cre* in each case. Analysis of the *pepQ-pepR1* region in the twin species *Lb. delbrueckii ssp. bulgaricus* confirmed the regulatory effect of PepR1 on *pepQ* expression (Morel et al. 2001). In the presence of glucose, PepR1 autoregulates its own expression via *cre* while it concomitantly confers activation of *pepQ* transcription presumably by stabilising the RNA polymerase–promoter complex (Morel et al. 2001). Sixteen genes of the proteolytic system of *L. lactis* were investigated for their response to carbon control (Guédon et al. 2001). Fusion of a reporter gene encoding luciferase to the respective promoter regions revealed that only *pepP* encoding an aminopeptidase showed a sugar-dependent expression, while *pepQ*, e.g., was not subject to carbon regulation. This is in agreement with the fact that the *pepP* region contains four *cre*, while no *cre* is found in the *pepQ-ccpA* intergenic region in *L. lactis*.

Conclusions and possible applications

Carbon regulation in low-GC Gram-positive bacteria seems to be mediated by a generally shared molecular mechanism of transcriptional control by CcpA/HPr-ser-P complex binding to *cre*. Depending on the location of *cre* with respect to the promoter this can lead to gene repression or activation. This global response depends on the metabolic state of the cell, which is sensed by the metabolite-activated HPrK/P that adjusts the level of HPr-ser-P. The process assures economic utilisation of a preferred carbon source by concomitant regulation of target systems: (i) catabolic operons will be repressed, (ii) genes of central pathways such as glycolysis and genes involved in secretion of metabolic end products will be activated, (iii) when required, transport of the preferred carbon source may be up- or down-regulated, and (iv) expression of *ccpA* itself may be autoregulated. The transcriptional control is supplemented by protein activity control via inducer exclusion. This is mediated by HPr-ser-P, which prevents the entry of less-preferred carbon sources but may also regulate the permease that imports the preferred carbon source, as was shown for LacS of *S. thermophilus*. It should be noted that inducer exclusion in low-GC Gram-positive bacteria has only been demonstrated so far in LAB. Therefore, it remains to be seen whether it exists in other members of this group.

The advancements in the field of carbon regulation in LAB will provide a molecular basis for the improvement of dairy producer strains. Any mutation in the genes encoding a PTS component, CcpA or HprK/P will change sugar fermentation via the regulatory network of carbon control. EI mutants are pleiotropically deficient in PTS-dependent sugar uptake and HPr mutants are further deficient in CcpA-mediated regulation and in inducer exclusion. Loss of CcpA or HPrK/P renders the cell insensitive to global carbon control. Accordingly, a *ccpA* mutant of *L. lactis* exhibited mixed acid fermentation characteristics, while the wild type was homolactic (Luesink et al. 1998). A set of *ccpA* alleles was introduced into *Lb. casei*, they could restore the reduced growth phenotype but the strains were still deficient in CCR, or resulted in altered CCR responses (Carlos Esteban and Kerstin Mahr, unpublished results). Variation of the expression of EI and HPr in *S. salivarius* showed that a threefold reduction of HPr caused derepression of several catabolic genes when the cells were grown in the presence of PTS sugars (Thomas et al. 2001). The construction of industrially relevant LAB strains deficient or modified in any of the genes involved in carbon regulation should therefore be considered as an approach towards production of new dairy products with potentially unique compositions of acid and flavour compounds.

Acknowledgements

We thank Stephan Parche for the design of figures. Work conducted by the authors was supported by grants of the EU project BIO4-CT96-0380, SFB473 of the *Deutsche Forschungsgemeinschaft* and the *Fond der chemischen Industrie*.

References

Bolotin A, Wincker P, Mauger S, Jaillon O, Malarme K, Weissenbach J, Ehrlich SD & Sorokin A (2001) The complete genome sequence of the lactic acid bacterium *Lactococcus lactis ssp. lactis* IL1403. Genome Res. 11: 731–753.

Brochu D & Vadeboncoeur C (1999) The HPr(Ser) kinase of *Streptococcus salivarius*: purification, properties, and cloning of the *hprK* gene. J. Bacteriol. 181: 709–717.

Brückner R & Titgemeyer F (2002) Carbon catabolite repression in bacteria: choice of the carbon source and autoregulatory limitation of sugar utilization. FEMS Microbiol. Lett. 209: 141–148.

Busby S & Ebright RH (1999) Transcription activation by catabolite activator protein (CAP). J. Mol. Biol. 293: 199–213.

Charrier V, Buckley E, Parsonage D, Galinier A, Darbon E, Jaquinod M, Forest E, Deutscher J & Claiborne A (1997) Cloning and sequencing of two enterococcal *glpK* genes and regulation of the encoded glycerol kinases by phosphoenolpyruvate-dependent, phosphotransferase system-catalyzed phosphorylation of a single histidyl residue. J. Biol. Chem. 272: 14166–14174.

Chaillou S, Postma PW & Pouwels PH (2001) Contribution of the phosphoenolpyruvate: mannose phosphotransferase system to carbon catabolite repression in *Lactobacillus pentosus*. Microbiology 147: 671–679.

Chauvaux S, Paulsen IT & Saier MH Jr. (1998) CcpB, a novel transcription factor implicated in catabolite repression in *Bacillus subtilis*. J. Bacteriol. 180: 491–497.

Colland F, Barth M, Hengge-Aronis R & Kolb A (2000) Sigma factor selectivity of *Escherichia coli* RNA polymerase: role for CRP, IHF and *lrp* transcription factors. EMBO J. 19: 3028–3037.

Deutscher J, Reizer J, Fischer C, Galinier A, Saier MH, Jr. & Steinmetz M (1994) Loss of protein kinase-catalyzed phosphorylation of HPr, a phosphocarrier protein of the phosphotransferase system, by mutation of the *ptsH* gene confers catabolite repression resistance to several catabolic genes of *Bacillus subtilis*. J. Bacteriol. 176: 3336–3344.

Deutscher J, Küster E, Bergstedt U, Charrier V & Hillen W (1995) Protein kinase-dependent HPr/CcpA interaction links glycolytic activity to carbon catabolite repression in gram-positive bacteria. Mol. Microbiol. 15: 1049–1053.

Djordjevic GM, Tchieu JH & Saier MH, Jr. (2001) Genes involved in control of galactose uptake in *Lactobacillus brevis* and reconstitution of the regulatory system in *Bacillus subtilis*. J. Bacteriol. 183: 3224–3236.

Dossonnet V, Monedero V, Zagorec M, Galinier A, Pérez-Martínez G & Deutscher J (2000) Phosphorylation of HPr by the bifunctional HPr Kinase/P-ser-HPr phosphatase from *Lactobacillus casei* controls catabolite repression and inducer exclusion but not inducer expulsion. J. Bacteriol. 182: 2582–2590.

Faires N, Tobisch S, Bachem S, Martin-Verstraete I, Hecker M & Stulke J (1999) The catabolite control protein CcpA controls ammonium assimilation in *Bacillus subtilis*. J. Mol. Microbiol. Biotechnol. 1: 141–148.

Fieulaine S, Morera S, Poncet S, Monedero V, Gueguen-Chaignon V, Galinier A, Janin J, Deutscher J & Nessler S (2001) X-ray structure of HPr kinase: a bacterial protein kinase with a P-loop nucleotide-binding domain. EMBO J. 20: 3917–3927.

Fisher SH, Strauch MA, Atkinson MR & Wray LV, Jr. (1994) Modulation of *Bacillus subtilis* catabolite repression by transition state regulatory protein AbrB. J. Bacteriol. 176: 1903–1912.

Galinier A, Haiech J, Kilhoffer MC, Jaquinod M, Stülke J, Deutscher J & Martin-Verstraete I (1997) The *Bacillus subtilis crh* gene encodes a HPr-like protein involved in carbon catabolite repression. Proc. Natl. Acad. Sci. U.S.A. 94: 8439–8444.

Galinier A, Kravanja M, Engelmann R, Hengstenberg W, Kilhoffer MC, Deutscher J & Haiech J (1998) New protein kinase and protein phosphatase families mediate signal transduction in bacterial catabolite repression. Proc. Natl. Acad. Sci. U.S.A. 95: 1823–1828.

Garrity LF, Schiel SL, Merrill R, Reizer J, Saier MH, Jr. & Ordal GW (1998) Unique regulation of carbohydrate chemotaxis in *Bacillus subtilis* by the phosphoenolpyruvate-dependent phosphotransferase system and the methyl-accepting chemotaxis protein McpC. J. Bacteriol. 180: 4475–4480.

Görke B & Rak B (1999) Catabolite control of *Escherichia coli* regulatory protein BglG activity by antagonistically acting phosphorylations. EMBO J. 18: 3370–3379.

Gosalbes MJ, Monedero V & Pérez-Martínez G (1999) Elements involved in catabolite repression and substrate induction of the lactose operon in *Lactobacillus casei*. J. Bacteriol. 181: 3928–3934.

Gosalbes MJ, Esteban CD & Pérez-Martínez G (2002) In vivo effect of mutations in the antiterminator LacT in *Lactobacillus casei*. Microbiology 148: 695–702.

Gösseringer R, Küster E, Galinier A, Deutscher J & Hillen W (1997) Cooperative and non-cooperative DNA binding modes of catabolite control protein CcpA from *Bacillus megaterium* result from sensing two different signals. J. Mol. Biol. 266: 665–676.

Guédon E, Renault P, Ehrlich SD & Delorme C (2001) Transcriptional pattern of genes coding for the proteolytic system of *Lactococcus lactis* and evidence for coordinated regulation of key enzymes by peptide supply. J. Bacteriol. 183: 3614–3622.

Gunnewijk MG & Poolman B (2000) Phosphorylation state of HPr determines the level of expression and the extent of phosphorylation of the lactose transport protein of *Streptococcus thermophilus*. J. Biol. Chem. 275: 34073–34079.

Gunnewijk Mg, van den Bogaard PT, Veenhoff LM, Heuberger EH, de Vos WM, Kleerebezem M, Kuipers OP & Poolman B (2001) Hierarchical control versus autoregulation of carbohydrate utilization in bacteria. J. Mol. Microbiol. Biotechnol. 3: 401–413.

Henkin TM (1996) The role of CcpA transcriptional regulator in carbon metabolism in *Bacillus subtilis*. FEMS Microbiol. Lett. 135: 9–15.

Henkin TM, Grundy FJ, Nicholson WL & Chambliss GH (1991) Catabolite repression of alpha-amylase gene expression in *Bacillus subtilis* involves a *trans*-acting gene product homologous to the *Escherichia coli* LacI and GalR repressors. Mol. Microbiol. 5: 575–584.

Hogema BM, Arents JC, Bader R, Eijkemans K, Yoshida H, Takahashi H, Aiba H & Postma PW (1998) Inducer exclusion in *Escherichia coli* by non-PTS substrates: the role of the PEP to pyruvate ratio in determining the phosphorylation state of enzyme IIA^{Glc}. Mol. Microbiol. 30: 487–498.

Hoskins J, Alborn WE, Jr., Arnold J, Blaszczak LC, Burgett S & DeHoff BS et al. (2001) Genome of the bacterium *Streptococcus pneumoniae* strain R6. J. Bacteriol. 183: 5709–5717.

Jones BE, Dossonnet V, Küster E, Hillen W, Deutscher J & Klevit RE (1997) Binding of the catabolite repressor protein CcpA to its DNA target is regulated by phosphorylation of its corepressor HPr. J. Biol. Chem. 272: 26530–26535.

Jourlin-Castelli C, Mani N, Nakano Mm & Sonenshein AL (2000) CcpC, a novel regulator of the LysR family required for glucose repression of the *citB* gene in *Bacillus subtilis*. J. Mol. Biol. 295: 865–878.

Kim JH, Voskuil MI & Chambliss GH (1998) NADP, corepressor for the *Bacillus* catabolite control protein CcpA. Proc. Natl. Acad. Sci. U.S.A. 95: 9590–9595.

Kravanja M, Engelmann R, Dossonnet V, Blüggel M, Meyer HE, Frank R, Galinier A, Deutscher J, Schnell N & Hengstenberg W (1999) The *hprK* gene of *Enterococcus faecalis* encodes a novel bifunctional enzyme: the HPr kinase/phosphatase. Mol. Microbiol. 31: 59–66.

Kunji ER, Mierau I, Hagting A, Poolman B & Konings WN (1996) The proteolytic systems of lactic acid bacteria. Antonie van Leeuwenhoek 70: 187–221.

Küster-Schöck E, Wagner A, Völker U & Hillen W (1999) Mutations in catabolite control protein CcpA showing glucose-independent regulation in *Bacillus megaterium*. J. Bacteriol. 181: 7634–7638.

Leboeuf C, Auffray Y & Hartke A (2000a) Cloning, sequencing and characterization of the *ccpA* gene from *Enterococcus faecalis*. Int. J. Food Microbiol. 55: 109–113.

Leboeuf C, Leblanc L, Auffray Y & Hartke A (2000b) Characterization of the *ccpA* gene of *Enterococcus faecalis*: identification of starvation-inducible proteins regulated by *ccpA*. J. Bacteriol. 182: 5799–5806.

Lokman BC, Heerikhuisen M, Leer RJ, van den Broek A, Borsboom Y, Chaillou S, Postma PW & Pouwels PH (1997) Regulation of expression of the *Lactobacillus pentosus xylAB* operon. J. Bacteriol. 179: 5391–5397.

Ludwig H, Homuth G, Schmalisch M, Dyka FM, Hecker M & Stülke J (2001) Transcription of glycolytic genes and operons in *Bacillus subtilis*: evidence for the presence of multiple levels of control of the *gapA* operon. Mol. Microbiol. 41: 409–422.

Luesink EJ, van Herpen RE, Grossiord BP, Kuipers OP & de Vos WM (1998) Transcriptional activation of the glycolytic *las* operon and catabolite repression of the *gal* operon in *Lactococcus lactis* are mediated by the catabolite control protein CcpA. Mol. Microbiol. 30: 789–798.

Luesink EJ, Beumer CM, Kuipers OP & De Vos WM (1999) Molecular characterization of the *Lactococcus lactis ptsHI* operon and analysis of the regulatory role of HPr. J. Bacteriol. 181: 764–771.

Lux R, Jahreis K, Bettenbrock K, Parkinson JS & Lengeler JW (1995) Coupling the phosphotransferase system and the methyl-accepting chemotaxis protein-dependent chemotaxis signaling pathways of *Escherichia coli*. Proc. Natl. Acad. Sci. U.S.A. 92: 11583–11587.

Mahr K, Hillen W & Titgemeyer F (2000) Carbon catabolite repression in *Lactobacillus pentosus*: analysis of the *ccpA* region. Appl. Environ. Microbiol. 66: 277–283.

Mahr K, Esteban CD, Hillen W, Titgemeyer F & Pérez-Martínez G (2002) Cross communication between components of carbon catabolite repression of *Lactobacillus casei* and *Bacillus megaterium*. J. Mol. Microbiol. Biotechnol.: in press.

Marasco R, Muscariello L, Varcamonti M, De Felice M & Sacco M (1998) Expression of the *bglH* gene of *Lactobacillus plantarum* is controlled by carbon catabolite repression. J. Bacteriol. 180: 3400–3404.

Miwa Y, Nagura K, Eguchi S, Fukuda H, Deutscher J & Fujita Y (1997) Catabolite repression of the *Bacillus subtilis gnt* operon exerted by two catabolite-responsive elements. Mol. Microbiol. 23: 1203–1213.

Miwa Y, Nakata A, Ogiwara A, Yamamoto M & Fujita Y (2000) Evaluation and characterization of catabolite-responsive elements (*cre*) of *Bacillus subtilis*. Nucleic Acids Res. 28: 1206–1210.

Monedero V, Gosalbes MJ & Pérez-Martínez G (1997) Catabolite repression in *Lactobacillus casei* ATCC 393 is mediated by CcpA. J. Bacteriol 179: 6657–6664.

Monedero V, Poncet S, Mijakovic I, Fieulaine S, Dossonnet V, Martin-Verstraete I, Nessler S & Deutscher J (2001a) Mutations lowering the phosphatase activity of HPr kinase/phosphatase switch off carbon metabolism. EMBO J. 20: 3928–3937.

Monedero V, Kuipers OP, Jamet E & Deutscher J (2001b) Regulatory functions of serine-46-phosphorylated HPr in *Lactococcus lactis*. J. Bacteriol. 183: 3391–3398.

Morel F, Frot-Coutaz J, Aubel D, Portalier R & Atlan D (1999) Characterization of a prolidase from *Lactobacillus delbrueckii subsp. bulgaricus* CNRZ 397 with an unusual regulation of biosynthesis. Microbiology 145: 437–446.

Morel F, Lamarque M, Bissardon I, Atlan D & Galinier A (2001) Autoregulation of the biosynthesis of the CcpA-like protein, PepR1, in *Lactobacillus delbrueckii subsp bulgaricus*. J. Mol. Microbiol. Biotechnol. 3: 63–66.

Moreno MS, Schneider BL, Maile RR, Weyler W & Saier MH, Jr. (2001) Catabolite repression mediated by the CcpA protein in *Bacillus subtilis*: novel modes of regulation revealed by whole-genome analyses. Mol. Microbiol. 39: 1366–1381.

Plumbridge J (2001) Regulation of PTS gene expression by the homologous transcriptional regulators, Mlc and NagC, in *Escherichia coli* (or how two similar repressors can behave differently). J. Mol. Microbiol. Biotechnol. 3: 371–380.

Posthuma CC, Bader R, Engelmann R, Postma PW, Hengstenberg W & Pouwels PH (2002) Expression of the xylulose 5-phosphate phosphoketolase gene, *xpkA*, from *Lactobacillus pentosus* MD363 is induced by sugars that are fermented via the phosphoketolase pathway and is repressed by glucose mediated by CcpA and the mannose phosphoenolpyruvate phosphotransferase system. Appl. Environ. Microbiol 68: 831–837.

Postma PW, Lengeler JW & Jacobson GR (1993) Phosphoenolpyruvate:carbohydrate phosphotransferase systems of bacteria. Microbiol. Rev. 57: 543–594.

Presecan-Siedel E, Galinier A, Longin R, Deutscher J, Danchin A, Glaser P & Martin-Verstraete I (1999) Catabolite regulation of the *pta* gene as part of carbon flow pathways in*Bacillus subtilis*. J. Bacteriol. 181: 6889–6897.

Reizer J, Novotny MJ, Panos C & Saier MH Jr. (1983) Mechanism of inducer expulsion in *Streptococcus pyogenes*: a two-step process activated by ATP. J. Bacteriol 156: 354–361.

Reizer J, Hoischen C, Titgemeyer F, Rivolta C, Rabus R, Stülke J, Karamata D, Saier MH Jr. & Hillen W (1998) A novel protein kinase that controls carbon catabolite repression in bacteria. Mol. Microbiol. 27: 1157–1169.

Reizer J, Bachem S, Reizer A, Arnaud M, Saier MH Jr. & Stülke J (1999) Novel phosphotransferase system genes revealed by genome analysis – the complete complement of PTS proteins encoded within the genome of *Bacillus subtilis*. Microbiology 145: 3419–3429.

Roseman S, Pettigrew DW & Remington SJ (1993) Structure of the regulatory complex of *Escherichia coli* III^{Glc} with glycerol kinase. Science 259: 673–677.

Saier MH Jr. (1989) Protein phosphorylation and allosteric control of inducer exclusion and catabolite repression by the bacterial phosphoenolpyruvate: sugar phosphotransferase system. Microbiol. Rev. 53: 109–120.

Saier MH Jr., Chauvaux S, Cook GM, Deutscher J, Paulsen IT, Reizer J & Ye JJ (1996) Catabolite repression and inducer control in Gram-positive bacteria. Microbiology 142: 217–230.

Saier MH Jr. & Ramseier TM (1996) The catabolite repressor/activator (Cra) protein of enteric bacteria. J. Bacteriol 178: 3411–3417.

Schick J, Weber B, Klein JR & Henrich B (1999) PepR1, a CcpA-like transcription regulator of *Lactobacillus delbrueckii subsp. lactis*. Microbiology 145: 3147–3154.

Seok YJ, Sun J, Kaback HR & Peterkofsky A (1997) Topology of allosteric regulation of lactose permease. Proc. Natl. Acad. Sci. U.S.A. 94: 13515–13519.

Seok YJ, Koo BM, Sondej M & Peterkofsky A (2001) Regulation of *E. coli* glycogen phosphorylase activity by HPr. J. Mol. Microbiol. Biotechnol. 3(3): 385–393..

Sondej M, Sun J, Seok YJ, Kaback HR & Peterkofsky A (1999) Deduction of consensus binding sequences on proteins that bind IIA^{Glc} of the phosphoenolpyruvate:sugar phosphotransferase system by cysteine scanning mutagenesis of *Escherichia coli* lactose permease. Proc. Natl. Acad. Sci. U.S.A. 96: 3525–3530.

Soutourina O, Kolb A, Krin E, Laurent-Winter C, Rimsky S, Danchin A & Bertin P (1999) Multiple control of flagellum biosynthesis in *Escherichia coli*: role of H-NS protein and the cyclic AMP-catabolite activator protein complex in transcription of the *flhDC* master operon. J. Bacteriol. 181: 7500–7508.

Stucky K, Schick J, Klein JR, Henrich B & Plapp R (1996) Characterization of pepR1, a gene coding for a potential transcriptional regulator of *Lactobacillus delbrueckii subsp. lactis* DSM7290. FEMS Microbiol. Lett. 136: 63–69.

Stülke J, Arnaud M, Rapoport G & Martin-Verstraete I (1998) PRD-a protein domain involved in PTS-dependent induction and carbon catabolite repression of catabolic operons in bacteria. Mol. Microbiol. 28: 865–874.

Stülke J & Hillen W (2000) Regulation of carbon catabolism in *Bacillus* species. Annu. Rev. Microbiol. 54: 849–880.

Titgemeyer F, Mason RE & Saier MH, Jr. (1994) Regulation of the raffinose permease of *Escherichia coli* by the glucose-specific enzyme IIA of the phosphoenolpyruvate:sugar phosphotransferase system. J. Bacteriol 176: 543–546.

Tobisch S, Zühlke D, Bernhardt J, Stülke J & Hecker M (1999) Role of CcpA in regulation of the central pathways of carbon catabolism in *Bacillus subtilis*. J. Bacteriol. 181: 6996–7004.

Turinsky AJ, Grundy FJ, Kim JH, Chambliss GH & Henkin TM (1998) Transcriptional activation of the*Bacillus subtilis ackA* gene requires sequences upstream of the promoter. J. Bacteriol. 180: 5961–5967.

Turinsky AJ, Moir-Blais TR, Grundy FJ & Henkin TM (2000) *Bacillus subtilis ccpA* gene mutants specifically defective in activation of acetoin biosynthesis. J. Bacteriol. 182: 5611–5614.

Ueguchi C, Misonou N & Mizuno T (2001) Negative control of *rpoS* expression by phosphoenolpyruvate:carbohydrate phosphotransferase system in *Escherichia coli*. J. Bacteriol. 183: 520–527.

van den Bogaard PT, Kleerebezem M, Kuipers OP & de Vos WM (2000) Control of lactose transport, beta-galactosidase activity, and glycolysis by CcpA in *Streptococcus thermophilus*: evidence for carbon catabolite repression by a non-phosphoenolpyruvate-dependent phosphotransferase system sugar. J. Bacteriol. 182: 5982–5989.

Viana R, Monedero V, Dossonnet V, Vadeboncoeur C, Pérez-Martínez G & Deutscher J (2000) Enzyme I and HPr from *Lactobacillus casei*: their role in sugar transport, carbon catabolite repression and inducer exclusion. Mol. Microbiol. 36: 570–584.

Wray LV Jr, Pettengill FK & Fisher SH (1994) Catabolite repression of the *Bacillus subtilis hut* operon requires a *cis*-acting site located downstream of the transcription initiation site. J. Bacteriol. 176: 1894–1902.

Ye JJ, Reizer J, Cui X & Saier MH Jr. (1994a) ATP-dependent phosphorylation of serine-46 in the phosphocarrier protein HPr regulates lactose/H^+ symport in *Lactobacillus brevis*. Proc. Natl. Acad. Sci. U.S.A. 91(8): 3102–3106.

Ye JJ, Reizer J, Cui X & Saier MH Jr. (1994b) Inhibition of the phosphoenolpyruvate:lactose phosphotransferase system and activation of a cytoplasmic sugar-phosphate phosphatase in *Lactococcus lactis* by ATP-dependent metabolite-activated phosphorylation of serine 46 in the phosphocarrier protein HPr. J. Biol. Chem. 269(16): 11837–11844.

Yebra MJ, Veyrat A, Santos MA & Pérez-Martínez G (2000) Genetics of L-sorbose transport and metabolism in *Lactobacillus casei*. J. Bacteriol 182(1): 155–163.

Yoshida K, Kobayashi K, Miwa Y, Kang CM, Matsunaga M, Yamaguchi H, Tojo S, Yamamoto M, Nishi R, Ogasawara N, Nakayama T & Fujita Y (2001) Combined transcriptome and proteome analysis as a powerful approach to study genes under glucose repression in *Bacillus subtilis*. Nucleic Acids Res. 29(3): 683–692.

Antonie van Leeuwenhoek **82:** 73–91, 2002.

Comparative genomics of phages and prophages in lactic acid bacteria

Frank Desiere*, Sacha Lucchini[1], Carlos Canchaya, Marco Ventura & Harald Brüssow
*Nestlé Research Center, Nestec Ltd., Vers-chez-les-Blanc, CH Lausanne 26, Switzerland (*Author for correspondence; E-mail: frank.desiere@rdls.nestle.com)*

Key words: phage genomics, prophages, classification

Abstract

Comparative phage genomics has become possible due to the availability of more than 100 complete phage genome sequences and the development of powerful bioinformatics tools. This technology, profiting from classical molecular-biology knowledge, has opened avenues of research for topics, which were difficult to address in the past. Now, it is possible to retrace part of the evolutionary history of phage modules by comparative genomics. The diagnosis of relatedness is hereby not uniquely based on sequence similarity alone, but includes topological considerations of genome organization. Detailed transcription maps have allowed *in silico* predictions of genome organization to be verified and refined. This comparative knowledge is providing the basis for a new taxonomic classification concept for bacteriophages infecting low G+C-content Gram-positive bacteria based on the genetic organization of the structural gene module. An Sfi21-like and an Sfi11-like genus of Siphoviridae is proposed. The gene maps of many phages show remarkable synteny in their structural genes defining a lambda super-group within Siphoviridae. A hierarchy of relatedness within the lambda super-group suggests elements of vertical evolution in Siphoviridae. Tailed phages are the result of both vertical and horizontal evolution and are thus fascinating objects for the study of molecular evolution. Prophage sequences integrated into the genomes of their bacterial host present theoretical challenges for evolutionary biologists. Prophages represent up to 10% of the genome in some LAB. In pathogenic streptococci prophages confer genes of selective value for the lysogenic cell. The lysogenic conversion genes are located between the lysin gene and the right phage attachment site. Non-attributed genes were found at the same genome position of prophages from lactic streptococci. These genes belong to the few prophage genes transcribed in the lysogen. Prophages from dairy bacteria might therefore also contribute to the evolutionary fitness of non-pathogenic LAB.

Introduction

Due to their economic importance for the dairy industry, there are currently more complete genome sequences for phages of *Streptococcus thermophilus* or *Lactococcus lactis* than for lambdoid coliphages. Dairy phages are currently the best-investigated phage system with respect to genome data (Brüssow & Hendrix 2002). The complete genome sequence of bacteriophage lambda has been extraordinarily important to study fundamental biological questions. Phage genomics has the potential to be as important for fundamental problems in genomics since phage genomics represents all problems that are currently discussed in bacterial genomics: unity or diversity of origin, vertical versus horizontal gene transfer, non-orthologous gene displacement, tree versus web-like phylogeny, synteny versus instability of gene order, gene-splitting versus domain accretion (Koonin et al. 2000). Bacteriophages have an important impact on bacterial genomes and their evolution. In fact, many sequenced bacterial genomes contain prophages, some of which contribute 3–10% of the chromosomal gene content of the host bacterium (Brüssow & Hendrix 2002). The scientific value of these prophage sequences is potentially great, but detailed analysis in the literature is scarce. Thorough analysis is needed to elucidate the role of prophages in the context of bacterial genomes. A major question is whether pro-

* Present address: Institute of Food Research, Norwich Research Park, Norwich, NR4 7UA, United Kingdom

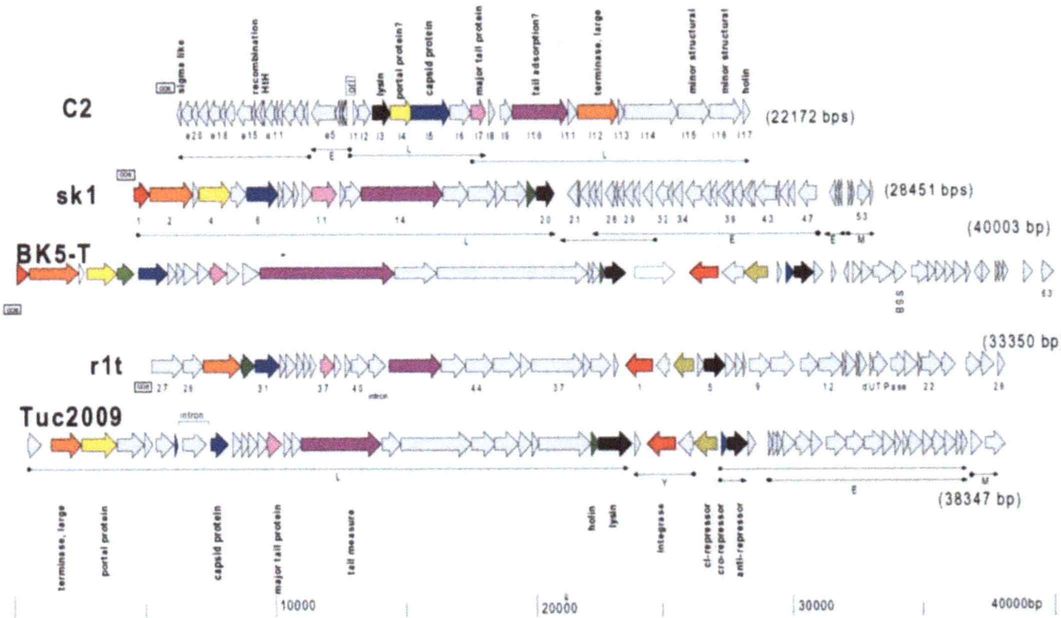

Figure 1. Alignment of the genetic maps of four distinct *L. lactis* phage species represented by phages c2, sk1, BK5-T and r1t. Transcription maps are indicated with arrows; E: early; M: middle; L: late transcripts. To facilitate comparison the genome maps from sk1, c2 and r1t are rearranged with respect to the Genbank entry. Corresponding genes are marked with the same colour.

phages contribute to the fitness of the bacterial host. Discussion of this subject can profit from the knowledge accumulated from pathogenic streptococci. Due to their clinical relevance, research regarding virulence in this group of organism has been extensive and prophages have been investigated for their potential to carry and disseminate virulence genes (Weeks & Ferretti 1984; Goshorn & Schlievert 1989; McShan et al. 1997). In the current contribution, we review phage and prophage genomics for lactic acid bacteria.

Genomes of lactic acid bacteriophages

Lactococcus lactis *phages*

Braun et al. (1989) differentiated five types of isometric head Siphoviridae according to head size and tail structure and two types of prolate headed phages. Most phages had DNA genomes ranging from 20 to 55 kb and defined seven slightly overlapping DNA homology groups. Interlaboratory phage comparisons led to the definition of twelve lactococcal phage species (Jarvis et al. 1991). However, industrial phage ecology is dominated by just three phage species (Prevots et al. 1990). The most abundant isolates were virulent small isometric head *Siphoviridae* of the 936-species ('sk1-like' phages , Figure 1) representing about half of all isolates. Another quarter of the isolates were classified into the c2-species of virulent lactococcal phages, Siphoviridae with prolate heads. A further quarter of the isolates is represented by small isometric phages of the P335 species. The P335 species is heterogeneous and comprises virulent and temperate phages and phages with two different DNA packaging mechanisms, i.e., *cos*-site and *pac*-site phages. The definition of this species can be questioned (see below).

Representatives from each of the three abundant lactococcal phage species have been completely sequenced. The phages sk1 and bIL170, members from the 936 phage species shared >80% nt sequence identity and differ from each other mainly by point mutations and numerous small insertion/deletions (Figure 2). Phage sk1 has a 28.4-kb long DNA genome with cohesive ends. The genome is subdivided into two segments (Chandry et al. 1997). The left most 16-kb encodes the structural gene cluster followed by the lysis module. The structural genes showed a very similar gene order as phage BK5-T (Figure 1), a temperate *cos*-site phage and the type phage of a minor lactococcal phage species. However, no sequence similarity linked the two structural gene clusters. The early phage genes are encoded on the opposite strand covering a major part of the right genome half of phage

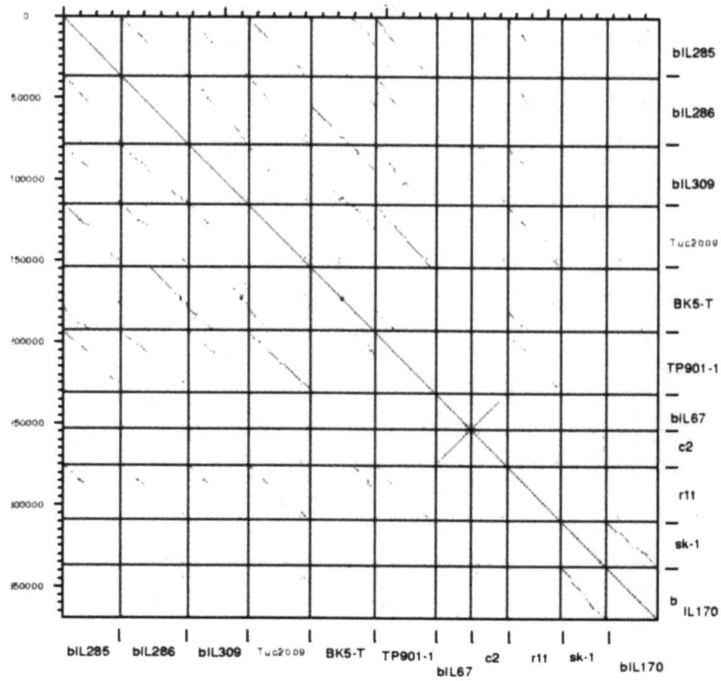

Figure 2. Dotplot matrix comparison calculated for the genome sequences of the lactococcal phages bIL285, bIL286, bIL309, Tuc2009, BK5-T, TP901-1, bIL67, c, r1t, sk-1 and bIL170. The left x-axis provides a scale in kb. The dotplot-matrix was calculated using Dotter (Sonnhammer & Durbin, 1995).

sk1. This head-to-head constellation of late and early genes is not found in any other dairy phage.

Eight proteins from phage sk1 shared sequence similarity with phage c2 proteins, the type phage of the c2-species of lactococcal phages. Despite that similarity, phage c2 showed a clearly distinct genome organization and a shorter genome size (22 kb) when compared to sk1 (Lubbers et al. 1995). The origin of replication, located 7 kb downstream of the left *cos*-site, divided the phage c2 genome in leftwardly transcribed early genes covering DNA replication and recombination functions and a larger rightward transcribed late gene cluster covering the lysis module and the structural genes. c2-like phages do not share nt sequence similarity with members of other lactococcal phage species. Phage bIL67 (Schouler et al. 1994), another member of the c2-species, could be aligned with c2 along the entire genome at the nt level (see also Figure 2). The c2-bIL67 comparison revealed gene replacements over the purported tail adsorption protein and three minor structural genes located at the right genome end (Lubbers et al. 1995). The two *pac*-site *L. lactis* phages, Tuc2009 (van Sinderen, van de Guchte, Seegers, and Fitzgerald, GenBank entry AF109874) and TP901-1 (Brondsted et al. 2001) are both members of the P335 species defined by phage taxonomists and could be aligned over three quarters of their genomes (Figure 2) and the nt sequence identity frequently exceeded 90%. The similarity was especially marked over the late gene cluster. Here the alignment was only interrupted at a few gene positions (downstream of a minor head gene, an intron in the major head gene from Tuc2009, insertions/deletions of two likely tail fiber genes, domain replacement in a minor tail gene). In contrast, over the early gene cluster the two lactococcal phages could only be aligned over small genome regions.

Phage r1t, another member of the P335 lactococcal phage species, has a 33.3-kb genome with cohesive ends (van Sinderen et al., 1996) (Figure 1). The structural genes are localized next to the *cos*-site, and were followed by the lysis genes, the lysogeny module, the DNA replication modules and transcriptional regulation region. The lysogeny genes were the only genes transcribed from the opposite strand. The structural gene cluster of r1t does not resemble in its organization any other group of dairy phages, while it shares significant protein sequence similarity with Siphoviridae from Mycobacteria (Lucchini et al. 1999a; Brüssow & Desiere 2001b). We propose that

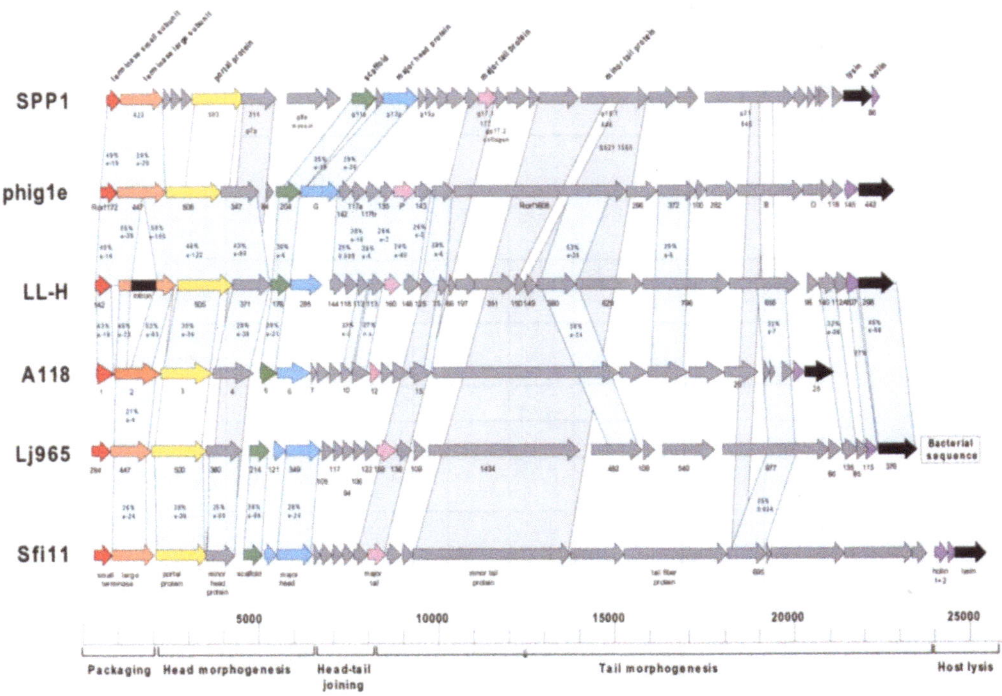

Figure 3. Comparison of the DNA packaging, head, tail, tail fiber and lysis genes in the *Lactobacillus johnsoni* prophage Lj965 with the corresponding genes in pac-site phages from Streptococcus thermophilus (Sfi11), L. delbrueckii (LL-H), L. plantarum (phig1e) and Bacillus subtilis (SPP1). Listeria phage A118 is given as a further reference (Loessner et al. 2000). Open reading frames are indicated by arrows. The number below the arrow gives the lengths of the deduced proteins in aa or the name attributed in the original publication. Corresponding genes in the five phages are indicated with the same colour code, unknown genes are indicated with gray arrows. Amino acid sequence identity between phage proteins is indicated with their percentage and BLAST E-value (expressed as decadic logarithmic exponent). Light blue shading links related proteins. Gray shading indicates similarities between proteins of bacteriophages, which are not depicted next to each other. The phage Sfi11 genes are marked with putative functions deduced from previous analysis (Lucchini et al. 1999a).

phage r1t be removed from the P335 species and that it be attributed to its own lactococcal phage species. As the current taxonomic phage system is primarily based on phage morphology, it seems logical to attribute a greater importance to the distinct structural gene module of phage r1t than to the fact that it shares early genes with lactococcal phages attributed to two distinct phage species (Tuc2009 and BK5-T).

The genetic diversity of the remaining lactococcal phage species is unexplored, but probably substantial. Phage ϕ111, for example, a member of the minor 949-species, is a *Siphovirus* with a large isometric head and a 110-kb genome (Prevots et al. 1990).

Lactobacillus phages

Lactobacillus phage sequence data are available from five distinct species of lactobacilli (*Lb. delbrueckii, gasseri, plantarum, casei and johnsoni*). No significant nucleotide sequence similarity was detected between *Lactobacillus* phages infecting distinct bacterial species (Desiere et al. 2000ab). Apparently, relatively tight barriers prevent the transfer of phage genes across *Lactobacillus* species. The overall genomic organization of *Lb. delbrueckii* phage LL-H was identical to that of *pac*-site *S. thermophilus* phages belonging to the proposed Sfi11-like genus of Siphoviridae. Similarly organized *pac*-site phages were identified in other lactobacilli (*Lb. plantarum* phage phi-g1e, *Lb. johnsoni* prophage Lj965). Related phages were identified in *Bacillus subtilis* (phage SPP1) and *Listeria monocytogenes* (phage A118, Figure 3). Sequence similarity of the structural proteins defined two sublineages of *pac*-site *Lactobacillus* phages. One group is constituted by *Lactobacillus* phages LL-H and phi-g1e. The second group of *pac*-site *Lactobacillus* phages is constituted by *Lb. johnsoni* prophage Lj965 whose closest relative is *S. thermophilus* phage Sfi11. The *pac*-site *B. subtilis* phage SPP1 is a linker between both subgroups since it alternatively shares

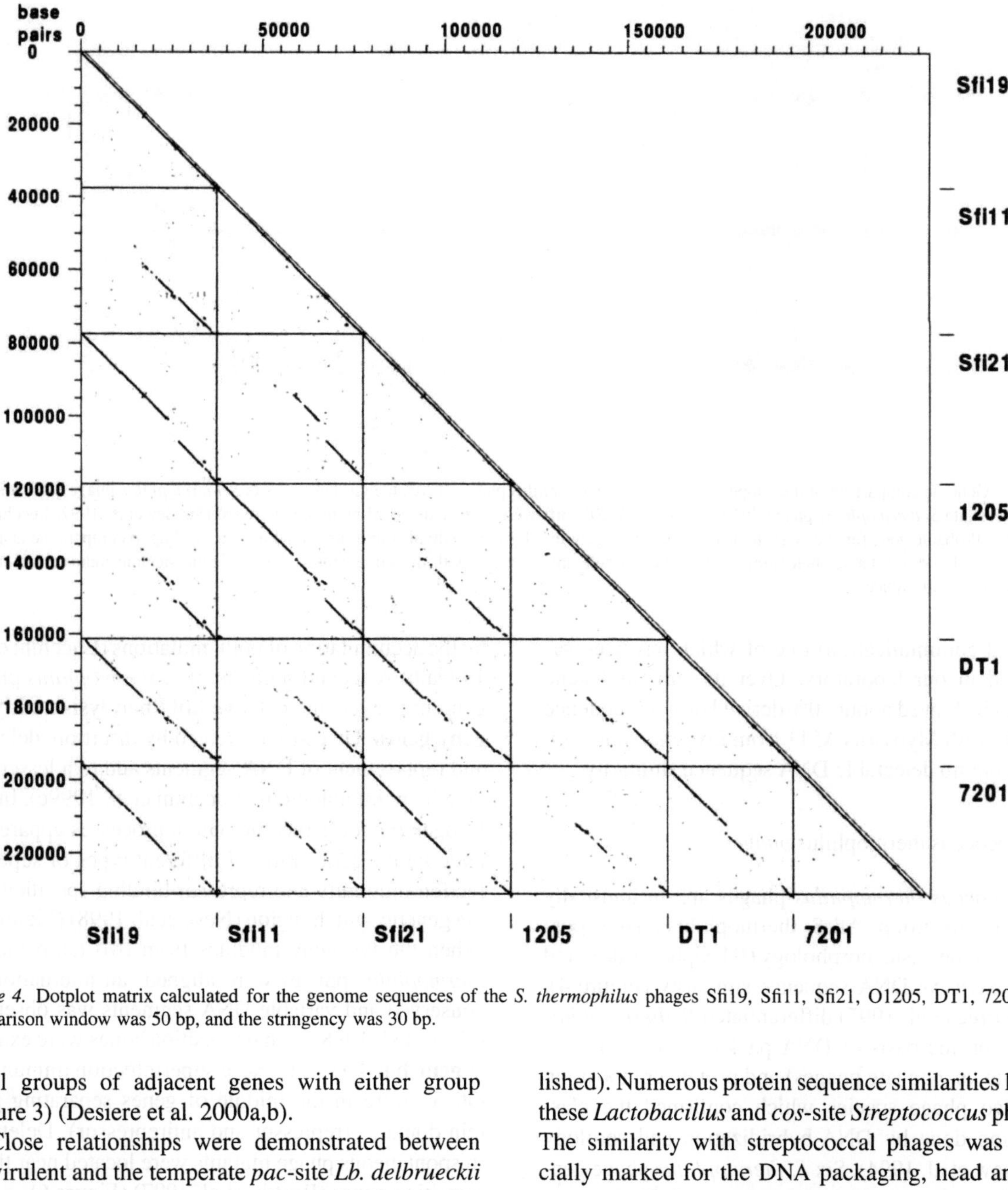

Figure 4. Dotplot matrix calculated for the genome sequences of the *S. thermophilus* phages Sfi19, Sfi11, Sfi21, O1205, DT1, 7201. The comparison window was 50 bp, and the stringency was 30 bp.

small groups of adjacent genes with either group (Figure 3) (Desiere et al. 2000a,b).

Close relationships were demonstrated between the virulent and the temperate *pac*-site *Lb. delbrueckii* phages LL-H and mv4. Over the structural genes the two phages shared up to 92% nt sequence identity (Vasala et al. 1993). Interestingly, the virulent phage LL-H still contained a remnant phage integrase and phage attachment sites demonstrating that it was recently derived from a temperate phage (Mikkonen et al. 1996a,b).

One complete and one partial genome sequence is currently available for *cos*-site *Lactobacillus* phages adh (Altermann et al. 1999) and a2 (Garcia, unpublished). Numerous protein sequence similarities linked these *Lactobacillus* and *cos*-site *Streptococcus* phages. The similarity with streptococcal phages was especially marked for the DNA packaging, head and tail genes of *Lb. gasseri* phage adh (Altermann et al. 1999) and the DNA replication genes of *Lb. casei* phage a2 (Moscoso & Suarez 2000).

Further genetic diversity exists in *Lactobacillus* phages. For example, group c *Lb. delbrueckii* phages contain prolate head Siphoviridae with peculiar knob-like appendages along the tail (phage JCL1032) (Ravin, Raisanen, Alatossava unpublished). In addition, Myoviridae (phages with contractile tails) have been isolated from *Lb. plantarum* (Chennoufi,

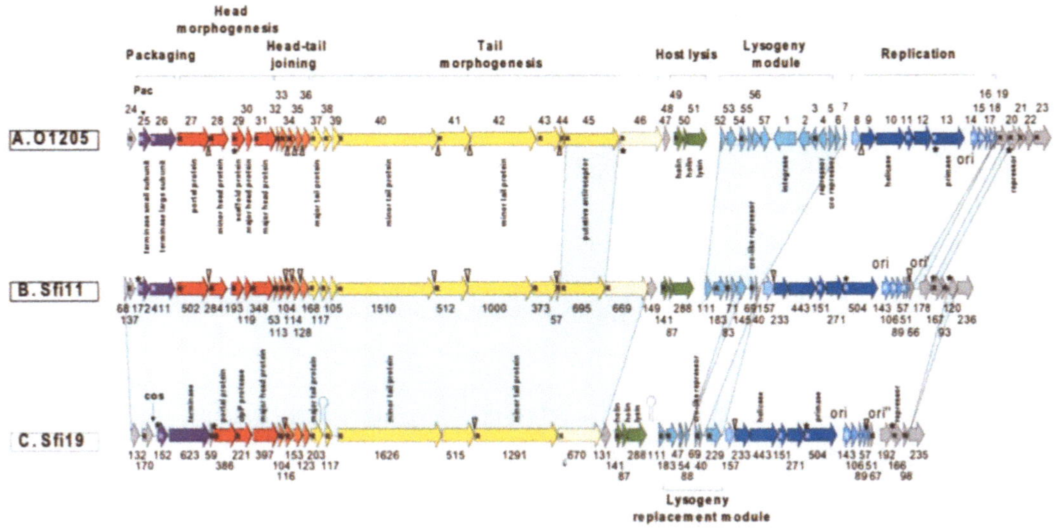

Figure 5. Genome comparison of the temperate pac-site *S. thermophilus* phage 01205, the virulent pac-site *S. thermophilus* phage Sfi11 and the virulent cos-site *S. thermophilus* phage Sfi19. For phage 01205, orfs were indicated according to the reference (Stanley et al. 1997; Lucchini et al. 1999c). Probable gene functions are indicated, the phage genomes have been divided into functional units according to comparative analysis (Lucchini et al. 1999c). Genes belonging to the same module are indicated with the same colour. Areas of blue shading indicate regions of major sequence difference.

personal communication) one of which has been sequenced in our laboratory. Over its structural gene clusters it showed about 40% derived protein sequence identity with Myovirus A511 from *Listeria monocytogenes*, but no detectable DNA sequence similarity.

Streptococcus thermophilus *phages*

Streptococcus thermophilus phages are an unusually homogenous group. All S. thermophilus phages possess the same basic morphology (B1 Siphoviridae) and share extensive DNA sequence similarity (Figure 4). (Le Marrec et al. 1997) differentiated *S. thermophilus* phages on the basis of DNA packaging mechanisms (*cos*-site vs. *pac*-site phages) and protein composition into two phage species which confirmed the classification done by DNA-hybridization and serology (Brüssow et al. 1994). Six *S. thermophilus* phage genomes have been completely sequenced. Based on a dotplot analysis the *S. thermophilus* phage genome can be subdivided into four large segments (Figure 4). One segment is the **late gene cluster** extending from the DNA packaging genes to the tail genes. This module is represented by two unrelated configurations: one is characteristic for *cos*-site phages (e.g., phage Sfi19), the other for *pac*-site phages (prototype: phage Sfi11) (Lucchini et al. 1999a) (Figure 5). The two structural gene clusters are not related to each other at the nucleotide or protein sequence level. Both clusters diversify by the accumulation of point mutations (Lucchini et al. 1999a,b) A second segment of *S. thermophilus* phage genomes covers the putative **tail fiber, lysis and lysogeny genes**. Diversity is created by insertion, deletion and replacement of DNA segments and to a lesser degree by point mutations (Lucchini et al. 1999c). In the lysogeny module recombination processes apparently underlie the acquisition of different types of superinfection immunity and repressor binding specificity in the genetic switch region (Neve et al. 1998) (Figure 6). When the lysogeny modules from two temperate *S. thermophilus* phages were aligned, an alternation of conserved and variable DNA segments was observed (Neve et al. 1998). Some transition zones were exactly at **gene borders** (integrase/ superinfection immunity). Others were in the middle of genes separating protein domains (repressor and antirepressor). Deletions in spontaneous phage mutants were located near these transition zones (Bruttin et al. 1997) (Figure 6).

The third genome segment is the putative DNA **replication module** represented by two distinct gene constellations: the Sfi21-like and the 7201-like DNA replication module. The Sfi21-like DNA replication module is present in the vast majority of the isolated phages and is unusually conserved (Desiere et al. 1997). Even at the third codon position $< 1\%$ sequence diversity was observed in independent phage isolates. The fourth *S. thermophilus* phage genome segment covers the rightmost 5 kb of the genome

Table 1. List of completely sequenced LAB bacteriophage genomes

	Phage	Host	Genome size	Temperate/ virulent	Phage type *	Reference
1	bIL67	*L. lactis*	22195	V	B2; cos: c2-species	Schouler et al. (1994)
2	c2	*L. lactis*	22172	V	B2; cos: c2-species	(Lubbers et al. (1995)
3	sk1	*L. lactis*	28451	V	B1; cos; 936-species	(Chandry et al. 1997)
4	bIL170	*L. lactis*	31754	V	B1; cos; 936-species	Crutz-Le Coq, A.M., Cesselin, B., Commissaire, J., Anba, J., Kyriakidis, S. and Chopin, M.C., (unpublished)
5	r1t	*L. lactis*	33350	T	B1; P335-species: cos	van Sinderen et al. (1996)
6	Tuc2009	*L. lactis*	38347	T	B1; P335-species: pac	van Sinderen, D., van de Guchte, M., Seegers, J.F.M.L. and Fitzgerald, G.F., (unpublished) Desiere et al. (2001)
7	TP901-1	*L. lactis*	37667	T	B1; P335-species: pac	Brondsted et al. (2001)
8	BK5-T	*L. lactis*	40003	T	B1; cos: BK5-T species	Mahanivong et al., (2001) Desiere et al. (2001)
9	LL-H	*Lb. delbrückii*	34 659	V	B1; pac group a	Mikkonen et al., (1996)
11	g1e	*Lactobacillus plantarum*	42259	T	B1; pac	(Kodaira et al. (1997)
12	adh	*Lactobacillus gasseri*	43785	T	B1; cos	Altermann et al. (1999)
13	01205	*S. thermophilus*	43075	T	B1; pac: Sfi11-species	(Stanley et al., (1997)
14	7201	*S. thermophilus*	35466	V	B1; cos: Sfi21-species	Stanley et al. (2000)
15	DT1	*S. thermophilus*	34820	V	B1; cos: Sfi21-species	Tremblay & Moineau (1999)
16	Sfi21	*S. thermophilus*	40739	T	B1; cos: Sfi21-species	Desiere et al., (1998)
17	Sfi19	*S. thermophilus*	37370	V	B1; cos: Sfi21-species	Desiere et al., (1998)
18	Sfi11	*S. thermophilus*	39807	V	B1; pac: Sfi11-species	Lucchini et al., (1999c)
19	Cp-1	*S. pneumoniae*	28451	V	Podovirus	Martin et al. (1996)
20	MM1	*S. pneumoniae*	40264	T	B1	Obregon, V., Garcia, P., Garcia, E., Lopez, R. and Garcia, J.L. (Unpublished)

* B1 – isometric head; B2, B3 – prolate head Siphoviridae (phages with long non-contractile tail) / cos – *cos*-site, genome with cohesive ends; *pac* – pac-site, DNA packaging by the head-full mechanism/phage species designation where available T – temperate phage; V – virulent phage.

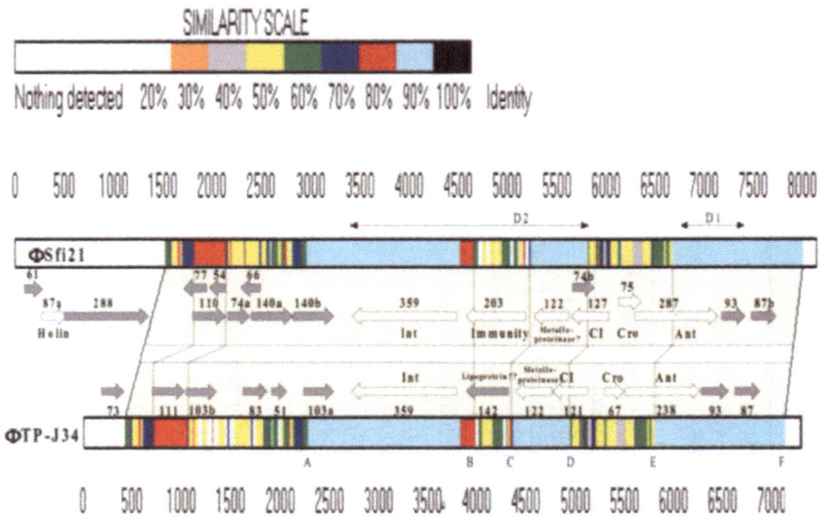

Figure 6. Alignment of the lysogeny modules from the cos-site phage Sfi21 and the pac-site phage TP-J34 from *S. thermophilus*. The colored bars give the percentage of nt sequence identity between both sequences as defined in the colour code at the top of the figure. The predicted open reading frames are indicated with their codon length. The vertical lines indicate the transition zones from high to low sequence similarity. D1, D2: extent of two spontaneous deletions

encoding early genes apparently involved in transcriptional regulation. Diversity is created by several insertions/ deletions, a duplication and a few gene replacements, while the overall DNA sequence is highly conserved (Lucchini et al. 1999c).

An unusual feature is the close genetic relationship between temperate and virulent *S. thermophilus* phages. Lytic phages dominate the *S. thermophilus* phage population. They are apparently derived from temperate phages by a combination of rearrangement and deletion events in the lysogeny module (Lucchini et al. 1999b). Recombination plays also a role in the creation of diversity in the putative tail fiber genes (Lucchini et al. 1998). Variable and conserved DNA segments were interspersed in the gene of the phage protein that probably interacts with the phage receptor on the bacterial cell (Lucchini et al. 1999c). Spontaneous deletions were observed that started and ended in DNA repeats of conserved regions encoding collagen-like protein motifs (Desiere et al. 1998). Phages that differed in host range showed completely unrelated variable regions, while phages with overlapping host ranges shared highly related variable regions. Swapping of the variable gene segments changed the host range of recombinant phages as predicted by the specificity of the variable region (Duplessis & Moineau 2001).

Genome comparison defines the Sfi21- and Sfi11-like genus of Siphoviridae

The comparison of *cos*-site temperate Siphoviridae infecting distinct genera of low GC content Gram-positive bacteria (Desiere et al. 2000) has yielded interesting insights into the evolution of the structural gene module. The compared sequences showed a **gradient of relatedness**, which, in an approximate way, reflected the relatedness of their bacterial hosts suggesting some co-evolution of phages with their host bacteria. Closest relatives to phage Sfi21 were other *cos*-site *S. thermophilus* phages which shared > 80% DNA sequence identity with Sfi21, followed by *Lactococcus* phage BK5-T, which shared 60% DNA identity over the DNA packaging and head morphogenesis genes with Sfi21 (Desiere et al. 2001a,b). *Lactobacillus* phage adh showed up to 61% aa identity over the DNA packaging, head and tail morphogenesis gene products but no detectable DNA identity (Desiere et al. 2000). *Bacillus* phage phi-105 and *Staphylococcus* phage PVL shared 26 to 36% protein sequence identity and an identical modular genome organization with the Sfi21 DNA packaging-head-tail-tail fiber-lysis-lysogeny-DNA replication-transcriptional regulation modules (Figure 7). Similarly an **SFi11-like genus** of Siphoviridae was proposed, based on the extensive sequence similarities of phage Sfi11 with a number of *pac*-site phages from low GC content Gram-positive bacteria including *Lactococcus* phage

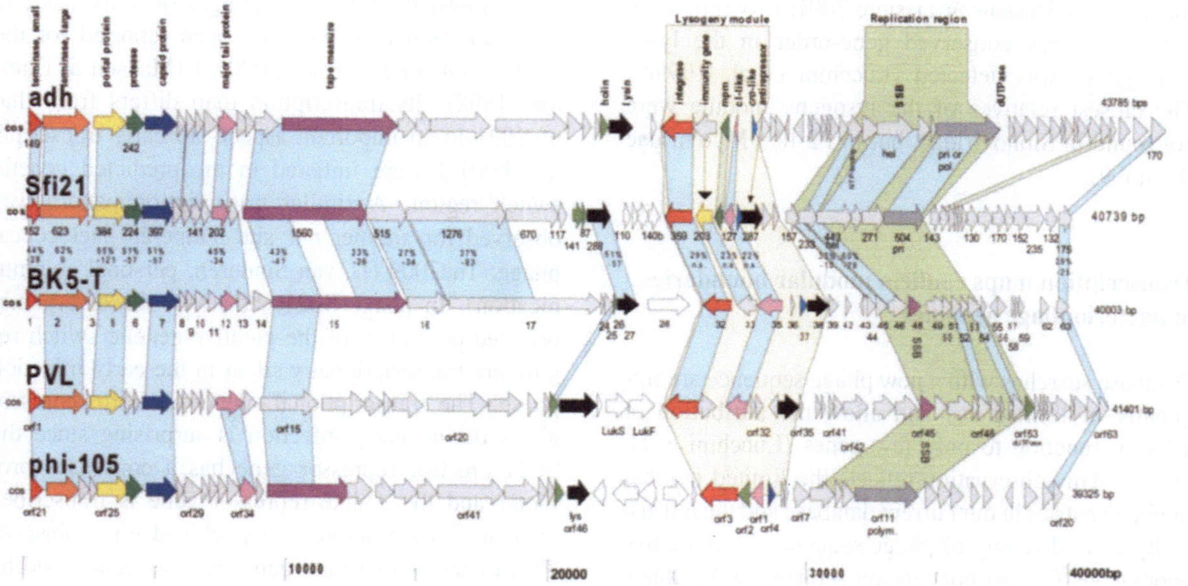

Figure 7. Alignment of the genetic maps of the *Lactobacillus* phage adh, *Streptococcus* phage Sfi21, *Lactococcus* phage BK5-T, *Staphylococcus* phage PVL, *Bacillus* phage phi-105 and *Staphylococcus aureus* phage PVL. Genes encoding proteins, which showed aa sequence similarity, are linked by blue shading. Corresponding genes are marked with the same colour. The location of the lysogeny and DNA replication modules are highlighted in yellow and green.

TP901-1, *Lactobacillus* phages phig1e, *Bacillus* phage SPP1 and *Listeria* phage A118 (Desiere et al. 2000).

The Sfi11-like phages have been shown to differ from the Sfi21-like phages by the possession of two instead of one major head proteins, the presence of a scaffold protein and the lack of proteolytic processing of the major head protein.

Many of the established genera of Siphoviridae share with Sfi21- and Sfi11-like phages a similarly organized structural gene cluster suggesting evolution from a **common ancestor module** (Figure 8). c2-like Siphoviridae (Jarvis et al. 1995) showed a distinct organization of their structural genes, but could still be aligned with the equivalent region of sk1-like phages if one allows for two rearrangement events (Desiere et al. 2001a,b). Sequence similarity was still detected between some phages from this super-group and allowed the distinction of at least five lineages of capsid genes represented by phages HK97/Sfi21/phiC31, λ/Sfi11, L5/TM4/r1t, sk1/c2 and ιM2. Notably, the archaeavirus ιM2 (Pfister et al. 1998) showed over its portal and major head proteins clear similarity with P2-like Myoviridae (contractile tail Caudovirales) from both Gram-negative and Gram-positive bacteria. P22-like Podoviridae (short tail Caudovirales) shared with the λ super-group of Siphoviridae a similarly organized capsid gene cluster (Eppler et al. 1991).

Siphoviridae genomes fall into two size classes: 121-134 kb (coliphage T5, *Bacillus* phage SPβ) and 22-56 kb (all other genera). The genome organization of phage SPβ (Lazarevic et al. 1999) shared no similarity with the smaller genomes of Siphoviridae, and thus may be polyphyletic. Most of the putative structural genes from SPβ are unrelated to the entries in the database, while part of the tail fiber genes shared similarity with defective prophages from *Bacillus*. Still other SPβ genes are present as a consequence of recent horizontal gene transfers from the *Bacillus* chromosome. This observation leads to the taxonomically unsatisfying situation that genomes of Siphoviridae from the λ super-group are more closely related to P22-like Podoviridae and P2-like Myoviridae (below) than to SPβ-like Siphoviridae. Apparently, the current taxonomic classification only partially reflects the natural relationships between tailed phages.

Since current phage taxonomy is based on the morphology of the phage particle, it seems logical to build a genomics-based phage taxonomy on the organization of the structural gene cluster. For other phage modules, a distinct picture of phage relatedness has been obtained. For example, at least four different lysogeny modules were observed within the lambda supergroup (Brüssow & Desiere 2001a). Within dairy phages, a very conserved gene-order in the lysogeny genes was detected (Lucchini et al. 1999b). The closest relatives of the lysogeny module were not found in Siphoviridae, but in P2-like Myoviridae (Figure 9).

Transcription maps confirm modular boundaries in bacteriophage genomes

Database searches with a new phage sequence are frequently of limited use and allow the attribution of a likely function to only few genes (Lucchini et al. 1999a). This observation reflects the limited number of phage entries in our current database and the intrinsically great diversity of phage sequences. Alternative approaches that do not rely on sequence relatedness like the comparison of phage gene maps with well-investigated model phages makes it possible to define the modular organization of dairy phage genomes. A 'reality check' for these *in silico* predictions was provided by **transcription mapping** of the temperate *cos*-site *S. thermophilus* phage Sfi21 (Ventura et al. 2002). Three classes of transcripts were observed based on the timing of their appearance (Figure 10). Early transcripts were identified in four different genome regions. Transcription mapping located early genes in the lysogeny module and between the DNA replication module and the *cos*-site (Desiere et al. 1997). Two middle transcripts were identified, one was initiated at the promoter of the *cro*-repressor gene, the other covered the DNA replication module (Foley et al. 1998). Four types of late transcripts were identified which covered the likely DNA packaging genes, the head morphogenesis module plus the major tail gene, the remainder of the tail genes, and the putative tail fiber and lysis genes, respectively. The Sfi21 transcription map confirmed the prediction of the Sfi21 modular structure established by comparative genome analysis with only one notable exception. Transcription mapping showed that the DNA replication module ended downstream of the primase gene with the origin of phage replication. The lack of database matches for the following five genes and the observation of a duplicated origin of replication downstream of these genes in virulent phages led to their erroneous attribution as DNA replication genes. This analysis showed that transcription mapping is a suitable tool that can verify *in silico* predictions about the modular genome organization of bacteriophages.

Amongst the temperate phages of dairy bacteria a transcription map has only been reported for the *Lactococcus lactis* phage TP901-1 (Madsen & Hammer 1998). Its transcription map differs from that of Sfi21 in an important aspect: all early transcripts in TP901-1 were initiated in the predicted genetic switch region. A similar transcription pattern was observed for another *pac*-site temperate lactococcal phage, Tuc2009 (D. van Sinderen, personal communication). In phage Sfi21, however, the divergently oriented promoters of the putative genetic switch region are transcriptionally silent in the early infection phase. The transcriptional silence of both repressor genes during early infection is surprising since the Sfi21 *cro*-like repressor gene has a consensus promoter and the *cI*-like repressor gene is transcribed from its own promoter when cloned on a plasmid. This indicates that both genes can be transcribed by an unmodified host RNA polymerase.

Prophages

About half of the sequenced bacterial genomes contain prophage sequences (Lawrence et al. 2001). The presence of such prophage DNA poses intriguing questions for evolutionary biologists. Prophages increase the metabolic burden of the host cell which has to replicate up to 8% extra phage DNA. In addition, prophages may lyse their host after prophage induction. To compensate for these disadvantages one has to presume that temperate phages encode functions that increase the fitness of the lysogen (Brüssow & Hendrix 2002). According to the selective value of these phage genes, the lysogenic cell will be maintained or even be over-represented in the bacterial population. An obvious selective advantage for the lysogenic host is the presence of immunity and super-infection exclusion genes of the prophage that protect the lysogen against phage infection. Further examples of phage-encoded genes that increase host fitness include diphtheria toxin, streptococcal erythrogenic toxin A, and the non-essential phage λ gene *bor* that confers serum-resistance to the *E. coli* lysogen. In these cases, the reproductive success of the lysogenic bacterium translates directly into an evolutionary success for the resident prophage. However, the host–parasite relationship is also a highly fragile and dynamic genetic equilibrium. Prophages can be considered dangerous molecular time bombs that can kill the lysogenic cell upon their eventual induction.

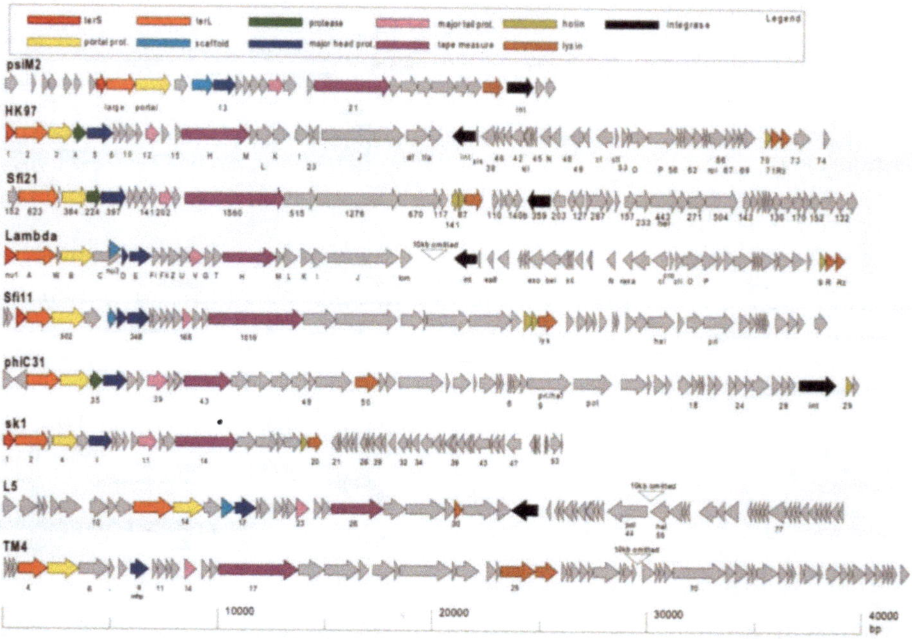

Figure 8. Comparison of the genomes from the established and proposed genera of Siphoviridae constituting the λ super-group of phages. Corresponding genes are indicated with the same colour.

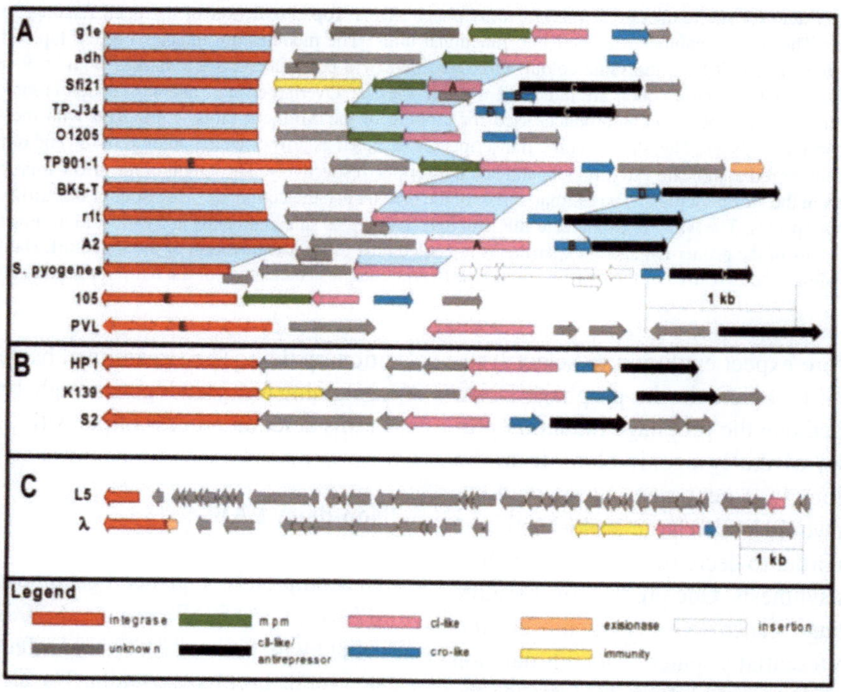

Figure 9. Comparative genetic organization of the lysogeny-related genes in Sfi21- and Sfi11-like Siphoviridae (A), P2-like Myoviridae infecting *Vibrio cholerae* (phage K139) and *Homophiles influenza* (phages S2 and HP1) (B) and L5-like and λ-like Siphoviridae (C; note the changed scale). Corresponding genes are indicated with the same colour. Blue shading or identical letters link genes of which the products share sequence similarity.

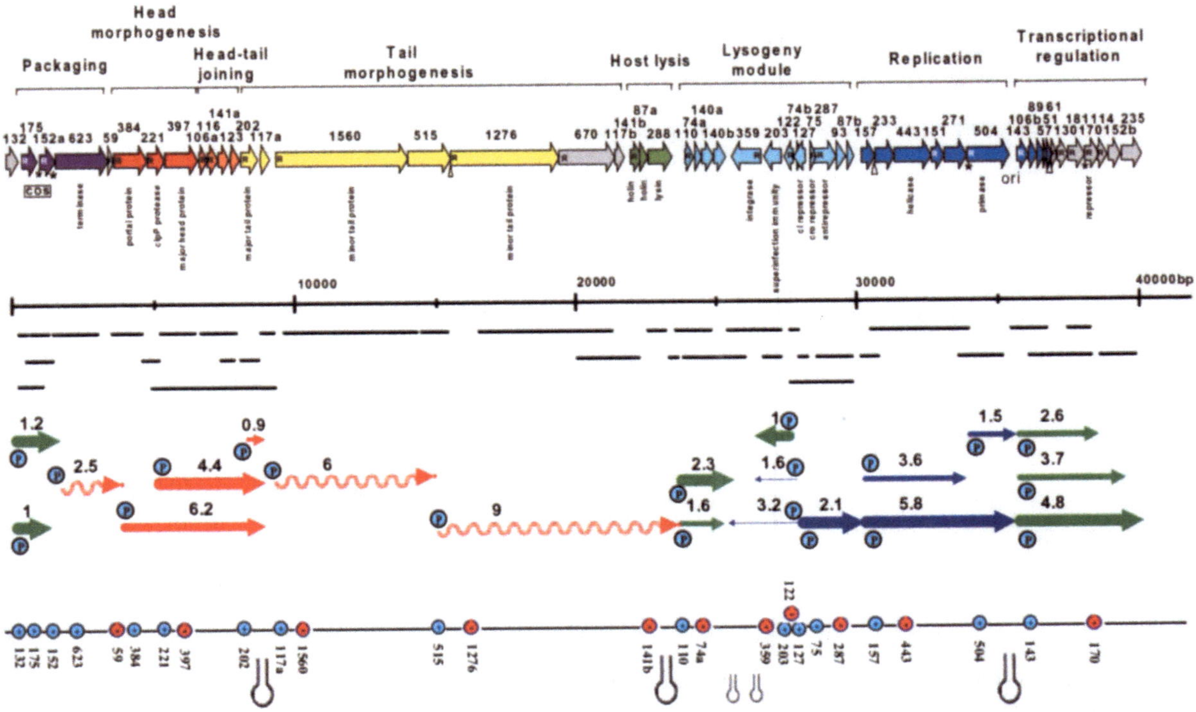

Figure 10. Transcription map for the temperate *S. thermophilus* phage Sfi21. **Top**: Prediction of the open reading frames in the complete genome of phage Sfi21. The phage genome is divided into functional units. The modules are indicated at the top of the gene map. Genes predicted to belong to the same unit have the same colour. Orfs preceded by a potential RBS are marked with an R inside the arrow. Orfs starting with an unconventional initiation codon are indicated with an asterisk. Overlap of start and stop codon is indicated with a triangle. **Middle**: The approximate position of the PCR products used for probing of the Northern blots is provided with the scale in bp. **Bottom**: Summary of the transcription analyses. The Sfi21-specific transcripts are depicted as arrows which point to the 3' end of the mRNA and which are coloured in green, blue or red to indicate early, middle and late transcripts, respectively. The length of the arrow is proportional to the length of the mRNA derived from the Northern blots. The estimated size of the mRNA is indicated in kb. The width of the arrows indicates the relative abundance of the mRNA species. The wavy lines indicate mRNAs that show up as smeared bands in hybridization. Positive (blue circles) and negative (red circles) results of the primer-extension experiments are noted next to the identification of the tested orf. Hairpins indicate possible rho-independent terminators, the two sizes of the hairpins refer to different energies calculated for the hairpin (Ventura et al. 2002).

One would therefore expect evolution to select lysogenic bacteria with mutations in the prophage DNA. Mutations that inactivate the prophage induction process avoid the loss of the lysogenic clone from the bacterial population. In a next step, one would expect that selection would lead to large-scale deletion of prophage DNA in order to decrease the metabolic burden of extra DNA synthesis. One predicts furthermore that useful prophage genes (e.g. lysogenic conversion genes) are preferentially spared from this deletion process since their loss would actually decrease the fitness of the cell. It was proposed that a high genomic deletion rate is instrumental in removing dangerous genetic parasites from the bacterial genome (Lawrence et al. 2001). Deletion processes could explain why bacterial genomes did not increase in size despite a constant bombardment with parasitic DNA over evolutionary time. The streamlined bacterial chromosome containing few pseudogenes might be the consequence of this deletion process of parasitic DNA.

Non-dairy LAB

The completely sequenced genome of the **S. pyogenes** *M1 strain SF370* (Ferretti et al. 2001) isolated from a patient with an invasive wound infection was investigated for its prophage content. The analysis was done to confirm the theoretical predictions of prophage–host interaction (Desiere et al. 2001a,b). It was already known that many *S. pyogenes* strains contain multiple prophage elements (Yu & Ferretti 1991; Hynes et al. 1995). In addition, it was established that some streptococcal exotoxins were phage-encoded (Nida & Ferretti 1982; Weeks & Ferretti 1984; Goshorn &

Schlievert, 1989). *S. pyogenes* strains encode further virulence factors like extracellular enzymes that contribute to the pathology of streptococcal disease. Some of them (e.g., hyaluronidase, DNase) are of phage origin (Hynes & Ferretti 1989; Marciel et al. 1997). Phages might thus be common vehicles of genetic exchange between *S. pyogenes* strains including the transfer of virulence factors.

The *S. pyogenes* **SF370** genome contains eight prophage elements (Figure 11). Only prophage SF370.1 could be induced by mitomycin C treatment. **SF370.1** possesses a 41-kb-long genome that closely resembles the genome of *pac*-site temperate Siphoviridae found in different lactic acid bacteria. Its closest relatives are prophage NIH1.1 from a Japanese *S. pyogenes* strain and to a lesser extent *S. pneumoniae* phage mm1. The possession of prophage NIH1.1 differentiated older from newly emerging Japanese *S. pyogenes* strains (Inagaki et al. 2000). Prophage acquisition might thus be a major mechanism of genetic variation in this highly dynamic and medically most relevant bacterial species (Smoot et al. 2002).

Prophage **SF370.3** has a 33-kb long genome of which the genome organisation closely resembles that of the *cos*-site temperate *L. lactis* phage r1t. Analysis of the prophage genome revealed mutations in the replisome organizer gene that may prevent the induction of the prophage. SF370.3 encodes a hyaluronidase and a DNase possibly implicated in the spreading of *S. pyogenes* through tissue planes of its human host. Notably, antibodies against both phage proteins are found in poststreptococcal autoimmune complications underlining a role for these phage proteins in the pathology of this human disease. Prophage **SF370.2** has a 43-kb long genome, the genome organization of which resembles that of *pac*-site temperate Siphoviridae infecting dairy bacteria and shows extensive protein sequence identity with structural proteins from *S. thermophilus* and *L. lactis* phages. SF370.2 showed two probable inactivating mutations: one in the replisome organizer gene and another in the gene encoding the portal protein.

All three SF370 and the NIH1.1 prophages possess likely lysogenic conversion genes between the phage lysin gene and the right attachment site (Figure 12). This region in prophage **SF370.1** genome encodes the pyrogenic exotoxin C (SpeC) and a mitogenic factor (mf2). Closely related proteins are encoded in prophages SF370.2 and SF370.3 at an identical phage genome position. The superantigens may contribute to the immune deregulation observed during invasive streptococcal infections. The lysogenic conversion genes in the prophages differ in their GC content from the surrounding prophage and bacterial DNA. Their location near the phage attachment site suggests a faulty phage excision process in an unusual bacterial host of a GC content lower than that of the genome from which the DNA originated. The horizontal transfer of these genes is also suggested by the presence of identical genes in a veterinary pathogen, *S. equi* (Harrington et al. 2002).

A clear trend for prophage genome inactivation was seen for the remaining five prophage sequences in *S. pyogenes* SF370 which all showed massive losses of prophage DNA. The largest prophage remnant SF370.4 showed a 13-kb long genome consisting of lysogeny, DNA replication and transcriptional regulation genes flanked by attachment sites. The remaining four prophage remnants were small (0.4–2.2-kb long), but all contained an integrase gene.

The sequencing of a second *S. pyogenes* strain, **Manfredo**, is close to completion. This allowed interesting interstrain comparisons of the prophage sequences. The comparison has revealed that prophages are hotspots for genetic recombination, which results not only in the reshuffling of genes between prophages, but apparently also in rearrangements of the bacterial genome (Brüssow & Hendrix 2002).

S. pyogenes infections are known for their variable clinical symptomatology ranging from symptom-less carriage and mild sore throat to scarlet fever and pyodermitis to include autoimmune diseases (rheumatic fever, glomerulonephritis) and life-threatening diseases (fasciitis, toxic shock syndrome). It is tempting to interpret the clinical potential of this protean bacterial pathogens, at least partly, as a function of the specific prophage combination that a given strain harbors (Brüssow & Hendrix, 2002). In fact, the possession of the prophage NIH1.1 has been found to be a genetic marker for newly emerging *S. pyogenes* isolates in Japan. Moreover, the recent publication about the genome sequence and the comparative microarray analysis of serotype M18 group A *Streptococcus* strains associated with acute rheumatics fever (Smoot et al. 2002) showed prophages and prophage-like elements were responsible for most of the genetic variation in 36 serotype M18 strains from diverse localities. This observation underlines the role that prophages might play for the ecological success of a given bacterial strain.

Pathogenic bacteria represent only a small fraction of LAB. Do the principles observed with *S. pyogenes*

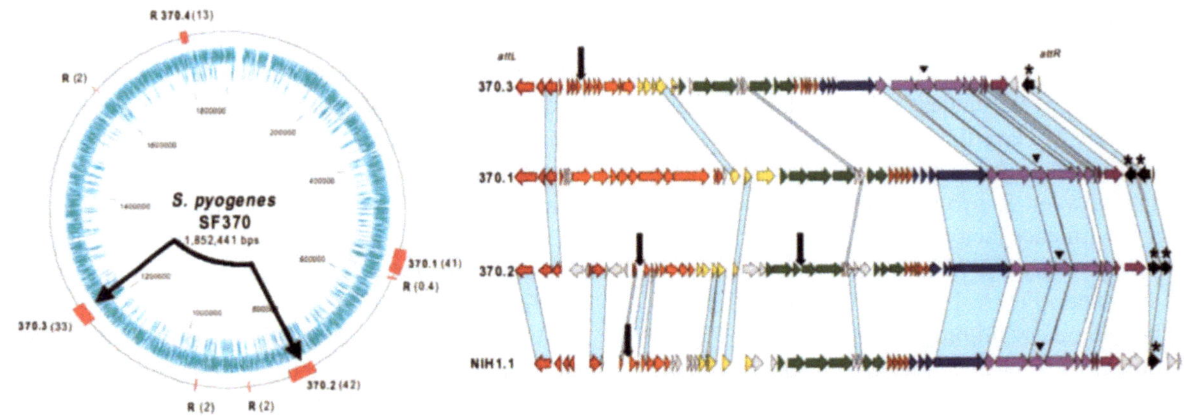

Figure 11. Aligned genome maps of three SF370 prophages and a prophage from *S. pyogenes* strain NIH1. The phage modules are colour-coded, (red, lysogeny; orange, DNA replication; yellow, transcriptional regulation; green, DNA packaging and head; brown, head-to-tail; blue, tail; mauve, tail fiber; dark pink, lysis, black; superantigen/mitogenic factor genes). Likely prophage inactivating mutations are indicated by large vertical arrows (replisome organizer, portal protein), an asterisk marks phage genes that potentially contribute to the virulence of the lysogenic host; the phage hyaluronidase is labeled by a triangle. Blue shading connects regions of DNA sequence similarity between the prophages

Figure 12. Compilation of currently available partial genome maps of temperate Siphoviridae from low GC content Gram-positive bacteria. The depicted genome regions are flanked on the left side by the phage lysin (blue) and on the right side by the phage integrase gene (green). Candidates for lysogenic conversion genes are highlighted in red. The code name of the prophages or temperate phages is indicated at the left side of the gene map, the host of the specified phage is indicated at the right side of the map (Canchaya et al. unpublished).

prophages also apply to prophages from other LAB occupying dramatically different ecological niches? Do some prophages increase the ecological fitness of dairy strains or commensals? Are phages then major drivers of the evolution of bacteria allowing the testing of many new genes and the continued maintenance of those genes that increase the competitiveness of the lysogen in its ecological niche?

Comparative phage genomics has demonstrated that **lysogenic conversion** genes might also be encoded by temperate phages infecting dairy streptococci. ORFs of no known phage-related function are fre-

Table 2. Prophages of lactic acid bacteria

Host	Prophage	Inducible	Size	Reference:
S. thermophilus	Sfi16	N	3kb	unpublished
S. mitis	SM1	Y	35 kb	Bensing 2001
L. lactis IL1403	bIL285	y	35 538	Chopin 2001
	bIL286	y	41 834	
	bIL309	y	36 949	
	bIL310	Y	14 957	
	bIL311	N	14 510	
	bIL312	Y	15 179	
Lb. johnsonii	Lj965	N	39 kb	Desiere 2000
	Lj928	N	39 kb	unpublished
	Lj771	N	42 kb	
Lb. plantarum	LP1, LP2a			unpublished
	LP2b, LP3			
S. pyogenes	SF370.1			Ferretti 2001
	SF370.2			Desiere 2001
	SF370.3			Canchaya 2002
	SF370.4			
E. faecalis V583	1	?	37 kb	unpublished
	2	2	?	33 kb
E. faecium	1			unpublished
Oenococcus oeni	fOg44			Parreira 1999

quently encoded in temperate phages and prophages from dairy bacteria at a genome position where pathogenic streptococci and staphylococci encode lysogenic conversion genes. Figure 12 provides a compilation of the genome maps of temperate phages from LAB between the lysin and the integrase genes. With a single exception (*S. pneumoniae* phage mm1), all phages encode at least one gene in this region. In all cases, the attachment site is located directly downstream of the integrase gene or it overlaps the 3'-end of the integrase gene. Due to this configuration, the extra-genes are located between the phage lysin gene and the right attachment site in the lysogen. Where investigated these genes have been shown to be transcribed in the lysogen (Ventura et al. 2002). Frequently, their transcription was found to be more prominent than the transcription of the phage repressor, for example. These observations suggest a physiological function for these prophage genes in the lysogen. The lack of biochemical or genetic evidence and of any bioinformatic leads for these phage genes makes it, however, difficult to speculate on possible lysogenic conversion phenotypes.

The close genetic relationships between phages from dairy bacteria and evolutionary related pathogenic bacteria have interesting scientific implications. Research in medical and dairy microbiology has been conducted with different priorities. For example, lysogenic conversion is a strong research area in medical bacteriology, while the basic aspects of phage replication have been investigated in more detail for dairy phages (Brüssow 2001). Medical microbiologists used phages for killing bacteria (phage therapy) (Nelson et al. 2001), while dairy microbiologists have developed phage-resistant starter strains based on the knowledge of the phage genome (e.g., Foley et al. 1998). Knowledge established in one field can be employed to complement research in the other.

Thus, the current knowledge on prophages in dairy species of LAB will be briefly summarized and an overview of the currently known LAB prophages is given in Table 2.

Phophages of dairy lab

Prophages of Lactobacillus johnsoni

Lactobacillus johnsoni strain La1 contains three prophage elements of 38–42 kb in length in its 1.8-Mb long genome. Each is flanked by attachment sites, but no phage particles could be induced by mitomycin C treatment. They share the genome organization of a widely distributed group of temperate *pac*-site phages from LAB (Desiere et al. 2000). Twelve kb upstream of the integrase gene, all three prophages contain a cluster of tRNA genes. Only two regions of the prophage genome are transcribed in the lysogenic cell. In all three prophages the strongest transcripts derive from a single open reading frame located between the lysin gene and the right phage attachment site. One of these genes shares sequence similarity with a plasmid maintenance system. The second transcribed region is the lysogeny module. In prophage Lj771 this transcript covers a conventional lysogeny module for dairy phages (integrase, superinfection exclusion, repressor genes). However, in prophages Lj965 and Lj928 additional genes were apparently inserted into the lysogeny module. Bioinformatic analysis suggests there is a further lysogenic conversion gene in this region in Lj965.

Prophages of Lactococcus lactis

Lactococcus lactis strain IL1403 is a major bacterial starter organism used in industrial cheese fermentation. Its 2.3-Mb genome contains six prophage elements; all were found to be flanked by the phage attachment sites, and all but one prophage remnant (which contained insertion elements) could be excised from the bacterial genome (Chopin et al. 2001). However, only two prophages gave rise to infectious phage particles after induction. Three prophages, ranging in size from 36 to 42 kb, showed a genome organization typical of temperate dairy phages. Prophages bIL286 and bIL309 share DNA sequence identity over the structural gene cluster with the temperate *cos*-site *L. lactis* phage BK5-T and protein sequence similarity with *S. thermophilus* phage Sfi21. Prophage bIL285 did not match the DNA sequence of any previously described lactococcal phage, but it still shared detectable protein sequence similarity with the BK5-T group of phages. Three prophage sequences of approximately 15 kb in length represent likely prophage remnants. All possessed the conserved part of the lysogeny module (integrase/ repressor) and, in variable numbers, DNA replication and a few structural genes.

Lactobacillus plantarum strain WCFS1 contains two about 43-kb long prophage genomes and several prophage remnants in its 3.3-Mb long genome (R. Siezen, M. Kleerebezem personal communication). The large prophage elements Lp1 and Lp2 resemble temperate *pac*-site dairy phages in their genome organization, the closest relative being *Lb. plantarum* phage phig1e. Both prophages contain genes unrelated to known phage functions near both attachment sites. In the case of prophage Lp1 these 'extra' genes at both genome ends yielded database matches suggesting lysogenic conversion functions (distant relatives of mitogenic factors from *S. pyogenes* prophages, and the lysogenic conversion gene *fun* from *E. coli* phage P1). The Lp2 prophage showed a point mutation in the large subunit terminase gene which was likely to cause gene inactivation. Further features make prophage Lp2 peculiar. It shares DNA sequence identity with prophage Lp1 over the entire DNA packaging/head/tail gene cluster and the lysis cassette. Directly downstream of the Lp2 integrase gene a 12 kb-long prophage remnant containing residual lysogeny, DNA replication, DNA packaging and head genes was found. Another 8-kb long prophage remnant carries lysogeny- and DNA replication-related phage genes.

In the future prophage genomics is likely to play an important role in bacterial genomics. Prophage DNA is apparently enriched for genes that play a crucial role in the phenotype of the host cell. This role is not only underlined by virulence factors encoded on the prophage genome but DNA microarrays have demonstrated that prophage genes are prominent within the group of host genes that experience a significantly changed transcription level after alterations in growth conditions (for example temperature changes in *S. pyogenes* (Smoot et al. 2001), planktonic vs. biofilm growth in *Pseudomonas* (Whiteley et al. 2001)). Focusing on prophage sequences in bacterial genomes might thus lead to genes influencing important phenotypic differences between otherwise closely related bacterial strains.

Outlook

Genomics started with the sequencing of phage genomes in the late 1970s. However, in the last decade only few laboratories were involved in phage genomics. More recently, the pace has quickened again so that now over 100 complete phage genomes are available and it can be expected that many more will be published in the near future. These sequences have been marvelously informative with respect to the biology of the individual phages. The real excitement for phage biology starts now, with the advent of high volume sequencing technology, as this allows for the first time at a whole genome resolution to analyze the sequences together and thereby address a set of fundamental biological questions related to populations, e.g., what is the structure of the global phage population, what are its dynamics, how do phages evolve? This is Comparative Genomics with a capital 'C' (Brüssow and Hendrix 2002).

There is now interest not only in sequencing new phage genomes but also in doing so from an array of phages as diverse as possible. Phage biologists are also beginning to explore methods of sampling phage sequences from environmental sources without introducing the severe bias of asking them to grow on culturable bacteria. Combining such an approach with techniques like DNA microarray analysis promises to give a better picture of the diversity of the phage population and also to provide tools to answer questions as to how that diversity changes over space and time. The ease and low cost of phage sequencing combined with the extensive knowledge of model phages could give phage genomics a lead role in population genetics, the evolution of simple DNA genomes and their genetic diversity (Brüssow & Hendrix 2002).

Acknowledgements

We thank our colleagues at Nestlé for their support of the project and the Swiss National Science Foundation for financial support in its Biotechnology Module (grant 5002-044545/1). We thank Michiel Kleerebezem, Roland Siezen and Willem de Vos, Wageningen Center for Food Sciences, the Netherlands for sharing data on *Lactobacillus plantarum* strain WCFS1 prophages.

References

Altermann E, Klein JR & Henrich B (1999) Primary structure and features of the genome of the *Lactobacillus gasseri* temperate bacteriophage phi adh. Gene 236: 333–346.

Botstein D (1980) A theory of modular evolution for bacteriophages. Ann. NY Acad. Sci. 354: 484–490.

Braun V, Hertwig S, Neve H, Geis A & Teuber M (1989) Taxonomic differentiation of bacteriophages of *Lactococcus lactis* by electron microscopy, DNA-DNA hybridization, and protein profiles. J. Gen. Microbiol. 135: 2551–2560.

Brondsted L, Ostergaard S, Pedersen M, Hammer K & Vogensen FK (2001) Analysis of the complete DNA sequence of the temperate bacteriophage TP901-1: evolution, structure, and genome organization of lactococcal bacteriophages. Virology 283: 93–109.

Bruttin A, Desiere F, Lucchini S, Foley S & Brüssow H (1997) Characterization of the lysogeny DNA module from the temperate *Streptococcus thermophilus* bacteriophage Sfi21. Virology 233: 136–148.

Brüssow H (2001) Phages of dairy bacteria. Annu. Rev. Microbiol. 55: 283–303.

Brüssow H & Desiere F (2001a) Comparative phage genomics and the evolution of Siphoviridae: insights from dairy phages. Mol. Microbiol. 39: 213–223.

Brüssow H & Desiere F (2001b) Comparative phage genomics and the evolution of Siphoviridae: insights from dairy phages. Mol. Microbiol. 39: 213–222.

Brüssow H & Hendrix R (2002) Phage genomics: Small is beautiful. Cell 108: 13–16. Brüssow H, Fremont M, Bruttin A, Sidoti J, Constable A & Fryder V (1994) Detection and classification of *Streptococcus thermophilus* bacteriophages isolated from industrial milk fermentation. Appl. Environ. Microbiol. 60: 4537–4543.

Casjens S, Hatfull GF & Hendrix R (1992) Evolution of dsDNA tailed-bacteriophage genomes. Sem. Virol. 3: 383–397.

Chandry PS, Moore SC, Boyce JD, Davidson BE & Hillier AJ (1997) Analysis of the DNA sequence, gene expression, origin of replication and modular structure of the *Lactococcus lactis* lytic bacteriophage sk1. Mol. Microbiol. 26: 49–64.

Chopin A, Bolotin A, Sorokin A, Ehrlich SD & Chopin M (2001) Analysis of six prophages in *Lactococcus lactis* IL1403: different genetic structure of temperate and virulent phage populations. Nucleic Acids Res. 29: 644–651.

Desiere F, Lucchini S, Bruttin A, Zwahlen MC & Brüssow H (1997) A highly conserved DNA replication module from *Streptococcus thermophilus* phages is similar in sequence and topology to a module from *Lactococcus lactis* phages. Virology 234: 372–382.

Desiere F, Lucchini S & Brüssow H (1998) Evolution of *Streptococcus thermophilus* bacteriophage genomes by modular exchanges followed by point mutations and small deletions and insertions. Virology 241: 345–356.

Desiere F, Pridmore RD & Brüssow H (2000) Comparative genomics of the late gene cluster from lactobacillus phages. Virology 275: 294–305.

Desiere F, Mahanivong C, Hillier A J, Chandry PS, Davidson BE & Brüssow H (2001) Comparative genomics of lactococcal phages: insight from the complete genome sequence of Lactococcus lactis phage BK5-T. Virology 283: 240–252.

Desiere F, McShan WM, van Sinderen D, Ferretti JJ & Brüssow H (2001b) Comparative genomics reveals close genetic relationships between phages from dairy bacteria and pathogenic streptococci: Evolutionary implications for prophage-host interactions. Virology 288: 325–341.

Duplessis M & Moineau S (2001) Identification of a genetic determinant responsible for host specificity in *Streptococcus thermophilus* bacteriophages. Mol. Microbiol. 41: 325–336.

Eppler K, Wyckoff E, Goates J, Parr R & Casjens S (1991) Nucleotide sequence of the bacteriophage P22 genes required for DNA packaging. Virology 183: 519–538.

Ferretti JJ, McShan WM, Ajdic D, Savic DJ, Savic G, Lyon K, Primeaux C, Sezate S, Suvorov AN, Kenton S, Lai HS, Lin SP, Qian Y, Jia HG, Najar FZ, Ren Q, Zhu H, Song L, White J, Yuan X, Clifton SW, Roe BA & McLaughlin R (2001) Complete genome sequence of an M1 strain of *Streptococcus pyogenes*. Proc. Natl. Acad. Sci. U.S.A. 98: 4658–4663.

Foley S, Lucchini S, Zwahlen MC & Brüssow H (1998) A short noncoding viral DNA element showing characteristics of a replication origin confers bacteriophage resistance to *Streptococcus thermophilus*. Virology 250: 377–387.

Goshorn SC & Schlievert PM (1989) Bacteriophage association of streptococcal pyrogenic exotoxin type C. J. Bacteriol. 171: 3068–3073.

Harrington DJ, Sutcliffe IC, Chanter N (2002) The molecular basis of *Streptococcus equi* infection and disease. Microbes Infect. 4(4): 501–105.

Hendrix RW, Smith MC, Burns RN, Ford ME & Hatfull GF (1999) Evolutionary relationships among diverse bacteriophages and prophages: all the world's a phage. Proc. Natl. Acad. Sci. U.S.A. 96: 2192–2197.

Hynes WL & Ferretti JJ (1989) Sequence analysis and expression in *Escherichia coli* of the hyaluronidase gene of *Streptococcus pyogenes* bacteriophage H4489A. Infect.Immun. 57: 533–539.

Hynes WL, Hancock L & Ferretti J J (1995) Analysis of a second bacteriophage hyaluronidase gene from *Streptococcus pyogenes*: evidence for a third hyaluronidase involved in extracellular enzymatic activity. Infect. Immun. 63: 3015–3020.

Inagaki Y, Myouga F, Kawabata H, Yamai S & Watanabe H (2000) Genomic differences in *Streptococcus pyogenes* serotype M3 between recent isolates associated with toxic shock-like syndrome and past clinical isolates. J. Infect. Dis. 181: 975–983.

Jarvis AW, Fitzgerald GF, Mata M, Mercenier A, Neve H, Powell IB, Ronda C, Saxelin M & Teuber M (1991) Species and type phages of lactococcal bacteriophages. Intervirology 32: 2–9.

Jarvis AW, Lubbers MW, Beresford TP, Ward LJ, Waterfield NR, Collins LJ & Jarvis BD (1995) Molecular biology of lactococcal bacteriophage c2. Dev. Biol. Stand. 85: 561–567.

Kodaira KI, Oki M, Kakikawa M, Watanabe N, Hirakawa M, Yamada K & Taketo A (1997) Genome structure of the *Lactobacillus* temperate phage phi g1e: the whole genome sequence and the putative promoter/repressor system. Gene 187: 45–53.

Koonin EV, Aravind L & Kondrashov AS (2000) The impact of comparative genomics on our understanding of evolution. Cell 101: 573–576.

Lawrence JG, Hendrix R & Casjens S (2001) Where are the pseudogenes in bacterial genomes? Trends Microbiol. 9(11): 535–540.

Lazarevic V, Dusterhoft A, Soldo B, Hilbert H, Mauel C & Karamata D (1999) Nucleotide sequence of the *Bacillus subtilis* temperate bacteriophage SPbetac2. Microbiology 145 (Pt 5): 1055–1067.

Le Marrec C, van Sinderen D, Walsh L, Stanley E, Vlegels E, Moineau S, Heinze P, Fitzgerald G & Fayard B (1997) Two groups of bacteriophages infecting *Streptococcus thermophilus* can be distinguished on the basis of mode of packaging and genetic determinants for major structural proteins. Appl. Environ. Microbiol. 63: 3246–3253.

Loessner MJ, Inman RB, Lauer P & Calendar R (2000) Complete nucleotide sequence, molecular analysis and genome structure of bacteriophage A118 of *Listeria monocytogenes*: implications for phage evolution. Mol. Microbiol. 35: 324–340.

Lubbers MW, Waterfield NR, Beresford TP, Le PR & Jarvis AW (1995) Sequencing and analysis of the prolate-headed lactococcal bacteriophage c2 genome and identification of the structural genes. Appl. Environ. Microbiol. 61: 4348–4356.

Lucchini S, Desiere F & Brüssow H (1998) The structural gene module in *Streptococcus thermophilus* bacteriophage phi Sfi11 shows a hierarchy of relatedness to Siphoviridae from a wide range of bacterial hosts. Virology 246: 63–73.

Lucchini S, Desiere F & Brüssow H (1999a) Comparative genomics of *Streptococcus thermophilus* phage species supports a modular evolution theory. J. Virol. 73: 8647–8656.

Lucchini S, Desiere F & Brüssow H (1999b) Similarly organized lysogeny modules in temperate Siphoviridae from low GC content Gram-positive bacteria. Virology 263: 427–435.

Lucchini S, Desiere F & Brüssow H (1999c) The genetic relationship between virulent and temperate *Streptococcus thermophilus* bacteriophages: Whole genome comparison of *cos*-site phages Sfi19 and Sfi21. Virology 260: 232–243.

Madsen PL & Hammer K (1998) Temporal transcription of the lactococcal temperate phage TP901-1 and DNA sequence of the early promoter region. Microbiology 144: 2203–2215.

Mahanivong C, Boyce JD, Davidson BE & Hillier AJ (2001) Sequence analysis and molecular characterization of the *Lactococcus lactis* temperate bacteriophage BK5-T. Appl Environ Microbiol. 67: 3564–3576.

Marciel AM, Kapur V & Musser JM (1997) Molecular population genetic analysis of a *Streptococcus pyogenes* bacteriophage-encoded hyaluronidase gene: recombination contributes to allelic variation. Microb. Pathog. 22: 209–217.

Martin AC, Lopez R & Garcia P (1996) Analysis of the complete nucleotide sequence and functional organization of the genome of *Streptococcus pneumoniae* bacteriophage Cp- 1. J. Virol. 70: 3678–3687.

McShan WM, Tang YF & Ferretti JJ (1997) Bacteriophage T12 of *Streptococcus pyogenes* integrates into the gene encoding a serine tRNA. Mol. Microbiol. 23: 719–728.

Mikkonen M, Dupont L, Alatossava T & Ritzenthaler P (1996) Defective site-specific integration elements are present in the genome of virulent bacteriophage LL-H of *Lactobacillus delbrueckii*. Appl. Environ. Microbiol. 62: 1847–1851.

Mikkonen M, Raisanen L & Alatossava T (1996) The early gene region completes the nucleotide sequence of *Lactobacillus delbrueckii* subsp. *lactis* phage LL-H. Gene 175: 49–57.

Moscoso M & Suarez JE (2000) Characterization of the DNA replication module of bacteriophage A2 and use of its origin of replication as a defense against infection during milk fermentation by *Lactobacillus casei*. Virology 273: 101–111.

Nelson D, Loomis L & Fischetti VA (2001) Prevention and elimination of upper respiratory colonization of mice by group A streptococci by using a bacteriophage lytic enzyme. Proc. Natl. Acad. Sci. U.S.A. 98: 4107–4112.

Neve H, Zenz KI, Desiere F, Koch A, Heller KJ & Brüssow H (1998) Comparison of the lysogeny modules from the temperate *Streptococcus thermophilus* bacteriophages TP-J34 and Sfi21: Implications for the modular theory of phage evolution. Virology 241: 61–72.

Nida SK & Ferretti JJ (1982) Phage influence on the synthesis of extracellular toxins in group A streptococci. Infect. Immun. 36: 745–750.

Prevots F, Mata M & Ritzenthaler P (1990) Taxonomic differentiation of 101 lactococcal bacteriophages and characterization of bacteriophages with unusually large genomes. Appl. Environ. Microbiol. 56: 2180–2185.

Schouler C, Ehrlich SD & Chopin MC (1994) Sequence and organization of the lactococcal prolate-headed bIL67 phage genome. Microbiology. 140: 3061–3069.

Smoot LM, Smoot JC, Graham MR, Somerville GA, Sturdevant DE, Migliaccio CA, Sylva GL & Musser JM (2001) Global differential gene expression in response to growth temperature alteration in group A *Streptococcus*. Proc. Natl. Acad. Sci. U.S.A. 98: 10416–10421.

Smoot JC, Barbian KD, Van Gompel JJ, Smoot LM, Chaussee MS, Sylva GL, Sturdevant DE, Ricklefs SM, Porcella SF, Parkins LD, Beres SB, Campbell DS, Smith TM, Zhang Q, Kapur V, Daly JA, Veasy LG & Musser JM (2002) Genome sequence and comparative microarray analysis of serotype M18 group A *Streptococcus* strains associated with acute rheumatic fever outbreaks. Proc. Natl. Acad. Sci. U.S.A. 99: 4668–4673.

Sonnhammer E L & Durbin R (1995) A dot-matrix program with dynamic threshold control suited for genomic DNA and protein sequence analysis. Gene 167: GC1–10.

Stanley, E., Fitzgerald, G. F., Le Marrec, C., Fayard, B., and van Sinderen, D. (1997). Sequence analysis and characterization of phi O1205, a temperate bacteriophage infecting *Streptococcus thermophilus* CNRZ1205. Microbiology. 143: 3417–3429.

Stanley E, Walsh L, van der ZA, Fitzgerald GF & van Sinderen D (2000) Identification of four loci isolated from two *Streptococcus thermophilus* phage genomes responsible for mediating bacteriophage resistance. FEMS Microbiol. Lett. 182: 271–277.

Tremblay DM & Moineau S (1999) Complete genomic sequence of the lytic bacteriophage DT1 of *Streptococcus thermophilus*. Virology 255: 63–76.

van Sinderen D, Karsens H, Kok J, Terpstra P, Ruiters MH, Venema G & Nauta A (1996) Sequence analysis and molecular characterization of the temperate lactococcal bacteriophage r1t. Mol. Microbiol. 19: 1343–1355.

Vasala A, Dupont L, Baumann M, Ritzenthaler P & Alatossava T (1993) Molecular comparison of the structural proteins encoding gene clusters of two related *Lactobacillus delbrueckii* bacteriophages. J. Virol. 67: 3061–3068.

Ventura M, Bruttin A, Canchaya C & Brüssow H (2002) Transcription analysis of *Streptococcus thermophilus* phages in the lysogenic state. (2002) Virology. (In Press).

Weeks CR & Ferretti JJ (1984) The gene for type A streptococcal exotoxin (erythrogenic toxin) is located in bacteriophage T12. Infect. Immun. 46: 531–536.

Whiteley M, Bangera MG, Bumgarner RE, Parsek MR, Teitzel GM, Lory S & Greenberg EP (2001) Gene expression in *Pseudomonas aeruginosa* biofilms. Nature 413: 860–864.

Yu CE & Ferretti JJ (1991) Molecular characterization of new group A streptococcal bacteriophages containing the gene for streptococcal erythrogenic toxin A (speA). Mol. Gen. Genet. 231: 161–168.

Antonie van Leeuwenhoek **82:** 93–112, 2001.

Gene regulation in *Lactococcus lactis*: the gap between predicted and characterized regulators

Eric Guédon, Emmanuel Jamet & Pierre Renault*
*Génétique Microbienne, INRA, Domaine de Vilvert, 78352 Jouy en Josas, France (*Author for correspondence; E-mail: renault@jouy.inra.fr)*

Key words: transcriptional control, gene expression, gene regulation, regulatory network, lactic acid bacteria

Abstract

The genome sequence of *Lactococcus lactis* IL1403 was previously determined with high quality, allowing a reliable determination of the potential ORFs present in the genome. It encodes 2310 proteins, and 138 of them were assigned as potential regulators, half of which being further classified by their similarity to known protein families. Among these regulators, most could have a direct role as transcriptional regulators, while the others may have less well defined functions in transcriptional regulation or more general functions, such as the GTP binding protein family. Current knowledge related to the regulators controlling gene expression in *L. lactis* will be confronted to data obtained in other bacteria. For example, comparison of the *L. lactis* regulators with those of *B. subtilis* reveals many orthologous regulators and also some clear differences in the type of regulator used in the two bacteria. Further comparison of the role and the effectors of orthologous regulators shows that direct transposition of a 'heterologous model' does not allow to build a reliable regulatory network in *L. lactis*. Moreover, many *L. lactis* regulators have functions that could not be proposed by transposition of the knowledge currently available in other bacteria. A considerable amount of work will be necessary to assess the function of *L. lactis* regulators and build a comprehensive model of the regulatory network. This would provide invaluable information on *L. lactis* biology and the way this bacterium interacts with the environment.

Introduction

The nucleotide sequences of entire genomes are now available for a continually growing number of microorganisms. The functional assignment of proteins encoded in these genomes is a major problem in post-genomic biology. Even in 'model microorganisms' such as *Escherichia coli* and *Bacillus subtilis*, about half of the proteins have unknown functions. Moreover, functions of many proteins are assigned on the basis of their 'relatedness' to functionally characterized proteins. Relatedness may include amino acid identity and genetic environment of the gene in the favourable cases, but some annotations are based on less pertinent similarities such as the presence of restricted motifs or potential similarities of threedimensional structures. More awkwardly, the function of a number of genes is assigned on the basis of similarities with proteins of which the function was assigned after a cascade of similar searches, obscuring access to the original data. Special care is thus needed to interpret and use *in silico* genome data. Nevertheless, the real potential of an organism will not be easily addressed until, at least, the function of most or all predicted ORFs is correctly assigned.

A sub-class of proteins encoded in genomes are those that have regulatory functions. Many of these have a direct role in modulating transcription and we will focus on them in this review. These proteins repress or activate the transcription of genes that ultimately will ensure particular cellular functions. The ability of the cell to respond quickly to environmental changes is essential for its survival. Moreover, the relevance of the functions selectively expressed in the cell will determine the fitness of the bacteria to resist adverse conditions or to optimally exploit the environment and proliferate. Characterization of the regulators, the signal governing their activity, the genes they control and the level of control is essential to understand how the bacteria respond to environmental changes. The data corresponding to these regulatory features could be defined as the regulatory network of the bacterial cell. Acting on this network may drastically affect the properties of the bacteria and have implications on bioprocesses in the case of industrial microorganism, survival in the environment after their release but, especially in the case of potential pathogens, eventually also on health.

In this paper we will review the current knowledge related to the regulators controlling gene expression in *Lactococcus lactis*, updating the last review on this topic (Kok 1996). *L. lactis* is widely used in the food industry and is consumed alive in many fermented foods such as cheeses and buttermilk. It is also present in other ecological niches, mainly on plant material from where they probably originated before being used in the industry. In addition, *L. lactis* has been isolated from niches as diverse as fishes, insects' guts, and more rarely diseased humans (Elliott et al. 1991; Pellizzer et al. 1996). The genome sequence of *L. lactis* IL1403 has been determined with high quality, allowing reliable detection of the potential ORFs present in the genome. Some 2310 proteins were reported and 138 of them were predicted to be potential regulators, with half of them being classified further by their similarity to known protein families. Most of these regulators have a direct role as transcriptional regulators, while the others have less well defined functions in transcriptional regulation or may have more general functions in cell life such as the GTP binding protein family. Comparisons of the *L. lactis* regulators with those of *B. subtilis* reveals some clear differences in the type of regulators used in the two bacterial species, probably reflecting differences in the environmental constraints these microorganisms have to face. For example, there are only three σ factors in *L. lactis*, while there are 18 in *B.subtilis*. Moreover, *L. lactis* has eight two-component signal transducers while *B. subtilis* contains 34. These two types of regulators are usually involved in the activation of arrays of genes in response to changing environmental conditions, suggesting that the *L. lactis* lifestyle is more sedate. A better understanding of the regulatory network of *L. lactis* should provide invaluable information on its biology and the way this bacterium interacts with the environment.

Sensing environmental signals

Sigma factors

In order to respond to drastic changes in the environment, bacteria may activate or repress the transcription of many genes. A particularly efficient way to modify transcriptional patterns is to directly interact with RNA polymerase and modify its properties. Cellular metabolites such as pppGpp produced during the stringent response probably interact with RNA polymerase to regulate gene expression at certain promoters during nutrient starvation (Cashel et al. 1996; Chatterji & Ojha 2001). During exponential growth, the vegetative sigma factor RpoD allows the binding of RNA polymerase, and thus the initiation of transcription, at the so-called 'vegetative promoters' whose general consensus is TTGACA-N_{16--19}-TATAAT. In many bacteria, but particularly in Gram-positive bacteria including *L. lactis*, this consensus could be reduced to a -10 extended box whose consensus is TGYTATAAT (Helmann 1995; Goupil-Feuillerat et al. 2000). Lastly, RpoD could be replaced by alternative sigma factors and extracytoplasmic function (ECF) RNA polymerase sigma factors (Missiakas & Raina 1998). In the presence of alternative sigma factors, new consensus motifs are recognized allowing the transcription of new sets of genes. Only two such factors have been found from the analysis of the proteins encoded by the *L. lactis* IL1403 chromosome compared to the 18 found in the *B. subtilis* genome (Bolotin et al. 2001). These two alternative sigma factors are annotated ComX and SigX and are localized rather closely on the genome (18-kb apart).

ComX function is likely to induce competence genes (Bolotin et al. 1999, 2001), whereas SigX function could not be suggested by the analysis of its sequence (closest homologues in other organisms or the genes situated in its vicinity). The closest homologue of ComX, with 26% identity, is the competence sigma factor comX from *S. pneumoniae* (Lee & Morrison 1999) and those from several other Streptococci. It is located upstream of an *rrn* operon, like the streptococcal homologues. Interestingly, several genes or operons, most of them having potential function in competence or DNA processing, are preceded by a motif close to the competence gene promoter consensus in *S. pneumoniae* (Bolotin et al. 1999, 2000). These data suggest that *L. lactis*, like many Streptococci, is able to develop natural competence. In other Streptococci, competence is induced by a peptide secreted by the cell in the medium that is later sensed *via* a two-component regulator relay. Search for a comparable system in the *L. lactis* IL1403 genome did not allow finding this competence induction system encoded by *comCDE* in *S. pneunomiae* (Havarstein et al. 1996; Pestova et al. 1996) or *S. mutans* (Li et al. 2001). The signal-inducing competence genes in *L. lactis* remain thus to be found since they may be different from the peptide-dependent system from Streptococci or may be even lost from the IL1403 genome.

Two-component systems

Microorganisms have developed several systems of detecting and responding to the changes in their environment. Two Component signal transduction Systems (TCS) represent one of the ways of accomplishing this task in the bacterial world. TCSs are composed of two main proteins (i) the Histidine-sensor Kinase (HK) and (ii) the Response Regulator (RR). HKs are generally integral membrane proteins that monitor a specific environmental parameter and RRs are DNA binding proteins that interacts with HK. Upon receipt of a specific stimulus, HK is activated by auto-phosphorylation on a conserved histidine residue and serves as phospho-donor for a conserved aspartate residue on its cognate RR. In its activated phosphorylated form, RRs lead to the appropriate cellular response. This frequently occurs by activation of transcription as many of the RRs are DNA binding proteins. Analysis of the complete genome sequence of *L. lactis* IL1403 revealed the presence of 8 RRs and 7 HKs (Bolotin et al. 2001). Each lactococcal TCS is located at a distinct locus containing a histidine kinase and response-regulator-encoding gene pair, with the exception of *llrH* and *kinE/llrE*. Indeed, *llrH* is an orphan RR gene without HK gene in its vicinity and the *kinE/llrE* pair is separated by a gene encoding a protein that has homology with a putative membrane protein of *Yersinia pestis* and a sodium/pantothenate symporter of *Rickettsia conorii*. The LlrA, LlrB, LlrC, LlrE, LlrF and LlrG proteins belong to the OmpR subfamily of bacterial RRs and LlrD and LlrH to the NarL subfamily. Finally, at least two other TCSs carried by genetic mobile elements were characterized in other *L. lactis* strains, the *Tn5276 nisR/nisK* and the plasmid-borne *lcoR/lcoS* systems. The *nisR/nisK* TCS has been extensively studied and controls the expression of the *nis* operon (Kuipers et al. 1995; de Ruyter et al. 1996). Transcription of *lcoR/lcoS* is induced by the presence of copper in the medium and may control copper resistance determinants (Khunajakr et al. 1999).

Among the chromosomal lactococcal TCSs, six (KinA/LlrA; KinB/LlrB; KinC/LlrC, KinD/LlrD; KinE/LlrE) have been analyzed in *L. lactis* subsp. *cremoris* MG1363 (O'Connell-Motherway et al. 2000). Transcriptional analyses of these six TCSs have shown that *kinA/llrA* and *kinC/llrC* are constitutively expressed at a high level, whilst *kinB/llrB*, *kinE/llrE*, *kinF/llrF* are induced during mid-log phase and *kinD/llrD* at the very end of exponential phase of growth.

Insertion mutagenesis and phenotypic analysis of the corresponding mutants suggest that KinA/LlrA is involved in arginine metabolism regulation in response to acid stress. Interestingly, the genes located downstream of *llrA* encode a potential amonium transporter and a nitrogen regulatory protein. The KinC/LlrC system might thus be involved in a global stress response in the lactococcal cell since the inactivation of *kinC* or *llrC* leads to an acid-sensitive phenotype, although less severe than inactivation of the *kinA* or *llrA* genes. Furthermore, this system might also be involved in a temperature-sensing pathway because LlrC was identified as a cold-induced protein (Wouters et al. 2001).

The KinD/LlrD system seems to be involved in salt or osmotic stress responses and the KinF/LlrF system in oxidative stress response. Mutant analysis suggested that KinD and KinF are necessary for the cell growth and/or survival since no mutant inactivated for the corresponding genes were obtained, despite repeated attempts (O'Connell-Motherway et al. 2000).

The KinE/LlrE system could play a role in the regulation of phosphatase activity in *L. lactis* as revealed by a plate assay with an X-P substrate as indicator. Lastly, the neighborhood of TCS genes in *L. lactis* IL1403 suggests that LlrH and LlrG/KinG could be involved in arginine and aromatic amino acids metabolism, respectively: (i) *llrH* is followed by *octA* and *yrfD*, which encode a potential ornithine carbamoyltransferase and an amino acid antiporter, respectively and (ii) *llrG-kinG* is in cluster with the *tyrA-aroA-aroK-pheA* genes. Finally, neither physiological data from analysis of MG1363 *kinB* nor the genetic environment indicates a function for the lactococcal KinB/LlrB TCS (O'Connell-Motherway et al. 2000).

To conclude, six out of the seven/eight TCSs of *L. lactis* were studied using classical phenotypic and genetic analysis, which gave partial indications on the function of some of them. A more extensive analysis with global methods, such as the use of DNA microarrays reported for *B. subtilis* (Kobayashi et al. 2001) and *S. pneumoniae* (Throup et al. 2000) might allow defining their regulons and their potential functions in response to environmental changes.

Osmolarity, ions, oxidative stress

In addition to the previously described systems, several additional regulators may sense environmental factors such as change in osmolarity (BusR, GadR), ions (PhoU, CopR, ZitR, FlpA, FlpB) or oxidants (Fur). It should be stressed that regulation of metal uptake and metabolism is often related to oxidative stress responses and that *L. lactis* contains 11 paralogous genes of the MarR family, the members of which are often involved in gene regulation of metal metabolism. Except for BusR, GadR and Flp, no published data are available yet to indicate the role of these regulators.

Tolerance to high osmolarity of *L. lactis* is related to the activity of a betaine transport system encoded by *busAA-busAB* genes. Transcription of the *busA* operon is strongly regulated by the external osmolarity of the medium, such as salt concentration (Obis et al. 1999). A regulator of this system, BusR, was characterized through a genetic screen as repressing the *busA* promoter in *E. coli*. The gene encoding BusR is located immediately upstream of the *busA/AB* operon. BusR is able to bind at a site that overlaps the *busA* promoter. In *L. lactis*, a *busR* mutant showed a severe growth defect at both low and high osmolarity. Since the BusAB proteins are present at a lower level in a *busR* mutant, which correlates with a reduced glycine betaine uptake activity, Romeo et al. (2002) concluded that BusR probably controls other genes in *L. lactis*.

A promoter specifically induced by chloride was also identified in *L. lactis*. It controls the transcription of the two genes *gadC* and *gadB*, which likely encode glutamate–γ-aminobutyrate antiporters and glutamate decarboxylase (Sanders et al. 1998). *L. lactis gad* mutants are more sensitive to low pH than the wild-type strain when NaCl and glutamate are present in the medium. Expression from this promoter increases in the presence of glutamate or chloride at low pH, suggesting that *gadBC* encode a glutamate-dependent acid resistance mechanism needed to maintain viability. A gene encoding GadR, a homologue of the activator Rgg from *Streptococcus gordonii*, is located immediately upstream of the chloride-dependent *gadCB* promoter. Analysis of a *gadR* mutant showed that GadR is the activator controlling sodium chloride induction of the *gadCB* operon.

Phosphate metabolism is known to play an important role in bacterial stress response (Duwat et al. 2000). *L. lactis phoU*, encoding a homologue of the *E. coli* PhoU phosphate regulon repressor, is located immediately downstream of the *ptsFEDCBA* operon. This operon encodes a high phosphate affinity transporter. A preliminary report suggests its role as phosphate regulon repressor (Cesselin et al. 2001). Mutants in *ptsA* have a higher transcriptional rate of *ptsF*, possibly due to the lower the expression of PhoU.

Aerotolerance and oxidative stress resistance genes have been found to be under the control of Fur (ferric uptake regulator) family regulators. *L. lactis* Fur is a 128-amino-acid protein sharing 35% identity with *S. aureus* Fur and *B. subtilis* PerR (peroxyde operon regulator), but only 28% identity with *B. subtilis* FurR. Multiple alignment of the regulators of this family clearly indicates that *L. lactis* Fur is related to *B. subtilis* PerR. In the latter bacterium, PerR controls the transcription of the genes for catalase, alkyl hyperoxyde reductase and MrgA, a protein involved in metalloregulation of an oxidative-stress protein (Bsat et al., 1998). It is not known if *L. lactis* Fur regulates the expression of other oxidative stress response genes such as *sodA* (Sanders et al. 1995), *ahpCF* and *dpsA*. It would be interesting to test whether *L. lactis*, like *S. pyogenes*, contains genes for a novel mechanism of managing peroxide stress that may be regulated by this PerR homologue (King et al. 2000). Recently, it has been shown that the *ohrR/ohrA* system may be involved in hydroperoxide response in *B. subtilis* (Fuangthong et al. 2001). *L. lactis* RmaJ, a MarR-family regulator, shares 45% identity with OhrR. The *rmaJ* gene is transcribed divergently from *yfiE*, which encodes a protein sharing 37% identity with *ohrA*. A motif with only one base mismatch from the proposed recognition site of OhrR (TACAATT-AATTGTA) is present in the intergenic region *rmaJ-yfiE*, suggesting a good conservation of the mechanism and function of these genes.

Other potential regulators for metal transport could be proposed, based on their homology and the genetic organization of their genes. The *L. lactis* genome encodes several potential metal transporters. For example, copper metabolism could be controlled by CopR, a regulator with 40% identity to the *Enterococcus hirae* repressor of the *copAB* genes, CopY (Strausak & Solioz 1997). CopR is likely encoded by the first gene of an operon that potentially encodes a metal reductase and a copper/phosphate ATPase transporter. Zinc transport could depend on *zitS*, previously characterized as *nlp3*, specifying an exported protein (Poquet et al. 1998). This gene is clustered in the *zitRSQP* operon, where *zitSQP* encode a potential zinc ABC transporter while *zitR* encodes a protein belonging to the MarR regulator family. ZitR contains a potential HTH DNA binding domain and seems to act as a repressor of the *zit* genes, as a function of the extracellular concentration of Zn^{2+} (Poquet, personal communication).

Interestingly, in addition to the potential zinc transporter encoded by *zitSPQ*, *L. lactis* MG1363 may possess high- and low affinity ATP-dependent Zn(II) uptake systems that could be encoded by two paralogous operons *orfXY-flpA/B* (Scott et al. 2000a,b). FlpA and FlpB share a high degree of identity, belong to the FNR-like protein family and control their own expression (Gostick et al. 1999). These regulators also share a high percentage of identity with FLP from *Lb. casei* (Scott et al. 2000a). Expression from the *orfXY-flpB* operon was shown to be activated by Cd(II) and Zn(II). A *flpA flpB* double mutant is hypersensitive to hydrogen peroxide and has a depleted intracellular Zn(II) pool suggesting that hypersensitivity to oxidative stress is due to Zn(II) requirement to protect vulnerable protein thiols from oxidation (Scott 2000b). FlpA from *L. lactis* MG1363 was purified as a homodimeric protein containing both Zn and Cu, a property similar to FLP from *Lb. casei* (Scott 2000a). However, FlpA recognizes FNR sites (TTGAT-N4-ATCAA) but not FLP sites (CCTGA-N4-TCAGG) in band-shift assays. Thus, it behaves more like the *E. coli* oxygen-responsive transcription factor FNR than the *Lb. casei* FLP, its closest characterized homologue. Studies of *E. coli* strains expressing *flpA* or *flpB* revealed that both homologues were able to activate expression of FNR-dependent promoters in vivo when the FNR site was positioned 61 base pairs upstream of the transcription start. However, FlpA was not able to restore the aerobic–anaerobic switch in the heterologous host, *E. coli*. Scanning the *L. lactis* IL1403 genome for FNR-like regulators did not allow finding true orthologs of FlpA and FlpB. However, the genome does encode two potential FNR-like proteins, RcfA and RcfB. RcfA is the closest homologue of the *L. lactis* MG1363 Flp regulators, sharing 55% identity. It is encoded in an operon and is preceded by *yuiA*, which encodes a potential metal transporting ATPase. Interestingly, a motif that perfectly matches the FNR binding site is present in the *yuiA-rcfA* promoter region, suggesting that RcfA may have the same binding properties as FlpA and may, thus, control the expression of its own operon. In addition, 22 motifs in the IL1403 genome perfectly match the FNR binding site consensus but only four are present in intergenic regions, namely upstream of *pbp2B*, *ygfC*, *fabZ2* and *yvfA* genes. FlpA may thus control the expression of the penicillin-binding protein 2B, a transcriptional regulator followed by an ABC transporter, proteins involved in biotin metabolism and an unknown protein, a hypothesis that could be tested experimentally.

Lastly RcfB, although belonging to the FNR type regulator family, is not a close homologue of the FLP proteins, but is closer to the *B. subtilis* type FNR (26% identity). The function of the *rcfB* gene has not been tested yet, and no hypothesis could be proposed for its function from its localization in the genome.

Heat and cold stresses

Among changes in environmental conditions, changes in temperature are very common and bacteria have developed systems to adapt to low or high temperatures. Among proteins known to allow these adaptations, chaperones were found to be induced almost ubiquitously. These proteins are very important for cell survival, helping to avoid protein denaturation, and are among the first to be induced upon a stress. This first response is followed by changes in cellular metabolism, leading to modifications in the composition of the cytoplasmic membrane and the cell wall, and probably also to other less well characterized reactions. The first response has been studied specifically in lactic acid bacteria, mostly because modification of temperature is an integral part of cheese making processes. For example, elevated temperatures are applied to accelerate whey expulsion and to select particular flora, while low temperatures are required for cheese ripening in the presence or not of molds. Proteins induced during stresses and their potential roles in cellular processes are detailed in another review in this issue (Maguin et al.).

L. lactis reacts to a lowering of temperature by the production of a set of proteins that includes the small so-called cold shock proteins (CSPs). *L. lactis* MG1363 produces at least five CSPs (CspA, B, C, D and E, Wouters et al. 1998) whereas *L. lactis* IL1403 contains only two *csp* genes (*cspD* and *cspE*). *L. lactis* IL1403 CspD is encoded by a gene located in a prophage, whereas CpsE might be the counterpart of *L.*

lactis MG1363 CspE. Members of this protein family were shown to bind to single-stranded DNA and RNA In model bacteria. It has been suggested that CSPs may be transcriptional regulators (Bae et al. 1999, 2000) or translational facilitators (Weber et al. 2001). Several proteins, including a number of cold-induced proteins, are induced in *L. lactis* upon overproduction of the CSPs. Inactivation of some CPSs in multiple disruption mutants lacking two or three CSPs was compensated by the overexpression of the remaining ones. Therefore, CSPs may play a regulatory role upon cold shock treatment and, likely, directly induce other factors involved in the cold adaptive response (Wouters et al. 2000, 2001). CspE may play the most important role in cryoprotection in *L. lactis* MG1363 (Wouters et al. 2001). The fact that CspD is dispensable for cold shock protection in *L. lactis* IL1403 upon excision of the prophage (Calero & Renault, in preparation) confirms that CspE may play a pivotal role in cold shock response, whereas the other CSPs may play accessory roles.

The heat shock response and the regulation of gene expression by heat shock stress have been well characterized in the two leading model microorganisms *E. coli* and *B. subtilis* (Narberhaus 1999). The alternative sigma factors σ^{32}, σ^{E} and σ^{54} positively regulate heat-shock-induced gene expression in *E. coli* (Bukau 1993). In *B. subtilis*, the heat-shock response involves at least four different classes of heat-inducible genes (Hecker et al. 1996; Kruger & Hecker 1998; Derré et al. 1999, 2000; Helmann et al. 2001; Petersohn et al. 2001). Class I genes are controlled by the HrcA repressor. The expression of Class II genes is dependent on the σ^{B} sigma factor whose synthesis and activity are increased under various stress conditions including heat shock. Class III genes were shown to be negatively regulated by CtsR. Class IV heat shock genes are defined as those that are expressed independently of σ^{B} and are devoid of CIRCE or CtsR operator sequences. In *L. lactis*, both the cellular response to heat shock and its regulation appear to be similar to those in other bacteria. Nevertheless, the number of heat-shock genes seems to be clearly less in *L. lactis* than in *B. subtilis*. Only two regulators for the heat shock response, HrcA and CtsR, have been identified in *L. lactis* so far (Kilstrup et al. 1997; Varmanen, et al. 2000).

HrcA represses the transcription of several heat shock genes in *B. subtilis* by its binding to the CIRCE elements (TTAGCACTC-N9-GAGTGCTAA; Hecker et al. 1996). Although the direct involvement of the HrcA repressor to CIRCE has not yet been really demonstrated in *L. lactis*, different data suggest a similar role during heat shock response: (i) the presence of CIRCE elements upstream of the *hrcA-gprE-dnaK* and *groES-groEL* operons and the *dnaJ* gene, all encoding classical chaperones, (ii) the heat shock induction of transcription of these genes and (iii) the functionality of the CIRCE element in heat shock dependent expression of *dnaJ* (Eaton et al. 1993; van Asseldonk et al. 1993; Kim & Batt 1993). No other CIRCE boxes (even allowing one mismatch) are present in the IL1403 chromosome sequence, suggesting that the *L. lactis* HcrA regulon is restricted to these three transcriptional units (unpublished data). Lastly, DnaK could be involved in the maturation of HrcA in *L. lactis* and, thus, modulate chaperone expression (Koch et al. 1998).

CtsR, which is present only in Gram-positive bacteria, represses gene expression by binding specifically to an operator sequence located in the promoter region of *clpP, clpC, clpE* of *B. subtilis* or *L. monocytogenes* and *clpP, clpC, clpE* and *groESL* of *S. pneumoniae* (Derré et al. 1999; Nair et al. 2000; Chastanet et al. 2001). This observation suggests that heat shock regulation by CtsR is highly conserved in Gram-positive bacteria. The regulatory role of CtsR in the control of several heat shock genes in *L. lactis*, such as *clpB*, *clpC*, *clpE* and *clpP*, has been demonstrated by constructing a *ctsR* deletion mutant and analyzing the transcription of most *clp* genes (Varmanen et al. 2000). Modulation of heat shock gene expression by CtsR seems to be dependent on the accumulation of denatured proteins, as described for HrcA of *B. subtilis* (Mogk et al. 1998; Frees & Ingmer 1999; Varmanen et al. 2000). Interestingly, potential CtsR binding motifs (RGTCARANWNRGTCAAA; Derre et al. 2000) are present upstream of the transcriptional units encoding Clp (*ctsR-clpC, yffA-clpE, clpP* and *clpB*), but also in the promoter regions of genes involved in different cellular processes. Among these are the *groES/EL* genes, where the CtsR motif is present 23 nt upstream of the CIRCE palyndrome, and the prophage-encoded *cspD* gene. Moreover, CtsR could also be involved in the control of genes involved in DNA metabolism, such as *dnaA* (initiation of replication), *ssbD* (single strand binding protein) and *parC* (topoisomerase IV subunitA), genes of nucleotide metabolism (*ndrHIEF*, encoding glutaredoxin-like protein and ribonucleotide reductase) and *glyA* (also controlled by PurR), L1 and L28 ribosome protein genes, envelop metabolism genes such as *acmB* (*N*-acetylmuramidase), *kdtB* (lipopolysaccharide core biosynthesis protein), potential cellular regulator YtaD (protein-tyrosine phosphatase), the metabolic genes *glnPQ* (Gln ABC transporter) and *adhE* (alcohol-acetaldehyde dehydrogenase), and several genes of unknown function (*ybjB, yjjG, yljD, ynfH, yqJA, ysbD, yueA* and *yscE*). Although the mere presence of a CtsR motif in a promoter region is not sufficient to predict binding of CtsR and, even less, its interaction with cognate promoters, it is likely that CtsR controls more genes than those experimentally described until now. The dual control probably exerted by CtsR and HrcA on *groES/EL* genes is interesting and a similar control was shown to exist in *S. pneumoniae* (Chastanet et al. 2001). The potential role of CtsR on CspD regulation is intriguing, while its possible involvement in the control of DNA replication, nucleotide metabolism and other cellular processes that might be affected by heat are promising research areas that should be examined in the future.

Regulation of carbon metabolism

Control of sugar catabolism

In contrast to the relatively low number of regulators sensing environmental conditions, *L. lactis* possesses a wide array of regulators related to sugar metabolism. Indeed, almost all operons involved in sugar catabolism may be controlled by the general regulator for catabolic repression, CcpA, and by specific regulators. CcpA is the DNA-binding protein responsible for the catabolite repression in Gram-positive bacteria (Hueck & Hillen 1995). It is well conserved in all bacteria from this group including *L. lactis,* where the overall scheme of regulation seems to be similar (Luesink et al. 1998; Aleksandrzak et al. 2000). A phosphorylated form of a small protein, Hpr, regulates the activity of CcpA. At high concentration of fructose 1,6-diphosphate, occuring when a well-assimilated sugar is catabolized, Hpr is phosphorylated by Hpr-kinase on its serine in position 46. For a detailed review of this topic, see the paper of Titgemyer & Hillen in this issue. The site recognized by CcpA, the so-called CRE box (catabolite responsive element; consensus TGNNANCGNTNNCA) is well conserved in Gram-positive bacteria. A search for this motif, allowing one to two mismatches except the central CG produces a few thousand hits in the *L. lactis* genome. Most of these motifs probably have no biological significance as many are not 'placed suitably' relatively to promoters or lack additional surrounding sequences that might be necessary for well recognized CRE boxes. Analysis of the 21 boxes that perfectly match the consensus could give a limited look of targets for CcpA-dependent catabolite repression. Seven of these boxes are properly placed to control the transcription of sugar catabolism genes, such as the ribose, mannitol, galactose, gluconate and a β-glucoside (*ybhE-celB-bglS*) operon, and a phospho-β-glucosidase gene (*yrcA*). Four other boxes may control genes involved in carbon metabolism, *cstA* (carbon starvation protein), *yxfA* (probable transporter of sugar), the *cit* operon (citrate lyase) and *ccpA*. Perfect CRE motifs are also present in the promoter regions of *prsA* (ribose-phosphate pyrophosphokinase), *octA* (ornithine carbamoyltransferase), *yhaA*, *ybjK*, *ygeD*, *yudB*, and the *arc* and *ilv* operons. The last two consensus CRE sites are within *ybgB* and *pip* genes.

The presence of CRE sites in the promoter regions of genes involved in carbon metabolism was expected and most other genes and operon involved in sugar catabolism are preceded by at least a CRE motif deviating from the consensus by one or two mismatches. Among genes known to be repressed by CcpA, only transcription of the *gal* (Luesink et al. 1999a,b), *xyl* (Jamet et al. in preparation) and aryl-β-D-glucosides (Monedero et al. 2001) operons, *pepP* (Guédon et al. 2001a) and *fbp* (Jamet et al. in preparation) has been measured. These results show that catabolic repression in *L. lactis* is similar to that in other Gram-positive bacteria such as *B. subtilis*. However, it is quite probable that in addition to the expected repressed genes described above, CcpA regulates many other genes containing less conserved CRE motifs. In particular, it is likely that CcpA regulates purine metabolism and arginine catabolism. The fact that glycolytic flux may influence purine synthesis is of particular interest and should motivate further studies, since purine pools are implied in many cellular processes, such as stress resistance (Rallu et al. 2000), cell division and transcription.

In addition to global carbon catabolism control, most sugar catabolic genes are also controlled specifically by secondary regulators that belong to different families of DNA binding proteins such as LacI, LysR, AraC, GntR, DeoR, BglG and possibly by regulators belonging to two protein families that contain many paralogous genes in *L. lactis* (see Conclusions below). About half of these regulators were characterized to control metabolism of salicine, maltose, xylose and fructose assimilation genes. In addition, plasmid encoded PTS-lactose and starch metabolism (Doman et al. 2000) and the Tn*5276* saccharose genes and their regulation have been studied in detail.

Lactose assimilation genes are clustered in the 8-kb lactose operon (*lacABCDFEGX*), which encodes the enzymes of the lactose phosphotransferase system and the tagatose 6-phosphate pathway (van Rooijen et al. 1990, 1991, 1992, 1993). The *lac* promoter sequence is located in a back-to-back configuration with the promoter of the divergently transcribed *lacR* gene, which encodes the LacR repressor. Transcriptional repression of the *lac* operon is mediated by the interaction between LacR, the *lac* promoter, and an operator sequence in the non-coding region between *lacR* and *lacA*. The LacR inducer is tagatose-6-phosphate produced by the cleavage of lactose-phosphate by phospho-β-galactosidase. Residues in the LacR repressor that are involved in the induction of *lacABCDFEGX* expression by tagatose-6-phosphate were determined, making this regulator one of the best studied in *L. lactis*. Transcription of the *lac* operon is also controlled by CcpA, which may bind at a site upstream of the promoter.

Several *L. lactis* strains utilize lactose via a chromosomally encoded permease and β-galactosidase (*lacZ*) (Vaughan et al. 1998). The genes are part of a *gal-lac* operon, *galKT-lacAZ-galE*. The *gal* genes encode enzymes of the Leloir pathway for galactose metabolism, and *lacA* encodes a galactoside acetyltransferase. This organization seems to be common in *L. lactis* subsp. *lactis* strains such as IL1403, but *lacAZ* genes are absent from strains of *L. lactis* subsp. *cremoris* such as MG1363 (Grossiord et al. 1998). Based on the high similarity to their Gram-negative homologues, the *lacA-lacZ* genes appear to be recently acquired by gene transfer and have engaged the promoters of the *gal* operon in order to direct and control their expression. No potential regulator involved in the lactose inducible expression of this operon is known as yet, in particular no gene encoding a GalR homologue (the galactose regulator present in several lactic acid bacteria such as *S. thermophilus*) is present either upstream of the operon (Grossiord et al. 1998; Vaughan et al. 2001), or elsewhere in the IL1403 chromosome.

The divergently transcribed *sacBK* and *sacAR* operons, involved in the utilization of sucrose by *L. lactis* NZ9800, are transcribed in three sucrose-inducible transcripts, containing *sacBK*, *sacAR*, and only *sacR*, respectively. The inactivation of *sacR* results in the constitutive transcription of the *sacBK* and *sacAR* operons in the presence of different carbon sources, indicating that SacR acts as a repressor of transcription (Luesink et al. 1999).

L. lactis contains several operons involved in β-glucoside assimilation. One of these, *ptbA-bglH*, encodes PTS-dependent enzyme II and a β-glucosidase, is preceded by an open reading frame denoted *bglR* (Bardowski et al. 1994). BglR shares 36–30% sequence identity with a family of regulatory proteins including BglG from *E. coli*, and SacT and SacY from *B. subtilis*. There is a transcription terminator upstream of *bglR*, whose 5′ end partly overlaps a 32-bp sequence highly homologous to the RNA-binding site that is conserved in regulatory systems of the BglG-SacY family. This suggests that BglR functions in a way similar to that of the other regulators of this family. Lastly, (i) increase of transcription of *bglR* in the presence of salicine, (ii) impaired growth of an *L. lactis* *bglR* mutant on aryl-β-D-glucosides such as esculin, salicin, and arbutin and (iii) positive autoregulation of *bglR* strongly suggest that BglR is involved in positive control of the utilization of aryl-β-D-glucosides by transcription antitermination. CRE sites are present in the *bglR* promoter region, suggesting that BglR activator is not produced in the presence of excess of rapidly metabolizable sugars, leading to the over 6-fold decrease of β-glucosidase expression (Monedero et al. 2001). Regulation of other β-glucoside operons has not been studied, but preliminary data suggest control by CcpA and by specific regulators (Aleksandrzak et al. 2000; Aleksandrzak, personal communication).

Particular strains of *L. lactis* are able to grow on xylose (Erlandson et al. 2000). The genes necessary for xylose assimilation are present in a large operon (*xylABM-xynNB-xyaX-xylT*), which also contains genes for xyloside/xylan utilization, although no *L. lactis* strains are known to use these large polymers (Bolotin et al. 2001; Erlandson et al. 2001). A potenial regulator gene, *xylR*, is transcribed divergently from the operon. Inactivation of *xylR* impairs growth on xylose, showing that XylR is an activator (Jamet et al. submitted). Xylose utilization genes present downstream of *xylA* are transcribed at lower level, because a potential terminator in the *xylA-xylB* intergenic region seems to attenuate transcription. Interestingly, xylose genes are transcribed from two and three promoters in IL1403 and NCDO2118 strains, respectively. In the latter strain the *pkt* gene encoding phosphoketolase (formerly *ypdE*) is separated from *xylT* by a 10-kb insertion encoding arabinose utilization genes (Jamet et al. in preparation). The three promoters are induced 100–1000-fold in the presence of xylose, and are completely repressed when glucose is added. Interestingly, glucose repression is CcpA independent, since only the second promoter of the operon possesses a CRE motif (Jamet et al. in preparation). It is likely that absence of transcription from the xylose inducible promoters is due to inducer expulsion/exclusion by a similar mechanism described for maltose and ribose (Monedero et al. 2001).

Regulation of the maltose assimilation genes has recently been studied (Andersson et al. pers comm). Inactivation of the gene present downstream of *malP* impairs maltose utilization, suggesting that this gene (designated *malR*, formerly *rliA*) encodes an activator. Interestingly, MalR belongs to the LacI family, the members of which are known to be repressors. The exact mechanism of activation is still not known, but it is suggested that MalR binds to a unique operator similar to MalR binding sites found in other bacteria. In *L. lactis*, only the *malEFG* genes, encoding the maltose transport system, might be controlled by MalR and maltose. Several CRE sites are present in the intergenic region between the *malE* and *dexC* diverging genes, and between *dexC-malA*, but only a 2-fold difference of maltose phosphorylase activity was found between cells growing in media containing maltose or glucose. This result suggests a low or insignificant effect of catabolic repression on the transcription of maltose genes.

Lastly, several sugar catabolic operons are in close proximity to regulator genes. Thus, fructose catabolism, depending on the transcription of *fruCA* (formerly *lacC-fruA*), is activated by FruR, a protein from the DeoR regulator family (formerly annotated as LacR, but only 38% identical to the LacR repressor of the *lac* PTS operon, C. Barrrière pers. comm.). A mannitol metabolism operon contains *mltR*, encoding a potential activator of the LysR family, while *rbsR, kdgR, gntR, rgrA*, and *rliB* encode proteins potentially controlling the expression of genes involved in the catabolism of ribose, mannonate, gluconate, and two unknown sugars, respectively.

Transcriptional regulation of glycolysis genes

Glycolysis is the central pathway to produce energy in lactic acid bacteria, but studies of regulation of glycolytic genes are relatively rare in *L. lactis*. The most relevant studies concern the role of CcpA that, in addition to its role as a repressor of sugar catabolic genes, also has been shown to activate the transcription of the *las* operon, encoding phosphofructo kinase, pyruvate kinase and L-lactate dehydrogenase (Luesink et al. 1998). In a *ccpA* mutant, the transcription of the *las* operon is reduced 4-fold. This decrease leads to a lower activity of pyruvate kinase and L-lactate dehydrogenase, resulting in a metabolic profile characteristic of mixed-acid fermentation rather than homolactic fermentation. Transcriptional activation is dependent on HPr(Ser-P) that functions here as a coactivator of CcpA (Luesink et al. 1999a,b). Systematic comparison of the transcription in glucose- and galactose-grown cells of *L. lactis* IL1403 have confirmed these results (Even et al. 2001) and further analysis of a *ccpA* mutant has shown that other glycolytic genes are controlled by glucose and CcpA (Jamet et al. submitted). In particular, the transcription of *pgi, gapA2* and *enoA* appears to increase in glucose grown cells. Only *pgi* and *las* activation seems to be under the direct control of CcpA, however, and degenerated CRE sites are present 13–14 bp upstream of the −35 sequence in the *pgi* and *las* promoters. Replacement of the CRE site in the *pgi* promoter leads to a 10-fold decrease of transcription rate and a 2-fold decrease of Pgi activity. The mechanism of CcpA-independent glucose-dependent activation of the other glycolytic genes remains unknown. It is worthwhile to mention that a systematic analysis of the transcription by DNA macroarrays has suggested that activation of the glyclytic genes in *B. subtilis* is CcpA-independent (Moreno et al. 2001) although CcpA could activate transcription of several genes such as acetate kinase gene (Grundy et al. 1993). Moreover, several genes activated or repressed by glucose are at least partially under CcpA-independent control (Moreno et al. 2001; Yoshida et al. 2001). It is likely that also in *L. lactis* specific regulators are involved, such as a possible functional equivalent of *B. subtilis* CggR, a protein that controls triose phosphate metabolism genes (Ludwig et al. 2001), although no true CggR homologue exists in *L. lactis* and novel regulators should thus be searched.

Control of organic acid metabolism

In addition to sugars, *L. lactis* actively metabolizes organic acids such as malate and citrate. The degradation of these two acids, which are also intermediary compounds in the few steps of the citric acid cycle in *L. lactis*, are encoded by the *mle* and *cit* operons. Upstream and divergently transcribed from the *cit* operon encoding citrate lyase, there is a gene encoding a potential repressor of the DeoR family. Although this gene may well be involved in the transcriptional control of citrate lyase expression, no experimental data have confirmed this hypothesis yet.

The transcription of the *mleSP* operon, encoding malolactic enzyme and the malate-lactate antiporter, has been well studied. In fact, MleR was the first positive regulator of the LysR-family characterized in Gram-positive bacteria (Renault et al. 1989). A mutant unable to ferment malate has been characterized (Renault et al. 1987). Its complementation by a DNA fragment carrying a gene with homology to a transcriptional activator of Gram-negative bacteria suggested that transcription of the gene for malolactic enzyme was under positive control of this regulator (Renault et al. 1989). This conclusion has been later confirmed by Northern blot analysis and transcriptional fusions, which showed that *mleSP* was induced 100–1000-fold in different strains by malate, in the presence of MleR (Ansanay et al. 1993; Guédon et al. unpublished).

Regulation of nitrogen metabolism

The proteolytic system

Proteolysis is essential for supplying *L. lactis* with amino acids during growth in milk. The transcription of 16 genes encoding 12 peptidases (*pepC, pepN, pepX, pepP, pepA, pepF2, pepDA1, pepDA2, pepQ, pepT, pepM*, and *pepO1*), P(I) and P(III) proteinases (*prtP1* and *prtP3*), and three transport systems (*dtpT, dtpP*, and *opp-pepO1*) of *L. lactis* MG1363 was analyzed in response to different environmental factors (Guédon et al. 2001a,b). Elevated temperature has no significant effect on the level of transcription of these genes whereas changes in the sugar source modulate *pepP* transcription only. The presence of potential CRE boxes in the *pepP* promoter region suggests that its expression is directly controlled by catabolic repression. Expression of *prtP1, prtP3, pepC, pepN, pepX, pepDA2* and the *opp-pepO1* operon is repressed 5–150-fold by addition of peptide sources such as casitone in a chemically defined medium, whereas that of *dtpT, pepP, pepA, pepF2, pepDA1, pepQ, pepT, pepM*, and the *dtpP* operon is not. However, another study has shown that the *dpp* operon (dipeptide uptake) is also controlled by nitrogen sources containing leucine or valine. (Sanz et al. 2001). These and mutant studies (Tynkkynen et al. 1993; Mierau et al. 1996) show that cell wall proteinase, the oligopeptide transport system and the three intracellular peptidases PepO1, PepN, PepC are the major components of the *L. lactis* proteolytic system.

Since it was previously shown that the dipeptide leucylproline exerts partial control on the transcription of *prtP* (Marugg et al. 1995), *pepC, pepN*, and the *opp-pepO1* operon (Guédon et al. 2001a), the signal controlling casitone-dependent repression was further searched for by analyzing the response of an *opp-lux* fusion to the addition of different dipeptides (Guédon et al. 2001b). Full correlation was found between the dipeptide content in branched-chain amino acids (BCAA; isoleucine, leucine or valine) and their ability to mediate repression of *opp-pepO1* expression. The repressive effect requires transport of the dipeptides and degradation into amino acids, showing that the signal was dependent on the BCAA pool in the cell. The role of BCAA in the regulation of the proteolytic system of *L. lactis* was confirmed by the study of mutants in one or more peptide transport systems (Opp, DtpT, and/or Dpp) (Sanz et al. 2001). The central role of BCAAs as a signal for the control of the proteolytic system in *L. lactis* was confirmed by the study of a mutant unable to degrade BCAAs (Guédon et al. 2001b).

The regulator sensing the BCAA signal was found by searching mutants constitutively derepressed for of *opp* expression (Guédon et al. 2001b). The protein shares 48% identity with CodY from *B. subtilis*, a pleiotropic repressor of the dipeptide permease operon (*dpp*) and several genes involved in amino acid degradation and competence induction. *B. subtilis* CodY was shown to bind to the promoter regions of *dpp, comK* and *srfA* (Serror & Sonenshein, 1996a,b). The almost perfect conservation of the HTH DNA binding domain in the C-terminal end of CodY of *L. lactis* and *B. subtilis* strongly suggests that *L. lactis* CodY is also a DNA binding protein. *B. subtilis* CodY was shown to be a GTP binding protein with increased affinity for its target promoter in presence of GTP *in vitro* (Ratnayake-Lecamwasam et al. 2001). Degenerated GTP binding motifs were proposed to be necessary for this binding. These motifs may also be present in *L. lactis* CodY, but their sequences are more distant from the general GTP binding consensus motif, making it difficult to predict whether or not *L. lactis* CodY has retained GTP binding properties. A model of regulation of the *L. lactis* CodY regulon has been proposed (Guédon et al. 2001b). After protein degradation and peptide assimilation, amino acid availability is not limiting in the cell and CodY represses functions whose expression is no longer required at high levels. The intracellular pool of BCAA was proposed to be used as a sensor of amino acid availability via CodY repressor and, thus, to lead to feed-back repression of genes involved in proteolysis. In *B. subtilis*, it was proposed that CodY activity is modulated by the GTP pool in the cell, and that this pool is dependent of the nitrogen state of the cell. Data obtained in *L. lactis* suggests that the BCAA pool is the most important factor to modulate CodY. It should be pointed out that neither the GTP pool nor the BCAA pool were measured *in vivo* in *B. subtilis* or *L. lactis*. Thus, further work will be necessary to determine whether (i) CodY responds to different signals in the two bacteria, (ii) the two signals may have a direct role on CodY activity or (iii) the BCAA pool modulates CodY activity indirectly by acting on the GTP pool in the cell. *L. lactis* contains a second gene, *codZ*, which encodes a homologue of CodY, but which lacks the C-terminal domain containing the HTH motif. The role of this gene, which is not present in *B. subtilis*, is still unknown. CodY may have a pleiotropic role in *L. lactis* as has been shown in *B. subtilis*. The use of DNA microarrays will be of great help to determine the complete set of genes modulated by CodY in *L. lactis*. Indeed, no CodY boxes or motifs have been described yet in any bacterium, although CodY DNA binding sites have been mapped for several *B. subtilis* promoters.

As stated above, several genes involved in the proteolytic system of *L. lactis* are not part of the CodY or CcpA regulons. The fact that no conditions were found to affect their expression does not imply that these genes are not regulated. For example, it is tempting to speculate that *pepDA1* transcription is regulated by YcfA, which is located head-to-head to *pepDA1*. Several genes encoding regulators of unknown function, such as *ysgA* and *rmeA*, are located in the vicinity of *pepO2* and *pepC*, respectively. However, location of a regulator gene near a possible target gene is only an indication of a functional link between the two. For example, *ccpA* is encoded back-to-back to *pepQ*, an organization conserved in many Gram-positive bacteria (Mahr et al. 2000). Although it has been suggested that *pepQ* transcription could be regulated by sugar source and catabolic repression in *Lactobacillus delbruekii* (Schick et al. 1999), the modulation measured is only minor and *pepDA* transcription seems not to be regulated by CcpA in *L. lactis* (Guédon et al. 2001a).

Amino acid biosynthesis and degradation

Although *L. lactis* requires the proteolytic system for optimal growth in milk, it is able to synthesize most amino acids in chemically defined medium (CDM), in which single amino acids are missing (Cocaign-Bousquet et al. 1995). Indeed, *L. lactis* possesses all the genes required for amino acid biosynthesis (Bolotin et al. 2001). Transcriptional regulation of the *trp*, *leu-ilv-ald*, *his* and *metC-cysK* operons has been characterized. Transcription is most often controlled at two levels: at initiation and at elongation. Transcriptional initiation of the *trp* operon is increased 4-fold in CDM lacking any amino acid (Raya et al. 1998). This unspecific response to amino acid starvation is reminiscent of the amino acid dependent stringent response in *E. coli* and other bacteria, although it remains possible that it depends on other factors such as the modification of RNA stability. The transcription of the *his* and *leu-ilv-ald* operons (from the *leu* and *ilv* promoters) is repressed 10- and 30-fold in the presence of histidine and isoleucine in the medium, respectively, suggesting the existence of a repressor (Delorme et al. 1992, 1999; Godon et al. 1992, 1993). However, the mere examination of the list of regulators, annotated mainly using homology scores to proteins present in databases, did not allow pinpointing possible candidate repressor genes. These might, therefore, either be unrelated to known histidine or isoleucine regulator genes, or be absent from the genome of the sequenced strain, which is auxotrophic for histidine and isoleucine. We propose that the first hypothesis is more likely, in particular for the isoleucine dependent regulation, as that is still operational in *L. lactis* IL1403 (Godon et al. 1993). Thus, the regulators modulating the initiation of transcription of many amino acid biosynthesis genes remain to be identified.

The regulation of the lactococcal *metC-cysK* operon, encoding a cystathionine beta-lyase (*metC*) and cysteine synthase (*cysK*) was studied in detail (Fernandez et al. 2002). Its expression is negatively affected by high concentrations of cysteine, methionine, or glutathione in the culture medium. *O*-acetyl-l-serine, the substrate of cysteine synthase induces the transcription of the operon, which is reminiscent of CysB dependent *metC* regulation in *E. coli*. The *L. lactis* regulator, encoded by *cmbR*, was identified by a random mutagenesis approach. CmbR belongs to the LysR family of regulatory proteins and likely activates transcription of *metC-cysK* in the presence of acetylserine. Surprisingly, *cmbR* (designated *fhuR* in

L. lactis IL1403), is the last gene of the *fhuCBGDR* operon, encoding a putative ferrichrome ABC transporter. This regulator also controls the transcription of the *fhu* operon, but the physiological role of this cross regulation is unclear (Fernandez et al. 2002).

Regulation of arginine synthesis has been well studied in model bacteria, allowing the characterization of ArgR and AhrC as transcriptional factors in *E. coli* and *B. subtilis*, respectively (Maas 1994; Miller et al. 1997). These proteins share homology and the ArgR/AhrC system seems to be universally conserved in different genomes. The recognition signal on DNA, a weak palindrome designated the ARG box, is also conserved between genomes, despite a very low degree of similarity between individual sites within a genome (Makarova et al. 2001). The role of ArgR and AhrC is slightly different. In *E. coli*, the ArgR regulator represses at least eight operons encoding arginine biosynthetic enzyme, whereas in *B. subtilis* AhrC has a dual role as repressor of the transcription of the arginine biosynthetic genes (four transcriptional units) and activator of the expression of genes involved in the arginase pathway (four transcriptional units). Two proteins, denoted ArgR and AhrC and sharing 35% and 36% of identity with *E. coli* ArgR and *B. subtilis* AhrC, respectively, are encoded by the *L. lactis* IL1403 genome. The occurrence of two potential arginine regulators seems not to be specific to *L. lactis*, since two homologues were identified in several Streptococci and Lactobacilli. Lactococcal *argR* is located downstream and transcribed divergently from the arginyl-tRNA synthetase gene, which is immediately followed by the *arcA-B-D1-C1-C2-T-D2* genes responsible for arginine catabolism. The second gene, *ahrC*, is not linked to genes involved in arginine metabolism but, interestingly, its genetic context is similar to that of *B. subtilis ahrC*. These data suggest that both ArgR and AhrC could play a role in the control of arginine metabolism of *L. lactis* and it is tempting to speculate that one is involved in the control of arginine anabolism while the other functions in arginine catabolism. As described above, arginine metabolism may also be controlled directly or indirectly by TCSs KinA/LlrA and LlrH. The possibility that *L. lactis* arginine metabolism is regulated by several regulators, specific (ArgR, AhrC) and global (stress related, TCS for example), would not be surprising, considering the possible role of arginine metabolism in acid stress response (Rallu et al. 2000), nitrogen metabolism and the complex control of this metabolic pathway in other bacteria, such as *B. subtilis* (Fisher 1999).

Glutamine synthetase (GS; L-glutamate:amonia ligase) is a major enzyme responsible for assimilation of ammonium ions to give glutamine, a donor of nitrogen for the synthesis of amino acids, bases and vitamins (Reitzer 1996). The control of GS is essential to provide nitrogen for cellular metabolism, but also to prevent ATP spilling. In *L. lactis*, *glnA* (encoding GS) is located dowstream of *glnR*, a gene encoding a potential regulator. This organization is similar to that of the *glnRA* operon in *B. subtilis*, where GlnR represses the operon. GlnR and GS are required to regulate *glnRA* transcription in function of nitrogen source, but the exact mechanism of this regulation is not known (Schreier et al. 1989). Lastly, GS activity is also regulated by PII/PII-UMP as a function of the nitrogen status of the cell (Reitzer 1996). In *L. lactis*, PII is encoded by GlnB, which is potentially regulated by LlrA/KinA (see section 'Two-component systems').

The regulation of a number of amino acid biosynthesis genes is still unknown. Concerning aromatic amino acids, it has already been mentioned that *tyrA-aroAK-pheA* is in one operon with *llrG/kinG*, but it is still not known to which signal it responds. Moreover, the regulation of the five transcriptional units *aroC, aroD, aroEB, aroF* and *aroH* was not studied. In the case of *aroH*, a terminator–antiterminator structure preceded by an ORF potentially encoding a phenylalanine-rich peptide could lead to modulation of transcription. If this is true, this gene and the *leu* operon (Godon et al. 1992) would be the only two units regulated by a leader-peptide-dependent attenuation mechanism. A systematic search for terminator–antiterminator structures allowed to identify the *alaS*, *ileS*, *trpS* and probably *hisS* genes, all encoding tRNA synthetases, as additional genes potentially regulated by antiterminating systems. However, all seem to belong to the Tbox-dependent system (Grundy & Henkin 1994), like the previously described *trp* and *his* operons (Raya et al. 1998, Delorme et al. 1999). Regulatory circuits for other genes involved in amino acid metabolism remain to be found, although not all genes seem to be regulated. For instance, the *hom-thrB* operon, encoding enzymes involved in threonine biosynthesis, is not subject to threonine-dependent regulation in *L. lactis* MG1614 (Madsen et al. 1996).

Nucleotide metabolism

Purine synthesis

The nucleotides ATP and GTP are both derived from IMP, which is synthesized *de novo* from 5-phosphoribosyl-1-pyrophosphate (PRPP), glycine, glutamine, aspartate, and C_1 units, by 10 enzymatic reactions (Zalkin & Dixon 1992). While the individual enzymatic steps in purine biosynthesis appear to be similar in all bacteria, the genetic organization and the regulation of the *pur* genes follow different rules in different bacteria. In *L. lactis*, the *pur* genes are scattered on the chromosome (Bolotin et al. 2000). Comparison of the *purC* and *purD* promoter regions revealed the presence of a sequence of three conserved motifs (AWWWCCGAACWWT) (PurBox), two of which are located precisely between −79 and −70 nucleotides upstream of the transcriptional start sites (Kilstrup et al. 1998a). By site-directed mutagenesis and deletion analysis, it was shown that mutations in the PurBox sequence resulted in low levels of transcription and the loss of purine-mediated regulation at the *purC* and *purD* promoters. These results support the notion that *purC* and *purD* transcription is regulated by an activator binding to the PurBox sequence.

A mutant of *L. lactis* MG1363 was obtained by transposon mutagenesis with pGh9:ISS1 and selection for purine auxotrophs (Kilstrup 1998b). The inactivated gene encodes the PurR protein, which is 51% identical to the *B. subtilis* PurR repressor. The *purR*::ISS1 mutation lowered the level of transcription from the purine-regulated *purD* promoter. A PurBox sequence overlaps the −35 sequence of the *purR* promoter, causing *purR* to be autoregulated. PurBox sequences, located almost identically with respect to the *L. lactis* PurR regulated promoter, were also identified in the promoter regions of the PurR-regulated genes of *B. subtilis*. However, PurR acts as a repressor in *B. subtilis*, whereas it is an activator in *L. lactis*. In addition to its HTH DNA binding domain that belongs to that of the LysR activator family, PurR shows a high degree of similarity with purine phosphoribosyltransferases. It binds specifically to a DNA sequence in the promoter region in response low concentrations of PRPP in *B. subtilis*, whereas it binds at low and high PRPP concentrations in *L. lactis* (Kilstrup & 1998). In the latter case, transcription is only activated at elevated concentrations of PRPP. It was suggested that the ancestral PurR protein could have evolved from an activator into a repressor in *B. subtilis*, regulating transcription initiation from the same promoters.

A search for perfect PurBoxes in *L. lactis* in IL1403 gave 16 hits, all but two being located in intergenic regions. This suggests that PurR may have a more extensive effect on transcription. In addition to the three, two and one PurBoxes found in the promoter regions of *purC*, *purD* and *purR*, respectively, single PurBoxes are present upstream of *purB* and *purH*, and at the end of *purM*, 262 bp upstream of the start codon of *purN*. If one mismatch is allowed, *hprT, xpt, purA, purF* (two boxes), *purM* (second box), *purN* and the *upp* genes are also precedeed by a PurBox. Moreover, PurBoxes are present in the promoter regions of genes involved in the metabolism of tetrahydrofolate, such as *glyA* and *fhs* and *folD* (with one mismatch for the latter), pyrimidine (*pyrZ-Db-F*) and upstream of several genes of unknown function (*yriD, yrjA, pppL, yjjG, bmpA*).

Activation of *pur* genes by PurR in the presence of PRPP is probably not the only regulation of purine synthesis in *L. lactis*. In *B. subtilis*, *pur* is also negatively regulated by an attenuation mechanism dependent on guanine concentration. Moreover, it has been suggested that YabJ could also participate in regulation of the *pur* operon, *purA* and *purR* (Rappu et al. 1999). The exact mechanism of this regulation is still not known. *L. lactis* AldR, encoded at the end of the *leu-ilv-ald* operon, is highly homologous to *B. subtilis* YabJ. *L. lactis aldR* mutants were shown to be affected in isoleucine metabolism (Goupil-Feuillerat et al. 2000). *L. lactis aldR* is also able to complement *yabJ* deletion in *B. subtilis* (Calero & Renault unpublished results). The *aldR yabJ* mutations seem to induce pleiotropic effects in *L. lactis* and *B. subtilis*, respectively (Calero et al. unpublished). The exact mechanism of action of these regulators remains to be elucidated in both bacteria. Lastly, there are regulators located in a region suggesting their role in the control of purine genes, such as *ypfD*, which is placed head-to-head with the *purDE-K* cluster.

Although central in cellular metabolism, both regulatory mechanisms and the signals controlling regulation of *pur* genes are different in *L. lactis, B. subtilis* and *E. coli*. In the last two bacteria, *pur* genes are regulated by classical *lacI*-type repressors with guanine and hypoxanthine acting as corepressors (Choi & Zalkin 1992).

Pyrimidine synthesis

The *de novo* synthesis of pyrimidines universally comprises six enzymatic steps, leading to the formation of UMP. The pyrimidine biosynthetic genes (*pyr*) constitute a single operon in many Gram-positive bacteria. The *B. subtilis pyr* operon is under the control of an attenuation mechanism involving RNA binding protein PyrR. PyrR forms a complex with UMP (Lu & Switzer 1996). Unlike in most Gram-positive bacteria, lactococcal *pyr* genes are scattered around the chromosome in small units (*pyrRPB-carA, pyrKDbF, pyrEC, carB, pyrDA*). With the exception of *pyrDA*, expression of the *pyr* genes in *L. lactis* is repressed in the presence of uracil. PyrR binding sites similar to those found in *B. subtilis* are present upstream of *carB, pyrKDbF* and the *pyrRPB-carA* putative attenuator. The *L. lactis* PyrR homologue, encoded as the first gene of the *pyrRPB-carA* cluster likely controls the expression of the *pyr* genes by an attenuator mechanism similar to that in *B. subtilis* (Martinussen et al. 2001).

NAD synthesis

L. lactis nadR, which encodes a homologue of the *E. coli* NadR regulator, is located downstream of *yugC*, encoding a probable organic acid dehydrogenase. In *E. coli*, NadR is a bifunctional protein controlling the expression of the *nadA, nadB* and *pncB* genes as NAD concentration increases in the cell and catalyses NAD synthesis from nicotinamide mononucleotide and ATP (Raffaelli et al. 1999). Repressor activity may be mediated through a HTH domain present at the NadR N-terminal end. However, *L. lactis* NadR does not contain the corresponding N-terminal HTH domain, suggesting that it may only posses the catalytic activity of NadR.

Conclusion

In this review, we have illustrated that many of the (putative) *L. lactis* regulators have counterparts in other bacteria. However, we have also observed intruiging differences, showing that gene regulation models cannot be transposed without proper verifications from one organism to another. Does *L. lactis* have other particularities? Since it contains about 400 specific genes, are some of them involved in the regulatory network of *L. lactis*? We have no answer to these questions yet. However, there are nine and eight potential regulators in *L. lactis* IL1403 belonging to two families, with EpsR and YebF as prototypes, respectively. These regulators, which have some loose homology to regulators present usually in not more than one copy in the genomes of other bacteria, form a large family of paralogous genes in *L. lactis*. None of these genes has been adequatly studied yet, so the question as to why they are that abundant in *L. lactis* has no experimentally supported answer at present.

Two large families of regulators, an L. lactis *specificity?*

EpsR is the prototype of the first family. It was annotated as the regulator for the *eps* cluster encoding an exopolysaccharide synthesis system present on plasmids in some *L. lactis* strains. Although a potential HTH DNA-binding motif is present in this small protein, no analytical data support its function. Interestingly, although the IL1403 genome does not contain an *epsR* ortholog, nine genes were identified that encode proteins sharing significant identity with EpsR, two of them being encoded by phages (Figure 1A). The only characterized homologues of this gene family are the two phage regulators Xre from the *B. subtilis* prophage PBSX (Wood et al. 1990) and RstR from *Vibrio cholerae* phage CTXphi (Waldor et al. 1997). Thus, the two prophage-located regulatory genes *ps115* and *ps205* form a subgroup in the EpsR family and are probably involved in the regulation of prophage gene expression. Another member of this family, *tagR*, is present upstream of a large operon encoding genes potentially involved in teichoic acid synthesis. This organization is reminiscent of the *espR-epsABCDEFG* operon, which contains other genes with a closely related function in the synthesis of extracellular polymer. TagR and EpsR are more closely related to each other than to the other proteins of the family and have similar sizes. It is tempting to speculate that they regulate *tag* and *eps* operons, respectively. However, their functionality should still be verified and the signal to which they respond be determined. The function of the remaining six paralogs remains even more mysterious and the potential targets of their regulation are difficult to predict from their genome organization although one of them, *ymcE*, is clustered with the *ymcEFG*, an operon of unknown function.

The second most important class of paralogous regulators contains products encoded by *yebF, gntR, yfeA, yugA, yidA, yecA, yleF* and *yliC*. These proteins do not have strong homologues in other Gram-positive bacteria although some of them share 20–30% identity

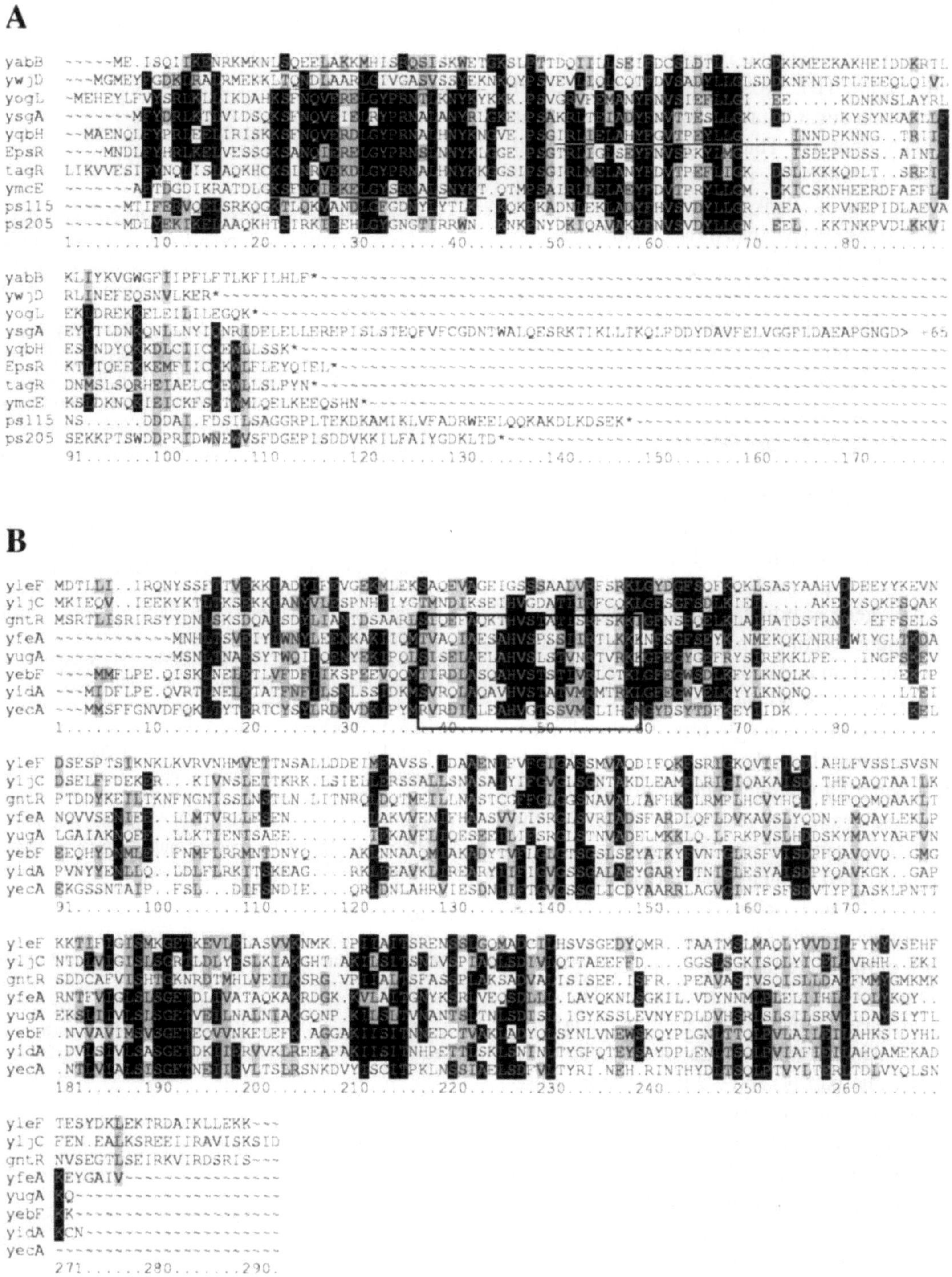

Figure 1. Multiple alignment of proteins belonging to EpsR (A) and YebF (B) regulator family. White letters on a black background and black letters on a gray background correspond to amino acids identical and similar in more than half of the sequences, respectively. Underlined or boxed regions are those predicted to be potential helix-turn-helix domains (Dodd & Egan, 1990) with a reliable level of confidence. These potential DNA-binding domains, which correspond to aligned regions in each regulator family, are shown by stars at the top of the sequences. The sequence of YsgA, which has a C-terminal significantly longer than the other proteins in the family, was truncated.

to RpiR, the repressor of ribose catabolism (Sorensen & Hove-Jensen 1996). These regulators have a highly probable HTH DNA-binding domain at their N terminal end (Figure 1B). Most of them are linked to genes involved in sugar catabolism, and in particular to β-glucoside operons. Thus, *gntR* is head-to-head to gluconate catabolic genes while *yidA, yebF, yecA, yleF* are upstream of, flanked by or downstream of β-glucoside genes, respectively. *yliC* and *yugA* are clustered and head-to-head with genes of unknown function while *yfeA* is a single gene. Preliminary data suggest that YebF could be involved in the activation of the transcription of cellobiose utilization genes (Aleksandrzak T. pers. comm).

Toward building a regulatory network for L. lactis

The analysis of the list of genes annotated in *L. lactis* leads to an estimate of some 111 transcriptional regulators. About 18 of them have been characterized, most often by the phenotyic study of null mutants constructed by gene insertion. In this review, we have proposed the function of 34 more regulators, based on genome organization and protein similarity. We were unable to do so for 59 potential regulators. Some of the latter are encoded in single transcriptional units, whereas others may be linked to genes of unknown function. A better knowledge of the biology of *L. lactis* will thus require further research to determine the function of these putative regulator genes, some of which are conserved in other microorganisms, while others are specific to *L. lactis*. In addition, to direct functional studies, placing these genes in the regulatory network of the bacteria will give indications as to the functions that they may be involved in. The combined use of classical techniques such as phenotypic tests, gene fusions, and post-genomic techniques, such as DNA microarrays and 2D gel electrophoresis will provide the frame to reveal the cell's regulatory network. Lastly, we have not mentioned in this review that other regulatory mechanisms are essential in cell life, including protein regulator-independent transcriptional controls such as attenuation, translational controls, post-transcriptional protein modification, as well as enzymatic controls.

Acknowledgements

We thank Charlotte Barrière, Isabelle Poquet, Peter Radstrom, Tamara Aleksandrzak, and Michel-Yves Mistou, who divulged some results prior to publication. We also thank Dusko Ehrlich for his useful comments on the manuscript.

References

Aleksandrzak T, Kowalczyk M, Kok J & Bardowski J (2000) Regulation of carbon catabolism in *Lactococcus lactis*. In: Bielecki S, Tramper J & Polak J (Eds.) Food Biotechnology – Progress in Biotechnology, Vol. 71 (pp 61–66). Elsevier Science, Amsterdam.

Ansanay V, Dequin S, Blondin B & Barre P (1993) Cloning, sequence and expression of the gene encoding the malolactic enzyme from *Lactoccus lactis*. FEBS Lett. 332: 74–80.

Bae W, Phadtare S, Severinov K & Inouye M (1999) Characterization of *Escherichia coli* cspE, whose product negatively regulates transcription of *cspA*, the gene for the major cold shock protein. Mol. Microbiol. 31: 1429–4141.

Bae W, Xia B, Inouye M & Severinov K (2000) *Escherichia coli* CspA-family RNA chaperones are transcription antiterminators. Proc. Natl. Acad. Sci. USA 97 (14): 7784–7789.

Bardowski J, Ehrlich SD & Chopin A (1994) BglR protein, which belongs to the BglG family of transcriptional antiterminators, is involved in beta-glucoside utilization in *Lactococcus lactis*. J. Bacteriol. 176 (18): 5681–5685.

Bolotin A, Mauger s, Malarme K, Ehrlich SD & Sorokin A (1999) Low-redundancy sequencing of the entire *Lactococcus lactis* IL1403 genome. Antonie Van Leeuwenhoek 76 (10-4): 27–76.

Bolotin A, Wincker P, Mauger S, Jaillon O, Malarme K, Weissenbach J, Ehrlich SD & Sorokin A (2001) The complete genome sequence of the lactic acid bacterium *Lactococcus lactis* ssp. *lactis* IL1403. Genome Res. 11 (5): 731–753.

Bsat N, Herbig A, Casillas-Martinez L, Setlow P & Helmann JD (1998) *Bacillus subtilis* contains multiple Fur homologues: identification of the iron uptake (Fur) and peroxide regulon (PerR) repressors. Mol. Microbiol. 29 (1): 189–198.

Bukau B (1993) Regulation of the *Escherichia coli* heat-shock response. Mol. Microbiol. 9 (4): 671–680.

Cashel M, Gentry DR, Hernandez VJ & Vinella D (1996) The stringent response. In: Neidhardt FC (Ed.) *Escherichia coli* and *Salmonella typhimurium*: Cellular and Molecular Biology, 2nd edn. Vol. 1 (pp 14580–1496). ASM Press, Washington, DC.

Cesselin B, Duwat P, Ali D, El Karoui M & Gruss A (2001) Analyse de mutants résistants au DTT chez *Lactococcus lactis*. 11ème Réunion Du Club Des Bactéries Lactiques, Bordeaux.

Chastanet A, Prudhomme M, Claverys JP & Msadek T (2001) Regulation of *Streptococcus pneumoniae clp* genes and their role in competence development and stress survival. J. Bacteriol. 183 (24): 7295–7307.

Chatterji D & Ojha AK (2001) Revisiting the stringent response, ppGpp and starvation signaling. Curr. Opin. Microbiol. 4 (2): 160–165.

Choi KY & Zalkin H (1992) Structural characterization and corepressor binding of the *Escherichia coli* purine repressor. J. Bacteriol. 174 (19): 6207–6214.

Cocaign-Bousquet M, Garrigues C, Novak L, Lindley ND & Loubiere P (1995) Rational development of a simple synthetic medium for the sustained growth of *Lactococcus lactis*. J. Appl. Bacteriol. 79: 108–116.

de Ruyter PGGA, Kuipers OP, Beerthuyzen MM, Vanalenboerrigter I & de Vos WM (1996) Functional analysis of promoters in the nisin gene cluster of *Lactococcus lactis*. J. Bacteriol. 178 (12): 3434–3439.

Delorme C, Ehrlich DS & Renault P (1992) Histidine biosynthesis genes in *Lactococcus lactis* subsp. *lactis*. J. Bacteriol. 174 (20): 6571–6579.

Delorme C, Ehrlich SD & Renault P (1999) Regulation of expression of the *Lactococcus lactis* histidine operon. J. Bacteriol. 181 (7): 2026–2037.

Derre I, Rapoport G & Msadek T (1999) CtsR, a novel regulator of stress and heat shock response, controls clp and molecular chaperone gene expression in gram-positive bacteria. Mol. Microbiol. 31 (1): 117–131.

Derre I, Rapoport G & Msadek T (2000) The CtsR regulator of stress response is active as a dimer and specifically degraded in vivo at 37 °C. Mol. Microbiol. 38 (2): 335–347.

Dodd IB & Egan JB (1990) Improved detection of helix-turn-helix DNA-binding motifs in protein sequences. Nucleic Acids Res. 18: 5019–5026.

Doman M, Czerniec E, Targonski Z & Bardowski J (2000) Production and genetic regulation of an amylase gene in *Lactococcus lactis*. In: Bielecki S, Tramper J & Polak J (Eds.) Food Biotechnology – Progress in Biotechnology, Vol. 17 (pp 67–72). Elsevier Science, Amsterdam.

Duwat P, Cesselin B, Sourice S & Gruss A (2000) *Lactococcus lactis*, a bacterial model for stress responses and survival. Int. J. Food Microbiol. 55 (1–3): 83–86.

Eaton T, Shearman CA & Gasson MJ (1993) Cloning and sequence analysis of the dnaK gene region of *Lactococcus lactis*. J. Gen. Microbiol. 139: 3253–3264.

Elliott JA, Collins MD, Pigott NE & Facklam RR (1991) Differentiation of *Lactococcus lactis* and *Lactococcus garvieae* from humans by comparison of whole-cell protein patterns. J. Clin. Microbiol. 29 (12): 2731–2734.

Erlandson KA, Park JH, Wissam, El Khal, Kao HH, Basaran P, Brydges S & Batt CA (2000) Dissolution of xylose metabolism in *Lactococcus lactis*. Appl. Environ. Microbiol. 66 (9): 3974–3980.

Erlandson KA, Delamarre SC & Batt CA (2001) Genetic evidence for a defective xylan degradation pathway in *Lactococcus lactis*. Appl. Environ. Microbiol. 67 (4): 1445–1452.

Even S, Lindley ND & Cocaign-Bousquet M (2001) Molecular physiology of sugar catabolism in *Lactococcus lactis* IL1403. J. Bacteriol. 183 (13): 3817–3824.

Fernandez M, Kleerebezem M, Kuipers OP, Siezen RJ & van Kranenburg R (2002) Regulation of the *metC-cysK* Operon, involved in sulfur metabolism in *Lactococcus lactis*. J. Bacteriol. 184 (1): 82–90.

Fisher SH (1999) Regulation of nitrogen metabolism in *Bacillus subtilis*: vive la difference! Mol. Microbiol. 32 (2): 223–232.

Frees D & Ingmer H (1999) ClpP participates in the degradation of misfolded protein in *Lactococcus lactis*. Mol. Microbiol. 31 (1): 79–87.

Fuangthong M, Atichartpongkul S, Mongkolsuk S & Helmann JD (2001) OhrR is a repressor of *ohrA*, a key organic hydroperoxide resistance determinant in *Bacillus subtilis*. J. Bacteriol. 183 (14): 4134–4141.

Godon JJ, Delorme C, Ehrlich SD & Renault P (1992) Divergence of genomic sequences between *Lactococcus lactis* subsp. *lactis* and *Lactococcus lactis* subsp. *cremoris*. Appl. Environ. Microbiol. 58 (12): 4045–4047.

Godon JJ, Delorme C, Bardowski J, Chopin MC, Ehrlich SD & Renault P (1993). Gene inactivation in *Lactococcus lactis* – branched-chain amino acid biosynthesis. J. Bacteriol. 175 (14): 4383–4390.

Gostick DO, Griffin HG, Shearman CA, Scott C, Green J, Gasson MJ & Guest JR (1999) Two operons that encode FNR-like proteins in *Lactococcus lactis*. Mol. Microbiol. 31 (5): 1523–1535.

Goupil-Feuillerat N, Corthier G, Godon JJ, Ehrlich SD & Renault P (2000) Transcriptional and translational regulation of alpha-acetolactate decarboxylase of *Lactococcus lactis* subsp. *lactis*. J. Bacteriol. 182 (19): 5399–5408.

Grossiord B, Vaughan EE, Luesink E & de Vos WM (1998) Genetics of galactose utilisation via the Leloir pathway in lactic acid bacteria. Lait 78: 77–84.

Grundy FJ & Henkin TM (1994) Conservation of a transcription antitermination mechanism in aminoacyl-tRNA synthetase and amino acid biosynthesis genes in Gram-positive bacteria. J. Mol. Biol. 235: 798–804.

Grundy FJ, Waters DA, Allen SH & Henkin TM (1993) Regulation of the *Bacillus subtilis* acetate kinase gene by CcpA. J. Bacteriol. 175 (22): 7348–7355.

Guedon E, Renault P, Ehrlich SD & Delorme C (2001a) Transcriptional pattern of genes coding for the proteolytic system of *Lactococcus lactis* and evidence for coordinated regulation of key enzymes by peptide supply. J. Bacteriol. 183 (12): 3614–3622.

Guedon E, Serror P, Ehrlich SD, Renault P & Delorme C (2001b) Pleiotropic transcriptional repressor CodY senses the intracellular pool of branched-chain amino acids in *Lactococcus lactis*. Mol. Microbiol. 40 (5): 1227–1239.

Havarstein LS, Gaustad P, Nes IF & Morrison DA (1996) Identification of the streptococcal competence-pheromone receptor. Mol. Microbiol. 21 (4): 863–869.

Hecker M, Schumann W & Volker U (1996) Heat-shock and general stress response in *Bacillus subtilis*. Mol. Microbiol. 19 (3): 417–428.

Helmann JD (1995) Compilation and analysis of *Bacillus subtilis* sigma(A)-dependent promoter sequences: Evidence for extended contact between RNA polymerase and upstream promoter DNA. Nucleic Acids Res. 23 (13): 2351–2360.

Helmann JD, Wu MF, Kobel PA, Gamo FJ, Wilson M, Morshedi MM, Navre M & Paddon C (2001) Global transcriptional response of *Bacillus subtilis* to heat shock. J. Bacteriol. 183 (24): 7318–7328.

Henkin TM (1996) The role of CcpA transcriptional regulator in carbon metabolism in *Bacillus subtilis*. FEMS Microbiol. Lett. 135 (11): 9–15.

Hueck CJ & Hillen W (1995) Catabolite repression in *Bacillus subtilis*: A global regulatory mechanism for the gram-positive bacteria? Mol. Microbiol. 15 (3): 395–401.

Khunajakr N, Liu CQ, Charoenchai P & Dunn NW (1999) A plasmid-encoded two-component regulatory system involved in copper-inducible transcription in *Lactococcus lactis*. Gene 229 (1–2): 229–235.

Kilstrup M & Martinussen J (1998) A transcriptional activator, homologous to the *Bacillus subtilis* PurR repressor, is required for expression of purine biosynthetic genes in *Lactococcus lactis*. J. Bacteriol. 180 (15): 3907–3916.

Kilstrup M, Jacobsen S, Hammer K & Vogensen FK (1997) Induction of heat shock proteins DnaK, GroEL, and GroES by salt stress in *Lactococcus lactis*. Appl. Environ. Microbiol. 63 (5): 1826–1837.

Kilstrup M, Jessing SG, Wichmand-Jorgensen SB, Madsen M & Nilsson D (1998). Activation control of pur gene expression in *Lactococcus lactis*: proposal for a consensus activator binding sequence based on deletion analysis and site-directed mutagenesis of purC and purD promoter regions. J. Bacteriol. 180 (15): 3900–3906.

Kim SG & Batt CA (1993) Cloning and sequencing of the *Lactococcus lactis* subsp. *lactis groESL* operon. Gene 127 (1): 121–126.

King KY, Horenstein JA & Caparon MG (2000) Aerotolerance and peroxide resistance in peroxidase and PerR mutants of *Streptococcus pyogenes*. J. Bacteriol. 182 (19): 5290–5299.

Kobayashi K, Ogura M, Yamaguchi H, Yoshida K, Ogasawara N, Tanaka T & Fujita Y (2001) Comprehensive DNA microarray analysis of *Bacillus subtilis* two-component regulatory systems. J. Bacteriol. 183 (24): 7365–7370.

Koch B, Kilstrup M, Vogensen FK & Hammer K (1998) Induced levels of heat shock proteins in a dnaK mutant of *Lactococcus lactis*. J. Bacteriol. 180 (15): 3873–3881.

Kok J (1996) Inducible gene expression and environmentally regulated genes in lactic acid bacteria. Antonie Van Leeuwenhoek 70 (2–4): 129–145.

Kruger E & Hecker M (1998) The first gene of the *Bacillus subtilis clpC* operon, *ctsR*, encodes a negative regulator of its own operon and other class III heat shock genes. J. Bacteriol. 180 (24): 6681–6688.

Kruger E, Zuhlke D, Witt E, Ludwig H & Hecker M (2001) Clp-mediated proteolysis in Gram-positive bacteria is autoregulated by the stability of a repressor. EMBO J. 20 (4): 852–863.

Kuipers OP, Beerthuyzen MM, Ruyter P, Luesink EJ & de Vos WM (1995) Autoregulation of nisin biosynthesis in *Lactococcus lactis* by signal transduction. J. Biol. Chem. 270: 1–6.

Lee MS & Morrison DA (1999) Identification of a new regulator in *Streptococcus pneumoniae* linking quorum sensing to competence for genetic transformation. J. Bacteriol. 181 (16): 5004–5016.

Li YH, Lau PC, Lee JH, Ellen RP & Cvitkovitch DG (2001) Natural genetic transformation of *Streptococcus mutans* growing in biofilms. J. Bacteriol. 183 (3): 897–908.

Lu Y & Switzer RL (1996) Transcriptional attenuation of the *Bacillus subtilis* pyr Operon by the PyrR regulatory protein and uridine nucleotides in vitro. J. Bacteriol. 178 (24): 7206–7211.

Ludwig H, Homuth G, Schmalisch M, Dyka FM, Hecker M & Stulke J (2001) Transcription of glycolytic genes and operons in *Bacillus subtilis*: evidence for the presence of multiple levels of control of the *gapA* operon. Mol. Microbiol. 41 (2): 409–422.

Luesink EJ, van Herpen RE, Grossiord BP, Kuipers OP & de Vos WM (1998) Transcriptional activation of the glycolytic las operon and catabolite repression of the gal operon in *Lactococcus lactis* are mediated by the catabolite control protein CcpA. Mol. Microbiol. 30 (4): 789–798.

Luesink EJ, Beumer CM, Kuipers OP & De Vos WM (1999a) Molecular characterization of the *Lactococcus lactis* ptsHI operon and analysis of the regulatory role of HPr. J. Bacteriol. 181 (3): 764–771.

Luesink EJ, Marugg JD, Kuipers OP & de Vos WM (1999b) Characterization of the divergent *sacBK* and *sacAR* operons, involved in sucrose utilization by *Lactococcus lactis*. J. Bacteriol. 181 (6): 1924–1926.

Maas WK (1994) The arginine repressor of *Escherichia coli*. Microbiol. Rev. 58 (4): 631–640.

Madsen SM, Albrechtsen B, Hansen EB & Israelsen H (1996) Cloning and transcriptional analysis of two threonine biosynthetic genes from *Lactococcus lactis* MG1614. J. Bacteriol. 178 (13): 3689–3694.

Mahr K, Hillen W & Titgemeyer F (2000) Carbon catabolite repression in *Lactobacillus pentosus*: analysis of the *ccpA* region. Appl. Environ. Microbiol. 66 (1): 277–283.

Makarova KS, Mironov AA & Gelfand MS (2001) Conservation of the binding site for the arginine repressor in all bacterial lineages. Genome Biol. 2 (4): RESEARCH0013.

Martinussen J, Schallert J, Andersen B & Hammer K (2001) The pyrimidine operon *pyrRPB-carA* from *Lactococcus lactis*. J. Bacteriol. 183 (9): 2785–2794.

Marugg JD, Meijer WC, Vankranenburg R, Laverman P, Bruinenberg PG & de Vos WM (1995) Medium-dependent regulation of proteinase gene expression in *Lactococcus lactis*: Control of transcription initiation by specific dipeptides. J. Bacteriol. 177 (11): 2982–2989.

Mierau I, Kunji ER, Leenhouts KJ, Hellendoorn MA, Haandrikman AJ, Poolman B & Kok J (1996) Multiple-peptidase mutants of *Lactococcus lactis* are severely impaired in their ability to grow in milk. J. Bacteriol. 178: 2794–2803.

Miller CM, Baumberg S & Stockley PG (1997) Operator interactions by the *Bacillus subtilis* arginine repressor/activator, AhrC: novel positioning and DNA-mediated assembly of a transcriptional activator at catabolic sites. Mol. Microbiol. 26 (1): 37–48.

Missiakas D & Raina S (1998) The extracytoplasmic function sigma factors: role and regulation. Mol. Microbiol. 28 (6): 1059–1066.

Mogk A, Volker A, Engelmann S, Hecker M, Schumann W & Volker U (1998) Non-native proteins induce expression of the *Bacillus subtilis* CIRCE regulon. J. Bacteriol. 180 (11): 2895–2900.

Monedero V, Kuipers OP, Jamet E & Deutscher J (2001) Regulatory functions of serine-46-phosphorylated HPr in *Lactococcus lactis*. J. Bacteriol. 183 (11): 3391–3398.

Moreno MS, Schneider BL, Maile RR, Weyler W & Saier Jr MH (2001) Catabolite repression mediated by the CcpA protein in *Bacillus subtilis*: novel modes of regulation revealed by whole-genome analyses. Mol. Microbiol. 39 (5): 1366–1381.

Nair S, Derre I, Msadek T, Gaillot O & Berche P. Feb. CtsR controls class III heat shock gene expression in the human pathogen *Listeria monocytogenes*. Mol. Microbiol. 35 (4): 800–811.

Narberhaus F (1999) Negative regulation of bacterial heat shock genes. Mol. Microbiol. 31 (1): 1–8.

O'Connell-Motherway M, van Sinderen D, Morel-Deville F, Fitzgerald GF, Ehrlich SD & Morel P (2000) Six putative two-component regulatory systems isolated from *Lactococcus lactis* subsp. *cremoris* MG1363. Microbiology 146 (4): 935–947.

Obis D, Guillot A, Gripon JC, Renault P, Bolotin A & Mistou MY (1999) Genetic and biochemical characterization of a high-affinity betaine uptake system (BusA) in *Lactococcus lactis* reveals a new functional organization within bacterial ABC transporters. J. Bacteriol. 181 (20): 6238–6246.

Pellizzer G, Benedetti P, Biavasco F, Manfrin V, Franzetti M, Scagnelli M, Scarparo C & de Lalla F (1996) Bacterial endocarditis due to *Lactococcus lactis* subsp. *cremoris*: case report. Clin. Microbiol. Infect. 2 (3): 230–232.

Pestova EV, Havarstein LS & Morrison DA (1996) Regulation of competence for genetic transformation in *Streptococcus pneumoniae* by an auto-induced peptide pheromone and a two-component regulatory system. Mol. Microbiol. 21 (4): 853–862.

Petersohn A, Brigulla M, Haas S, Hoheisel JD, Volker U & Hecker M (2001) Global analysis of the general stress response of *Bacillus subtilis*. J. Bacteriol. 183 (19): 5617–5631.

Poquet I, Ehrlich SD & Gruss A (1998) An export-specific reporter designed for gram-positive bacteria: application to *Lactococcus lactis*. J. Bacteriol. 180 (7): 1904–1912.

Raffaelli N, Lorenzi T, Mariani PL, Emanuelli M, Amici A, Ruggieri S & Magni G (1999) The *Escherichia coli* NadR regulator is endowed with nicotinamide mononucleotide adenylyltransferase activity. J. Bacteriol. 181 (17): 5509–5511.

Rallu F, Gruss A, Ehrlich SD & Maguin E (2000) Acid- and multistress-resistant mutants of *Lactococcus lactis*: identification of intracellular stress signals. Mol. Microbiol. 35 (3): 517–528.

Rappu P, Shin BS, Zalkin H & Mantsala P (1999) A role for a highly conserved protein of unknown function in regulation of *Bacillus subtilis purA* by the purine repressor. J. Bacteriol. 181 (12): 3810–3815.

Ratnayake-Lecamwasam M, Serror P, Wong KW & Sonenshein AL (2001) *Bacillus subtilis* CodY represses early-stationary-phase genes by sensing GTP levels. Genes Dev. 15 (9): 1093–1103.

Raya R, Bardowski J, Andersen PS, Ehrlich SD & Chopin A (1998) Multiple transcriptional control of the *Lactococcus lactis trp* operon. J. Bacteriol. 180 (12): 3174–3180.

Reitzer LJ (1996) Ammonia assimilation and the biosynthesis of glutamine, glutamate, aspartate, arsparagine, L-alanine, and D-alanine. In: Neidhardt FC (Ed.) *Escherichia coli* and *Salmonella typhimurium*: Cellular and Molecular Biology, 2nd Edn., Vol. 1 (pp 391–407). ASM Press, Washington, DC.

Renault P, Gaillardin C & Heslot H (1989) Product of the *Lactococcus lactis* gene required for malolactic fermentation is homologous to a family of positive regulators. J. Bacteriol. 171: 3108–3114 .

Renault P & Heslot H (1987) Selection of mutants of *Streptococcus lactis* defective for malolactic fermentation. Appl. Environ. Microbiol. 53: 320–324.

Rodrigues UM, Aguirre M, Facklam RR & Collins MD (1991) Specific and intraspecific molecular typing of lactococci based on polymorphism of DNA encoding rRNA. J. Appl. Bacteriol. 71 (6): 509–516.

Romeo Y, Obis D, Guillot A, Fourçans A, Bouvier I, Guitierrez C & Mistou M-Y (2002) Osmoregulation in *Lactococcus lactisl*: BusR, a transcriptional repressor of the glycine betaine uptake sustem BusA (Submitted).

Sanders JW, Leenhouts KJ, Haandrikman AJ, Venema G & Kok J (1995) Stress response in *Lactococcus lactis*: cloning, expression analysis, and mutation of the lactococcal superoxide dismutase gene. J. Bacteriol. 177 (18): 5254–5260.

Sanders JW, Leenhouts K, Burghoorn J, Brands JR, Venema G & Kok J (1998) A chloride-inducible acid resistance mechanism in *Lactococcus lactis* and its regulation. Mol. Microbiol 27 (2): 299–310.

Sanz Y, Lanfermeijer FC, Renault P, Bolotin A, Konings WN & Poolman B (2001) Genetic and functional characterization of *dpp* genes encoding a dipeptide transport system in *Lactococcus lactis*. Arch. Microbiol. 175 (5): 334–343.

Schick J, Weber B, Klein JR & Henrich B (1999) PepR1, a CcpA-like transcription regulator of *Lactobacillus delbrueckii* subsp. *lactis*. Microbiology 145 (11): 3147–3154.

Schreier HJ, Brown SW, Hirschi KD, Nomellini JF & Sonenshein AL (1989) Regulation of *Bacillus subtilis* glutamine synthetase gene expression by the product of the *glnR* gene. J. Mol. Biol. 210 (1): 51–63.

Scott C, Guest JR & Green J (2000) Characterization of the *Lactococcus lactis* transcription factor FlpA and demonstration of an in vitro switch. Mol. Microbiol. 35 (6): 1383–1393.

Scott C, Rawsthorne H, Upadhyay M, Shearman CA, Gasson MJ, Guest JR & Green J (2000) Zinc uptake, oxidative stress and the FNR-like proteins of *Lactococcus lactis*. FEMS Microbiol. Lett. 192 (1): 85–89.

Serror P & Sonenshein AL (1996) CodY is required for nutritional repression of *Bacillus subtilis* genetic competence. J. Bacteriol. 178 (20): 5910–5915.

Serror P & Sonenshein AL (1996) Interaction of CodY, a novel *Bacillus subtilis* DNA-binding protein, with the *dpp* promoter region. Mol. Microbiol. 20 (4): 843–852.

Sorensen KI & Hove-Jensen B (1996) Ribose catabolism of *Escherichia coli*: characterization of the *rpiB* gene encoding ribose phosphate isomerase B and of the *rpiR* gene, which is involved in regulation of *rpiB* expression. J. Bacteriol. 178 (4): 1003–1011.

Strausak D & Solioz M (1997) CopY is a copper-inducible repressor of the *Enterococcus hirae* copper ATPases. J. Biol. Chem. 272 (14): 8932–8936.

Throup JP, Koretke KK, Bryant AP, Ingraham KA, Chalker AF, Ge YG, Marra A, Wallis NG, Brown JR, Holmes DJ, Rosenberg M & Burnham MKR (2000) A genomic analysis of two-component signal transduction in *Streptococcus pneumoniae*. Mol. Microbiol. 35 (3): 566–576.

Tynkkynen S, Buist G, Kunji E, Kok J, Poolman B, Venema G, Haandrikman A (1993) Genetic and biochemical characterization of the oligopeptide transport system of *Lactococcus lactis*. J. Bacteriol. 175: 7523–7532.

van Asseldonk M, Simons A, Visser H, de Vos WM & Simons G (1993) Cloning, nucleotide sequence, and regulatory analysis of the *Lactococcus lactis dnaJ* gene. J. Bacteriol. 175 (6): 1637–1644.

van Rooijen RJ & de Vos WM (1990) Molecular cloning, transcriptional analysis, and nucleotide sequence of *lacR*, a gene encoding the repressor of the lactose phosphotransferase system of *Lactococcus lactis*. J. Biol. Chem. 265 (30): 18499–18503.

van Rooijen RJ, van Schalkwijk S & de Vos WM (1991) Molecular cloning, characterization, and nucleotide sequence of the tagatose 6-phosphate pathway gene cluster of the lactose operon of *Lactococcus lactis*. J. Biol. Chem. 266 (11): 7176–7181.

van Rooijen RJ, Gasson MJ & de Vos WM (1992) Characterization of the *Lactococcus lactis* lactose operon promoter: contribution of flanking sequences and LacR repressor to promoter activity. J. Bacteriol. 174 (7): 2273–2280.

van Rooijen RJ, Dechering KJ, Niek C, Wilmink J & de Vos WM (1993) Lysines 72, 80 and 213 and aspartic acid 210 of the *Lactococcus lactis* LacR repressor are involved in the response to the inducer tagatose-6-phosphate leading to induction of *lac* operon expression. Protein Eng. 6 (2): 201–206.

Varmanen P, Ingmer H & Vogensen FK (2000) *ctsR* of *Lactococcus lactis* encodes a negative regulator of clp gene expression. Microbiology 146 (6): 1447–1455.

Vaughan EE, Pridmore RD & Mollet B (1998) Transcriptional regulation and evolution of lactose genes in the galactose-lactose operon of *Lactococcus lactis* NCDO2054. J. Bacteriol. 180 (18): 4893–4902.

Vaughan EE, van den Bogaard PT, Catzeddu P, Kuipers OP & de Vos WM (2001) Activation of silent *gal* genes in the *lac-gal* regulon of *Streptococcus thermophilus*. J. Bacteriol. 183 (4): 1184–1194.

Waldor MK, Rubin EJ, Pearson GD, Kimsey H & Mekalanos JJ (1997) Regulation, replication, and integration functions of the *Vibrio cholerae* CTXphi are encoded by region RS2. Mol. Microbiol. 24 (5): 917–926.

Weber MH, Beckering CL & Marahiel MA (2001) Complementation of cold shock proteins by translation initiation factor IF1 in vivo. J. Bacteriol. 183 (24): 7381–7386.

Wood HE, Devine KM & McConnell DJ (1990) Characterisation of a repressor gene (*xre*) and a temperature-sensitive allele from the *Bacillus subtilis* prophage, PBSX. Gene 96 (1): 83–88.

Wouters JA, Mailhes M, Rombouts FM, de Vos WM, Kuipers OP & Abee T (2000) Physiological and regulatory effects of controlled overproduction of five cold shock proteins of *Lactococcus lactis* MG1363. Appl. Environ. Microbiol. 66 (9): 3756–3763.

Wouters JA, Frenkiel H, de Vos WM, Kuipers OP & Abee T (2001) Cold shock proteins of *Lactococcus lactis* MG1363 are involved in cryoprotection and in the production of cold-induced proteins. Appl. Environ. Microbiol. 67 (11): 5171–5178.

Yoshida K, Kobayashi K, Miwa Y, Kang CM, Matsunaga M, Yamaguchi H, Tojo S, Yamamoto M, Nishi R, Ogasawara N, Nakayama T & Fujita Y (2001) Combined transcriptome and proteome analysis as a powerful approach to study genes under glucose repression in *Bacillus subtilis*. Nucleic Acids Res. 29 (3): 683–692.

Zalkin H & Dixon JE (1992) De novo purine nucleotide biosynthesis. Prog. Nucleic Acid Res. Mol. Biol. 42: 259–287.

Antonie van Leeuwenhoek **82:** 113–122, 2002.

Transcriptome analysis and related databases of *Lactococcus lactis*

Oscar P. Kuipers*, Anne de Jong, Richard J.S. Baerends, Sacha A.F.T. van Hijum, Aldert L. Zomer, Harma A. Karsens, Chris D. den Hengst, Naomi E. Kramer, Girbe Buist & Jan Kok
Molecular Genetics, University of Groningen, Groningen Biomolecular Sciences and Biotechnology Institute, PO Box 14, 9750 AA Haren, The Netherlands (Author for correspondence; E-mail: o.p.kuipers@biol.rug.nl)

Key words: Lactococcus lactis, genome sequence, transcriptome, DNA-microarrays, gene regulation, functional genomics, bioinformatics

Abstract

Several complete genome sequences of *Lactococcus lactis* and their annotations will become available in the near future, next to the already published genome sequence of *L. lactis* ssp. *lactis* IL1403. This will allow intra-species comparative genomics studies as well as functional genomics studies aimed at a better understanding of physiological processes and regulatory networks operating in lactococci. This paper decribes the initial set-up of a DNA-microarray facility in our group, to enable transcriptome analysis of various Gram-positive bacteria, including a ssp. *lactis* and a ssp. *cremoris* strain of *Lactococcus lactis*. Moreover a global description will be given of the hardware and software requirements for such a set-up, highlighting the crucial integration of relevant bioinformatics tools and methods. This includes the development of MolGenIS, an information system for transcriptome data storage and retrieval, and LactococCyc, a metabolic pathway/genome database of *Lactococcus lactis*.

Introduction

Lactococcus lactis ssp. *lactis* IL1403 was the first lactic acid bacterium for which the complete genome sequence and its annotation have been published (Bolotin et al. 2001). This important achievement has set the stage for various other genome sequencing projects on *L. lactis* and on a wide variety of other lactic acid bacteria (see also Klaenhammer et al. 2002 in this issue). The completion of the genome sequence of *L. lactis* ssp. *cremoris* MG1363 by an English/Irish/Dutch consortium is in progress and is expected to be completed in 2002. Moreover, in the USA an effort is going on to sequence the chromosome and plasmids of *L. lactis* ssp. *cremoris* SK11 (D. Mills et al., personal communication), the sequence of which is also expected to be finished in 2002. It is very likely that the genomes of even more strains will be sequenced in the near future, enabling comparative genomics within this species that is of such great interest to the dairy industry. Next to *Bacillus subtilis*, which is the paradigm for genetic, physiological and biochemical research in Gram-positive bacteria, *L. lactis* has evolved to be a model species for Gram-positive cocci.

The availability of genomic information on lactococcal and other bacterial species has opened the way for a number of analytical and experimental approaches, which were impossible to perform without this data. First, genome mining and comparison studies yield valuable information on the presence or absence of certain features in the strains studied, as well as on evolutionary phenomena. For instance, the IL1403 sequence has revealed the presence of some unexpected pathways, e.g., on competence development and some expected but interesting traits, such as respiration ability (Duwat et al. 2001), as well as on horizontal gene transfer (Bolotin et al. 2001). Apart from the mining approaches, which in essence will primarily place genes in functional categories and allow to compare gene organizations in various related and unrelated species, it is exciting to see how fast a number of experimental approaches have evolved to enable transcriptome analysis, genomotyping, proteome analysis, high-throughput screening and high-throughput structural biology in microbial systems (Kuipers 2000; Paton et al. 2000; Sebaihia et al. 2001; Tomita 2001; Ye et al. 2001; Zheng et al. 2001; Oliver et al. 2002). Since there is a huge interest to identify lactococci with desired properties or to construct such strains by genetic modification, it is essential to know how the function of each individual gene/protein is related to its expression/production during growth and in stationary phase in various environments. For this reason the elucidation of gene regulatory networks and identification of protein–protein interactions involved in important industrial processes is of crucial importance. Fortunately, these issues can now be addressed in a systematic way, by using the genomics tools that have evolved over the past years. Research using these tools will undoubtedly be directed towards studying important industrial phenomena like acidification rate, various stress responses, flavour formation, effects of metabolic engineering, assessment of the safety of GMOs, use of lactococci as delivery vehicles in a host or food-product, or for improved biopreservation.

One of the most powerful approaches is to use proteomics to determine the full complement of proteins in a cell under a specific condition and at a defined time. Technology development in this field is taking place rapidly, e.g., directed to the development of protein- or antibody arrays and improved and high-throughput protein separation techniques. However, the most commonly used methodology is to combine separation of proteins by 2D-electrophoresis, followed by trypsinolysis and protein identification by MALDI-TOF. Some drawbacks of this approach are the difficulty to detect low-abundancy proteins, and the problems in analysing membrane proteins and unstable proteins.

Another very powerful tool in genomic research is provided by DNA-microarrays, which offer a near-complete view of the relative abundance of all messenger RNAs in a cell in a spatio-temporal way. Various technologies for the production and analysis of DNA-microarrays have been described (Diehl et al. 2001; Kamb & Ramaswami 2001; Lucchini et al. 2001; Wei et al. 2001; Ball & Trevors 2002), but it is beyond the scope of this paper to further describe them. The purpose of this paper is to describe in some detail how DNA-microarray analysis has been set-up in our group, including a description of the infrastructure and the bioinformatics tools that are extremely important for being able to extract useful information from primary transcriptome data. The set-up that will be described has already been implemented for microarrays of *Bacillus subtilis* 168 and for *Streptococcus pneumoniae* TIGR4 (collaboration with Dr. P.W.M. Hermans, Pediatrics, Rotterdam), while arrays for a *Lactobacillus* species and another *Bacillus* species will be developed in the near future in various collaborations with groups having in-depth expertise with these organisms.

DNA-microarray production

Because standardisation of DNA-microarray production and analysis is of crucial importance to be able to create robust and validated data, a series of steps in the whole procedure was carefully analysed and optimised. Amplicons of all genes of the genome studied are synthesised by PCR, have a maximal length of 800 bp and are commonly, if gene size allows, longer than 500 bp. The 5′ and 3′ specific primers are synthesized with generic tags (15 nucleotides forward and reverse, respectively), allowing reamplification of the amplicons obtained in the first round of PCR (original amplicons). These amplicons can be reamplified with an aminated forward primer and a non-aminated reverse primer-based on the generic tag sequences, which yield single-aminated amplicons after PCR. This approach has the advantage of only detecting cDNAs originating from gene transcripts and not those originating from antisense RNA, produced by read-through from genes/operons into downstream genes in a back-to-back orientation.

The purity, size-correctness and amounts of the amplicons are routinely checked by running them on agarose gels. Subsequently, amplicons are purified by the Millipore 96 colums PCR purification system. Amplicons are transferred to 384-well microtiter plates and are diluted 1:1 with Array-It spotting solution to a final concentration of 0.25–1.0 mg/ml. Using Telechem stealth pins (SMP5) 0.2 μl is sampled and 1–3 nL are spotted in duplicate onto aldehyde-coated slides (Cell Associates) and further handled using standard protocols for aldehyde slides. Before and after boiling, slides are incubated with either SyberGreenI, for detection of double stranded DNA, or SyberGreenII, for detection of single stranded DNA (Molecular Probes) and subsequently scanned with a Gene TAC LS IV (Genomics Solutions) to check the transition of double stranded DNA before boiling and single stranded DNA after boiling treatment.

DNA-microarray analysis: the hardware and methods

During the validation of the transcriptome analysis

procedure the following steps have been optimised: RNA isolation, cDNA labeling, and hybridisation.

RNA isolation

RNA isolation is performed by first growing cells in 50 ml GM17 medium and harvesting cells at the desired optical density (OD_{600} ranging from 0.6 to 1.6). From 50 ml MG1363 culture usually 200–300 μg total RNA is routinely obtained. Analysis of isolated RNA is performed using agarose gels in order to assess that the material is of good quality (no degradation, and a 16S to 23S rRNA ratio of approximately 1:2). Alternatively, an Agilent bioanalyzer will be used in the near future, for improved standardisation.

cDNA labeling

Single-strand reverse transcription (amplification) and direct labeling of 25–50 μg of isolated total RNA with Cy3-dCTP or Cy5-dCTP is done either with the Amersham CyScribe First-Strand cDNA Labeling kit or the Invitrogen FluoroScript cDNA labeling system. Both kits give good label incorporation when the labeled cDNA is subsequently used for hybridisation experiments. However, since Cy5 incorporation in cDNA is less efficient than that of Cy3 (due to preference of the reverse transcriptase) and to improve on the sensitivity by enhancing the absolute fluorescence signals, the Amersham CyScribe Post Labelling Kit is currently being tested. Checking the quality of the cDNA is done by hybridising a series of dilutions on a polylysine slide and subsequent analysis with the GeneTAC LS IV.

Hybridisation and scanning

Slides are pre-hybridised in Ambion SlideHyb buffer I for 15 min at 40 °C in a Genomic Solutions Hybstation. After removal of the pre-hybridisation buffer, 5–10 μl of the Cy3/Cy5-labeled cDNA mix in 100 μl Ambion SlideHyb buffer I is added and incubated for 1 h at 42 °C and finally 16 h at 40 °C. Subsequently, the hybridised slides are washed for 1 min in 2× SSC, 0.5% SDS and 5 min in 1×SSC, 0.25% SDS. The slides are scanned using a confocal laser scanner (GeneTAC LS IV).

Signal analysis

After scanning of the slides with the GeneTAC LS IV, spot intensities are determined. The raw data, along with the scanning image are stored in the Molecular Genetics Information System (MolGenIS), which is described below. A grid definition was made to enable the spot analysis software, i.e., Array Pro (Phoretix), to produce tables containing gene names and signal intensities. The program Genespring (Silicon Genetics) is used for further statistical analyses and clustering. The whole procedure for microarray production and analysis is schematically depicted in Figure 1.

DNA-microarray analysis: bioinformatics

Primer design

Large scale production of PCR products of genes or gene fragments is a crucial step in the production of spotted DNA-microarrays. In order to use a high-throughput approach for the production of these PCR products (amplicons) we have developed a software program for the design of primer pairs, named 'GenomePrimer'. A tab-delimited text file or Excel file with the nucleotide sequences of all putative open reading frames (ORFs) is needed as input. GenomePrimer runs under Windows and was written in Borland Delphi 4.0. Various selection parameters for primers, amplicons and annealing temperatures (T_m) can be set within the user-friendly interface.

The parameters to be defined for primer selection are length, GC content, G or C at the 3′ end and inclusion of start or stop codon. For amplicon selection, length, location within an ORF, and flexibility in choice of the length for optimal adjustment of the T_m of the primer can be set. Distribution of G and C and palindromic check of the primer pairs is routinely performed. Calculation of the melting temperature can be performed according to the rules of Suggs et al. (1981) T_m=4(GC)+2(AT) or via the method of Sugimoto et al. (1996), $T_m = 62.3 + 0.41(GC) - 500/\text{length}$. Desired tags, for instance for the amplification of all amplicons using one single oligonucleotide pair directed against the 3′ and 5′ tags, can be added to the designed oligonucleotides.

Thousands of primers can be selected within seconds. General result-statistics such as success rate, average primer size, number of nucleotides to synthesize and specific characteristics such as the number of failed primers, (short and/or low GC) amplicons, or number of failed amplicons are given. If primers have not been selected for all ORFs, the setting can easily be adjusted and a new selection run can be performed

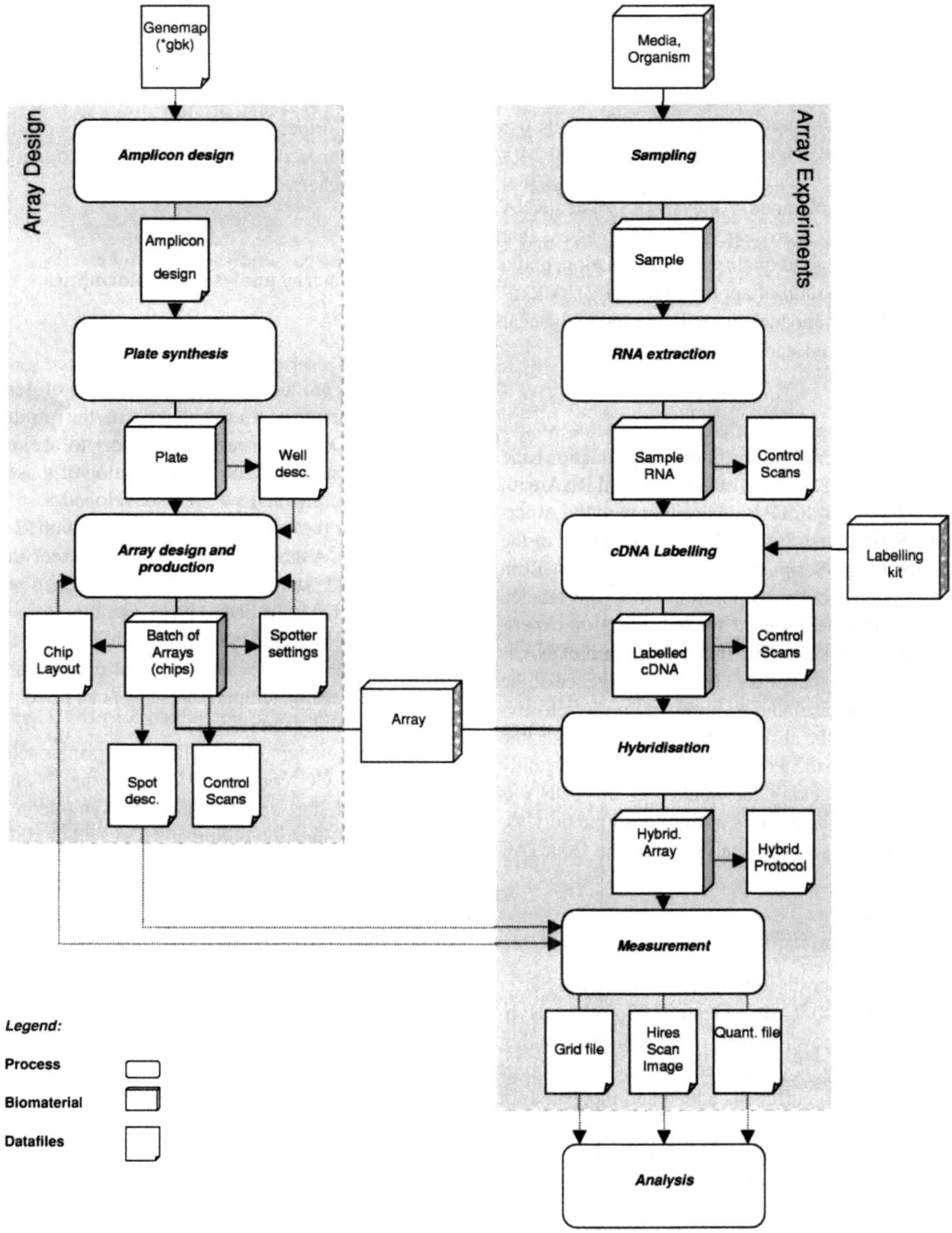

Figure 1. Process overview of a DNA-microarray experiment.

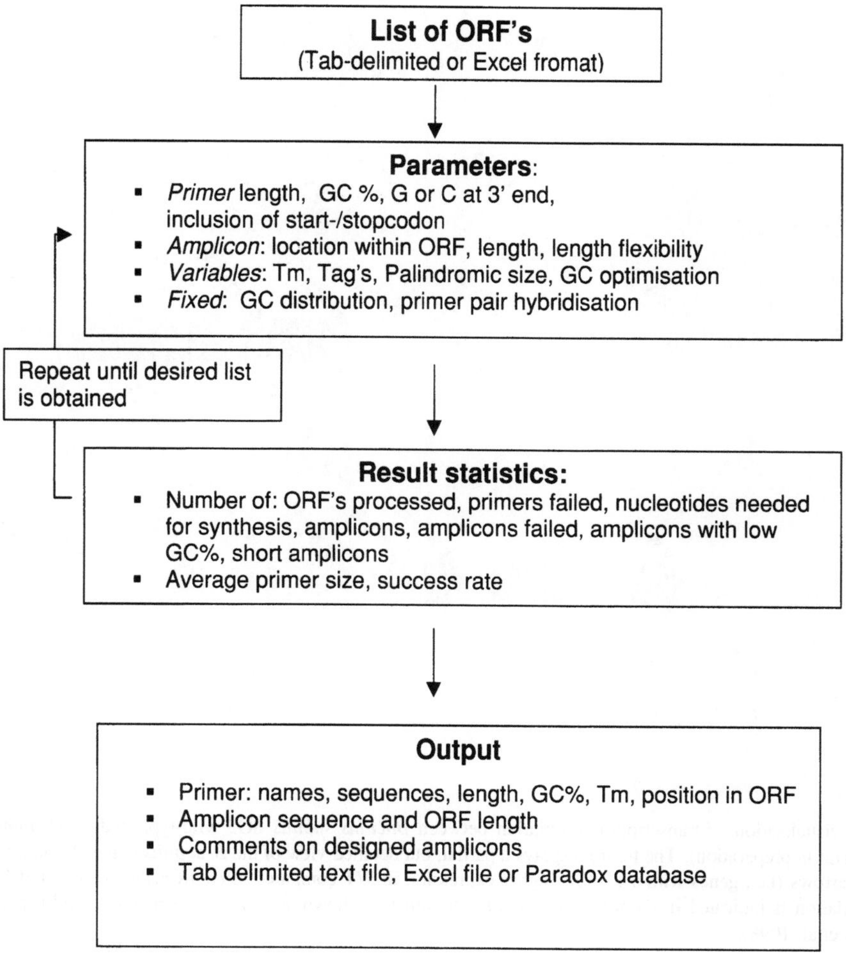

Figure 2. Flow chart of GenomePrimer.

until all primers/amplicons fit within the desired range. The procedure is schematically depicted in Figure 2. The primer pairs and amplicons of the output file can be used as input for the 'Genome2D' program (see below) to indicate their positions within the ORFs.

GenomePrimer was first tested on coding sequences of the lactococcal phage r1t (Table 1). A high-throughput approach, using 96-well microtiterplates, is used for amplicon production using a pipetting robot in combination with a PCR machine for 96-well plates. After obtaining a 100% success rate for amplicon production with r1t, the program was used to select primers for the complete genome of *S. pneumoniae* TIGR4 and *L. lactis* IL1403, which also gave good results. Simultaneously, a similar software tool for design of primer pairs for DNA microarray construction named 'PrimeArray' was designed by researchers of the Max-Planck Institute of Biology in Switserland (Raddatz et al. 2001). Both GenomePrimer and PrimeArray need the input of coding sequences. Recently the program GST-PRIME was published which can be used to design a large number of primer pairs starting from a list of protein sequences (Varotto et al. 2000). DNA sequences corresponding to the amino acid sequence of the proteins are retrieved, extracted and assembled into gene sequences with and without introns followed by design of primers.

Genome2D: Visualisation of transcript profiles on a linear chromosome map

To visualise a bacterial genome with all its individual genes on a computer screen, we developed a program called Genome2D. Full-screen visualisation of a complete genome enables quick identification of biologically relevant information (Figure 3). For example, it can illustrate which genes have the same orientation and could be transcriptionally linked. Furthermore,

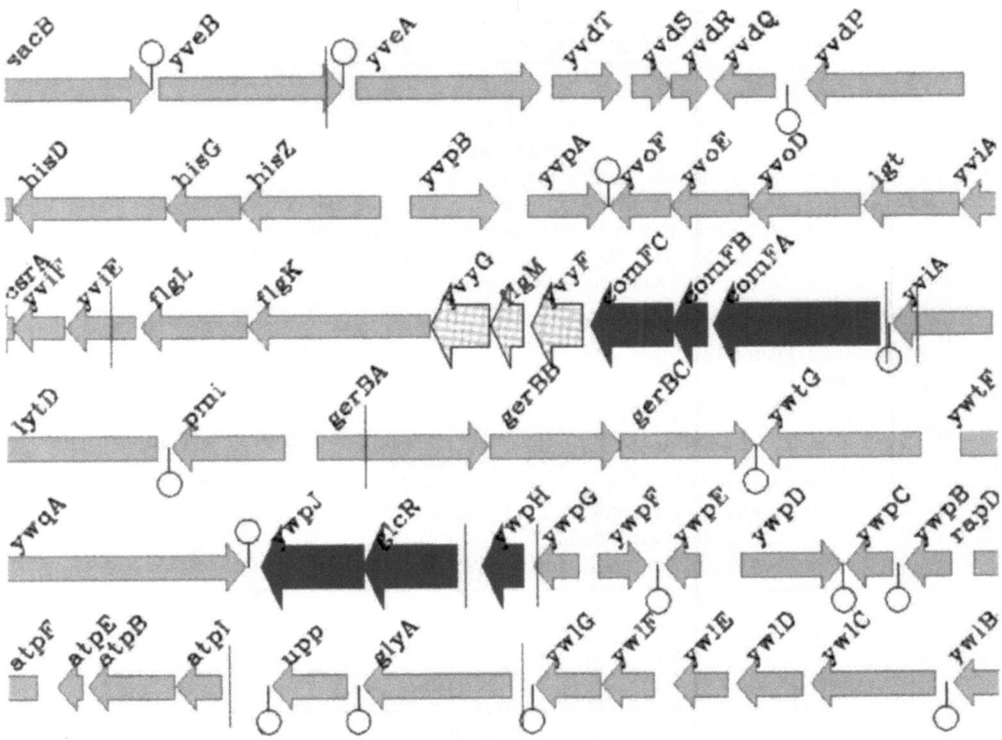

Figure 3. Genome2D visualisation of transcriptional variation between *Bacillus subtilis* 8G5 wild-type and *comK* mutant cells (data from Hamoen et al. manuscript in preparation). The figure displays a partial, but detailed view of the *B. subtilis* genes. Variations in gene expression are indicted by larger arrows (i.e., genes with a 1.5–2× higher expression in WT compared to *comK* cells are coloured *lighter gray*, whereas more than 2× up-regulation is indicated in *darker gray*). Putative terminators shown as stem-loop structures, and predicted *comK* boxes as vertical lines (Hamoen et al. 1998).

Table 1. Overview of primer selection using GenomePrimer and results obtained for amplicon production (GenomePrimer software settings are indicated in grey)

	Phage r1t	*L. lactis* IL1403	S. pneumoniae TIGR4
Nr ORF's	51	2126	2229
GC% ORF's	35.5%	35.5%	40.6%
Primer length (in nt)	17 or 20	18 to 22	18 to 22
Primer GC%	6 to 16	9 to 11	9 to 11
Amplicon length (in bp)	80 to 2000	80 to 800	200 to 800
Amplicon location	Most unique	Most unique	150 bp from ORF ends
Minimal T_m	52 °C	52 °C	56 °C
T_m calculation	T_m=62.3+0.41 (GC)-500/length	T_m=62.3+0.41 (GC)-500/length	T_m=62.3+0.41 (GC)-500/length
Success primer selection	100%	96.5%	98%
Amplicon length (in bp)	89 to 1842	83 to 798	79 to 800
Success rate original PCR	100%	100%	99.5%
Success rate re-amplification	100%	96.2%	100%

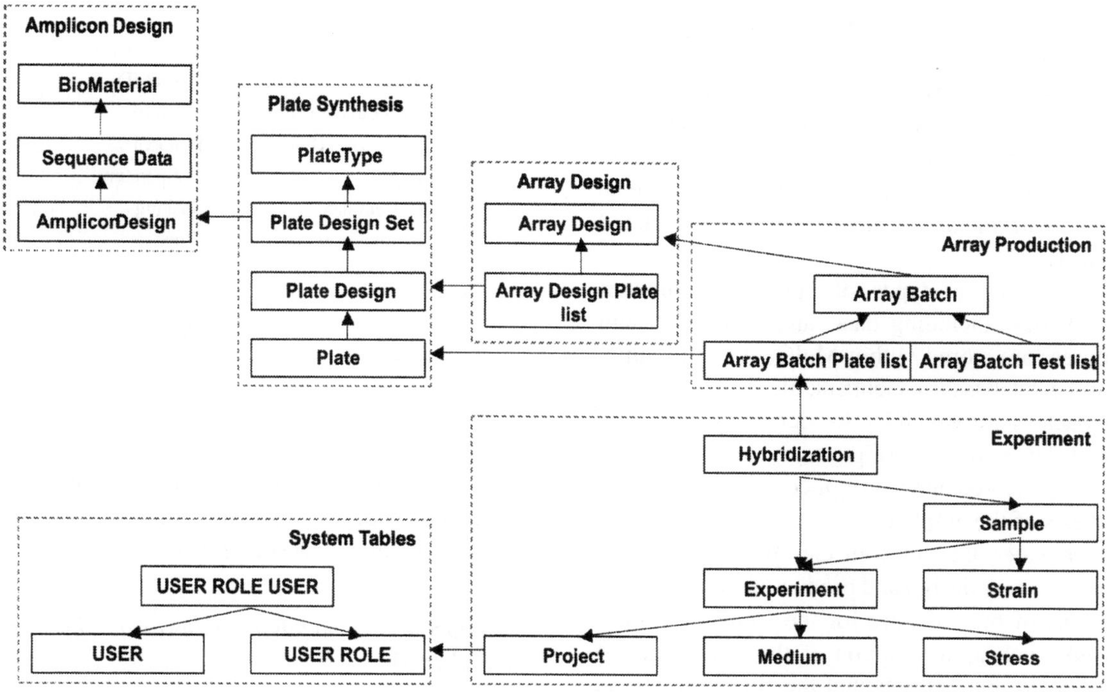

Figure 4. Schematic overview of the MolGenIS information system.

transcriptional terminators or regulator binding sites can be indicated to help in the discovery of operons. Using a simple input file (i.e., a tab-delimited text file, comprising one column with the genes to be colored and one column indicating their colors), subsets of genes can be differentially visualised for DNA-microarray analysis. The input file is based on the differences in transcript levels. This feature enables easy and rapid identification of genes which are transcriptionally linked. In a multiple transcriptome analysis experiment (e.g., measurements during a time course), all data-sets can be loaded as separate input files and subsequently shown in animation. Thus, changes in gene expression levels can be readily recognised.

Genome2D toolbox

Apart from visualisation properties, Genome2D includes a large number of utilities. These comprise tools to: (1) convert gene lists (tab-delimited text files) to FastA format, (2) perform a BLAST analysis at a location of choice (local, NCBI, etc.), (3) search in the genome for binding sites of transcriptional regulators (with or without a trained set), (4) calculate codon usage (of a complete genome or a subset of genes), (5) randomise the genome for statistical purposes (gene fragments or single basepairs), and (6) extract non-coding regions from a genome. Additional utilities can be easily included in the program.

Genome2D development

Genome2D is programmed in Borland Delphi 4. It runs on PCs with Microsoft Windows 98 and higher. To enable the 2D visualisation of a genome, Delphi makes use of the CADSys 4 library version 4.2 (obtained from Piero Valagussa). This library extends the Delphi vectorial graphics support to include 2D/3D CAD-like functions in applications.

MolGenIS: an information system for transcriptome data storage and retrieval

The result of high-throughput DNA-microarray production and analyses will result in large numbers of data. In order to control the processing and storage of these data a dedicated database system, named MolGenIS, has been set-up within the Molecular Genetics group. The MolGenIS system is implemented using the Invengine product of Inventory Netherlands (www.inventory.nl). The setup of an Invengine system is always based on a schematic design file and a user interface design file. MolGenIS is built up of tables that have been grouped in five functional modules: Amplicon design, Plate synthesis, Array design,

Array production and Array experiment (Figure 4). A module for users and user roles has also been added. Access at different restricted levels is given to each user via the use of passwords. The tables of each module have to be completed via a user-friendly interface. Design of the DNA-microarrays and the experiments performed with the arrays are schematically depicted in Figure 1.

Various different types of apparatus/computer-generated files, including data ranging from design to analyses, are stored within MolGenIS. The initial input of the database is a Genbank file map of the genome of bacteria for which DNA-microarrays are being produced. The output file of primers selected by GenomePrimer is stored under 'Amplicon design'. Results of synthesis of the amplicons and their position in the microtiter plates are stored in the 'Well description' file. Under 'Array design and production' the different output files of the software for the GenPack spotter are saved, such as description of the grid-layout of the spotter, spotter settings and the spot description, linking a spot to a microtiterplate well. In addition, a scan of one of the slides from a batch is made after staining with SyberGreen to check the slide quality.

The section 'Array experiments' stores images of the isolated RNA analysed by gel electrophoresis, a scan used to control the labeling of the cDNA by dilution series, and the hybridisation protocol. During measurement of the signal, an ArrayPro grid file with the description of the spot position, which is specific for each type of DNA-microarray, is generated. The results of the scan image obtained with the confocal laser scanner and the ArrayPro software package (image processing solutions), and quantification output files are subsequently generated and stored. The final analysis is performed using the Genespring software package that links the initial input data (Genbank file) with the ArrayPro grid and quantification files. These output results can be linked to pathway data. In addition to the stored files, specific experimental details have to be filled in by the user for each step of the various processes. For instance, within the process of DNA-microarray design and array experiments the following information has to be given: *See table top right.*

Array Design ID	Experiment ID
Spotter Settings (File)	Project ID
Grid Layout (File)	Medium
Spot Notes (File)	Protocol
Log	Temperature
Current date	Shaking
Current user	Stress
	Log
	Current date
	Current user

In a similar way, fields also have to be completed for the description of the project, strain, medium, stress condition, sample size and hybridisation conditions within the DNA-microarray experiment section. In this way all generated data are stored within the MolGenIS database in a highly structured manner, such that it can be easily accessed to compare data obtained from different studies and be made available to other users in any desired format.

LactococCyc: the metabolic pathway database for L. lactis IL1403

The recent availability of the complete chromosomal DNA sequence of *L. lactis* ssp. *lactis* IL1403 enables us to get insight in the numerous properties encoded in this genome (Bolotin et al. 2001; Kuipers 2001). Some 10 years ago, Peter Karp and co-workers started building a knowledge base concerning chemical compounds of intermediary metabolism (Karp 1992). This initiative gradually evolved into the currently well-known encyclopaedia of *Escherichia coli* genes and metabolism (EcoCyc) (Karp et al. 1996). Besides exploration of the metabolic routes of various (at present 11) bacteria in the EcoCyc package (http://biocyc.org/), its computational analysis tool PathoLogic enables users to generate new databases of any sequenced micro-organism of interest. The EcoCyc PathoLogic program uses a genome sequence and annotation in GenBank file format as input. Gene or protein names are extracted/searched (text recognition) from this annotation file and compared to those in the reference database, MetaCyc. Subsequently, when components/enzymes of metabolic routes in the reference database are found, PathoLogic will predict the presence of this metabolic pathway in the analysed organism (Karp et al. 2002).

In a research program on Gram-positive bacteria, we and our collaborators at the Wageningen Center for Food Sciences (Wageningen, The Netherlands) and the Center of Molecular and Biomolecular Informatics (Nijmegen, The Netherlands) aim to reconstruct their metabolic pathways and gene regulatory networks by performing genome-wide transcriptome analysis of *L. lactis*, *B. subtilis*, *Streptococcus pneumoniae* and *B. cereus* using DNA microarrays. With the aid of the EcoCyc program, in which transcription profiles can be overlayed onto metabolic maps, we expect to be able to identify metabolic pathway-related changes in gene expression (i.e., data filtering).

We have set-up an EcoCyc Pathway/Genome Data-Base (PGDB) of *L. lactis* ssp. *lactis* IL1403 by using the EcoCyc PathoLogic program, and called it the LactococCyc PGDB. The program was fed with the GenBank accession-file AE005176. PathoLogic predicted 134 metabolic routes in the genome of IL1403 (whereas 162 pathways are described in EcoCyc for *E. coli*). Out of the 134 predicted metabolic pathways, 88 pathways contained at least 50% of the contributing/participating enzymes in *L. lactis* (for the remaining 46 pathways less than 50% but at least one of the enzymes involved were present). In more detail, from 566 enzymes involved in the various pathways, 408 are present in the genome of IL1403, of which 71% are possibly involved in multiple routes. The MetaCyc reference database (version 5.7) contains 445 pathways composed of 4221 enzymatic and transport reactions. The LactococCyc PGDB comprises 669 enzymatic and transport reactions, while 579 metabolic conversions are missing. This means that either the corresponding enzymes are not present at all in the genome of *L. lactis* IL1403, or the program failed to recognise them, or the genes were not correctly annotated in the GenBank input file. An initial requirement for successful building of a PGDB is that the genome has been adequately annotated, as it is the primary and only input of the PathoLogic program (Ouzounis & Karp 2002). Secondly, since the PathoLogic program produces a prediction of the metabolic content of an organism, the generated PGDB has to be updated with knowledge from literature.

Although the generated *L. lactis* IL1403 PGDB is not fully complete, this does not hamper visualisation of its metabolic pathways and exploration of these with transcriptome data. Reactions missing in pathways are easily recognised, since the corresponding genes (differently colored) are absent in the graphs. Furthermore, the PathoLogic program generates hypertext markup language files displaying the missing reactions or absent genes or gene products. To enable refining and curation of the database an Editor program is included. We are currently curating and refining the PGDB of *L. lactis* IL1403, and building a PGDB of *L. lactis* ssp. *cremoris* strain MG1363.

Concluding remarks

To be able to generate transcriptome data that are both reliable and reproducible, all procedures for production and analysis have to be standardized and validated. Only when this process has been successfully completed, does it make sense to start doing the actual experiments, and to begin to answer biologically relevant questions. Subarrays and complete arrays of IL1403, as well as subarrays containing 96 amplicons of MG1363 are currently being analysed to generate the first reference sets. The full arrays of MG1363 will become available in autumn 2002, providing us with the unique opportunity of doing comparative transcriptome analyses of these two model strains of *L. lactis*. Moreover, methods for genomotyping using DNA–DNA hybridisation are currently being implemented, to be able to analyse the gene content of other lactococcal species and to be able to determine correction factors for the subsequent transcriptome analysis. Moreover, available and future sequence information on lactococcal plasmids, transposons and phages will be used to further enlarge the set of amplicons already spotted on the two master slides.

First experiments will concentrate on the analysis of gene transcription during growth on different media, providing reference sets for the analysis of modified strains or strains grown under different conditions. Moreover, cross hybridisations between IL1403 and MG1363 will be performed to further assess the possibilities and problems associated with using model slides for analysis of non-isogenic strains.

Great challenges now lie ahead to use transcriptome analysis, if possible in conjunction with proteome analysis, for instance (i) to unravel and visualize the complex gene regulatory networks underlying relevant physiological processes, (ii) to study adaptation to changing or completely different environments, (iii) to investigate possible side-effects of genetic engineering and (iv) to optimize the design of metabolic engineering experiments.

Acknowledgements

We wish to express our gratitude to various funding agencies (NWO-ALW, NWO-Bioinformatics program, NWO-STW, BTS-program, IOP-Genomics program, EU-Express Fingerprints FW5 program) for financial support, and our academic, institutional and industrial collaborators for valuable advice, stimulating discussions and other ways of support. We thank Wiep-Klaas Smits for making available results as an example of Genome2D in Figure 3.

References

Ball KD & Trevors JT (2002) Bacterial genomics: the use of DNA microarrays and bacterial artificial chromosomes. J. Microbiol. Methods 49: 275–284.

Bolotin A, Wincker P, Mauger S, Jaillon O, Malarme K, Weissenbach J, Ehrlich SD & Sorokin A (2001) The complete genome sequence of the lactic acid bacterium *Lactococcus lactis* ssp. *lactis* IL1403. Genome Res. 11: 731–753.

Diehl F, Grahlmann S, Beier M & Hoheisel JD (2001) Manufacturing DNA microarrays of high spot homogeneity and reduced background signal. Nucleic Acids Res. 29: e38.

Duwat P, Sourice S, Cesselin B, Lamberet G, Vido K, Gaudu P, Le Loir Y, Violet F, Loubiere P & Gruss A (2001) Respiration capacity of the fermenting bacterium *Lactococcus lactis* and its positive effects on growth and survival. J. Bacteriol. 183: 4509–4516.

Kamb A & Ramaswami M (2001) A simple method for statistical analysis of intensity differences in microarray-derived gene expression data. BMC Biotechnol. 1: 1–8.

Karp PD (1992) A knowledge base of the chemical compounds of intermediary metabolism. Comput. Appl. Biosci. 8: 347–357.

Karp PD, Riley M, Paley SM & Pelligrini-Toole A (1996) EcoCyc: an encyclopedia of *Escherichia coli* genes and metabolism. Nucleic Acids Res. 24: 32–39.

Karp PD, Riley M, Saier M, Paulsen IT, Collado-Vides J, Paley SM, Pellegrini-Toole A, Bonavides C & Gama-Castro S (2002) The EcoCyc Database.Nucleic Acids Res. 30: 56–58.

Klaenhammer TR et al. (2002) Discovering lactic acid bacteria by genomics. Antonie van Leeuwenhoek, in press.

Kuipers OP (2001) Complete DNA sequence of *Lactococcus lactis* adds flavor to genomics. Genome Res. 11: 673–674.

Kuipers OP, Buist G & Kok J (2000) Current strategies for improving food bacteria. Res. Microbiol. 151: 815–822.

Lucchini S, Thompson A & Hinton JC (2001) Microarrays for microbiologists. Microbiology 147: 1403–1414.

Oliver DJ, Nikolau B & Wurtele ES (2002) Functional genomics: high-throughput mRNA, protein, and metabolite analyses. Metab. Eng. 4: 98–106.

Ouzounis CA & Karp PD (2002) The past, present and future of genome-wide re-annotation. Genome Biology 3: comment2001.

Paton NW, Khan SA, Hayes A, Moussouni F, Brass A, Eilbeck K, Goble CA, Hubbard SJ & Oliver SG (2000) Conceptual modelling of genomic information. Bioinformatics 16: 548–517.

Raddatz G, Dehio M, Meyer TF & Dehio C (2001) PrimeArray: genome-scale primer design for DNA-microarray construction. Bioinformatics 17: 98–99.

Sebaihia M, Thomson N, Holden M & Parkhill J (2001) Microbial genomics. Dynamic duos. Trends Microbiol. 9: 579.

Suggs SV, Wallace RB, Hirose T, Kawashima EH & Itakura K (1981) Use of synthetic oligonucleotides as hybridization probes: isolation of cloned cDNA sequences for human beta 2-microglobulin. Proc. Natl. Acad. Sci. U.S.A. 78: 6613–6617.

Sugimoto N, Nakano S, Yoneyama M & Honda K (1996) Improved thermodynamic parameters and helix initiation factor to predict stability of DNA duplexes. Nucleic Acids Res. 24: 4501–4505.

Tomita M (2001) Whole-cell simulation: a grand challenge of the 21st century. Trends Biotechnol. 19: 205–210.

Varotto C, Richly E, Salamini F & Leister D (2000) GST-PRIME: a genome-wide primer design software for the generation of gene sequence tags. Nucleic Acids Res. 29: 4373–4377.

Wei Y, Lee JM, Richmond C, Blattner FR, Rafalski JA & LaRossa RA (2001) High-density microarray-mediated gene expression profiling of *Escherichia coli*. J. Bacteriol. 183: 545–56.

Ye RW, Wang T, Bedzyk L & Croker KM (2001) Applications of DNA microarrays in microbial systems. J. Microbiol. Methods 47: 257–272.

Zheng M, Wang X, Templeton LJ, Smulski DR, LaRossa RA & Storz G (2001) DNA microarray-mediated transcriptional profiling of the *Escherichia coli* response to hydrogen peroxide. J. Bacteriol. 183: 4562–4570.

Antonie van Leeuwenhoek **82:** 123–132, 2002.

Genome plasticity in *Lactococcus lactis*

Nathalie Campo, Miguel J. Dias, Marie-Line Daveran-Mingot, Paul Ritzenthaler* & Pascal Le Bourgeois
*Laboratoire de Microbiologie et Génétique Moléculaire du CNRS (UMR5100), Université Paul Sabatier, 31062 Toulouse, France. (*Author for correspondence; Email: ritzenth@ibcg.biotoul.fr)*

Key words: Lactococcus lactis, Genome analysis, comparative genomics

Abstract

Comparative genome analyses contribute significantly to our understanding of bacterial evolution and indicate that bacterial genomes are constantly evolving structures. The gene content and organisation of chromosomes of lactic acid bacteria probably result from a strong evolutionary pressure toward optimal growth of these microorganisms in milk. The genome plasticity of *Lactococcus lactis* was evaluated at inter- and intrasubspecies levels by different experimental approaches. Comparative genomics showed that the lactococcal genomes are not highly plastic although large rearrangements (a.o. deletions, inversions) can occur. Experimental genome shuffling using a new genetic strategy based on the Cre-*loxP* recombination system revealed that two domains are under strong constraints acting to maintain the original chromosome organisation: a large region around the replication origin, and a smaller one around the putative terminus of replication. Future knowledge of the rules leading to an optimal genome organisation could facilitate the definition of new strategies for industrial strain improvement.

Introduction

During the last decade, the increasing number of chromosome maps (more than 100 species) and sequenced genomes (74 at the end of 2001) has led us away from a view of bacterial genomes as static, monolithic structures toward the view that they are relatively variable, fluid structures (Casjens 1998). Comparative genomics at the intraspecies level tends to show that bacteria split into two groups according to their genomic plasticity. The first group contains eubacteria in which the overall genomic organisation (i.e. gene order) is strongly preserved, as for *Bacillus subtilis* (Anagnostopoulos 1990), *Bordetella pertussis* (Stibitz & Yang 1997), *Escherichia coli* (Perkins et al. 1993), *Salmonella enterica* sv. Typhimurium (Liu & Sanderson 1995), *Clostridium perfringens* (Canard et al. 1992), *Lactococcus lactis* (Davidson et al. 1996), *Mycoplasma galliseptum* (Tigges & Minion 1994), *Streptococcus salivarius* subsp.*thermophilus* (Roussel et al. 1997), and *Streptomyces ambofaciens* (Leblond & Decaris 1998), although chromosome rearrangements do occasionally occur. The second group comprises species in which the gene order is highly rearranged, such as *Salmonella enterica* sv. Typhi (Liu & Sanderson 1996), *Rhodobacter capsulatus* (Nikolskaya et al. 1995), *Bacillus cereus* (Carlson et al. 1992), *Leptospira interrogans* (Zuerner et al. 1993), *Pseudomonas aeruginosa* (Römling et al. 1995), and *Vibrio cholerae* (Nandi et al. 1997). Despite the different extent of chromosome rearrangement among bacterial species, there is now evidence that physical constraints or selective pressures exist to retain the bacterial genome organisation, such as preservation of nearly equal lengths for the two replication arms (Bergthorsson & Ochman 1998), maintenance of gene position and orientation relative to the chromosomal replication origin (Rocha et al. 1999), control of exogenous gene transfer in preferred chromosomal regions (Janssen et al. 2001), and conservation of the relative order of some loci even in distantly related bacterial genomes (Watanabe et al. 1997; Glaser et al. 2001). Conceivably, some evolutionary processes act to optimise the genome organisation of a particular species allowing the best adaptative response of the cell within its natural environment. However, there is still a lack of knowledge about the constraints on bacterial genome plasticity: to what extent is a sequenced genome representative for the entire bacterial species and what is the degree of genome rearrangements within a species? Which mechanisms are usually involved in this plasticity (homologous recombination, transposition, ...)? Are the constraints acting to preserve the genome organisation similar in all bacteria? And what is the nature (structural or functional) of these constraints?

At the evolutionary point of view, the dairy bacterium *Lactococcus lactis* can be considered a specialised microorganism because it possesses a reduced genome of approximately 2.5 Mb and has a very narrow ecological niche (its growth medium is usually milk). Thus, exploration of lactococcal genome plasticity can provide relevant data on fundamental aspects of bacterial genome dynamics, and can guide industrial strain improvement strategies. For instance, chromosome engineering without following the rules for optimised chromosome organisation could lead to genetic instability of the biotechnological traits. In this review, we present state-of-the-art knowledge of lactococcal genome plasticity, using recent comparative genome analyses derived from both physical mapping methods and sequencing projects.

Identification of spontaneous genome rearrangements in *Lactococcus lactis*

Genome rearrangements are provoked by various genetic elements, such as temperate bacteriophages and conjugative transposons (which have uncontested roles in acquisition of foreign DNA by horizontal gene transfer), or insertion sequences (IS). Except for few cases of transpositional rearrangements (DeBoy & Craig 1996), the largest contribution of IS elements to genome plasticity is to provide target DNA for homologous recombination. The nature of the rearrangements generated by homologous recombination depends on the relative orientation of the target sequences: interchromosomal recombination leads to deletion or duplication if the repeated elements are in direct orientation, and intrachromosomal recombination leads to inversion if they are in inverse orientation (for a review of the mechanisms see Roth et al. 1996). During the last few years, a large number of comparative analyses have demonstrated that, among the different repetitive elements found in the bacterial chromosomes, the primary target sequences associated with genome plasticity by homologous recombination are the ribosomal operons (*rrn*) (Liu & Sanderson 1996) and the IS elements (Syvanen 1997).

Characterisation of the spontaneous chromosomal rearrangements occurring in the two subspecies of *L. lactis* used in dairy industry offers the intriguing possibility of measuring the extent of lactococcal genome plasticity in natural conditions or during biotechnological processes.

Genome comparison at the intersubspecies level

Different methods have been used to compare the genomes of the *L. lactis* strains IL1403 and MG1363, which constitute the reference strains of the subspecies *lactis* and *cremoris,* respectively, although MG1363 can not be considered an authentic *cremoris* strain since it was formerly classified as a *lactis* strain from its phenotypic characteristics. Physical and genetic mapping of their chromosomes by pulse field gel electrophoresis revealed an overall similar genetic organisation, although the physical maps were very different (Le Bourgeois et al. 1995). This observation, corroborated by other comparative studies of lactococcal strains belonging to the same subspecies (Le Bourgeois et al. 1992; Davidson et al. 1996), led us to postulate that the main factors responsible for the restriction fragment length polymorphism (RFLP) between strains belonging to the same subspecies were genome rearrangements. On the contrary, the RFLP observed between strains of different subspecies were mostly due to sequence divergence, in addition to genome rearrangements. From the genome plasticity point of view, the most striking result of the comparative analysis of the IL1403 and MG1363 chromosomes by PFGE was the identification of a large inversion covering 50% of the chromosome, with borders located near the *uncD* and *rpoD* genes (Le Bourgeois et al. 1995). This inversion did not disturb the length of the replication arms. However, since the position of the *oppDFBCA-pepO* gene cluster, which is located close to one side of the inversion, was not conserved between the two chromosomes, this probably indicated that the genome rearrangement observed between the two strains was more complex than a single inversion.

Using the data of the complete genome sequence of the IL1403 chromosome (Bolotin et al. 2001), a co-linear scaffolding of the IL1403 and MG1363 chromosomes at the nucleotide level has been recently performed by Bolotin and co-workers (Bolotin et al.

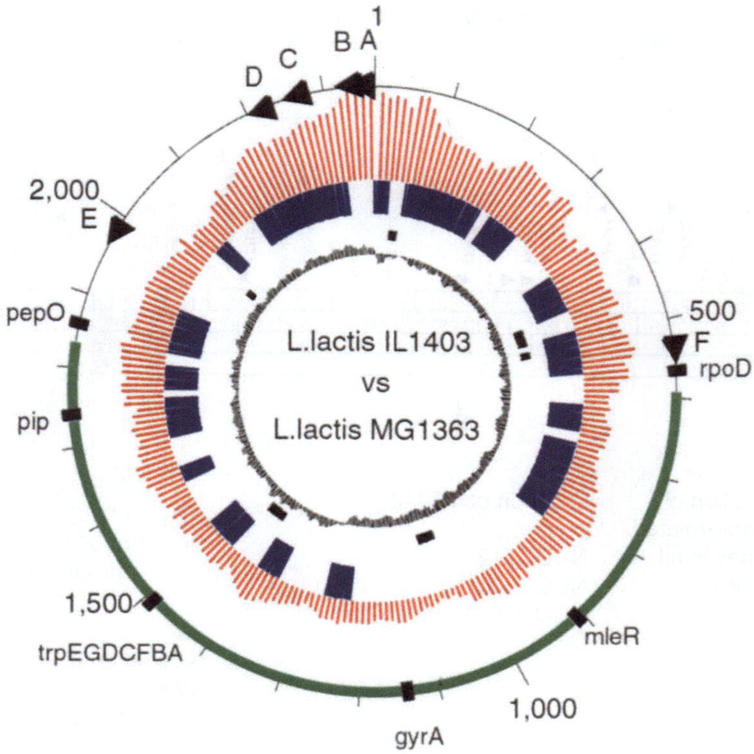

Figure 1. Co-linear scaffolding of the MG1363 and IL1403 chromosomes. The outer circle shows the IL1403 chromosome with a number of loci. Triangles indicate ribosomal operons. Scale on the outer circle shows coordinates (kbp) over the IL1403 chromosome. The IL1403/MG1363 genome inversion is indicated in green. Moving inward, the second circle (red) indicates MG1363 protein hits on the IL1403 chromosome, the third circle (blue) indicates the co-linear regions established by long range PCR. Starting from nucleotide 1, the clockwise order of co-linear regions (CR) is: CR19–CR03–CR08–CR06–CR04–CR12–CR02–CR14–CR10–CR09–CR15–CR07–CR11–CR05–CR16–CR01. The fourth circle corresponds to the location of the six prophages in the IL1403 chromosome, and the fifth circle shows GC-skew distribution (modified from Bolotin et al. 2002, Elsevier 2002).

2002). For that purpose, the DNA sequence of approximately 300 kb of the MG1363 chromosome was obtained by sequencing 513 random chromosomal fragments. Computer-based comparison at the DNA level allowed an estimate of a percentage of identity ranging from 70 to 98%, with an average value of 85%. At the protein sequence level, the distribution of significant hits over the IL1403 chromosome revealed a gradient of protein hits from the replication origin toward the terminus region (Figure 1), suggesting a higher conservation of the genome organisation in the vicinity of *oriC*. A comparable situation has been observed in *Pseudomonas aeruginosa* (Römling et al. 1995). The extent of the synteny between the two chromosomes was performed using a long range PCR approach (LR-PCR). Using a total of 160 PCR primers, these authors were able to identify 16 co-linear regions (CR) longer than 20 kb. These co-linear regions are distributed all along the chromosome, with the exception of a 400-kb region located near the replication terminus (Figure 1). It is noteworthy that this region contains strain-specific genetic mobile elements, such as the 36-kb prophage pi2 (Bolotin et al. 2001) for IL1403, and the 60-kb conjugative sex factor for MG1363 (Le Bourgeois et al. 1995). Two small non co-linear regions, one located between CR04 and CR12, and the second between CR11 and CR05 (Figure 1), should correspond to the borders of the large inversion described above. The total size of the co-linear regions was estimated to be 1072 kb for MG1363 and 1172 kb for IL1403, indicating that small genome rearrangements (e.g. location of IS elements) arise even in regions of synteny, although insertion of genes specific for each subspecies cannot be ruled out. However, it is not known whether the numerous non co-linear regions (which represent 55% of the genomes) are caused by occurrence of subspecies specific genes or genetic elements or by DNA sequence divergence, in addition to

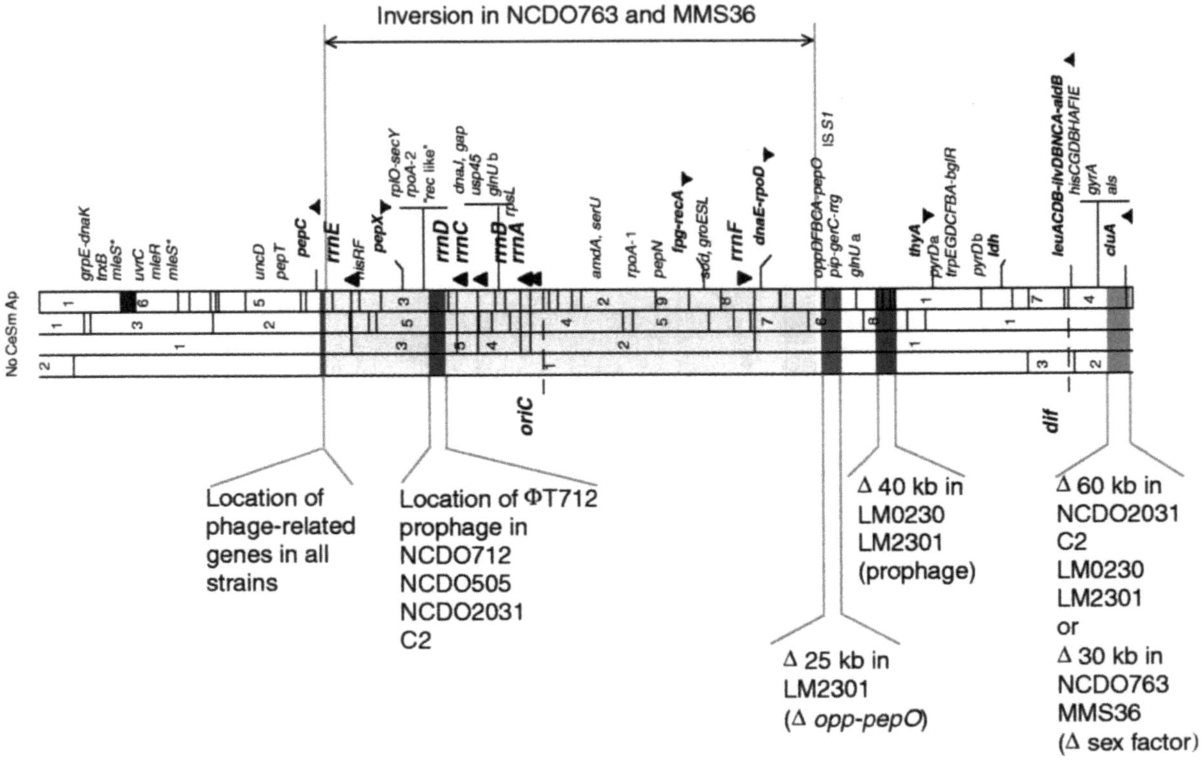

Figure 2. Location of the polymorphic regions in the MG1363 chromosome. Prophage locations, excisable elements and the inverted region are shown in shades of grey. Abbreviations: No, *Not*I; Ce, I-*Ceu*I; Sm, *Sma*I; Ap, *Apa*I.

probable genome rearrangements. The availability of the complete genome sequence of the MG1363 strain should answer this uncertainty. In conclusion, it seems realistic to state that the genetic content of the different *L. lactis* subspecies can differ substantially from each other.

Genome comparison at the intrasubspecies level: the case of the L. lactis *subsp cremoris NCDO712 family*

The most informative comparisons, in terms of evaluating genome dynamics, are those made between closely related strains of the same subspecies. We have recently identified the genetic events involved in spontaneous chromosomal rearrangements of nine strains genetically related to *L. lactis* subsp. *cremoris* strain NCDO712 (Davies et al. 1981), and characterised the genetic elements associated with these rearrangements (Le Bourgeois et al. 2000). The strains of the NCDO712 family differ from each other by some genetic traits, such as the content in plasmid or prophage DNA (McKay et al. 1976; Davies et al. 1981), the location of the conjugative sex factor (Gasson & Davies 1980; Walsh & McKay 1981) or the oligopeptide transport system (Tynkkynen et al. 1993; Yu et al. 1996). Comparison of their genome fingerprints revealed several macro-RFLPss that were located into five discreet regions on the physical map of the MG1363 chromosome, and assigned to a particular genetic event (Figure 2). Several strains of the NCDO712 family gave indistinguishable PFGE-fingerprints, suggesting that they correspond to strains differing either by their plasmid content (NCDO712 and NCDO505) or by point mutations (NCDO763 and MMS36) and very small rearrangements (NCDO2031 and C2). One striking chromosomal rearrangement found was a large inversion covering half of the genome of strain NCDO763. This inversion restored a genome organisation similar to that observed for the strain IL1403 (Le Bourgeois et al. 1995), and was provoked by a homologous recombination event between two inactive copies of the IS*905* element (Daveran-Mingot et al. 1998). Unexpectedly, several phage-related elements were found in the chromosome of these strains, even for those described as prophage-

Table 1. Analysis of IS elements involved in *L. lactis* subsp. *lactis* IL1403 chromosome rearrangements

Name	Family	Location[a]	Predicted length of the DR[b]	Upstream flanking sequence	Downstream flanking sequence
IS*983*A	IS*30*	682	12/14	TTGTATTCTTTTTAGGAC	GTCCACACGTGAAACAAA
IS*983*D	"	1338	12/14	TTTACTTAATTTTTTAAC	TTCTTGCACTGTAAGTTG
IS*981*B	IS*3*	94	3	TTTATTG	CTTTAGT
IS*981*C	"	650	"	CAATAAG	CTTTTTT
IS*981*D	"	728	"	ATTTGAG	ATATGTT
IS*981*E	"	1216	"	CATGTTT	ACTATTT
IS*981*F	"	1221	"	GAAGTAA	GTTTTTT

[a] In kb, as according to Bolotin et al. (2001).
[b] As determined by alignment of all the copies of each IS using the MULTALIGN program (Corpet 1988).

free. In addition to the characterised prophage in strains NCDO712, NCDO505, NCDO2031 and C2 (Davies et al. 1981), all strains contain a defective phage-related region (Daveran-Mingot et al. 1998), whereas strains NCDO712, NCDO505, NCDO763, NCDO2031, C2 and MG1363 contain a 40-kb excisable element probably corresponding to an uncharacterised defective prophage (Le Bourgeois et al. 2000). A similar situation was observed when analysing the IL1403 chromosome (Chopin et al. 2001). This comparative genome analysis revealed that two isogenic laboratory strains (MG1363 and LM2301) can differ from each other by more than 5% of their genome content simply by the presence of three large excisable elements: the 40-kb uncharacterised prophage, the 60-kb conjugative sex factor, and the 25-kb transposon-like structure carrying the *oppDFBCA-pepO* operon. This feature needs to be taken into account when performing chromosome engineering or systematic deletion analysis.

Identification of remnant rearrangements in the L. lactis *subsp.* lactis *IL1403 chromosome*

One of the major forces leading to chromosomal rearrangements is homologous recombination between repetitive elements, such as IS sequences in *E. coli* (Papadopoulos et al. 1999), REP sequences in *Mycobaterium* (Cole et al. 2001), or AUD elements in *Streptomyces* (Redenbach et al. 1993; Leblond et al. 1996). IS elements are grouped into 17 families on the basis of their genetic organisation, similarity of their transposases, similar features of their ends (inverted repeats), and their target sites (Mahillon & Chandler 1998). The IL1403 chromosome contains six different IS elements present at various copy numbers: IS*904* (9 copies), IS*905* (1 copy), IS*981* (10 copies), IS*982* (1 copy), IS*983* (15 copies), and IS*1077* (7 copies). Since IS transposition events are generally signed by the duplication of the target site (Mahillon & Chandler 1998), it is possible to detect remnant rearrangements provoked by homologous recombination between IS elements. We investigated the DNA sequences surrounding each IS element present in the IL1403 chromosome, assuming that absence of the transposition signature corresponded to a genome rearrangement between two copies of the same IS sequence (unpublished results). If the recombination led to inversions or translocations, one copy of the duplicated target site should be found associated with a similar IS located elsewhere in the chromosome. In the case of a deletion event, no direct repeat should be found. Each copy of IS*1077* was flanked by a 3-bp duplicated repeat, indicating that no homologous recombination had taken place between the seven copies of this element. In addition, IS*904* elements were never found alone, but were always inserted in one of the IS*1077* inverted repeats. These results indicate that IS*904* and IS*1077* elements are unlikely to represent a source of chromosome rearrangements in the lactococcal strain IL1403.

Among the fifteen copies of the IS*983*, two (IS*983*A and IS*983*D) are not flanked by direct repeats (Table 1). As no similarity could be found between the sequences bordering these two copies, it is likely that IS*983*A and IS*983*D correspond each to a deletion event between two ancestral IS elements. Similarly, six out of the ten IS*981*-copies were remnant products of a deletion event between tandemly repeated IS elements (Table 1). A computer-based comparison (BLASTp) of the ORFs surrounding these IS strongly corroborated the deletion hypothesis. For example, IS*983*A is flanked by two putative ORFs similar to

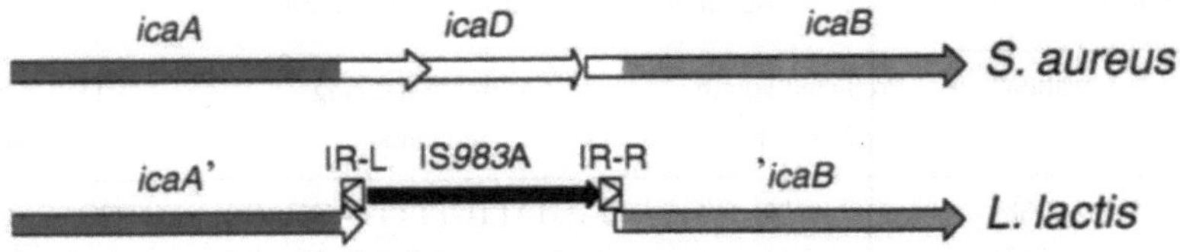

Figure 3. Comparison of the genetic organisation of the *ica* operons in *S. aureus* ATCC 35556 (Cramton et al. 1999) and *L. lactis* IL1403. Sequence similarities of *icaA* and *icaB* are shown in grey.

icaA and *icaB*, two genes belonging to the *ica* operon (intercellular adhesion) required for biofilm formation in *Staphylococcus aureus* (Cramton et al. 1999). Comparative analysis of the *ica* locus in *L. lactis* and *S. aureus* clearly indicated that *icaD* gene was lost by deletion between two copies of IS*983* in *L. lactis* (Figure 3). Thus, it appears that the actual organisation of the IL1403 chromosome does not reveal IS-mediated extensive genome rearrangements such as translocation or inversions of DNA segments, but rather small deletions between proximal copies of IS*981* or IS*983*.

Characterisation of constraints in the lactococcal genome

One challenge for understanding the dynamic state of the lactococcal chromosome is to characterise the constraints that could act to preserve its organisation: are there constrained regions in the lactococcal chromosome, and where are they located? Are these constraints structural (i.e. mechanical barriers in the nucleoid that could prevent intrachromosomal recombination processes) or functional (i.e. rearrangements are mechanistically possible but lethal)? A direct approach to these questions consists of experimental chromosome shuffling by construction of artificial large inversions, which disrupt the organisation of the bacterial chromosome without altering its genetic content, and studying the phenotypic consequences of these inversions.

A site-specific recombination system to construct large chromosomal inversions

To date, large genome inversions in bacteria have been constructed in *E. coli*, *S. enterica* sv. Typhimurium, and to a lesser extent in *B. subtilis*, by homologous recombination between two inverse-order inactivated copies of a selectable marker (Rebollo et al. 1988; Segall et al. 1988; Toda et al. 1996). These recombination events were phenotypically selected because they led to the reconstitution of one functional copy of the marker. Since this strategy was not easily applicable to *L. lactis*, we recently devised a new site-specific recombination strategy to obtain large chromosomal inversions in the chromosome of *L. lactis* (Campo et al. 2002). This strategy is based on the use of the recombinase Cre from the *E. coli* phage P1, which catalyses the recombination between two identical 34-bp sites called *loxP* (Abremski et al. 1983). The chromosomal inversions were obtained in *L. lactis* by a three-step procedure: (i) insertion of two inverse-order *loxP* cassettes (Figure 4A) into the chromosome of a lactococcal strain. These insertions can be either at random (via IS*S1* replicative transposition) or at specific sites (via homologous recombination), depending on the integrative plasmid used; (ii) controlled recombination between the two *loxP* sites by transient expression of the Cre recombinase provided *in trans* by a thermosensitive replication plasmid derived from the pGhost plasmid (Maguin et al. 1996). In permissive conditions for plasmid replication, the Cre recombinase catalyses recombination between the two *loxP* sites with an efficiency ranging from 5 to 60% of the cell population, depending on the location of the sites; (iii) recovery of cells harbouring either parental- or inverted-type chromosome structures by removal of the replication thermosensitive plasmid. Cells that had acquired the inverted-type chromosome could be screened for their ability to grow in the presence of tetracycline, because the recombination between the *loxP1* and *loxP2* sites generated a transcriptional fusion between the P59 promoter and the *tetL* ORF (Figure 4A). This Cre/*loxP* recombination system appears to be a powerful tool for creating various chromosome rearrangements (inversions, deletions, and transpositions) in a variety of Gram-positive bacteria, such as lactic acid bacteria.

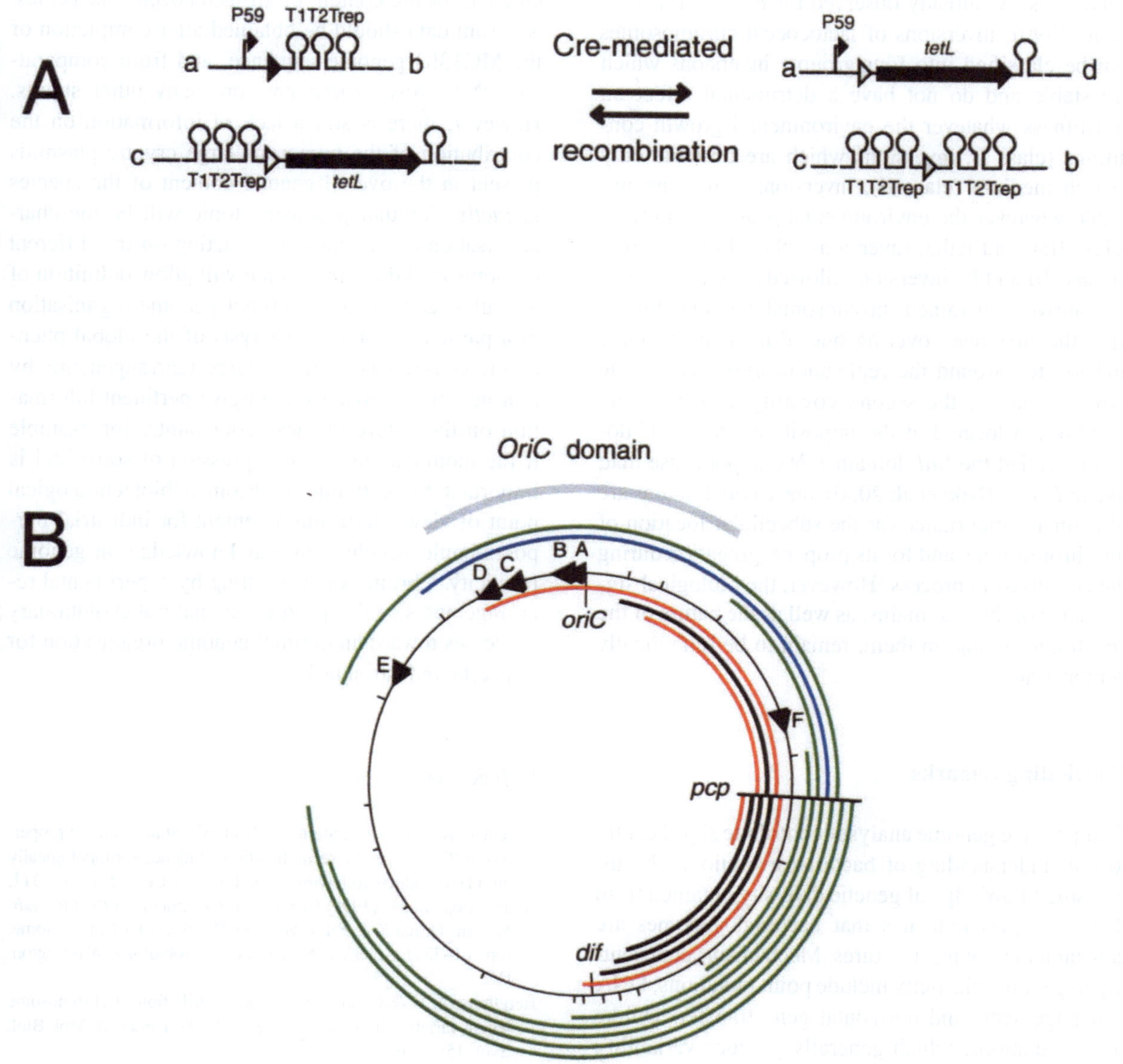

Figure 4. (A) Structure of the *loxP* cassettes used in the Cre/*loxP* recombination system. The *loxP* sites are represented by triangles. The recombined *loxP* sites are indicated in grey.The letters a, b, c, and d represent the chromosome order. (B) Invertible (outside circular segments) and non-invertible (inside circular segments) regions of the MG1363 chromosome. Triangles indicate ribosomal operons. The classes of inversions are represented as follows: Green, Class I; blue, class II; red, class III; black, class IV. The *dif* site is defined as the location of the shift of GC skew.

Identification of several constrained domains in the genome of strain MG1363

Several large inversions of the chromosome of the MG1363 strain were constructed by site-specific recombination (unpublished results). These inversions, which vary from 140- to 1200 kb in length, modify to various extents either the length of the two replication arms, or the orientation of genes relative to the origin of replication (Figure 4B). Each inversion has been analysed for its stability and its phenotypic consequence on cell fitness under different environmental conditions. As already observed for *E. coli* (François et al. 1990), inversions of lactococcal chromosomes can be classified into four groups: inversions which are stable and do not have a detrimental effect on cell fitness whatever the environmental growth conditions (class I), inversions which are unstable only in rich medium (class II), inversions which are unstable whatever the environmental growth conditions (class III), and lethal inversions (class IV). Location of class III and IV inversions allowed the characterisation of two constrained chromosomal domains (Figure 4B), the first one covering one fifth of the genome and located around the replication origin (called the '*oriC* domain'), the second covering approximately 100 kb and located at the opposite of the '*oriC* domain' (called the '*dif* domain'). We hypothesise that, like in *E. coli* (Niki et al. 2000), these two domains are of primary importance for the subcellular location of the chromosome, and for its proper segregation during the cell division process. However, the biological significance of these domains, as well as the nature of the constraints acting on them, remain to be functionally demonstrated.

Concluding remarks

Comparative genome analyses contribute significantly to our understanding of bacterial evolution. The increasing knowledge of genetic diversity obtained from these analyses indicates that bacterial genomes are constantly evolving structures. Mechanisms contributing to genome plasticity include point mutations, DNA rearrangements and horizontal gene transfer. Unlike point mutations, which generally produce variability in only small regions of the genome, chromosomal rearrangements generating variation across large molecular distances appear to be a major driving force for rapid evolution. Conceivably, the bacterial cell carries on a fast exploration of the landscape to which it has to adapt using alternative genome configurations. Thus, actual genome organisations are the result of an ongoing process of genome optimisation. Many of these 'genomic experiments' do not lead to an increase in cell fitness, but an unknown fraction may prove to be beneficial (Papadopoulos et al. 1999). Our knowledge of lactococcal genome flexibility has largely increased during the last few years, partly thanks to genome sequencing projects. Comparative genomics of strains IL1403 and MG1363 has clearly demonstrated that the IL1403 genetic content is probably not representative for the content of all lactococcal subspecies. Relevant data should be obtained after completion of the MG1363 genome sequence, and from comparative DNA array experiments on many other strains. However, there is still a lack of information on the contribution of the numerous large cryptic plasmids present in the overall genetic content of the species *L. lactis*. Another promising topic will be the characterisation of the constraints acting on the different chromosomal domains, which will allow definition of the rules leading to an optimal genome organisation in a particular context. Analysis of the global phenotypic consequences of the large rearrangements by a proteomic approach should give pertinent information on the nature of these constraints, for example if the modification of the expression of some loci is important for cell fitness. From a biotechnological point of view, strain improvement for industrial purposes could benefit from our knowledge on genome plasticity. Chromosome shuffling by experimental rearrangements could speed up the natural evolutionary processes toward an optimal genome organisation for a specific industrial task.

References

Abremski K, Hoess R & Sternberg N (1983) Studies on the properties of P1 site-specific recombination: evidence for topologically unlinked products following recombination. Cell 32: 1301–1311.

Anagnostopoulos C (1990) Genetic rearrangements in *Bacillus subtilis*. In: Drlica K & Riley M (Eds) The Bacterial Chromosome. (pp 361–371). American Society For Microbiology, Washington, DC.

Bergthorsson U & Ochman H (1998) Distribution of chromosome length variation in natural isolates of *Escherichia coli*. Mol. Biol. Evol. 15: 6–16.

Bolotin A, Ehrlich D & Sorokin A (2002) Studies of genomes of dairy bacteria *Lactococcus lactis*. Science des Aliments. *In press*.

Bolotin A, Wincker P, Mauger S, Jaillon O, Malarme K, Weissenbach J, Ehrlich SD & Sorokin A (2001) The complete genome sequence of the lactic acid bacterium *Lactococcus lactis* ssp. *lactis* IL1403. Genome Res. 11: 731–753.

Campo N, Daveran-Mingot M-L, Leenhouts KJ, Ritzenthaler P & Le Bourgeois P (2002) A Cre/*loxP* recombination system for large genome rearrangements in *Lactococcus lactis*. Appl. Environ. Microbiol. 68: 2359–2367.

Canard B, Saint-Joanis B & Cole ST (1992) Genomic diversity and organization of virulence genes in the pathogenic anaerobe *Clostridium perfringens*. Mol. Microbiol. 6: 1421–1429.

Carlson CR, Gronstad A & Kolsto AB (1992) Physical maps of the genomes of three *Bacillus cereus* strains. J. Bacteriol. 174: 3750–3756.

Casjens S (1998) The diverse and dynamic structure of bacterial genomes. Annu. Rev. Genet. 32: 339–377.

Chopin A, Bolotin A, Sorokin A, Ehrlich SD & Chopin MC (2001) Analysis of six prophages in *Lactococcus lactis* IL1403: different genetic structure of temperate and virulent phage populations. Nucleic Acids Res. 29: 644–651.

Cole ST, Supply P & Honore N (2001) Repetitive sequences in *Mycobacterium leprae* and their impact on genome plasticity. Lepr. Rev. 72: 449–461.

Corpet F (1988) Multiple sequence alignment with hierarchical clustering. Nucleic Acids Res. 16: 10881–10890.

Cramton SE, Gerke C, Schnell NF, Nichols WW & Gotz F (1999) The intercellular adhesion (*ica*) locus is present in *Staphylococcus aureus* and is required for biofilm formation. Infect. Immun. 67: 5427–5433.

Daveran-Mingot M-L, Campo N, Ritzenthaler P & Le Bourgeois P (1998) A natural large chromosomal inversion in *Lactococcus lactis* is mediated by homologous recombination between two insertion sequences. J. Bacteriol. 180: 4834–4842.

Davidson BE, Kordias N, Dobos M & Hillier AJ (1996) Genomic organization of lactic acid bacteria. In: Venema G, Huis in't Veld JHJ & Hugenholtz J (Eds) Lactic Acid Bacteria: Genetics, Metabolism and Applications (pp 65–87). Kluwer Academic Publishers, Dordrecht.

Davies FL, Underwood HM & Gasson MJ (1981) The value of plasmid profile for strain identification in lactic streptococci and the relationship between *Streptococcus lactis* 712, ML3 and C2. J. Appl. Bacteriol. 51: 325–337.

DeBoy RT & Craig NL (1996) Tn*7* transposition as a probe of *cis* acting interaction between widely separated (190 kilobases apart) DNA sites in the *Escherichia coli* chromosome. J. Bacteriol. 178: 6184–6191.

François V, Louarn J, Rebollo JE & Louarn JM (1990) Replication termination, nondivisible zones, and structure of the *Escherichia coli* chromosome. In: Drlica K & Riley M (Eds) The Bacterial Chromosome (pp 351–359). American Society for Microbiology, Washington, DC.

Gasson MJ & Davies FL (1980) High-frequency conjugation associated with *Streptococcus lactis* donor cell aggregation. J. Bacteriol. 143: 1260–1264.

Glaser P, Frangeul L, Buchrieser C, Rusniok C, Amend A, Baquero F, Berche P, Bloecker H, Brandt P, Chakraborty T, Charbit A, Chetouani F, Couve E, de Daruvar A, Dehoux P, Domann E, Dominguez-Bernal G, Duchaud E, Durant L, Dussurget O, Entian KD, Fsihi H, Portillo FG, Garrido P, Gautier L, Goebel W, Gomez-Lopez N, Hain T, Hauf J, Jackson D, Jones LM, Kaerst U, Kreft J, Kuhn M, Kunst F, Kurapkat G, Madueno E, Maitournam A, Vicente JM, Ng E, Nedjari H, Nordsiek G, Novella S, de Pablos B, Perez-Diaz JC, Purcell R, Remmel B, Rose M, Schlueter T, Simoes N, Tierrez A, Vazquez-Boland JA, Voss H, Wehland J & Cossart P (2001) Comparative genomics of *Listeria* species. Science 294: 849–852.

Janssen PJ, Audit B & Ouzounis CA (2001) Strain-specific genes of *Helicobacter pylori*: distribution, function and dynamics. Nucleic Acids Res. 29: 4395–4404.

Le Bourgeois P, Daveran-Mingot ML & Ritzenthaler P (2000) Genome plasticity among related *Lactococcus* strains: identification of genetic events associated with macrorestriction polymorphisms. J. Bacteriol. 182: 2481–2491.

Le Bourgeois P, Lautier M, Mata M & Ritzenthaler P (1992) Physical and genetic map of the chromosome of *Lactococcus lactis* subsp. *lactis* IL1403. J. Bacteriol. 174: 6752–6762.

Le Bourgeois P, Lautier M, van den Berghe L, Gasson MJ & Ritzenthaler P (1995) Physical and genetic map of the *Lactococcus lactis* subsp. *cremoris* MG1363 chromosome: comparison with that of *Lactococcus lactis* subsp. *lactis* IL1403 reveals a large genome inversion. J. Bacteriol. 177: 2840–2850.

Leblond P & Decaris B (1998) Chromosome geometry and intraspecific genetic polymorphism in Gram-positive bacteria revealed by pulsed-field gel electrophoresis. Electrophoresis 19: 582–588.

Leblond P, Fischer G, Francou FX, Berger F, Guérineau M & Decaris B (1996) The unstable region of *Streptomyces ambofaciens* includes 210 kb terminal inverted repeats flanking the extremities of the linear chromosome. Mol. Microbiol. 19: 261–271.

Liu SL & Sanderson KE (1995) I-*Ceu*I reveals conservation of the genome of independent strains of *Salmonella typhimurium*. J. Bacteriol. 177: 3355–3357.

Liu SL & Sanderson KE (1996) Highly plastic chromosomal organization in *Salmonella typhi*. Proc. Natl. Acad. Sci. U.S.A. 93: 10303–10308.

Maguin E, Prévots H, Ehrlich SD & Gruss A (1996) Efficient insertional mutagenesis in Lactococci and other Gram-positive bacteria. J. Bacteriol. 178: 931–935.

Mahillon J & Chandler M (1998) Insertion sequences. Microbiol. Mol. Biol. Rev. 62: 725–774.

McKay LL, Baldwin KA & Efstathiou JD (1976) Transductional evidence for plasmid linkage of lactose metabolism in *Streptococcus lactis* C2. Appl. Environ. Microbiol. 32: 45–52.

Nandi S, Khetawat G, Sengupta S, Majumder R, Kar S, Bhadra RK, Roychoudhury S & Das J (1997) Rearrangements in the genomes of *Vibrio cholerae* strains belonging to different serovars and biovars. Int. J. Syst. Bacteriol. 47: 858–862.

Niki H, Yamaichi Y & Hiraga S (2000) Dynamic organization of chromosomal DNA in *Escherichia coli*. Genes Dev. 14: 212–223.

Nikolskaya T, Fonstein M & Haselkorn R (1995) Alignment of a 1.2 Mb chromosomal region from three strains of *Rhodobacter capsulatus* reveals a significantly mosaic structure. Proc. Natl. Acad. Sci. U.S.A. 92: 10609–10613.

Papadopoulos D, Schneider D, Meier-Eiss J, Arber W, Lenski RE & Blot M (1999) Genomic evolution during a 10,000-generation experiment with bacteria. Proc. Natl. Acad. Sci. U.S.A. 96: 3807–3812.

Perkins JD, Don Heath J, Sharma BR & Weinstock GM (1993) *Xba*I and *Bln*I genomic cleavage maps of *Escherichia coli* K-12 strain MG1655 and comparative analysis of other strains. J. Mol. Biol. 232: 419–445.

Rebollo JE, François V & Louarn JM (1988) Detection and possible role of two large nondivisible zones on the *Escherichia coli* chromosome. Proc. Natl. Acad. Sci. U. S. A. 85: 9391–9395.

Redenbach M, Flett F, Piendl W, Glocker I, Rauland U, Wafzig O, Kliem R, Leblond P & Cullum J (1993) The *Streptomyces lividans* 66 chromosome contains a 1 Mb deletogenic region flanked by two amplifiable regions. Mol. Gen. Genet. 241: 255–262.

Rocha EP, Danchin A & Viari A (1999) Universal replication biases in bacteria. Mol. Microbiol. 32: 11–16.

Roth JR, Benson N, Galitsky T, Haack K, Lawrence JG & Miesel L (1996) Rearrangements of the bacterial chromosome: formation and applications. In: Neidhardt FC, Curtiss III R, Ingraham JL, Lin ECC, Low KB, Magasanik B, Resnikoff WS, Riley M, Schaechter M & Umbarger HE (Eds) *Escherichia coli* and *Salmonella:* Cellular and Molecular Biology. (pp 2256–2276). ASM Press, Washington, DC.

Roussel Y, Bourgoin F, Guedon G, Pebay M & Decaris B (1997) Analysis of the genetic polymorphism between three *Streptococcus thermophilus* strains by comparing their physical and genetic organization. Microbiology 143: 1335–1343.

Römling U, Greipel J & Tümmler B (1995) Gradient of genomic diversity in the *Pseudomonas aeruginosa* chromosome. Mol. Microbiol. 17: 323–332.

Segall AM, Mahan MJ & Roth JR (1988) Rearrangement of the bacterial chromosome: forbidden inversions. Science 241: 1314–1318.

Stibitz S & Yang MS (1997) Genomic fluidity of *Bordetella pertussis* assessed by a new method for chromosomal mapping. J. Bacteriol. 179: 5820–5826.

Syvanen M (1997) Insertion sequences and their evolutionary role. In: de Bruijn FJ, Lupski JR & Weinstock GM (Eds) Bacterial Genomes: Physical Structure and Analysis (pp 213–220). Chapman & Hall, NY.

Tigges E & Minion FC (1994) Physical map of *Mycoplasma galliseptum*. J. Bacteriol. 176: 4157–4159.

Toda T, Tanaka T & Itaya M (1996) A method to invert DNA segments of the *Bacillus subtilis* 168 genome by recombination between homologous sequences. Biosci. Biotech. Biochem. 60: 773–778.

Tynkkynen S, Buist G, Kunji ERS, Kok J, Poolman B, Venema G & Haandrikman AJ (1993) Genetic and biochemical characterization of the oligopeptide transport system of *Lactococcus lactis*. J. Bacteriol. 175: 7523–7532.

Walsh PM & McKay LL (1981) Recombinant plasmid associated with cell aggregation and high-frequency conjugation in *Streptococcus lactis* ML3. J. Bacteriol. 146: 937–944.

Watanabe H, Mori H, Itoh T & Gojobori T (1997) Genome plasticity as a paradigm of eubacteria evolution. J. Mol. Evol. 44 (Suppl 1): S57–S64.

Yu W, Gillies K, Kondo JK, Broadbent JR & McKay LL (1996) Loss of plasmid-mediated oligopeptide transport system in Lactococci: another reason for slow milk coagulation. Plasmid 35: 145–155.

Zuerner RL, Herrmann JL & Saint Girons I (1993) Comparison of genetic maps for two *Leptospira interrogans* serovars provides evidence for two chromosomes and intraspecies heterogeneity. J. Bacteriol. 175: 5445–5451

Antonie van Leeuwenhoek **82:** 133–145, 2002.

Regulation of antimicrobial peptide production by autoinducer-mediated quorum sensing in lactic acid bacteria

Luis E.N. Quadri
Department of Microbiology and Immunology, Medical College of Cornell University, 1300 York Avenue, Box 62, Room W-706, New York, NY 10021, USA (Tel.: +1-212-746-4497; Fax: +1-212-746-4028; E-mail: leq2001@med.cornell.edu)

Key words: Quorum sensing, antimicrobial peptide, autoinducer signaling, signal transduction

Abstract

Several lactic acid bacteria produce peptides with antimicrobial activity. During the last few years, cell–cell communication has emerged as the key regulatory mechanism that controls the production of many of these antimicrobial peptides via a regulatory strategy denominated quorum sensing. Quorum sensing allows population-wide synchronised production of antimicrobial peptides as a function of cell density. The cell–cell communication phenomenon required for sensing of the cell density is mediated by secreted signalling molecules. These 'molecular messengers' accumulate in the environment as the cell density increases and activate signal transduction cascades that result in the production of antimicrobial peptides by the stimulated bacterial cell.

Introduction

Many lactic acid bacteria (LAB) produce ribosomally synthesized antimicrobial peptides (AMPs) usually referred to as bacteriocins. These peptides are diverse in terms of structure, mode of action, spectrum of antimicrobial activity and potency. Because of their antimicrobial properties, the peptides are of relevance for the food and pharmaceutical industries, and therefore their production has been investigated. In recent years, several research laboratories have reported examples of LAB in which the production of AMPs is an inducible phenotype dependent on the presence of 'inducing peptides' (referred to herein as autoinducers, AIs) in the culture supernatant (for a recent review see Kleerebezem & Quadri 2001). These examples contrast the more commonly found situation, where AMPs are apparently produced in a constitutive fashion and without the need for AIs. The first inducible systems were identified by serendipity. For example, production of AMPs by *Carnobacterium piscicola* was observed to be mysteriously lost from time to time. The explanation for this phenomenon was not as simple as it was first thought to be, that is, the head of the laboratory, who decided to work at the bench that summer, inoculated the wrong strain. Rather, the AMP$^-$ phenotype was eventually correlated with small-size inocula utilized to start the cultures and the lack of AIs in the culture supernatants (Saucier et al. 1995, 1997; Quadri et al. 1997a,b; Franz et al. 2000a,b; Kleerebezem & Quadri 2001; Kleerebezem et al. 2001). Inducible production of the AMP nisin A by strains of *Lactococcus lactis* was also discovered by chance. It was observed that a 4-bp deletion in the *nisA* gene not only abrogated nisin A production, but also suppressed the transcription of the mutant allele Δ*nisA* (Kuipers et al. 1993). It was subsequently discovered that addition of nisin A to the culture supernatant of the mutant restored Δ*nisA* transcription, indicating that nisin A had AI activity (Kuipers et al. 1995; Dodd et al. 1996; Van Kraaij et al. 1997).

Today, 7 years after the first reported example of AI-mediated induction of AMPs in LAB, it is widely recognized that not only many LAB, but also other Gram-positive bacteria have evolved mechanisms to control production of AMPs via a phenomenon called quorum sensing (de Vos et al. 1997; Dunny & Leonard 1997). Quorum sensing, in its broadest definition, is a cell–cell communication strategy that enables unicellular organisms to behave in a multicellular manner by allowing population-wide synchronized behavioural responses as a function of cell density. All quorum sensing systems utilize AIs, however of different chemical natures, as communication signals (Fuqua et al. 1996, 2001; Dunny & Leonard 1997; Fuqua & Greenberg 1998; Kleerebezem et al. 1997a). In addition to the production of AMPs, examples of behavioural responses modulated by quorum sensing are production of antibiotics and toxins, sporulation and cell differentiation, development of genetic competence, bioluminescence, conjugative plasmid transfer, biofilm formation, and virulence response (for review see Dunny & Winans 1999).

In the first part of this review, the highlights of different systems where production of AMPs by LAB is regulated by AI-mediated quorum sensing will be presented. The emergent models for these regulatory mechanisms will be discussed in the second part of the review.

Studies on regulation of class II AMP production

Class II AMPs are small, heat-stable, unmodified secreted peptides that are produced by many Gram-positive bacteria, including several LAB (Diep et al. 2000; Ennahar et al. 2000; Franz et al. 2000a). Class II AMPs are produced as precursors containing a leader peptide, which, in most cases, is believed to be cleaved concomitantly with export by a dedicated *sec*-independent export/processing machinery (Havarstein et al. 1995). Most of these leaders can be identified by the presence of two glycine residues preceding the cleavage site (Nes et al. 1996). Since 1995, production of several class II AMPs has been shown to be inducible. In these inducible systems, small secreted peptides, with or without antimicrobial activity, function as AIs. The AIs are involved in the induction of their own structural genes and other genes required for production of and immunity to the class II AMPs; e.g., AMP genes, immunity genes, secretion genes, and regulatory genes (Kleerebezem & Quadri 2001). AI-mediated regulation of class II AMP production has been investigated in many LAB, including *Carnobacterium piscicola*, *Lactococcus lactis*, *Lactobacillus plantarum* and *Lactobacillus sakei*. These systems are described in the sections below.

Class II AMP production in Carnobacterium piscicola *LV17*

The first clear example of AI-mediated regulation of class II AMP production was reported in 1995 (Saucier et al. 1995). The phenomenon was observed in strains of *C. piscicola*. These strains were *C. piscicola* LV17 (carrying the plasmids pCP49, pCP40 and pCP9) and its derivatives, strains LV17A and LV17B (carrying only the plasmids pCP49 and pCP40, respectively). A third derivative, the plasmidless strain LV17C, was shown to lack AMP production. Three class II AMPs are known to be produced by these strains: carnobacteriocins A, B2 and BM1. Carnobacteriocin A and carnobacteriocins B2 and BM1 were purified from cultures of strains LV17A and LV17B, respectively (Quadri et al. 1994; Worobo et al. 1994). It was observed that cultures of strains LV17, LV17A and LV17B started with diluted inocula would grow well but would not produce AMPs (Saucier et al. 1995). The AMP$^-$ phenotype persisted during subsequent inoculations, unless supernatant from an AMP$^+$ culture of the strains LV17, LV17A or LV17B was added to the AMP$^-$ culture before the cultures would reach the stationary phase of growth. Growing the cells on solid medium was also shown to restore the AMP$^+$ phenotype. Notably, addition of purified carnobacteriocins A or B2 also rescued the AMP$^+$ phenotype in the three strains, whereas carnobacteriocin BM1 only rescued the phenotype in LV17 and LV17A (Saucier et al. 1995). Recombinant carnobacteriocin B2 and five carnobacteriocin B2 variants also rescued the AMP$^+$ phenotype in LV17B (Quadri et al. 1997a). Since a Phe33-Ser variant had no detectable antimicrobial activity but retained its AI capacity, it has been hypothesized that the AI function does not require the antimicrobial activity of the peptide (Quadri et al. 1997a). However, a C-terminally truncated variant had neither AI nor antimicrobial activity, suggesting that the C-terminus of carnobacteriocin B2 is important for both activities. Notably, recombinant precarnobacteriocin B2 (with marginal antimicrobial activity) also had AI activity, suggesting that the free N-terminus of the mature peptide is not essential for its AI function (Quadri et al. 1997a). Overall, these results indicated that the carnobacteriocins act as AIs.

Further analysis of the carnobacteriocin systems revealed the existence of two additional AIs. These AIs are termed CS and CbnX, and are encoded as precursors with double-glycine-type leaders by the genes *cbaX* and *cbnS*, respectively (Quadri et al. 1997b;

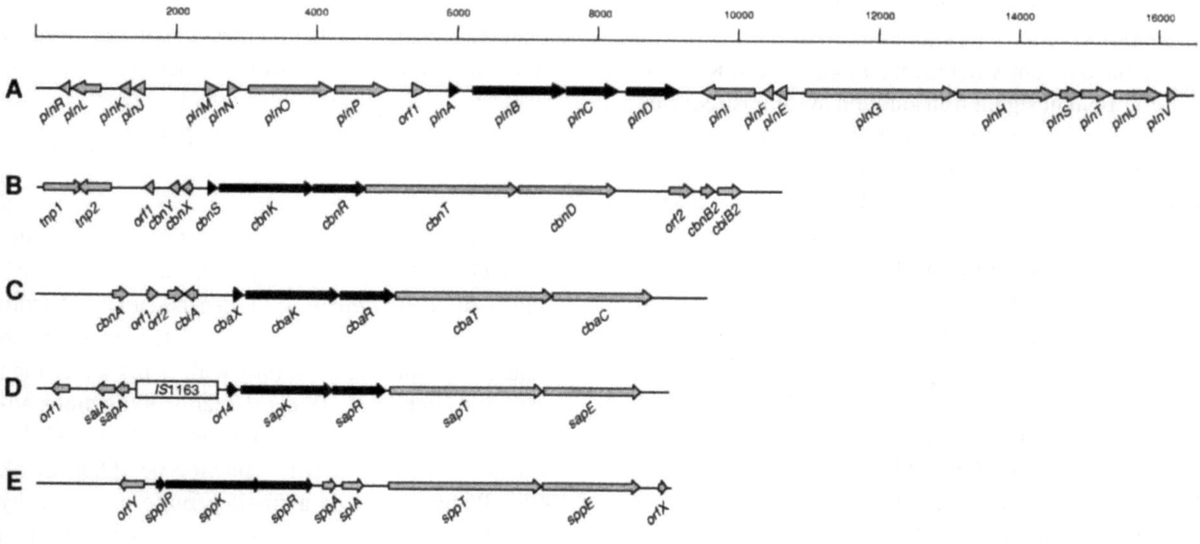

Figure 1. Gene clusters involved in regulated production of and immunity to various class II AMPs in LAB. (A) Plantaricin A gene cluster of *Lactobacillus plantarum* C11; (B) carnobacteriocin B2 gene cluster of *Carnobacterium piscicola* LV17B; (C) carnobacteriocin A gene cluster of *Carnobacterium piscicola* LV17A; (D) sakacin A gene cluster of *Lactobacillus sake* Lb706; and (E) sakacin P gene cluster of *Lactobacillus sake* Lb674 and *Lactobacillus sake* LTH673. Genes forming the three-component regulatory systems are highlighted in black. The scale is given in base pairs.

Franz et al. 1999, 2000b; Kleerebezem et al. 2001). The gene *cbaX* is located in the carnobacteriocin A gene cluster of pCP49, whereas *cbnS* resides on the carnobacteriocin B2 gene clusters of pCP40 (see below). Synthetic CbnX or sterile culture supernatants of homologous and heterologous hosts genetically engineered to produce CbnX induce AMP production in LV17A (Franz et al. 1999, 2000b). Similarly, synthetic CS triggers AMP production in LV17B (Kleerebezem et al. 2001). Interestingly, we have recently analyzed the AI activity of several CS variants, some including an Ala substitution at one of two cysteine residues conserved in several AIs (Kleerebezem & Quadri 2001a). Our preliminary analysis suggests that both cysteine residues are involved in the AI activity of the CS peptide (Rohde et al. 2001). Like carnobacteriocin B2, NMR analysis reveals that CS displays α-helix structure in membrane-mimicking environments, although the CS peptide has no detectable antimicrobial activity (Rohde et al. 2001).

Genetic analysis suggested that the AI-mediated regulation of carnobacteriocin production takes place via a signal transduction mechanism. The genetic determinant of carnobacteriocins A, B2 and BM1 are located on pCP49, pCP40 and the chromosome, respectively (Quadri et al. 1994; Worobo et al. 1994). The carnobacteriocin B2 gene cluster of pCP40 includes three well defined operons: *cbnXY*, *cbnSKRTD* and *cbnB2-cbiB2* (Figure 1B) (Quadri et al. 1994, 1995, 1997b). *cbnB2* and *cbiB2* encode precarnobacteriocin B2 and carnobacteriocin B2 immunity protein, respectively. Based on sequence analysis, *cbnK* and *-R* encode a histidine protein kinase (HPK) and a response regulator (RR) of a two-component signal transduction system, respectively, whereas *cbnT* and *-D* encode a *sec*-independent export/processing system. The products of *cbnX* and *-Y* resemble class II AMP precursors with double-glycine-type leaders. *cbnS* encodes the precursor of the CS AI mentioned above (Kleerebezem et al. 2001). Frameshift mutations in *cbnR*, *-T* and *-D* yield AMP^- mutants with reduced immunity to both carnobacteriocins. However, a frameshift mutation in *cbnK* results in a AMP^+ strain, probably because of the elimination of a phosphatase activity present in CbnK (Quadri et al. 1997b). The carnobacteriocin A gene cluster of pCP49 has also been investigated (Figure 1C) (Worobo et al. 1994; Franz et al. 2000b). It contains the genes *cbaA* and *cbiA*, encoding precarnobacteriocin A and its immunity proteins, respectively, and the genes *cbaX, -K, -R, -T* and*-D,* which are homologs of*cbnS, -K, -R, -T* and*-D,* respectively. Deletion analysis suggested that the *cbaKRTD* region is required for carnobacteriocin A production; however further analysis is needed to ascertain the relevance of individual genes (Franz et al. 2000b).

We have recently determined that the CbnK/CbnR two-component signal transduction system is required for the mechanism via which carnobacteriocin B2 and CS induce AMP production in the strain LV17B (Kleerebezem et al. 2001; Rohde et al. 2001). This was accomplished using a two-plasmid system, where plasmid one constitutively expressed *cbnK* and *cbnR* (as wild-type or site-directed mutants) and plasmid two expressed the β-glucuronidase reporter gene (*gusA*) under the control of the promoters of *cbnB2*, *cbnBM1* or *cbnS* (PcbnB2, PcbnBM1 and PcbnS, respectively). All plasmid-one plasmid-two pairs and control vector combination were introduced into *C. piscicola* LV17C (plasmidless derivative of strain LV17) and *L. lactis* MG1163, and the glucuronidase activity of transformants subjected to different treatments was determined. Transformants containing wild-type *cbnK* and *cbnR* and PcbnB2-, PcbnBM1- or PcbnS-*gusA* fusions showed CS- and carnobacteriocin B2-dependent induction of glucuronidase activity. However, transformants containing residue substitutions at predicted phosphorylation sites in CbnK (His 239 to Ala) and CbnR (Asp 10 or Asp 62 to Asn) had no carnobacteriocin B2- or CS-dependent activation of *cbn* promoters, as indicated by glucuronidase assays. These and early results strongly suggest that both *cbnK* and *cbnR* are necessary and sufficient for reconstruction of the functional signal-transduction pathway required for AI-dependent activation of *cbn* promoters in homologous and heterologous hosts (Kleerebezem et al. 2001; Rohde et al. 2001).

Using the two-plasmid system, the promoters PcbnBM1, PcbnB2 and PcbnS were functionally characterized. The promoters have comparable carnobacteriocin B2- and CS-dependent induction patterns (Kleerebezem et al. 2001; Rohde et al. 2001), supporting the earlier observation that transcripts originated from these promoters are only detected in AMP$^+$ or induced cultures of *C. piscicola* LV17B (Quadri et al. 1997b). The common regulation of the promoters could take place via two conserved direct repeat sequences present in the promoters. The repeats have been suggested to be sites for binding of the RR CbnR (Diep et al. 1996; Quadri et al. 1997b). Similar repeats are present in inducible promoters of the carnobacteriocins A gene cluster (Diep et al. 1996). Overall, the analysis of the carnobacteriocin B system supports a complex regulatory model of double-autoinduction in which expression of the genes required for AMP production and immunity is controlled by a CbnK-CbnR dependent signal transduction mechanism triggered by at least two distinct AIs, carnobacteriocin B2 and CS.

Class II AMP production in Lactobacillus plantarum *C11*

Elegant work has been done to investigate the production of class II AMPs by *L. plantarum* C11. AMP$^-$ cultures of *L. plantarum* C11 can be obtained from AMP$^+$ cultures by growing the bacterium on solid media or by starting a culture with a highly diluted inoculum (Diep et al. 1995). Notably, exogenous addition of a peptide called plantaricin A to AMP$^-$ cultures rescues the AMP$^+$ phenotype (Diep et al. 1995). Plantaricin A is a small peptide produced by *L. plantarum* C11 and was originally identified as a class II AMP (Nissen-Meyer et al. 1993). Plantaricin A also induces transcription of a 3.3-kb operon of *L. plantarum* C11. The operon contains the genes *plnA*, encoding the precursor of plantaricin A, *plnB*, encoding a predicted HPK, and *plnC* and *plnD*, encoding two predicted RRs (Figure 1A) (Diep et al. 1994). The report of this operon is the first example of genetic linkage between genes encoding a two-component signal transduction system and a gene encoding a class II AMP-like precursor. The studies on the plantaricin A system clearly indicate that plantaricin A acts as an AI not only of AMP production but also of its own production. During the study in which the AI activity of plantaricin A was discovered, the investigators failed to detect the previously reported AMP activity of plantaricin A and suggested that the peptide was an AI rather than an AMP (Nissen-Meyer et al. 1993). However, analysis of the activity of N-terminally truncated variants of plantaricin A demonstrated that the peptides have strain-specific antimicrobial activity and the ability to dissipate the membrane potential (Hauge et al. 1998). Notably, synthetic L- and D-enantiomers of plantaricin A exhibit antimicrobial and membrane potential dissipation activities and both adopt helical secondary structures in a membrane-like environment. However, only the L-enantiomer has AI activity, suggesting that unlike the antimicrobial activity of the peptide, its AI function requires a stereospecific interaction, possibly with PlnB (Hauge et al. 1998).

Further exploration of the *pln* locus unveiled four additional operons induced by plantaricin A (Figure 1A) (Diep et al. 1996). Collectively, the operons (*plnEFI* and *plnJKLR*, *plnMNOP and plnGHSTUV*) contain genes encoding: (i) five predicted class II

AMP-like precursors with double-glycine-type leaders (*plnE, -F, -J, –K* and– *N*), (ii) four predicted class II AMP immunity proteins (*plnI, –L, -M* and *–P*), and (iii) a putative ABC transporter and its partner accessory protein (*plnG* and *-H*, respectively) possibly involved in processing of double-glycine leaders and secretion of the Pln peptides encoded in the locus. Formal demonstration is required to ascertain the function of the predicted immunity and transport/processing genes. However, production of the mature peptides derived from the products of *plnE, -F, -J, -K and -N* has been validated by mass spectrometry and amino acid sequencing and, using synthetic peptides, it has been shown that PlnEF and PlnJK constitute two-peptide AMP systems (Anderssen et al. 1998).

In vitro and *in vivo* analysis of the plantaricin A system revealed that the RRs PlnC and -D modulate the activity of plantaricin A-induced promoters by binding to conserved direct repeat sequences in the target promoters (Risoen et al. 1998). Using DNase I footprinting, it was demonstrated that PlnC and -D bind to and protect the repeats of the *plnA* promoter. Furthermore, electrophoretic mobility gel shift assays showed that PlnC and -D bind to the repeats in the promoter of *plnA, -E* and *-J*. The efficiency of binding increased in the presence of acetyl phosphate, suggesting that the phosphorylation state of the RRs determines their binding affinity. Moreover, the assays indicated that the RRs bind as homodimers to the repeats and that elimination of one of the repeats, alterations of the spacing between them, or single nucleotide substitutions in the consensus sequence reduce binding. Using a glucuronidase reporter gene, the integrity of the repeats and the spacing between repeats was shown to be required for promoter activity *in vivo*. Overall, the *in vitro* and *in vivo* analysis demonstrated that the repeats are sites of binding for PlnC and -D and function as *cis*-regulatory elements involved in the regulation of AMP production (Risoen et al. 1998).

Class II AMP production in Lactobacillus sakei *Lb706*

Excellent work has been done to show that production of the class II AMP sakacin A by *L. sakei* Lb706 is temperature-dependent and regulated by a peptide AI (Diep et al. 2000). *L. sakei* Lb706 produces sakacin A at 25 or 30 °C, but production is reduced or abolished at 33.5–35 °C. AMP production in the AMP^- cultures can be restored by reducing the growth temperature to 30 °C (Diep et al. 2000). AMP can also be induced in AMP^- cultures by addition of a peptide AI to the growth medium. By analogy with the plantaricin A system, a 23-residue peptide derived from a class II AMP-like precursor encoded by *orf4*, a gene located on the sakacin A gene cluster (see below), was hypothesised to be an AI in the sakacin A system. A synthetic peptide corresponding to the predicted AI was indeed shown to rescue the AMP^+ phenotype in AMP^- cultures. This observation supports the proposed function of the product of *orf4*. Further studies showed that a mutation in *orf4* abolishes sakacin A production and that the mutation can be complemented by exogenous addition of AI (Diep et al. 2000). Plantaricin A, the AI of *L. plantarum* C11, failed to rescue the AMP^+ phenotype in *L. sakei* Lb706, demonstrating specificity for the cognate AI (Diep et al. 2000). Sakacin A production by *Lactobacillus curvatus* LTH1174 showed a similar response to temperature and AI addition (Diep et al. 2000).

The 8.7-kb gene cluster required for sakacin A production has been characterized (Figure 1D) (Axelsson et al. 1995). It is organized in two divergently transcribed operons. The first operon encompasses *sapA* (previously named *sakA*; Holck et al. 1992) and *saiA*, encoding the sakacin A precursor and sakacin A immunity protein, respectively. The second operon contains the AI precursor gene, *orf4*, followed by *sapK* (previously designated *sakB*; Axelsson et al. 1993), *-R, -T* and *-E*. *sapK* and *-R* encode an HPK and an RR of a predicted two-component signal transduction system, respectively. The *sapT* and *-E* products share similarity with *sec*-independent export/processing systems. Transcriptional analysis confirmed that the predicted operon organization of the locus. The operons are separated by a region that contains the insertion element IS*1163*, whose integrity is not required for AMP production or immunity (Axelsson et al. 1995). On each side of IS*1163*, but not being part of it, there is one of two 35-bp sequence motifs (named LIR and RIR) that are inverted versions of each other. Deletion analysis indicated that LIR, located upstream of the *sapA–saiA* operon, is essential for AMP production, AMP immunity, and transcription of the *sapA–saiA* operon. However, deletion of RIR results in immune but AMP^- mutants (Axelsson et al. 1995). The requirement of RIR for transcription of the downstream operon has not been clearly demonstrated. Mutational analysis suggested that the SapK–SapR regulatory system is necessary for AMP production and immunity, whereas the SapT–SapE export/processing system is only required for AMP production (Axelsson et al. 1995). Moreover, Northern blot analysis indicated that mutations in *sapK* or *-R* eliminates the transcription of the two *sap* operons (Axelsson et al. 1995). Based on the overall analysis of the sakacin A system, and by analogy with other inducible class II AMP production systems, it has been proposed that the SapK/SapR two-component system is involved in regulation of transcription of the *sap* operons and that LIR and RIR are target sites for SapR-dependent transcriptional activation (Axelsson et al. 1995).

Class II AMP production in Lactobacillus sakei *LTH673 and* Lactobacillus sakei *Lb674*

A great deal of research has been done to investigate the cell-density regulated production of the class II AMP sakacin P by *L. sakei* LTH673. That sakacinP production in *L. sakei* LTH673 was lost upon dilution of the culture into fresh culture broth or by plating the cells on solid media was reported in 1996 (Eijsink et al. 1996). As seen in analogous inducible AMP production systems, loss of sakacin P production can be prevented by conditioning the media with the addition of a small volume of culture supernatant from the cognate AMP$^+$ culture. Colonies grown on conditioned solid medium also retain the AMP$^+$phenotype. Characterization of the AI present in the supernatant of the AMP$^+$ cultures led to the identification of a peptide of 19 amino acids. Both the peptide purified from the culture supernatant and the chemically synthesized peptide have comparable AI activity (Eijsink et al. 1996). The AI peptide exhibited no antimicrobial activity and circular dichroism studies showed that it adopts a helical structure in membrane-mimicking environments (Eijsink et al. 1996). Northern blot analysis confirmed that the AI induces expression of the structural genes of sakacin P and the predicted cognate immunity protein. The induced transcripts, which are not expressed in AMP$^-$ cultures, can be detected after 30 min of addition of the AI. Not only AMP activity, but also AI activity appear in the culture supernatant 2 h after addition of the AI, indicating that the peptide also induces its own production (Eijsink et al. 1996). A similar regulatory strategy is expected in *L. sakei* Lb674, which contains a gene cluster for sakacin P production identical to that of *L. sakei* LTH673 (see below).

Clues to the mechanism controlling transcription of the operons required for sakacin P production and immunity were provided by sequence, mutational and transcription analysis of the 8.9-kb locus that contains the gene cluster involved production of and in immunity to sakacin P in *L. sakei* LTH673 and *L. sakei* Lb674 (Figure 1E) (Holck et al. 1994; Tichaczek et al. 1994; Huhne et al. 1996; Brurberg et al. 1997). The gene cluster includes three main operons. One operon contains the gene *sppIP*, encoding the precursors (with a double-glycine-type leader) of the 19-residue AI mentioned above, and *sppK* and *-R*, encoding an HPK and an RR, respectively, of a predicted two-component signal-transduction system. The second operon encompasses *sppA* and *spiA*, encoding sakacin P precursor and sakacin P immunity protein, respectively. The third operon contains *sppT* and *-E*, encoding a predicted ATP-dependent transporter and an accessory protein, respectively, of a *sec*-independent export/processing system postulated to be involved in processing and secretion of sakacin P and the AI. Deletion and frameshift mutation analysis strongly suggests that *sppK*, *-T* and *-E* are essential for sakacin P production (Huhne et al. 1996). The cluster contains two additional open reading frames, designated *orfX* and *-Y*. The predicted product of *orfX* resembles a class II AMP-like precursor, whereas the product of *orfY* has similarity to class II AMP immunity proteins.

Transcription analysis has shown that addition of the AI to AMP$^-$ cultures triggers the appearance of several transcripts, including one encompassing *sppA* and *spiA*, and a second containing *sppIP*, *-K* and *-R* (Brurberg et al. 1997). The start sites of the AI-triggered transcripts initiated directly upstream of *sppIP*, *-A* or *-T* have been determined by primer extension. The mapped promoters have similar direct repeat sequences in the -40 and -80 regions, suggesting that the promoters are under the same regulatory control. These repeats are similar to those noticed in the inducible promoters of the plantaricin A system of *L. plantarum* C11 (Diep et al. 1996). When plasmids containing the promoters of *sppIP* or *-A* fused to the promoterless chloramphenicol acetyltransferase (CAT) gene are introduced in *L. plantarum* C11, CAT activity can be induced with the AI of the plantaricin A system, plantaricin A (Brurberg et al. 1997). Further analysis of *spp* promoter-CAT fusions in *L. plantarum C11* revealed that the *sppA* promoter is tightly controlled, and with a dynamic range of more than 1000-fold upon activation. Constitutive basal expression for the transcripts *sppIPKR* and *sppTE* from promoters upstream of the conserved direct repeats was also demonstrated (Risoen et al. 2000). During

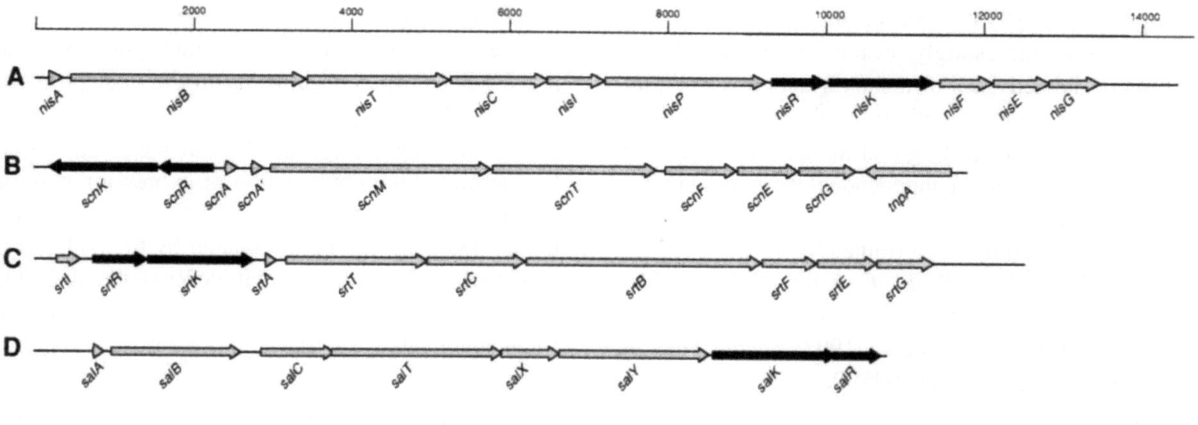

Figure 2. Gene clusters involved in regulated production of, and immunity to, various class I AMPs in LAB. (A) Nisin gene cluster of *Lactococcus lactis*; (B) streptococcin A-FF22 gene cluster of *Streptococcus pyogenes* FF22; (C) gene cluster of an uncharacterized AMP of *Streptococcus pyogenes* BL-T; and (D) salivaricin gene cluster of *Streptococcus salivarius* 20P3. The regulatory genes are highlighted in black. The scale is given in base pairs.

the same study, electrophoretic mobility shift assays showed that the RR SppR of*L. sakei* LTH673, as well as its homologs PlnC and PlnD of *L. plantarum* C11, bind to the conserved repeats of the *sppIP* and *-A* promoters. It has been proposed that promoter activation is mediated by binding of SppR to the repeats (Risoen et al. 2000).

Studies on regulation of class I AMP production

Class I AMPs (often referred to as lantibiotics) are small post-translationally modified peptides produced by a large variety of Gram-positive microorganisms, including several LAB (van Kraaij et al. 1999; Guder et al. 2000; Jack et al. 2000; Sablon et al. 2000; McAuliffe et al. 2001). The precursors of the class I AMPs are often extensively modified to include uncommon residues (such as dehydroalanine and dehydrobutyrine) and intramolecular lanthionine and ß-methyl-lanthionine bridges. As observed for the production of many AMPs of class II, production of several class I AMPs has been shown to be regulated by an AI-mediated quorum-sensing strategy (Kleerebezem & Quadri 2001). Although AI-regulated production of class I AMPs is well documented in *Lactococcus lactis*, production of this class of AMPs by several other bacteria, including species of the genera *Bacillus*, *Streptococcus* and *Staphylococcus,* is believed to be regulated as well. The studies on regulation of class I AMP production by LAB are summarized in the following sections.

Class I AMP production in Lactococcus lactis

The most extensively studied class I AMP is nisin, produced by strains of *L. lactis*. Two natural nisin variants have been characterized (called nisin A and nisin Z), which differ in a single amino acid (Buchman et al. 1988; Mulders et al. 1991). The genes *nisA* and *nisZ,* encoding the precursors of nisin A (Buchman et al. 1988) and nisin Z (Mulders et al. 1991), respectively, have been characterized. The nisin gene encodes a precursor peptide of 57 residues which undergoes proteolytic cleavage of a leader peptide of 23 residues (de Vos et al. 1995). Two mechanisms of antimicrobial activity have been attributed to nisin. One mechanism is the inhibition of peptidoglycan biosynthesis, thought to take place via binding to lipid II, whereas the other is based on dissipation of membrane potential through the formation of pores in the cell membrane (Breukink & de Kruijff 1999; Breukink et al. 1999; Wiedemann et al. 2001). In addition to its antimicrobial activity, nisin has been shown to have AI function and stimulate its own synthesis. Transcription analysis of an AMP$^-$ mutant strain bearing a deletion of 4 base pairs in the nisin precursor gene *nisA* (Δ*nisA* allele) demonstrated that Δ*nisA* transcription was abrogated in the strain (Kuipers et al. 1993). Notably, addition of trace amounts of nisin to the culture was sufficient to induce transcription of Δ*nisA* (Kuipers et al. 1995). Further analysis of the AI activity of various non-natural occurring nisin variants and nisin fragments indicated that the N-terminal and C-terminal residues were relevant for AI function; however, the N-terminus of the peptide appeared more important

(Kuipers et al. 1995; Dodd et al. 1996; Van Kraaij et al. 1997). Interestingly, evaluation of the AI and antimicrobial activities of nisin variants revealed that variants with poor AI activity retained good antimicrobial activity, suggesting that the antimicrobial and AI activities of nisin are independent from one another (Kuipers et al. 1995).

Nisin production is determined by the gene cluster *nisA, -B, -T, -C, -I, -P, -R, -K, -F, -E* and *-G* (Figure 2A) (Buchman et al. 1988; Kaletta & Entian 1989; Engelke et al. 1992, 1994; Kuipers et al. 1993; van der Meer et al. 1993; Siegers & Entian 1995; Ra et al. 1999). In addition to *nisA*, the nisin precursor structural gene, the cluster encompasses genes encoding proteins (thought to be) involved in: (i) post-translational modification of the nisin precursor (*nisB* and *nisC*; van der Meer et al. 1994; Siezen et al. 1996; Ra et al. 1999), (ii) secretion of the modified pre-nisin (*nisT*; Engelke et al. 1992; Kuipers et al. 1993; Siegers et al. 1996), (iii) proteolytic cleavage of the leader peptide (*nisP*; van der Meer et al. 1993, 1994; Siezen et al. 1995), (iv) immunity to nisin (*nisI* and *nisF*, *-E* and *-G*; Kuipers et al. 1993; Siegers & Entian 1995); Ra et al. 1999, and (v) regulation of nisin production (*nisK* and *-R*; Kuipers et al. 1993, 1995; van der Meer et al. 1993; Engelke et al. 1994; de Ruyter et al. 1996a,b; Kleerebezem et al. 1997b, 1999). The genes *nisK* and *-R* encode an HPK and an RR, respectively, of a predicted two-component signal-transduction system required for nisin production. Addition of trace amounts of nisin to the culture of Δ*nisA* mutants with additional gene disruptions in *nisR* and/or *nisK* fails to induce Δ*nisA* transcription. It has been suggested that NisK is the sensor of nisin (Kuipers et al. 1995; Kleerebezem et al. 1999).

Three promoters have been identified within the nisin gene cluster and their activities have been studied using transcriptional fusions to *gusA* (Kuipers et al. 1995; de Ruyter et al. 1996b). Transcription of *gusA* from the *nisA* and *nisF* promoters is nisin-induced and *nisK* and *-R* dependent. The activity of these promoters directly correlates to the concentration of added nisin (de Ruyter et al. 1996a,b). The third promoter (upstream of *nisR*) directs nisin-independent *gusA* expression, indicating that the promoter constitutively drives the expression of the *nisRK* operon (de Ruyter et al. 1996b). A direct repeat (PDR, 5 nucleotides) separated by 6 nucleotides (in place of a canonical -35 region) has been noted in the *nisA* and *nisF* promoters. Mutation analysis showed that the integrity of the PDR in the *nisA* promoter is required for nisin-induced transcription (Kleerebezem et al. 1999). The PDR sequences are the proposed binding position for the predicted RR NisR. Preliminary data indicate that NisR can bind specifically to the *nisA* promoter *in vitro* (Kleerebezem et al. 1999). The deconvolution of the regulatory elements in the nisin system has permitted the construction of expression systems for Gram-positive bacteria in which the expression of the cloned gene can be modulated by the addition of nisin to the culture supernatant (de Ruyter et al. 1996a, 1997; Kuipers et al. 1997; Kleerebezem et al. 1997b; Eichenbaum et al. 1998).

Class I AMP production in other LAB

Streptococcus pyogenes FF22 produces the class I AMP streptococcin A-FF22. The locus involved in the production of this AMP has been recently identified (McLaughlin et al. 1999) (Figure 2B). In addition to the AMP precursor structural gene (*scnA*; Hynes et al. 1993), the cluster encodes genes predicted to be involved in pre-AMP modification (*scnM*), pre-AMP transport and processing (*scnT*), AMP immunity (*scn, -F, -E* and *-G*), and regulation of AMP production (*scnK* and *scnR*). The genes *scnK* and *scnR*, encoding an HPK and an RR of a predicted two-component regulatory system, have been shown to be essential for streptococcin A-FF22 production (McLaughlin et al. 1999). The cluster also contains a small gene named *scnA′*, which is located downstream of the AMP structural gene and resembles the gene encoding the class I AMP salivaricin (see below; McLaughlin et al. 1999). Based on the similarities with the nisin system, it is likely that streptococcin A-FF22 production is regulated and that streptococcin A-FF22 and/or the predicted product of *scnA′* have AI function. This speculation remains to be validated experimentally.

Two other loci involved in class I AMP production by streptococci have been sequenced. Analysis of the loci suggests that AI-mediated quorum sensing regulates production of the AMPs. One locus contains the genes for production of salivaricin in *S. salivarius* strain 20P3 (Figure 2D) (Genbank accession nr: AY005472). This locus contains the previously identified salivaricin structural gene (*salA*; Ross et al. 1993) and genes encoding proteins predicted to be involved in the modification of AMP precursor (*salB* and *salC*), precursor transport and processing (*salT* and *salX*), and regulation of production via a two-component regulatory system (*salK* and *salR*). The second locus is from *S. pyogenes* strain BL-T (Genbank accession number AB030831; Fig-

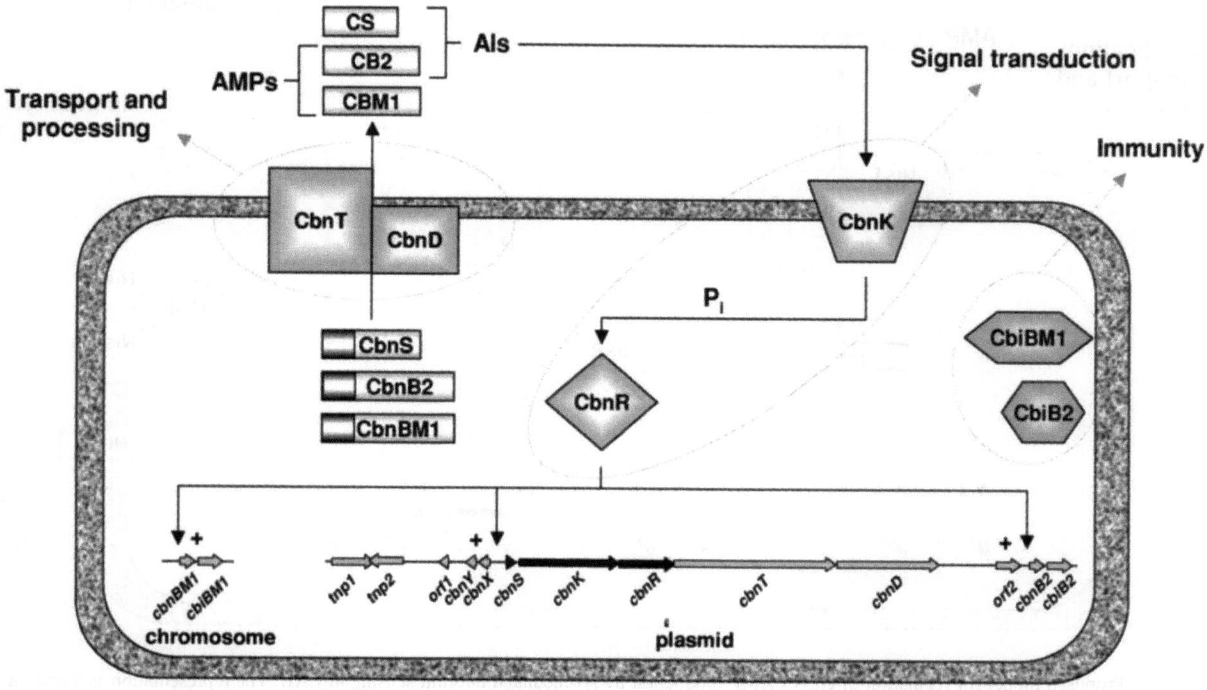

Figure 3. Proposed model for regulation of class II AMP production by AI-mediated quorum sensing in LAB. The representation is based on the production of carnobacteriocins by *Carnobacterium piscicola* LV17B. In this bacterium both a dedicated AI peptide, part of the three-component regulatory system, and the AMP carnobacteriocin B2 act as AIs. In other systems, only the dedicated AI of the three-component regulatory system has been shown to induce AMP production. The genes forming the three-component regulatory system are highlighted in black. For gene nomenclature and further information, see text.

ure 2C). The locus contains genes thought to encode the AMP precursor peptide (*srtA*), precursor modification enzymes (*srtB* and *srtC*), AMP export (*srtT*), AMP immunity (*srtI*, *srtFEG*), and regulation of AMP production via a two-component signal transduction system (*srtR* and *srtK*). The AMP produced by *S. pyogenes* strain BL-T remains uncharacterized. Overall, the sequence analysis of the two streptococci gene clusters suggests an AI-mediated regulation of AMP production similar to that present in the nisin system. The existence of such regulatory strategy awaits experimental demonstration.

Emerging models for regulation of AMP production via AI-mediated quorum sensing

From the analysis of the systems in which production of class I and class II AMPs is regulated in a cell-density dependent manner, two respective working models for AI-mediated quorum-sensing regulation have emerged (Figures 3 and 4). In class I AMP production systems, the AIs are the AMPs themselves, whereas class II AMP production systems have dedicated AIs as part of a 'three-component regulatory system' (see below). These dedicated AIs are unmodified small cationic peptides with or without antimicrobial activity and α-helix propensity. The dedicated AIs are synthesized as class II AMP-like precursors with double-glycine-type leader peptides, which are cleaved off upon export by the same *sec*-independent machinery involved in possessing and secretion of class II AMPs. Notably, production of the class II AMPs called carnobacteriocins is not only controlled by dedicated AIs, but also by the carnobacteriocins themselves. The dual (AI and antimicrobial) activity of the carnobacteriocins and of the dedicated AI plantaricin A raises interesting questions as to the evolutionary relatedness of dedicated AIs and AMPs.

The loci involved in regulated production of class I or class II AMPs encode two-component signal transduction systems composed of an HPK and an RR required for regulation. The regulatory protein pairs in class I and class II AMP systems have similarity to the EnvZ–OmR and AgrC–AgrD signal transduction systems, respectively. Notably, in class II AMP

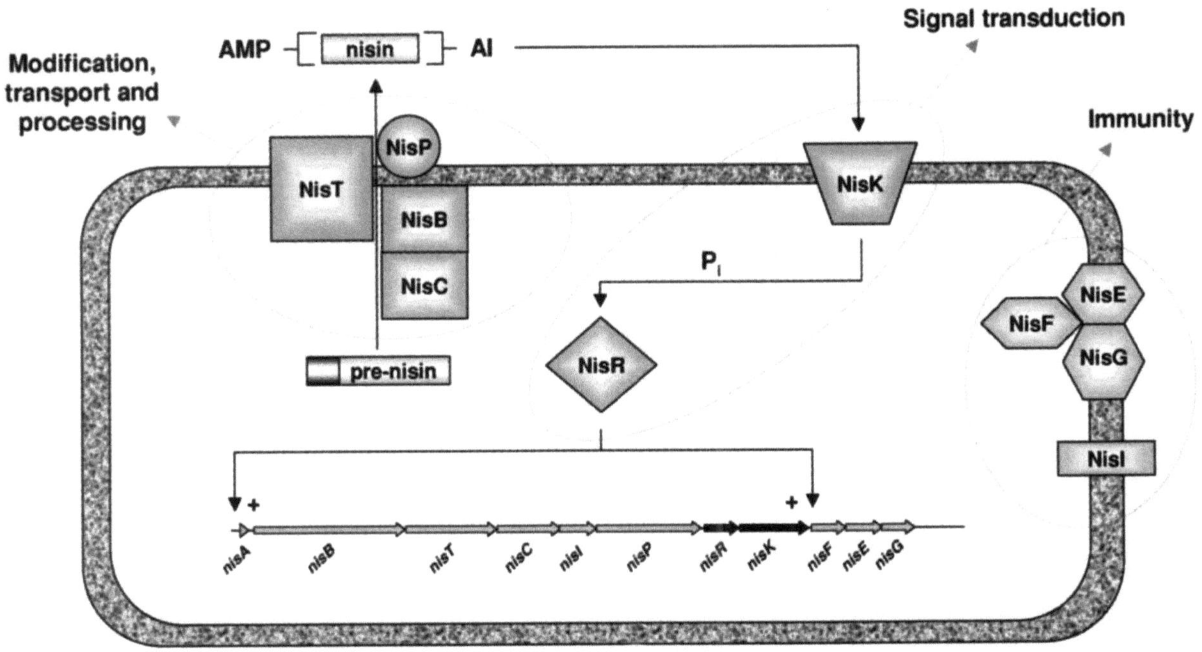

Figure 4. Proposed model for regulation of class I AMP production by AI-mediated quorum sensing in LAB. The representation is based on the production of nisin by *Lactococcus lactis*. The regulatory genes are highlighted in black. For gene nomenclature and further information, see text.

production loci the genes of the dedicated AIs are followed by the regulatory genes. The AI-HPK-RR regulatory triad is frequently referred to as a 'three component regulatory system' (Figure 3). It is believed that the HPKs are the sensors of the peptides (AMPs or dedicated AIs) with AI activity. However, the multiplicity of AIs acting via the same sensor HPK in the carnobacteriocin B system raises questions about the nature of the stimuli sensed by and the specificity of sensor HPKs. In this regard, direct contact between HPKs and AIs awaits demonstration. Nevertheless, the HPKs are believed to modulate the activity of cognate RRs via phosphorylation in an AI-dependent manner. The activated (phosphorylated) RRs in turn control transcription of: (i) AI precursor and regulatory genes, and (ii) all the genes required for AMP production and immunity through binding to conserved repeat sequences in the regulated promoters of the AMP gene clusters (Figures 3 and 4).

Concluding remarks and future directions

A clear view of only some aspects of the of regulatory mechanisms controlling AMP production via AI-mediated quorum-sensing has emerged in recent years, and the information has been incorporated in the current models for class I and class II AMP production. However, many aspects such as AI-sensor recognition and control of the AI autoinduction loop require further analysis. Furthermore, by analogy with other cell-density dependent regulatory models, AIs in AMP production systems should be constitutively produced at low levels to permit accumulation until a threshold concentration required for triggering AMP production is reached. However, some of the systems described here do not appear to fulfil this requirement. It is possible that additional yet unidentified environmental factors play a role in modulating the production of AIs or the sensitivity of the cells to the AIs. Interestingly, the production of sakacin A appears to depend on the growth temperature and the presence of the AI. Finally, the characterization of other systems where cell-density dependent regulation of AMP production is suspected and comparison with peptide-AI regulated systems involved in other behavioural responses will be required to develop a comprehensive picture of the AI-mediated quorum-sensing phenomena in LAB and other Gram-positive bacteria.

Acknowledgements

We thank the Niarchos Foundation for financial support to L.E.N. Quadri, Stavros S. Niarchos Scholar.

References

Anderssen EL, Diep DB, Nes IF, Eijsink VGH & Nissen-Meyer J (1998) Antagonistic activity of *Lactobacillus plantarum* C11: Two new two-peptide bacteriocins, plantaricins EF and JK, and the induction factor plantaricin A. Appl. Environ. Microbiol. 64: 2269–2272.

Axelsson L & Holck A (1995) The genes involved in production of and immunity to sakacin A, a bacteriocin from *Lactobacillus sake* Lb706. J. Bacteriol. 177: 2125–2137.

Axelsson L, Holck A, Birkeland SE, Aukrust T & Blom H (1993) Cloning and nucleotide sequence of a gene from *Lactobacillus sake* Lb706 necessary for sakacin A production and immunity. Appl. Environ. Microbiol. 59: 2868–2875.

Breukink E & de Kruijff B (1999) The lantibiotic nisin, a special case or not? Biochim. Biophys. Acta 1462: 223–234.

Breukink E, Wiedemann I, van Kraaij C, Kuipers OP, Sahl H & de Kruijff B (1999) Use of the cell wall precursor lipid II by a pore-forming peptide antibiotic. Science 286: 2361–2364.

Brurberg MB, Nes IF & Eijsink VG (1997) Pheromone-induced production of antimicrobial peptides in *Lactobacillus*. Mol. Microbiol. 26: 347–360.

Buchman GW, Banerjee S & Hansen JN (1988) Structure, expression, and evolution of a gene encoding the precursor of nisin, a small protein antibiotic. J. Biol. Chem. 263: 16260–16266.

de Ruyter PG, Kuipers OP & de Vos WM (1996a) Controlled gene expression systems for *Lactococcus lactis* with the food-grade inducer nisin. Appl. Environ. Microbiol. 62: 3662–3667.

de Ruyter PG, Kuipers OP, Beerthuyzen MM, van Alen-Boerrigter I & de Vos WM (1996b) Functional analysis of promoters in the nisin gene cluster of *Lactococcus lactis*. J. Bacteriol. 178: 3434–3439.

de Ruyter PG, Kuipers OP, Meijer WC & de Vos WM (1997) Food-grade controlled lysis of *Lactococcus lactis* for accelerated cheese ripening. Nat. Biotechnol. 15: 976–979.

de Vos WM, Kuipers OP, van der Meer JR & Siezen RJ (1995) Maturation pathway of nisin and other lantibiotics: Post-translationally modified antimicrobial peptides exported by Gram-positive bacteria. Mol. Microbiol. 17: 427–437.

de Vos WM, Kleerebezem M & Kuipers OP (1997) Expression systems for industrial Gram-positive bacteria with low guanine and cytosine content. Curr. Opin. Biotechnol. 8: 547–553.

Diep DB, Havarstein LS, Nissen-Meyer J & Nes IF (1994) The gene encoding plantaricin A, a bacteriocin from *Lactobacillus plantarum* C11, is located on the same transcription unit as an *agr*-like regulatory system. Appl. Environ. Microbiol. 60: 160–166.

Diep DB, Havarstein LS & Nes IF (1995) A bacteriocin-like peptide induces bacteriocin synthesis in *Lactobacillus plantarum* C11. Mol. Microbiol. 18: 631–639.

Diep DB, Havarstein LS & Nes IF (1996) Characterization of the locus responsible for the bacteriocin production in *Lactobacillus plantarum* C11. J. Bacteriol.178: 4472–4483.

Diep DB, Axelsson L, Grefsli C & Nes IF (2000) The synthesis of the bacteriocin sakacin A is a temperature-sensitive process regulated by a pheromone peptide through a three-component regulatory system. Microbiology 146: 2155–2160.

Dodd HM, Horn N, Chan WC, Giffard CJ, Bycroft BW, Roberts GC & Gasson MJ (1996) Molecular analysis of the regulation of nisin immunity. Microbiology 142: 2385–2392.

Dunny GM & Leonard BA (1997) Cell-cell communication in Gram-positive bacteria. Annu. Rev. Microbiol. 51: 527–564.

Dunny GM & Winans SC (Eds) (1999) Cell-Cell Signaling in Bacteria. ASM Press, Washington, D.C.

Eichenbaum Z, Federle MJ, Marra D, de Vos WM, Kuipers OP, Kleerebezem M & Scott JR (1998) Use of the lactococcal *nisA* promoter to regulate gene expression in Gram-positive bacteria: Comparison of induction level and promoter strength. Appl. Environ. Microbiol. 64: 2763–2769.

Eijsink VG, Brurberg MB, Middelhoven PH & Nes IF (1996) Induction of bacteriocin production in *Lactobacillus sake* by a secreted peptide. J. Bacteriol.178: 2232–2237.

Engelke G, Gutowski-Eckel Z, Hammelmann M & Entian KD (1992) Biosynthesis of the lantibiotic nisin: Genomic organization and membrane localization of the *nisB* protein. Appl. Environ. Microbiol. 58: 3730–3743.

Engelke G, Gutowski-Eckel Z, Kiesau P, Siegers K, Hammelmann M & Entian KD (1994) Regulation of nisin biosynthesis and immunity in *Lactococcus lactis* 6F3. Appl. Environ. Microbiol. 60: 814–825.

Ennahar S, Sashihara T, Sonomoto K & Ishizaki A (2000) Class IIa bacteriocins: Biosynthesis, structure and activity. FEMS Microbiol. Rev. 24: 85–106.

Franz CM, Worobo RW, Quadri LE, Schillinger U, Holzapfel WH, Vederas JC & Stiles ME (1999) Atypical genetic locus associated with constitutive production of enterocin B by *Enterococcus faecium* BFE900. Appl. Environ. Microbiol. 65: 2170–2178.

Franz CM, Stiles ME & van Belkum MJ (2000a) Simple method to identify bacteriocin induction peptides and to auto-induce bacteriocin production at low cell density. FEMS Microbiol. Lett. 186: 181–185.

Franz CM, van Belkum MJ, Worobo RW, Vederas JC & Stiles ME (2000b) Characterization of the genetic locus responsible for production and immunity of carnobacteriocin A: The immunity gene confers cross-protection to enterocin B. Microbiology 146: 621–631.

Fuqua C & Greenberg EP (1998) Self perception in bacteria: Quorum sensing with acylated homoserine lactones. Curr. Opin. Microbiol. 1: 183–189.

Fuqua C, Winans SC & Greenberg EP (1996) Census and consensus in bacterial ecosystems: The *luxR-luxI* family of quorum-sensing transcriptional regulators. Annu. Rev. Microbiol. 50: 727–751.

Fuqua C, Parsek MR & Greenberg EP (2001) Regulation of gene expression by cell-to-cell communication: acyl-homoserine lactone quorum sensing. Annu. Rev. Genet. 35: 439–468.

Guder A, Wiedemann I & Sahl HG (2000) Posttranslationally modified bacteriocins – the lantibiotics. Biopolymers. 55: 62–73.

Hauge HH, Mantzilas D, Moll GN, Konings WN, Driessen AJ, Eijsink VG & Nissen-Meyer J (1998) Plantaricin A is an amphiphilic alpha-helical bacteriocin-like pheromone which exerts antimicrobial and pheromone activities through different mechanisms. Biochemistry 37: 16026–16032.

Havarstein LS, Diep DB & Nes IF (1995) A family of bacteriocin ABC transporters carry out proteolytic processing of their substrates concomitant with export. Mol. Microbiol. 16: 229–240.

Holck A, Axelsson L, Birkeland SE, Aukrust T & Blom H (1992) Purification and amino acid sequence of sakacin a, a bacteriocin from *Lactobacillus sake* Lb706. J. Gen. Microbiol. 138: 2715–2720.

Holck AL, Axelsson L, Huhne K & Krockel L (1994) Purification and cloning of sakacin 674, a bacteriocin from *Lactobacillus sake* Lb674. FEMS Microbiol. Lett. 115: 143–149.

Huhne K, Axelsson L, Holck A & Krockel L (1996) Analysis of the sakacin P gene cluster from *Lactobacillus sake* Lb674 and its expression in sakacin-negative *Lb. sake* strains. Microbiology 142: 1437–1448.

Hynes WL, Ferretti JJ & Tagg JR (1993) Cloning of the gene encoding streptococcin A-FF22, a novel lantibiotic produced by *Streptococcus pyogenes*, and determination of its nucleotide sequence. Appl. Environ. Microbiol. 59: 1969–1971.

Jack RW & Jung G (2000) Lantibiotics and microcins: polypeptides with unusual chemical diversity. Curr. Opin. Chem. Biol. 4: 310–317.

Kaletta C & Entian KD (1989) Nisin, a peptide antibiotic: cloning and sequencing of the nisa gene and posttranslational processing of its peptide product. J. Bacteriol. 171: 1597–1601.

Kleerebezem M, Quadri LE, Kuipers OP & de Vos WM (1997a) Quorum sensing by peptide pheromones and two-component signal-transduction systems in Gram-positive bacteria. Mol. Microbiol. 24: 895–904.

Kleerebezem M, Beerthuyzen MM, Vaughan EE, de Vos WM & Kuipers OP (1997b) Controlled gene expression systems for lactic acid bacteria: transferable nisin-inducible expression cassettes for *Lactococcus*, *Leuconostoc*, and *Lactobacillus* spp. Appl. Environ. Microbiol. 63: 4581–4584.

Kleerebezem M, de Vos WM & Kuipers OP (1999) The lantibiotics nisin and subtilin act as extracellular regulators of their own biosynthesis. In: Dunny GM &Winans SC (eds) Cell-Cell Signaling in Bacteria. (pp 159–174). ASM Press, Washington DC.

Kleerebezem M & Quadri LE (2001) Peptide pheromone-dependent regulation of antimicrobial peptide production in Gram-positive bacteria: A case of multicellular behavior. Peptides 22: 1579–1596.

Kleerebezem M, Kuipers OP, de Vos WM, Stiles ME & Quadri LE (2001b) A two-component signal-transduction cascade in Carnobacterium piscicola LV17B: Two signaling peptides and one sensor-transmitter. Peptides 22: 1597–1601.

Kuipers OP, Beerthuyzen MM, Siezen RJ & De Vos WM (1993) Characterization of the nisin gene cluster *nisABTCIPR* of *Lactococcus lactis*. Requirement of expression of the *nisA* and *nisI* genes for development of immunity. Eur. J. Biochem. 216: 281–291.

Kuipers OP, Beerthuyzen MM, de Ruyter PG, Luesink EJ & de Vos WM (1995) Autoregulation of nisin biosynthesis in *Lactococcus lactis* by signal transduction. J. Biol. Chem. 270: 27299–27304.

Kuipers OP, de Ruyter PG, Kleerebezem M & de Vos WM (1997) Controlled overproduction of proteins by lactic acid bacteria. Trends Biotechnol. 15: 135–140.

McAuliffe O, Ross RP & Hill C (2001) Lantibiotics: Structure, biosynthesis and mode of action. FEMS Microbiol. Rev. 25: 285–308.

McLaughlin RE, Ferretti JJ & Hynes WL (1999) Nucleotide sequence of the streptococcin A-FF22 lantibiotic regulon: model for production of the lantibiotic SA-FF22 by strains of *Streptococcus pyogenes*. FEMS Microbiol. Lett. 175: 171–177.

Mulders JW, Boerrigter IJ, Rollema HS, Siezen RJ & de Vos WM (1991) Identification and characterization of the lantibiotic nisin Z, a natural nisin variant. Eur. J. Biochem. 201: 581–584.

Nes IF, Diep DB, Havarstein LS, Brurberg MB, Eijsink V & Holo H (1996) Biosynthesis of bacteriocins in lactic acid bacteria. Antonie Van Leeuwenhoek. 70: 113–128.

Nissen-Meyer J, Larsen AG, Sletten K, Daeschel M & Nes IF (1993) Purification and characterization of plantaricin A, a *Lactobacillus plantarum* bacteriocin whose activity depends on the action of two peptides. J. Gen. Microbiol. 139: 1973–1978.

Quadri LE, Sailer M, Roy KL, Vederas JC & Stiles ME (1994) Chemical and genetic characterization of bacteriocins produced by *Carnobacterium piscicola* LV17B. J. Biol. Chem. 269: 12204–12211.

Quadri LE, Sailer M, Terebiznik MR, Roy KL, Vederas JC & Stiles ME (1995) Characterization of the protein conferring immunity to the antimicrobial peptide carnobacteriocin B2 and expression of carnobacteriocins B2 and BM1. J. Bacteriol. 177: 1144–1151.

Quadri LE, Yan LZ, Stiles ME & Vederas JC (1997a) Effect of amino acid substitutions on the activity of carnobacteriocin B2. Overproduction of the antimicrobial peptide, its engineered variants, and its precursor in *Escherichia coli*. J. Biol. Chem. 272: 3384–3388.

Quadri LE, Kleerebezem M, Kuipers OP, de Vos WM, Roy KL, Vederas JC & Stiles ME (1997b) Characterization of a locus from *Carnobacterium piscicola* LV17B involved in bacteriocin production and immunity: Evidence for global inducer-mediated transcriptional regulation. J. Bacteriol. 179: 6163–6171.

Ra R, Beerthuyzen MM, de Vos WM, Saris PE & Kuipers OP (1999) Effects of gene disruptions in the nisin gene cluster of *Lactococcus lactis* on nisin production and producer immunity. Microbiology 145: 1227–1233.

Risoen PA, Havarstein LS, Diep DB & Nes IF (1998) Identification of the DNA-binding sites for two response regulators involved in control of bacteriocin synthesis in *Lactobacillus plantarum* C11. Mol. Gen. Genet. 259: 224–232.

Risoen PA, Brurberg MB, Eijsink VG & Nes IF (2000) Functional analysis of promoters involved in quorum sensing-based regulation of bacteriocin production in *Lactobacillus*. Mol. Microbiol. 37: 619–628.

Rohde BH, Eliezer D & Quadri LEN (2001) Unpublished results.

Ross KF, Ronson CW & Tagg JR (1993) Isolation and characterization of the lantibiotic salivaricin a and its structural gene *salA* from *Streptococcus salivarius* 20P3. Appl. Environ. Microbiol. 59: 2014–2021.

Sablon E, Contreras B & Vandamme E (2000) Antimicrobial peptides of lactic acid bacteria: mode of action, genetics and biosynthesis. Adv. Biochem. Eng. Biotechnol. 68: 21–60.

Saucier L, Poon A & Stiles ME (1995) Induction of bacteriocin in *Carnobacterium piscicola* LV17. J. Appl. Bacteriol. 78: 684–690.

Saucier L, Paradkar AS, Frost LS, Jensen SE & Stiles ME (1997) Transcriptional analysis and regulation of carnobacteriocin production in *Carnobacterium piscicola* LV17. Gene 188: 271–277.

Siegers K & Entian KD (1995) Genes involved in immunity to the lantibiotic nisin produced by *Lactococcus lactis* 6F3. Appl. Environ. Microbiol. 61: 1082–1089.

Siegers K, Heinzmann S & Entian KD (1996) Biosynthesis of lantibiotic nisin. Posttranslational modification of its prepeptide occurs at a multimeric membrane-associated lanthionine synthetase complex. J. Biol. Chem. 271: 12294–12301.

Siezen RJ, Rollema HS, Kuipers OP & de Vos WM (1995) Homology modelling of the *Lactococcus lactis* leader peptidase NisP and its interaction with the precursor of the lantibiotic nisin. Protein. Eng. 8: 117–125.

Siezen RJ, Kuipers OP & de Vos WM (1996) Comparison of lantibiotic gene clusters and encoded proteins. Antonie Van Leeuwenhoek. 69: 171–184.

Tichaczek PS, Vogel RF & Hammes WP (1994) Cloning and sequencing of *sakP* encoding sakacin P, the bacteriocin produced by *Lactobacillus sake* LTH673. Microbiology 140: 361–367.

van der Meer JR, Polman J, Beerthuyzen MM, Siezen RJ, Kuipers OP & De Vos WM (1993) Characterization of the *Lactococcus lactis* nisin a operon genes *nisP*, encoding a subtilisin-like serine protease involved in precursor processing, and *nisR*, encoding a regulatory protein involved in nisin biosynthesis. J. Bacteriol. 175: 2578–2588.

van der Meer JR, Rollema HS, Siezen RJ, Beerthuyzen MM, Kuipers OP & de Vos WM (1994) Influence of amino acid substitutions in the nisin leader peptide on biosynthesis and secretion of nisin by *Lactococcus lactis*. J. Biol. Chem. 269: 3555–3562.

Van Kraaij C, Breukink E, Rollema HS, Siezen RJ, Demel RA, De Kruijff B & Kuipers OP (1997) Influence of charge differences in the C-terminal part of nisin on antimicrobial activity and signaling capacity. Eur. J. Biochem. 247: 114–120.

van Kraaij C, de Vos WM, Siezen RJ & Kuipers OP (1999) Lantibiotics: Biosynthesis, mode of action and applications. Nat. Prod. Rep. 16: 575–587.

Wiedemann I, Breukink E, van Kraaij C, Kuipers OP, Bierbaum G, de Kruijff B & Sahl HG (2001) Specific binding of nisin to the peptidoglycan precursor lipid II combines pore formation and inhibition of cell wall biosynthesis for potent antibiotic activity. J. Biol. Chem. 276: 1772–1779.

Worobo RW, Henkel T, Sailer M, Roy KL, Vederas JC & Stiles ME (1994) Characteristics and genetic determinant of a hydrophobic peptide bacteriocin, carnobacteriocin a, produced by *Carnobacterium piscicola* LV17A. Microbiology 140: 517–526.

Antonie van Leeuwenhoek **82:** 147–164, 2002.

Transporters and their roles in LAB cell physiology

Bert Poolman
Department of Biochemistry, Groningen Biomolecular Sciences and Biotechnology Institute, University of Groningen, Nijenborgh 4, 9747 AG Groningen, The Netherlands (e-mail: B.Poolman@chem.rug.nl)

Key words: phosphotransferase system, metabolic capacity, transport system

Abstract

For most metabolic pathways, the uptake of the substrate into the cell represents the first step. This transport reaction can exert a large control on the flux through the pathway, in particular when the substrate concentration becomes limiting. Besides serving a role in the uptake of nutrients and the excretion of metabolic (end)products or drugs, transport systems can have one or more other functions in the physiology of the cell. Two of these functions, control of carbohydrate utilization and regulation of cell volume, have been well established in lactic acid bacteria (LAB). The first example concerns the phospho*enol*pyruvate-dependent phosphotransferase system (PTS), which serves a role in the transport of sugars into the cell but also regulates the activity of metabolic pathways, either through regulation of transcription and/or (in)activation of transporters and key enzymes already present. The regulation by the PTS results in a hierarchy in the utilization of sugars and/or adjustment of the first step(s) of a metabolic pathway to the metabolic capacity of the cell and the availability of a particular substrate. The second example relates to the activation of transporters (and mechanosensitive channels), which represents the first mechanism of defence against osmotic stress. The activation by osmotic-upshift of the ATP-binding Cassette (ABC) transporter OpuA from *Lactococcus lactis* is compared with the activation by osmotic-downshift of mechanosensitive channels. The mechanosensitive channels have been best studied in organisms other than LAB, but the presence of similar systems in LAB, and their conservation of structure, suggest that the postulated functions and mechanisms generally hold.

Introduction

The first step in the metabolism of almost any substrate is the transport of the molecule into the cell. In bacteria substrates are taken up by *primary* or *secondary transport* systems or group translocation systems. Primary transporters are driven by ATP, whereas secondary transporters utilize the free energy difference stored in the electrochemical gradient(s) of the translocated solute(s) across the membrane (Poolman & Konings 1993). The by far most abundant class of primary transport systems in lactic acid bacteria (LAB) is that of the ATP-binding cassette transporters, and this type of mechanism is used to accumulate substrates and compatible solutes but also to excrete unwanted products (xenobiotics, drugs) (Figure 1A). Among the secondary transport systems one can distinguish *symporters* (cotransport of two or more solutes), *uniporters* (transport of one molecule) and *antiporters* (countertransport of two or more solutes) (Figure. 1B–D). Symporters usually couple the uphill movement of the substrate to the downhill movement of a proton (or sodium ion), i.e., the electrochemical proton (or sodium ion) gradient drives the accumulation of substrate. Antiporters use the electrochemical ion gradient to excrete a (end)-product, whereas uniporters do not use a coupling ion. Substrate transport by group translocation is restricted to carbohydrates and alditols and involves phospho*enol*pyruvate-dependent phosphotransferase systems (PTSs) (Postma et al. 1993). The PTS catalyzes the uptake of carbohydrate or alditol concomitant with its phosphorylation (Fig. 2).

The phosphoryl group is transferred from phospho*enol*pyruvate (PEP) via the general energy coupling proteins Enzyme I and HPr, and the substrate-

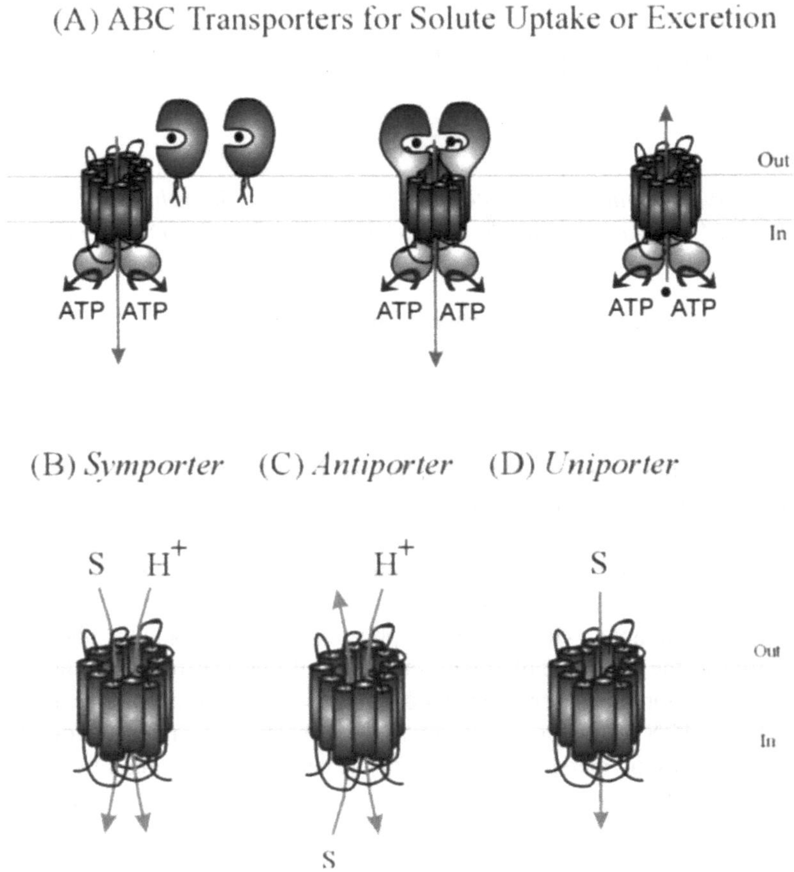

Figure 1. Transport mechanisms. (A) ATP-binding cassette (ABC) transporters for uptake and excretion of solutes. The domain organization of the first, second and third system are that of the oligopeptide transporter Opp, the osmoregulated glycine betaine transporter OpuA from *L. lactis*, and the drug efflux system LmrA, respectively. (B–D). Subdivision of secondary transport mechanisms into symporter (B), antiporter (C) and uniporter (D), respectively. S and H^+ refer to solute and proton, respetively. The coupling ion shown is H^+ (proton motive force-driven uptake or efflux) but in other systems this can be Na^+ (sodium motive force-driven).

specific phosphoryl transfer proteins/domains IIA and IIB. IIB~P transfers the phosphoryl group to the sugar or alditol that is translocated via the substrate-specific IIC protein/domain. IIA, IIB and IIC can be separate proteins, domains in a single polypeptide or linked as pairs in any possible combination (Robillard & Lolkema 1988; Saier & Reizer 1992; Lengeler et al. 1994).

In this review, the regulatory roles of transporters in the cell physiology of LABs are discussed, with emphasis on the regulation of carbohydrate utilization (Section 2) and cell volume control (Section 3). Although the review focuses on lactic acid bacteria, key findings originally made in other organisms (e.g., *Bacillus subtilis* or *Escherichia coli*) but generally true for low-GC Gram-positive bacteria, or prokaryotes in general, are described for sake of completeness.

Regulation of carbohydrate utilization

Most bacterial cells have the capacity to utilize several carbohydrates as carbon and energy source and possess various transport proteins and catabolic enzymes for the metabolism of the different carbohydrates. In addition, different mechanisms that control the transport and the first steps of metabolism of a particular carbohydrate have evolved. These mechanisms generally result in sequential uptake and metabolism of the available carbohydrates and/or a tuning of the metabolic rate to the needs of the cell. These two regulatory phenomena, hereafter referred to as *hierarchical control* and *autoregulation*, are universal and have been reported for many bacteria. Hierarchical control of carbohydrates has been explained by (i) inhibition of expression of genes encoding enzymes that are in-

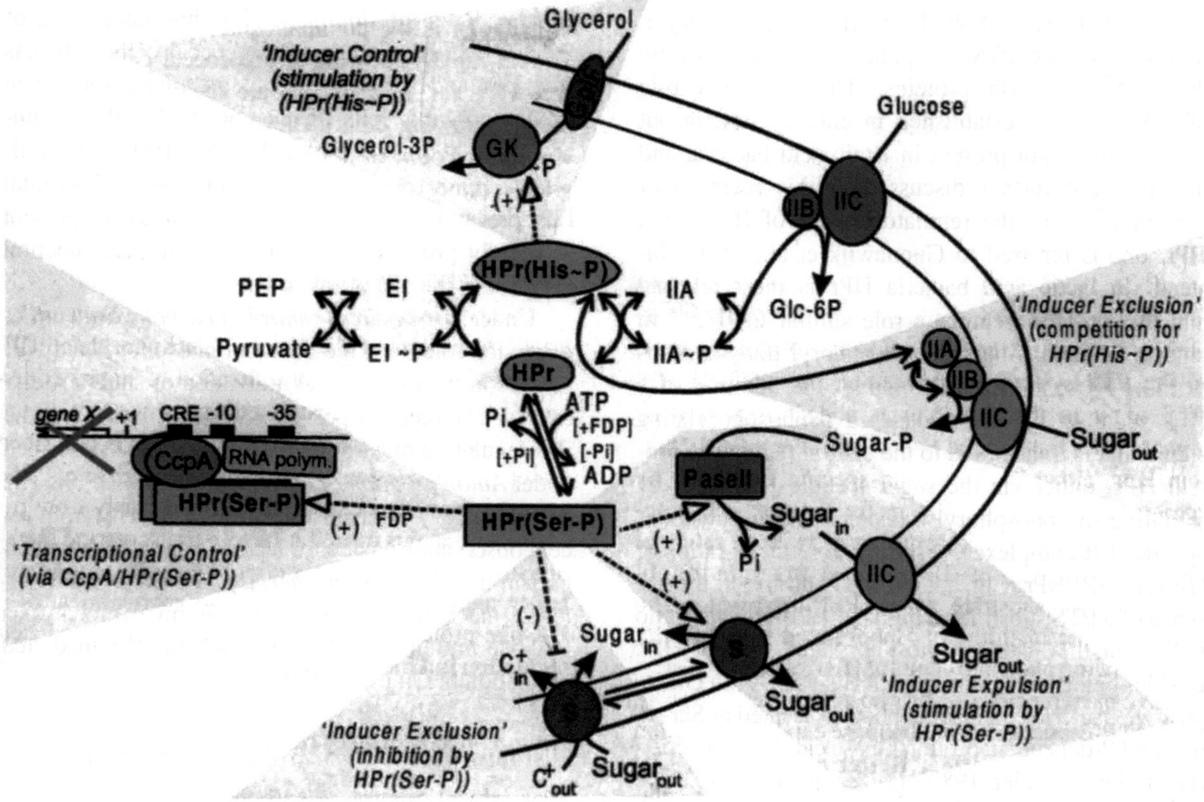

Figure 2. Schematic representation of the phospho*enol*pyruvate:glucose phosphotransferase system of Gram-positive bacteria. The central role of the HPr species in controlling transcription, inducer exclusion, inducer expulsion and inducer control is illustrated. EI, Enzyme I; Glc-6P, glucose-6P; Pase II, sugar phosphatase II CcpA, catabolite control protein A; RNA polym., RNA polymerase; C^+, cation; S, secondary transport protein; GK, glycerol kinase; GlpF, glycerol facilitator; CRE, catabolite responsive element; FDP, fructose-1,6-bisphosphate; and P_i, free phosphate.

volved in transport and metabolism of less preferred carbohydrates (Cohn & Horibata 1959); (ii) inhibition of activity of enzymes that effect the uptake or production of the transcriptional inducer (hereafter referred to as *inducer exclusion*; Mcginnis & Paigen 1969; Dills et al. 1980); and (iii) stimulation of efflux of intracellular inducer, that is, the carbohydrate or the phosphorylated derivative (this phenomenon is referred to as *inducer expulsion*, Reizer & Panos 1980; Thompson & Saier 1981; Romano et al. 1987). The inducer exclusion and expulsion mechanisms result in a lowering of the intracellular inducer concentration, and, thereby, indirectly affect gene expression. Autoregulatory control of carbohydrate utilization, on the other hand, occurs via adjustment of the rate of transport of a particular carbohydrate to the rate of its metabolism and the availability of the substrate, hereby providing a feedback or feedforward control to the pathway. Like with hierarchical control, autoregulation involves both control of gene transcription and control of enzyme/transporter activity. The two regulatory mechanisms differ in the sense that autoregulation of carbohydrate utilization controls the catabolic activities within a specific metabolic pathway, whereas hierarchical control involves the metabolic pathway of the preferred carbohydrate as well as that of the less preferred carbohydrate.

The mechanisms underlying the regulation of the initial steps of carbohydrate metabolism have been best studied in Gram-negative enteric bacteria, such as *Escherichia coli* and *Salmonella typhimurium*, and Gram-positive low-GC bacteria, such as *Bacillus subtilis* and several streptococcal, lactococcal and lactobacillus species. It has been established that the PTS system plays a crucial role both in the hierarchical control and auotoregulation of carbohydrate utilization.

How does the PTS exert all these regulatory functions? The PTS is able to sense the availability of carbohydrates and the metabolic capacity of the cell to metabolize these carbohydrates via the phosphorylation state of the PTS components; key players are the IIA^{Glc} and HPr proteins. The regulatory role of IIA^{glc} is well established in enteric bacteria but this protein is not present in lactic acid bacteria and therefore not further discussed in this review (for a comparison of the regulatory roles of IIA^{glc} and HPr, one is referred to Gunnewijk et al. 2001). Instead, in lactic acid bacteria HPr is most relevant and this protein assumes a role similar to IIA^{Glc} in enteric bacteria. Since the phosphoryl transfer steps in the PTS system are reversible, the addition of a PTS sugar to the cell induces a dephosphorylating signal that is transmitted to the central regulatory protein HPr, either via the sugar-specific IICBA or by rerouting the phosphoryl transfer to other substrate-specific EII complexes, resulting in reduced ratios of HPr(His~P)/HPr (Fig. 2). On the contrary, in the absence of a PTS substrate, a high [PEP]/[pyruvate] ratio will favor the histidine-phosphorylated state of HPr. Besides being phosphorylated at His-15, HPr in Gram-positive bacteria can also be phosphorylated at Ser-46 in an ATP-dependent protein kinase catalyzed reaction (Deutscher & Saier 1983; Reizer et al. 1984). The reverse reaction, the hydrolysis of HPr(Ser-P), is catalyzed by a cytosolic HPr(Ser-P)phosphatase, which is stimulated by high concentrations of phosphate (Deutscher et al. 1985). HPr(Ser)kinases from several Gram-positive bacteria, including *B. subtilis*, *Streptococcus pyogenes*, *Lactobacillus brevis* and *Lactococcus casei*, are stimulated by early glycolytic intermediates, in particular fructose 1,6-bisphosphate (FDP) (see references cited in Gunnewijk et al. 2001). The HPr kinases from *S. salivarius*, *S. mutans* Ingbritt and *E. faecalis* do not seem to be stimulated by FDP or other glycolytic intermediates, but instead, these enzymes are controlled by the cellular ATP and P_i levels (Brochu et al., 1999, Kravanja et al., 1999). Recent experiments on the HPr(Ser)kinase from *B. subtilis* showed that stimulation by FDP essentially occurred at low ATP and enzyme concentrations, and that positive cooperativity for FDP binding is related to oligomerization of the enzyme (Jault et al., 2000). Overall, the formation of HPr(Ser-P) is proposed to be governed by the relative cellular concentrations of ATP, P_i and/or FDP, which are indicators of the energy status of the cells (Mason et al., 1981; Thompson and Torchia, 1984). This suggestion is supported by measurements of the relative levels of HPr, HPr(Ser-P), HPr(His~P) and the doubly phosphorylated species HPr(Ser-P/His~P) in the cell. It has been shown that HPr(Ser-P) is the dominant phosphorylated form of HPr in rapidly growing streptococcal cells, whereas free HPr and HPr(His~P) are the major species in slowly growing cells (Thevenot et al. 1995; Gunnewijk & Poolman 2000a). HPr(Ser-P/His~P) is always a minor species (ranging from 5 to 30% of total HPr present in different streptococci) and only present in rapidly growing cells; the physiological function HPr(Ser-P/His~P) is unknown.

Under *Hierarchical control of carbohydrate utilization*, the role(s) of the different phosphorylated HPr species in the hierarchical utilization of sugars is discussed. The recently postulated role of the PTS in the autoregulation of carbohydrate utilization is described under *Autoregulation of carbohydrate utilization*. The autoregulatory mechanism follows primarily from recent observations made for lactose transport and metabolism in *Streptococcus thermophilus*. This autoregulatory mechanism is unique for its involvement of a IIA-like protein, which is unusual for PTS-mediated regulation in Gram-positive bacteria.

Hierarchical control of carbohydrate utilization

Hierarchical control of carbohydrate utilization was first described by Monod in 1942. Monod demonstrated that on a mixture of carbohydrates the growth of *E. coli* is biphasic as a result of the sequential use of the carbohydrates. For instance, glucose is used first when present in combination with lactose, melibiose, maltose and/or raffinose. The molecular basis for hierarchical carbohydrate utilization is well understood, in particular in enteric bacteria, and is generally referred to as catabolite repression. Catabolite repression is defined as the inhibitory effect of a preferred carbohydrate on the expression of other (catabolic) genes. Catabolite repression also includes inducer exclusion and inducer expulsion, regulatory mechanisms via which the cellular concentrations of inducer are reduced. The inducer exclusion and expulsion mechanisms thus indirectly affect gene expression. Most often the activity of the transport protein is modified such that the uptake of the inducer is prevented or the accumulated inducer is expelled from the cell. In some cases the first step(s) of the metabolism that produce(s) the transcriptional inducer is inhibited. The different mechanisms of inducer exclusion and inducer expulsion are described under *Catabolite repression by inducer exclusion or inducer expulsion*. In the control of gene expression not only 'inducer-specific' transcription factors, but also 'general' ones are involved. In enteric bacteria, general transcriptional control is mediated by CRP (cAMP receptor protein), which requires cAMP as cofactor. The synthesis of cAMP is catalyzed by adenylate cyclase, which is activated by IIA^{Glc} ~P and thereby under the control of the PTS. In Gram-positive bacteria, cAMP is not present and CcpA, the equivalent of CRP, is regulated by HPr(Ser-P).

Catabolite repression by inducer exclusion or inducer expulsion

Inducer exclusion is established by different mechanisms in Gram-negative and Gram-positive bacteria, and involves different PTS proteins. In Gram-negative enteric bacteria, inducer exclusion is determined by the phosphorylation state of IIA^{Glc}, a mechanism not to be discussed here. In Gram-positive bacteria, on the other hand, inducer exclusion involves allosteric control of transporters by HPr(Ser-P) or control via HPr(His~P)-dependent phosphorylation (Figure 2). In *L. brevis*, HPr(Ser-P) has been implicated in the control of uptake of non-PTS carbohydrates such as glucose, lactose and ribose (Ye et al. 1994a,b). Recent studies with *ptsH* and *hprK* mutants in *L. casei* showed that HPr(Ser-P) can also act in inducer exclusion by inhibiting the uptake of the non-PTS carbohydrate maltose (Dossonet et al. 2000; Viana et al. 2000). In some Gram-positive bacteria, like *E. faecalis, E. casseliflavus* and *B. subtilis*, the concentration of the transcriptional inducer of the *glp*-operon is controlled via the activity of glycerol kinase. The enzyme is stimulated via HPr(His~P)-dependent phosphorylation (Charrier et al. 1997). The phosphorylation of glycerol kinase by HPr is reversible, and the dephosphorylated, less active form of glycerol kinase, is dominant when a PTS-substrate is present in the medium (Deutscher et al. 1993). The net result of this regulation is equivalent to the allosteric control of transporters by HPr(Ser-P), as in both cases the intracellular inducer concentration is lowered. HPr also affects the hierarchical utilization of PTS carbohydrates via competition for HPr(His~P), a general mechanism that is operative in both Gram-negative and Gram-positive bacteria. As the affinity of HPr(His~P) varies for the carbohydrate-specific IIA proteins/domains, competition for HPr(His~P) leads to hierarchical uptake of PTS carbohydrates.

In some low-GC Gram-positive bacteria, catabolite repression is achieved by a mechanism in which the inducer is expelled from the cell. Two types of inducer expulsion mechanisms have been postulated (Figure 2). In homofermentative lactic acid bacteria like *E. faecalis, S. pyogenes, S. bovis* and *L. lactis*, lactose and glucose accumulate in the cytoplasm in their phosphorylated forms (Reizer & Panos 1980; Ye et al. 1996). Addition of a rapidly metabolizable sugar results in dephosphorylation of the accumulated sugar-P, which is followed by a rapid efflux of the free sugar from the cell (Reizer et al., 1983). A sugar-P phosphatase (Pase II) has been identified in *E. faecalis, S. pyogenes, S. bovis* and *L. lactis* and this enzyme seems to be absent in *S. aureus, S. mutans, S. salivarius* or *B. subtilis*, organisms that do not exhibit the sugar-P hydrolysis dependent expulsion phenomenon. Based on *in vitro* studies with toluenized vesicles or purified Pase II, it has been suggested that HPr(Ser-P) stimulates PaseII (Ye et al. 1994c). A second type of inducer expulsion has been observed in heterofermentative lactobacilli such as *L. brevis* and *L. buchneri* These bacteria transport lactose and glucose via proton symport mechanisms and accumulate these substrates as free (non-phosphorylated) sugars. Binding of HPr(Ser-P) to the glucose/H^+ and lactose/H^+ symporters is thought to alter the energy coupling mechanism, resulting in a conversion of the systems from carbohydrate-proton symport into carbohydrate uniport. Consequently, the accumulated sugars leave the cell down their concentration gradients, and thereby the inducer levels are lowered (Romano et al. 1987; Ye et al. 1994a,b).

CcpA-mediated catabolite repression

In many Gram-positive bacteria the general transcription factor, CcpA, mediates the repression of a range of catabolic genes (Hueck & Hillen 1995) (Figure 2). HPr(Ser-P) has been shown to interact with CcpA, allowing the latter protein to bind specifically to a *cis*-acting sequence. This sequence, named catabolite-responsive element (*cre*), is present in or near the promoter regions of many catabolite repression-sensitive operons (Weickert & Chambliss 1990). HPr(Ser-P)/CcpA forms a ternary complex with *cre*, consisting of two molecules of HPr(Ser-P), the CcpA dimer and the *cre* sequence (Jones et al. 1997). Both the formation of the HPr(Ser-P)/CcpA-complex and its binding to *cre* sequences is stimulated by FDP (Deutscher et al. 1995; Kim et al. 1998). The histidine residue at position 15 in the HPr protein, the active site for PEP-dependent Enzyme I phosphorylation, is important for CcpA-mediated repression, as mutation or phosphorylation of His-15 blocks the interaction of HPr(Ser-P) with CcpA. This observation suggests a direct link between catabolite repression and PTS-mediated carbohydrate transport (Deutscher et al. 1995; Reizer et al. 1996). Indeed, the uptake of glucose or other rapid metabolizable PTS carbohydrates leads to dephosphorylation of the PTS proteins and to an increase of the concentrations of glycolytic intermediates that activate the HPr(Ser)kinase. As a result the levels of HPr(Ser-P) rise and CcpA-dependent genes become less efficiently transcribed. This regulatory mechanism leads to a hierarchy in the utilization of carbohydrates.

Catabolite control by CcpA not only involves repression but also activation of genes and operons. In *Lactococcus lactis*, CcpA was found to be a transcriptional activator of the *las* operon, thereby controlling the production of the three key glycolytic enzymes, that are, phosphofructokinase, pyruvate kinase and lactate dehydrogenase (Luesink et al., 1998).

Autoregulation of carbohydrate utilization

Besides hierarchical control, the PTS also mediates autoregulation of carbohydrate utilization. The mechanistic concepts of the autoregulatory control circuits are emerging and, in a few cases, it has been shown that the rate of carbohydrate uptake is tuned to the metabolic capacity of the cell and the carbohydrate availability in the medium. In *S. thermophilus* evidence was obtained for autoregulation of the transport of the non-PTS carbohydrate lactose. This involved the tuning of the uptake to the rate of sugar metabolism. In the following sections, the regulation of transport and metabolism of lactose at the level of protein activity (Section *Regulation of lactose transport in S. thermophilus*) and gene transcription (Section *Transcriptional control of the lac operon in S. thermophilus*) are described. The underlying mechanism involves HPr(His~P)-mediated phosphorylation of the IIA-like domain of the non-PTS transport protein lactose transporter (LacS) from *S. thermophilus* and HPr(Ser-P)-dependent binding of CcpA to a *cre* site in the *lacS* promoter region.

S. thermophilus has a very limited capacity to utilize carbohydrates. Lactose and sucrose are fermented most rapidly, glucose is used very slowly, and only one or few other carbohydrates can be used by most strains. *S. thermophilus* co-metabolizes sucrose and lactose, a PTS and a non-PTS substrate, respectively, indicating that the utilization of these carbohydrates is not (strongly) hierarchically controlled. Instead, it has been proposed that HPr(His~P)-mediated regulation of the lactose uptake rate serves to control the flux of glycolysis. This mechanism is based on studies of the kinetic properties of phosphorylated and unphosphorylated lactose transporter (LacS) from *S. thermophilus*, and on the LacS levels of the cell as a function of the phosphorylation state of HPr. The various species of HPr present in lactose-growing *S. thermophilus* cells have been quantified at different stages of growth (Gunnewijk & Poolman 2000). HPr(Ser-P) appears to be the dominant phosphorylated species in the exponentional phase of growth, whereas HPr(His~P) dominates in the stationary phase. Similar results were obtained when *S. thermophilus* cells were grown on sucrose. The fact that the levels of HPr(Ser-P), HPr(His~P) and HPr are similar in sucrose- and lactose-growing cells suggest that the rate of glycolysis of both carbohydrates is sufficiently high to keep Ser-46 phosphorylated and that the drain of phosphoryl transfer to sucrose is minor compared to the phosphorylation activity of Enzyme I. The similar HPr(His~P)/HPr ratios also suggest that the PEP/pyruvate ratios are comparable in sucrose and lactose growing cells. Although PEP levels have not been measured in *S. thermophilus*, it has been firmly established for other lactic acid bacteria that concentrations of PEP are relatively low in rapidly metabolizing cells, whereas PEP concentrations increase under conditions of carbohydrate limitation (Mason et al. 1981; Thompson & Torchia, 1984; Konings et al. 1989). The increase in HPr(His~P) in *S. thermophilus* at later stages of growth would thus correlate with increased PEP levels and a decreased metabolic activity.

Regulation of lactose transport in S. thermophilus

In *S. thermophilus*, lactose is taken up via the secondary transport protein LacS (Poolman et al., 1996). LacS catalyzes two modes of transport, solute-H^+ symport, driven by the proton motive force (Δp) and lactose/galactose exchange, which is driven by the concentration gradients of lactose and galactose across the membrane (Foucaud & Poolman 1992). The lactose/galactose exchange reaction via LacS is the most relevant transport mode *in vivo* as it is much faster than the lactose/H^+ symport reaction (Knol et al. 1996). In addition, the galactose moiety of lactose cannot be metabolized in most *S. thermophilus* strains, and therefore galactose has to be expelled from the cell. Kinetic studies have revealed that the affinity of LacS for galactose and lactose at the cytoplasmic

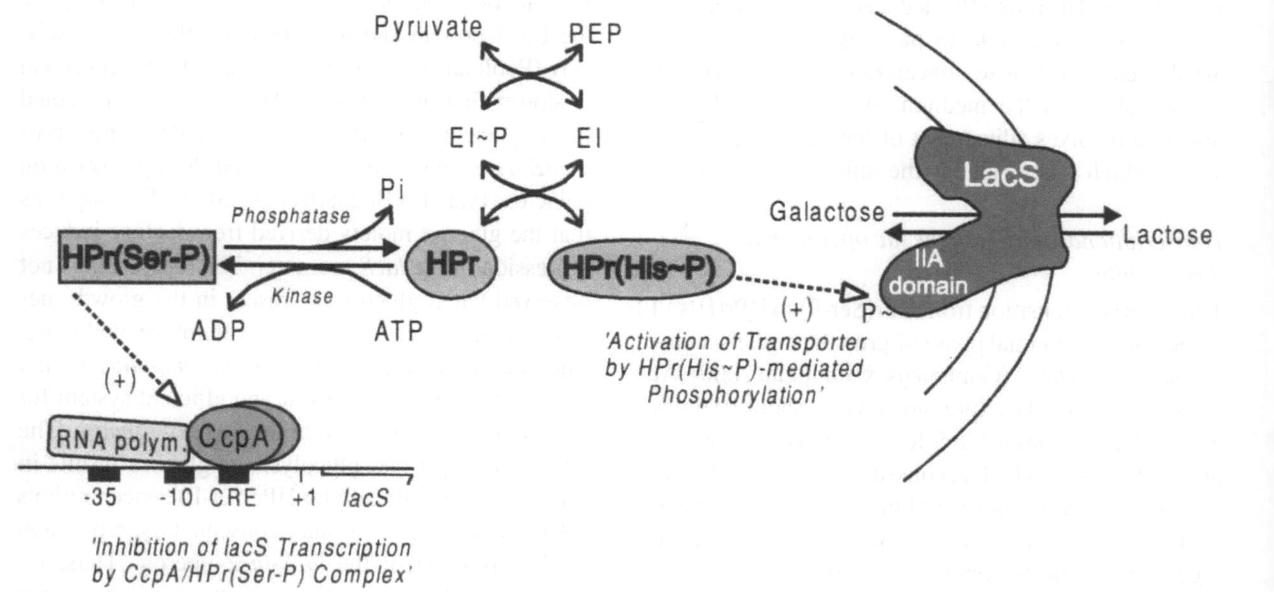

Figure 3. Regulation of lactose transport in *Streptococcus thermophilus*. Schematic representation of HPr(His~P)-mediated phosphorylation of LacS (stimulation of lactose transport activity), and HPr(Ser-P)/CcpA-mediated regulation of lacS transcription. The depicted symbols are described in the legend of Figure 2.

binding site is 20-fold higher than at the extracellular binding site, and that, in this conformation, galactose is preferred over lactose (Veenhoff & Poolman 1999). These observations are consistent with the view that LacS has evolved into an efficient lactose/galactose exchanger.

Although LacS is not a PTS transport system, the protein has a carboxyl-terminal hydrophilic domain of about 160 amino acids (Figure 3), which is homologous to IIA proteins/domains of various PTS systems. This so-called IIA^{LacS} domain has evolved into a regulatory element, whose main function is not to transfer phosphoryl groups rapidly but rather to control the transport activity. The effect of HPr(His~P)-mediated phosphorylation on lactose transport has been studied *in vitro* using an artificial membrane system, in which purified LacS protein was incorporated into liposomes with the IIA^{LacS} domain facing outwards (Gunnewijk & Poolman, 2000b). This system allowed phosphorylation and manipulation of LacS activity by adding PEP, Enzyme I and/or HPr to the outside medium. Upon phosphorylation of LacS the maximal rate of lactose exchange transport is increased, whereas the rate of Δp-driven lactose uptake is not affected. In line with a range of kinetic studies (Foucaud & Poolman 1992; Poolman et al., 1995b), it has been proposed that phosphorylation affects the rate constants for the reorientation of the ternary complex (LacS with bound lactose plus proton), which is rate-determining for exchange transport but not for Δp-driven uptake. Since the lactose/galactose exchange reaction and not the Δp-driven uptake is most relevant in lactose (glycolysing)-metabolizing cells of *S. thermophilus*, HPr(His~P)-mediated phosphorylation of LacS evokes maximal activity of the lactose transport protein in vivo by increasing the V_{max} of the lactose/galactose exchange reaction. This condition is met in cells at the late-exponentional and stationary phase of growth, when HPr(His~P) is the dominant species of HPr (Gunnewijk & Poolman 2000a).

The transition from HPr(Ser-P) to HPr(His~P) at the late-exponentional phase of growth parallels an increase in the extent of LacS phosphorylation, a decrease in lactose and an increase in galactose concentration in the growth medium. Since both lactose and galactose are substrates of LacS (Veenhoff & Poolman 1999), the decrease in lactose/galactose ratio in the medium will reduce the lactose uptake (and galactose excretion) capacity as growth proceeds. This will at some point during growth be reflected in a reduced glycolytic activity, to which the HPr(Ser-P)/HPr(His~P) ratio is very sensitive (Reizer et al. 1984,1989b; Deutscher et al. 1985; Deutscher & Engelman, 1984). By increasing the specific transport activity via HPr(His∼P)-mediated phosphorylation, *S. thermophilus* is able to partially compensate for the decrease in lactose concentration (and galactose accumulation) in the medium. Another, but slower, response involves adjustment of the LacS expression levels, which is described in the following section.

Transcriptional control of the lac operon in S. thermophilus

The observed transition from HPr(Ser-P) to HPr(His∼P) at the late-exponential phase of growth parallels an increase in LacS level (Gunnewijk & Poolman 2000). At stationary phase, the expression level is about 10 times higher than the basal LacS level at early-exponential phase of growth, which is consistent with the idea that HPr(Ser-P) is a corepressor of the lac operon Figure 3.

Direct evidence for HPr(Ser-P)/CcpA-mediated regulation of the *lac* operon came from studies with a *ccpA* disruption mutant. Disruption of the *ccpA* gene impaired the growth of *S. thermophilus* on several sugars as has been observed for other Gram-positive bacteria (Hueck et al. 1995; Egeter & Brückner 1996; Monedero et al. 1997). The *lacS* promoter contains a *cre* site, overlapping the −10 box and the transcriptional start site (Poolman 1993), suggesting that expression of the *lacS-lacZ* operon is under control of CcpA (Henkin 1996). In accordance, disruption of the *ccpA* gene in *S. thermophilus* CNRZ302 resulted in derepression of *lacSZ* transcription during exponential growth on lactose (van den Bogaard et al. 2000). Moreover, the rates of lactose uptake and galactose excretion were at least 4-fold increased in the *ccpA* disruption strain relative to wild-type cells. The increased lactose uptake and hydrolysis does not result in an increased growth rate on lactose, but leads to massive expulsion of glucose into the fermentation medium. Apparently, loss of a functional CcpA in *S. thermophilus* uncouples the control of metabolism over transport and *vice versa*, as glycolysis can no longer keep up with the massive lactose intake. In other words, the *S. thermophilus ccpA* disruption mutant has a lactose transport capacity that exceeds the maximal glycolytic rate. The data indicate that the concerted activity of HPr(Ser-P) and CcpA results in fine-tuning of lactose transport and hydrolysis capacity in order to accommodate maximal glycolytic flux.

Although most *S. thermophilus* strains cannot use galactose as a carbon source, galactose-fermenting mutants of strain CNRZ302 are readily obtained (Hutkins et al. 1985; Vaughan et al. 2001). In the Gal^+ variants of *S. thermophilus*, galactose is taken up by the LacS protein and fermented via the Leloir pathway (Poolman 1993; Vaughan et al. 2001), but growth is slower than with lactose. When *ccpA* is disrupted in the galactose-fermenting variants, derepression of *lacSZ* transcription is not observed during growth on galactose (van den Bogaard et al., 2000). This suggests that the glucose moiety derived from lactose induces repression of the *lacS* promoter. The repression is not observed when glucose is present in the growth medium, which is due to the low rate of uptake of glucose. The LacS transport protein of *S. thermophilus*, on the other hand, constitutes a fast and efficient system for lactose uptake, leading to high intake of glucose. The accompanying rapid glycolysis of glucose results in relatively high intracellular HPr(Ser-P) concentrations (Deutscher et al. 1995) and, consequently, repression of the promoter of the *lacS-lacZ* operon. Thus, repression of the lac operon in *S. thermophilus* is not carbon-source dependent but determined by the rate of glycolysis. Probably, FDP, PEP and/or ATP function as the intracellular indicators of the glycolytic flux, as has been suggested for other Gram-positive bacteria.

Model for autoregulation of lactose transport and metabolism in S. thermophilus

A model has been proposed for the control of lactose transport and metabolism in *S. thermophilus*, which accomodates the knowledge of the kinetic properties of the transporter, the regulation of transporter activity, the regulation of expression of the *lacS-lacZ* operon, and the metabolic status of the cells (Figure 3). The rate of lactose transport via LacS is susceptible to the lactose/galactose ratio in the growth-medium as the transporter has a higher affinity for galactose (end-product of the fermentation) than for the substrate lactose. This implies that the transport capacity will decrease when galactose accumulates in the medium even with millimolar concentrations of lactose available. At some point during growth this will be reflected in a reduced glycolytic activity, which affects the concentrations of different glycolytic intermediates to which the HPr(Ser-P)/HPr(His∼P) ratio is very sensitive. ATP is an effector of HPr(Ser) kinase, whereas P_i is an inhibitor. In addition, the HPr(Ser-P)phosphatase is stimulated by P_i and inhibited by ATP (Deutscher & Saier 1983; Reizer et al. 1984,1989b; Deutscher et al. 1985). The intracellular concentrations of ATP, PEP and P_i vary in response to the carbohydrate availability as has been firmly established for other lactic acid bacteria. ATP levels are relatively high in rapidly

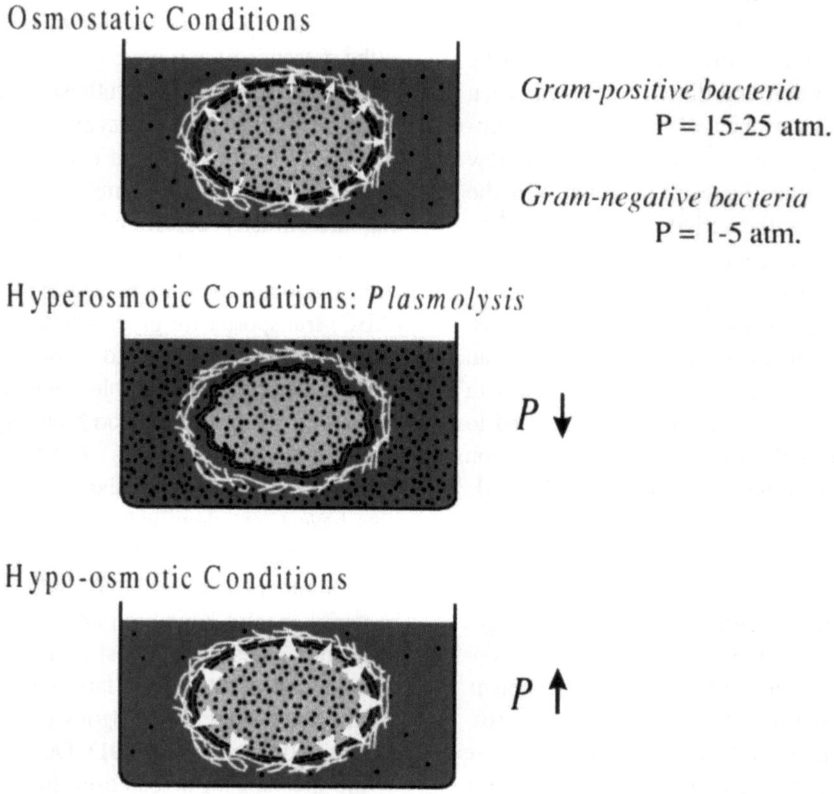

Figure 4. Schematic representation of the osmotic conditions experienced by a microbial cell. P, cell turgor.

metabolizing cells, whereas P_i and PEP are low under these conditions. These latter compounds become high at the end of the exponentional phase of growth and remain high in the stationary phase (Mason et al., 1981; Thompson & Torchia 1984; Konings et al. 1989).

The concentrations of the metabolites FDP, P_i, PEP and ATP reflect the metabolic status of the cell and determine the phophophyrylation state of HPr. Increasing PEP concentrations result in a rise in the HPr(His~P) concentration due to PEP-dependent Enzyme I phosphorylation of HPr. HPr(His~P) phosphorylates the LacS protein and this modification increases the maximal rate of transport rate. At the same time, the modified activities of HPr(Ser)kinase and HPr(Ser-P)phosphatase will decrease the HPr(Ser-P) concentration. Accordingly, a relief of the HPr(Ser-P)/CcpA-mediated repression of the *lacS* promoter will result in the synthesis of more LacS and β-galactosidase, which in turn will provide more glucose for glycolysis. In this way the uptake of lactose is tuned to the lactose/galactose ratio in the medium as well as the glycolytic capacity of the cell. The autoregulatory mechanism for the transport and metabolism of the non-PTS sugar lactose in *S. thermophilus* also allows the organism to co-metabolize lactose and sucrose.

Regulation of cell volume

Introduction

Enzymes and other macromolecules are not only sensitive to physical parameters such as pH, temperature and solute composition but also to water activity. This factor is often ignored in studies of enzyme activities and their regulation. Estimates of reaction rates and equilibria are generally made in dilute solutions with water activities close to one. The cytoplasm of a bacterial cell, however, is highly concentrated with 300–400 g/l of biological macromolecules (predominantly protein and RNA) and ~100 g/l of low molecular weight osmolytes (incl. salts, amino acids, compatible solutes, etc.), resulting in low water activities. The fraction of macromolecules occupies 20–30% of the cellular volume, and, in this crowded and viscous environment, the diffusion of solutes and macromolecules is impaired (Ellis 2001). The rate of any biochemical process that is diffusion-limited will be reduced in such a milieu and directly affected when the cytoplasmic volume changes upon osmotic up- or downshift. An increase in crowdedness will slow down diffusion and negatively effect rates. On the other hand, it increases the activity coefficients, i.e., the ratio of effective concentration to actual concentration, of enzymes by favoring the association of molecules. Knowledge of these omotic stress-related factors is not only crucial in understanding how an organism works, but also allows one to interfere with its metabolic activity. Osmotic challenge imposed to microorganisms forms the basis of food preservation strategies for several millennia, but, generally, little attention is paid to the actual consequences of this type of stress.

In their natural environment, the majority of microbial cells experience from time to time changes in extracellular water activity, which has direct consequences for the water activity of the cytoplasm. Following an increase in external water activity (osmotic downshift), passive influx of water will increase the turgor and eventually the cells will lyse if there are no mechanisms to counteract the stress (Figure 4); for the definition of cell turgor see Section *Other parameters relevant for cell volume control and osmoregulation*. Similarly, upon osmotic upshift water will flow out of the cell, the turgor will decrease, and in the end the cells will plasmolyse. To keep turgor within a specific range, and to prevent cells from lysing or plasmolysing, microorganisms adjust their intracellular osmolyte concentration. Both Gram-positive and Gram-negative bacteria prefer particular zwitterionic organic co-solvents such as glycine betaine, carnitine or ectoine as osmoprotectants (Wood 1999). These compounds, generally referred to as compatible solutes, can be accumulated to molar levels without negative effects on macromolecular structure or function. In fact, several compatible solutes have been shown to stabilize enzyme structure (Arakawa & Timasheff, 1985). The stabilization of native protein structures by these compounds involves preferential exclusion of the compatible solutes from the protein's surface. The preferential exclusion implies that the interaction between protein and compatible solutes is thermodynamically unfavorable. Because more protein surface is exposed in the unfolded than in the native state, the free energy difference for the transfer from water to compatible solute solution is largest for unfolded protein. This large positive Gibbs energy effect renders proteins thermodynamically more stable in the presence of compatible solutes.

Because the *de novo* synthesis of additional transporters and biosynthetic enzymes would be too slow to be effective against a rapid external osmotic change, cells need transport systems for compatible solutes that are directly controlled by osmotic conditions. To accumulate compatible solutes upon osmotic upshift, *Lactococcus lactis* uses an ATP-binding cassette (ABC)-transporter for glycine betaine (OpuA); equivalent systems can be found in other lactic acid bacteria. To excrete compatible solutes, that is, in the event the turgor becomes too high, organisms activate mechanosensitive channels. The molecular identities of three proteins that contribute to these channel activities have been identified in *E. coli* (Sukharev et al. 1994; Levina et al. 1999). BLAST searches of the non-redundant database NCBI indicate that lactic acid bacterial species have homologues of one or more of these molecules. By far, best studied of the three is the protein responsible for the largest conducting activity, MscL, and a gene homologous to *mscL* is present in *L. lactis* (Bolotin et al. 2001). Our current knowledge of the mechanism of osmosensing and regulation of the osmotic upschift-activated OpuA and downshift-activated MscL is presented in the following sections. Other osmotic stress-related parameters, perhaps not directly relevant for these systems, but important for a complete understanding of cell volume control are summarized.

Mechanisms of Oosmosensing

Osmotic activation of membrane proteins may be signaled via: (i) a change in cell turgor; (ii) macroscopic change in membrane structure; (iii) mechanical stimulus originating within the exo- or cytoskeleton of the cell; (iv) a change in the hydration state of the protein (internal or external osmolality); (v) alterations in the physicochemistry of the membrane bilayer (protein–lipid interactions); (vi) a change in cytoplasmic ion concentration or ionic strength; (vii) specific molecule interacting directly with the protein. Of these physicochemical parameters, the ones relating to changes in cytoplasmic osmolality, ionic strength, specific molecular stimulus, and protein interactions with putative cytoskeleton, are also relevant for cytoplasmic enzymes and other macromolecules. The relevance of each of the signaling parameters for membrane-embedded osmoregulated proteins has re-

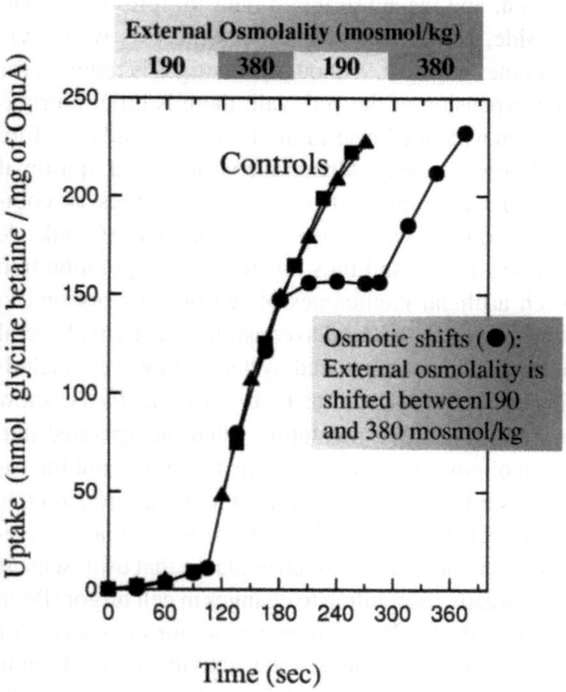

Figure 5. Kinetics and reversibility of the osmotic activation of OpuA. Uptake of [^{14}C]glycine betaine was assayed in 100 mM KP_i, pH 7.0, corresponding to 190 mosmol/kg. At 105 (●, ■, ▲) and 285 s (●) the proteoliposomes were subjected to hyperosmotic conditions by the addition of 100 mM KCl (final osmolality corresponding to 380 mosmol/kg). Isoosmotic conditions were restored at 180 s (●). Modified after van der Heide et al. 2001.

cently been evaluated (Poolman et al. 2002). In the *Molecular Microbiology* review, the focus is on transporters, channels and sensor kinases that have been well studied in intact cells and in vitro model systems, allowing detailed evaluation of the bacterial osmosensing mechanisms. Except for the ABC transporter OpuA from *L. lactis*, the systems are from organisms other than LABs, but the main conclusions may represent a firm basis for the situation in LABs, that is, if one believes in the conservation of biological structure and function.

To summarize the *Molecular Microbiology* review, the data from *in vivo* and *in vitro* experiments indicate that external ionic and non-ionic osmolytes activate a range of osmosensing devices, provided the compounds do not rapidly equilibrate (cross the membrane). Under these conditions, the osmotic stress causes the cytoplasmic or liposomal volume to decrease, resulting in an increased concentration of intracellular osmolytes, of which the ionic ones are critical for the activity of the OpuA transporter. *In vivo*, not only the decrease in cell volume, but also the accumulation of potassium ions in the initial response to osmotic upshift, may contribute to the increase in cytoplasmic ionic osmolyte concentration. The ionic osmolytes (or ionic strength) are sensed not only by the ABC-transporter OpuA from *L. lactis* but also by a number of other upshift-activated systems, among which sensor kinases of two-component regulatory systems (Poolman et al. 2002). Figure 5 shows that activation of OpuA by osmotic upshift is instantaneous and reversible upon returning to iso-osmotic conditions, which is a prerequisite for an effective response mechanism. Another important message here is that the effect of the osmotic upshift is indirect, that is, through the increase in concentration of ionic osmolytes on the inside (van der Heide et al. 2001). This condition is met with ionic (salts) and non-ionic (sugar, polyols) osmolytes added to the outside medium, provided the molecules do not equilibrate across the membrane on the time scale of the transport measurements (van der Heide & Poolman 2000b). Membrane permeant osmolytes such as glycerol cause a transient activation because these molecules diffuse across bacterial membranes with half times of seconds.

The equivalence of salts and sugars in the activation of an essential osmoresponsive system is important to emphasize as this point is not always clearly resolved in published literature. Salts and sugars do not always have the same osmotic effect, and often the stronger inhibition by salts is ascribed to additional 'electrolyte stress'. But what is actually meant by (additional) electrolyte stress? *Lb. plantarum* is dependent of an ATP-dependent glycine betaine transporter (QacT) for growth at high medium osmolalities. Like OpuA from *L. lactis*, *in vivo* the *Lb. plantarum* QacT system is activated by KCl, NaCl, sucrose and lactose (Glaasker et al., 1996). However, the growth defect elicited by the salts is much more severe than with the sugars (Glaasker et al. 1998a). It turns out that lactose and sucrose, although essentially membrane-impermeable in proteoliposomal systems, slowly enter the cell via some low-affinity sugar transport system. On the minute time-scale, the sugars cause osmotic stress, which can be observed as an activation of the QacT transporter. On the longer time-scale, the external and internal sugars equilibrate, that is, the carrier-mediated influx of sugar can keep up with the growth of the organism. Consequently, growth inhibition does not occur at medium osmolalities at which equiosmolar concentrations are already inhibitory. Thus, salts and sugars have the same 'osmotic effect' in terms of activation of the transporter but, for rather trivial reasons, the two types of osmolytes have different effects when growth is monitored. These observations, together with the transient activation of OpuA by glycerol, indicate that the increase in internal osmolality upon salt or sugar stress does not represent the signal for activation of these systems. What matters is the increase in intracellular ion concentration (ionic strength). In case of OpuA, there is strong evidence that the ionic signal is transduced to the protein complex via alterations in the protein–lipid interactions rather than direct sensing of the ion concentration or ionic strength by the protein (van der Heide et al. 2001).

The osmotic signal and regulation of downshift-activated mechanosensitive channels, including MscL of *L. lactis*, is different from that of OpuA, but the primary activation signal is also transduced via the membrane. The MscL channel opens, and jettisons solutes with little discrimination (except for size), when the tension in the membrane reaches a certain threshold value. The critical value for MscL gating is close to the tension at which the membrane ruptures and the cell lyses. The system thus serves as pressure valve that opens when the difference in internal and external osmolality becomes too high. An analogy with a balloon under high pressure may be useful to explain this concept. Upon further inflation with air, the balloon breaks since it does not have a means (equivalent of MscL) to release the pressure. Too high membrane tension may occur in nature when, for instance, a soil bacterium, after a period of drought, is suddenly confronted with rainfall. Such osmotic downshift conditions lead to an increase in turgor and tension in the membrane, and, depending on the mechanic and elastic properties of the cytoplasmic membrane and peptidoglycan layer, the cell may lyse.

Other parameters relevant for cell volume control and osmoregulation

Cell turgor

Cell turgor (ΔP) is the hydrostatic pressure difference that balances the difference in internal and external osmolyte concentration. In equation:

$$\Delta P = (RT/V_w)\ln(a_o/a_i) \sim RT(c_i - c_o)$$

in which V_w is the partial molal volume of water, 'a' is the water activity, 'c' is the total osmolyte concentration, and the subscripts 'i' and 'o' refer inside and outside, respectively. A cell plasmolyses when ΔP becomes negative. Although cell turgor is required for the expansion of the cell wall, there is little information on what the lower limit of turgor should be before cell growth ceases. Membrane vesicles and liposomal systems can only withstand low pressures as compared to cells with a peptidoglycan layer (Csonka & Hansen 1991), and thus functional incorporation into such artificial membranes of sensors that respond to low turgors should lead to constitutive activity. Several classes of osmoregulated systems, however, including the ABC transporter OpuA from *L. lactis*, show normal functional regulation when incorporated into proteoliposomes, suggesting that turgor is not the salient stimulus. In fact, contrary to the impression one may get from published literature on bacterial osmoregulation, there is little or no evidence that osmosensors are directly responding to changes in cell turgor (Poolman et al. 2002). A change in turgor does lead to a change in intracellular water activity, ionic strength and crowdedness (depending on the elasticity of the cell envelope) and membrane properties, and one or more of these parameters are more likely candidates to which osmoregulated systems respond.

Hydration state

Substrate and ligand binding to enzymes and transporters is typically associated with changes in the conformation of the proteins. Since it is likely that different protein conformations sequester different amounts of water, osmotic stress could potentially affect a system's activity through a change in the hydration state of the protein. Hexokinase is a classic example of water activity as regulator of enzyme activity (Parsegian et al. 1995). The dissociation constant (*Kd*) for glucose binding to hexokinase decreases with increasing osmotic pressure of the assay medium, when varied with either low or high molecular weight polyethylene glycol (PEG) in the solution (Reid & Rand 1997). The smaller effects of the low molecular weight PEGs are explained by their less effective steric exclusion from a cleft in the surface of the enzyme. Similarly for the channel-forming peptide alamethicin, it has been shown that the open probability decreases with increasing concentrations of high molecular weight PEGs (Vodyanoy et al. 1993). Parsegian and colleagues (1995) have formulated a thermodynamic hypothesis for these observations. In the transition from the closed to the open state, the channel will require additional water as the open state is most likely more hydrated. Solutes too big to enter the channel such as high molecular weight PEGs will compete with the protein for water. Consequently, the excluded solute will cause the channel to perform extra osmotic work, which will lower the probability of the open conformation. The extra amount of work is less with solutes that are able to enter the channel, e.g., low molecular weight PEGs. The low molecular weight PEGs give rise to a smaller excluded volume and are thermodynamically less unfavorable than high molecular weight PEGs. In other words, the low molecular weight PEGs have a smaller dehydrating effect on the protein than high molecular weight ones. These two examples illustrate how osmotic stress could affect cytosolic and membrane-bound enzymes (and other macromolecules, e.g., protein-DNA interactions), but for LABs this is largely an unexplored area of research.

Macromolecular crowding

Cellular volume changes as a result of osmotic stress will result in changes in cytoplasmic protein concentration (macromolecular crowding), which affect the equilibrium of oligomeric enzymes and thereby their function. Although macromolecular crowding may not directly affect the function of systems embedded in the cytoplasmic membrane, membrane proteins that have a tendency to associate with soluble macromolecules may be influenced (Minton et al. 1992). As stated before, the crowdedness of the cytoplasm influences the diffusion of molecules. Three independently acting factors have been identified that account for a slowed diffusion (Verkman 2002): (i) the viscosity of the fluid-phase; (ii) the binding of molecules to other components; (iii) the collision of molecules with other cell components. The larger a molecule, the greater the contribution from viscosity and collision is. For a typical enzyme with a mass of 50 kDa, the diffusion may be slowed 10-fold relative to a dilute aqueous medium. With regard to factor (ii), it is worth noting that rod shaped bacteria possess genes coding for actin-like filaments, and in case of *Bacillus subtilis* it has been shown that these filamentous helical structures ('bacterial equivalent of a cytoskeleton') have a actin-like role in cell morphogenesis (Jones et al., 2001). Although homologues of these actin-like proteins are absent in round-shaped organisms (*Streptococcus* and *Enterococcus* sp.), they are likely to be present in lactobacilli. Whether or not these proteins affect the diffusion of cytoplasmic or membrane-bound proteins, the elasticity of the cell envelope, etc., is unknown, but, certainly, it represents an area of research that is relevant for a basic understanding of the stress response of (lactic acid) bacteria.

Physicochemical properties of the membrane

The role of the membrane as signal transducer of ionic osmolyte- (ionic strength) or bilayer tension-stress has already been mentioned in the context of the osmotic regulation of the OpuA transporter and the MscL channel. Ionic strength and bilayer tension, however, are not the only osmotic signals to be transduced via the membrane. On the assumption that the lipid bilayer behaves as an elastic solid, the intrinsic mechanical properties of the membrane can be described by four elasticity moduli that describe the response of a unit area of bilayer to volume compression, area expansion, bending/curvature and extension/shear (Evans & Skalak 1980). The responses of membranes to these elastic deformations have been recently reviewed by Hamill & Martinac (2001), and their main conclusions relevant for this paper are summarized here. Firstly, the bilayer is at least 10-fold more compressible in area than in volume during mechanical deformations encountered under physiological conditions. Osmotic downshifts will thus lead to relative increases in membrane area and concomitant decreases in membrane thickness. Secondly, the bending rigidity of the bilayer is dependent upon the lipid composition and area of each monolayer, and this parameter determines, amongst others, the shape of lipid vesicles. Thirdly, above the phase transition temperature, the membrane behaves like a fluid in response to extension. Of the elastic deformations, the ones that lead to thinning of the membrane are thought to have major impact on protein conformations and may thus signal activity changes. For the mechanosensitive channel MscL, it is thought that thinning of the membrane upon osmotic downshifts contributes to the ability of the protein to sense membrane tension (Hamill & Martinac 2001).

Bilayer thickness is obviously important for any membrane protein, as mismatch would result in exposure of hydrophobic surfaces of either the protein or lipid to an aqueous environment. The membrane–water interface of the bilayer comprises a chemically complex environment, which offers multiple possibilities for interactions with protein side-chains (Killian & von Heijne, 2000). If the bilayer thickness is suboptimal for these interactions, the protein or lipid may react to prevent hydrophobic mismatch that may lead to alterations in protein conformation and activity. When OpuA from *Lactococcus lactis* was incorporated in membranes in which 50% of the lipid fraction was varied in terms of acyl chain length (from C14 to C22), the protein displayed a clear optimum in maximal activity at C18 (dioleoyl lipids), but the threshold for osmotic activation was the same in all cases (van der Heide et al., 2001). Thus, changes in bilayer thickness upon osmotic up- or downshift are not of prime importance for the regulation of OpuA activity. On the other hand, in one model for gating the mechanosensitive channel MscL, the tilts of the transmembrane α-helices are predicted to increase as they move away from the axis of the pore, that is, when the membrane expands (Sukharev et al. 2001ab). The osmotic downshift-induced thinning of the membrane and the resulting hydrophobic mismatch could thus provide at least part of the energy required for the transit from the closed to the open state.

The build up of membrane tension upon osmotic downshift has been compared with the increase in tension in a balloon upon inflation, but the biological equivalent of the balloon is clearly more complex and can be described in further detail. The different local intermolecular forces between lipid molecules in a fluid membrane originate from steric hindrance, hydration, electrostatic charge and/or hydrogen bonding in the headgroup region, interfacial tension and acyl chain pressure. The differences in the components of the interactions as a function of membrane depth lead to enormous local transverse pressures that correspond to bulk pressures of several hundreds of atmospheres (Marsh 1996; Cantor 1999). A protein embedded in such membrane will thus not experience uniform tension as a function of membrane depth. The local pressure as a function of membrane depth is referred to as the lateral pressure profile.

Statistical thermodynamic calculations of the equilibrium pressure profiles of membranes predict large redistributions of lateral pressure when the acyl chain length, the degree, position and configuration of unsaturation, or headgroup repulsion are varied (Cantor 1999). Similarly, the incorporation into a lipid membrane of cholesterol or interfacial active solutes such as anesthetics are predicted to increase the lateral pressure selectively near the aqueous interfaces, resulting in a compensating decrease in lateral pressure near the center of the bilayer. Also an osmotic up- or downshift or the application of food preservatives with amphipatic properties (e.g., parabens) will affect the lateral pressure and such changes may influence protein conformation and activity. Because the osmotic activation profile of the OpuA transporter from *L. lactis* is not affected by variations in acyl chain length, position and configuration of the unsaturation, and/or headgroup repulsion as long as the charge of the lipid headgroups is kept constant (van der Heide et al. 2001), it seems unlikely that osmotic stress is detected as a perturbation of the lateral pressure profile. On the other hand, given the membrane tension-dependent gating of MscL, it is likely that changes in the lateral pressure profile influence the opening and closing of this and related mechanosensitive channel proteins.

Specific stimulus

A shift in medium osmolality causes a change in the cytoplasmic volume. In principle, the change in concentration of one particular molecule could be sensed and serve as a specific stimulus to switch on or off a particular system. A few such examples are described in the literature on osmotic regulation of transporters and sensor kinases (reviewed in Poolman et al. 2002). A special case of such a regulatory mechanism concerns the *trans*-inhibition of transport proteins, which is the equivalent of product-inhibition in metabolic pathways. For *Listeria monocytogenes* and *Lactobacillus plantarum*, it has been shown that pre-accumulated (*trans*) substrate inhibits the corresponding osmoregulated transport systems (Verheul et al., 1996; Glaasker et al., 1996b). The *trans*-inhibition may serve as a control mechanism to prevent the accumulation of these compatible solutes to too high levels and thereby the turgor pressure from becoming too high.

In the case of *L. monocytogenes*, carnitine is taken up via an ABC transporter that is specific for this substrate but is inhibited by high intracellular concentrations of both carnitine and glycine betaine (and perhaps other compatible solutes) depending on the osmotic status of the cells. In kinetic terms, the osmotic activation of the system parallels an increase in apparent inhibition constant (K_I) for glycine betaine and carnitine at the inner surface of the membrane. Apparently, as a consequence of the water efflux following an osmotic upshift, the internal binding site for glycine betaine and carnitine is altered. Binding of compatible solutes to an internal site thus seems to represent a key step in the activation–inactivation mechanism of some, but not all, osmoregulated transporters. We have recently shown that the *trans*-inhibition mechanism does not play a role in the osmotic regulation of the OpuA transporter from *L. lactis* (J. Patzlaff, unpublished).

Response of cells to osmotic up- and downshift

In the previous sections, the various osmotic stress-related signals to which a system, and thus a bacterial cell, could respond have been summarized. Despite the variety of possible signals and osmosensing mechanisms, we propose that not all of these are used as primary mechanism. In line with a strong believe in the conservation of biological mechanisms, the majority of osmotic upshift-activated systems (transporters and sensor kinases) may respond to changes in cytoplasmic concentrations of ionic osmolytes, whereas the downshift-activated channels sense tension in the membrane (Poolman et al. 2002).

Why would the cell use ionic osmolytes rather than intracellular osmolality (affecting protein hydration) or a specific signaling molecule (specific regulatory site on the protein)? When the medium osmolality is raised, the initial change in cytoplasmic water activity depends on the elasticity of the cell wall. Contrary to what is often thought, the cell wall of bacteria is not rigid but actually quite elastic (Csonka & Hansen 1991; Doyle & Marquis 1994). Consequently, even at turgors above zero, the cytoplasmic volume decreases with increasing external osmolality, and the ion (osmolyte) concentrations increase accordingly. The increase in ionic strength accompanying the volume decrease is undesirable as too high concentrations of electrolytes interfere with macromolecular functioning in eubacteria as well as higher organisms (Yancey et al. 1982). Most eubacteria expel ionic compounds in the event the electrolyte concentration becomes too high and replace these molecules with neutral ones such as glycine betaine to balance the cellular osmolality. The increase in electrolyte concentration (or ionic strength) upon a modest decrease in turgor pressure would thus represent an excellent trigger ('osmotic signal') for the activation of any osmoregulated transporter of neutral compatible solutes. Actually, it would prevent the osmotic stress from turning into 'cytoplasmic electrolyte stress'. Why, then, is the increase in intracellular osmolality less suitable as osmotic signal? In order to maintain a relatively constant turgor at different external osmolalties, the cell must be able to switch on transporters and take up organic compatible osmolytes with maximal activity at different internal osmolalities. In other words, the ability of (the majority of) microorganisms to grow at their maximal rate over a wide range of medium osmolalities implies that cellular processes function optimally over a wide range of intracellular osmolalities. Finally, the cell could use the osmotic upshift-dependent change in concentration of a specific molecule as signal, but ionic strength or collective ion concentration seems to be a more general signal for osmoresponsive systems. Specific signals may be used to tune the activity of a system as examplified by the *trans*-inhibition mechanisms in *L. monocytogenes* and *Lb. plantarum*.

Why does membrane tension, or more specifically a change in the lateral pressure profile of the lipid bilayer, represent a sensible mechanism to respond to osmotic downshifts? Following an osmotic downshift, the increase in turgor can be to some extent sustained by the cell envelope. The elastic properties of the cell membrane and peptidoglycan layer are such that some expansion is tolerated, but at some point the membrane or wall will rupture and the cell lyses. In this regard, it is worth noting that the cell turgor of a Gram-positive bacterium is in the range of 20–30 atm, which is more than 10 times the pressure a car tire has to resist. We propose that the decrease in cytoplasmic ion concentration, upon a severe osmotic downshift is much less detrimental to the cell than the accompanying increase turgor. Given the elasticity of the cell envelope, the increase in turgor will result in an increase in membrane tension. This tension within the membrane will be experienced by any membrane-embedded protein, and, for pressure valves such as MscL, this parameter represents an ideal gating signal. At present, we cannot exclude the possibility that the decrease in cytoplasmic ionic strenth or osmolality has some modulatory role in the gating of the mechanosensitive channels.

Concluding remarks

The emerging picture is that intracellular ionic solutes (or ionic strength) serve as a signal for the activation of the upshift-activated bacterial transporters and sensor kinases. For at least one system from a lactic acid bacterium, there is strong evidence that the signal is transduced to the protein complex via alterations in the protein-lipid interactions rather than direct sensing of ion concentration or ionic strength by the protein(s). The osmotic downshift-activated mechanosensitive channels, on the other hand, sense tension in the membrane but also here protein–lipid interactions serve a direct role in the transduction of the osmotic signal. This membrane-mediated transduction of the osmotic signal offers, in principle, additional means of control via alterations in lipid composition in the process of osmotic adaptation. Such potential regulation is suggested by the *in vitro* studies with the OpuA ABC transporter from *L. lactis*, for which the ionic set point for activation could be shifted to higher concentrations of intracellular ionic osmolytes by increasing the fraction of ionic lipids in the membrane.

Acknowledgements

This research was supported by grants from The Netherlands Foundation of Life Sciences (ALW) and the Foundation for Chemcial Sciences (CW), which are subsidized by the Netherlands Organization for Scientific Research (NWO), and the National Leading Research Institute 'Materials Science Center^{plus}'. I would like to thank the previous and present group members who have contributed to the research described in this chapter. These are F.J.M. Detmers, M.K. Doeven, R. Duurkens, J. Folgering, R.H.E. Friesen, E.R. Geertsma, E. Glaasker, M.G.W. Gunnewijk, A. Hagting, S.A. Henstra, E.H. Heuberger, T. van der Heide, J. Knol, W.N. Konings, E.R.S. Kunji, F.C. Lanfermeijer, J.S. Lolkema, E. Oldehinkel, J. Patzlaff, A. Picon, Y. Sanz, G. Schuurman-Wolters, L.M. Veenhoff, many M.Sc. students and many collaborators from various laboratories.

References

Arakawa T & Timasheff SN, (1985) The stabilization of proteins by osmolytes. Biophys. J., 47: 411–414.

Blount P & Moe PC (1999) Bacterial mechanosensitive channels: integrating physiology, structure and function. Trends Microbiol. 7: 420–424.

Bolotin A, Wincker P, Mauger S, Jaillon O, Malarme K, Weissenbach J, Ehrlich SD & Sorokin A (2001) The complete genome sequence of the lactic acid bacterium *Lactococcus lactis* ssp. *lactis* IL1403. Genome Res. 11: 731–753.

Botsford JL & Harman JG (1992) Cyclic AMP in prokaryotes. Microbiol. Rev. 56: 100–122.

Brochu D, and Vadeboncoeur C (1999). The HPr(Ser) kinase of Streptococcus salivarius: purification, properties, and cloning of the *hprK* gene. J. Bacteriol. 181: 709–717.

Cantor RS (1999) Lipid composition and the lateral pressure profile in bilayers. Biophys. J. 76: 2625–2639.

Chang G, Spencer RH, Lee AT, Barclay MT & Rees DC (1998) Structure of the MscL homolog from *Mycobacterium tuberculosis*: a gated mechanosensitive ion channel. Science 282: 2220–2226.

Charrier V, Buckley E, Parsonage, D, Galinier A, Darbon E, Jaquinod M, Forest E, Deutscher J & Claiborne A (1997) Cloning and sequencing of two enterococcal glpK genes and regulation of the encoded glycerol kinases by phospho*enol*pyruvate-dependent, phosphotransferase system-catalyzed phosphorylation of a single histidyl residue. J. Biol. Chem. 272: 14166–14174.

Cohn M & Horibata K (1959) Physiology of the inhibition by glucose of the induced synthesis of β-galactosidase-enzyme system of *Escherichia coli*. J. Bacteriol. 78: 624–635.

Csonka LN & Hanson AD (1991) Prokaryotic osmoregulation: genetics and physiology. Annu. Rev. Microbiol. 45: 569–606.

Deutscher J & Engelman R (1984) Purification and characterization of an ATP-dependent protein kinase from *Streptococcus feacalis*. FEMS Microbiol. Lett. 23: 57–162.

Deutscher J & Saier MH Jr (1983) ATP-dependent protein kinase-catalyzed phosphorylation of a seryl residue in HPr, a phosphate carrier protein of the phosphotransferase system in *Streptococcus pyogenes*. Proc. Natl. Acad. Sci. U.S.A. 80: 6790–6794.

Deutscher J & Sauerwald H (1986) Stimulation of dihydroxyacetone and glycerol kinase activity in *Streptococcus faecalis* by phospho*enol*pyruvate-dependent phosphorylation catalyzed by enzyme I and HPr of the phosphotransferase system. J. Bacteriol. 166: 829–836.

Deutscher J, Bauer B & Sauerwald H (1993) Regulation of glycerol metabolism in *Enterococcus faecalis* by phospho*enol*pyruvate-dependent phosphorylation of glycerol kinase catalyzed by enzyme I and HPr of the phosphotransferase system. J. Bacteriol. 175: 3730–3733.

Deutscher J, Kessler U & Hengstenberg W (1985) Streptococcal phospho*enol*pyruvate: sugar phosphotransferase system: purification and characterization of a phosphoprotein phosphatase which hydrolyzes the phosphoryl bond in seryl-phosphorylated histidine- containing protein. J. Bacteriol. 163: 1203–1209.

Deutscher J, Kuster E, Bergstedt U, Charrier V & Hillen W (1995) Protein kinase-dependent HPr/CcpA interaction links glycolytic activity to carbon catabolite repression in gram-positive bacteria. Mol. Microbiol. 15: 1049–1053.

Dills SS, Apperson A, Schmidt MR & Saier MH. Jr (1980) Carbohydrate transport in bacteria. Microbiol. Rev. 44: 385–418.

Dossonnet V, Monedero V, Zagorec M, Galinier A, Perez-Martinez G & Deutscher J (2000) Phosphorylation of HPr by the bifunctional HPr Kinase/P-ser-HPr phosphatase from *Lactobacillus casei* controls catabolite repression and inducer exclusion but not inducer expulsion. J. Bacteriol. 182: 2582–2590.

Doyle RJ & Marquis RE (1994) Elastic, flexible peptidoglycan and bacterial cell wall properties. Trends Microbiol, 2: 57–60.

Egeter O & Brückner R (1996) Catabolite repression mediated by the catabolite control protein CcpA in *Staphylococcus xylosus*. Mol. Mocrobiol. 21: 739–749.

Ellis RJ (2001) Macromolecular crowding: obvious but underappreciated. Trends Biochem. Sci. 26: 597–604.

Evans E & Skalak R (1980) Mechanics and thermodynamics of membranes. CRC Crit. Rev. Bioeng. 3: 181–418.

Foucaud C & Poolman B (1992) Lactose transport system of *Streptococcus thermophilus*. Functional reconstitution of the protein and characterization of the kinetic mechanism of transport. J. Biol. Chem. 267: 22087–22094.

Galinier A, Kravanja M, Engelmann R, Hengstenberg W, Kilhoffer MC, Deutscher J & Haiech J (1998) New protein kinase and protein phosphatase families mediate signal transduction in bacterial catabolite repression. Proc. Natl. Acad. Sci. U.S.A. 95: 1823–1828.

Glaasker E, Konings WN & Poolman B (1996) Glycine-betaine fluxes in *Lactobacillus plantarum* during osmostasis and hyper- and hypoosmotic shock. J. Biol. Chem. 271: 10060–10065.

Glaasker E, Heuberger EHML, Konings WN & Poolman B (1998) Mechanism of osmotic activation of the quaternary ammonium compound transporter (QacT) of *Lactobacillus plantarum*. J. Bacteriol, 180: 5540–5546.

Gunnewijk MGW & Poolman B (2000a) Phosphorylation state of HPr determines the level of expression and the extent of phosphorylation of the lactose transport protein of *Streptococcus thermophilus*. J. Biol. Chem. 275: 34073–34079.

Gunnewijk MGW & Poolman B (2000b) HPr(His~P)-mediated phosphorylation differently affects counterflow and proton motive force-driven uptake via the lactose transport protein of *Streptococcus thermophilus*. J. Biol. Chem. 275: 34080–34085.

Gunnewijk MGW, Sulter G, Postma PW & Poolman B (1997) Molecular Mechanisms of Signaling and Membrane Transport. Springer, Berlin, Heidelberg

Gunnewijk MGW, Postma PW & Poolman B (1999) Phosphorylation and functional properties of the IIA domain of the lactose transport protein of *Streptococcus thermophilus*. J. Bacteriol. 181: 632–641.

Hamill OP & Martinac B (2001) Molecular basis of mechanotransduction in living cells. Physiol. Rev. 81: 685–740.

Heide van der T & Poolman B (2000a) Osmoregulated ABC-transport system of *Lactococcus lactis* senses water stress via changes in the physical state of the membrane. Proc. Natl. Acad. Sci. USA, 97: 7102–7106.

Heide van der T & Poolman B (2000b) Glycine betaine transport in *Lactococcus lactis* is osmotically regulated at the level of expression and translocation. J. Bacteriol, 182: 203–206.

Heide van der T, Stuart MCA & Poolman B. (2001) On the osmotic signal and osmosensing mechanism of an ABC transporter for glycine betaine. EMBO J. 20: 7022–7032.

Hueck CJ & Hillen W (1995) Catabolite repression in *Bacillus subtilis*: a global regulatory mechanism for the gram-positive bacteria? Mol. Microbiol. 15: 395–401.

Hutkins R, Morris HA & McKay LL (1985) Galactose transport in *Streptococcus thermophilus*. Appl. Environ. Microbiol. 50: 772–776.

Jault JM, Fieulaine S, Nessler S, Gonzalo P, Di Pietro A, Deutscher J & Galinier A (2000) The HPr kinase from *Bacillus subtilis* is a homo-oligomeric enzyme which exhibits strong positive cooperativity for nucleotide and fructose 1,6-bisphosphate binding. J. Biol. Chem. 275: 1773–1780.

Jones BE, Dossonnet V, Kuster E, Hillen W, Deutscher J & Klevit RE (1997) Binding of the catabolite repressor protein CcpA to its DNA target is regulated by phosphorylation of its corepressor HPr. J. Biol. Chem. 272: 26530-26535.

Jones LJ, Carballido-Lopez R & Errington J (2001) Control of cell shape in bacteria: helical, actin-like filaments in *Bacillus subtilis*. Cell. 104: 913–922.

Killian JA & von Heijne G (2000) How proteins adapt to a membrane-water interface. Trend Biochem. Sci, 25: 429–434.

Kim JH, Guvener ZT, Cho JY, Chung KC & Chambliss GH (1995) Specificity of DNA binding activity of the *Bacillus subtilis* catabolite control protein CcpA. J. Bacteriol. 177: 5129–5134.

Knol J, Veenhoff L, Liang WJ, Henderson PJF, Leblanc G & Poolman B (1996) Unidirectional reconstitution into detergent-destabilized liposomes of the purified lactose transport system of *Streptococcus thermophilus*. J. Biol. Chem. 271: 15358–15366.

Konings WN, Poolman B & Driessen AJ (1989) Bioenergetics and solute transport in lactococci. Crit Rev.Microbiol. 16: 419–476.

Kravanja M, Engelmann R, Dossonnet V, Bluggel M, Meyer HE, Frank R, Galinier A, Deutscher J, Schnell N & Hengstenberg W (1999) The hprK gene of *Enterococcus faecalis* encodes a novel bifunctional enzyme: the HPr kinase/phosphatase. Mol. Microbiol. 31: 59–66.

Levina L, Totemeyer S, Stokes NR, Louis P, Jones MA & Booth IR (1999) Protection of *Escherichia coli* cells against extreme turgor by activation of MscS and MscL mechanosensitive channels: identification of genes required for MscS activity. EMBO Journal, 18: 1730–1737.

Luesink EJ, van Herpen RE, Grossiord BP, Kuipers OP & de Vos WM (1998) Transcriptional activation of the glycolytic *las* operon and catabolite repression of the gal operon in *Lactococcus lactis* are mediated by the catabolite control protein CcpA. Mol. Microbiol. 30: 789–798.

Marsh D (1996) Lateral pressure in membranes. Biochim. Biophys. Acta 1286: 183–223.

Mason PW, Carbone DP, Cushman RA & Waggoner AS (1981) The importance of inorganic phosphate in regulation of energy metabolism of *Streptococcus lactis*. J. Biol. Chem. 256: 1861–1866.

McGinnis JF & Paigen K (1969) Catabolite inhibition: a general phenomenon in the control of carbohydrate utilization. J. Bacteriol. 100: 902–913.

Minton AP, Colclasure GG & Parker JC (1992) Model for the role of macromolecular crowding in regulation of cellular volume. Proc. Natl. Acad. Sci. U.S.A. 89: 10504–10506.

Monedero V, Gosalbes MJ & Perez-Martinez G (1997) Catabolite repression in *Lactobacillus casei* ATCC 393 is mediated by CcpA. J. Bacteriol. 179: 6657–6664.

Monod J (1942) Rescherches sur la croissance des cultures bacteriennes, 2nd ed. Hermann, Paris.

Obis D, Guillot A, Gripon, J-C, Renault P, Bolotin A & Mistou M-Y (1999) Genentic and biochemical characterization of a high-affinity betaine uptake system (BusA) in *Lactococcus lactis* reveals a new functional organization within bacterial ABC transporters. J. Bacteriol, 181; 6238–6246.

Parsegian VA, Rand RP & Rau DC (1995) Macromolecules and water: probing with osmotic stress. Methods Enzymol. 259: 43–94.

Perlman RL, De Crombrugghe B & Pastan I (1969) Cyclic AMP regulates catabolite and transient repression in *E. coli*. Nature 223: 810–812.

Poolman B (1993) Energy transduction in lactic acid bacteria. FEMS Microbiol. Rev. 12: 125–148.

Poolman B and Glaasker E (1998) Regulation of compatible solute accumulation in bacteria. Mol. Microbiol, 29: 397–407.

Poolman B & Konings WN (1993) Secondary solute transport in bacteria. Biochim. Biophys. Acta 1183: 5–39.

Poolman B, Royer TJ, Mainzer SE & Schmidt BF (1989) Lactose transport system of *Streptococcus thermophilus*: a hybrid protein with homology to the melibiose carrier and enzyme III of phospho*enol*pyruvate-dependent phosphotransferase systems. J. Bacteriol. 171: 244–253.

Poolman B, Knol J, Mollet B, Nieuwenhuis B & Sulter G (1995) Regulation of bacterial sugar-H^+ symport by phospho*enol*pyruvate-dependent enzyme I/HPr-mediated phosphorylation. Proc. Natl. Acad. Sci. U.S.A. 92: 778–782.

Poolman B, Knol J, van de DC, Henderson PJ, Liang WJ, Leblanc G, Pourcher T & Mus-Veteau I (1996) Cation and sugar selectivity determinants in a novel family of transport proteins. Mol. Microbiol. 19: 911–922.

Poolman B, Blount P, Folgering JHA, Friesen RHE, Moe P & van der Heide T. (2002) How do membrane proteins sense water stress? Mol. Microbiol, 44, 889–902

Reid C & Rand RP (1997) Probing protein hydration and conformational states in solution. Biophys J., 72, 1022–1030.

Reizer J & Panos C (1980) Regulation of beta-galactoside phosphate accumulation in *Streptococcus pyogenes* by an expulsion mechanism. Proc. Natl. Acad. Sci. U.S.A. 77: 5497–5501.

Reizer J, Novotny MJ, Hengstenberg W & Saier MH Jr (1984) Properties of ATP-dependent protein kinase from *Streptococcus pyogenes* that phosphorylates a seryl residue in HPr, a phosphocarrier protein of the phosphotransferase system. J. Bacteriol. 160: 333–340.

Reizer J, Sutrina SL, Saier MH, Stewart GC, Peterkofsky A & Reddy P (1989) Mechanistic and physiological consequences of HPr(ser) phosphorylation on the activities of the phosphoenolpyruvate:sugar phosphotransferase system in gram-positive bacteria: studies with site-specific mutants of HPr. EMBO J. 8: 2111–2120.

Reizer J, Bergstedt U, Galinier A, Kuster E, Saier MH Jr, Hillen W, Steinmetz M & Deutscher J (1996) Catabolite repression resistance of gnt operon expression in *Bacillus subtilis* conferred by mutation of His-15, the site of phospho*enol*pyruvate-dependent phosphorylation of the phosphocarrier protein HPr. J. Bacteriol. 178: 5480–5486.

Robillard GT & Lolkema JS (1988) Enzymes II of the phospho*enol*pyruvate-dependent sugar transport systems: a review of their structure and mechanism of sugar transport. Biochim. Biophys. Acta 947: 493–519.

Romano AH, Brino G, Peterkofsky A & Reizer J (1987) Regulation of beta-galactoside transport and accumulation in heterofermentative lactic acid bacteria. J. Bacteriol. 169: 5589–5596.

Sukharev S, Durell SR & Guy HR (2001b) Structural models of the mscl gating mechanism. Biophys. J. 81: 917–936.

Sukharev S, Betanzos M, Chiang C-S & Guy HR (2001a) The gating mechanism of the large mechanosensitive channel MscL. Nature 409: 720–724.

Sukharev SI, Blount P, Martinac B, Blattner FR & Kung C. (1994) A large-conductance mechanosensitive channel in *E. coli* encoded by *mscL* alone. Nature 368: 265–268.

Sukharev SI, Sigurdson WJ, Kung C & Sachs F (1999) Energetic and spatial parameters for gating of the bacterial large conductance mechanosensitive channel, MscL. J. Gen. Physiol. 113: 525–540.

Thompson J & Torchia DA (1984) Use of ^{31}P nuclear magnetic resonance spectroscopy and ^{14}C fluorography in studies of glycolysis and regulation of pyruvate kinase in *Streptococcus lactis*. J. Bacteriol. 158: 791–800.

van den Bogaard PTC, Kleerebezem M, Kuipers OP & de Vos WM (2001) Control of lactose transport, β-galactosidase activity and glycolysis by CcpA in *Streptococcus thermophilus*: evidence for carbon catabolite repression by a non-phospho*enol*pyruvate-dependent phosphotransferase system sugar. J. Bacteriol., 182: 5982–5989.

Vaughan EE, van den Bogaard PTC, Catzeddu P, Kuipers OP & de Vos WM (2001) Activation of silent gal genes in the *lac-gal* regulon of *Streptococcus thermophilus*. J. Bacteriol. 183: 1184–1194.

Veenhoff LM & Poolman B (1999) Substrate recognition at the cytoplasmic and extracellular binding site of the lactose transport protein of *Streptococcus thermophilus*. J. Biol. Chem. 274: 33244–33250.

Veenhoff LM, Heuberger EHML & Poolman B (2001) The lactose transport protein is a cooperative dimer with two sugar translocation pathways. EMBO J. 20: 3056–3062.

Verheul A, Glaasker E, Poolman B & Abee T (1997) Betaine and L-carnitine transport in response to osmotic signals in *Listeria monocytogenes* Scott A. J. Bacteriol., 179: 6979–6985.

Verkman, A.S. (2002) Solute and macromolecule diffusion in cellular agueous compartments. Trends Biochem. Sci. 2: 27–33.

Viana R, Monedero V, Dossonnet V, Vadeboncoeur C, Perez-Martinez G & Deutscher J (2000) Enzyme I and HPr from *Lactobacillus casei*: their role in sugar transport, carbon catabolite repression and inducer exclusion. Mol. Microbiol. 36: 570–584.

Vodyanoy I, Bezrukow SM & Parsegian VA (1993) Probing alamethicin channels with water-soluble polymers. Size-modulated osmotic action. Biophys. J. 65: 2097–3105.

Wehtje C, Beijer L, Nilsson RP & Rutberg B (1995) Mutations in the glycerol kinase gene restore the ability of a *ptsGHI* mutant of *Bacillus subtilis* to grow on glycerol. Microbiology 141: 1193–1198.

Weickert MJ & Chambliss GH (1990) Site-directed mutagenesis of a catabolite repression operator sequence in *Bacillus subtilis*. Proc. Natl. Acad. Sci. U.S.A. 87: 6238–6242.

Wood JM (1999) Osmosensing by bacteria: signals and membrane-based sensors. Microbiol. Mol. Biol. Rev. 63: 230–262.

Yancey PH, Clark ME, Hand SC, Bowlus RD & Somero GN (1982) Living with water stress: evolution of osmolyte systems. Science 217: 1214–1222.

Ye JJ & Saier MH Jr. (1995) Allosteric regulation of the glucose:H^+ symporter of *Lactobacillus brevis*: cooperative binding of glucose and HPr(ser-P). J. Bacteriol. 177: 1900–1902.

Ye JJ & Saier MH Jr. (1996) Regulation of sugar uptake via the phospho*enol*pyruvate-dependent phosphotransferase systems in *Bacillus subtilis* and *Lactococcus lactis* is mediated by ATP-dependent phosphorylation of seryl residue 46 in HPr. J. Bacteriol. 178: 3557–3563.

Ye JJ, Neal JW, Cui X, Reizer J & Saier MH Jr. (1994a) Regulation of the glucose:H^+ symporter by metabolite-activated ATP-dependent phosphorylation of HPr in *Lactobacillus brevis*. J. Bacteriol. 176: 3484–3492.

Ye JJ, Reizer J, Cui X & Saier MH Jr. (1994b) ATP-dependent phosphorylation of serine-46 in the phosphocarrier protein HPr regulates lactose/H^+ symport in Lactobacillus brevis. Proc. Natl. Acad. Sci. U.S.A. 91: 3102–3106.

Ye JJ, Reizer J, Cui X & Saier MH Jr. (1994c) Inhibition of the phospho*enol*pyruvate:lactose phosphotransferase system and activation of a cytoplasmic sugar-phosphate phosphatase in *Lactococcus lactis* by ATP-dependent metabolite-activated phosphorylation of serine 46 in the phosphocarrier protein HPr. J. Biol. Chem. 269: 11837–11844.

Antonie van Leeuwenhoek **82:** 165–185, 2002.

Lantibiotics produced by lactic acid bacteria: structure, function and applications

Denis Twomey[1,3], R. P. Ross[3], Maire Ryan[1,3], Billy Meaney[3] & C. Hill[1,2,*]
[1]*Department of Microbiology, University College Cork, Co. Cork, Ireland;* [2]*National Food Biotechnology Centre, University College Cork, Co. Cork, Ireland;* [3]*Teagasc, Dairy Products Research Centre, Moorepark, Fermoy, Co. Cork, Ireland (*Author for correspondence; E-mail: c.hill@ucc.ie)*

Abstract

Lantibiotics are a diverse group of heavily modified antimicrobial and/or signalling peptides produced by a wide range of bacteria, including a variety of lactic acid bacteria. Based on their diverse structures and mode of action, at least six separate lantibiotic subgroups can be suggested, but all subgroups are characterized by significant post-translational modifications, which include the formation of (β-methyl)lanthionines, among other unusual alterations. These small peptides are produced, modified, exported, sensed and combated by a complex set of proteins encoded by (usually) co-ordinately regulated operons. In some instances, the production and immunity have been shown to be auto-regulated by the mature lantibiotic. Since their discovery, interest in lantibiotics has been fuelled by their obvious potential as food-grade antimicrobials to improve food safety and quality; a potential which, to date, has been realised only by the longest characterised molecule, nisin. In addition, these peptides are often mooted as alternatives to antibiotics for some biomedical applications. The purpose of this paper is to review recent developments in our understanding of lantibiotic structure, molecular genetics and applications for this unusual class of bacteriocins.

Introduction

Lactic acid bacteria (LAB) are capable of producing a wide range of proteins and peptides which have an associated antimicrobial activity. The larger proteins (>20 kDa) often possess enzymatic activity directed against the cell envelope of target bacteria; for example, lysins and muramidases. Antimicrobial peptides (bacteriocins) also target the cell envelope, but use non-enzymatic mechanisms to disrupt the integrity of the target cell membrane and/or inhibit cell wall synthesis. These peptides vary in structure from essentially linear, unmodified bacteriocins (which may contain one or more disulphide bridges) to complex post-translationally modified peptides which may contain multiple rings bridged by lanthionine (Lan) or β-methyllanthionine (MeLan) residues. Bacteriocins containing either Lan or MeLan are referred to as lantibiotics and constitute a complex set of molecules which have come under intense scrutiny in recent years, as they offer an attractive addition to 'conventional' antimicrobials, particularly in an era where antibiotic resistance compromises many current applications. The genetics and structural characteristics of many of these lantibiotics have been extensively studied and has been recently reviewed (Jack et al., 1998; Sahl & Bierbaum; 1998; McAuliffe et al. 2000). This review will focus on the genetics and characterisation of a number of lantibiotics produced by members of the LAB and will give a brief overview of some of their potential applications. Those characteristics which make lantibiotics suitable candidates for the control of food pathogens and food spoilage bacteria will also be discussed, in addition to some of their potential as biomedical antimicrobial agents.

Classification of lantibiotics produced by LAB

Lantibiotics are distinguishable from other bacteriocins by the presence of internal ring structures resulting from the formation of either Lan or MeLan bridges. These unusual amino acids are formed following the dehydration of either serine or threonine to form the α,β-unsaturated amino acids didehydroalanine (Dha) or 2,3-didehydrobutyrine (Dhb), re-

Table 1. Lantibiotics produced by members of the lactic acid bacteria

Lantibiotic	Producing Strain	Reference
Nisin A	*Lactococcus lactis* NIZOR5, 6F3, NCFB894, ATCC11454	Gross & Morell (1971)
Nisin Z	*Lactococcus lactis* N8, NIZO22186	Mulders et al. (1991); de Vos (1993)
Lacticin 481	*Lactococcus lactis* CNRZ481, ADRIA85LO30	Piard et al. (1993), van den Hooven et al. (1996)
Lacticin 3147	*Lactococcus lactis* DPC3147, IFLP105	Ryan et al. (1996), Martinez-Cuesta
Lactocin S	*Lactobacillus sake* L45	Mortvedt et al. (1991)
Plantaricin C	*Lactobacillus plantarum*	Turner et al. (1999)
Plantaricin W	*Lactocbacillus plantarum* LMG 2379	Holo et al. (2001)
Cytolysin	*Enterococcus faecalis* FA22	Booth et al (1996)

spectively. These residues are subsequently covalently linked to a free cysteine within the same peptide leading to the formation of either Lan or MeLan. The relative positions of the dehydrated amino acid and its specific target cysteine dictates the size and position of the resulting ring. To date, the presence of Lan and/or MeLan residues has been confirmed in eight different lantibiotics produced by LAB (Table 1) and the producer strains include lactococci, lactobacilli and enterococci. In addition, a number of antimicrobial peptides closely related to known lantibiotics have also been identified in LAB but the presence of Lan or MeLan remains to be confirmed. These include a lactococcal peptide designated lacticin J46 (previously referred to as bacteriocin J46; Huot et al. 1996) which appears to be a closely related derivative of lacticin 481 and has just four amino changes in the deduced mature peptide. It is likely that J46 will represent a further addition to the lantibiotic 'family' of inhibitors. While the presence of either Lan or MeLan is the principal diagnostic feature of lantibiotics, other post-translationally modified residues are also encountered, including free Dha or Dhb residues which have no cysteine partner with which to interact; 'D' isomers of amino acids such as D-alanine derived from an original L-serine residue (Skaugen et al. 1994; Ryan et al. 1999); and deaminated amino acids such as 2-oxobutyryl and 2-oxopropionyl (Table 1).

Structural similarities and characteristics of LAB lantibiotics

A number of LAB lantibiotics have been characterised to the point where their complete structure has been determined, and extensive knowledge has been acquired about the mechanistic basis of the antimicrobial activity. The most comprehensive studies have concerned nisin, a molecule whose antimicrobial properties were first discovered in the 1920s by Rogers & Whittier (1928). It is produced by certain strains of *Lactococcus lactis,* and has been approved as a food preservative in over 50 countries worldwide (Delves-Broughton 1990). While many LAB lantibiotics are single peptides, some require the complementary activity of two peptides, including cytolysin ($CylL_l$ and $CylL_s$; Gilmore et al. 1994), lacticin 3147 (LtnA1 and LtnA2; Ryan et al. 1999) and plantaricin W (Plwα and Plwβ; Holo et al. 2001).

While the presence of Lan and MeLan is a unifying feature of lantibiotics these post-translationally modified antimicrobial peptides are not a homogenous group and can be sub-grouped based on a number of other criteria, including the structure of the mature peptide, the structure of the prepeptide, the biosynthetic machinery and lastly, the mode of action (Jung 1991; Vos et al. 1995; Guder et al. 1998). Jung (1991) suggested that the lantibiotics could be separated into two groups, type A and type B. The classical view of the type A lantibiotics is that they are elongated flexible molecules with a positive charge, and were generally thought to act by depolarising the cytoplasmic membranes, leading to the formation of pores and the leakage of essential cell constituents. In these lantibiotics, the formation of Lan or MeLan occurs following the reaction of the sulphydryl group of cysteine with a proximal unsaturated amino acid (Dha or Dhb). The prototype Type A lantibiotic is nisin, which has a linear corkscrew like structure with a series of five ring structures and a flexible central hinge region (Figure 1). In contrast, Type B lantibiotics are globular in structure, and have a negative or no net charge and interfere with enzymatic reactions

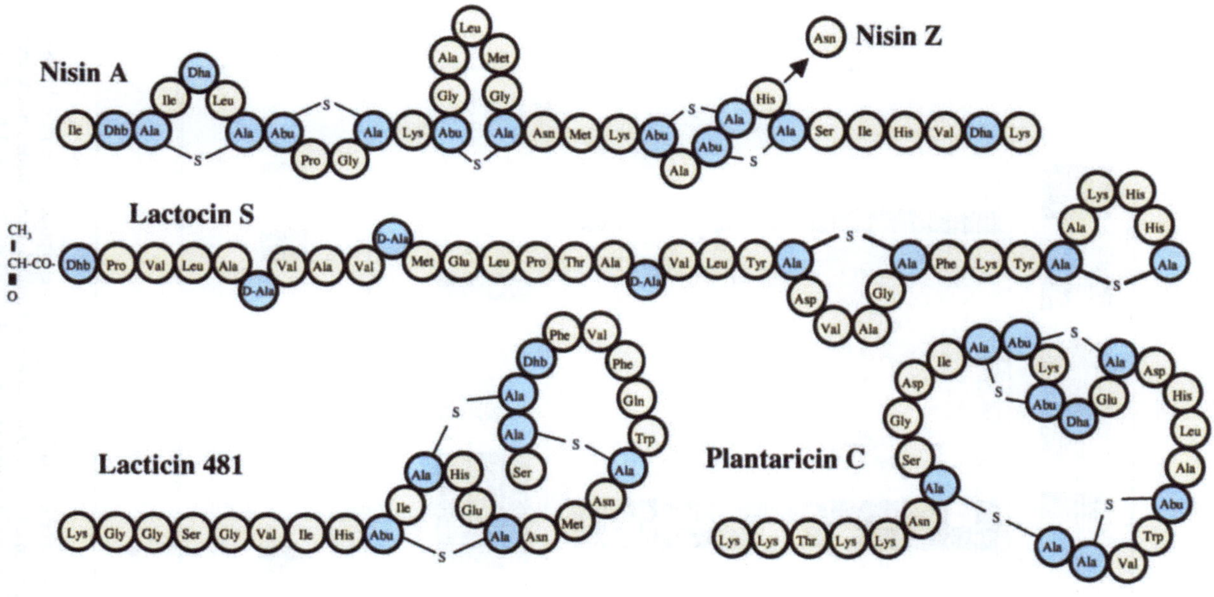

Figure 1. Structure of a selection of lantibiotic peptides representative of different lantibiotic groupings. Modified amino acids are coloured blue. Nisin A and nisin Z are representative of the nisin subgroup and have an elongated, corkscrew-like linear structure. Lactocin S also has an elongated structure but due to its lack of primary sequence similarity with other lantibiotics, it is in a group by itself. Lacticin 481 has a linear N-terminal end with a globular cross-bridged C-terminus.

essential to the growth and survival of the target bacterium. In these peptides, Lan or MeLan formation involves the interaction between a cysteine sulphydryl group and a distal unsaturated amino acid, resulting in a biomolecule with a compact globular structure. Members include mersacidin and actagardine (Figure 1). While the majority of lantibiotics appear to be derived from either of these two lineages, lantibiotics with intermediate properties have now also been characterised, and it is clear that as more lantibiotics are characterised new classification schemes will have to be devised. In addition, the identification of two component lantibiotics may make it more difficult to characterise these lantibiotics as strictly type A or B as they may have one peptide from each lineage. A further complication is that, in addition to pore formation, some type A lantibiotics also effect enzymatic reactions essential to the cell (i.e., nisin interferes with peptidogylcan biosynthesis), a feature normally associated with the type B lantibiotics. Recently, lantibiotics have been further subgrouped into groups of related peptides based on a number of functional characteristics (Sahl & Bierbaum 1998; Guder et al. 2000). As this review focuses on lantibiotics from LAB, we have attempted to group the LAB lantibiotics into their appropriate subgroup and to compare them with their related peptides. As the exact bridging pattern in all LAB lantibiotics is not known, groupings were based on primary sequence comparisons of the mature peptides. Sequences of all confirmed LAB lantibiotics were compared to database sequences using BLASTP and similar sequences were aligned using the CLUSTALW algorithm. This permitted sub-grouping based principally on primary sequence comparisons and does not take into consideration the known (or proposed) bridging pattern, or the presence of associated biosynthetic processing or transport genes.

Based on alignment of mature peptides, lantibiotic peptides from LAB were subdivided into six subgroups (Figure 2): Nisin A and Z belong to the nisin group; lacticin 481 (Piard et al. 1992; Rince et al. 1994), lacticin J46 (Huot et al. 1996) belong to the lacticin 481 group; plantaricin C (Turner et al. 1999), LtnA1 (Ryan et al. 1999), and Plwα (Holo et al. 2001) belong to the mersacidin group; LtnA2 (Ryan et al. 1999), and Plwβ (Holo et al. 2001) belong to the LtnA2 group; the two peptides from the two component cytolysin ($CylL_l$ and $CylL_S$; Gilmore et al. 1994) form a group of their own; while lactocin S (Skaugen et al. 1994) is grouped separately. Included in these alignments are a number of putative lantibiotic peptides from other genera whose identity as lantibiotics remains to be confirmed. Their inclusion is based

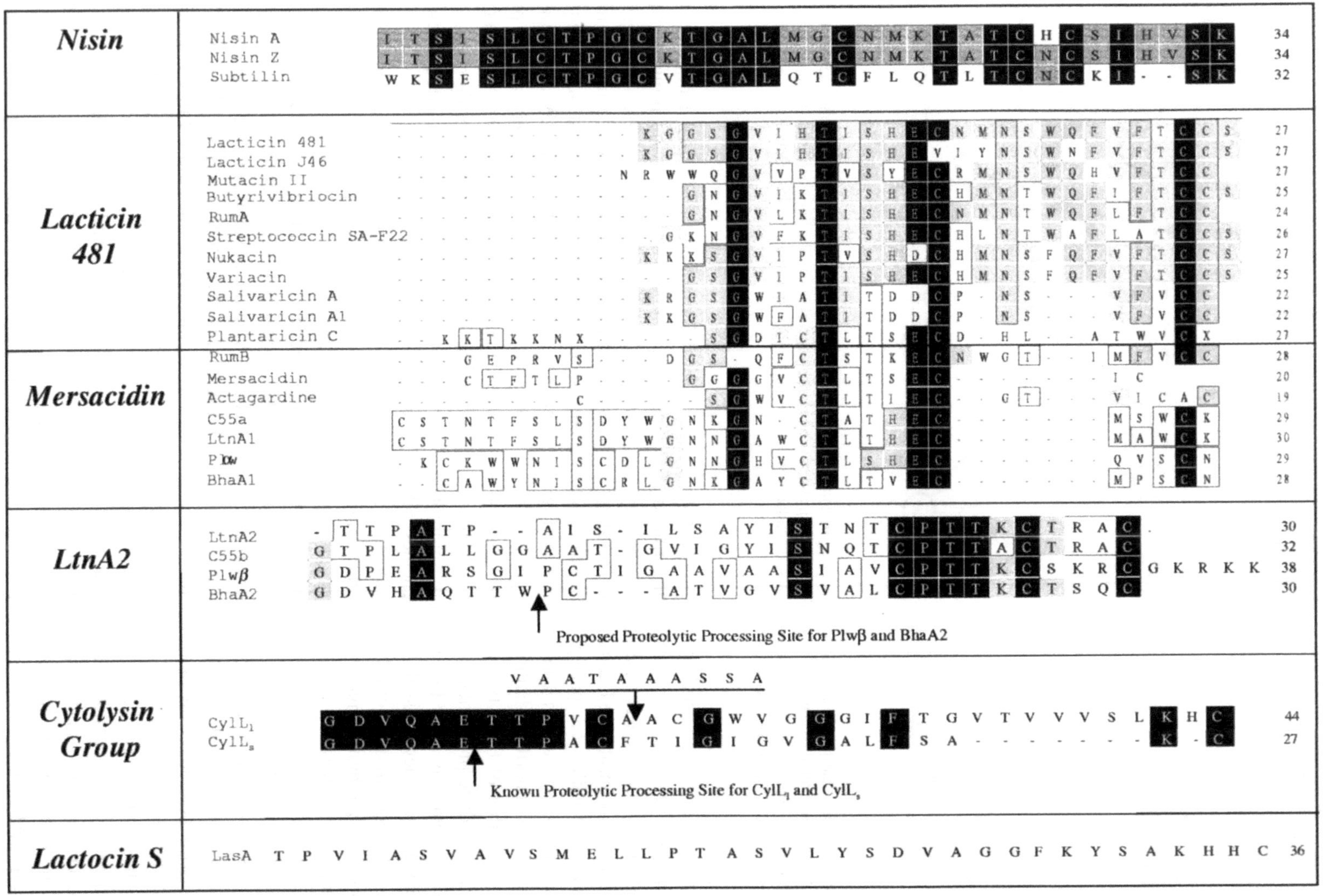

Figure 2. Subgrouping and alignment of LAB lantibiotic peptides and related peptides. Alignments are based on primary sequence comparisons of the mature peptides. The lacticin 481 and mersacidin subgroups have been aligned with each other as they share several conserved residues. Identical residues are highlighted with highly conserved amino acids shaded black and less conserved residues shaded lighter or boxed. The known proteolytic cleavage site in the cytolysin peptides is indicated. A proposed proteolytic processing site for Plwβ and BhaA2 is also

on their similarity with known lantibiotics and the location of their structural determinants among genes involved in immunity, modification, and/or transport of other lantibiotics. These include RumB from the strictly anaerobic bacterium, *Ruminococcus gnavus* which exhibits similarity to members of the lacticin 481 and mersacidin groups and is located immediately upstream of an apparently truncated-lantibiotic modification gene (EMBL AF320327). Immediately downstream of these genes are three almost identical copies of the structural gene for the recently described lantibiotic, ruminococcin A (Dabard et al. 2001; Gomez et al. 2002). Both components of a putative two-component lantibiotic from the completely sequenced genome of the alkaliphilic *Bacillus halodurans* have also been included (Takami et al. 2000). The deduced peptides have been designated BhaA1 and BhaA2, and have been included in the mersacidin and LtnA2 groups, respectively. Assigning the putative cleavage site of the precursor molecules was based on similarities with similar peptides. Given new recent insights into the proposed mode of action of a variety of lantibiotics this subgrouping scheme may help shed light on the mode of action of some LAB lantibiotics.

The Nisin subgroup

The nisin subgroup includes nisin A and its natural variant nisin Z which has a single amino acid change with Asn^{27} substituted by His^{27} as a consequence of a single nucleotide change. This substitution has little effect on the antimicrobial activity of nisin, although the increased solubility of nisin Z at pH 7.0 does offer greater potential applications at neutral pH. Subtilin, produced by *Bacillus subtilis,* is also included in the nisin group. Subtilin has ~60% identity to nisin and shares a pentacyclic structure. Members of this group have been extensively characterised and in particular, the precise structure of each molecule have been established. Nuclear magnetic resonance (NMR) studies have led to 3-D structures of nisin A and subtilin, both in aqueous solution and also bound to detergent micelles which mimic a membrane environment (van de Ven et al. 1991; Lian et al. 1992; van den Hooven et al. 1996a; Lavanant et al. 1998). The nisin molecule is composed of two structural domains, the N-terminus contains three lanthionine rings (A, B and C), while the C-terminus contains two intertwined rings (D and E). The Met^{21} residue acts as a flexible hinge region between the two domains and seems to be necessary for the mode of action (see below). Both nisin and subtilin are highly flexible in aqueous solution (although Ring B and the overlapping Rings D and E do confer some degree of localised rigidity) whereas, in lipophilic solutions they adopt a more rigid, helical structure. The N-terminus of both nisin and subtilin is essential for full activity, given that changes introduced at this site result in peptides with reduced activity. In particular, the Dha5 residue and an intact Ring A both appear to be essential for full activity (Kuipers et al. 1992). The nisin group bears similarity to two further lantibiotic clusters (Guder et al. 2000), the epidermin group (epidermin, Leu6-epidermin, Val1-Leu6-epidermin, mutacin B-Ny266, mutacin 1140 and mutacin III) and the pep5 group (pep5, epilancin K7 and epicidin 280). The N-terminal segment of the epidermin group shows significant structural relatedness to the nisin sub-group and the second ring of epidermin is identical to the second ring of nisin. The solution structures of Leu6-epidermin (also known as gallidermin) (Freund et al. 1991) and pep5 (Kellner et al. 1991) have been determined in a lipophilic environment and demonstrated that they too adopt an overall helical conformation with a corkscrew like structure. The charged residues appear to align to one side of the helix while the hydrophobic residues align along the opposite face. Similar to nisin and subtilin, the central region of gallidermin behaves as a flexible hinge region. In pep5, the presence of two didehydroamino acids tends to add structural rigidity to this central region (Kellner et al. 1991). These structural similarities indicate that nisin, subtilin, epidermin, gallidermin and all their variants constitute a structurally related subgroup of the lantibiotics.

The Lacticin 481 and Mersacidin subgroups

Lacticin 481 has been typically associated with type A lantibiotics while mersacidin was more associated with the globular type B lantibiotics. However, conserved residues are evident among these and related peptides indicating a relatively large group of related lantibiotics which can be subdivided into two related subgroups. The N-termini of the lacticin 481 cluster and the C-termini of the mersacidin cluster share the motif GXXXTX(T/S)EC. This motif includes several residues implicated in the bridging of several molecules (van den Hooven et al. 1996b; Krull et al. 2000) and also a Glu-Cys motif which has been proposed to be responsible for the inhibition of murein biosynthesis in mersacidin and actagardin (Zimmermann & Jung 1997; Sahl & Bierbaum 1998). While lacticin 481 was previously classified as a Type A lantibiotic, this bacteriocin and closely related peptides display similarities with mersacidin and actagardine which are more characteristic of the Type B lantibiotics. Studies on lacticin 481 (Piard et al. 1993; van den Hooven et al. 1996b) and the closely related lantibiotics streptococcin A-FF22 (Sahl & Bierbaum 1998) and mutacin II (Krull et al. 2000) indicated that a thioether bridge spans half the length of each peptide, thereby resulting in a more compact molecule with a bicyclic ring structure towards the C-terminal end of the molecule and an N-terminal linear conformation (Figure 1). These proposed structures for lacticin 481, streptococcin A-FF22 and mutacin II raises questions about their designation as linear Type A lantibiotics and indeed the similarities with mersacidin and actagardine, whose solution structures are also known, suggests they may have features more associated with the classical Type B grouping. The solution structure of plantaricin C has also been determined (Turner et al. 1999) and reveals a 27-amino acid peptide with a linear N-terminal end and a globular C-terminus. This overall structure bears similarity to the lacticin 481 structure and its designation as either a member of the lacticin 481 or mersacidin subgroups is difficult since it has features intermediate to both. Other lantibiotics within the lacticin 481 subgroup include salivaricin A (Ross et al. 1993), salivaricin A1 (Lys2, Phe7-salivaricin A, Ferretti et al. 2001), streptococcin A-M49 (Hynes et al. 1994), variacin (Pridmore et al. 1996), nukacin A (Kimura et al. 1998), mutacin II (Woodruff et al. 1998; Krull et al. 2000), RumB (EMBL AF320327) and butyrivibriocin OR79A (Kalmokoff et al. 1999). The similarities between lacticin 481 and lacticin J46 (also called bacteriocin J46) from *L. lactis* (Huot et al. 1996) suggest that these are highly related peptides. Twenty of the 24 residues of bacteriocin J46 are identical to those of lacticin 481. However, among the four differences observed is a Cys^{11} to Val change in lacticin J49. The corresponding Cys in lacticin 481 is involved in thioether bridge formation and forms part of the first ring structure which is highly conserved among other members of this group and forms part of the proposed active site of the mersacidin-like molecules. Peptides within the lacticin 481 subgroup contain a number of conserved regions, particularly among the amino acids comprising the ring structures, thereby suggesting a conserved bridging pattern. This is reflected in Figure 2 whereby all the residues known to be involved in thioether formation in lacticin 481 and mutacin II are highly conserved amongst their closest relatives and many are also conserved in mersacidin, actagardine and related molecules.

The best studied members of the mersacidin subgroup are mersacidin itself (Prasch et al. 1997) and actagardine (Zimmerman & Jung 1997), both of which are globular rigid structures with overlapping bridging structures (Figure 1). Both mersacidin and actagardine are tetracyclic lantibiotics. Ring B of actagardine is almost identical to Ring C of mersacidin, and thus may explain their similar activities in inhibiting cell wall biosynthesis at the level of transglycosylation by binding to lipid II (Brotz et al. 1998, 1997, 1995). A feature of actagardine, as shown by NMR, is the presence of two putative binding pockets. A region of high sequence similarity with mersacidin is located at the position of the second pocket in actagardine, suggesting that this is the site responsible for the mode of action (Zimmerman & Jung 1997). A number of peptides from two component LAB lantibiotics belong to the mersacidin subgroup, the lacticin 3147 LtnA1 peptide (Ryan et al. 1999) and the plantaricin W Plwα component (Holo et al. 2001). Further members of the mersacidin group include C55a (Navaratna et al. 1999) and the proposed lantibiotic peptide, BhaA1 (Takami et al. 2000), both of which are associated with two-component lantibiotics. Even though LtnA1, C55a and Plwα are found as components of a two-peptide lantibiotic, each has intrinsic antimicrobial properties (Ryan et al. 1999; Navaratna et al. 1999; Holo et al. 2001). The antimicrobial activity of BhaA1 or BhaA2 has not been reported. The four members of the two component systems are more closely related to each other than they are to other members of the group. It has previously been reported that LtnA1, C55a and Plwα form a novel group of lantibiotics (Holo et al. 2001), however their similarities with mersacidin and related peptides indicate they form part of this group which also has similarities to the larger lacticin 481 group (Edel et al. unpublished results). The thioether bridging patterns have not been determined for any of the two component lantibiotics. Based on the assumption that Plwα was a type A lantibiotic, Holo et al. (2001) proposed a bridging pattern consistent with an elongated, linear molecule. However, based on the similarities with the mersacidin group, Edel et al. (unpublished results) proposed a thioether bridging patterns between distal N-terminal and C-terminal residues suggesting a more globular conformation.

An unusual feature of the mature lacticin 3147 peptides is that they both contain the unusual amino acid D-alanine. For example, while the *ltnA1* structural gene dictates that the seventh amino acid in LtnA1 must be ribosomally synthesised as a serine, amino acid sequencing indicates the presence of an alanine residue which chiral phase GC analysis confirms as a D-isoform. It has been proposed that these D isoforms derive from the stereospecific hydrogenation of Dha formed following the dehydration of Ser to Dha (Ryan et al. 1999). The only other occurrence of D-alanine residues in ribosomally synthesised peptides is in the lantibiotic, lactocin S (Skaugen et al. 1994), where it was first discovered. The exact mechanism of D-Ala generation and the enzymes involved remain to be established. However, a candidate gene in the lacticin 3147 operon is *ltnD* located after the second modification gene and which bears some sequence similarity to dehydrogenase enzymes. A similar gene encoding a protein with almost 47% identity to LtnD is also present in the staphylococcin c55-encoding plasmid pETB (Yamaguchi et al. 2001). However, even though c55a and LtnA1 are very closely related peptides with just four amino acid differences predicted from their amino acid sequences, the published amino acid sequence of c55a (derived from Edman degradation following treatment with mercaptoethanol) indicates the presence of a Dha residue at position 7 rather than an alanine residue (Navaratna et al. 1999).

The LtnA2 group

The LtnA2 group of lantibiotics comprises four members, all of which are part of two component lantibiotics. In general, the two peptides of the two-component lantibiotics act synergistically to mediate their effect. However, while both of the plantaricin W peptides have intrinsic antimicrobial properties, both LtnA2 or C55b require the complimentary action of their partner peptide (Navaratha et al. 1999; Ryan et al. 1999; Holo et al. 2001). The structure or thioether bridging pattern has not been determined for any of these molecules but proposed models suggest they have a central lanthionine with two overlapping thioether bridges close to the C-terminus (Holo et al. 2001; Edel et al. unpublished results). This would give these molecules a general linear conformation similar to the type A lantibiotics. However, this proposed conformation would have a rigid central segment unlike the flexible hinge-like region of the type A lantibiotics, nisin and subtilin. Similar to other lantibiotic groups, greatest sequence conservation is seen among residues implicated in thioether bridge formation.

Holo et al. (2001) proposed that the N-terminus of Plwβ undergoes proteolytic attack after transport due to the presence of a consensus leader peptide consistent with a Gly-Ala cleavage site located upstream of the mature N-terminal residue, as happens in plantaricin A (Diep et al. 1994). This type of proteolytic cleavage has been demonstrated for both of the peptides of the two-component lantibiotic, cytolysin (see below; Gilmore et al. 1994) Considering the similarities between the N-termini of Plwβ, BhaA2 and the two cytolysin peptides (Figure 2), it is likely that all of these may undergo proteolytic processing after cleavage at the 'double-Gly' motif. By analogy with other lantibiotics including Pep5 (Kellner et al. 1991), the N-terminal hydroxy amino acids of LtnA2 and Plwβ have been proposed to contain a number of unusual residues which cause them to be blocked during amino acid sequencing via Edman degradation. LtnA2 is proposed to contain a 2-oxobutyryl group following the dehydration of the N-terminal threonine residue and subsequent deamination of Dhb (Ryan et al. 1999). Similarily, the N-terminus of Plwβ is proposed to contain a pyruvic acid residue, derived from the dehydration and subsequent deamination of Dha (Holo et al. 2001).

Lactocin S

Lactocin S is a 37-amino acid lantibiotic produced by *Lactobacillus sake* (Mortvedt, et al. 1991; Skaugen et al. 1994), whose primary sequence shares little similarity with other lantibiotic peptides and is therefore not grouped with any other lantibiotic. Although lactocin S shares little similarity with other lantibiotics, PCR analysis has indicated that the lasA gene is present in a variety of bacteriocin-producing lactobacilli strains isolated from fermented meat (Rodriguez et al. 1995). Lactocin S is notable as being the first procaryotic ribosomally synthesised peptide discovered containing D-Ala (Skaugen et al. 1994). Three D-Ala residues are located in this relatively long lantibiotic. The solution structure of lactocin S remains to be established, however, the proposed structure suggests a linear type A lantibiotic with two ring structures, both of which are located at the C-terminus. In addition to Lan and D-Ala, it has been inferred that lactocin S also contains either an N-terminal 2-oxo or 2-hydroxypropionyl group (Skaugen et al. 1994).

Mode of action studies indicated that lactocin S acts bactericidally in a pH-dependent fashion. The proposed linear structure of lactocin S suggests that it is more like the type A lantibiotics than the type B lantibiotics. However unlike the nisin subgroup, lactocin S is predicted to have a neutral p*I* (predicted pI = 7.1; Skaugen et al. 1994). Lactocin S is stable over a pH range of 4.5–9.0 but has no biological activity above pH 6.0. This pH dependent activity may be explained in that below pH 6.0 the effective charge contribution of the two histidine residues in the C-terminal ring are considered essential for the activity of this molecule, permitting the bacteriocin to interact with the cytoplasmic membrane (Sahl et al. 1987; Mortvedt et al. 1995). The importance of these residues and the majority of the other charged residues, which are also located towards the C-terminus, awaits confirmation of the tertiary structure and detailed mode of action studies.

Cytolysin

Cytolysin is a two component lantibiotic produced by some enterococcci which is unique in it that it has both bactericidal and cytotoxic activity. It has been shown to represent a significant virulence factor in enterococcal endocarditis (Chow et al. 1993) and endophthalmitis (Jett et al. 1992) and is encoded by as many as 60% of clinical enteroccocal isolates (Ike et al. 1987). Both peptides are required for hemolytic and bactericidal activity and are of unequal length, with $CylL_L$ containing 38 amino acids and $CylL_s$ containing 21 amino acids (Gilmore et al. 1994). There is a highly conserved 26-amino acid sequence in the prepeptides which may represent the signal that directs each of the subunits to the secretion machinery (Gilmore et al. 1994). After cleavage of the leader sequence, the peptides share a conserved N-terminus with an identical first 10 residues, while the C-termini share significantly less similarity. The cytolysin peptides have been grouped together in a separate group due to the lack of similarity with other lantibiotics. Following cleavage of the leader sequence at the 'double-Gly' motif, the N-termini of these peptides are similar to the proposed N-terminus of BhaA2 when its' leader is cleaved, with six identical residues in the first eight. While the cytolysin peptides and BhaA2 share a common motif, BhaA2 has been grouped with the LtnA2 group due to the greater similarity with this group of peptides, especially with Plwβ, and the presence of the highly conserved CPTT-motif. A segment of $CylL_L$ (AATAAAS), which is not present in $CylL_s$, is also present in another LAB bacteriocin, plantaricin 1.25β produced by *Lactobacillus plantarum* (Reminger et al. 1999). Although Plantaricin 1.25β is not a lantibiotic, the presence of the conserved motif suggests that it may be important for the functionality of these antimicrobial peptides. The solution structure of cytolysin remains to be determined and the bridging structure resolved. $CylL_L$ has the potential to form three thioether-bridges and a number of free dehydroamino acids and CylLs has been proposed to contain two thioether-bridges and three free dehydroamino acids (Booth et al. 1996; Jack et al. 1998). It is also speculated that both peptides possess a 2-oxobutyryl group at the N-terminus, similar to Pep5 and LtnA2 (Jack et al. 1998; Ryan et al. 1999).

In addition to cytotoxic activity, a further unusual feature of the cytolysin peptides is the double-processing step involved in peptide maturation. Most lantibiotics are processed once, when the leader sequence is cleaved off during export. However, after modified $CylL_L$and$CylL_S$ are secreted by an ABC-transporter (during which the leader is proteolytically removed), a further cleavage is performed by CylP (previously CylA) (Coburn et al. 1999). It was initially thought that this second processing step was responsible for self-protection, however, immunity has recently been ascribed to a novel open reading frame immediately adjacent to *cylP* (Coburn et al. 1999). It has recently been proposed that the N-terminus of Plwβ also undergoes proteolysis after transport due to the present of a consensus leader peptide consistent with Gly-Ala cleavage site located upstream of the known N-terminal residue (Holo et al. 2001). Considering the similarities between the N-termini of Plwβ, BhaA2 and the two cytolysin peptides, it is likely that all of these lantibiotics may undergo proteolysis after cleavage at the 'double-Gly' motif in order to become active.

The immense strides which have been made in structural analysis of lantibiotics using technologies such as NMR and MS have allowed researchers to propose functions for the different regions of the molecules and hence, build up a greater understanding of the functional nature of lantibiotics. For example, structural studies on Type A lantibiotics have shown that they are more flexible in aqueous environment than in lipophilic solution, which is also proposed to occur in the membrane of sensitive cells. This is in contrast to the globular Type B molecules which do not change their conformation and reflects their

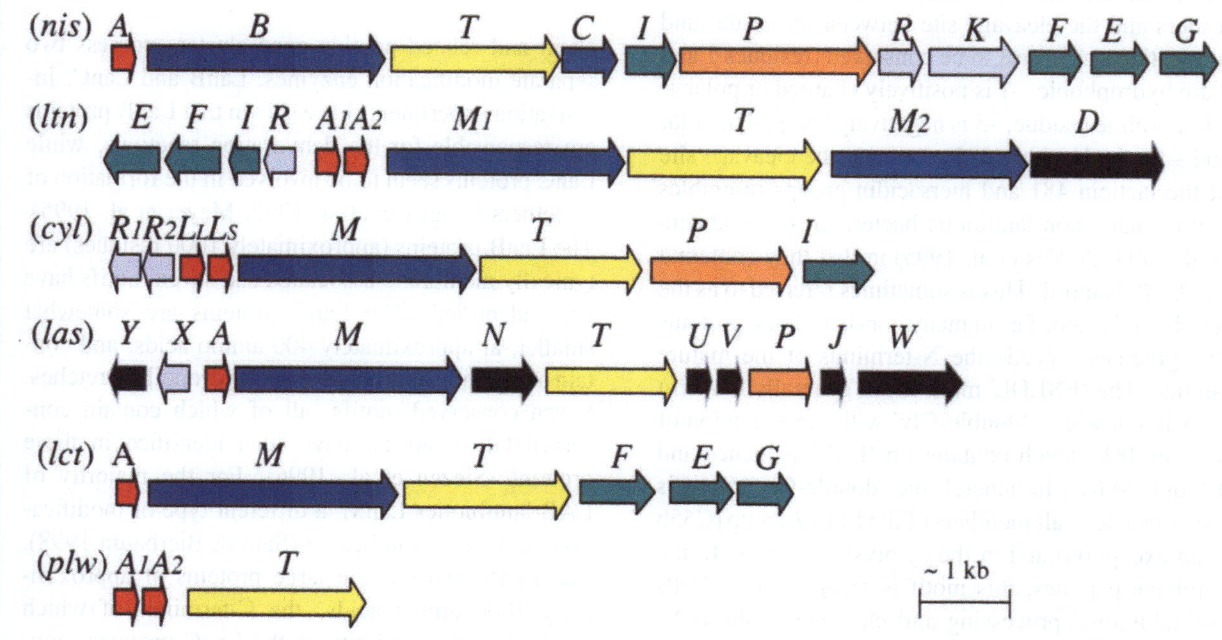

Figure 3. Schematic representation of the known genetic organisation of LAB lantibiotics; nisin (*nis*), lacticin 3147 (*ltn*), cytolysin (*cyl*), lactocin S (*las*), lacticin 481 (*lct*), and plantaracin W (*plw*). Genes with similar functions are highlighted in the same colour; red for structural genes; blue for modification; yellow for transport; green for immunity; orange for proteolytic processing; light blue for regulation. Genes with unassigned functions are highlighted in black.

activity of interacting with precise active sites of enzymes. Furthermore, the amphiphilic nature of Type A lantibiotics also fits into the model whereby molecules adopt a helical structure in the membrane with hydrophobic residues in contact with the lipid tails of the membrane and the charged residues interacting with charged phospholipid head groups. Conformational analysis may also be used to deduce the role of the biosynthetic intermediates in order to gain some insight into the role of leader peptides during biosynthesis (van den Hooven et al. 1997).

Genetics of Lantibiotics from LAB

The genetic determinants of several lantibiotics from LAB have been either partially or fully determined. Complete sequences are considered to be available for nisin, lacticin 3147, cytolysin, lactocin S and lactocin 481, while only partial sequence is available for plantaricin W. The only LAB bacteriocin which is known to be a lantibiotic and for which no nucleotide sequence has been published is plantaricin C (Turner et al. 1999). Lantibiotics are derived from ribosomally synthesised prepeptides, composed of an inactive propeptide with an N-terminal extension or leader peptide. This short-lived prepeptide, encoded by the structural gene, generically termed *lanA*, undergoes a series of post-translational modification reactions in which the unusual amino acid changes are introduced. Once these modifications have been completed, processing and export occurs, leading to the active mature peptide. The structural, modification, processing, transport, immunity, regulatory and associated genes are located in a number of transcriptional units. The genetic organisation of known LAB lantibiotic gene clusters are presented in Figure 3.

Structural genes

Lantibiotic prepeptides are difficult to study as they are relatively short-lived. The leader peptides generally do not contain cysteine residues, whereas they may contain serine and threonine, the substrates for the majority of modifications (de Vos et al. 1995; Jack et al. 1998). Additionally, the different sub-groups of lantibiotics appear to be distinguished by characteristic leaders. For example, nisin-type leaders (including subtilin, epidermin and variants, epicidin 280, epilancin K7 and Pep5) have a hydrophilic nature, contain a large number of charged residues and a net negative or slightly positive charge. A conserved motif, 'FNLDV', was observed between positions –20 and –15 in these leaders and the cleavage site between the leader and propeptide also seems to be conserved (residues 1 and 2 are hydrophobic, –1 is positively charged or polar, –2 is a proline residue, –3 is negatively charged or polar and –4 is hydrophobic). In contrast, the cleavage site of the lacticin 481 and mersacidin groups resembles that of many non-lantibiotic bacteriocins (Havarstein et al. 1995; de Vos et al. 1995) in that they contain a GA/GS/GG motif. This is sometimes referred to as the 'double-Gly' motif as in many non-lantibiotic proteins two glycines precede the N-terminus of the mature peptide. The 'FNLDL' motif is not generally found in association with a 'double-Gly' with the exception of lacticin 481, which contains an 'FNL' sequence and a 'double-Gly'. In general, the 'double-Gly' motif is recognisable in all members of the LtnA2 group (C55b is an exception) and in the cytolysin peptides. In the cytolysin peptides, this motif is recognised by CylB and additional processing and cleavage of the 6 N-terminal amino acids is mediated by CylP (previously referred to as CylA; Segarra et al. 1991; Booth et al. 1996). Although Plwβ possesses a 'double-Gly' motif, it precedes the known N-terminus by 6 amino acids residues suggesting that it may also undergo further proteolytic attack following cleavage of the leader peptide (Holo et al. 2001). BhaA2 is also likely to undergo similar processing as the six residues after the 'double Gly' motif are similar to those of Plwβ and the cytolysin peptides, however, this needs to be determined experimentally. These four peptides (BhaA2, Plwβ, $CylL_l$ and $CylL_s$) possess an invariant GDXXA motif after the 'double Gly' motif which may be a common protease recognition domain. The cleavage site of lactocin S differs from all other LAB lantibiotics in that its processing site resembles neither of the conserved types (Skaugen et al. 1994).

In general, the lantibiotic leader sequences are not typical of the signal peptides of the sec-dependent transport systems and hence, probably have different functions. In the case of nisin which is processed extracellularly, the leader was found to keep the prepeptide inactive, thereby potentially protecting the producer from the antimicrobial action of the intracellular bacteriocin. Furthermore, there is also evidence obtained from site-directed mutagenesis experiments, that the 'FNLDL' motif may have a role in modification and/or transport functions of Pep5 and nisin (van der Meer et al. 1994; Neis et al. 1997).

Post-translational modifications of lantibiotics

Nisin and related peptide gene clusters possess two separate modification enzymes, LanB and LanC. Inactivation experiments have shown that LanB proteins are responsible for the dehydration reactions, while LanC proteins seem to be involved in the formation of thioethers (Augustin et al. 1992; Meyer et al. 1995). The LanB proteins (approximately 1000 residues) are typically membrane-associated and seven motifs have been identified. The LanC proteins are somewhat smaller, at approximately 400 amino acids, and contain alternating hydrophobic and hydrophilic stretches. Seven conserved motifs, all of which contain conserved Gly residues, have been identified in these proteins (Siezen et al. 1996). For the majority of LAB lantibiotics LanM, a different type of modification enzyme, is implicated (Sahl & Bierbaum 1998). The LanM proteins are large proteins of approximately 1000 amino acids, the C-terminus of which displays some similarity to the LanC proteins, containing the same seven conserved motifs. However, the N-terminus bears no similarity to the LanB proteins (Siezen et al. 1996). It has been proposed that the C-terminal part of the protein is responsible for folding and thioether formation, whereas the N-terminus performs the dehydration reactions, but this remains to be verified. In both lactocin S and the two-component cytolysin systems a single *lanM* gene is present. However, in the two-component bacteriocin lacticin 3147, two *lanM* genes are present (Dougherty et al. 1998). Mutational analysis of genes in the lacticin 3147 system supported the assertion that the *lanM* genes are implicated in post-translational modification of the peptides and that LtnM1 modifies LtnA1, while LtnM2 modifies LtnA2 (McAuliffe et al. 2000). Two *lanM* genes are also present in the recently completed sequences of the two-component lantibiotics, staphylococcin c55 (Yamaguchi et al. 2001) and in the proposed lantibiotic system encoded by *B. halodurans* (Takami et al. 2000). With the exception of cytolysin, all of the two-component lantibiotic systems contain two *lanM* genes. The fact that the two cytolysin peptides share more similarity with each other than is common in other two-component systems, most likely permits a single modification protein to modify both. As the peptides of the plantaricin W system belong to two different groups (i.e. the mersacidin group and LtnA2 group), it is likely that two *lanM* genes are also present in this system.

In addition to these dehydrated and lanthionine residues, a number of other modified amino acids have been identified or proposed to be present in lantibiotics from LAB. These include D-Ala in lactocin S and both of the lacticin 3147 peptides. It remains to confirmed that these unusual amino acids are enzymatically generated and the putative genetic determinants responsible remain to be identified. Other unusual amino acids found in LAB lantibiotics include an N-terminal 2-oxopropionyl or alternatively, a 2-hydroxypropionyl group in lactocin S (Jack et al. 1998), a 2-oxobutyryl at the N-terminus of both of the cytolysin peptides (Booth et al. 1996), LtnA2, (and possibly BhaA2), and a pyruvic acid residue at the N-terminus of Plwβ. By analogy with other lantibiotic systems these N-terminal modifications are thought to occur following dehydration of the N-terminal threonine or serine residue, which is then spontaneously hydrated and deaminated (Kellner et al. 1991; Jack et al. 1998). Therefore, it is possible that the LanM protein and the proteins implicated in cleavage of the leader protein are ultimately responsible for the generation of these unusual amino acids.

Proteolytic processing and transport of lantibiotics from LAB

Once post-translational modification has taken place, the leader peptide is proteolytically removed by a gene-encoded protease (LanP), whereupon the lantibiotic is activated. For nisin and many lantibiotics with an 'FNLDL' motif in their leaders, a subtilisin-like serine protease that recognises the cleavage site is present. The other leader sequences containing the characteristic 'double-Gly' motif are usually recognised by a protease which appears to be present as a domain towards the N-terminus of the ATP binding cassette (ABC)-transporter protein encoded by *lanT* (Havarstein et al. 1995; Siezen et al. 1996). These dual protease and transport proteins include LctT (lacticin 481; Rince et al. 1994), LtnT (lacticin 3147; Dougherty et al. 1998), PlwT (plantarin W; Holo et al. 2001) and CylT (cytolysin). Cytolysin is unusual in that, even though CylT has an N-terminal protease domain, it also has a further protease CylP, which is required for the dual processing this lantibiotic undergoes during maturation (Booth et al. 1996). The first cleavage is carried out during export by CylT and the second processing occurs extracellularly by the subtilisin-type protease, CylP, resulting in cleavage of the six amino acid peptide 'GDVQAE' from both peptides. It is interesting to speculate that since Plwβ has been proposed to undergo proteolytic attack after cleavage of the leader peptide (Holo et al. 2001), that plantaricin W should encode a LanP homolog in addition to the N-terminal protease domain which is known to be present in PlwT. The lactocin S exporter (LasT) is similar to NisT (nisin transporter) in that it does not contain an N-terminal protease domain but has a separate subtilisin-like serine protease (LasP). The relative size of the subtilisin-like serine proteases associated with lantibiotics from LAB present differs and depends on whether an N-terminal signal sequence and/or a C-terminal extension is attached. The nisin protease, NisP, contains both a signal sequence, which is thought to act as a chaperone to direct the enzyme to the membrane, as well as a C-terminal extension which contains a membrane-anchor consensus motif, thereby indicating that NisP attaches to peptidoglycan (van der Meer et al. 1993). CylP does not contain the C-terminal extension, although it does have a signal sequence (Booth et al. 1996; Geissler et al. 1996). Protease activity has been identified in the supernatant of cultures encoding CylP and hence can be considered to act extracellularly (Segarra et al. 1991; Siezen et al. 1996). In contrast, LasP (Skaugen et al. 1997), probably acts intracellularly, as indicated by the absence of both a signal sequence and a membrane anchor motif.

All lantibiotics are exported *via* ABC-transporter proteins, which are encoded by *lanT* genes. Energy for transport comes from hydrolysis of ATP and interestingly, the mechanism by which transport occurs differs to the usual sec-dependent pathway as the leaders lack the typical signal sequence. A functional transporter usually consists of two subunits, each of which contains a N-terminal, membrane-spanning domain and a C-terminal intracellular ATP-binding domain. The membrane-spanning domain contains six hydrophobic regions while the C-terminus is characterised by two highly conserved ATP-binding 'Walker motifs' (Siezen et al. 1996; Jack et al. 1998). Some transport proteins have been found to be essential for export of the lantibiotic, e.g., NisT (Qiao et al. 1996) and LasT (Skaugen et al. 1997), whereas others seem to be dispensable such as EpiT (Peschel et al. 1997) and PepT (Meyer et al. 1995). In these cases, host cell transporters probably complement the specific transporter deficiency.

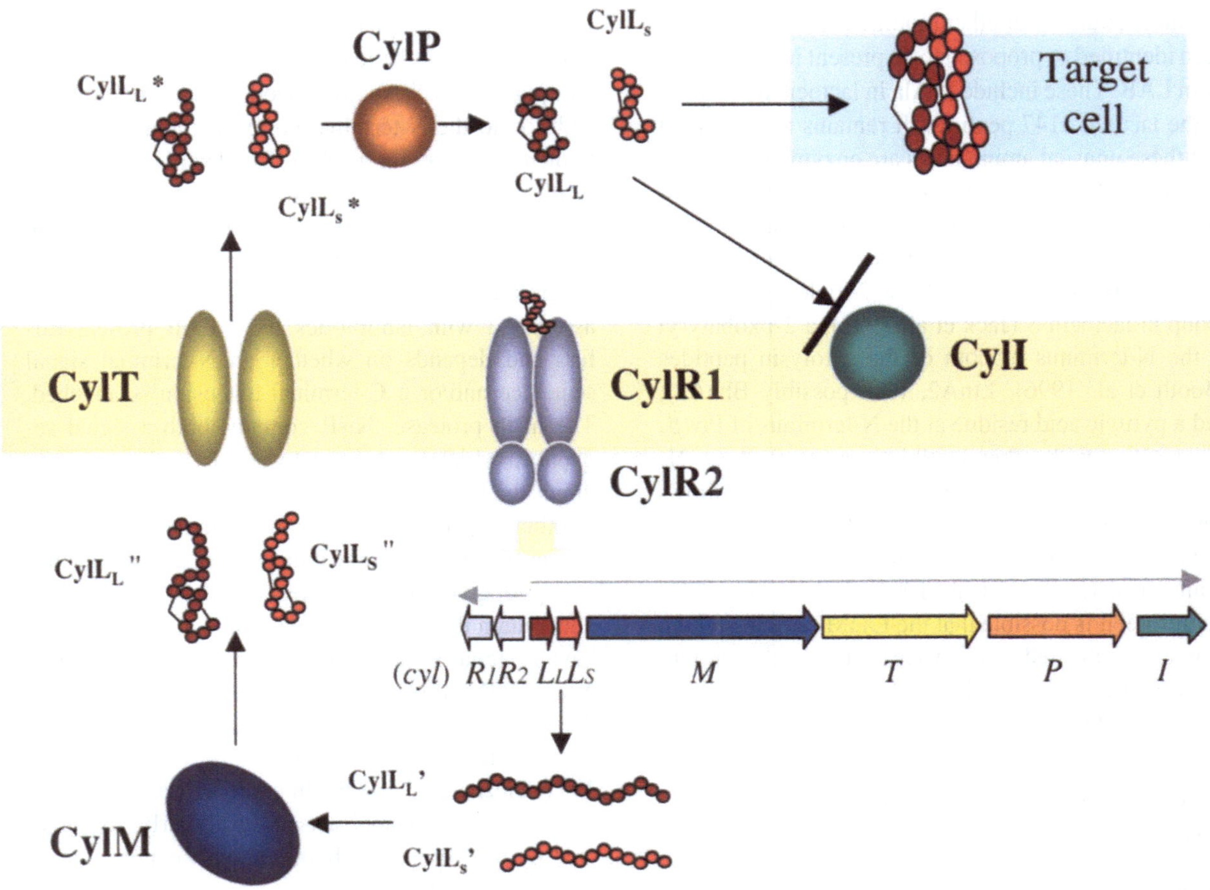

Figure 4. Proposed model for the maturation and regulation of cytolysin (adapted from Haas et al. 2002). The peptide subunits $CylL_L'$ and $CylL_S'$ are ribosomally synthesised and modified post-translationally by CylM. The modified peptides $CylL_L''$ and $CylL_S''$ are proteolytically cleaved and secreted by CylT , followed by extracellular activation by the protease CylP generating the two mature peptides $CylL_L$ and $CylL_S$. The two mature peptides are required for lysis of prokaryotic and eucaryotic cells. The cytolysin producing cell is protected by CylI by an unknown immunity mechanism which is proposed to be membrane associated. $CylL_S$ induces expression of the cytolysin operon by a mechanism involving the regulatory proteins CylR1 and CylR2. It is proposed that $CylL_S$ interacts with CylR1 at the membrane surface and in turn CylR1 interacts with the DNA binding protein, CylR2 to repress transcription of *cylL*$_L$*lL*$_s$*MTPI* via promoter PL when an extracellular threshold level of $CylL_S$ is reached.

Regulation of lantibiotic biosynthesis

Lantibiotic biosynthesis has been shown to be regulated in a number of instances and genes encoding regulatory proteins have been identified. In the case of nisin (Engelke et al. 1994) and subtilin (Klein et al. 1993), two-component regulatory systems have been identified. These regulatory molecules have been designated LanR and LanK. LanK is a membrane-bound histidine-kinase and acts as a sensor. Once activated by an extracellular signal, LanK undergoes autophosphorylation of a His residue, which subsequently phosphorylates an Asp residue in LanR. This results in a conformational change in LanR, which then induces transcriptional activation of the biosynthetic genes. In the case of nisin regulation, nisin itself has been identified as the induction factor. Nisin is produced at low levels during log phase and when a certain extracellular nisin threshold concentration is achieved, transcription of the biosynthesis and immunity genes is triggered (Kuipers et al. 1995; De Ruyter et al. 1996). This type of regulation is termed quorum sensing and suggests that nisin has a dual role, both as a signal molecule as well as an antimicrobial peptide.

Two recently identified regulatory genes designated CylR1 and CylR2 are located upstream of the promoter of the cytolysin operon and are divergently transcribed with respect to the cytolysin operon (Haas et al. 2002; Figure 4). CylR1 has no known homologues and is predicted to be a transmembrane protein whereas CylR2 has characteristics of a helix-turn-helix DNA binding protein. It has been proposed that both of these proteins combine to form a novel two-component regulator which incompletely represses transcription of the cytolysin operon. Once an appropriate extracellular concentration threshold of fully mature $ClyL_S$ is reached, the right-facing promoter is activated. The other component of the cytolysin system, $CylL_l$ (or partially processed $CylL_S$) is incapable of inducing the same promoter. This proposed quorum-sensing mechanism is unusual and is different to that seen for the nisin system or any other system involved in the control of lantibiotic expression. In the proposed model for cytolysin regulation (Haas et al. 2002; Figure 4), the dual functioning CylR1 sits in the membrane sensing external $CylL_S$ and also interacts with the DNA binding protein, CylR1, which in turn interacts with the cytolysin operator site and represses transcription when insufficient $CylL_S$ is present in the environment.

The first evidence of a negative regulator associated with a lantibiotic system was provided by the lacticin 3147 system (McAuliffe et al. 2001a),. LtnR is encoded by the first gene of a short operon which also contains *ltnIFE*, genes implicated in lacticin 3147 immunity. This operon is counter-transcribed to the lacticin 3147 structural, modification and transport genes (McAuliffe et al. 2001b). LtnR has similarities to the PBSx (Xre) family of transcriptional regulators and was shown to autoregulate its own transcription. Overexpression of *ltnR* resulted in an increase in the sensitivity to lacticin 3147, which agrees with the earlier proposal that the downstream genes (*ltnIFE*) are involved in lacticin 3147 immunity (McAuliffe et al. 2000). When *ltnR* was cloned in *cis* with the promoter of the lacticin 3147 biosynthetic genes, no significant reduction in the reporter gene expression was observed leading to the conclusion that LtnR does not down regulate the biosynthetic genes. The occurrence of a transcriptional attenuator within one of the lacticin 3147 modification genes contributes to the regulation of these biosynthetic genes and was proposed to maintain the correct ratio of structural prepeptide mRNA to mRNA for the biosynthetic gene (McAuliffe et al. 2000).

In the lactocin S system, a transcriptional regulator is located upstream of the divergently transcribed biosynthetic gene cluster (Skaugen et al. 2002). The regulator protein, designated LasR, (originally designated LasX) bears similarity to the Rgg-like transcriptional regulators which are found exclusively in Gram-positive bacteria. Mutations in *lasR* led to a reduced level of *lasA* mRNA and *lasA-W* mRNA and was associated with a bacteriocin-deficient phenotype. Unlike the regulators associated with the cytolysin and lacticin 3147 systems which are transcriptional repressors, the LasR regulator is proposed to function as a transcriptional activator in agreement with observations with other Rgg-like regulators. An Rgg-like regulator has also been identified in the mutacin II system which is responsible for the transcription of the mutacin II structural gene, *mutA* (Chen et al. 1999). The *lasR* gene is co-transcribed with *lasY*, which encodes an ABC transporter homologue, the role of which remains to be determined. Based on similarities to the *E. coli* regulator (WrbA), a further putative regulator encoded by *lasN* (previously referred to as *orf239*, Skaugen et al. 1997) is also present in the lactocin S system. However, the function of this putative regulator remains to be determined.

To date no known regulatory protein has been linked to lacticin 481 expression. The genetic organisation of lacticin 481 has been determined in three identical or closely related plasmids and in each case the *lctAMTFEG* genes are flanked by two copies of IS*1675* which has the potential to form a composite transposon designated Tn*5721* (Dufour et al. 2000). In each case, the putative transposon is flanked by similar genes which are not implicated in regulation. These data suggest that unlike related operon structures (i.e., streptococcin A-FF22 and mutacin II) lacticin 481 may lack a specific regulator. This is reinforced by the fact that cloning *lctAMTFEG* in *L. lactis* IL1403 permitted high levels of bacteriocin production (Rince et al. 1997). It has been proposed that global regulators influence the expression of lacticin 481 and indeed its production can be stimulated when exposed to the stress of increasing osmolarity (Uguen et al. 1999). Potential regulators of plantaricin W, lacticin J46 and plantaricin C remain to be identified.

Immunity to LAB lantibiotics

Lantibiotic producers need to protect themselves against the action of their own antimicrobials and accordingly they encode immunity mechanisms which are genetically linked to the production of the bacteriocin. LAB lantibiotic immunity mechanisms involve either individual proteins capable of conferring resistance alone (LanI) or individual domains of ABC-transporters (LanEFG), which in the case of epicidin, has been demonstrated to transport the bacteriocin out of the cell, lowering the concentration below an inhibitory threshold (Otto et al. 1998). In some instances, (i.e., nisin) both LanI and LanEFG mechanisms exist within the same lantibiotic gene cluster.

The three *lanI* genes which have been identified in LAB do not share any significant homology with each other nor with any of the other known immunity peptides. The absence of homology among the LanI peptides makes it relatively difficult to identify these proteins based on sequence similarity, and indeed no immunity determinant has yet been identified in the relatively well characterised lactocin S system. Interestingly, although the nisin and subtilin lantibiotics are very similar to each other (Figure 1), their immunity proteins, NisI (Kuipers et al. 1993) and SpaI (Klein & Entian 1994) do not share any homology with each other and do not confer cross immunity. NisI is predominantly hydrophilic, with an N-terminal lipoprotein signal sequence and has been shown to be covalently attached to the outside of the cytoplasmic membrane (Qiao et al. 1995). However, the exact mechanism of nisin immunity mediated by NisI remains to be established. The *nisI* gene is regulated in a quorum-sensing dependent manner and its expression involves the two component regulator NisRK and mature nisin (Kuipers et al. 1995; Dodd et al. 1996; Ra et al. 1996). In lacticin 3147, LtnI conferred wild-type levels of immunity on its host when its genetic determinant was cloned downstream of a strong promoter (McAuliffe et al. 2000). LtnI is proposed to have a cytoplasmic membrane location which may provide immunity by preventing lacticin 3147 gaining access to its target molecules. LtnI may also interact with LtnR to regulate expression of the lacticin 3147 determinants in a manner similar to the quorum-sensing mechanism proposed for cytolysin (Haas et al. 2002). The *ltnRIFE* genes are co-transcribed and their expression is known to be controlled by the self-regulating DNA-binding protein, LtnR. While LtnF and LntE are similar to components of type B ABC transporters, implicated in immunity in other lantibiotic systems, their role in lacticin 3147 immunity or any other function remains to be established. While hyperexpression of *lntI* conferred immunity at a level comparable to the wild-type strain, other factors including LtnF and LntE may also be involved. In the lacticin 481 system, the level of immunity was related to factors influencing gene expression including copy number, promoter strength and regulatory factors (Rince et al. 1997). The cytolysin immunity gene, *cylI*, also encodes a protein which does not share significant similarity with other known proteins. This predicted protein is predominantly hydrophobic suggesting that it has a membrane location and it does not appear to have a lipoprotein signal sequence (Coburn et al. 1999). Although, *cylI* is co-transcribed with *cylA*, CylA is not implicated in the immunity phenotype as was previously proposed (Segarra et al. 1991; Coburn et al. 1999).

In addition to the NisI immunity proteins, nisin producers also encode transport systems as a second mechanism of immunity. Indeed, this mechanism appears to be solely responsible for immunity to lacticin 481 (Rince et al. 1997). This type of immunity mechanism is present in a number of lantibiotic systems including mutacin II (Qi et al. 1999), epidermin (Peschel et al. 1997) and mersacidin (Sahl & Bierbaum 1998). Unlike the dedicated Type A ABC-transporters encoded by the *lanT* genes, the transporters involved in immunity are composed of multiple gene products and hence, are Type B. In the case of nisin and lacticin 481, the system is composed of three components; LanF is the ATP-binding domain while LanE and LanG are membrane spanning regions (Siegers & Entian 1995; Rince et al. 1997). Since many ABC transporters consist of four domains comprising two hydrophobic domains and two ATPase subunits, it is hypothesised that one molecule each of LanE and LanG together with two molecules of LanF comprise the immunity transporter (Peschel & Gotz 1996; Jack et al. 1998).

An additional mechanism of immunity/resistance not genetically linked to production, has also been reported. A nisin immunity gene has been linked to lactococcal plasmid DNA and is commonly associated with other traits such as phage resistance. The *nsr* gene (nisin resistance) on the lactococcal plasmid, pNP40, encodes a 35-kDa protein (Froseth & McKay 1991). Although the mechanism of protection is unknown, the hydrophobic N-terminus suggests a cell membrane location. Specific nisin-inactivating enzymes or nisinases have also been reported for a variety of Gram-positive strains including *Streptococcus thermophilus* (Harris et al. 1991) and several *Bacillus* subsp. (Jarvis & Farr 1971). The only type of nisinase to be well characterised from these species is that produced by *B. cereus* where it was described as possessing dehydroalanine reductase activity (Jarvis & Farr 1971).

Mode of action of LAB lantibiotics

Until recently the traditional view of the mode of action was that Type A lantibiotics acted mainly by transient pore formation in the cytoplasmic membrane of sensitive cells, whereas Type B targeted an essential enzymatic component of the cells metabolism. However, recent insights into the mode of action of some lantibiotics, particularly nisin and related type A lantibiotics, has meant that this view has had to be revised. It has long been known that nisin inhibited peptidoglycan biosynthesis (Linnet & Strominger 1973) and that it interacted with cell wall precursors, either lipid I or lipid II (Reisinger et al. 1980). Later studies indicated that nisin was capable of pore formation (Benz et al. 1981; Ruhr & Sahl 1985; Sahl et al. 1987). The two aspects of the mode of action were united recently in a model which demonstrated that nisin uses lipid II as an integral part of the pore formation process (Brotz et al. 1998). Lipid II appears to be the sole docking molecule for the high affinity binding of nisin (Breukink et al. 1999) and in the absence of lipid II, significantly higher concentrations of nisin are required to form target independent pores (Wiedemann et al. 2001). Mutations in the N-terminal rings of nisin indicated that these rings are involved in the specific binding to lipid II (Breukink et al. 1999) and complemented the earlier observation that a peptide comprising the 12 N-terminal amino acids was antagonistic to nisin activity (Chan et al. 1996). Some nisin mutants incapable of pore formation (i.e. peptides mutated in the flexible hinge region) still retain antimicrobial activity which correlates with their role as antagonists of cell wall biosynthesis (Wiedemann et al. 2001). Therefore, while binding to lipid II promotes pore formation, it can also independently block cell wall biosynthesis. In this respect these mutant peptides behave in a manner analagous to the antibiotics vancomycin and ramoplanin, and the type B lantibiotic, mersacidin (Brotz et al. 1998). Unlike many pore-forming bacteriocins, the specific targeted affinity for a cell wall component combined with pore forming ability enables nisin to act in the nanomolar range as opposed to the micromolar range for many other pore formers such as the eucaryotic peptide magainin (Breukink et al. 1999; Wiedemann et al. 2001). However, since death by cell wall inhibition is a comparatively slow process in contrast to death by pore formation, pore formation ultimately leads to death when both mechanisms are capable of killing a cell. The absence of lipid II in cells such as yeasts or fungi may explain the low activity of nisin against these micorganisms (Wiedemann et al. 2001). Lipid II also serves as the docking molecule for the Type A lantibiotic epidermin, but not pep5 or epilancin which are proposed to use an alternative, as yet unidentified docking molecule (Brotz et al. 1998).

The mode of action of other LAB lantibiotics has also been investigated, but not to the same extent as nisin. Plantaricin C has been shown to be a pore-forming bacteriocin which targets the cytoplasmic membrane leading to the dissipation of the proton motive force and the release of intracellular molecules (i.e., glutamate and ATP) in sensitive cells (Gonzalez et al. 1996). As a result, many vital cellular functions are blocked and the glycolytic rate is diminished presumably due to the lowering of the internal pH. In addition, the release and hydrolysis of internal ATP all contribute to death of the sensitive cell. Electron microscopy indicated that plantaricin C also has an effect on the peptidoglycan layer in *Lactobacillus fermentum*, as observed by an electron-transparent layer between the plasma membrane and the outer cell wall layer. It is interesting to speculate that this destruction of the peptidoglycan layer may be linked to the similarities observed between plantaricin C and the mersacidin group of lantibiotics, some of which are known antagonists of peptidoglycan synthesis.

Plantaricin C has been grouped here with the lacticin 481 group of lantibiotics. The mode of action of lacticin 481 and the closely related lacticin J46 remain to be determined but its' similarities to SA-FF22, a known pore former (Jack et al. 1994), suggests that it may act in a similar fashion. However, lacticin 481 is also very similar to variacin (Figure 2) which has been proposed not to act *via* a pore-forming mechanism but through the inhibition of an essential enzyme involved in metabolic energy generation (Chikindas et al. 1995).

The mode of action of lacticin 3147 has also been investigated (McAuliffe et al. 1998) where it has been shown to cause selective leakage of certain ions but not larger metabolites such as ATP. The selectivity of the lacticin 3147 pores causes the efflux of K^+ ions leading to intracellular ATP hydrolysis and the immediate dissipation of the membrane potential which ultimately leads to death of the cell. The similarities between one of the lacticin 3147 peptides (LtnA1) and mersacidin suggests that this peptide may interact with a peptidoglycan component of the cell and disrupt synthesis of the cell wall. Recent results indicate that when LtnA1 is added to sensitive cells which are then washed and exposed to LtnA2, significant killing occurs. However, when the sequence of addition of these peptides is reversed, the level of kill is dramatically reduced (Morgan et al. 2002, unpublished results). This suggests that LtnA1 may serve as the docking molecule which targets a particular peptidoglycan component, whereas LtnA2, which has a proposed elongated linear-type structure, may facilitate pore formation. The partial activity of LtnA1 alone and the ability of lacticin 3147 to act at nanomolar concentrations correlates with this model (Morgan et al. unpublished results). In this respect lacticin 3147 and other similar two component bacteriocins (i.e., plantaricin W, staphylococcin c55) may have a similar mode of action as nisin except that the dual killing effect is separated across two peptides.

Applications of LAB lantibiotics

There are a number of features of the LAB-derived lantibiotics which make them particularly attractive for both food and biomedical applications. Many of these post-translationally modified peptides have a broad spectrum of activity and are relatively thermostable and resistant to proteolytic breakdown. For example nisin and lacticin 3147 inhibit a wide range of Gram-positive bacteria, including many food spoilage and pathogenic bacteria. The lantibiotic can be delivered to a food in a number of different ways; for example, a lantibiotic-producing strain can be used as a starter or as an adjunct strain in a fermented food. Equally, a producer strain could be used as a protective culture on the surface of a suitable foodstuff, to protect against temperature abuse induced growth of undesirable bacteria. These delivery systems have the advantage of being extremely cost effective, in that the bacteriocin producer can simply be substituted for the normal starter and the lantibiotic is naturally produced *in situ*. Alternatively, the lantibiotic can be added as a purified or concentrated compound, depending again on the specific food, and the degree of added value such an addition might represent. Nisin, in particular, has been used for decades in the food industry and remains the only lantibiotic which is commercially produced. It was discovered more than 70 years ago (Rogers & Whittier 1928, Whitehead 1933) and was first sold in England in 1953. Currently, it is sold in more than 40 countries and was added to the positive list of food additives by the EU as additive number E234 (EEC 1983). Nisin applications have been the focus of numerous extensive reviews, and these will not be reviewed here (see Delves-Broughton et al. 1990; De Vuyst & Vandamme 1994; Delves-Broughton et al. 1996). As a simple guide to its effectiveness in foods, it has been utilised in foods as varied as dairy products: processed cheese and cheese spreads, direct acidified cheeses, pasteurised dairy desserts, fresh and recombined milk; liquid egg, dressings and sauces, high moisture/reduced fat foods, canned foods and crumpets (http://www.aplin-barrett.co.uk/key_benefits.htm). Using nisin as a benchmark, we can predict that other lantibiotics produced by LAB also have commercial potential. An example is lacticin 3147, which has been shown to be effective for inhibition of certain food spoilage and pathogenic bacteria in a variety of food systems (Ryan et al. 1996; Ross et al. 1999; Morgan et al. 2001). A particularly useful feature of this bacteriocin is that its genetic determinants are encoded on a large self-transmissible plasmid which has now been transferred to over 30 different lactococcal hosts, many of which are derivatives of commercial starter strains (Ryan et al. 1996; Coakley et al. 1997). These strains are suitable as alternative starters for cheese manufacture where they can offer the product an 'in built' protection against undesirable flora. Indeed, use of one such starter strain resulted in the killing of at least 99.9% of *Listeria monocytogenes* deliberately added to Cottage cheese (McAuliffe et al. 1999).

Lantibiotics produced by LAB have also potential biomedical applications where they may provide valuable alternatives to antibiotics for the treatment of human and animal infections. Indeed, a recent study demonstrated that many antibiotic resistant human pathogens such as MRSA (methicillin-resistant *Staphylococcus aureus*) or VRE (vancomycin-resistant *Enterococcus*) are sensitive to lacticin 3147 (Galvin et al. 1999). As an example of the potential for LAB lantibiotics in animal health, both nisin and lacticin 3147 have also been investigated for the treatment and/or prevention of mastitis in cattle (Ryan et al. 1998; Twomey et al. 2000). This disease is clinically defined as an inflammation of the udder and results in major economic losses in dairy production worldwide. Consequently, animals are commonly administered antibiotics prophylactically, particularly when they are most prone to contracting the disease such as when they are approaching the end of the lactating period. Studies have shown, for example, that administration of lacticin 3147 preparations contained in a Teat seal^{TM} formulation can protect animals from

Table 2. The effectiveness of teat seal plus lacticin 3147 in eliminating *Staphylococcus aureus* DPC5246 in artificially infected teats of lactating cows when compared with untreated controls. Shedding was evaluated 18 h. post-*S. aureus* inoculation

Trial	Inoculum (cfu per 100 μl)	[Lacticin 3147] AU/4g of teat seal	Treatment	Teats inoculated (*n*)	Teats shedding *S. aureus* (*n*)
1	1.5×10^3	16 384	Untreated	20	20 (100%)
			Teat seal + lacticin 3147	20	20 (100%)
2	1.7×10^3	32 768	Untreated	29	19 (66%)^a
			Teat seal + lacticin 3147	29	4 (14%)^b
			Teat seal + 1% Tween 80	1	1 (100%)
3	6.8×10^3	32 768	Untreated	20	16 (80%)
			Teat seal + lacticin 3147	20	11 (55%)

^{ab}Values with different superscripts within an experiment are significant $P < 0.001$.

deliberate infection with mastitic pathogens such as *Streptococcus dysgalactiae* or *Staph. aureus* (Table 2; Twomey et al. 2000). These types of experiments highlight the potential for lantibiotics to prevent or treat some animal infections especially in cases where antibiotic treatment is either undesirable or ineffective.

The proposed dual use of lantibiotics in both food and biomedical applications raises some interesting questions about their 'food grade' status. Can the same antimicrobial (or even the same class of antimicrobial) be used both to prevent and/or treat disease, and also as a food preservative? It is notable that nisin is fulfilling both roles currently, in that it is used as a mastitis dips and wipes, and of course in foods. Therefore, we can ask if the proposed uses are indeed mutually exclusive and, if so, is one type of application a more appropriate use of this valuable biological resource than the other? This question remains academic in the absence of significant biomedical applications, but it is a topic which will require careful thought and planning by lantibiotic researchers.

Conclusions

As the list of lantibiotics produced by members of the LAB continues to grow, it is clear that there is significant diversity among these post-translationally modified peptides in terms of their structure, genetics and mode of action. Indeed, the proposed classification of these complex peptides suggested in this paper will undoubtedly need to be changed to accommodate new bacteriocins as they continue to be discovered. This class of antimicrobial peptides offer exciting possibilities in terms of protein engineering if their unusual mechanisms for introducing post-translation modifications can be harnessed and exploited in the construction of novel lantibiotics.

It is also fair to say that, as of yet, these peptides have promised more than they have delivered in terms of commercial potential, be it as an alternative to chemical preservatives in foods or to antibiotics used to treat or prevent human and animal diseases. However, the lantibiotics have not disappointed those research groups who have decided to devote their attention to these fascinating and rewarding molecules. Whether the interest is at the fundamental level (gene regulation, peptide export, post-translational modifications, protein engineering, structure–function studies, protein–membrane interactions, protein–protein interactions, cell–cell signalling) or in applications (elimination of pathogens, control of spoilage organisms, starter technology, biomedicine, development of resistance), the lantibiotics offer a wealth of interesting phenomena – and with a readily scorable phenotype. If LAB lantibiotics did not exist, we would crave such interesting and diverse molecules for their obviously rewarding research and commercial opportunities. Given that they do exist, it is up to those of us fortunate to be in the LAB lantibiotic community to ensure that these multi-functional peptides deliver on their full potential, from both basic and applied perspectives.

References

Abel T & Maniatis T (1989) Gene regulation. Action of leucine zippers. Nature 341: 24–25.

Augustin J, Rosenstein R, Wieland B, Schneider U, Schnell N, Engelke G, Entian K-D & Götz F (1992) Genetic analysis of biosynthetic genes and epidermin-negative mutants of *Staphylococcus epidermidis*. Eur. J. Biochem. 204: 1149–1154.

Benz R, Jung G & Sahl H-G (1991) Mechanism of channel-forming lantibiotics in black lipid membranes. In: Jung G & Sahl, H-G (Eds) Nisin and Novel Lantibiotics (pp 359–372) ESCOM, Leiden, The Netherlands.

Booth MC, Bogie CP, Sahl H-G, Siezen RJ, Hatter KL & Gilmore MS (1996) Structural analysis and proteolytic activation of *Enterococcus faecalis* cytolysin, a novel lantibiotic. Mol. Microbiol. 21: 1175–1184.

Breukink E, van Kraaij C, Demel RA, Siezen RJ, Kuipers OP & de Kruijff B (1997) The C-terminal region of nisin is responsible for the initial interaction of nisin with the target membrane. Biochemistry 36: 6968–6976.

Breukink E, Wiedemann I, van Kraaij C, Kuipers OP, Sahl H-G & de Kruijff B (1999) Use of the cell wall precursor lipid II by a pore-forming peptide antibiotic. Science 286: 2361–2364.

Brotz H, Bierbaum G, Markus A, Molitor E. & Sahl H-G (1995) Mode of action of the lantibiotic mersacidin: inhibition of peptidoglycan biosynthesis via a novel mechanism. Antimicrob. Agents Chemother. 39: 714–719.

Brotz H, Bierbaum G, Reynolds PE & Sahl H-G (1997) The lantibiotic mersacidin inhibits peptidoglycan biosynthesis at the level of transglycosylation. Eur. J. Biochem. 246: 193–199.

Brotz H, Josten M, Wiedemann I, Schneider U, Gotz F, Bierbaum G & Sahl H-G (1998) Role of lipid-bound peptidoglycan precursors in the formation of pores by nisin, epidermin and other lantibiotics. Mol. Microbiol. 30: 317–327.

Coakley M, Fitzgerald G, Ross RP, (1997) Application and evaluation of the phage resistance and bacteriocin encoding plasmid pMRC01 for the improvement of dairy starter cultures. Appl. Environ. Microbiol. 63: 1434–1440.

Chan WC, Bycroft BW, Lian L-Y & Roberts GCK (1989) Isolation and characterization of two degradation products derived from the peptide antibiotic nisin. FEBS Lett. 252: 29–36.

Chan WC, Dodd HM, Horn N, Maclean K, Lian L-Y, Bycroft BW, Gasson MJ & Roberts GCK (1996) Structure-activity relationships in the peptide antibiotic nisin: role of dehydroalanine 5. Appl. Environ. Microbiol. 62: 2966–2969.

Chen P, Qi F, Novak J & Caufield PW (1999) The specific genes for lantibiotic mutacin II biosynthesis in *Streptococcus mutans* T8 are clustered and can be transferred *en bloc*. Appl. Environ. Microbiol. 65: 1356–1360.

Chinkindas ML, Novak J, Driessen AJM, Konings WN, Schilling KM & Caulfield PW (1995) Mutacin II, a bactericidal lantibiotib form *Streptococcus mutans* Antimicrobiol. Agents Chemother. 39: 2656–2660.

Chow JW, Thal LA, Perri MB, Vasquez JA, Donabedian SM, Clewell DB & Zervos MJ (1993) Plasmid-associated haemolysin and aggregation substance production contributes to virulence in experimental enterococcal endocarditis. Antimicrob. Agents Chemother. 37: 2474–2477.

Coburn PS, Hancock LE, Booth MC & Gilmore MS (1999) A novel means of self-protection unrelated to the bactericidal effects of the *Enterococcus faecalis* cytolysin. Infect. Immun. 67: 3339–3347.

Dabard J, Bridonneau C, Phillipe C, Anglade P, Molle D, Nardi M, Ladiré M, Girardin H, Marcille F, Gomez A & Fons M (2001) Ruminococcin A, a new lantibiotic produced by a *Ruminococcus gnavus* strain isolated form human faeces. Appl. Environ. Microbiol. 67: 4111–4118.

de Ruyter PGGA, Kuipers OP & de Vos WM (1996) Controlled gene expression systems for *Lactococcus lactis* with the food-grade inducer nisin. Appl. Environ. Microbiol. 62: 3662–3667.

de Vos WM, Kuipers OP, van der Meer JR & Siezen RJ (1995) Maturation pathway of nisin and other lantibiotics: post-translationally modified antimicrobial peptides exported by gram-positive bacteria. Mol. Microbiol. 17: 427–437.

Delves-Broughton J (1990) Nisin and its uses as a food preservative. Food Tech. 44: 100–117.

Delves-Broughton J, Blackburn P, Evans RJ, Hugenholtz J (1996) Applications of the bacteriocin, nisin. Antonie Van Leeuwenhoek 2: 193–202.

Diep DB, Håvarstein LS, Nissen-Meyer J & Nes IF (1994) The gene encoding plantaricin A, a bacteriocin from *Lactobacillus plantarum* C11, is located on the same transcription unit as an *agr*-like regulatory element. Appl. Environ. Microbiol. 60: 160–166.

Dodd HM, Horn N, Chan WC, Giffard CJ, Bycroft BW, Roberts GCK & Gasson MJ (1996) Molecular analysis of the regulation of nisin immunity. Microbiology 142: 2385–2392.

Dougherty BA, Hill C, Weidman JF, Richardson DR, Venter JC & Ross RP (1998) Sequence and analysis of the 60 kb conjugative, bacteriocin-producing plasmid pMRC01 from *Lactococcus lactis* DPC3147. Mol. Microbiol. 29: 1029–1038.

Dufour A, Rincé A, Uguen P & le Pennec J-P (2000) IS*1675*, a novel lactococcal insertion element, forms a transposon-like structure including the lacticin 481 lantibiotic operon. J. Bacteriol. 182: 5600–5605.

EEC (1983) EEC Commission Directive 83/463/EEC.

Engelke G, Gutowski-Eckel Z, Hammelmann M. & Entian K-D (1992) Biosynthesis of the lantibiotic nisin: genomic organization and membrane localization of the NisB protein. Appl. Environ. Microbiol. 58: 3730–3743.

Ferretti JJ, McShan WM, Adjic D, Savic D, Savic G, Lyon K, Primeaux C, Sezate SS, Surorov AN, Kenton S, Lai H, Lin S, Qian Y, Jia HG, Najar FZ, Ren Q, Zhu H, Song L, White J, Yuan X, Clifton SW, Roe BA & McLaughlin RE (2001) Complete genome sequence of an M1 strain of *Streptococcus pyogenes*. Proc. Natl. Acad. Sci. U.S.A. 98: 4658–4663.

Freund S, Jung, G, Gibbons WA, & Sahl H-G (1991) NMR and circular dichroism studies on Pep5. In: Sahl H-G, & Jung, J, (Eds) Nisin and Novel Lantibiotics (pp 103–112). ESCOM Scientific Publishers BV, Leiden.

Froseth BR & McKay LL (1991) Molecular characterisation of the nisin resistance region of *Lactococcus lactis* subsp. *lactis* DRC3. Appl. Environ. Microbiol. 54: 2636–2639.

Galvin M, Hill C, Ross RP (1999) Lacticin 3147 displays activity in buffer against Gram-positive bacterial pathogens which appear insensitive in standard plate assays. Lett. Appl. Microbiol. 28: 355–358.

Geissler S, Götz F & Kupke T (1996) Serine protease EpiP from *Staphylococcus epidermidis* catalyses the processing of the epidermin precursor peptide. J. Bacteriol. 178: 284–288.

Gilmore MS, Segarra RA, Booth MC, Bogie, CP, Hall, LR & Clewell DB (1994) Genetic structure of the *Enterococcus faecalis* plasmid pAD1-encoded cytolytic toxin system and its relationship to lantibiotic determinants. J. Bacteriol. 176: 7335–7344.

Gomez A, Ladiré M, Marcille F & Fons M (2002) Trypsin mediates growth phase-dependent transcritional regulation of genes involved in biosynthesis of Ruminococcin A, a lantibiotic produced by a *Ruminococcus gnavus* strain from a human intestinal microbiota. J Bacteriol. 184: 18–28.

Gonzalez B, Glaasker E, Kunji ERS, Driessen AJM, Suarez JE & Konings WN (1996) Bactericidal mode of action of plantaricin C. Appl. Environ. Microbiol. 62: 2701–2709.

Guder A, Wiedemann I & Sahl H-G (2000) Post-translationally modified bacteriocins – the lantibiotics. Biopolymers 55: 62–73.

Haas W, Shephard BD & Gilmore MS (2002) Two-component regulator of *Enterococcus faecalis* cytolysin responds to quorum sensing auto-induction. Nature 415: 84–87.

Harris LJ, Fleming HP & Klaenhammer TR (1991) Sensitivity and resistance of *Listeria monocytogenes* ATCC 19115, Scott A and UAL500 to nisin. J. Food. Prot. 54: 836–840.

Håvarstein LS, Diep DB & Nes IF (1995) A family of bacteriocin ABC transporters carry out proteolytic processing of their substrates concomitant with export. Mol. Microbiol. 16: 229–240.

Holo H, Jeknic Z, Daeschel M, Stevanovic S & Nes I (2001) Plantaricin W from *Lactobacillus plantarum* belongs to a new family of two-component lantibiotics. Microbiology 147: 643–651.

Huot E, Meghrous J, Barrena-Gonzalez C & Petitdemange H (1996) Bacteriocin J46, a new bacteriocin produced by *Lactococcus lactis* subsp. *cremoris* J46: isolation and characterization of the protein and its gene. Anaerobe 2: 137–145.

Hynes WL, Friend VL & Ferretti JJ (1994) Duplication of the lantibiotic structural gene in M-type 49 group A streptococcus strains producing streptococcin A-M49. Appl. Environ. Microbiol. 60: 4207–4209.

Ike Y, Hashimoto H & Clewell DB (1987) High incidence of hemolysin production by *Enterococcus* (*Streptococcus*) *faecalis* strains associated with human parenteral infections. J Clin. Microbiol. 25: 1524–1528

Ike Y, Clewell DB, Segarra RA & Gilmore MS (1990) Genetic analysis of the pAD1 hemolysin/bacteriocin determinant in *Enterococcus faecalis*: Tn917 insertional mutagenesis and cloning. J. Bacteriol. 172: 155–163.

Jack RW, Benz R, Tagg JR & Sahl H-G (1994) The mode of action of SA-FF22, a lantibiotic from *Streptococcus pyogenes* strain FF22. Eur. J. Biochem. 219: 699–705.

Jack RW, Bierbaum G. & Sahl H-G (1998) Lantibiotics and Related Peptides. Springer, New York.

Jarvis B, Farr J. Partial purification, specificity and mechanism of action of the nisin-inactivating enzyme from *Bacillus cereus*. Biochim. Biophys. Acta. 1971. 227: 232–240.

Jett DD, Jenson HG, Nordquist RE & Gilmore MS. (1992) Contribution of the pAD1-encoded cytolysin to the severity of experimental *Enterococcus faecalis* endophthalmilis. Infect. Immun. 60: 2445–2452.

Kalmokoff ML & Teather RM (1999). Isolation and characterisation of a bacteriocin, Butyrivibriocin AR10, from the ruminal anaerobe *Butyrivibrio fibrisolvens* AR10: evidence in support of the widespread occurrence of bacteriocin-like activity among ruminal isolates of *B. fibrisolvens*. Appl. Environ. Microbiol. 63: 394–402.

Kellner R, Jung G & Sahl H-G (1991) Structure elucidation of the tricyclic lantibiotic Pep5 containing eight positively charged amino acids. In: Sahl H-G & Jung J (Eds) Nisin and Novel (pp 141–158). ESCOM Scientific Publishers BV, Leiden.

Kimura H, Matsusaki H, Sashihara T, Sonomoto K & Ishizaki A (1998) Purification and partial identifiaction of bacteriocin ISK-1, a new lantibiotic produced by *Pediococcus* sp. ISK-1. Biosci. Biotechnol. Biochem. 62: 2341–2345.

Klein C & Entian K-D (1994) Genes involved in self-protection against the lantibiotic subtilin produced by *Bacillus subtilis* ATCC6633. Appl. Environ. Microbiol. 60: 2793–2801.

Klein C, Kaletta C & Entian K-D (1993) Biosynthesis of the lantibiotic subtilin is regulated by a histidine kinase/response regulator system. Appl. Environ. Microbiol. 59: 296–303.

Krull RE, Chen P, Novak J, Kirk M, Barnes S, Baker J, Krishna NR & Caufield PW. (2000) Biochemical structural analysis of the lantibiotic mutacin II. J. Biol. Chem. 26: 15845–15850.

Kuipers OP, Beerthuyzen MM, Siezen RJ & de Vos WM (1993) Characterization of the nisin gene cluster *nisABTCIPR* of *Lactococcus lactis*: requirement of expression of the *nisA* and *nisI* genes for development of immunity. Eur. J. Biochem. 216: 281–291.

Kuipers OP, Rollema HS, Yap G. J, Boot H, Siezen RJ & de Vos WM (1992) Engineering dehydrated amino acid residues in the antimicrobial peptide nisin. J. Biol. Chem. 267: 24340–24346.

Kuipers OP, Beerthuyzen MM, De Ruyter PG, Luesink EJ, & De Vos WM (1995) Autoregulation of nisin biosynthesis in *Lactococcus lactis* by signal transduction. J. Biol. Chem. 270: 27299–27304.

Lavanant H, Derrick PJ, Heck AJ & Mellon FA (1998) Analysis of nisin A and some of its variants using Fourier transform ion cyclotron resonance mass spectrometry Anal. Biochem. 255: 74–89.

Lian L-Y, Chan WC, Morlaey SD, Roberts GCK, Bycroft, BW & Jackson D (1991) Solution structures of nisin and its two major degradation products determined by NMR. Biochem. J. 283: 413–420.

Linnet PE & Strominger JL (1973) Additional antibiotic inhibitors of peptidoglycan biosynthesis. Antimicrobiol. Agents Chemother. 4: 231–236.

McAuliffe O, Hill C, Ross RP (1999) Inhibition of *Listeria monocytogenes* in cottage cheese manufactured with a lacticin 3147-producing starter culture. J. Appl. Microbiol. 86: 251–256.

McAuliffe O, Ryan MP, Ross RP, Hill C, Breeuwer P & Abee T (1998) Lacticin 3147, a broad-spectrum bacteriocin which selectively dissipates the membrane potential. Appl. Environ. Microbiol. 64: 439–445.

McAuliffe, O, Hill C & Ross R P (2000) Identification and overexpression of *ltnI*, a novel gene which confers immunity to the two-component lantibiotic, lacticin 3147. Microbiology 146: 129–138.

McAuliffe O, O'Keeffe T, Hill C & Ross RP (2001a) Regulation of immunity to the two-component lantibiotic, lacticin 3147, by the transcriptional repressor LtnR. Mol. Microbiol. 39: 982–993.

McAuliffe O, Ross RP & Hill C (2001b) Lantibiotics: structure, biosynthesis and mode of action. FEMS Microbiol. Rev. 25: 285–308.

Meyer C, Bierbaum G, Heidrich C, Reis M, Suling J, Iglesias-Wind MI, Kempter C, Molitor E & Sahl HG. Nucleotide sequence of the lantibiotic Pep5 biosynthetic gene cluster and functional analysis of PepP and PepC. Evidence for a role of PepC in thioether formation. Eur. J. Biochem. 1995. 232: 478–489.

Mortvedt C I, Nissen-Meyer J, Sletten K, & Nes IF (1991) Purification and amino acid sequence of lactocin S, a bacteriocin produced by *Lactobacillus sake* L45. Appl. Environ. Microbiol. 57: 1829–1834.

Mortvedt-Abilgaard CI, Nissen-Meyer J, Jelle B, Grenov B, Skaugen M & Nes IF (1995) Production and pH-dependent bacericidal activity of lactocin S, a lantibiotic from Lactobacillus sake L45. Appl. Environ. Microbiol. 61: 175–179.

Navaratna, MADB, Sahl H-G & Tagg JR (1998) Two-component anti-*Staphylococcus aureus* lantibiotic activity produced by *Sta-*

phylococcus aureus C55. Appl. Environ. Microbiol. 64: 4803–4808.

Navaratna MA, Sahl H-G & Tagg JR (1999) Identification of genes encoding two-component lantibiotic production in *Staphylococcus aureus* strain C55 and other phage group II *S. aureus* strains and demonstration of an association with the exfoliative toxin B gene. Infect. Immun. 67: 4268–4271.

Neis S, Bierbaum G, Josten C, Pag MU, Kempter C, Jung G & Sahl H-G (1997) Effect of leader peptide mutations on biosynthesis of the lantibiotic Pep5. FEMS Microbiol. Lett. 149: 249–255.

Nes I F, Diep DB, Håvarstein LS, Brurberg MB, Eijsink V & Holo H (1996) Biosynthesis of bacteriocins in lactic acid bacteria. Antonie van Leeuwenhoek 70: 113–128.

Otto M, Peschel A & Götz F (1998) Producer self-protection against the lantibiotic epidermin by the ABC transporter EpiFEG of *Staphylococcus epidermidis* Tu3298. FEMS Microbiol. Lett. 166: 203–211.

Peschel A & Gotz F. Analysis of the Staphylococcus epidermidis genes epiF, -E, and -G involved in epidermin immunity. J. Bacteriol. 1996. 178: 531–536.

Peschel A, Schnell N, Hille M, Entian K-D & Götz F (1997) Secretion of the lantibiotics epidermin and gallidermin: sequence analysis of the genes *gdmT* and *gdmH*, their influence on epidermin production and their regulation by EpiQ. Mol. Gen. Genet. 254: 312–318.

Piard, J-C, Muriana PM, Desmazeaud MJ & Klaenhammer TR (1992) Purification and partial characterisation of lacticin 481, a lanthionine-containing bacteriocin produced by *Lactococcus lactis* subsp. *lactis* CNRZ481. Appl. Environ. Microbiol. 58: 279–284.

Piard J-C, Kuipers OP, Rollema H S, Desmazeuad M J & de Vos WM (1993) Structure, organization and expression of the *lct* gene for lacticin 481, a novel lantibiotic produced by *Lactococcus lactis*. J. Biol. Chem. 268: 16361–16368.

Prasch T, Naumann T, Markert RL, Sattler M, Schubert W, Schaal S, Bauch M, Kogler H & Griesinger C. Constitution and solution conformation of the antibiotic mersacidin determined by NMR and molecular dynamics. Eur. J. Biochem. 1997. 224: 501–512.

Pridmore D, Rekhif N, Pittett A-C, Suri B, & Mollet B (1996) Variacin, a new lanthionine-containing bacteriocin produced by *Micrococcus varians*: comparison to lacticin 481 of *Lactococcus lactis*. Appl. Environ. Microbiol. 62: 1799–1802.

Qi F, Chen P & Caufield PW (1999) Functional analyses of the promoters in the lantibiotic mutacin II biosynthetic locus in *Streptococcus mutans*. Appl. Environ. Microbiol. 65: 652–658.

Qiao M & Saris PE J (1996) Evidence for a role of NisT in transport of the lantibiotic nisin produced by *Lactococcus lactis* N8. FEMS Microbiol. Lett. 144: 89–93.

Qiao M, Immonen T, Koponen O & Saris PE J (1995) The cellular location and effect on nisin imunity of the NisI protein from *Lactococcus lactis* N8 expressed in *Escherichia coli* and *L. lactis*. FEMS Microbiol. Lett. 131: 75–80.

Rogers LA & Whittier EO (1928) Limiting factors in lactic fermentation. J. Bacteriol. 16: 211–14.

Reisinger P, Seidel H, Tschesche H, & Hammes WP (1980) The effect of nisin on murein synthesis. Arch. Microbiol. 127: 187–193.

Reminger A, Eijsink VGH, Ehrmann MA, Sletten K, Nes IF & Vogel RF (1999) Purification and partial amino acid sequence of plantaricin 1.25a and 1.25b, two bacteriocins produced by *Lactobacillus plantarum* TMW1.25. J. Appl. Microbiol. 86: 1053–1058.

Rince A, Dufour A, Le Pogam S, Thuault D, Bourgeois C M & Le Pennec JP (1994) Cloning, expression and nucleotide sequence of genes involved in production of lactococcin DR, a bacteriocin from *Lactococcus lactis* subsp. *lactis*. Appl. Environ. Microbiol. 60: 652–1657.

Rince A, Dufour A, Uguen P, Le Pennec J-P & Haras D (1997) Characterization of the lacticin 481 operon: the *Lactococcus lactis* genes *lctF*, *lctE*, and *lctG* encode a putative ABC transporter involved in bacteriocin immunity. Appl. Environ. Microbiol. 63: 4252–4260.

Rodríguez JM, Cintas LM, Casaus P, Suárez A & Hernández PE (1995) PCR detection of the lactocin S structural gene in bacteriocin-producing lactobacilli from meat. Appl. Environ. Microbiol. 61: 2802–2805.

Rogers LA & Whittier ED (1928) Limiting factors in lactic fermentation. J. Bacteriol. 16: 211–229.

Ross KF, Ronson CW & Tagg JR (1993) Isolation and characterisation of the lantibiotic salivaricin A and its structural gene *salA* from *Streptococcus salivarius* 20P3. Appl Environ. Microbiol. 59: 2014–2021.

Ross RP, Galvin M, McAuliffe O, Morgan SM, Ryan MP, Twomey DP, Meaney WJ & Hill C (1999) Developing applications for lactococcal bacteriocins. Antonie van Leeuwenhoek 76: 337–346.

Ruhr E & Sahl H-G (1985) Mode of action of the peptide antibiotic nisin and influence on the membrane potential of whole cells and on artificial membrane vesicles. Antimicrobiol. Agents Chemother. 27: 841–845.

Ryan MP, Rea MC, Hill C & Ross RP (1996) An application in Cheddar cheese manufacture for a strain of *Lactococcus lactis* producing a novel broad spectrum bacteriocin, lacticin 3147. Appl. Environ. Microbiol. 62: 612–619.

Ryan MP, Meaney WJ, Ross RP & Hill C (1998) Evaluation of lacticin 3147 and a teat seal containing bacteriocin for the inhibition of mastitis pathogens. Appl. Environ. Microbiol. 64: 2287–2290.

Ryan MP, Jack R, Josten W, Sahl H-G, Jung G, Ross, RP & Hill C (1999) Extensive -translational modification, including a serine to D-alanine conversion, in the two-component lantibiotic, lacticin 3147. J. Biol. Chem. 274: 37544–37550.

Sahl H-G, Kordel M & Benz R (1987) Voltage-dependent depolarization of bacterial membranes and artificial lipid bilayers by the peptide antibiotic nisin. Arch. Microbiol. 149: 120–124.

Sahl H-G & Bierbaum G (1998) Lantibiotics: biosynthesis and biological activities of uniquely modified peptides from gram-positive bacteria. Annu. Rev. Microbiol. 52: 41–79.

Segarra RA, Booth MC, Morales DA, Huycke MM & Gilmore MS (1991) Molecular characterization of the *Enterococcus faecalis* cytolysin activator. Infect. Immun. 59: 1239–1246.

Siegers K & Entian K-D (1995) Genes involved in immunity to the lantibiotic nisin produced by *Lactococcus lactis* 6F3. Appl. Environ. Microbiol. 61: 1082–1089.

Siezen RJ, Kuipers OP & de Vos WM (1996) Comparison of lantibiotic gene clusters and encoded proteins. Antonie van Leeuwenhoek 69: 171–184.

Skaugen M, Nissen-Meyer J, Jung G, Stevanovic S, Sletten K, Abildgaard CIM & Nes I F (1994) *In vivo* conversion of L-serine to D-alanine in a ribosomally synthesized polypeptide. J. Biol. Chem. 269: 27183–27185.

Skaugen M, Abildgaard CIM & Nes I F (1997) Organisation and expression of a gene cluster involved in the biosynthesis of the lantibiotic, lactocin S. Mol. Gen. Genet. 253: 674–686.

Skaugen M, Anderson EL, Christie VH & Nes I (2002) Identification, characterisation and expression of a second bicistronic operon involved in the production of lactocin S in *Lactocbacillus sake* L45. Appl. Environ. Microbiol. (in press).

Takami H, Nakasone K, Takaki Y, Maeno G, Sasaki R, Masui N, Fuji F, Hirama C, Nakamura Y, Ogasawara N, Kuhara S & Horikoshi K (2000) Complete genome sequence of the alkaliphilic bacterium *Bacillus halodurans* and genomic sequence comparison with *Bacillus subtilis*. Nucleic Acids Res. 28: 4317–4331.

Turner DL, Brennan L, Meyer HE, Lohaus C, Siethoff C, Costa HS, Gonzalez B, Santos H & Suárez JE (1999) Solution structure of plantaricin C, a novel lantibiotic. Eur. J. Biochem. 264: 833–839.

Twomey DP, Wheelock AI, Flynn J, Meaney WJ, Hill C & Ross RP (2000) Protection against *Staphylococcus aureus* mastitis in dairy cows using a bismuth-based teat seal containing the bacteriocin, lacticin 3147. J. Dairy Sci. 83: 1981–1988.

Uguen P, Hamelin J, le Pennec J-P & Blanco C (1999) Influence of osmolarity and the presence of an osmoprotectant on *Lactococcus lactis* growth and bacteriocin production. Appl. Environ. Microbiol. 65: 291–293.

van de Ven FJM, van den Hooven HW, Konings RNH & Hilbers CW (1991) NMR-studies of lantibiotics: the structure of nisin in aqueous solution. Eur. J. Biochem. 202: 1181–1188.

van den Hooven HW, Doeland CC, van den Kamp M, Konings RN, Hilbers CW & van den Ven FJ (1996a) Three-dimensional structure of the lantibiotic nisin in the presence of membrane mimetic micelles of dodecylphosphocholine and of sodium dodecysulphate. Eur. J. Biochem. 235: 382–393.

van den Hooven HW, Lagerwerf FM, Heerma W, Haverkamp J, Piard JC, Hilbers CH, Siezen RJ, Kuipers OP & Rollema HS (1996b) The structure of the lantibiotic lacticin 481 produced by *Lactococcus lactis*: location of the thioether bridges. FEBS Lett. 391: 317–322.

van den Hooven HW, Rollema HS, Siezen RJ, Hilbers CW & Kuipers OP (1997) Structural features of the final intermediate in the biosynthesis of the lantibiotic nisin. Influence of the leader peptide. Biochemistry 36: 14137–14145.

van der Meer JR, Polman J, Beerthuyzen MM, Siezen RJ, Kuipers OP & de Vos WM (1993) Characterization of the *Lactococcus lactis* nisin A operon: genes *nisP*, encoding a subtilisin-like serine protease involved in precursor processing, and *nisR*, encoding a regulatory protein involved in nisin biosynthesis. J. Bacteriol. 175: 2578–2588.

van der Meer JR, Rollema HS, Siezen RJ, Beerthuyzen MM, Kuipers OP & de Vos WM (1994) Influence of amino acid substitutions in the nisin leader peptide on biosynthesis and secretion of nisin by *Lactococcus lactis*. J. Biol. Chem. 269: 3555–3562.

Walker JE, Saraste M, Runswick MJ & Gay NJ (1982) Distantly related sequences in the α- and β-subunits of ATP synthase, myosin, kinases and other ATP requiring enzymes and a common nucleotide binding fold. EMBO J. 1: 945–951.

Whitehead HR (1933) A substance inhibiting bacterial growth, produced by certain strains of lactic streptococci. Biochem. J. 27: 1793–1800.

Wiedemann I, Breukink E, van Kraaij C, Kuipers O, Bierbaum G, de Kruijff B & Sahl H-G (2001) Specific binding of nisin to the peptidoglycan precursor lipid II combines pore formation and the inhibition of cell wall biosynthesis for potent antibiotic activity. J. Biol. Chem. 276: 1772–1779.

Woodruff WA, Novak J & Caufield PW (1998) Sequence analysis of *mutA* and *mutM* genes involved in the biosynthesis of the lantibiotic mutacin II in *Streptococcus mutans*. Gene 206: 37–43.

Yamaguchi T, Hayashi T, Takami H, Ohnishi M, Murata T, Nakayama K, Asakawa M, Ohara M, Komatsuzawa H & Sugai M (2001) Complete nucleotide sequence of a *Staphylococcus aureus* exfoliative toxin B plasmid and identification of a novel ADP-ribosyltransferase, EDIN-C. Infect. Immun. 69: 7760–7771.

Zimmermann N & Jung G (1997) The three dimensional solution structure of the lantibiotic murein-biosynthesis-inhibitor actagardine determined by NMR. Eur. J. Biochem. 246: 809–819.

Antonie van Leeuwenhoek **82:** 187–216, 2002.

Stress responses in lactic acid bacteria

Maarten van de Guchte, Pascale Serror[1], Christian Chervaux[2], Tamara Smokvina[2], Stanislav D. Ehrlich & Emmanuelle Maguin*
*Génétique Microbienne, INRA, Domaine de Vilvert, 78352 Jouy en Josas Cedex, France (*Author for correspondence; E-mail: Emmanuelle.Maguin@jouy.inra.fr)*

Key words: lactic acid bacteria, stress response, adaptive response, cross protection, gene regulation.

Abstract

Lactic acid bacteria (LAB) constitute a heterogeneous group of bacteria that are traditionally used to produce fermented foods. The industrialization of food bio-transformations increased the economical importance of LAB, as they play a crucial role in the development of the organoleptique and hygienic quality of fermented products. Therefore, the reliability of starter strains in terms of quality and functional properties (important for the development of aroma and texture), but also in terms of growth performance and robustness has become essential. These strains should resist to adverse conditions encountered in industrial processes, for example during starter handling and storage (freeze-drying, freezing or spray-drying). The development of new applications such as life vaccines and probiotic foods reinforces the need for robust LAB since they may have to survive in the digestive tract, resist the intestinal flora, maybe colonize the digestive or uro-genital mucosa and express specific functions under conditions that are unfavorable to growth (for example, during stationary phase or storage). Also in nature, the ability to quickly respond to stress is essential for survival and it is now well established that LAB, like other bacteria, evolved defense mechanisms against stress that allow them to withstand harsh conditions and sudden environmental changes. While genes implicated in stress responses are numerous, in LAB the levels of characterization of their actual role and regulation differ widely between species. The functional conservation of several stress proteins (for example, HS proteins, Csp, etc) and of some of their regulators (for example, HrcA, CtsR) renders even more striking the differences that exist between LAB and the classical model micro-organisms. Among the differences observed between LAB species and *B. subtilis*, one of the most striking is the absence of a σ^B orthologue in *L. lactis* ssp. *lactis* as well as in at least two streptococci and probably *E. faecalis*. The overview of LAB stress responses also reveals common aspects of stress responses. As in other bacteria, adaptive responses appear to be a usual mode of stress protection in LAB. However, the cross-protection to other stress often induced by the expression of a given adaptive response, appears to vary between species. This observation suggests that the molecular bases of adaptive responses are, at least in part, species (or even subspecies) specific. A better understanding of the mechanisms of stress resistance should allow to understand the bases of the adaptive responses and cross protection, and to rationalize their exploitation to prepare LAB to industrial processes. Moreover, the identification of crucial stress related genes will reveal targets i) for specific manipulation (to promote or limit growth) , ii) to develop tools to screen for tolerant or sensitive strains and iii) to evaluate the fitness and level of adaptation of a culture. In this context, future genome and transcriptome analyses will undoubtedly complement the proteome and genetic information available today, and shed a new light on the perception of, and the response to, stress by lactic acid bacteria.

Lactic acid bacteria (LAB) constitute a heterogeneous group of bacteria which are found in diverse environments from the human and animal body to plants. These bacteria have been used for long to produce various fermented foods from products de-

[1] Present address: Unité de Recherche Laitière et Génétique Appliquée, INRA, Domaine de Vilvert, 78352 Jouy en Josas Cedex, France.
[2] Present address: Danone Vitapole, 15 avenue Galilée, 92350 Le Plessis Robinson, France.

rived from animals (milk, meat, fish, etc.) or plants (vegetables, wine, olives, etc.) (Stiles 1996). The industrialization of food bio-transformations increased the economical importance of LAB. Although LAB are a low cost ingredient of the food transformation processes, they play a crucial role in the development of the organoleptique and hygienic quality of fermented products. Therefore, the reliability of starter cultures in terms of quality and functional properties (important for the development of aroma and texture), but also in terms of growth performance and robustness has become essential for successful fermentations. Therefore LAB strains were selected for resistance against bacteriophages, for fast growth and acidification, for proteolytic properties, for bacteriocin resistance, etc. However, in addition these strains must also resist the adverse conditions encountered in industrial processes, for example during starter handling and storage (freeze-drying, freezing or spray-drying). The development of new applications such as live vaccins (Wells et al. 1996; Mercenier et al. 2000) and probiotic foods (Schiffrin et al. 2001) reinforces the need for robust LAB since they may have to survive in the digestive tract, resist the intestinal flora, eventually colonize the digestive or uro-genital mucosa and express specific functions in conditions unfavorable to growth (for example, during stationary phase or storage). Except probiotic strains for which high tolerance to acid and bile was used as a selection criteria (Dunne et al. 1999), LAB have seldomly been selected for stress resistance.

However, bacteria are not only submitted to potentially stressful environmental changes in industrial processes, but also in nature where the ability to quickly respond to stress is essential for survival (Stortz et al. 2000) It is now well established that LAB, like other bacteria, evolved stress-sensing systems and defenses against stress which allow them to withstand harsh conditions and sudden environmental changes. Although a micro-organism could, in theory, have specific regulators tailored to each of its regulated genes and adapt their expression according to its environment, this would represent a tremendous genetic burden. Instead, regulators usually control several genes and sometimes even control other regulators (VanBogelen et al. 1999). Stress defenses are good examples of such integrated regulation systems. Bacterial stress responses rely on the coordinated expression of genes which alter different cellular processes (cell division, DNA metabolism, housekeeping, membrane composition, transport, etc.) and act in concert to improve the bacterial stress tolerance (Stortz et al. 2000). The integration of these stress responses is accomplished by networks of regulators which allow the cell to react to various and complex environmental shifts.

Identifying regulators and regulatory networks is essential if the goal is to control, predict or engineer LAB behavior (in given conditions). The knowledge of regulators and a better understanding of LAB stress responses could constitute a basis of comparison with the well known model micro-organisms, *E. coli* and *B. subtilis*. Such comparisons should reveal the specificity of LAB stress responses which may have evolved and been selected indirectly to fit the specific constraints of a given substrate and/or process (for example, milk and milk fermentation). The current knowledge on the environmental stress responses in LAB varies between species and depending on the type of stress. The best studied are acid, heat and cold stress, although for the latter most of the studies focused on a specific family of proteins instead of the whole response. Therefore, the comparison between the classical model bacteria and LAB remains partial. Nevertheless, this comparison already suggests that regulatory circuits differ.

The acid stress response

The growth of lactic acid bacteria (LAB) is characterized by the generation of acidic end products of fermentation, that accumulate in the extracellular environment. The pronounced organic acid production of these bacteria creates an environment unfavorable for many other organisms. This characteristic is the basis of numerous methods of food preservation by fermentation. These bacteria can also encounter an acidic environment in the stomach after consumption, and the development of probiotics renewed the interest for LAB survival in the digestive tract. The cariogenicity of oral LAB such as streptococci and lactobacilli is directly related to their acidogenicity (ability to produce acid at low pH) and acidurance (capacity to function at low pH; Harper et al. 1984; Quivey et al. 2000b).

Except for some species of the genera *Lactobacillus*, *Leuconostoc* and *Oenococcus*, LAB are neutrophiles (i.e., optimal pH for growth between 5 and 9). The effects of acid stress on bacterial physiology are not known in detail. It is well established, however, that acids can passively diffuse through the cell membrane and after entry into the cytoplasm, rapidly dissociate into protons and charged derivatives to which the cell membrane is impermeable (Presser et al. 1997). The intracellular accumulation of protons may lower the intracellular pH (pH_i) and thus affects the transmembrane ΔpH which contributes to the proton motive force (pmf), that is used as an energy source in numerous transmembrane transport processes. The internal acidification also reduces the activity of acid-sensitive enzymes and damages proteins and DNA. The accumulation in the cytoplasm of the anionic moiety of the dissociated organic acids also has a detrimental effect on cellular physiology (Presser et al. 1997) perhaps through a chelating interaction with essential elements.

In LAB, acid tolerance (AT) increases in at least two distinct physiological states. (i) during logarithmic growth an adaptive response referred to as L-ATR, can be induced by incubation at a non-lethal acidic pH; (ii) after entry in the stationary phase, AT increases as a result of the induction of a general stress response (GSR, see the section concerning starvation) (Hartke et al. 1996). The latter response is usually independent of the external pH (pH_e), although in *L. acidophilus* CRL639 a low pH_e is needed (Lorca et al. 2001). It is not known whether these responses are independent or overlap. The growth in biofilms may be a third state which improves AT, but it was only demonstrated for *S. mutans* (Li et al. 2001; Zhu et al. 2001).

Most of the LAB species tested (except for many *L. lactis* ssp. *cremoris*; Kim et al. 1999) possess an L-ATR which improves the bacterial survival to a lethal acid challenge compared to untreated cells (Belli et al. 1991; Flahaut et al. 1996b; Hartke et al. 1996; Rallu et al. 1996; O'Sullivan et al. 1999; Takahashi et al. 1999; Kim et al. 2001). The induction of the L-ATR often protects LAB not only from an acid challenge but also from other stress such as heat, osmotic or oxidative shocks. This broad protective effect of L-ATR varies between species and does not always protect from the same set of stress (Quivey et al. 1995; Flahaut et al. 1996b; Svensater et al. 2000; Lorca et al. 2001; E.M. Lim and E. Maguin, pers. comm.). Numerous proteomic studies performed on L-ATR showed that numbers of proteins are induced during acid adaptation in LAB (Champomier-Verges et al. 2002). However, only few of these putative acid-resistance proteins have so far been characterized (Hartke et al. 1996; Guzzo et al. 1997; Lim et al. 2000; De Angelis et al. 2001; Giard et al. 2001; Wilkins et al. 2001).

In order to unravel the basis of AT, genetic studies were carried out in *L. lactis* and *S. mutans* (i) to obtain acid-sensitive mutants (Yamashita et al. 1993; Gutierrez et al. 1996), (ii) to find acid-regulated promoters (Israelsen et al. 1995; Sanders et al. 1998a; Cvitkovitch et al. 2000) and (iii) to select for acid-resistant mutants (Rallu et al. 2000). Altogether, biochemical, proteomic and genetic analyses indicate that LAB acid responses are intricate processes that involve the synthesis of a variety of proteins and several mechanisms. Although our understanding of these responses is not yet complete, it is already possible to enlighten some resistance mechanisms which address negative effects of the acid stress.

ATP dependent expulsion of protons

As mentioned above, the dissipation of the transmembrane pH gradient (ΔpH) would be an expected consequence of an acid shock. LAB can maintain a ΔpH over a wide range of low pH_e values (Hutkins et al. 1993). Some LAB, including lactococci maintain their internal pH (pH_i) close to neutrality until a threshold value of pH_e below which the pH_i starts to decrease (Breeuwer et al. 1996). In contrast other LAB, for example *L. delbrueckii* maintain a constant ΔpH of about 1 and let decrease pH_i in parallel with pH_e (Siegumfeldt 1999).

The F_0F_1-ATPase

The F_0F_1-ATPase is ubiquitous among bacteria and its molecular architecture and operation have been unraveled (Rastogi et al. 1999; Sambongi et al. 1999; Stock et al. 1999). This multimeric enzyme can either synthesize ATP using protons or conversely expulse protons out of the cell with the energy provided by the ATP hydrolysis. In LAB, the latter activity increases at low pH and is crucial to maintain the ΔpH (Futai et al. 1989; Nannen et al. 1991). Mutants affected in the F_0F_1-ATPase activity were obtained in various LAB, all of them exhibited growth defects at low pH (*E. faecalis*, Suzuki et al. 1988; *L. lactis*, Yokota et al. 1995; *Lactobacillus helveticus*, Yamamoto et al. 1996; and *O. oeni*, Tourdot-Marechal et al. 1999). In *O. oeni*, an acid-tolerant mutant was isolated which displayed an increased H^+-ATPase activity (Drici-Cachon et al. 1996). Finally, it was recently shown that the F_0F_1-ATPase was essential for the growth of *L. lactis* (Koebmann et al. 2000) except when an electron transfer chain is active (Blank et al. 2001; Duwat et al. 2001). Nevertheless, the F_0F_1-ATPase activity has a drawback: its energetic cost may ultimately lead to growth arrest at low pH (Bond et al. 1996).

Several operons encoding this enzyme (*atp* operon) have been characterized (Shibata et al. 1992; Fenoll et al. 1994; Smith et al. 1996; Kullen et al. 1999; Koebmann et al. 2000; Martin-Galiano et al. 2001) and an increasing number of other *atp* sequences are becoming available as a result of LAB genome sequence projects. The *atp* loci contain all the genes encoding the five subunits (α, β, δ, γ, ε) of the cytoplasmic F_1 complex and the three subunits (a, b, c) forming the F_0 membrane proton channel. In LAB, the *atp* operons have a genetic organization which differs from other bacteria as reviewed by Quivey et al. (2001) and Martin-Galiano et al. (2001); however, the significance (if any) of this feature is yet unknown. The stimulation of the F_0F_1-ATPase activity at low pH may result from (i) the stimulation of transcription of the *atp* locus as reported for *L. acidophilus* (with a 2-fold induction, Kullen et al. 1999) and for oral streptococci (Belli et al. 1991) and/or (ii) the stabilization of the subunits that constitute the enzyme as observed in *E. hirae* (Arikado et al. 1999). In all LAB studied, the 2DE analysis of L-ATR always revealed all or some of the subunits of the cytoplasmic F_1 complex among the upregulated proteins (Champomier-Verges et al. 2002). Nevertheless, in *O. oeni* no significant increase of ATPase activity was measured at low pH although as mentioned above, mutants of the F_1F_0-ATPase are altered in their aciduránce. Preliminary results suggested the existence of several ATPases with different optimum pHs (Guzzo et al. 2000); a K^+-ATPase may be involved (see below).

K^+-ATPase

Cation transport ATPases such as a K^+-ATPase can contribute to pH homeostasis. The exchange of K^+ for H^+ converts the transmembrane potential ($\Delta\Psi$) generated by the K^+-ATPase, into a transmembrane pH gradient (ΔpH). This ion exchange allows to establish the ΔpH and can participate in pH homeostasis (Kashket 1984). For example in *S. mutans* grown at a pH_e of 5.0, the pH_i of glucose energized cells equals 5.5 or 6.14 in the absence or presence of 25 mmol/l of K^+, respectively (Dashper 1992). Similar observations were reported for *L. lactis* (Kashket et al. 1977) and *E. hirae* (Bakker et al. 1980; Kobayashi et al. 1982).

Production of basic compounds

Arginine deiminase

Another mechanism for pH homeostasis is the arginine deiminase pathway (ADI, Cunin et al. 1986). Three enzymes, arginine deiminase, ornithine cabamoyltransferase and carbamate kinase, constitute this pathway and catalyze the conversion of arginine into ornithine, ammonia (two per arginine) and carbon dioxide with the formation of 1 mol of ATP per mol of arginine consumed. The resulting NH_3 reacts with H^+ and helps to alkalize the environment and the generated ATP can enable extrusion of cytoplasmic protons by the F_0F_1-ATPase. An arginine/ornithine antiporter completes the system and allows the exchange of these two molecules at no energy cost. Marquis et al. (1987) showed that in the presence of 45 mM of arginine, *L. lactis* increased the pH_e from 4 to 6.5 within a few hours. The ADI pathway has been detected in several LAB (Table 1, reviewed in Sanders et al. 1995) but its direct involvement in AT was not always demonstrated. Moreover, the factors involved in ADI regulation appeared to be a combination of arginine availability, energy depletion, catabolite repression and oxygenation rather than low pH (Cunin et al. 1986; Zuniga et al. 1998; Champomier Verges et al. 1999; Tonon et al. 2000). Therefore, although ADI activity can alkalize the environment, its importance for the acidurance of LAB may vary between species.

Urease

The urease activity was detected both in the dairy *S. thermophilus* (J. Anba, P. Renault, pers. comm.) and in the oral bacteria *S. salivarius* in which it has been mostly studied (Quivey et al. 2000b). The nickel metalloenzyme urease catalyses the hydrolysis of urea to CO_2 and ammonia (two per urea molecule) which can alkalize the pH. In the presence 25 mM urea, cells at pH 3 raised the pH_e to 7.6 and improved their survival 1000-fold compared to a control without urea (Chen et al. 2000). The complex urease operon (Chen et al. 1996) is derepressed at low pH and with an excess of carbohydrates (Chen et al. 1998). Since urea is present in the saliva, ureolysis is probably especially important for the survival of *S. salivarius* under the extreme acidic conditions that can be generated by the more aciduric lactobacilli and other oral streptococci. Note that in the dairy *S. thermophilus*, the regulation of urease expression seems different from *S. salivarius* (J. Anba, P. Renault, pers. comm.).

Table 1. Different systems that may contribute to acid tolerance and/or energy generation at low pH

System	Basic compound	Electrogenic transporter	Energy production	Organism	References
ADI	NH_3	Antiport	ATP	*L. lactis* (ssp. *cremoris* mostly ADI-minus)	Crow et al. (1982),
				Streptococci (some sp. are ADI-minus)	Abdelal (1979)
				Lactobacilli sp.?	Abdelal (1979), Arena et al. (1999)
				L. sakei	Champomier Verges et al. (1999)
				O. oeni	Arena et al. (1999), Tonon et al. (2000)
Urease	NH_3			*S. salivarius*	Chen et al. (1996)
				S. thermophilus	
Glu/GABA (Gad)	GABA	Antiport	ND	*L. lactis*	Sanders et al. (1998a)
			ATP	*Lactobacillus* sp.	Higuchi et al. (1997)
Asp/Ala	Ala	ND	ATP	*Lactobacillus* sp.	Abe et al. (1996)
His/Histamine	Histamine	Antiport	ATP	*L. buchneri*	Molenaar et al. (1993)
MLF	lactacte	Antiport	ATP	*L. lactis*	Poolman et al. (1991), Renault et al. (1988)
		uniport		*L. sakei*	Champomier-Verges et al. (2001)
		uniport		*L. plantarum*	Olsen et al. (1991)
				O oeni, Lactobacilli, Pediococcus	Salema et al. (1996)
Citrate/lactate	Lactate	antiport CitP	ND	*L. lactis diacetylactis*	Garcia-Quintans et al. (1998)
Citrolactique fermentation	Lactate		ATP	*Leuconostoc mesenteroides*	Marty-Teysset et al. (1996)
End product efflux		Lactate uniport		*L. lactis*	ten Brink et al. (1985), Konings et al. (1997)

ND: not determined

Decarboxylation reactions and electrogenic transport

Several systems based on a decarboxylation and an electrogenic transporter found in LAB may contribute to pH homeostasis. In these reactions, a carboxylic acidic compound (for example an amino acid) is transported into the cell to be decarboxylated. A proton is consumed in the reaction and the product is exported from the cell via a transporter. The net effect of this reaction is to increase the alkalinity of the cytoplasm. Moreover, the coupling of the decarboxylation to an electrogenic transporter (antiport or uniport) allows to generate ATP via the pmf. The different systems identified in LAB are presented in Table 1 (for reviews: Sanders et al. 1999; Konings et al. 1997). A direct link between these activities and the AT was only established for the *L. lactis* Gad system. In the presence of chloride and Glu needed for Gad activity, a *gad* mutant is 1000-fold more acid-sensitive than the wild-type strain (Sanders et al. 1998a). The Glu decarboxylase is often found in prokaryotes that either colonize or pass through the gut. It was proposed that this enzyme might strengthen bacteria during their transit through the stomach (Small et al. 1998).

Malolactic fermentation (MLF) is the conversion of the dicarboxylic malic acid to the monocarboxylic lactic acid. MLF has a major role for the wine bacteria *O. oeni*, *Lactobacillus* and *Pediococcus* sp. (Salema et al. 1996) and has also been identified in *L. sakei* (Champomier-Verges et al. 2001), *L. plantarum* (Olsen et al. 1991) and *L. lactis* (Renault et al. 1988). Depending on the bacterium the L-lactate produced is excreted via (i) a lactate–malate antiporter (*L. lactis*, Poolman et al. 1991) or (ii) an electrogenic uniport (*O. oeni*, *L. plantarum*) which both allow the synthesis of ATP at low pH. This ATP production is probably the basis for the implication of MLF in *O. oeni* acid tolerance (Tourdot-Marechal et al. 1999).

Electrogenic transport of end products

In *L. lactis*, the excretion of lactic acid has been studied at high and low pH. At high pH values, the excretion process is electrogenic (i.e., more than one proton is translocated with the lactate) and can contribute to the generation of the pmf, whereas at low pH values this transport is electroneutral and cannot contribute to the generation of a membrane potential (Konings et al. 1997). Nevertheless, Koebman et al. (2000) showed that an *L. lactis* strain defective for F_0F_1-ATPase was not able to form colonies on medium buffered at pH 7. This observation suggests that the lactic acid efflux at neutral pH does not generate enough energy to ensure cellular growth.

In *L. lactis* ssp. *diacetylactis*, bacterial growth at low pH is favored by the availability of both citrate and lactate compared to each of the substrates alone (Garcia-Quintans et al. 1998) and the citrate-lactate antiporter is induced. It is likely that the generation of energy via the electrogenic transport and the alkalization of the medium are responsible for this stimulation of growth (Marty-Teysset et al. 1996).

Cell envelope and acid tolerance

The cell envelope is probably among the first targets of physicochemical stress. A change in fatty acids (FA) composition of the membrane is a usual response to environmental stress, and in *E. coli* FA modifications play a major role in the protection from acid stress (Chang et al. 1999). The adaptation of *S. mutans* to acidification includes an enrichment in mono-unsaturated and longer chain FA in the membrane (Quivey et al. 2000a) which could cause the reduction of permeability to protons observed for acid adapted cells (Ma & Marquis 1997). The cell wall may also be involved in AT. In *S. mutans*, the inactivation of the *dlt* operon, responsible for the D-alanine esterification of lipotechoïc acids (Delcour et al. 1999), resulted in growth alteration and an increased acid sensitivity correlated with a higher permeability to protons than in the wild-type (Boyd et al. 2000). In *L. lactis*, the inactivation of certain penicillin-binding proteins involved in peptidoglycan synthesis increases the acid sensitivity (F. Rallu, A. Budin, pers. comm.). These data suggest that cell wall composition could have a role in acidurance, possibly through alteration of the cell surface properties. Finally, other systems might also help to maintain cell envelope integrity in acid stress conditions. Firstly, in *O. oeni* a small heat-shock protein (smHSP), Lo18, is induced after heat, ethanol and acid stress (Guzzo et al. 1997). The prokaryotic smHSPs share a conserved sequence with the α-crystallin protein of the vertebrate eye lens and are probably involved in the refolding of denatured proteins with the help of other chaperones. Interestingly, Lo18 became mostly membrane associated after heat stress and its expression was induced by membrane fluidization (Delmas et al. 2001). These data suggest that Lo18 might be involved in the stabilization of membrane and/or envelope proteins after thermal or acid stress. Secondly, the Ffh protein of *S. mutans* has a role in AT. In *E. coli*, Ffh participates in the SRP translocation pathway which handles the transport of membrane and extracellular proteins. In *S. mutans*, the *ffh* gene is upregulated by low pH. An *ffh* mutant is acid-sensitive and is not able to rapidly increase F_0F_1-ATPase activity in the membrane upon acid shock (Gutierrez et al. 1999; Kremer et al. 2001) This phenotype might reflect a role of the SRP system in the translocation of the F_0F_1-ATPase and possibly other acid-induced proteins into the membrane.

Repair of damaged proteins

In addition to *O. oeni* Lo18 and *S. mutans* Ffh which may also function as chaperones, the proteomic analysis of L-ATR in several LAB revealed that heat-shock chaperones were always upregulated presumably to repair acid-induced damaged proteins and/or to facilitate the folding of neosynthesized proteins. However, the subset of acid-induced heat-shock proteins varied between species although DnaK and GroEL were often identified (Champomier-Verges et al. 2002). Specific studies of the *dnaK* and *groE* operons in *S. mutans* (Jayaraman et al. 1997; Lemos et al. 2001) and the *groE* and *clpP* operons in *L. lactis* (Hartke et al. 1997; Frees et al. 1999) suggested that the induction by low pH was mediated by their usual heat-shock regulators, HrcA and CtsR (see the section concerning heat-shock).

Repair of DNA damage

The intracellular acidification causes depurination and depyrimidination of DNA. This mechanism involves protonation of the base followed by cleavage of the glycosyl bond. The residues left at sites of base loss are called abasic or AP (apurinic, apyrimidique) sites (Lindahl et al. 1972). In *L. lactis*, a moderate UV irradiation induced a tolerance against several stresses including acid shock (Hartke et al. 1995), and four proteins upregulated during acid adaptation were also induced by DNA damaging treatments (Hartke et al. 1996) suggesting that L-ATR might include DNA repair systems. This hypothesis was confirmed in *S. mutans* where L-ATR was shown to improve the resistance to DNA damage via the induction of an acid-inducible RecA-independent DNA repair system (Quivey et al. 1995). Further studies showed that at low pH, wild-type and *recA* strains of *S. mutans* exhibited an endonuclease activity targeted to AP sites (Hahn et al. 1999), and that the *uvrA* gene was induced (Hanna et al. 2001). An *uvrA* mutant has a reduced ability to mount a complete L-ATR protection, and its chromosomal DNA is more extensively degraded during exposure to low pH than in a wild-type strain (Hanna et al. 2001). The Uvr system is activated by helical distortion (Sancar 1996), whereas the AP endonuclease activity usually handles the repair of small DNA damaged regions by base excision. It is therefore possible that these two systems work together during an acid shock.

Regulation of the AT systems

Two component systems (2CS) are often involved in the adaptation of bacteria to their environment (E. Guédon et al. this issue). In *L. lactis* and *L. sakei*, several 2CS mutants displayed AT related phenotypes (Morel-Delville 1998; O'Connell-Motherway et al. 2000). This possible involvement of 2CS in AT suggests that an external signal of acidity may exist. A recent study suggests that *S. mutans* has an external signaling system related to quorum sensing which stimulates its AT. Although the competence stimulating factor participates in the signaling, another factor is needed to obtain a complete induction of AT (Li et al. 2001).

Some intracellular pools could also play a role in the control of the tolerance to acid and probably also to other stress. *L. lactis* mutants defective in the high affinity phosphate transporter (*pst* operon) have a high tolerance to H_2O_2 and acidic pH (Rallu et al. 2000). It is likely that, as documented in *B. subtilis*, phosphate limitation induces a regulon which includes stress resistance genes (Eymann et al. 1996). Since it was previously shown that low pH decreases the activity of the phosphate transporter (Poolman et al. 1987a), the phosphate controlled regulon may have a role in L-ATR. In *L. lactis*, the inactivation of genes (*guaA*, GMP synthetase, and *relA*, (p)ppGpp synthetase) involved in guanine nucleotides metabolism leads to acid, heat-shock and glucose starvation constitutive tolerance (Rallu et al. 2000). In *S. mutans*, the protein SGP (*Streptococcus* GTP-Binding protein) homologous to the universal Era subfamily of GTPases, may be involved in acid resistance (Yamashita et al. 1993; Baev et al. 1999). One of the functions proposed for SGP is the regulation of the GTP/GDP pool. It is noteworthy that both the amount of SGP and its association with membranes increase in stress conditions (acidic pH, heat, starvation). The GTPases are considered as necessary for either ribosome function or for the transmission of information from the ribosome to specific targets in order to generate specific cellular responses (Caldon et al. 2001). It is tempting to propose that (i) ribosomes which can act as stress sensors (VanBogelen et al. 1990), (ii) RelA which binds to the ribosomes and modulates its (p)ppGpp synthetase activity according to their activity (Cashel et al. 1996), and (iii) SGP which senses and may modulate the GTP/GDP ratio (Baev et al. 1999) are the players of an integrated regulatory circuit.

Oxidative stress resistance

The food-associated lactic acid bacteria are facultative anaerobic micro-organisms that have in common that, in order to regenerate NAD^+ from NADH formed during glycolysis, most or an important part of the pyruvate produced is reduced to lactate. They do not require oxygen for growth and in fact, a negative effect of oxygen on the growth of these bacteria has often been observed. It was generally believed that these bacteria could under no condition use oxygen as the terminal electron acceptor although in the past one report had suggested that *Lactococcus lactis* could respire when heme was present in the culture medium (Sijpesteijn 1970). Recently, this observation has been further documented (Duwat et al. 2001), and analysis of the genome sequence of *L. lactis* IL1403 revealed the presence of all genes necessary for aerobic respiration, as well as the genes involved in the late steps of heme synthesis (Bolotin et al. 2001). In spite of the observation made by Duwat et al. (2001) that respiration (in the presence of heme) results in a remarkably improved long-term survival in *L. lactis*, and the beneficial effects of microaerobic conditions for the growth of *L. lactis* in a glucose-limited medium observed by Jensen et al. (2001), oxygen is generally associated with negative effects in the lactic acid bacteria, including *L. lactis*.

The toxicity of oxygen is generally attributed to reactive oxygen species like O_2^- (superoxide), and $OH^{\cdot}$ (hydroxyl radical), that attack proteins, lipids and nucleic acids, thereby constituting one of the major causes of aging and cell death. Living cells from prokaryotes to eukaryotes have developed more or less successful ways to cope with oxygen toxicity by either preventing the formation of these reactive oxygen species, eliminating them (by enzymatic degradation or scavenging), rendering their possible targets less vulnerable, or repairing the damaged caused.

Table 2. Proteins or genes mentioned in the text that play a role in oxidative stress resistance, and their presence in *L. lactis* IL1403 according to the annotation by Bolotin et al. (1999). x, not present

Protein	Gene	*L. lactis* gene
Glutathione reductase	*gor*	*gshR*
Thioredoxin	*trxA*	*trxA, trxH*
Thioredoxin reductase	*trxB*	*trxB1, trxB2*
NADH oxidase		*noxC, noxD, noxE*
Catalase	*kat*	x
Pseudocatalase		x
Superoxide dismutase	*sodA*	*sodA*
FLP (FNR-like protein)	*flpA, flpB*	*flpA, flpB*
RecA	*recA*	*recA*
Phosphate ABC transporter	*pstS*	*pstS*

For the food-associated lactic acid bacteria a still fragmented picture of the resistance mechanisms present emerges. Representatives of the different mechanisms have been described in different lactic acid bacteria, but the spectrum of mechanisms present in one species is not always clear, and the presence of enzymatic and scavenging activities may even differ between strains. We expect that this situation will change rapidly in the near future as data of the genome sequencing projects of a large number of lactic acid bacteria will become available (see Table 2 for an example). Therefore, in this review examples of the different mechanisms found in lactic acid bacteria will be presented, rather than an exhaustive listing per species or per genus.

Apart from the toxic effects of oxygen, aeration can induce important changes in the sugar metabolism of lactic acid bacteria. For these aspects, that are beyond the scope of this review, we refer to a review by Condon (1987) and a more recent report on the role of redox balance in pyruvate metabolism by Lopez de Felipe et al. (1998).

The reducing intracellular environment

In *Escherichia coli*, the importance of a reducing intracellular environment for key cellular functions has been well established. The cells require either of two partially overlapping disulfide-reducing pathways, the thioredoxin or the glutathione/glutaredoxin pathway, for normal aerobic growth (Stewart et al. 1998). The latter pathway also seems to exist in *Streptococcus thermophilus* where the *gor* gene, encoding glutathione reductase, was identified (Pebay et al. 1995).

The reductase activity was doubled upon aeration. The gene *trxB* encoding the thioredoxin pathway enzyme thioredoxin reductase has been identified in *Lb. bulgaricus*. While in *E. coli* the transcription of *trxB* depends on OxyR (Prieto-Alamo et al. 2000), one of the key regulators in oxidative stress response which is activated by H_2O_2, no indication could be obtained for transcriptional regulation of *L. bulgaricus trxB* expression upon addition of H_2O_2 or catalase in the culture medium (M. van de Guchte and P. Serror, unpublished). In contrast, in *Oenococcus oeni* the gene *trxA*, encoding a thioredoxin, was found to be induced in the presence of H_2O_2 or after heat-shock (Jobin et al. 1999).

In *Lactobacillus fermentum*, an L-cystine uptake system has been directly implicated in the resistance to oxidative stress (Turner et al. 1999). The underlying mechanism would involve the intracellular breakdown of cystine to produce the reducing free sulfhydryl compound thiocysteine. As most of this sulfhydryl compound is subsequently exported, it is not known whether it would primarily protect the cell from internal or external oxidizing agents.

Prevention of reactive oxygen species formation

One way to limit the formation of the reactive oxygen species mentioned above is to eliminate free oxygen as much as possible. Guerzoni et al. (2001) showed that the fatty acid composition in the cell membrane of *Lactobacillus helveticus* changed in response to oxidative stress. This change is explained in terms of an increased activity of the oxygen consuming fatty acid desaturase system which would serve to reduce free radical damage to the cell. *Lactobacillus delbrueckii* ssp. *bulgaricus* presumably eliminates oxygen in a reaction catalyzed by an NADH oxidase (Yi et al. 1998; Marty-Teysset et al. 2000). The contribution of this reaction to NAD^+ regeneration was found to be relatively unimportant, and therefore the main role for this constitutively produced enzyme seems to be the elimination of O_2. Paradoxically, this reaction yields H_2O_2 which in itself is toxic to the cells (Condon 1987; Marty-Teysset et al. 2000; Van de Guchte et al. 2001) as *L. bulgaricus* does not possess a catalase that could render the H_2O_2 harmless, and eventually causes growth arrest.

Elimination of reactive oxygen species

Many other lactobacilli do produce H_2O_2 (Engesser et al. 1994). In contrast to *Lb. bulgaricus*, some of these can produce a catalase (when grown in a medium containing heme) or pseudocatalase (non-heme catalase) to detoxify H_2O_2 (Engesser et al. 1994). An example of a true catalase is the catalase produced by *Lactobacillus sakei* (Knauf et al. 1992), one of the most prevalent organisms in meat fermentations where ample heme is available. Transcription of the *katA* gene encoding this enzyme was shown to be induced by aeration or addition of H_2O_2 (Hertel et al. 1998). One of the best studied examples of a non-heme catalase is the Mn-containing pseudocatalase of *Lb. plantarum* ATCC14431 (Kono et al. 1983; Igarashi et al. 1996). The fact that *Lactobacillus casei*, like *Lb plantarum*, contains an extremely high concentration of Mn (about 10–20 mM) (Archibald et al. 1981) as opposed to *Lb. bulgaricus* (about 0.06 mM) might explain why this enzyme could be functionally expressed in *Lb. casei* while no activity was detected after cloning in *Lb. bulgaricus* (M. van de Guchte, unpublished).

The extremely high Mn content of *Lb. plantarum, Lb. casei, Lb. fermentum* and *Leuconostoc mesenteroides* serves as an efficient scavenger of O_2^-, thereby compensating for the lack of a superoxide dismutase in these bacteria (Archibald et al. 1981). *Lb. bulgaricus* and *Lb. acidophilus* that are among the least oxygen-tolerant lactobacilli do not contain high levels of Mn nor a superoxide dismutase (Archibald et al. 1981).

In contrast, *L. lactis* has been reported to possess a superoxide dismutase which eliminates O_2^- in a reaction that produces H_2O_2 (Sanders et al. 1995). The transcription of the *L. lactis sodA* gene encoding this enzyme was found to be induced by aeration, while the gene had originally been identified because of its induction at low pH. As *L. lactis* does not possess catalase activity to subsequently remove H_2O_2, Miyoshi et al. (2002) studied the incidence of production of heterologous catalases on the oxidative stress resistance of *L. lactis*. Two catalases originating from *Bacillus subtilis* (katE) and *Salmonella typhimurium* (KatN), respectively, were produced and shown to improve the resistance to H_2O_2 100- to 1000-fold.

No enzymes are known to degrade the highly reactive hydroxyl radicals, but scavenging activity for these reactive oxygen species was demonstrated in the cell-free extracts of several lactic acid bacteria examined (Lin et al. 1999).

Target protection

An example illustrating how defense against oxidative damage may function by rendering the potential targets less vulnerable, is found in *L. lactis*. In *L. lactis*, FlpA and FlpB (FNR-like proteins, see below) appear to regulate the uptake of Zn(II). Inactivation of *flpA* and *flpB* leads to the depletion of the intracellular Zn(II) pool, and the mutant strain shows increased sensitivity to H_2O_2. This result suggests that *L. lactis* uses Zn(II) as a defence mechanism against oxidative stress, possibly by protecting thiol groups in proteins from oxidative conversion to disulfide bonds (Scott et al. 2000b).

Repair of oxidative damage

Damage repair seems to be the ultimate mechanism of resistance against oxidative and other stresses. Duwatt et al. (1995) showed that a *recA* mutant of *L. lactis* was markedly more sensitive to oxygen than the *wt* strain. The authors speculate that the *recA* gene product could alleviate oxidative stress directly through its role in DNA repair, or indirectly via a regulatory role on other genes required for the repair of oxygen damage. Notably the amount of chaperone proteins was found to be reduced in the *recA* strain.

General stress resistance mechanisms may also confer resistance to oxidative stress. Rallu et al. (2000) isolated several acid-resistant mutants of *L. lactis* that appeared also more resistant to oxidative stress (H_2O_2). One of these mutants that was particularly resistant to oxidative stress was affected in the *pstS* gene encoding a phosphate ABC transporter. The authors evoke a putative role of a decreased intracellular phosphate concentration as the internal stress signal, that elicits a stress response conferring protection to multiple stress conditions.

Although for some of the above-mentioned defense mechanisms an induction of transcription or enzyme activity has been observed upon aeration, in general little is known about the regulatory circuits involved in oxidative stress resistance in the food-associated lactic acid bacteria. In *L. lactis* a two-component system has been suggested to be involved (O'Connell-Motherway et al. 2000). In *L. casei* a gene encoding an FNR-like protein (FLP) has been identified (Irvine et al. 1993), while in *L. lactis* two FNR-like protein encoding genes, *flpA* and *flpB*, are present (Gostick et al. 1999). Although both resembling the *E. coli* oxygen-responsive transcription factor FNR, FlpA and FLP have basically different architectures that result in different mechanisms of activation, and recognize slightly different DNA binding sites (Scott et al. 2000a). Interestingly, FlpA recognizes the same sequence as FNR (Scott et al. 2000a). Potential FlpA/B binding sites are present in the promoter regions of the *flpA* and *flpB* operons, at positions that suggest that the *flpA* operon would be repressed and the *flpB* operon activated by the FlpA/B regulators (Scott et al. 2000b).

Cold stress in lactic acid bacteria

Considering their optimal growth temperatures, lactic acid bacteria are either mesophilic or thermophilic. During industrial processes, like frozen storage of starter cultures, low temperature fermentation during cheese ripening and refrigerated storage of fermented products, LAB are exposed to temperatures far below their optimal growth temperature. Optimal survival of LAB during freezing and at low temperatures contributes to the industrial performance of the strains. A better understanding of the responses to low temperatures and freezing may contribute to the optimization of the fermentation processes, the storage of the products and the conservation conditions.

When living cells are exposed to a temperature downshift they undergo important physiological changes such as a decrease in membrane fluidity and a stabilization of secondary structures of RNA and DNA resulting in a reduced efficiency of translation, transcription and DNA replication. To overcome these effects, microorganisms have developed a transient adaptive response, termed the cold-shock response, during which a number of cold-induced proteins (CIPs) are synthesized in order to maintain (i) membrane fluidity by increasing the proportion of shorter and/or unsaturated fatty acids in the lipids, (ii) DNA supercoiling by reducing the negative supercoiling, and (iii) transcription and translation needed for cellular adaptation to low temperature (Phadtare et al. 2000). The most strongly induced proteins include a family of closely related low-molecular weight ($\sim$7.5 kDa) proteins termed cold-shock proteins (Csp). These proteins share a high degree of sequence identity (>45%) and orthologs have been found in multiple copies (from two to nine) in many Gram-positive and Gram-negative bacteria (Wouters et al. 2000c). The CSPs are expressed at different growth conditions, in *E.*

coli four of the nine Csp proteins are cold inducible (Yamanaka et al. 1998; Wang et al. 1999). Cold induced *csp* genes have an unusually long 5′ untranslated region (5′-UTR) which, in the case of *E. coli cspA*, plays a role in mRNA stability and translation efficiency (Jiang et al. 1996; Fang et al. 1997; Yamanaka et al. 1999). Cold induction of Csp proteins is complex but appears to be controlled mainly at the post-transcriptional level (Brandi et al. 1996; Goldenberg et al. 1996; Fang et al. 1997; Kaan et al. 1999). Single deletions of *B. subtilis* and *E. coli* cold-induced *csp* genes caused no distinct phenotype at either normal or low temperature (Bae et al. 1997; Willimsky et al. 1992). However the level of remaining CSPs increased suggesting that CSPs can compensate the loss of each other (Graumann et al. 1996; Bae et al. 1997; Xia et al. 2001). Furthermore, multiple deletion analysis showed that at least one CSP is required for the viability of *B. subtilis* indicating that CSPs play an important role, not only during cold-shock adaptation but also during active growth (Graumann et al. 1997). Csp proteins are β-barrel proteins with two RNA-binding motifs, able to bind to single-stranded DNA and to RNA with little specificity and to destabilize secondary RNA structures (Graumann et al. 1997; Jiang et al. 1997; Hanna et al. 1998; Phadtare et al. 1999). Thereby, CSPs have been proposed to act as RNA chaperones to facilitate transcription and translation at low temperature by preventing formation of mRNA secondary structures (Graumann et al. 1997; Jiang et al. 1997). Although *E. coli* CSPs were shown to act as transcription antiterminators of cold induced genes (Bae et al. 2000), *B. subtilis* CSPs were found to colocalize with ribosomes at cellular sites of translation suggesting that CSPs might also be involved in translation (Mascarenhas et al. 2001; Weber et al. 2001b). Very recently, Marahiel and co-workers have proposed that *B. subtilis* CSPs may act as alternative translation initiation factors (Weber et al. 2001a).

Adaptive response of LAB to temperature downshift

In contrast to *E. coli* (Jones et al. 1987), LAB can rapidly adapt to a temperature downshift. Exponentially growing cells of *L. lactis* sp. *lactis* (Panoff et al. 1994), *L. lactis* sp. *cremoris* (Kim et al. 1997; Wouters et al. 1999a), *S. thermophilus* (Perrin et al. 1999; Wouters et al. 1999b), *L. acidophilus* (Baati et al. 2000), *L. bulgaricus* (Panoff et al. 2000), *L. plantarum* (Mayo et al. 1997) and *E. faecalis* (Thammavongs et al. 1996) continue to grow at a reduced rate after a temperature downshift to about 20 °C below the optimal growth temperature. However more important temperature downshifts lead to growth arrest for *S. thermophilus* (Wouters et al. 1999b), *L. lactis* sp. *cremoris* (Wouters et al. 1999a), and *L. acidophilus* (Baati et al. 2000). Using the two-dimensional electrophoresis (2DE) approach this adaptability to cold-shock was correlated to the synthesis of CIPs, at least in *L. lactis* (Panoff et al. 1994; Wouters et al. 1999a), *S. thermophilus* (Wouters et al. 1999b) and *E. faecalis* (Panoff et al. 1997) in which a maximum of 22, 24 and 11 CIPs, respectively, were revealed. Identification of several *L. lactis* CIPs suggests that they might be involved in a large diversity of cellular processes such as sugar metabolism (Hpr, CcpA, and β-PGM; β-phosphoglucomutase), chromosome structuring (HU-like protein HslA), signal transduction (LlrC, a response regulator of a two-component signal transduction system homologous to YycF, an essential gene for *B. subtilis* (Fabret et al. 1998), and stress adaptation (OsmC, an ortholog of OsmC of *E. coli* (Gordia et al. 1996) and YkzA of *B. subtilis* (Volker et al. 1998)) which both belong to stress regulons (Wouters et al. 2000a, b, 2001). Recently, an unidentified cold-inducible 45-kDa protein of *L. lactis* (Wouters et al. 1999) has been proposed to correspond to the ClpX ATPase, (Skinner et al. 2001) which might be involved in the proteolysis of misfolded (damaged) proteins after cold-shock.

As part of the cold-adaptive response, CSP proteins or corresponding genes were detected and/or identified in several LAB such as *L. lactis* sp. *cremoris*, *L. lactis* sp. *lactis*, *S. thermophilus*, *Pediococcus pentosaceus*, *L. helveticus*, *E. faecalis*, *L. acidophilus*, *B. animalis*, *S. pyogenes*, *L. casei*, *S. dysgalactiae*, *L. plantarum* and *S. pyogenes* (Chapot-Chartier et al. 1997; Kim et al. 1997, 1998; Mayo et al. 1997; Francis et al. 1998; Wouters et al. 1998; Ferretti et al. 2001). As pointed out previously, the number of CSP genes may vary between subspecies and strains (Champomier-Verges et al. 2002). The *L. lactis* sp. *lactis* strain IL1403 (Bolotin et al. 2001) has only two *csp* genes, whereas genetic and biochemical analysis revealed seven CSPs (CspA to CspG) in *L. lactis* sp. *cremoris* strain MG1363 (Wouters et al. 2000b). Similarly, strains C3.8 and NC8 of *L. plantarum* might contain two and three *csp* genes, respectively (Mayo et al. 1997; Derzelle et al. 2000).

Elements for a regulation of CSPs in LAB

During the past few years, progress has been made in the characterization of the cold-adaptive response in *L. lactis* sp. *cremoris*. Transcriptional and 2DE analysis have shown that at least five (CspA, CspB, CspC, CspD, CspF) out of the seven *L. lactis* Csp proteins were induced upon cold-shock. Maximal induction efficiency of the CSPs was obtained at 10 °C, whereas lower induction was observed at 20 and 4 °C, indicating an optimal temperature for efficient induction. CspE appears to be the major CSP present at 30 °C, and is slightly induced after cold-shock at 10 °C, whereas the expression of CspB, CspD and CspF was at least 50-fold induced (Wouters et al. 1998, 1999a). Double and triple deletions of *cspA cspB* and *cspA cspB cspE*, respectively, did not affect growth, presumably because of an increased level of remaining CSPs upon deletion of their counterparts. The impossibility to inactivate *cspC* and *cspD* genes in a *cspA cspB cspE* strain (Wouters et al. 2001) suggest a CSP dependent regulation of CSP family members and the necessity for a minimal CSP level in the cell. Furthermore, overproduction or deletion of CSPs modified the levels of OsmC, HslA, LlrC and three other unidentified CIPs indicating that similarly to *E. coli* and *B. subtilis* CSPs of *L. lactis* somehow regulate proteins involved in cold adaptation (Wouters et al. 2000b, 2001).

As to regulation of CSPs of other lactic acid bacteria, transcriptional studies revealed that one (*cspL*) out of the three *L. plantarum* NC8 *csp* genes is significantly induced upon cold-shock at 8 °C (Derzelle et al. 2000). A common feature of the non-cold-induced *csp* genes of *L. lactis* (*cspE*) and *L. plantarum* (*cspP* and *cspC*) consists in a longer 5′-UTR compared to their cold-induced counterparts. This difference might be responsible for higher instability of the transcripts. Recently, six putative CSPs (CSPA to CSPF) have been detected in *S. thermophilus*. Five of them appeared to be induced after cold-shock, whereas the last one was expressed at low level both at 42 and 20 °C (Wouters et al. 1999b). So far, a single cold-inducible *csp* gene (*cspA*) has been identified in *L. bulgaricus* strain ATCC11842. Transcription of *L. bulgaricus cspA* was detectable during growth at normal temperature (42 °C) and was maximally induced upon cold-shock at 25 °C (P. Serror et al., pers. comm.). Like in *L. lactis*, the induction of cold-inducible *csp* genes or proteins in *L. plantarum*, *L. bulgaricus* and *S. thermophilus* appears to respond to an optimal induction temperature below which induction is delayed. Inefficient induction of the CSPs below this threshold would compromise the cold adaptive response and might cause the arrest of growth. Determination of the optimal temperature for maximal induction of CSPs will allow to optimize the adaptive response for industrial applications.

CSPs and cryotolerance

A correlation between cryotolerance and cold-shock response has been established for the first time in *B. subtilis* with the observation that deletion of *cspB* was responsible for an increased sensitivity to freezing (Willimsky et al. 1992). Acquisition of cryotolerance after adaptation at suboptimal temperatures has been documented for several LAB including *L. lactis* (Panoff et al. 1995; Kim et al. 1997; Broadbent et al. 1999; Wouters et al., 1999a) *L. acidophilus* (Lorca et al. 1999; Baati et al. 2000), *L. bulgaricus* (Panoff et al. 2000), *S. thermophilus* (Wouters et al. 1999b) and *E. faecalis* (Thammavongs et al. 1996). So far, only few studies aiming to link CSP levels and resistance to freezing have been carried out, mainly in *L. lactis*. First, an increased survival rate (~100-fold) to freezing of cold adapted *L. lactis* sp. *cremoris* exponentially growing cells (4 h, 10 °C) coincided with an increase of CSP expression. However, the level of freeze protection induced by cold-shock did not directly correlate with the expression level of the *csp* genes (Wouters et al. 1999a). Secondly, deletions of *cspA cspB* and of *cspA cspB cspE* did not affect the survival to freezing of a culture grown at normal temperature. However, the cold-adaptation of the *cspA cspB cspE* strain is less efficient than in wild-type, probably due to the absence of CspE and the low total CSPs amount (Wouters et al. 2001). Thirdly, overproduction of CspB, CspD and CspE seems to increase the survival to freezing 2–10-fold compared to wild-type cells (Wouters et al. 1999a, 2000b). All these data suggest that CspB, CspD and CspE play a role in the cold-induced cryotolerance of *L. lactis*. According to their putative function as RNA chaperones or alternative translation initiation factors, CSP proteins might either protect nucleic acids by non-specific binding or participate to the synthesis of factors needed for cryotolerance. Although stationary phase cells had no increased CSP levels compared to cold-induced cells, they exhibited 20-fold higher survival rates after freezing than exponentially growing cells. Among the 19 proteins induced during stationary phase, three were also induced upon cold-shock and might constitute important factors for cryotolerance other than CSPs (Wouters et al. 1999a).

Interestingly, cold-shock has also been found to enhance thermotolerance of *L. lactis* sp. *lactis* IL1403 (Panoff et al. 1995) indicating a probable overlap between the cold-shock and the heat-shock regulons. This hypothesis is supported by indirect observations, such as the increased level of *L. mesenteroides* DnaK and GroEL homologues upon cold-shock (Salotra et al. 1995) and a correlation between the induction of the heat-shock response and a slight cross-protection to freezing for *L. johnsonii* (Walker et al. 1999) and *L. lactis* sp. *lactis* (Broadbent et al. 1999). As previously speculated, heat-shock proteins might act as macromolecular stabilizers (Komatsu et al. 1990). However, heat-shock did not improved cryotolerance of *L. lactis* sp. *cremoris* strain MG1363 (Wouters et al. 1999a), suggesting a strain specificity regarding cross-protection between cold-shock and heat-shock responses.

CSPs appear to be involved in cold adaptation of LAB by an overlapping regulatory role on the expression of both CSPs and several CIPs. They probably also play an important role during growth at optimal temperature as suggested by the requirement for a minimal CSP level in the cell. Although progress has been made in the characterization of the cold-shock response and its role in the induction of increased survival after freezing, further identification of CIPs and their connection with CSPs is still required for a better understanding of the low-temperature adaptation of *L. lactis*. The cold-induced response regulator (LlrC) recently identified (Wouters et al. 2001) is part of a two-component system which might be involved in thermosensing and signal transduction at low temperature.

Osmotic stress resistance

For active metabolism to occur, the intracellular conditions must remain relatively constant with respect to ionic composition, pH, and metabolite levels (Csonka et al. 1991). In addition, the maintenance of constant positive turgor is generally considered as the driving force for cell expansion. As the bacterial cytoplasmic membrane is permeable to water but forms an effective barrier for most solutes, a change in the osmolality of the environment could, therefore, rapidly compromise essential cell functions, and bacteria need to adapt to such a change in their environment in order to survive. In general, they can do so by accumulating compatible solutes (by uptake or synthesis) under hyperosmotic conditions, and releasing (or degrading) them under hypoosmotic conditions. Apart from their effect on the osmotic balance, compatible solutes may also stabilize enzymes and thereby provide protection not only against osmotic stress but also against high temperature, freeze-thawing and drying (Kets et al. 1996; Poolman et al. 1998; Panoff et al. 2000).

In their various applications in the food and feed industry, lactic acid bacteria can be exposed to osmotic stress when important quantities of salt or sugar are added to the product. Unlike the enteric bacteria and *B. subtilis*, they have limited or no possibilities to synthesize compatible solutes (Poolman et al. 1998), and primarily rely on the uptake of such compounds from the culture medium. This process, which ensures a rapid response to changing osmotic conditions, has been studied in detail in *Lactobacillus plantarum* and *Lactococcus lactis*.

It was observed that KCl and NaCl inhibit the growth of *L. plantarum* and *L. lactis* much more than equiosmolar concentrations of sucrose or lactose, and a stimulatory effect of glycine-betaine (see below) was found only in the case of salt stress (KCl or NaCl). As shown by Glaasker et al. (1998b), these observations may be explained by the fact that sucrose and lactose only impose a transient osmotic stress because the internal and external sugar concentrations rapidly equilibrate as a result of sugar uptake. Unlike *E. coli*, where the rapid accumulation of K^+ constitutes the first response to hyperosmotic stress (Csonka et al. 1991), *Lb. plantarum* is unable to respond adequately to osmotic stress by accumulation of K^+ or Na^+ to levels beyond those already existing in the unstressed cell (Glaasker et al. 1998b). Glaasker et al. (1996b) determined pool sizes of a number of compatible solutes in *Lb. plantarum* grown in the presence or absence of 0.8 M KCl. They showed that especially the amino acids glutamate and proline had accumulated in the cells grown in the high-osmolality medium. Like many other bacteria, *Lb. plantarum* preferentially accumulated the quaternary ammonium compound glycine-betaine when provided in the medium. In that case, glutamate and proline levels were much lower than in the absence of glycine-betaine. Whereas proline and glycine-betaine uptake were activated by an osmotic upshift, no activation of glutamate uptake was observed. Osmotic downshift caused a rapid efflux of a limited number of compatible solutes, among which proline, glycine-betaine and glutamate. Interestingly, a basal flux (without net uptake or efflux) of glycine betaine, a compound that *L. plantarum* is unable to synthesize or metabolize (Glaasker et al. 1998a), can be observed at steady state conditions (Glaasker et al. 1996a). A single semiconstitutive ATP-dependent uptake system, QacT, appears to be responsible for the uptake of glycine-betaine, its analogue carnitine (both with high affinity), and proline (with low affinity) (Glaasker et al. 1998a). The efflux of glycine-betaine upon osmotic downshift is mediated by a separate system, most probably a mechanosensitive channel protein. So far, neither of these systems has been described at the genetic level.

Like in *Lb. plantarum*, glycine-betaine has been shown to augment osmotolerance in *Lactobacillus acidophilus* (Hutkins et al. 1987) and *L. lactis* (Obis et al. 1999; Van der Heide et al. 2000a). In contrast to what was found in *Lb. plantarum*, glycine-betaine uptake in *L. lactis* is subject to osmoregulation at two levels, i.e., gene expression and transport activity (Obis et al. 1999; Bouvier et al. 2000; Van der Heide et al. 2000a). The genes encoding this transport system, an ABC transporter, have been identified (Obis et al. 1999; Bouvier et al. 2000; Van der Heide et al. 2000b), as well as a transcriptional regulator of their expression, BusR (Roméo et al. 2002). The transporter, OpuA or BusA, not only acts as osmoregulator, but also functions as osmosensor. At present it is believed that a change in intracellular ionic strength serves as a primary signal of osmotic stress, which is transmitted to OpuA via its effect on interactions between membrane lipids and the protein (Van der Heide et al. 2001). The change in membrane fatty acid composition observed after growth in high osmolality media (Guillot et al. 2000) may equally affect transport activity.

In a genetic approach to identify genes involved in osmotic stress response in *L. lactis*, Sanders et al. (1998b) used a promoter screening assay, and detected a promoter that was induced during growth in the presence of 0.5 M NaCl. After further testing, the authors concluded that this promoter was specifically induced by the presence of chloride ions rather than by the osmolality or ionic strength of the medium, however.

Kilstrup et al. (1997) used two-dimensional electrophoresis to identify proteins induced upon osmotic upshift in *L. lactis*. They observed that largely the same proteins were induced after salt stress and heat-shock, although the induction factors varied. Among the proteins identified were the general stress proteins GroES, GroEL, and DnaK. One should keep in mind, however, that the induction of a membrane protein like OpuA would probably not have been detected because of the limitations inherent to this technique. Two other membrane-associated proteins, the intracellular protease FtsH (Nilsson et al. 1994) and the extracellular housekeeping protease HtrA (Foucaud-Sceunemann et al. 2002), have been reported to have an effect on osmotic stress resistance in *L. lactis*. Mutants of either protease showed increased sensitivity to NaCl stress, and *htrA* transcription was transiently induced under NaCl stress.

Apart from effects on growth rate and membrane fatty acid composition, growth medium osmolality was also reported to affect exopolysaccharide (Liu et al. 1998) and bacteriocin (Uguen et al. 1999) production in *L. lactis*.

Heat shock tolerance

The major problem encountered by cells at high temperature is the denaturation of proteins and their subsequent aggregation (Somero 1995). In addition, destabilization of macromolecules as ribosomes and RNA, and alterations of membrane fluidity were also described (Earnshaw et al. 1995; Teixera et al. 1997; Hansen et al. 2001). The heat-shock (HS) response has been studied notably in *E. coli* and *B. subtilis*, the Gram-positive paradigm. In *B. subtilis*, HS inducible genes are classified according to their regulators. Genes of class I are regulated by the HrcA repressor which binds to the palindromic operator sequence CIRCE (controlling inverted repeat of chaperone expression) (Zuber et al. 1994). Class II comprises the genes which are under the control of the alternate sigma factor σ^B, which governs the expression of a large set of genes involved in stress tolerance (Petersohn et al. 2001; Price et al. 2001). Class III genes are controlled by the class three stress gene repressor CtsR which binds to a specific direct repeat referred to as CtsR-box (Derre et al. 1999). Finally, the heat-shock genes of unknown regulation are grouped in class IV.

Physiological studies demonstrated that LAB elicit heat-shock responses similar to that of other Gram-positive bacteria (*L. lactis*, Whitaker et al. 1991; Auffray et al. 1992; Kilstrup et al. 1997; *Leuconostoc mesenteroides*, Salotra et al. 1995; *E. faecalis*, Flahaut et al. 1996b; *O. oeni*, Guzzo et al. 1997; *L. bulgaricus*, Gouesbet et al. 2002). Examination of

LAB HS responses using 2DE revealed variable numbers of induced proteins: 34 in *E. faecalis* (Flahaut et al. 1996b), 17 in *L. lactis* (Kilstrup et al. 1997) and 40 in *S. mutans* (Svensater et al. 2000). Among the latter 40, only six proteins were specifically upregulated by heat (Svensater et al. 2000). In *L. lactis* the 12 proteins induced by NaCl addition as detected on 2DE, all belong to the HS stimulon (Kilstrup et al. 1997). This striking overlap between heat-shock and osmotic-stress responses may also exist in *S. mutans* as 21 heat-inducible proteins also belong to the osmotic stress response. Among the HS proteins, well conserved chaperones (DnaK, DnaJ, GrpE, GroES and GroEL) and proteases (Clp, HtrA, FtsH) have often been identified.

Heat-inducible chaperones and HrcA/CIRCE regulation

Chaperones are involved in the folding of nascent proteins and the re-folding of denatured proteins. It is assumed that LAB, like most bacteria, possess the two chaperones complexes constituted by DnaK-GrpE-DnaJ and GroES-GroEL since some or all of these proteins and/or encoding genes were identified in LAB. In several LAB, putative CIRCE sequences (TTAGCAGTC-N9-GAGTGCTAA), the operator recognized by HrcA in *B. subtilis*, were found upstream of the *dnaJ* gene (*L. lactis*, van Asseldonk et al. 1993), *hrcA-grpE-dnaK* (*L. lactis*, Eaton et al. 1993; *S. mutans*, Jayaraman et al. 1997; *L. sakei*, Schmidt et al. 1999; and *S. pyogenes*, Laport et al. 2001). Note that in these last two species, the *dnaK* operon also comprises *dnaJ* and *groES-groEL* (*L. lactis*, Kim et al. 1993; *S. mutans*, Lemos et al. 2001; *L. johnsonii*, Walker et al. 1999; *L. helveticus*, Broadbent et al. 1998; *S. pneumonia*, Chastanet et al. 2001; and *S. pyogenes*, Laport et al. 2001) operons. The only evidence for the conservation of the HrcA–CIRCE regulatory function in LAB results from the three following studies. Firstly in *L. lactis*, the activity of native or CIRCE-truncated *dnaJ* promoters was monitored at normal and high temperatures. A complete CIRCE appeared to be required for the thermal regulation of the gene (van Asseldonk et al. 1993). Secondly, a biochemical study performed with *S. thermophilus* cell extracts showed that a protein recognized by HrcA-specific antiserum bound to a CIRCE oligonucleotide (Martirani et al. 2001). Thirdly, in *S. mutans* the inactivation of *hrcA* lead to the derepression of the *groESL* and the *hrcA-grpE-dnaK* promoters (Lemos et al. 2001) suggesting that HrcA repressed these operons. But, the *hrcA* mutant also exhibited a 50% reduction of DnaK amount (most probably due to a polar effect of the *hrcA* mutation) which complicates the interpretation of data and might explain the acid-sensitivity of the mutant.

In *L. lactis*, a C-terminal deletion of DnaK lead to the upregulation of the putative HrcA–CIRCE regulon and of HS proteins of unknown function called Hps84, Hsp85 and Hsp100 while the transcription of *hflB*, a HS protease (see below) was not modified. These observations have been interpreted as evidence for the involvement of DnaK in the regulation of HrcA activity (Koch et al. 1998), a role which is attributed to the GroEL chaperone in *B. subtilis* (Mogk et al. 1997). It is not known whether Hsp84, 85 and 100 encoding genes (yet unidentified) have CIRCE operators.

Heat-inducible proteases and CtsR/CtsR-box regulation

HtrA/DegP protease

HtrA belongs to a widely distributed family of serine proteases (Pallen 1997) which can display both chaperone and protease functions. In *E. coli*, HtrA acts as a chaperone at low temperature, whereas at normal and high temperature the protease activity directed against misfolded proteins predominates (Spiess et al. 1999). HtrA orthologs have been characterized in *L. helveticus* (Smeds et al. 1998), *L. lactis* (Poquet et al. 2000) and *S. mutans* (Diaz-Torres et al. 2001). In all these species, *htrA* mutants have reduced ability to withstand high temperature, and in *S. mutans* the mutated strain is also sensitive to low pH and H_2O_2. Transcription of *htrA* was only monitored in *L. helveticus*, where heat, ethanol, puromycin and NaCl treatments gave rise to 2–8-fold induction. The role of HtrA in the degradation of native or heterologous proteins was unequivocally established in *L. lactis* whereas in *S. mutans*, *htrA* inactivation had no effect on the degradation of native extracellular proteins.

FtsH/HflB protease

The FtsH membrane-bound ATP-dependent metalloprotease is essential in *E. coli*. It is heat inducible and is involved in the regulation of σ^{32}, a sigma factor involved in the control of the HS response in *E. coli* (Yura et al. 2000). An ortholog of the *ftsH* gene was found in *L. lactis* (Nilsson et al. 1994) and shown to be thermally induced (Duwat et al. 1995; Arnau et al. 1996). When FtsH (695 aa) was truncated at 163 residues from its C-terminus, the strain displayed heat, salt and cold sensitivity (Nilsson et al. 1994). This work indicated that FtsH may not be essential in *L. lactis* but is involved in the tolerance to several stress conditions including heat. In addition, the study of a *recA* mutant of *L. lactis* suggested that this protease could be a negative regulator of the HS response as in *E. coli* (Duwat et al. 1995; see below) but further experiments are needed to conclusively establish that function. Orthologs of *ftsH* were detected by Southern hybridization in several Lactobacillus and Leuconostoc species illustrating the likely conservation of this protease in LAB. *Clp protease*. The caseinolytic protease Clp has been extensively studied in *E. coli*. The ClpP multimeric complex is a serine protease which degrades peptides less than seven amino acids length. Bacteria also contain a number of Clp-ATPases (four in *L. lactis*) which can have chaperone activity. When the proteolytic component ClpP associates with a Clp-ATPase, the resulting complex displays a protease activity against specific substrates determined by the Clp-ATPase subunit. After HS, the Clp protease degrades damaged proteins which cannot be properly refolded by the chaperones. In *L. lactis*, a *clpP* and four *clp*-ATPase (*clpB*, *clpC*, *clpE* and *clpX*) genes have been found (Ingmer et al. 1999; Bolotin et al. 2001; Skinner et al. 2001). A *clpP* mutated strain was more sensitive than wild-type *L. lactis* to HS and puromycin, an aminoacyl–tRNA analogue which leads to truncated and misfolded proteins (Frees et al. 1999). The abnormal polypeptides generated by the addition of puromycin were less efficiently degraded in *clpP*, confirming that ClpP has a role in the housekeeping of proteins in *L. lactis* (Frees et al. 1999). During growth at normal temperature (30 °C) of the *clpP* strain, four HS proteins and 11 non-heat inducible proteins were overexpressed. The latter may correspond to ClpP degradation targets. Except for *clpX*, all *clp* genes exhibited a sequence homologous to the CtsR box (Derre et al. 1999; $^{A}/_{G}$GTCAAANAN$^{A}/_{G}$GTCAAA) in their promoter regions. In *L. lactis*, the *ctsR* ortholog was found upstream of *clpC* as in *B. subtilis* and *S. pneumoniae*. The transcription of *clpB, clpC, clpE* and *clpP* increased about 4-fold in a *ctsR* mutant. However, the *clpB, clpC* and *clpE* genes still remained heat inducible to some extent, suggesting that another regulator may be involved in their control (Varmanen et al. 2000). The TrmA protein may be a candidate. A *trmA* mutation alleviates the thermosensitivity of a *clpP* mutant *via* an increase of the cellular proteolytic capacity. Since the proteolysis stimulating effect of *trmA* inactivation was stronger in the wild-type than in a *clpP* mutant, it has been proposed that TrmA partly controls the cellular Clp-dependent proteolysis activity as well as other proteolytic functions. Note that the *htrA* and *hflB* expression were not modified in a *trmA* background (Frees et al. 2001). TrmA may also act by modulating the interaction between protease(s) and their substrates which may require adaptor protein(s) (Becker et al. 1999; Persuh et al. 1999). In other LAB, CtsR operators were also found upstream of several *clp* genes (*L. sakei* and *S. salivarius*; Derre et al. 1999) and *clpP, clpE, clpL, cstR-clpC* operon of *S. pneumoniae* (Chastanet et al. 2001) and other Hsp encoding genes (*Lo18* of *O. oeni* and *hsp16* of *S. thermophilus* (Derre et al. 1999), *groESL* of *S. pneumoniae* (Chastanet et al. 2001). In *S. pneumoniae*, gel mobility shift assays demonstrated the specific interaction of purified CtsR with the operators located upstream of *clpP, clpE, ctsR-clpC* and *groESL* and these genes were overexpressed in a *ctsR* deleted strain (Chastanet et al. 2001). The same study also revealed the presence of a CIRCE operator upstream of *groESL* suggesting that this operon is under dual control by CtsR and HrcA.

Temporal expression of heat shock genes

In *L. lactis*, the induction of several HSP has been monitored at both transcriptional (Arnau et al. 1996) and translational (Kilstrup et al. 1997) levels. In both studies, the temporal pattern of induction suggested that the HS proteins fall in more than one induction class. Analysis of mRNA was focused on the *ftsH* gene and the CIRCE regulated genes (*dnaJ, groESL* and *hrcA-grpE-dnaK*). These latter genes were rapidly induced to high levels (10–100-fold) 10–15 min after HS, but were repressed at 20 min. *ftsH* mRNA was detected in similar amounts from 10 min to more than 20 min. (Note that this study revealed extended transcript processing for the *dnaK* operon.) The proteomic study also distinguished two temporal classes for HS induction: HrcA, DnaK, GroES, GroEL, Hsp 85 (ClpE), Hsp84, Hsp100 and Hsp26 were induced at least 10-fold in the first 10–15 min after HS (Kilstrup et al. 1997). In contrast, ClpP (Hsp23, Frees et al. 1999) and eight uncharacterized Hsp were induced 2–8-fold during the 25 min following HS (Kilstrup et al. 1997). These observations suggest that one or several partners of HS regulation in *L. lactis* remain to be identified.

Global regulation of the HS response

No global regulator of the HS response has yet been identified in LAB. Except for HrcA and CtsR, the only information concerning HS regulation results from studies of the function of RecA in *L. lactis*. As expected a *recA* mutant was recombination deficient and sensitive to DNA damaging agents including H_2O_2 but strikingly, it also appeared to be thermosensitive. Western blot experiments showed that a *recA* mutant displayed diminished levels of DnaK, GroEL and GrpE chaperones and an increased amount of FtsH. It was proposed that, in *L. lactis*, RecA had a role in HS regulation *via* the control of FtsH (Duwat et al. 1995). In an attempt to better understand the regulatory circuit involved, suppressors of *recA* thermosensitivity were isolated after insertional mutagenesis (Duwat et al. 1999). These thermoresistant mutants (trm) affected (i) purine metabolism (*deoB*, *guaA*, and *tktA*), (ii) high affinity phosphate uptake (*pstB*, *pstS*), (iii) mRNA stability (*pnpA*, polynucleotide phosphorylase) and (iv) the *trmA* gene which was shown since to affect proteolytic activity(ies) (see above, *clpP*). The amounts of DnaK and FtsH were assessed by immunodetection in the *trm recA* mutants. In the *pnpA recA* and *trmA recA* mutants, the amounts of DnaK and FtsH were similar to that observed in *recA*. In contrast, for the *trm* mutations affecting the purine metabolism or the phosphate uptake, DnaK amounts in the *trm recA* mutants were increased up to the level observed in the wild-type strain, while HflB remained elevated as in *recA*. Furthermore, alteration of purine and phosphate metabolism also improved the tolerance of the *recA* and wild-type strains to several stress (Duwat et al. 1999; Rallu et al. 2000; see acid stress response). Altogether, these data suggest that the guanine nucleotide (GP) and phosphate pools control multi-stress tolerance responses which can suppress the *recA* thermosensitivity. The *pnpA* and *trmA* mutations seemed to have more specific effects although their involvement in HS response is probably distinct from the classical chaperone machinery.

The analysis of the complete genome of *L. lactis* did not reveal any counterpart of the σ^B factor of *B. subtilis* which has a major role in the regulation of stress responsive genes and similar observations have been drawn for *S. pyogenes* and *S. pneumoniae*. This observation suggests that lactococci, streptococci and possibly other LAB have developed different stress regulatory networks than *B. subtilis* which is often regarded as the model for all Gram-positive bacteria.

Bile tolerance

Bile is a complex digestive secretion that plays a role in the dispersion and absorption of fats. Bile acids (also often referred to as bile salts), the major constituents of bile, are derivatives of cholic acid (CA) that is itself synthesized from cholesterol. Some of the bile acids are conjugated to either glycine or taurine in the liver (Russell et al. 1992). Although the toxicity of bile acids for bacterial cells is not completely understood, bile acids are surface active, amphipatic molecules with potent antimicrobial activity since they act as detergents and disassemble biological membranes. Gram-positive bacteria that lack an outer membrane which constitutes a first permeability barrier, are often more sensitive to the toxic agents present in their environment than Gram-negative bacteria. However, it was shown for Gram-negative bacteria that bile penetrates through porins (Thanassi et al. 1997) and that the outer membrane only retards the penetration of the uncharged forms of bile salts (Ma et al. 1994). Therefore, both Gram-negative and Gram-positive bacteria evolved specific mechanisms to resist to the toxic action of bile and to be able to live in a bile-rich environment such as the digestive tract (Gunn 2000). Resistance to bile stress has been mostly studied in two categories of bacteria: the enteric pathogens which survive in the digestive tract and the food-associated lactic acid bacteria (LAB) or bifidobacteria which are considered as potential probiotics and were often selected to resist to the digestive stress.

Metabolism of bile salts by lactic acid bacteria

Several microorganisms of the intestinal flora including some lactic acid bacteria can metabolize bile acids and this ability was thought to contribute to the protection against bile. In lactobacilli, bile salt hydrolases (BSH) deconjugate bile acids, hydrolysing the amino acid glycine or taurine from the steroid core. This hydrolysis alters the properties of bile and notably decreases its solubility at low pH and its detergent activity (Adamowicz et al. 1991; De Smet et al. 1995). In *Eubacterium* sp. (Mallonee et al. 1996) and *Lactobacillus johnsonii* 100–100 (Elkins et al. 1998) the BSH locus appeared to be linked to genes encoding transporters involved in the active import of bile acids. However, Moser and Savage (2001) recently showed that the BSH activity and the resistance to bile were unrelated properties in lactobacilli. In several species of lactobacilli, an active uptake of CA, dependent on the ΔpH and the presence of glucose, was also observed. This activity which is unrelated to BSH gives rise to an intracellular accumulation of CA. The role of this system in lactobacilli and its relation to CA or bile tolerance is unknown (Kurdi et al. 2001).

Adaptation to bile stress

Adaptive responses to bile, non-conjugated bile acids (such as cholic acid, CA) and/or conjugated bile acids (glycocholic acid (GCA) or taurodeoxycholic acid (TDCA)) were observed in *E. faecalis* (Flahaut et al. 1996a), *L. acidophilus* (Kim et al. 2001), *L. bulgaricus* (T. Smokvina, E. Maguin, pers. comm.) and *L. lactis* ssp. *lactis* but not in the three strains of *L. lactis* ssp. *cremoris* tested (Kim et al. 1999). In both *E. faecalis* (Flahaut et al. 1996b) and *L. bulgaricus* (T. Smokvina, E. Maguin, pers. comm.), bile adaptation was shown to increase the tolerance not only to bile salts but also to heat-shock. Inversely, heat adaptation protected from a lethal bile challenge. In contrast, bile adaptation of *L. acidophilus*, protected it from heat stress but heat adaptation did not improve its tolerance to bile (Kim et al. 2001). All together, these data suggest that the mechanisms of bile tolerance vary between LAB although bile adaptation probably involves mechanisms similar or identical to those involved in heat-shock response.

The proteins synthesized during bile adaptation were investigated using a proteomic approach based on 2DE. Although the membrane proteins that may be induced during adaptation in response to the membrane damaging effect of bile would not be detected with this technique, bile adaptation was found to upregulate 23 cytoplasmic proteins in *L. bulgaricus* (T. Smokvina, E. Maguin., pers. comm.) and 45 proteins in *E. faecalis*. Among the latter, 12 were also induced during detergent (SDS) stress (Flahaut et al. 1996a, b; Rince et al. 2000). For the majority of these proteins, the involvement in the protection of bacteria from bile remains to be proven. In *E. faecalis*, the conserved chaperones DnaK and GroEL were identified. Among the other proteins of unknown function, three proteins (Gsp62, Gsp65 and Gls24) were further studied. The expression of the Gsp62 encoding gene appeared to be stimulated by bile, SDS, heat and several other stress treatments, but gene disruption had no effect on the bacterial tolerance to these stress conditions (Rince 2002). Gsp65 is homologous to the family of organic hydroperoxide resistance (Ohr) proteins. Inactivation of the corresponding gene did not affect bile tolerance while the sensitivity to other stress increased (Rince 2001). Gls24 was upregulated by bile stress, glucose starvation and oligotrophic conditions (Hartke et al. 1998). A *gls24* mutant was severely affected in growth and cell shape, and was more sensitive to bile salt after 24 h starvation (Giard et al. 2000). However, the exact role of this gene is still poorly understood.

MDR efflux systems and bile resistance

The only characterized mechanism responsible for bile acid resistance is the extrusion of bile from Gram-negative bacteria, notably *E. coli* (Lomovskaya et al. 1992; Thanassi et al. 1997; Nishino et al. 2001). This extrusion is mediated by efflux systems belonging to the family of multidrug resistance (MDR) transporters. MDR systems are responsible for the resistance to numerous toxic compounds including antibiotics, dyes, organic solvents, detergents, bile acids and many other hydrophobic agents. The activity of the majority of bacterial MDR transporters depends on the proton motive force (pmf) or on ATP hydrolysis for MDR systems that are members of the ABC transporters superfamily (Higgins 1992). These systems are widespread among both Gram-negative and Gram-positive bacteria (for review see: Putman et al. 2000; Markham et al. 2001). MDR efflux systems have been well studied in several Gram-positive bacteria including *B. subtilis, S. pneumoniae* and *L. lactis* (Putman et al. 2000) but the role of efflux in bile resistance has not yet been established. In *L. lactis* two MDR transporters, LmrP (pmf-dependent) and LmrA (ATP-dependent), (Bolhuis et al. 1994) were studied in detail (for a review, see Konings et al. 1997; van Veen et al. 1999) and another system (which is ATP-dependent) has recently been shown to be involved in the efflux and resistance to CA, one of the non-conjugated bile acids (Yokota et al. 2000). So far, the HorA ABC transporter of *Lactobacillus brevis* is the only MDR efflux system which has been genetically characterized in lactobacilli (Sami et al. 1997; Sakamoto et al. 2001). HorA is involved in the resistance to hop and several toxic agents but an eventual role in bile acids efflux has not been described (Sami et al. 1997). In *L. bulgaricus*, a cross protection between TDCA, a conjugated bile acid and other toxic compounds, such as ethidium bromide (EtBr) and acriflavine, has been observed, suggesting that an MDR system(s) could be responsible for the resistance to these compounds. The active efflux of EtBr indicated that at least one

MDR efflux system should exist in *L. bulgaricus*. Although this system could contribute to the resistance to TDCA, it did not represent the major mechanism of bile resistance in *L. bulgaricus* (T. Smokvina and E. Maguin, pers. comm.).

Regulation of bile resistance

The understanding of regulatory pathways controlling the response to bile stress in bacteria is still very poor. In Gram-negative bacteria some general virulence regulators such as PhoQ-PhoP in *S. thyphimurium* (van Velkinburgh et al. 1999) or TorR in *Vibrio cholerae* (Provenzano et al. 2000) seem to be involved in the control of bile resistance. The regulation of MDR systems often involves specific transcriptional regulators and in addition some efflux pumps are also controlled by a general stress response regulator (Grkovic et al. 2001). In LAB, the strong cross-protection between bile and heat-shock which was observed in some species, suggests overlapping regulatory pathways. However, the possible interplay between adaptive responses and MDR transporters has not been studied and the regulator(s) of these systems remains to be identified.

Starvation response and stress tolerance

Bacteria spend most of their time in stationary phase. Growth arrest and entry into stationary phase can be provoked by numerous stress conditions like cold, heat, osmotic, oxidative or acid stress, or starvation. Among these stresses, nutrient starvation is one of the most frequent and bacterial growth itself contributes to nutrient exhaustion and subsequent starvation for one or several compounds. Moreover, some extreme environmental stress conditions may provoke a deprivation of one or several components, apart from their direct effects on the cells constituents. For example, extreme acidic conditions can decrease the activity of some transporters, thereby diminishing the availability of essential substrates. Therefore, stress conditions can indirectly provoke starvation or energy depletion, irrespective of the extracellular amount of the substrate (Poolman et al. 1987b; Konings et al. 1997). These conditions of energy or essential elements depletion could be deleterious for long-term cell viability. It has been established, however, that many bacteria are well adapted to survive long-term starvation. Some bacteria can enter a stress-resistant spore-forming process. Others, including LAB, do not have this capacity and have developed other strategies. In these bacteria, nutrient starvation leading to growth arrest is generally associated with the modification of cell morphology. Cell division at the entry into stationary phase, leading to a diminution of cell size, has been described for many non-spore forming bacteria like *E. coli* (Lange et al. 1991), and has also been reported for some enterococci and lactococci (Giard et al. 1996; Hartke et al. 1998).

LAB constitute an heterogeneous group of bacteria that grow in very different media, and therefore do not encounter identical starvation conditions. Therefore, starvation surviving mechanisms developed by different LAB are may be diverse.

In order to better understand the starvation response, research has generally been performed in conditions of limitation for a unique compound. Responses to three types of limiting compounds have largely been studied in bacteria: carbohydrate (sugar) starvation leading to cell energy depletion, phosphate starvation which can be deleterious for both energy supply and DNA/RNA synthesis, and nitrogen (amino acids) starvation which primarily results in the limitation of protein synthesis. In LAB, these tree types of starvation conditions have also been studied, but the majority of the results concerns carbohydrate starvation.

Glucose-starved cultures of *L. lactis* sp. *lactis*, *E. faecalis* and *S. pyogenes* show bi-phasic survival curves. The first phase corresponds to a fast decrease of the number of viable cells and the second one to a stabilisation of the number of viable cells in the culture (Hartke et al. 1994; Giard et al. 1997; Trainor et al. 1999). In contrast, for *L. bulgaricus* a complete loss of viability is observed after 48 h in lactose starvation conditions (Chervaux et al. unpublished data) as it is observed for cultures in non-limited rich medium. Although these studies were not performed in exactly the same media and conditions, they suggest that the responses to starvation could be diverse among LAB.

An important common aspect to survival in stationary phase for most of LAB seems to be the capacity to maintain an active metabolic state. Firstly, Poolman et al. (1987) and Kunji et al. (1993) pointed out that survival capacity of lactococci upon carbohydrate starvation is related to the maintenance of glycolytic capacity, and that, therefore, the regulation of metabolism is important for survival. Other studies indicated that amino acid catabolism plays a role in survival for *Lactococcus lactis* (Stuart et al.

1999), *Oenococcus oeni* (Tonon et al. 2000) and *Lactobacillus sakei* (Champomier Verges et al. 1999). In each case, arginine catabolism was used for energy production and was linked to survival. It has been reported that bacteria can prevent total energy depletion by the synthesis and accumulation of energetic polymeric compounds at the entry into stationary phase. Polyphosphate accumulation is observed after phosphate or nitrogen starvation in *E. coli* and there is evidence that polyphosphate synthesis is important for long term survival (Kornberg et al. 1999). Polyphosphate may represent a 'high energy' stock convertible to ATP, or promote the degradation of ribosomal proteins (Rao et al. 1999; Kuroda et al. 2001). Likewise, glycogen and trehalose are described as energy reservoirs in many organisms. So far, however, the synthesis or accumulation of polyphosphate, trehalose or glycogen has not been observed in LAB.

It is well established that bacteria that enter into stationary phase develop a general stress-resistant state. In *E. coli* and *B. subtilis* general stress proteins (GSP) are induced in numerous stress conditions and notably in glucose-starvation conditions (Hengge-Aronis 1993, 1999; Bernhardt et al. 1997). The alternative sigma factor σ^B is known to control the expression of the majority of GSP and activation of this regulon can be induced by distinct pathways depending on the stress conditions (Voelker et al. 1995). In *B. subtilis* at least 75 GSP were identified by 2D-gel analysis (Bernhardt et al. 1997) and at least 125 genes have been shown to be regulated by σ^B using transcriptome analysis (Petersohn et al. 2001). Some of the GSP may have a role in survival during stationary phase and some may have a preventive role to protect against potential environmental stress.

In the case of LAB, it is known that glucose starvation induces an increased resistance to many stress conditions (heat, oxidative, ethanol, acid and osmotic stress) in both *E. faecalis* and *L. lactis* (Hartke et al. 1994; Giard et al. 1996). Heat, acid and bile stress resistance were also increased for lactose-starved *L. bulgaricus* cells (Chervaux et al. unpublished data). In order to identify sugar starvation-induced proteins, 2D-electrophoresis analysis was performed in *L. lactis* (Kunji et al. 1993) *E. faecalis* (Giard et al. 1996; Hartke et al. 1998) and *S. mutans* (Svensater et al. 2000). Fourteen, 42 and 58 proteins, respectively, were found to be induced by sugar starvation in these bacteria. In *E. faecalis*, some of these glucose-starvation induced proteins were characterized (Giard et al. 2001). The majority of them are involved in carbon metabolism (triose phosphate isomerase, a putative dihydroxi-acetone kinase, the Gls24 protein which is probably involved in the regulation of pyruvate metabolism). Other proteins (carbamate kinase, a putative glycine cleavage system, and L-serine dehydratase) are involved in amino acid catabolism. Only one characterized glucose starvation-induced protein, a manganese superoxide dismutase, could play a direct role in protection against oxidative stress. Similar studies should be done in other LAB in order to establish whether starvation-induced proteins are generally involved in the modulation of metabolism in LAB as it was observed in *E. faecalis*.

In order to detect the induction of GSPs in LAB in different stress conditions, 2D-gel analysis of proteins was performed in *E. faecalis* and *S. mutans* (Svensater et al. 2000; Giard et al. 2001). Comparison of these proteins indicated that many proteins are commonly induced by more than one stress, but only few proteins are common to all stress. In contrast to what was found in *B. subtilis*, sugar starvation does not induce a large class of GSPs that could explain the general stress resistance state observed.

As mentioned in the 'heat-shock' section, no σ^B homologue has so far been found in LAB and the regulation of starvation-induced proteins in these bacteria is not yet understood. However, some observations indicate possible pathways for general stress response regulation, and particularly for induction during amino acids starvation. Similarly to *E. coli* (Cashel et al. 1996) and *B. subtilis* (Wendrich et al. 1997), amino acid starvation has been shown to induce the stringent response in *S. equisimilis* (Mechold et al. 1996) and *S. pyogenes* (Whitehead et al. 1998). This response is characterized by the accumulation of guanine nucleotides (ppGpp) synthesized by RelA (the ppGpp synthetase). Accumulation of ppGpp in response to different starvation conditions has been reported *E. coli* (Kvint et al. 2000), *Staphylococcus aureus* (Crosse et al. 2000) and Corynebacteria (Wehmeier et al. 1998). ppGpp has been shown to be involved in the control of many stress induced genes in *E. coli* (Loewen et al. 1998) and ppGpp is now considered as a major signal involved in saving the cells energy (for a review, see Chatterji et al. 2001). In *L. lactis*, a role for ppGpp synthesis in acid-stress resistance has been proposed (Rallu et al. 2000) and, more generally, pools of guanine and phosphate have been implicated in stress response in lactococci (Duwat et al. 1999; Rallu et al. 2000). Moreover, stringent response in *S. pyogenes* is reported to strongly support the cells' survival under nutritional stress conditions (Steiner et al. 2000). All these observations suggest that ppGpp can play a role in stress response, and could represent a global regulator in starvation conditions. Recently, a *relA*-independent response to amino acid starvation has been described in *S. pyogenes*. This response controls both stress genes and virulence genes (Steiner et al. 2001). In *L. lactis*, a *relA*-independent response to amino acid starvation has been shown to involve the *codY* gene which regulates several genes of the proteolysis pathway (Guedon et al. 2001). Although it remains to be demonstrated, it was recently suggested that the *relA* independent response of *S. pyogenes* may involve a *codY* ortholog (Steiner et al. 2001).

Emerging data from studies on LAB strongly suggest that the starvation-induced mechanisms are different from those of the two well-studied bacterial models *E. coli* and *B. subtilis*. The data presented above suggest that hitherto unknown regulation mechanisms may be present in LAB compensating for the absence of a σ^B homologue.

First proteome analyses of *E. faeca*lis and *S. mutans*, indicates that only a small overlap exists between stress-specific and starvation regulons, and that general stress proteins seem to be rare. It would be important to confirm such primary tendencies in other LAB, and especially to compare these results with the response of some species found in radically different environments that show different long-term starvation survival capacity.

Concluding remarks

Genes implicated in stress responses are numerous and in LAB the levels of characterization of their actual role and regulation differ widely between species. The studies concerning stress responses in LAB sometimes benefit from the knowledge already acquired in other bacteria. For example, part of the studies on heat response focused on specific genes because of their major role demonstrated in other micro-organisms. The functional conservation of several stress proteins (for example, HS proteins, Csp, etc.) and of some of their regulators (for example, HrcA, CtsR) renders even more striking the differences existing between LAB and the classical model micro-organisms. However, two points should be keep in mind. First, comparison of the phylogenetic distribution of various stress-related genes suggests that only very few of them are really indispensable (Koonin et al. 2000). Second, comparative genomics revealed that horizontal gene transfer and lineage-specific gene loss are important forces of evolution among prokaryotes (Fitzgerald et al. 2001; Lawrence et al. 2001). With these conclusions in mind, the differences between 'model bacteria' and LAB appear less surprising, emphasizing the caution that should be employed when extrapolating the observations made in one species to others.

Among the differences observed between LAB species and *B. subtilis*, one of the most striking is the absence of a σ^B orthologue in *L. lactis* ssp. *lactis* (Bolotin et al. 2001) as in at least two streptococci (Ferretti et al., 2001; Hoskins et al. 2001; Tettelin et al. 2001) and probably *E. faecalis* (A. Hartke, JC. Giard, pers. comm.). The observation that only few proteins seem to be induced by 'all' stress (see the section on starvation response) suggests that the σ^B function is not carried out by a distantly related or an unrelated protein (i.e., non-orthologous gene displacement, Galperin et al. 1998) but rather is absent from *L. lactis* and *E. faecalis*. Nevertheless, general stress resistance is induced by starvation in these species, and the question of its regulation remains to be answered. The increasing number of LAB genome sequences becoming available (Klaenhammer et al. this issue) should allow to determine in a near future how many LAB are devoid of σ^B orthologues.

The information presented in the preceding chapters also reveals that, not unexpectedly, the cell membrane plays an important role in stress resistance. First of all, the membrane itself can change in adaptation to environmental conditions and these changes contribute to the protection of the bacteria (see the sections focusing on acid stress, thermal shocks and oxidative stress). In addition, many processes involved in stress resistance that do not directly alter the structure of the lipid bilayer itself take place at the cell membrane. A multitude of proteins (for example, transport systems, sensors, housekeeping proteases, etc.), is physically linked to the cell membrane. In spite of the importance of these proteins, they are rarely catalogued in studies aiming at the identification of clusters of stress or adaptation induced proteins. Two obvious reasons may account for this discrepancy. First, the global identification of stress-induced proteins in a given organism is usually based on the analysis of protein profiles after 2DE. Due to technical limitations, membrane proteins are rarely detected by this method. Secondly, it may be that changes in membrane protein composition result from long-term adaptation processes, while short-term responses may primarily be accounted for by the activation (and/or stabilization) of proteins already present. The latter hypothesis is valid especially in the case of transport systems, although for some of the systems studied a transcriptional induction has also been observed.

The overview of LAB stress responses also reveals common aspects of stress responses. As in other bacteria, adaptive responses appear to be a usual mode of stress protection in LAB. The cross-protections often induced by the expression of a given adaptive response can be advantageous when cells are exposed to a combination of stress. However, the cross protections associated to the adaptive response to a given stress appear to vary between species (for example, see section focusing on bile resistance, etc.). This observation suggests that the molecular bases of adaptive responses are, at least in part, species (or even ssp.) specific. Further studies of the genes induced in adaptation will be needed to determine how many are species specific, and if this specificity correlate with a life-style.

Still, the induction of adaptive responses can undoubtedly increase the bacterial tolerance to stress and can *per se* be used to prepare strains to harsh conditions. Among the successful examples of applications is the heat adaptation of *O. oeni* starter cultures prior to storage. Heat treatment leads to higher pH and ethanol tolerance, and improves the efficiency of direct inoculation of *O. oeni* in wine (Guzzo et al. 1994). In many LAB, several treatments (incubation at cold temperature or in lactose (de Urraza et al. 1997), osmotic stress) were also reported to improve freezing-thawing or freeze-drying survival. A better understanding of the mechanisms of stress resistance should allow to understand the bases of the adaptive responses and cross protections and to exploit them better to prepare LAB to industrial processes. Moreover, the identification of crucial stress-related genes will reveal targets (i) for specific manipulation (to promote or limit growth), (ii) to develop tools to screen for tolerant or sensitive strains and (iii) to evaluate the fitness of a culture which can indicate whether cells are fully adapted and can be used in harsh conditions or, inversely, are stressed and will not optimally perform in a process. This can also help to optimize procedures according to cells fitness. In this context, future genome and transcriptome analyses will undoubtedly complement the proteome and genetic information available today, and shed a new light on the perception of, and the response to, stress by lactic acid bacteria.

References

Abdelal A T (1979) Arginine catabolism by microorganisms. Annu Rev Microbiol. 33: 139–168.

Abe K, Hayashi H & Maloney P C (1996) Exchange of aspartate and alanine: machanism for development of a proton-motive force in bacteria. J. Biol. Chem. 271: 3079–3084.

Adamowicz M, Kelley P M & Nickerson K W (1991) Detergent (sodium dodecyl sulfate) shock proteins in *Escherichia coli*. J. Bacteriol. 173: 229–233.

Archibald F S & Fridovich I (1981) Manganese, superoxide dismutase, and oxygen tolerance in some lactic acid bacteria. J Bacteriol. 146: 928–936.

Arena M E, Saguir F M & Manca de Nadra M C (1999) Arginine, citrulline and ornithine metabolism by lactic acid bacteria from wine. Int. J. Food Microbiol. 52: 155–161.

Arikado E, Ishihara H, Ehara T, Shibata C, Saito H, Kakegawa T, Igarashi K & Kobayashi H (1999) Enzyme level of enterococcal F1Fo-ATPase is regulated by pH at the step of assembly. Eur. J. Biochem. 259: 262–268.

Arnau J, Sorensen K I, Appel K F, Vogensen F K & Hammer K (1996) Analysis of heat shock gene expression in Lactococcus lactis MG1363. Microbiology 142: 1685–1691.

Auffray Y, Gansel X, Thammavongs B & Boutibonnes P (1992) Heat-shock induced protein synthesis in *Lactococcus lactis* subsp. *lactis*. Curr. Microbiol. 24: 281–284.

Baati L, Fabre-Gea C, Auriol D & Blanc P J (2000) Study of the cryotolerance of *Lactobacillus acidophilus*: effect of culture and freezing conditions on the viability and cellular protein levels. Int. J. Food Microbiol. 59: 241–247.

Bae W, Jones P G & Inouye M (1997) CspA, the major cold shock protein of *Escherichia coli*, negatively regulates its own gene expression. J. Bacteriol. 179: 7081–7088.

Bae W, Xia B, Inouye M & Severinov K (2000) *Escherichia coli* CspA-family RNA chaperones are transcription antiterminators. Proc. Natl. Acad. Sci. U.S.A. 97: 7784–7789.

Baev D, England R & Kuramitsu H K (1999) Stress-induced membrane association of the *Streptococcus mutans* GTP-binding protein, an essential G protein, and investigation of its physiological role by utilizing an antisense RNA strategy. Infect Immun. 67: 4510–4516.

Bakker E P & Harold F M (1980) Energy coupling to potassium transport in *Streptococcus faecalis*. J. Biol. Chem. 255: 433–440.

Becker G, Klauck E & Hengge-Aronis R (1999) Regulation of RpoS proteolysis in *Escherichia coli*: the response regulator RssB is a recognition factor that interacts with the turnover element in RpoS. Proc. Natl. Acad. Sci. U.S.A. 96: 6439–6444.

Belli W A & Marquis R E (1991) Adapation of *Strepococcus mutans* and *Enteroccus hirae* to acid stress in continuous culture. Appl. Environ. Microbiol. 57: 1134–1138.

Bernhardt J, Volker U, Volker A, Antelmann H, Schmid R, Mach H & Hecker M (1997) Specific and general stress proteins in *Bacillus subtilis* – a two-dimensional protein electrophoresis study. Microbiology 143: 999–1017.

Blank L M, Koebmann B J, Michelsen O, Nielsen L K & Jensen P R (2001) Hemin reconstitutes proton extrusion in an H(+)-ATPase-negative mutant of *Lactococcus lactis*. J. Bacteriol. 183: 6707–6709.

Bolhuis H, Molenaar D, Poelarends G, van Veen H W, Poolman B, Driessen A J & Konings W N (1994) Proton motive force-driven and ATP-dependent drug extrusion systems in multidrug-resistant *Lactococcus lactis*. J Bacteriol. 176: 6957–6964.

Bolotin A, Mauger S, Malarme K, Ehrlich S D & Sorokin A (1999) Low-redundancy sequencing of the entire *Lactococcus lactis* IL1403 genome. Antonie Van Leeuwenhoek 76: 27–76.

Bolotin A, Wincker P, Mauger S, Jaillon O, Malarme K, Weissenbach J, Ehrlich S D & Sorokin A (2001) The complete genome sequence of the lactic acid bacterium *Lactococcus lactis* ssp. *lactis* IL1403. Genome Res. 11: 1–23.

Bond D R & Russell J B (1996) A role for fructose 1,6-diphosphate in the ATPase-mediated energy spilling reaction of *Streptococcus bovis*. Appl. Environ. Microbiol. 62: 2095–2099.

Bouvier J, Bordes P, Romeo Y, Fourcans A, Bouvier I & Gutierrez C (2000) Characterization of OpuA, a glycine–betaine uptake system of *Lactococcus lactis*. J. Mol. Microbiol. Biotechnol. 2: 199–205.

Boyd D A, Cvitkovitch D G, Bleiweis A S, Kiriukhin M Y, Debabov D V, Neuhaus F C & Hamilton I R (2000) Defects in D-alanyl-lipoteichoic acid synthesis in *Streptococcus mutans* results in acid sensitivity. J. Bacteriol. 182: 6055–6065.

Brandi A, Pietroni P, Gualerzi C O & Pon C L (1996) Post-transcriptional regulation of CspA expression in *Escherichia coli*. Mol. Microbiol. 19: 231–240.

Breeuwer P, Drocourt J-L, Rombouts F M & Abee T (1996) A novel method for coninuous determination of the intracellular pH in bacteria with the internally conjugated fluorescent probe (and 6) carboxyfluorescein succinimidyl ester. Appl. Environ. Microbiol. 62: 178–183.

Broadbent J R & Lin C (1999) Effect of heat shock or cold shock treatment on the resistance of lactococcus lactis to freezing and lyophilization. Cryobiology 39: 88–102.

Broadbent J R, Oberg C J & Wei L (1998) Characterization of the *Lactobacillus helveticus* groESL operon. Res. Microbiol. 149: 247–253.

Caldon C E, Yoong P & March P E (2001) Evolution of a molecular switch: universal bacterial GTPases regulate ribosome function. Mol. Microbiol. 41: 289–297.

Cashel M, Gentry D, Hernandez V J & Vinella D (1996) The stringent response. *In*: Neidhardt F C, Ingraham J L, Low K B, Magasanik B, Schaechter M & Umberger H E (eds.), *Escherichia coli* and *Salmonella typhimurium*: Cellular and Molecular Biology (pp 1458–1496). American Society for Microbiology Press, Washington, DC.

Champomier-Verges M C, Zuniga M, Morel-Deville F, Perez-Martinez G, Zagorec M & Ehrlich S D (1999) Relationships between arginine degradation, pH and survival in *Lactobacillus sakei*. FEMS Microbiol. Lett. 180: 297–304.

Champomier-Verges M C, Chaillou S, Cornet M & Zagorec M (2001) *Lactobacillus sakei*: recent developments and future prospects. Res. Microbiol. 152: 839–848.

Champomier-Verges M C, Maguin E, Mistou M Y, Anglade P & Chich J F (2002) Lactic acid bacteria and proteomics: current knowledge and perspectives. J. Chromatogr. (in press).

Chang Y Y & Cronan Jr. J E, (1999) Membrane cyclopropane fatty acid content is a major factor in acid resistance of *Escherichia coli*. Mol. Microbiol. 33: 249–259.

Chapot-Chartier M P, Schouler C, Lepeuple A S, Gripon J C & Chopin M C (1997) Characterization of cspB, a cold-shock-inducible gene from *Lactococcus lactis*, and evidence for a family of genes homologous to the *Escherichia coli* cspA major cold shock gene. J. Bacteriol. 179: 5589–5593.

Chastanet A, Prudhomme M, Claverys J P & Msadek T (2001) Regulation of *Streptococcus pneumoniae* clp genes and their role in competence development and stress survival. J. Bacteriol. 183: 7295–7307.

Chatterji D & Ojha A K (2001) Revisiting the stringent response, ppGpp and starvation signaling. Curr. Opin. Microbiol. 4: 160–165.

Chen Y Y & Burne R A (1996) Analysis of *Streptococcus salivarius* urease expression using continuous chemostat culture. FEMS Microbiol. Lett. 135: 223–229.

Chen Y Y, Weaver C A, Mendelsohn D R & Burne R A (1998) Transcriptional regulation of the *Streptococcus salivarius* 57.I urease operon. J. Bacteriol. 180: 5769–5775.

Condon S (1987) Responses of lactic acid bacteria to oxygen. FEMS Microbiol. Rev. 46: 269–280.

Chen Y Y, Weaver C A & Burne R A (2000) Dual functions of *Streptococcus salivarius* urease. J. Bacteriol. 182: 4667–4669.

Crosse A M, Greenway D L & England R R (2000) Accumulation of ppGpp and ppGp in *Staphylococcus aureus* 8325-4 following nutrient starvation. Lett. Appl. Microbiol. 31: 332–337.

Crow V L & Thomas T D (1982) Arginine metabolism in lactic Streptococci. J. Bacteriol. 150: 1024–1032.

Csonka L N & Hanson A D (1991) Prokaryotic osmoregulation: genetics and physiology. Annu. Rev. Microbiol. 45: 569–606.

Cunin R, Glansdorff N, Pierard A & Stalon V (1986) Biosynthesis and metabolism of arginine in bacteria. Microbiol. Rev. 50: 314–352.

Cvitkovitch D G, Gutierrez J A, Behari J, Youngman P J, Wetz J E, Crowley P F, Hillman J D, Brady L J & Bleiweis A S (2000) Tn917-lac mutagenesis of *Streptococcus mutans* to identify environmentally regulated genes. FEMS Microbiol. Lett. 182: 149–154.

De Angelis M, Bini L, Pallini V, Cocconcelli P S & Gobbetti M (2001) The acid-stress response in *Lactobacillus sanfranciscensis* CB1. Microbiology 147: 1863–1873.

de Urraza P & de Antoni G (1997) Induced cryotolerance of *Lactobacillus delbrueckii* subsp. *bulgaricus* LBB by preincubation at suboptimal temperature with a fermentable sugar. Cryobiology 35: 159–164.

Delcour J, Ferain T, Deghorain M, Palumbo E & Hols P (1999) The biosynthesis and functionality of the cell-wall of lactic acid bacteria. Antonie Van Leeuwenhoek 76: 159–184.

Delmas F, Pierre F, Coucheney F, Divies C & Guzzo J (2001) Biochemical and physiological studies of the small heat shock protein Lo18 from the lactic acid bacterium *Oenococcus oeni*. J. Mol. Microbiol. Biotechnol. 3: 601–610.

Derre I, Rapoport G & Msadek T (1999) CtsR, a novel regulator of stress and heat shock response, controls clp and molecular chaperone gene expression in gram-positive bacteria. Mol. Microbiol. 31: 117–131.

Derzelle S, Hallet B, Francis K P, Ferain T, Delcour J & Hols P (2000) Changes in cspL, cspP, and cspC mRNA abundance as a function of cold shock and growth phase in *Lactobacillus plantarum*. J. Bacteriol. 182: 5105–5113.

Diaz-Torres M L & Russell R R (2001) HtrA protease and processing of extracellular proteins of *Streptococcus mutans*. FEMS Microbiol. Lett. 204: 23–28.

Drici-Cachon Z, Guzzo J, Cavin J-F & Diviès C (1996) Acid tolerance in *Leuconostoc oenos*. Isolation and characterization of an acid-resistant mutant. Appl. Microbiol. Biotechnol. 44: 785–789.

Dunne C, Murphy L, Flynn S, O'Mahony L, O'Halloran S, Feeney M, Morrissey D, Thornton G, Fitzgerald G, Daly C, Kiely B, Quigley E M, O'Sullivan G C, Shanahan F & Collins J K (1999) Probiotics: from myth to reality. Demonstration of functionality in animal models of disease and in human clinical trials. Antonie Van Leeuwenhoek 76: 279–292.

Duwat P, Ehrlich S D & Gruss A (1995) The recA gene of *Lactococcus lactis*: characterization and involvement in oxidative and thermal stress. Mol. Microbiol. 17: 1121–1131.

Duwat P, Ehrlich S D & Gruss A (1999) Effects of metabolic flux on stress response pathways in *Lactococcus lactis*. Mol. Microbiol. 31: 845–858.

Duwat P, Sourice S, Cesselin B, Lamberet G, Vido K, Gaudu P, Le Loir Y, Violet F, Loubiere P & Gruss A (2001) Respiration capacity of the fermenting bacterium *Lactococcus lactis* and its positive effects on growth and survival. J. Bacteriol. 183: 4509–4516.

Earnshaw R G, Appleyard J & Hurst R M (1995) Understanding physical inactivation processes: combined preservation opportunities using heat, ultrasound and pressure. Int. J. Food Microbiol. 28: 197–219.

Eaton T, Shearman C & Gasson M (1993) Cloning and sequence analysis of the dnaK gene region of *Lactococcus lactis* subsp. *lactis*. J. Gen. Microbiol. 139: 3253–3264.

Elkins C A & Savage D C (1998) Identification of genes encoding conjugated bile salt hydrolase and transport in *Lactobacillus johnsonii* 100-100. J. Bacteriol. 180: 4344–4349.

Engesser D M & Hammes W P (1994) Non-heme catalase activity of lactic acid bacteria. Syst. Appl. Microbiol. 17: 11–19.

Eymann C, Mach H, Harwood C R & Hecker M (1996) Phosphate-starvation-inducible proteins in *Bacillus subtilis*: a two-dimensional gel electrophoresis study. Microbiology 142: 3163–3170.

Fabret C & Hoch J A (1998) A two-component signal transduction system essential for growth of *Bacillus subtilis*: implications for anti-infective therapy. J. Bacteriol. 180: 6375–6383.

Fang L, Jiang W, Bae W & Inouye M (1997) Promoter-independent cold-shock induction of cspA and its derepression at 37 degrees C by mRNA stabilization. Mol. Microbiol. 23: 355–364.

Fenoll A, Munoz R, Garcia E & de la Campa A G (1994) Molecular basis of the optochin-sensitive phenotype of pneumococcus: characterization of the genes encoding the F0 complex of the *Streptococcus pneumoniae* and *Streptococcus oralis* H(+)-ATPases. Mol. Microbiol. 12: 587–598.

Ferretti J. J, McShan W M, Ajdic D, Savic D J, Savic G, Lyon K, Primeaux C, Sezate S, Suvorov A N, Kenton S, Lai H S, Lin S P, Qian Y, Jia H G, Najar F Z, Ren Q, Zhu H, Song L, White J, Yuan X, Clifton S W, Roe B A & McLaughlin R (2001) Complete genome sequence of an M1 strain of *Streptococcus pyogenes*. Proc. Natl. Acad. Sci. U.S.A. 98: 4658–4663.

Fitzgerald J R & Musser J M (2001) Evolutionary genomics of pathogenic bacteria. Trends Microbiol. 9: 547–553.

Flahaut S, Frere J, Boutibonnes P & Auffray Y (1996a) Comparison of the bile salts and sodium dodecyl sulfate stress responses in *Enterococcus faecalis*. Appl. Environ. Microbiol. 62: 2416–2420.

Flahaut S, Hartke A, Giard J C, Benachour A, Boutibonnes P & Auffray Y (1996b) Relationship between stress response toward bile salts, acid and heat treatment in *Enterococcus faecalis*. FEMS Microbiol. Lett. 138: 49–54.

Foucaud-Sceunemann C & Poquet I (2002) The *Lactococcus lactis* HtrA protease is induced and essential for cell survival under stress conditions. In preparation.

Francis K P, Mayr R, von Stetten F, Stewart G S & Scherer S (1998) Discrimination of psychrotrophic and mesophilic strains of the *Bacillus cereus* group by PCR targeting of major cold shock protein genes. Appl. Environ. Microbiol. 64: 3525–3529.

Frees D & Ingmer H (1999) ClpP participates in the degradation of misfolded protein in *Lactococcus lactis*. Mol. Microbiol. 31: 79–87.

Frees D, Varmanen P & Ingmer H (2001) Inactivation of a gene that is highly conserved in Gram-positive bacteria stimulates degradation of non-native proteins and concomitantly increases stress tolerance in *Lactococcus lactis*. Mol. Microbiol. 41: 93–103.

Futai M, Noumi T & Maeda M (1989) ATP synthase (H+-ATPase): results by combined biochemical and molecular biological approaches. Annu. Rev. Biochem. 58: 111–136.

Galperin M Y, Walker D R & Koonin E V (1998) Analogous enzymes: independent inventions in enzyme evolution. Genome Res. 8: 779–790.

Garcia-Quintans N, Magni C, de Mendoza D & Lopez P (1998) The citrate transport system of *Lactococcus lactis* subsp. *lactis* biovar *diacetylactis* is induced by acid stress. Appl. Environ. Microbiol. 64: 850–857.

Giard J C, Hartke A, Flahaut S, Benachour A, Boutibonnes P & Auffray Y (1996) Starvation-induced multiresistance in *Enterococcus faecalis* JH2-2. Curr. Microbiol. 32: 264–271.

Giard J C, Hartke A, Flahaut S, Boutibonnes P & Auffray Y (1997) Glucose starvation response in *Enterococcus faecalis* JH2-2: survival and protein analysis. Res. Microbiol. 148: 27–35.

Giard J C, Rince A, Capiaux H, Auffray Y & Hartke A (2000) Inactivation of the stress- and starvation-inducible gls24 operon has a pleiotrophic effect on cell morphology, stress sensitivity, and gene expression in *Enterococcus faecalis*. J. Bacteriol. 182: 4512–4520.

Giard J C, Laplace J M, Rince A, Pichereau V, Benachour A, Leboeuf C, Flahaut S, Auffray Y & Hartke A (2001) The stress proteome of *Enterococcus faecalis*. Electrophoresis 22: 2947–2954.

Glaasker E, Konings W N & Poolman B (1996a) Glycine betaine fluxes in *Lactobacillus plantarum* during osmostasis and hyper- and hypo-osmotic shock. J. Biol. Chem. 271: 10060–10065.

Glaasker E, Konings W N & Poolman B (1996b) Osmotic regulation of intracellular solute pools in *Lactobacillus plantarum*. J. Bacteriol. 178: 575–582.

Glaasker E, Heuberger E H, Konings W N & Poolman B (1998a) Mechanism of osmotic activation of the quaternary ammonium compound transporter (QacT) of *Lactobacillus plantarum*. J Bacteriol. 180: 5540–5546.

Glaasker E, Tjan F S, Ter Steeg P F, Konings W N & Poolman B (1998b) Physiological response of *Lactobacillus plantarum* to salt and nonelectrolyte stress. J. Bacteriol. 180: 4718–4723.

Goldenberg D, Azar I & Oppenheim A B (1996) Differential mRNA stability of the cspA gene in the cold-shock response of *Escherichia coli*. Mol. Microbiol. 19: 241–248.

Gordia S & Gutierrez C (1996) Growth-phase-dependent expression of the osmotically inducible gene osmC of *Escherichia coli* K-12. Mol. Microbiol. 19: 729–736.

Gostick D O, Griffin H G, Shearman C A, Scott C, Green J, Gasson M J & Guest J R (1999) Two operons that encode FNR-like proteins in *Lactococcus lactis*. Mol. Microbiol. 31: 1523–1535.

Gouesbet G, Jan G & Boyaval P (2002) Two-dimensional electrophoresis study of *Lactobacillus delbrueckii* subsp. *bulgaricus* thermotolerance. Appl. Environ. Microbiol. 68: 1055–1063.

Graumann P, Schroder K, Schmid R & Marahiel M A (1996) Cold shock stress-induced proteins in *Bacillus subtilis*. J. Bacteriol. 178: 4611–4619.

Graumann P, Wendrich T M, Weber M H, Schroder K & Marahiel M A (1997) A family of cold shock proteins in *Bacillus subtilis* is essential for cellular growth and for efficient protein synthesis at optimal and low temperatures. Mol. Microbiol. 25: 741–756.

Grkovic S, Brown M H & Skurray R A (2001) Transcriptional regulation of multidrug efflux pumps in bacteria. Semin. Cell Dev. Biol. 12: 225–237.

Guedon E, Serror P, Ehrlich S D, Renault P & Delorme C (2001) Pleiotropic transcriptional repressor CodY senses the intracellular pool of branched-chain amino acids in *Lactococcus lactis*. Mol. Microbiol. 40: 1227–1239.

Guerzoni M E, Lanciotti R & Cocconcelli P S (2001) Alteration in cellular fatty acid composition as a response to salt, acid, oxidative and thermal stresses in *Lactobacillus helveticus*. Microbiology 147: 2255–2264.

Guillot A, Obis D & Mistou M Y (2000) Fatty acid membrane composition and activation of glycine–betaine transport in *Lactococcus lactis* subjected to osmotic stress. Int. J. Food Microbiol. 55: 47–51.

Gunn J S (2000) Mechanisms of bacterial resistance and response to bile. Microbes Infect. 2: 907–913.

Gutierrez J A, Crowley P J, Brown D P, Hillman J D, Youngman P & Bleiweis A S (1996) Insertional mutagenesis and recovery of interrupted genes of *Streptococcus mutans* by using transposon Tn917: preliminary characterization of mutants displaying acid sensitivity and nutritional requirements. J. Bacteriol. 178: 4166–4175.

Gutierrez J A, Crowley P J, Cvitkovitch D G, Brady L J, Hamilton I R, Hillman J D & Bleiweis A S (1999) *Streptococcus mutans* ffh, a gene encoding a homologue of the 54 kDa subunit of the signal recognition particle, is involved in resistance to acid stress. Microbiology 145: 357–366.

Guzzo J, Cavin J & Divies C (1994) Induction of stress proteins in *Leuconostoc oenos* to perform direct inoculation of wine. Biotechnol. Lett. 16: 1189–1194.

Guzzo J, Delmas F, Pierre F, Jobin M P, Samyn B, Van Beeumen J, Cavin J F & Divies C (1997) A small heat shock protein from *Leuconostoc oenos* induced by multiple stresses and during stationary growth phase. Lett. Appl. Microbiol. 24: 393–396.

Guzzo J, Jobin M-P, Delmas F, Fortier L-C, Garmyn D, Tourdot-Maréchal R, Lee B & Diviès C (2000) Regulation of stress response in *Oenococcus oeni* as a function of environmental changes and growth phase. Int. J. Food Microbiol. 55: 27–31.

Hahn K, Faustoferri R C & Quivey Jr. R G (1999) Induction of an AP endonuclease activity in *Streptococcus mutans* during growth at low pH. Mol. Microbiol. 31: 1489–1498.

Hanna M M & Liu K (1998) Nascent RNA in transcription complexes interacts with CspE, a small protein in *E. coli* implicated in chromatin condensation. J. Mol. Biol. 282: 227–239.

Hanna M N, Ferguson R J, Li Y H & Cvitkovitch D G (2001) uvrA is an acid-inducible gene involved in the adaptive response to low pH in *Streptococcus mutans*. J. Bacteriol. 183: 5964–5973.

Hansen M C, Nielsen A K, Molin S, Hammer K, Kilstrup M, Palmer Jr. R J, Udsen C & White D C (2001) Changes in rRNA levels during stress invalidates results from mRNA blotting: fluorescence in situ rRNA hybridization permits renormalization for estimation of cellular mRNA levels. J. Bacteriol. 183: 4747–4751.

Harper D S & Loesche W J (1984) Growth and acid tolerance of human dental plaque bacteria. Arch. Oral Biol. 29: 843–848.

Hartke A, Bouché S, Gansel X, Boutibonnes P & Auffray Y (1994) Starvation-induced stress resistance in *Lactococcus lactis* subsp. *lactis* IL1403. Appl. Environ. Microbiol. 60: 3474–3478.

Hartke A, Bouche S, Laplace J-M, Benachour A, Boutibonnes P & Auffray Y (1995) UV-inducible proteins and UV-induced cross-protection against acid, ethanol, H_2O_2 or heat treatments in *Lactococcus lactis* subsp. *lactis*. Arch. Microbiol. 163: 329–336.

Hartke A, Bouché S, Giard J C, Benachour A, Boutibonnes P & Auffray Y (1996) The lactic acid stress response of *Lactococcus lactis* subsp. *lactis*. Curr. Microbiol. 33: 194–199.

Hartke A, Frere J, Boutibonnes P & Auffray Y (1997) Differential induction of the chaperonin GroEL and the Co-chaperonin GroES by heat, acid, and UV-irradiation in *Lactococcus lactis* subsp. *lactis*. Curr. Microbiol. 34: 23–26.

Hartke A, Giard J C, Laplace J M & Auffray Y (1998) Survival of *Enterococcus faecalis* in an oligotrophic microcosm: changes in morphology, development of general stress resistance, and analysis of protein synthesis. Appl. Environ. Microbiol. 64: 4238–4245.

Hengge-Aronis R (1993) Survival of hunger and stress: the role of rpoS in early stationary phase gene regulation in *E. coli*. Cell 72: 165–168.

Hengge-Aronis R (1999) Interplay of global regulators and cell physiology in the general stress response of *Escherichia coli*. Curr. Opin. Microbiol. 2: 148–152.

Hertel C, Schmidt G, Fischer M, Oellers K & Hammes W P (1998) Oxygen-dependent regulation of the expression of the catalase gene katA of *Lactobacillus sakei* LTH677. Appl. Environ. Microbiol. 64: 1359–1365.

Higgins C F (1992) ABC transporters: from microorganisms to man. Annu. Rev. Cell Biol. 8: 67–113.

Higuchi T, Hayashi H & Abe K (1997) Exchange of glutamate and gamma-aminobutyrate in a *Lactobacillus* strain. J. Bacteriol. 179: 3362–3364.

Hoskins J, Alborn Jr. W E, Arnold J, Blaszczak L C, Burgett S, DeHoff B S, Estrem S T, Fritz L, Fu D J, Fuller W, Geringer C, Gilmour R, Glass J S, Khoja H, Kraft A R, Lagace R E, LeBlanc D J, Lee L N, Lefkowitz E J, Lu J, Matsushima P, McAhren S M, McHenney M, McLeaster K, Mundy C W, Nicas T I, Norris F H, O'Gara M, Peery R B, Robertson G T, Rockey P, Sun P M, Winkler M E, Yang Y, Young-Bellido M, Zhao G, Zook C A, Baltz R H, Jaskunas S R, Rosteck Jr. P R, Skatrud P L & Glass J I (2001) Genome of the bacterium *Streptococcus pneumoniae* strain R6. J. Bacteriol. 183: 5709–5717.

Hutkins R W & Nannen N L (1993) pH homeostasis in lactic acid bacteria. J. Dairy Sci. 76: 2354–2365.

Hutkins R W, Ellefson W L & Kashket E R (1987) Betaine transport imparts osmotolerance on a strain of *Lactobacillus acidophilus*. Appl. Environ. Microbiol. 53: 2275–2281.

Igarashi T, Kono Y & Tanaka K (1996) Molecular cloning of manganese catalase from *Lactobacillus plantarum*. J. Biol. Chem. 271: 29521–29524.

Ingmer H, Vogensen F K, Hammer K & Kilstrup M (1999) Disruption and analysis of the clpB, clpC, and clpE genes in *Lactococcus lactis*: ClpE, a new Clp family in gram-positive bacteria. J. Bacteriol. 181: 2075–2083.

Irvine A S & Guest J R (1993) *Lactobacillus casei* contains a member of the CRP-FNR family. Nucleic Acids Res. 21: 753.

Israelsen H, Madsen S M, Vrang A, Hansen E B & Johansen E (1995) Cloning and partial characterization of regulated promoters from *Lactococcus lactis* Tn917-lacZ integrants with the new promoter probe vector, pAK80. Appl. Environ. Microbiol. 61: 2540–2547.

Jayaraman G C, Penders J E & Burne R A (1997) Transcriptional analysis of the *Streptococcus mutans* hrcA, grpE and dnaK genes and regulation of expression in response to heat shock and environmental acidification. Mol. Microbiol. 25: 329–341.

Jensen N B, Melchiorsen C R, Jokumsen K V & Villadsen J (2001) Metabolic behavior of *Lactococcus lactis* MG1363 in microaerobic continuous cultivation at a low dilution rate. Appl. Environ. Microbiol. 67: 2677–2682.

Jiang W, Fang L & Inouye M (1996) The role of the 5′-end untranslated region of the mRNA for CspA, the major cold-shock protein of *Escherichia coli*, in cold-shock adaptation. J. Bacteriol. 178: 4919–4925.

Jiang W, Hou Y & Inouye M (1997) CspA, the major cold-shock protein of *Escherichia coli*, is an RNA chaperone. J. Biol. Chem. 272: 196–202.

Jobin M P, Garmyn D, Divies C & Guzzo J (1999) Expression of the *Oenococcus oeni* trxA gene is induced by hydrogen peroxide and heat shock. Microbiology 145: 1245–1251.

Kaan T, Jurgen B & Schweder T (1999) Regulation of the expression of the cold shock proteins CspB and CspC in *Bacillus subtilis*. Mol. Gen. Genet. 262: 351–354.

Kashket E R (1984) Bioenergetics of lactic acid bacteria: cytoplasmic pH and osmotolerance. FEMS Microbiol. Rev. 46: 233–244.

Kashket E R & Barker S L (1977) Effects of potassium ions on the electrical and pH gradients across the membrane of *Streptococcus lactis* cells. J. Bacteriol. 130: 1017–1023.

Kets E P W, Teunissen P J M & De Bont J A M (1996) Effect of compatible solutes on survival of lactic acid bacteria subjected to drying. Appl. Environ. Microbiol. 62: 259–261.

Kilstrup M, Jacobsen S, Hammer K & Vogensen F K (1997) Induction of heat shock proteins DnaK, GroEL, and GroES by salt stress in *Lactococcus lactis*. Appl. Environ. Microbiol. 63: 1826–1837.

Kim S G & Batt C A (1993) Cloning and sequencing of the *Lactococcus lactis* subsp. *lactis* groESL operon. Gene. 127: 121–126.

Kim W S & Dunn N W (1997) Identification of a cold shock gene in lactic acid bacteria and the effect of cold shock on cryotolerance. Curr. Microbiol. 35: 59–63.

Kim W S, Khunajakr N & Dunn N W (1998) Effect of cold shock on protein synthesis and on cryotolerance of cells frozen for long periods in *Lactococcus lactis*. Cryobiology 37: 86–91.

Kim W S, Ren J & Dunn N W (1999) Differentiation of *Lactococcus lactis* subspecies *lactis* and subspecies *cremoris* strains by their adaptive response to stresses. FEMS Microbiol. Lett. 171: 57–65.

Kim W S, Perl L, Park J H, Tandianus J E & Dunn N W (2001) Assessment of stress response of the probiotic *Lactobacillus acidophilus*. Curr. Microbiol. 43: 346–350.

Knauf H J, Vogel R F & Hammes W P (1992) Cloning, sequence, and phenotypic expression of katA, which encodes the catalase of *Lactobacillus sake* LTH677. Appl. Environ. Microbiol. 58: 832–839.

Kobayashi H, Murakami N & Unemoto T (1982) Regulation of the cytoplasmic pH in *Streptococcus faecalis*. J. Biol. Chem. 257: 13246–13252.

Koch B, Kilstrup M, Vogensen F K & Hammer K (1998) Induced levels of heat shock proteins in a *dnaK* mutant of *Lactococcus lactis*. J. Bacteriol. 180: 3873–3881.

Koebmann B J, Nilsson D, Kuipers O P & Jensen P R (2000) The membrane-bound H(+)-ATPase complex is essential for growth of *Lactococcus lactis*. J. Bacteriol. 182: 4738–4743.

Komatsu Y, Kaul S C, Iwahashi H & Obuchi K (1990) Do heat shock proteins provide protection against freezing? FEMS Microbiol. Lett. 60: 159–162.

Konings W N, Lolkema J S, Bolhuis H, van Veen H W, Poolman B & Driessen A J (1997) The role of transport processes in survival of lactic acid bacteria. Energy transduction and multidrug resistance. Antonie Van Leeuwenhoek. 71: 117–128.

Kono Y & Fridovich I (1983) Isolation and characterization of the pseudocatalase of *Lactobacillus plantarum*. J. Biol. Chem. 258: 6015–6019.

Koonin E V, Aravind L & Glaperin M Y (2000) A comparative-genomic view of the microbial stress response. *In*: Stortz G & Hengge-Aronis R (Eds.) Bacterial Stress Responses (pp 417–444). ASM Press, Washington, DC.

Kornberg A, Rao N N & Ault-Riche D (1999) Inorganic polyphosphate: a molecule of many functions. Annu. Rev. Biochem. 68: 89–125.

Kremer B H, van der Kraan M, Crowley P J, Hamilton I R, Brady L J & Bleiweis A S (2001) Characterization of the sat operon in *Streptococcus mutans*: evidence for a role of Ffh in acid tolerance. J. Bacteriol. 183: 2543–2552.

Kullen M J & Klaenhammer T R (1999) Identification of the pH-inducible, proton-translocating F1F0-ATPase (atpBEFHAGDC) operon of *Lactobacillus acidophilus* by differential display: gene structure, cloning and characterization. Mol. Microbiol. 33: 1152–1161.

Kunji E R, Ubbink T, Matin A, Poolman B & Konings W N (1993) Physiological responses of *Lactococcus lactis* ML3 to alternating conditions of growth and starvation. Arch. Microbiol. 159: 372–379.

Kuroda A, Nomura K, Ohtomo R, Kato J, Ikeda T, Takiguchi N, Ohtake H & Kornberg A (2001) Role of inorganic polyphosphate in promoting ribosomal protein degradation by the Lon protease in *E. coli*. Science 293: 705–708.

Kvint K, Farewell A & Nystrom T (2000) RpoS-dependent promoters require guanosine tetraphosphate for induction even in the presence of high levels of sigma(s). J. Biol. Chem. 275: 14795–14798.

Lange R & Hengge-Aronis R (1991) Growth phase-regulated expression of bolA and morphology of stationary-phase *Escherichia coli* cells are controlled by the novel sigma factor sigma S. J. Bacteriol. 173: 4474-4481.

Laport M S, de Castro A C, Villardo A, Lemos J A, Bastos M C & Giambiagi-deMarval M (2001) Expression of the major heat shock proteins DnaK and GroEL in *Streptococcus pyogenes*: a comparison to *Enterococcus faecalis* and *Staphylococcus aureus*. Curr. Microbiol. 42: 264–268.

Lawrence J G, Hendrix R W & Casjens S (2001) Where are the pseudogenes in bacterial genomes? Trends Microbiol. 9: 535–540.

Lemos J A, Chen Y Y & Burne R A (2001) Genetic and physiologic analysis of the groE operon and role of the HrcA repressor in stress gene regulation and acid tolerance in *Streptococcus mutans*. J. Bacteriol. 183: 6074–6084.

Li Y H, Hanna M N, Svensater G, Ellen R P & Cvitkovitch D G (2001) Cell density modulates acid adaptation in *Streptococcus mutans*: implications for survival in biofilms. J. Bacteriol. 183: 6875–6884.

Lim E M, Ehrlich S D & Maguin E (2000) Identification of stress-inducible proteins in *Lactobacillus delbrueckii* subsp. *bulgaricus*. Electrophoresis 21: 2557–2561.

Lin M Y & Yen C L (1999) Antioxidative ability of lactic acid bacteria. J. Agric. Food Chem. 47: 1460–1466.

Lindahl T & Nyberg B (1972) Rate of depurination of native deoxyribonucleic acid. Biochemistry 11: 3610–3618.

Liu S, Asmundson R V, Gopal P K, Holland R & Crow V L (1998) Influence of reduced water activity on lactose metabolism by *lactococcus lactis* subsp. *cremoris* at different pH values. Appl. Environ. Microbiol. 64: 2111–2116.

Loewen P C, Hu B, Strutinsky J & Sparling R (1998) Regulation in the rpoS regulon of *Escherichia coli*. Can. J. Microbiol. 44: 707–717.

Lomovskaya O & Lewis K (1992) Emr, an *Escherichia coli* locus for multidrug resistance. Proc. Natl. Acad. Sci. U.S.A. 89: 8938–8942.

Lopez de Felipe F, Kleerebezem M, de Vos W M & Hugenholtz J (1998) Cofactor engineering: a novel approach to metabolic engineering in *Lactococcus lactis* by controlled expression of NADH oxidase. J. Bacteriol. 180: 3804–3808.

Lorca G L & de Valdez G F (1999) The effect of suboptimal growth temperature and growth phase on resistance of *Lactobacillus acidophilus* to environmental stress. Cryobiology 39: 144–149.

Lorca G L & Valdez G F (2001) A low-pH-inducible, stationary-phase acid tolerance response in *Lactobacillus acidophilus* CRL 639. Curr. Microbiol. 42: 21–25.

Ma Y & Marquis R E (1997) Thermophysiology of *Streptococcus mutans* and related lactic-acid bacteria. Antonie Van Leeuwenhoek 72: 91–100.

Ma D, Cook D N, Hearst J E & Nikaido H (1994) Efflux pumps and drug resistance in gram-negative bacteria. Trends Microbiol. 2: 489–493.

Ma Y, Curran T M & Marquis R E (1997) Rapid procedure for acid adaptation of oral lactic-acid bacteria and further characterization of the response. Can. J. Microbiol. 43: 143–148.

Mallonee D H & Hylemon P B (1996) Sequencing and expression of a gene encoding a bile acid transporter from *Eubacterium* sp. strain VPI 12708. J. Bacteriol. 178: 7053–7058.

Markham P N & Neyfakh A A (2001) Efflux-mediated drug resistance in Gram-positive bacteria. Curr. Opin. Microbiol. 4: 509–514.

Marquis R E, Bender G R, Murray D R & Wong A (1987) Arginine deiminase system and bacterial adaptation to acid environments. Appl. Environ. Microbiol. 53: 198–200.

Martin-Galiano A J, Ferrandiz M J & de la Campa A G (2001) The promoter of the operon encoding the F0F1 ATPase of *Streptococcus pneumoniae* is inducible by pH. Mol. Microbiol. 41: 1327–1338.

Martirani L, Raniello R, Naclerio G, Ricca E & De Felice M (2001) Identification of the DNA-binding protein, HrcA, of *Streptococcus thermophilus*. FEMS Microbiol. Lett. 198: 177–182.

Marty-Teysset C, Posthuma C, Lolkema J S, Schmitt P, Divies C & Konings W N (1996) Proton motive force generation by citrolactic fermentation in *Leuconostoc mesenteroides*. J. Bacteriol. 178: 2178–2185.

Marty-Teysset C, de la Torre F & Garel J (2000) Increased production of hydrogen peroxide by *Lactobacillus delbrueckii* subsp. *bulgaricus* upon aeration: involvement of an NADH oxidase in oxidative stress. Appl. Environ. Microbiol. 66: 262–267.

Mascarenhas J, Weber M H & Graumann P L (2001) Specific polar localization of ribosomes in *Bacillus subtilis* depends on active transcription. EMBO Rep. 2: 685–689.

Mayo B, Derzelle S, Fernandez M, Leonard C, Ferain T, Hols P, Suarez J E & Delcour J (1997) Cloning and characterization of cspL and cspP, two cold-inducible genes from *Lactobacillus plantarum*. J. Bacteriol. 179: 3039–3042.

Mechold U, Cashel M, Steiner K, Gentry D & Malke H (1996) Functional analysis of a relA/spoT gene homolog from *Streptococcus equisimilis*. J. Bacteriol. 178: 1401–1411.

Mercenier A, Muller-Alouf H & Grangette C (2000) Lactic acid bacteria as live vaccines. Curr. Issues Mol. Biol. 2: 17–25.

Miyoshi A, Gratadoux J J, Azevedo V, Rochat T, Duwat P, Sourice S, Oliveira S C, Gruss A & Langella P (2002) Expression of heterologous catalases confers high-level resistance to oxidative stress in *Lactococcus lactis*. In preparation.

Mogk A, Homuth G, Scholz C, Kim L, Schmid F X & Schumann W (1997) The GroE chaperonin machine is a major modulator of the CIRCE heat shock regulon of *Bacillus subtilis*. EMBO J. 16: 4579–4590.

Molenaar D, Bosscher J S, ten Brink B, Driessen A J & Konings W N (1993) Generation of a proton motive force by histidine decarboxylation and electrogenic histidine/histamine antiport in *Lactobacillus buchneri*. J. Bacteriol. 175: 2864–2870.

Moser S A & Savage D C (2001) Bile salt hydrolase activity and resistance to toxicity of conjugated bile salts are unrelated properties in lactobacilli. Appl. Environ. Microbiol. 67: 3476–3480.

Nannen N L & Hutkins R W (1991) Proton translocating adenosine triphosphatase activity in lacic acid bacteria. J. Dairy Sci. 74: 747–751.

Nilsson D, Lauridsen A A, Tomoyasu T & Ogura T (1994) A *Lactococcus lactis* gene encodes a membrane protein with putative ATPase activity that is homologous to the essential *Escherichia coli* ftsH gene product. Microbiology 140: 2601–2610.

Nishino K & Yamaguchi A (2001) Analysis of a complete library of putative drug transporter genes in *Escherichia coli*. J. Bacteriol. 183: 5803–5812.

Obis D, Guillot A, Gripon J C, Renault P, Bolotin A & Mistou M Y (1999) Genetic and biochemical characterization of a high-affinity betaine uptake system (BusA) in *Lactococcus lactis* reveals a new functional organization within bacterial ABC transporters. J. Bacteriol. 181: 6238–6246.

O'Connell-Motherway M, van Sinderen D, Morel-Deville F, Fitzgerald G F, Ehrlich S D & Morel P (2000) Six putative two-component regulatory systems isolated from *Lactococcus lactis* subsp. cremoris MG1363. Microbiology 146: 935–947.

Olsen E B, Russell J B & Henick-Kling T (1991) Electrogenic L-malate transport by *Lactobacillus plantarum*: a basis for energy derivation from malolactic fermentation. J. Bacteriol. 173: 6199–6206.

O'Sullivan E & Condon S (1999) Relationship between acid tolerance, cytoplasmic pH, and ATP and H^+-ATPase levels in chemostat cultures of *Lactococcus lactis*. Appl. Environ. Microbiol. 65: 2287–2293.

Panoff J M, Legrand S, Thammavongs B & Boutibonnes P (1994) The cold shock response in *Lactococcus* subsp. *lactis*. Curr. Microbiol. 29: 213–216.

Panoff J M, Thammavongs B, Laplace J M, Hartke A, Boutibonnes P & Auffray Y (1995) Cryotolerance and cold adaptation in *Lactococcus lactis* subsp. *lactis* IL1403. Cryobiology 32: 516–520.

Panoff J M, Corroler D, Thammavongs B & Boutibonnes P (1997) Differentiation between cold shock proteins and cold acclimation proteins in a mesophilic gram-positive bacterium, *Enterococcus faecalis* JH2-2. J. Bacteriol. 179: 4451–4454.

Panoff J M, Thammavongs B & Gueguen M (2000) Cryoprotectants lead to phenotypic adaptation to freeze–thaw stress in *Lactobacillus delbrueckii* ssp. *bulgaricus* CIP 101027T. Cryobiology 40: 264–269.

Pebay M, Holl A C, Simonet J M & Decaris B (1995) Characterization of the gor gene of the lactic acid bacterium *Streptococcus thermophilus* CNRZ368. Res. Microbiol. 146: 371–383.

Perrin C, Guimont C, Bracquart P & Gaillard J L (1999) Expression of a new cold shock protein of 21.5 kDa and of the major cold shock protein by *Streptococcus thermophilus* after cold shock. Curr. Microbiol. 39: 342–347.

Persuh M, Turgay K, Mandic-Mulec I & Dubnau D (1999) The N- and C-terminal domains of MecA recognize different partners in the competence molecular switch. Mol. Microbiol. 33: 886–894.

Petersohn A, Brigulla M, Haas S, Hoheisel J D, Volker U & Hecker M (2001) Global analysis of the general stress response of *Bacillus subtilis*. J. Bacteriol. 183: 5617–5631.

Phadtare S & Inouye M (1999) Sequence-selective interactions with RNA by CspB, CspC and CspE, members of the CspA family of *Escherichia coli*. Mol. Microbiol. 33: 1004–1014.

Phadtare S, Yamanata K & Inouye M (2000) The cold-shock response. *In*: Stortz G & Hengge-Aronis R (Eds.) Bacterial Stress Response (pp 33–45). ASM Press, Washington, DC.

Poolman B, Nijssen R M & Konings W N (1987a) Dependence of *Streptococcus lactis* phosphate transport on internal phosphate concentration and internal pH. J. Bacteriol. 169: 5373–5378.

Poolman B, Smid E J, Veldkamp H & Konings W (1987b) Bioenergetic consequences of lactose starvation for continuously cultured *Streptococcus cremoris* and *Streptococcus lactis*. J. Bacteriol. 169: 1460–1468.

Poolman B, Molenaar D, Smid E J, Ubbink T, Abee T, Renault P P & Konings W N (1991) Malolactic fermentation: electrogenic malate uptake and malate/lactate antiport generate metabolic energy. J. Bacteriol. 173: 6030–6037.

Poolman B & Glaasker E (1998) Regulation of compatible solute accumulation in bacteria. Mol. Microbiol. 29: 397–407.

Poquet I, Saint V, Seznec E, Simoes N, Bolotin A & Gruss A (2000) HtrA is the unique surface housekeeping protease in *Lactococcus lactis* and is required for natural protein processing. Mol. Microbiol. 35: 1042–1051.

Presser K A, Ratkowsky D A & Ross T (1997) Modelling the growth rate of *Escherichia coli* as a function of pH and lactic acid concentration. Appl. Environ. Microbiol. 63: 2355–2360.

Price C W, Fawcett P, Ceremonie H, Su N, Murphy C K & Youngman P (2001) Genome-wide analysis of the general stress response in *Bacillus subtilis*. Mol. Microbiol. 41: 757–774.

Prieto-Alamo M J, Jurado J, Gallardo-Madueno R, Monje-Casas F, Holmgren A & Pueyo C (2000) Transcriptional regulation of glutaredoxin and thioredoxin pathways and related enzymes in response to oxidative stress. J. Biol. Chem. 275: 13398–13405.

Provenzano D & Klose K E (2000) Altered expression of the ToxR-regulated porins OmpU and OmpT diminishes *Vibrio cholerae* bile resistance, virulence factor expression, and intestinal colonization. Proc. Natl. Acad. Sci. U.S.A. 97: 10220–10224.

Putman M, van Veen H W & Konings W N (2000) Molecular properties of bacterial multidrug transporters. Microbiol. Mol. Biol. Rev. 64: 672–693.

Quivey Jr. R G, Faustoferri R C, Clancy K A & Marquis R E (1995) Acid adaptation in *Streptococcus mutans* UA159 alleviates sensitization to environmental stress due to RecA deficiency. FEMS Microbiol. Lett. 126: 257–261.

Quivey Jr. R G, Faustoferri R, Monahan K & Marquis R (2000a) Shifts in membrane fatty acid profiles associated with acid adaptation of *Streptococcus mutans*. FEMS Microbiol. Lett. 189: 89–92.

Quivey Jr. R G, Kuhnert W L & Hahn K (2000b) Adaptation of oral streptococci to low pH. Adv. Microb. Physiol. 42: 239–274.

Quivey R G, Kuhnert W L & Hahn K (2001) Genetics of acid adaptation in oral streptococci. Crit. Rev. Oral Biol. Med. 12: 301–314.

Rallu F, Gruss A & Maguin E (1996) *Lactococcus lactis* and stress. Antonie van Leeuwenhoek 70: 243–251.

Rallu F, Gruss A, Ehrlich S D & Maguin E (2000) Acid- and multistress-resistant mutants of *Lactococcus lactis*: identification of intracellular stress signals. Mol. Microbiol. 35: 517–528.

Rao N N & Kornberg A (1999) Inorganic polyphosphate regulates responses of *Escherichia coli* to nutritional stringencies, environmental stresses and survival in the stationary phase. Prog. Mol. Subcell. Biol. 23: 183–195.

Rastogi V K & Girvin M E (1999) Structural changes linked to proton translocation by subunit c of the ATP synthase. Nature 402: 263–268.

Renault P, Gaillardin C & Heslot H (1988) Role of malolactic fermentation in lactic acid bacteria. Biochimie 70: 375–379.

Rince A, Flahaut S & Auffray Y (2000) Identification of general stress genes in *Enterococcus faecalis*. Int. J. Food Microbiol. 55: 87–91.

Roméo Y, Gotierrez C & Mistou M Y (2002) Osmoregulation in *Lactococcus lactis*. BusR, a transcriptional repressor of the glycine betaine uptake system BusA. Submitted for publication.

Russell D W & Setchell K D (1992) Bile acid biosynthesis. Biochemistry 31: 4737–4749.

Sakamoto K, Margolles A, van Veen H W & Konings W N (2001) Hop resistance in the beer spoilage bacterium *Lactobacillus brevis* is mediated by the ATP-binding cassette multidrug transporter HorA. J. Bacteriol. 183: 5371–5375.

Salema M, Lolkema J S, San Romao M V & Lourero Dias M C (1996) The proton motive force generated in *Leuconostoc oenos* by L-malate fermentation. J. Bacteriol. 178: 3127–3132.

Salotra P, Singh D K, Seal K P, Krishna N, Jaffe H & Bhatnagar R (1995) Expression of DnaK and GroEL homologs in *Leuconostoc esenteroides* in response to heat shock, cold shock or chemical stress. FEMS Microbiol. Lett. 131: 57–62.

Sambongi Y, Iko Y, Tanabe M, Omote H, Iwamoto-Kihara A, Ueda I, Yanagida T, Wada Y & Futai M (1999) Mechanical rotation of the c subunit oligomer in ATP synthase (F0F1): direct observation. Science 286: 1722–1724.

Sami M, Yamashita H, Hirono T, Kadokura H, Kitamoto K, Yoda K & Yamasaki M (1997) Hop-resistant *Lactobacillus brevis* contains a novel plasmid harboring a multidrug resistance-like gene. J. Ferment. Bioeng. 84: 1–6.

Sancar A (1996) DNA excision repair. Annu. Rev. Biochem. 65: 43–81.

Sanders J W, Leenhouts K J, Haandrikman A J, Venema G & Kok J (1995) Stress response in *Lactococcus lactis*: cloning, expression analysis, and mutation of the lactococcal superoxide dismutase gene. J. Bacteriol. 177: 5254–5260.

Sanders J W, Leenhouts K, Burghoorn J, Brands J R, Venema G & Kok J (1998a) A chloride-inducible acid resistance mechanism in *Lactococcus lactis* and its regulation. Mol. Microbiol. 27: 299–310.

Sanders J W, Venema G, Kok J & Leenhouts K (1998b) Identification of a sodium chloride-regulated promoter in *Lactococcus lactis* by single-copy chromosomal fusion with a reporter gene. Mol. Gen. Genet. 257: 681-685.

Sanders J W, Venema G, Kok J (1999) Environmental stress responses in *Lactococcus lactis*. FEMS Microbiol. Rev. 23: 483–501.

Schiffrin E J & Blum S (2001) Food processing: probiotic microorganisms for beneficial foods. Curr. Opin. Biotechnol. 12: 499–502.

Schmidt G, Hertel C & Hammes W P (1999) Molecular characterisation of the dnaK operon of *Lactobacillus sakei* LTH681. Syst. Appl. Microbiol. 22: 321–328.

Scott C, Guest J R & Green J (2000a) Characterization of the *Lactococcus lactis* transcription factor FlpA and demonstration of an in vitro switch. Mol. Microbiol. 35: 1383–1393.

Scott C, Rawsthorne H, Upadhyay M, Shearman C A, Gasson M J, Guest J R & Green J (2000b) Zinc uptake, oxidative stress and the FNR-like proteins of *Lactococcus lactis*. FEMS Microbiol. Lett. 192: 85–89.

Shibata C, Ehara T, Tomura K, Igarashi K & Kobayashi H (1992) Gene structure of *Enterococcus hirae* (*Streptococcus faecalis*)

F1F0-ATPase, which functions as a regulator of cytoplasmic pH. J. Bacteriol. 174: 6117–6124.

Sijpesteijn A K (1970) Induction of cytochrome formation and stimulation of oxidative dissimilation by hemin in *Streptococcus lactis* and *Leuconostoc mesenteroides*. Antonie Van Leeuwenhoek 36: 335–348.

Skinner M M & Trempy J E (2001) Expression of clpX, an ATPase subunit of the Clp protease, is heat and cold shock inducible in *Lactococcus lactis*. J. Dairy Sci. 84: 1783–1785.

Small P L & Waterman S R (1998) Acid stress, anaerobiosis and gadCB: lessons from *Lactococcus lactis* and *Escherichia coli*. Trends Microbiol. 6: 214–216.

Smeds A, Varmanen P & Palva A (1998) Molecular characterization of a stress-inducible gene from *Lactobacillus helveticus*. J. Bacteriol. 180: 6148–6153.

Smith A J, Quivey Jr. R G & Faustoferri R C (1996) Cloning and nucleotide sequence analysis of the *Streptococcus mutans* membrane-bound, proton-translocating ATPase operon. Gene 183: 87–96.

Somero G N (1995) Proteins and temperature. Annu. Rev. Physiol. 57: 43–68.

Spiess C, Beil A & Ehrmann M (1999) A temperature-dependent switch from chaperone to protease in a widely conserved heat shock protein. Cell 97: 339–347.

Steiner K & Malke H (2000) Life in protein-rich environments: the relA-independent response of *Streptococcus pyogenes* to amino acid starvation. Mol. Microbiol. 38: 1004–1016.

Steiner K & Malke H (2001) relA-Independent amino acid starvation response network of *Streptococcus pyogenes*. J. Bacteriol. 183: 7354–7364.

Stewart E J, Aslund F & Beckwith J (1998) Disulfide bond formation in the *Escherichia coli* cytoplasm: an in vivo role reversal for the thioredoxins. EMBO J. 17: 5543–5550.

Stiles M E (1996) Biopreservation by lactic acid bacteria. Antonie Van Leeuwenhoek 70: 331–345.

Stock D, Leslie A G & Walker J E (1999) Molecular architecture of the rotary motor in ATP synthase. Science 286: 1700–1705.

Stortz G & Hengge-Aronis R (Eds.) (2000) Bacterial Stress Responses (p 485). ASM Press, Washington, DC.

Stuart M R, Chou L S & Weimer B C (1999) Influence of carbohydrate starvation and arginine on culturability and amino acid utilization of *Lactococcus lactis* subsp. *lactis*. Appl. Environ. Microbiol. 65: 665–673.

Suzuki T, Unemoto T & Kobayashi H (1988) Novel streptococcal mutants defective in the regulation of H^+-ATPase biosynhesis and in F0 complex. J. Biochem. Chem. 263: 11840–11843

Svensater G, Sjogreen B & Hamilton I R (2000) Multiple stress responses in *Streptococcus mutans* and the induction of general and stress-specific proteins. Microbiology 146: 107–117.

Takahashi N & Yamada T (1999) Acid-induced acid tolerance and acidogenicity of non-mutans streptococci. Oral Microbiol. Immunol. 14: 43–48.

Teixera P, Castro H, Mohacsi-Farkas C & Kirby R (1997) Identification of sites of injury in *Lactobacillus bulgaricus* during heat stress. J. Appl. Microbiol. 83: 219–226.

ten Brink B, Otto R, Hansen U P & Konings W N (1985) Energy recycling by lactate efflux in growing and nongrowing cells of *Streptococcus cremoris*. J. Bacteriol. 162: 383–390.

Tettelin H, Nelson K E, Paulsen I T, Eisen J A, Read T D, Peterson S, Heidelberg J, DeBoy R T, Haft D H, Dodson R J, Durkin A S, Gwinn M, Kolonay J F, Nelson W C, Peterson J D, Umayam L A, White O, Salzberg S L, Lewis M R, Radune D, Holtzapple E, Khouri H, Wolf A M, Utterback T R, Hansen C L, McDonald L A, Feldblyum T V, Angiuoli S, Dickinson T, Hickey E K, Holt I E, Loftus B J, Yang F, Smith H O, Venter J C, Dougherty B A, Morrison D A, Hollingshead S K & Fraser C M (2001) Complete genome sequence of a virulent isolate of *Streptococcus pneumoniae*. Science 293: 498–506.

Thammavongs B, Corroler D, Panoff J M, Auffray Y & Boutibonnes P (1996) Physiological response of *Enterococcus faecalis* JH2-2 to cold shock: growth at low temperatures and freezing/thawing challenge. Lett. Appl. Microbiol. 23: 398–402.

Thanassi D G, Cheng L W & Nikaido H (1997) Active efflux of bile salts by *Escherichia coli*. J. Bacteriol. 179: 2512–2518.

Tonon T & Lonvaud-Funel A (2000) Metabolism of arginine and its positive effect on growth and revival of *Oenococcus oeni*. J. Appl. Microbiol. 89: 526–531.

Tourdot-Marechal R, Fortier L C, Guzzo J, Lee B & Divies C (1999) Acid sensitivity of neomycin-resistant mutants of *Oenococcus oeni*: a relationship between reduction of ATPase activity and lack of malolactic activity. FEMS Microbiol. Lett. 178: 319–326.

Trainor V C, Udy R K, Bremer P J & Cook G M (1999) Survival of *Streptococcus pyogenes* under stress and starvation. FEMS Microbiol. Lett. 176: 421–428.

Turner M S, Woodberry T, Hafner L M & Giffard P M (1999) The bspA locus of *Lactobacillus fermentum* BR11 encodes an L-cystine uptake system. J. Bacteriol. 181: 2192–2198.

Uguen P, Hamelin J, Le Pennec J P & Blanco C (1999) Influence of osmolarity and the presence of an osmoprotectant on *Lactococcus lactis* growth and bacteriocin production. Appl. Environ. Microbiol. 65: 291–293.

van Asseldonk M, Simons A, Visser H, de Vos W M & Simons G (1993) Cloning, nucleotide sequence, and regulatory analysis of the *Lactococcus lactis* dnaJ gene. J. Bacteriol. 175: 1637–1644.

Van de Guchte M, Ehrlich S D & Maguin E (2001) Production of growth-inhibiting factors by *Lactobacillus delbrueckii*. J. Appl. Microbiol. 91: 147–153.

Van der Heide T & Poolman B (2000a) Glycine betaine transport in *Lactococcus lactis* is osmotically regulated at the level of expression and translocation activity. J. Bacteriol. 182: 203–206.

Van der Heide T & Poolman B (2000b) Osmoregulated ABC-transport system of *Lactococcus lactis* senses water stress via changes in the physical state of the membrane. Proc. Natl. Acad. Sci. U.S.A. 97: 7102–7106.

Van der Heide T, Stuart M C & Poolman B (2001) On the osmotic signal and osmosensing mechanism of an ABC transport system for glycine betaine. EMBO J. 20: 7022–7032.

van Veen H W, Putman M, Margolles A, Sakamoto K & Konings W N (1999) Structure-function analysis of multidrug transporters in *Lactococcus lactis*. Biochim. Biophys. Acta 1461: 201–206.

van Velkinburgh J C & Gunn J S (1999) PhoP-PhoQ-regulated loci are required for enhanced bile resistance in *Salmonella* spp. Infect. Immun. 67: 1614–1622.

VanBogelen R A, Greis K D, Blumenthal M, Tani T H & Matthews R (1999) Mapping regulatory networks in microbial cells. Trends Microbiol. 7: 320–328.

VanBogelen R A & Neidhardt F (1990) Ribosomes as sensors of heat and cold shock in *Escherichia coli*. Proc. Natl. Acad. Sci. U.S.A. 87: 5589–5593.

Varmanen P, Ingmer H & Vogensen F K (2000) ctsR of *Lactococcus lactis* encodes a negative regulator of clp gene expression. Microbiology 146: 1447–1455.

Voelker U, Voelker A, Maul B, Hecker M, Dufour A & Haldenwang W G (1995) Separate mechanisms activate sigma B of *Bacillus subtilis* in response to environmental and metabolic stresses. J. Bacteriol. 177: 3771–3780.

Volker U, Andersen K K, Antelmann H, Devine K M & Hecker M (1998) One of two osmC homologs in *Bacillus subtilis* is part of the sigmaB- dependent general stress regulon. J. Bacteriol. 180: 4212–4218.

Walker D C, Girgis H S & Klaenhammer T R (1999) The groESL chaperone operon of *Lactobacillus johnsonii*. Appl. Environ. Microbiol. 65: 3033–3041.

Wang N, Yamanaka K & Inouye M (1999) CspI, the ninth member of the CspA family of *Escherichia coli*, is induced upon cold shock. J. Bacteriol. 181: 1603–1609.

Weber M H, Beckering C L & Marahiel M A (2001a) Complementation of cold shock proteins by translation initiation factor IF1 in vivo. J. Bacteriol. 183: 7381–7386.

Weber M H, Volkov A V, Fricke I, Marahiel M A & Graumann P L (2001b) Localization of cold shock proteins to cytosolic spaces surrounding nucleoids in *Bacillus subtilis* depends on active transcription. J. Bacteriol. 183: 6435–6443.

Wehmeier L, Schafer A, Burkovski A, Kramer R, Mechold U, Malke H, Puhler A & Kalinowski J (1998) The role of the *Corynebacterium glutamicum* rel gene in (p)ppGpp metabolism. Microbiology 144: 1853–1862.

Wells J M, Robinson K, Chamberlain L M, Schofield K M & Le Page R W (1996) Lactic acid bacteria as vaccine delivery vehicles. Antonie Van Leeuwenhoek 70: 317–330.

Wendrich T M & Marahiel M A (1997) Cloning and characterization of a relA/spoT homologue from *Bacillus subtilis*. Mol. Microbiol. 26: 65–79.

Whitaker R D & Batt C A (1991) Characterization of the heat shock response in *Lactococcus lactis* subsp. *lactis*. Appl. Environ. Microbiol. 57: 1408–1412.

Whitehead K E, Webber G M & England R R (1998) Accumulation of ppGpp in *Streptococcus pyogenes* and *Streptococcus rattus* following amino acid starvation. FEMS Microbiol. Lett. 159: 21–26.

Wilkins J C, Homer K A & Beighton D (2001) Altered protein expression of *Streptococcus oralis* cultured at low pH revealed by two-dimensional gel electrophoresis. Appl. Environ. Microbiol. 67: 3396–3405.

Willimsky G, Bang H, Fischer G & Marahiel M A (1992) Characterization of cspB, a *Bacillus subtilis* inducible cold shock gene affecting cell viability at low temperatures. J. Bacteriol. 174: 6326–6335.

Wouters J A, Frenkiel H, de Vos W M, Kuipers O P & Abee T (2001) Cold shock proteins of *Lactococcus lactis* MG1363 are involved in cryoprotection and in the production of cold-induced proteins. Appl. Environ. Microbiol. 67: 5171–5178.

Wouters J A, Sanders J W, Kok J, de Vos W M, Kuipers O P & Abee T (1998) Clustered organization and transcriptional analysis of a family of five csp genes of *Lactococcus lactis* MG1363. Microbiology 144: 2885–2893.

Wouters J A, Jeynov B, Rombouts F M, de Vos W M, Kuipers O P & Abee T (1999a) Analysis of the role of 7 kDa cold-shock proteins of *Lactococcus lactis* MG1363 in cryoprotection. Microbiology 145: 3185–3194.

Wouters J A, Rombouts F M, de Vos W M, Kuipers O P & Abee T (1999b) Cold shock proteins and low-temperature response of *Streptococcus thermophilus* CNRZ302. Appl. Environ. Microbiol. 65: 4436–4442.

Wouters J A, Kamphuis H H, Hugenholtz J, Kuipers O P, de Vos W M & Abee T (2000a) Changes in glycolytic activity of *Lactococcus lactis* induced by low temperature. Appl. Environ. Microbiol. 66: 3686–3691.

Wouters J A, Mailhes M, Rombouts F M, de Vos W M, Kuipers O P & Abee T (2000b) Physiological and regulatory effects of controlled overproduction of five cold shock proteins of *Lactococcus lactis* MG1363. Appl. Environ. Microbiol. 66: 3756–3763.

Wouters J A, Rombouts F M, Kuipers O P, de Vos W M & Abee T (2000c) The role of cold-shock proteins in low-temperature adaptation of food-related bacteria [In Process Citation]. Syst. Appl. Microbiol. 23: 165–173.

Xia B, Ke H & Inouye M (2001) Acquirement of cold sensitivity by quadruple deletion of the cspA family and its suppression by PNPase S1 domain in *Escherichia coli*. Mol. Microbiol. 40: 179–188.

Yamamoto N, Masujima Y & Takano T (1996) Reduction of membrane bound ATPase activity in a *Lactobacillus helveticus* strain with slower growth at low pH. FEMS Microbiol. Rev. 138: 179–184.

Yamanaka K, Fang L & Inouye M (1998) The CspA family in *Escherichia coli*: multiple gene duplication for stress adaptation. Mol. Microbiol. 27: 247–255.

Yamanaka K, Mitta M & Inouye M (1999) Mutation analysis of the 5′ untranslated region of the cold shock cspA mRNA of *Escherichia coli*. J. Bacteriol. 181: 6284–6291.

Yamashita Y, Takehara T & Kuramitsu H K (1993) Molecular characterization of a *Streptococcus mutans* mutant altered in environmental stress responses. J. Bacteriol. 175: 6220–6228.

Yi X, Kot E & Bezkorovainy A (1998) Properties of NADH oxidase from *Lactobacillus delbrueckii* ssp. *bulgaricus*. J. Sci. Food Agric. 78: 527–534.

Yokota A, Amachi S, Ishii S & Tomita F (1995) Acid sensitivity of a mutan of *Lactococcus lactis* subsp. *lactis* C2 with reduced membrane bound ATPase activity. Biosci. Biotech. Biochem. 59: 2004–2007.

Yokota A, Veenstra M, Kurdi P, van Veen H W & Konings W N (2000) Cholate resistance in *Lactococcus lactis* is mediated by an ATP- dependent multispecific organic anion transporter. J. Bacteriol. 182: 5196–5201.

Yura T, Kanemori M & Morita T M (2000) The heat shock response: regulation and function. *In*: Storz G & Hengge-Aronis R (Eeds.) Bacterial Stress Responses (pp 3–18). ASM Press, Washington, DC.

Zhu M, Takenaka S, Sato M & Hoshino E (2001) Influence of starvation and biofilm formation on acid resistance of *Streptococcus mutans*. Oral Microbiol. Immunol. 16: 24–27.

Zuber U & Schumann W (1994) CIRCE, a novel heat shock element involved in regulation of heat shock operon dnaK of *Bacillus subtilis*. J. Bacteriol. 176: 1359–1363.

Zuniga M, Champomier-Verges M, Zagorec M & Perez-Martinez G (1998) Structural and functional analysis of the gene cluster encoding the enzymes of the arginine deiminase pathway of *Lactobacillus sake*. J. Bacteriol. 180: 4154–4159.

Antonie van Leeuwenhoek **82:** 217–235, 2002.

Metabolic engineering of lactic acid bacteria for the production of nutraceuticals

Jeroen Hugenholtz[1,2,*], Wilbert Sybesma[1], Masja Nierop Groot[1], Wouter Wisselink[1,6], Victor Ladero[4], Kay Burgess[3], Douwe van Sinderen[3], Jean-Christophe Piard[5], Gerrit Eggink[1,6], Eddy J. Smid[2], Graciela Savoy[7], Fernando Sesma[7], Tanja Jansen[2], Pascal Hols[4] & Michiel Kleerebezem[1,2]
[1]*Wageningen Centre for Food Sciences, the Netherlands;* [2]*NIZO Food Research, Ede, The Netherlands;* [3]*University College Cork, Ireland;* [4]*Universite Catholic de Louvain-la-Neuve, Belgium;* [5]*INRA Jouy-en Josas, France;* [6]*ATO-DLO, Wageningen, The Netherlands;* [7]*CERELA, Tucuman, Argentina (*Author for correspondence; Tel: +31-318-659540; Fax: +31-318-650400; E-mail: hugenhol@nizo.nl)*

Key words: folate, galactose, lactic acid bacteria, *Lactococcus lactis*, low-calorie sugars, mannitol, metabolic engineering, nutraceuticals, oligosaccharides, riboflavin, sorbitol, tagatose

Abstract

Lactic acid bacteria display a relatively simple and well-described metabolism where the sugar source is converted mainly to lactic acid. Here we will shortly describe metabolic engineering strategies on the level of sugar metabolism, that lead to either the efficient re-routing of the lactococcal sugar metabolism to nutritional end-products other than lactic acid such as L-alanine, several low-calorie sugars and oligosaccharides or to enhancement of sugar metabolism for complete removal of (undesirable) sugars from food materials. Moreover, we will review current metabolic engineering approaches that aim at increasing the flux through complex biosynthetic pathways, leading to the production of the B-vitamins folate and riboflavin. An overview of these metabolic engineering activities can be found on the website of the Nutra Cells 5th Framework EU-project (www.nutracells.com). Finally, the impact of the developments in the area of genomics and corresponding high-throughput technologies on nutraceutical production will be discussed.

Lactic acid bacteria as cell-factories

Lactic acid bacteria (LAB) are industrially important microbes that are used all over the world in a large variety of industrial food fermentations. Their contribution in these fermentation processes primarily consists of the formation of lactic acid from the available carbon source resulting in a rapid acidification of the food raw-material, which is a critical parameter in the preservation of these products. However, besides their lactic acid forming capacity, LAB also have the ability to contribute to other product characteristics like flavour, texture and nutrition. Next to their most important application, which is undoubtedly in the dairy industry, LAB are also applied at an industrial scale in the fermentation of other food-raw materials like meat and vegetables.

Lactococcus lactis is by far the most extensively studied lactic acid bacterium, and over the last decades elegant and efficient genetic tools have been developed. These tools are of critical importance in metabolic engineering strategies that aim at inactivation of undesired genes and/or (controlled) overexpression of existing or novel ones. Especially, the nisin controlled expression (NICE) system for controlled heterologous and homologous gene expression in Gram-positive bacteria (de Ruyter et al. 1996; Kleerebezem et al. 1997; Kuipers et al. 1998) was proven to be very valuable and has been employed in most of the metabolic engineering strategies discussed here (Hols et al. 1999; Hugenholtz et al. 2000; Boels et al. 2001; Sybesma et al. 2002).

Initial metabolic engineering of *Lactococcus lactis* has focussed on rerouting of pyruvate metabol-

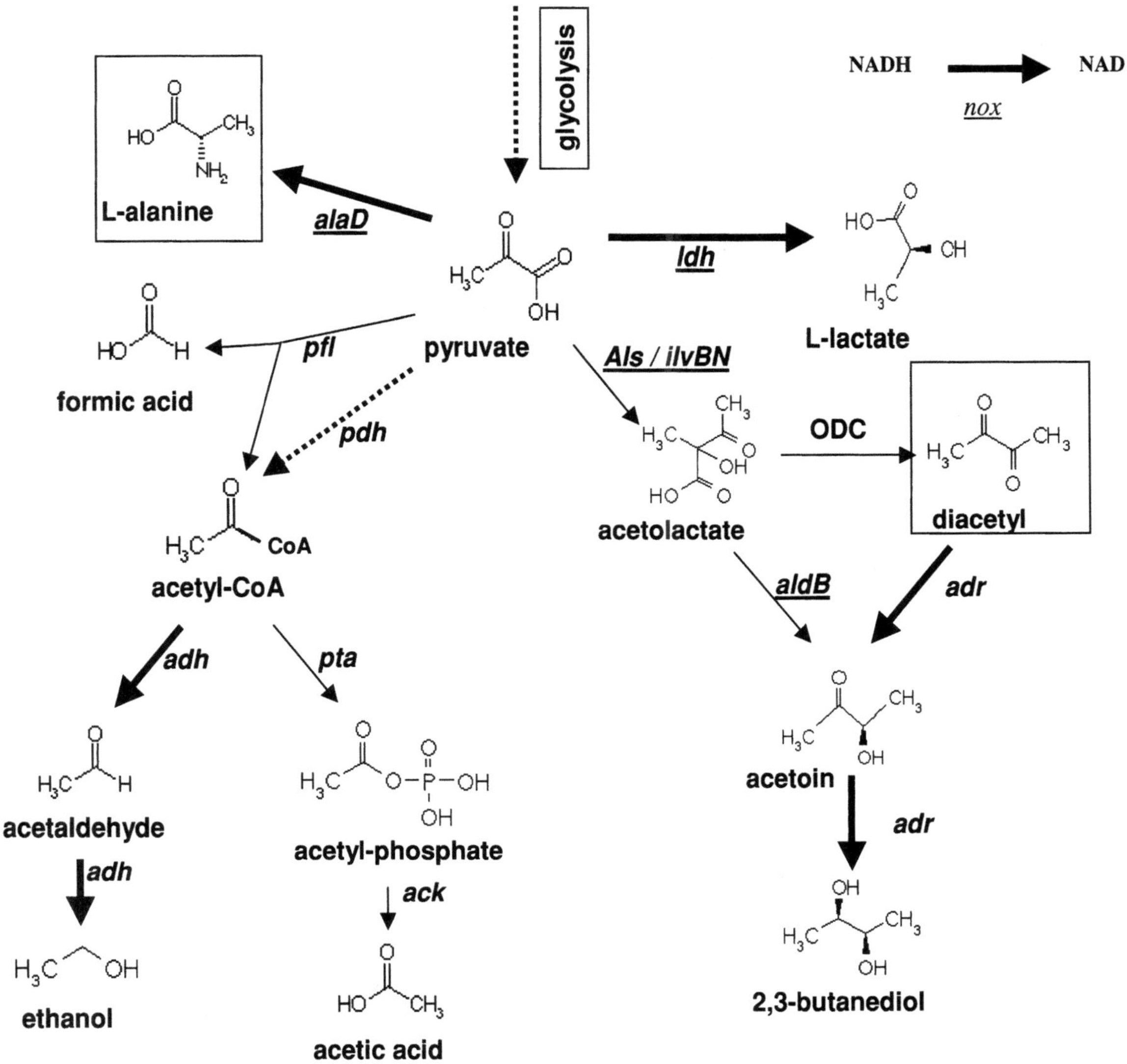

Figure 1. Schematic representation of pyruvate metabolism in *Lactococcus lactis*. NADH consuming (bold arrows) and producing (dotted arrows) and redox-neutral (normal arrows) conversions are indicated. The genes encoding the enzymes involved in individual conversions are indicated using abbreviations based on established genetic nomenclature. The heterologously introduced alanine dehydrogenase catalysed conversion (see text for further explanation) is added (top-left) and the conversion of acetolactate to diacetyl does not require an enzyme, but involves a chemical conversion (oxidative decarboxylation; ODC). Moreover, the reaction catalysed by the water-forming NADH-oxidase (see text for further explanation) is indicated. The metabolites that have been the target of metabolic engineering strategies (for more detail see text) are boxed, while the names of the genes that were targeted by genetic manipulations during these metabolic engineering procedures are underlined (for more detail see the text).

ism (Figure 1). Sugar metabolism was diverted away from lactate production towards production of α-acetolactate, the precursor of diacetyl, by either disruption of lactate dehydrogenase or overproduction of NADH oxidase. By combining this strategy with disruption of the *aldB* gene encoding for α-acetolactate decarboxylase, very effective diacetyl production from glucose and lactose was achieved. This example shows how a minor pathway in *Lactococcus lactis* could be boosted to become the major metabolic pathway, and how powerfull metabolic engineering can be in these, relatively simple, lactic acid bacteria. The extremely efficient production of L-alanine from sugar, was a subsequent achievement in metabolic engineering on the level of pyruvate metabolism. The (heterologous) *Bacillus sphaericus alaD* gene, encoding for alanine dehydrogenase enzyme, was cloned in *L. lactis*, using the NICE system (Hols et al. 1999). The alanine dehydrogenase enzyme converts pyruvate to L-alanine in the presence of ammonium (Figure 1). Upon introduction of this system in lactate dehydrogenase deficient lactococcal cells, a complete conversion of the pyruvate pool to alanine was obtained when cells were incubated under the appropriate, high ammonium, conditions (Hols et al. 1999). The alanine produced by these cells consisted of a mixture of both stereoisomers (D- and L-alanine) as a consequence of the endogenous alanine racemase activity of *L. lactis* (encoded by the *alr* gene). To obtain a stereo-specific L-alanine producing bioreactor, the *alr* gene was functionally disrupted in the lactate dehydrogenase deficient lactococcal strain. High level expression of alanine dehydrogenase in the resulting cells indeed led to stereo-specific L-alanine production as the only end-product of fermentation (Hols et al. 1999).

These two examples illustrate how the, basically simple, homofermentative metabolism of *Lactococcus lactis*, centred around lactate as sole fermentation product, can be redirected into efficient production of other, industrially important, (food) ingredients. In the remainder of this overview, we will present a number of examples where metabolic engineering of homofermentative lactic acid bacteria have led or will lead to efficient production of food ingredients (nutraceuticals) or food products with a beneficial effect for the consumer.

Nutraceuticals

The term 'Nutraceuticals', launched by Stephen DeFelici in the 1980s, defines a wide range of foods and food components with a claimed medical or health benefit (Pszczola 1992). An increasing number of food components are being labelled as such, often without proper scientific basis. In this overview, we will discuss the (increased) production of a range of metabolites produced by lactic acid bacteria, most of them definitely adding to the health status of (fermented) foods. Clear examples are the B-vitamins, riboflavin (B2), folate (B11) and cobalamin (B12). There is no doubt that these vitamins are essential in the human diet. The only discussion, especially for folate and cobalamin, centres on the protective role against a series of diseases, of increased dosage of these vitamins. Other examples of true nutraceuticals that can be produced by lactic acid bacteria are the low-calorie sweeteners. Production of alanine, a component with comparable sweetness as glucose and fructose was already discussed above. Production of sugars such as mannitol, sorbitol, and tagatose, as a result of metabolic engineering of lactic acid bacteria, will be described below. These sugars have similar sweetening power as sucrose and glucose, are all poorly degraded by the human body and, as such, can be considered low-calorie sugars. Oligosaccharides have a slightly weaker basis as true nutraceuticals. It is well established that some of these polysugars can actually stimulate growth of specific intestinal microorganisms, such as bifidobacteria, but a direct beneficial effect on the human health in more difficult to show. This holds even more true for food components such as antioxidants. Lactic acid bacteria are able to produce components such as glutathione and thioredoxin and overproduction is certainly feasible using metabolic engineering. However, the beneficial effect of increased intake of anti-oxidants on the human health is rather doubtful. In the final part of this overview, bacterial activities will be discussed that definitely contribute to the health benefit of food materials, but, as such, cannot really be considered nutraceuticals. We describe the use of the lactic acid bacteria to remove undesirable sugars present in the foods. Milk is a healthy and nutritious component of the human diet for only that part of the human population that is tolerant to the milk sugar, lactose. By removal of lactose (and galactose) from milk and liquid fermented dairy products, these food products would become suitable for the lactose-intolerant people. Strategies and actual examples will be shown where lactic acid bacteria are successfully used to achieve this goal. Similar solutions are considered for the raffinose and stachyose containing food materials, such as soybeans. Both these sugars are poorly metabolised by humans, and cause rapid growth of, often undesirable microorganisms, in the human intestine (Khane et al. 1994). Removal of these sugars from soy products would avoid intestinal problems after consumption.

Engineering of B-vitamin production

Folate

Folates are essential components in the human diet. They are involved, as cofactor, in many metabolic reactions, including the biosynthesis of nucleotides, the building blocks of DNA and RNA. The daily recommended intake for an adult is 200 μg. For pregnant women a double dose is recommended, since folates are known to prevent neural-tube defect in newly borns (Wald et al. 1991). Moreover, folates are reported to protect against some forms of cancer (Ames 1999). In contrast, a low folate level in the diet is associated with high homocysteine levels in the blood and, consequently, with coronary diseases (Boushey et al. 1996; Brattstrom 1996). Folates are produced in different (green) plants (folium (latin)=leaf) and by some micro-organisms. Therefore, vegetables and dairy products are the main source of folates for humans. Milk is a well-known source of folate. It contains between 20 and 50 μg/L folate and thus contributes significantly to the daily requirement of the average human. Fermented milk products, especially yoghurt, are reported to contain even higher amounts of folate (Alm 1980).

The term 'folate' is a non-specific term referring to any folate compound with vitamin activity. In the general metabolism, folates act as acceptors or donors of C1-residues. Folate is isolated from natural sources in many different forms, in which the basic structure consists of a pteridine moiety, a *p*-aminobenzoic acid residue and one or more γ-linked L-glutamic acid residues. However, variation of the C1-residues (including methyl-, methenyl-, methylene-, formimino- or formyl-) substitutions at either the N-5 or N-10 position, combined with the variable lengths of the poly-γ-L-glutamate 'tail' (ranging from a single glutamyl residue to more than 10 residues) results in a large variety of chemical variants that are all clustered under the term 'folate' (Hamm-Alvarez et al. 1989). Finally, the pyrazine rings can be reduced to the so-called dihydro- and tetrahydro-form of folate. The polyglutamyl 'tail' is required for the intracellular retention of folate derivatives and plays a role in the binding efficiency of folate and substrates to enzymes involved in one-carbon transfer or inter-conversion of folic acid derivatives (McBurney & Whitmore 1974). Most mammals are not able to synthesise folates. Plants and more particular leafy vegetables and fruits and fermented dairy products are the main source of folates, but animal products like meat and milk also contain folate. Folates are taken up by mammals, including humans, in the small intestines. Only monoglutamyl-folate derivatives can directly be absorbed in the human gut. Polyglutamyl-folate derivatives with two or more glutamate residues have to be processed in the intestines by a mammalian enzyme deconjugase to release monoglutamyl-folate. The activity of the deconjugase enzymes is susceptible to inhibition by various constituents found in some foods (Rosenberg & Goldwin 1971; Bhandari & Gregory 1990; Seyoum & Selhub 1998). The level of monoglutamyl folates (MGF), and hence the bio-availability of folate, varies considerably between food products, i.e. egg yolk (72% MGF), cow liver (56% MGF), orange juice (21% MGF), cabbage (6% MGF), lima beans (5% MGF), and lettuce (less than 1% MGF, Seyoum & Selhub 1998).

Compared to vegetable sources, dairy and other fermented products have the advantage that at least part of the folate is stored intracellularly in (lactic acid) bacteria. As a result, the intracellularly stored folate – up to mM concentrations in the cytoplasm (Sybesma, personal communication) – may be better protected against the acid-peptic environment of the stomach, leading to a higher effective folate consumption.

Many fermented dairy products contain higher folate levels compared to raw milk. The explanation for this is that some of the starter bacteria used in dairy fermentation execute *de novo* biosynthesis of folates and secrete a surplus of these compounds into the growth medium. Examples are the yoghurt bacterium *Streptococcus thermophilus* and the cheese and butter(milk) bacterium *Lactococcus lactis*. This property of LAB offers the possibility to fortify fermented dairy products with folate by natural means, i.e. without the addition of food supplements. The natural diversity amongst dairy starter culture with respect to their capacity to produce folate can be exploited to design new complex starter cultures, which yield fermented dairy products with elevated folate levels. Using this strategy, production of experimental yoghurt containing up to 150 μg/L folate has been reported (Smid et al. 2001). Besides selecting vitamin-producing lactic acid bacteria for inclusion in starter cultures, these lactic acid bacteria can also be supplied as probiotics or health-promoting additives in food or pharmaceutical products. These new applications of lactic acid bacteria will be boosted even more if folate production can be improved through optimisation of fermentation conditions or if LAB-variants (resulting from classical mutagenesis) with improved folate production capacities can be selected.

Total folate production in *L. lactis* strain MG1363 is approximately 100 ng/ml. Most of the folate (90%) is accumulated intracellularly, probably due to polyglutamylation of these intracellular folates. Only a small amount (10%) is released into the environment (Sybesma et al. 2002). Folate biosynthesis involves a multi-enzyme pathway, which is schematically presented in Figure 2. The genes involved in folate biosynthesis in *L. lactis* MG1363 have re-

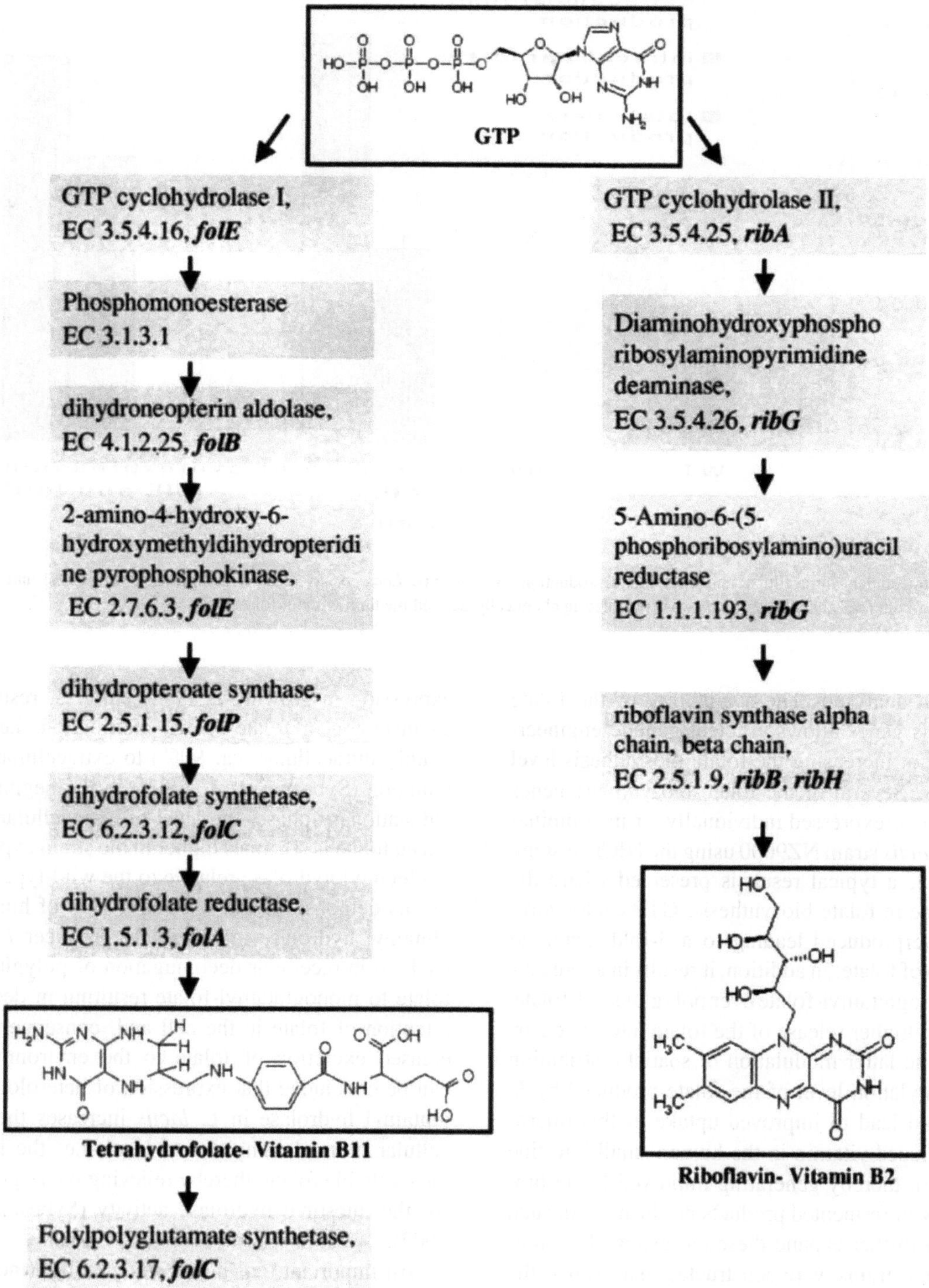

Figure 2. Biosynthetic pathway for folate (Vitamin B11) and for riboflavin (vitamin B2) (see text for more detail).

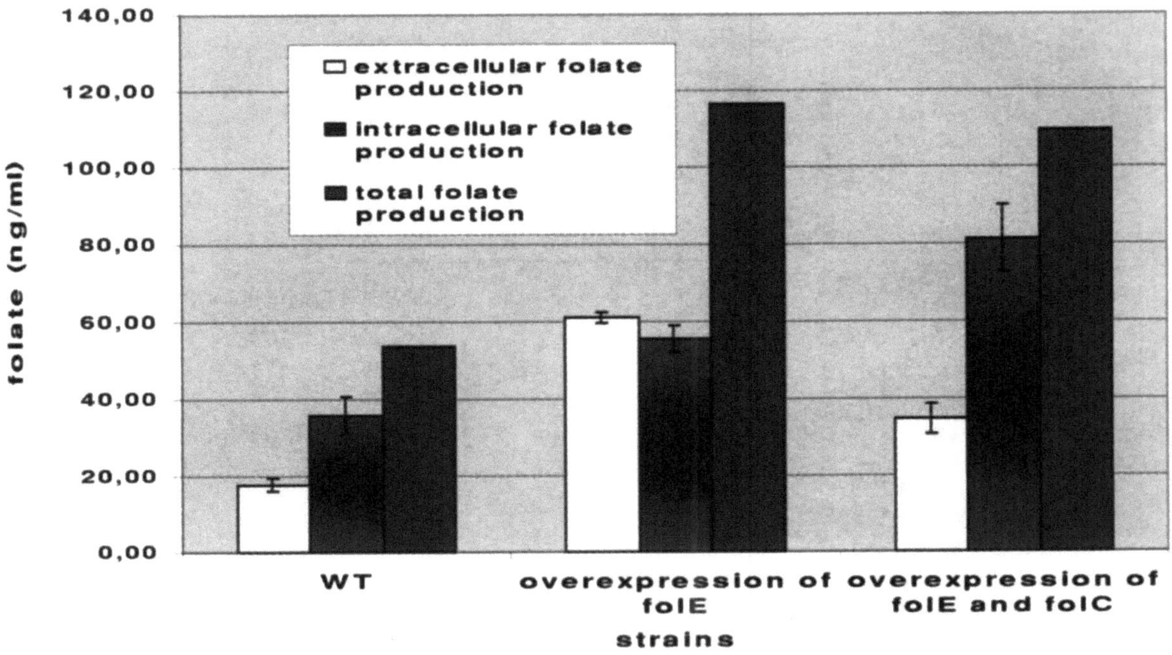

Figure 3. Extracellular, intracellular and total folate production by wild-type *Lactococcus lactis* overexpressing *folE (gch)* and *L. lactis* overexpressing both *folE* and *folC*. *L. lactis* was cultivated in chemically defined medium without folate.

cently been analysed. The availability of the folate biosynthesis genes allows efficient genetic engineering aiming at increasing the folate biosynthesis level in *L. lactis*. Several of the folate biosynthetic genes have been overexpressed individually, or in combination, in *L. lactis* strain NZ9000 using the NICE system. In Figure 3, a typical result is presented where the first enzyme in folate biosynthesis, GTP cyclohydrolase, is overproduced leading to a 3-fold increased production of folate. In addition, it results in altered ratios of monoglutamyl-folate over polyglutamyl-folate, resulting in higher release of the folate into the environment. The latter modulation of spatial distribution and glutamylation level of the folate produced by *L. lactis* would lead to improved uptake of this microbially produced vitamin in the human small intestine (see above), thereby generating improved folate bioavailability in fermented products produced with such strains. To further expand these bio-availability studies, *L. lactis* strains were constructed that express the γ-glutamyl hydrolase enzyme derived from rat or human origin (Yao et al. 1996). cDNA clones encoding these enzymes were cloned in the NICE system and introduced in *L. lactis* and it could be shown that both the human and rat enzymes could be expressed functionally in this organism. The γ-glutamyl hydrolase expression in growing *L. lactis* cultures, resulted in an inversion of folate spatial distribution, i.e. from mainly intracellular (ca. 90%) to extracellular accumulation (Sybesma et al. 2002). In the beginning of the stationary phase, the level of extracellularly produced folate is six times higher in the strain expressing γ-glutamyl hydrolase relative to the wild-type. These results demonstrate that the expression of human γ-glutamyl hydrolase in the folate producer *L. lactis* leads to intracellular deconjugation of polyglutamyl-folate to monoglutamyl-folate resulting in decreased retention of folate in the cell and, consequently, increased excretion of folate to the environment. It can be concluded that expression of heterologous γ-glutamyl hydrolase in *L. lactis* increases the extracellular monoglutamyl folate level, i.e. the level of bio-avalaible folate, thereby relieving the requirement for the intestinal hydrolase activity (Sybesma et al. 2002).

An important tool in the study of the (increased) bioavailability of folate, is a good methodology to separate and quantify the different folate molecules. HPLC-methods are now available to distinguish folates with different bound C1-moieties and with varying lengths of the polyglutamate-tail (Figure 4). Formyl-tetrahydrofolate and methylene-

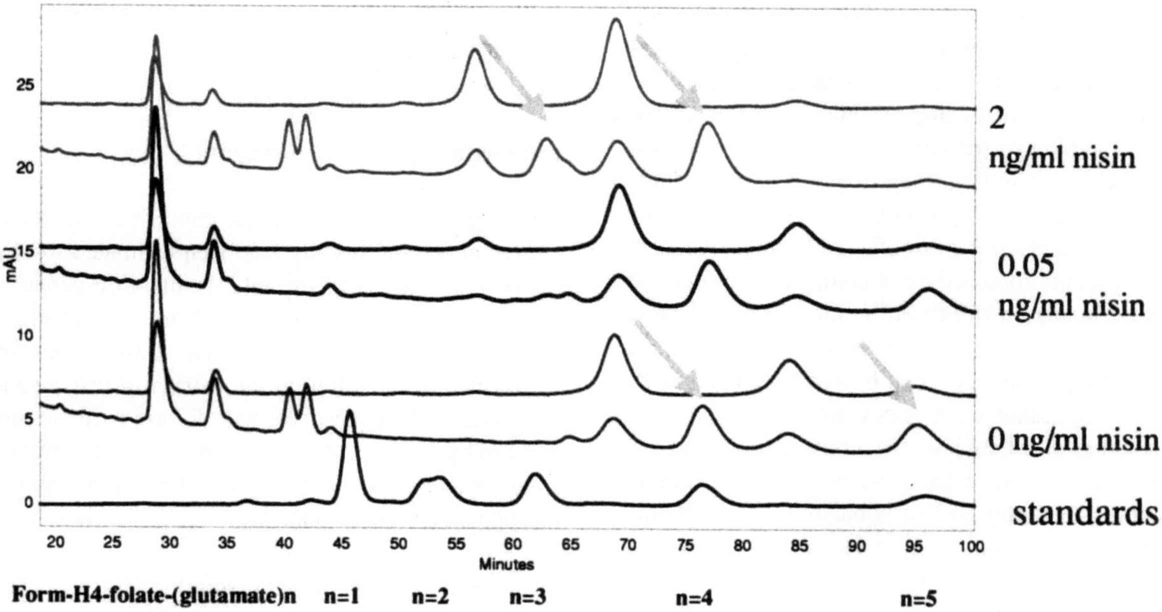

Figure 4. HPLC-analysis of intracellular folates produced by wild-type cells and cells producing increasing amounts of γ-glutamyl hydrolase. Analysis at 280 nm shows formyl-tetrahydrofolates with different polyglutamyl-tails and at 360 nm shows methenyl-tetrahydrofolates.

tetrahydrofolate represent the two major groups of folate in *Lactococcus lactis*. Preliminary results indicate that the major bound C1-moieties is not the same in all lactic acid bacteria, but that also the methyl-tetrahydrofolate form can be the major component as also found in many plant materials. The effect of the cloning strategies described above on the length of the polyglutamate-tail can be directly visualised via HPLC in for instance a clear change from predominantly tetra- and pentaglutamate folates into mostly monoglutamate-folates (Figure 4).

This work represents an important first step in the development of novel, functional foods with increased levels of bio-available folate. The identification of γ-glutamyl hydrolases from (food grade) bacterial origin may enhance the market introduction of a food grade micro-organism that has improved bio-available-folate production capacities. Moreover, metabolic engineering of the lactococcal folate biosynthesis pathway itself by changing expression levels of the relevant biosynthetic enzymes is expected to result in significant increases in folate production. These approaches could eventually be combined to generate starter bacteria that can be used for the production of novel fermented dairy products providing 100% instead of the current 15–20% of the human daily requirement for folate intake.

Other B-vitamins

Recent population studies performed all over the world have revealed that malnutrition is a frequently occurring phenomenon, also in the well-developed Western world. This is sometimes a direct result of medical treatment, but often also a result of social deprivation especially for the young and old. Malnutrition is specifically observed in deficiencies in B-vitamin content of the diet and/or the human blood. Besides deficiencies in folate, deficiencies in riboflavin (vitamin B2) and in cobalamin (vitamin B12) are most commonly observed. Riboflavin-deficiency can lead to various physiological effects such as liver- (Ross & Klein 1990) and skin-disorders (Lakshimi 1998), disturbed metabolism of the red blood cells (Hassan & Thurnham 1977) and, in general, reduced performance during physical exercise (Belko et al. 1983; Bates 1987). Some lactic acid bacteria, such as *Lactococcus lactis*, are able to produce and excrete small amounts of riboflavin into the surrounding medium. Based on this idea, fermented products could serve as valuable source of this vitamin for the human diet. The riboflavin is synthesised from the precursor (GTP) as is the case for folate (Figure 2), and the initial step in its biosynthesis involves a similar enzyme (GTP cyclohydrolase II) as present in the folate biosynthetic pathway (see above). In *Bacillus subtilis*, this first reaction in riboflavin biosynthesis has been demonstrated to be rate limiting (Humbelin et al. 1999). The gene coding for this enzyme, *ribA*, has been brought to overexpression in *L. lactis* using the NICE-system. Similar as described for folate production, this metabolic engineering strategy resulted in a 3-fold overproduction of riboflavin. Further studies in this respect are focussed on obtaining natural mutants showing increased riboflavin-production by selecting for strains or variants that are resistant to analogues of purine and riboflavin, or their biosynthetic precursors, such as 8-azaguanine and decoyinine.

Another B-vitamin that is conspicuously low in many diets and repeatedly leads to deficiencies in humans, is vitamin B12 or cobalamin. This vitamin is produced by only a limited number of micro-organisms and, related to food fermentations, only very few food-grade micro-organisms are known to produce this vitamin. Up to now, no lactic acid bacteria have been described producing this essential co-factor in metabolism. In recent studies on *Lactobacillus reuteri*, some interesting observations were made on the production of a cobalamin-like substance, especially in the presence of glycerol which is converted in this organisms to propanediol (and to hydroxypropionaldehyde, reuterin) through a cobalamin dependent enzymatic step. Culture supernatant, containing this substance, was shown to complement vitamin B12-deficiency in auxotrophic *Lactobacillus delbrueckii* ATCC7830 and *Escherichia coli* 113 (*metE*-negative), indicating that we are dealing with a cobalamin or a related component. Molecular analysis of *Lactobacillus reuteri* has revealed the presence of at least five genes involved in the biosynthesis of vitamin B12. This is the first report of production of vitamin B12 by a lactic acid bacterium. Currently, studies are on the way to analyse the exact nature of the chromophore and to investigate the potential to produce more via metabolic engineering. The potential to produce a corrin-containing component such as cobalamin, should also open the way to produce other corrin-containing (bio)molecules such as cytochromes. Earlier studies have indicated that addition of heme, the precursor for cytochromes, to *Lactococcus lactis* cells leads to more efficient growth, presumably via the introduction of, normally lacking, an aerobic electron transport chain. Inducing production of cytochromes in lactic acid bacteria could be a whole new strategy to improve growth rates and growth yields of these important starter and probiotic bacteria.

Production of low-calorie sugars

Body-weight control is a major concern in our Western countries and obesity has been estimated to cost between 2 and 5% of the total health-care expenses of various developed countries. New food products containing low-calorie sugars and/or fat-replacers are in constant progression on the market in response to the consumer's request. We will discuss here the production by lactic acid bacteria of mannitol, sorbitol, tagatose and possibly some other low-calorie sugars through metabolic engineering.

Production of mannitol and sorbitol

Polyols, such as mannitol and sorbitol, are low-calorie sugars that could replace sucrose, lactose, glucose or fructose in food products as they display equivalent sweetness and taste (Dwivedi 1978; Debord et al. 1987). Mannitol can also serve as anti-oxidant in biological cells (Shen et al. 1997) as shown by increased survival of cells during freezing and/or drying in the presence of mannitol (Efiuvwevwere et al. 1999). Heterofermentative lactic acid bacteria such as *Leuconostoc mesenteroides* are known for their ability to produce mannitol in the fermentation of fructose (Soetaert et al. 1995). These bacteria convert part of the fructose for energy generation via the usual heterofermentative pathway while the other part of fructose is reduced directly to mannitol. In this way, high yield of mannitol is found, especially when fructose is supplied with f.i. glucose. Mannitol production, at low levels, has also been reported for homofermentative lactic acid bacteria under unusual conditions. In both *Lactobacillus plantarum* (Ferain et al. 1996) and *Lactococus lactis* (Neves et al. 2000), disruption of lactate dehydrogenase (LDH) resulted in production of a whole new range of metabolites, including mannitol. The mannitol is, most probably, produced by reduction of fructose-6-P to mannitol-1-P, which is subsequently dephosphorylated to yield mannitol (Figure 5). With this in mind, metabolic engineering strategies can be designed to induce significant mannitol production in these lactic acid bacteria. Overproduction of the mannitol-P dehydrogenase (MPDH) in a LDH-deficient *L. lactis* strain has already resulted in strong increase in, mostly, intracellular mannitol pro-

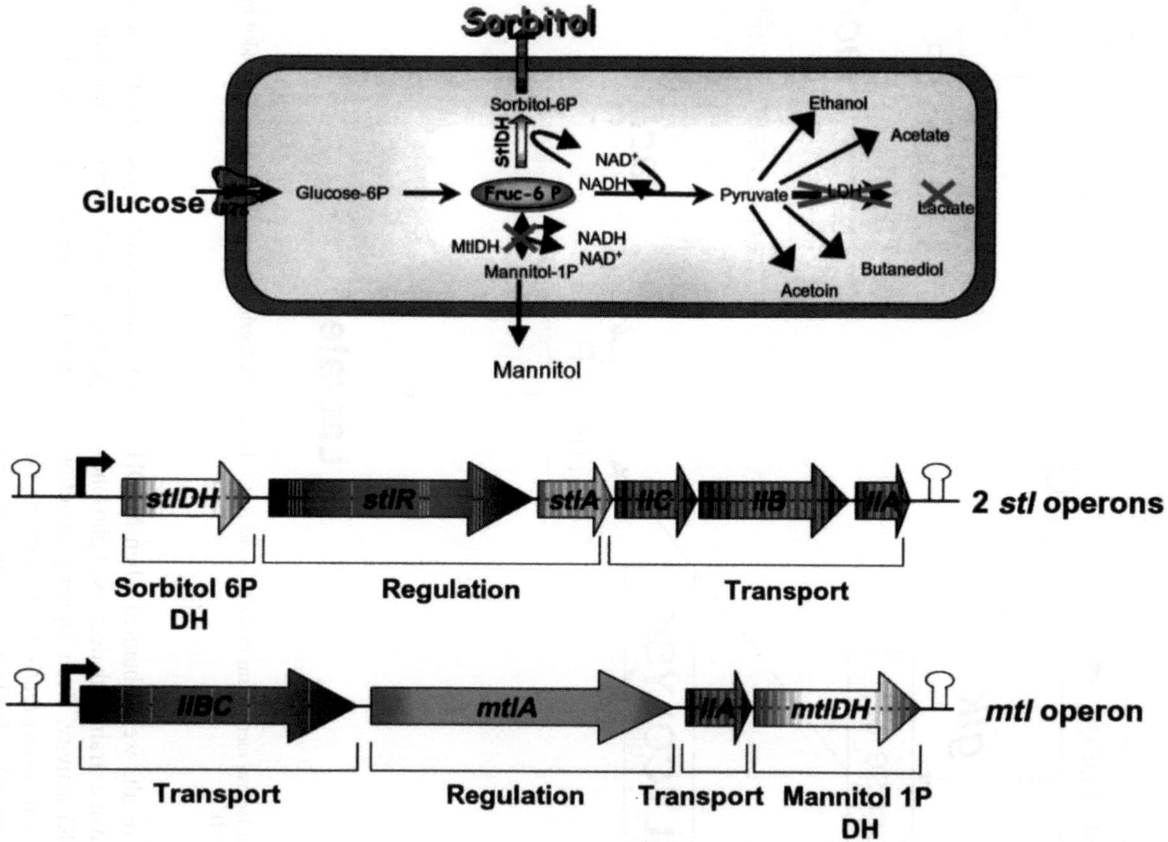

Figure 5. Schematic view of a *Lactobacillus plantarum* cell engineered for sorbitol production and organisation of sorbitol and mannitol genes.

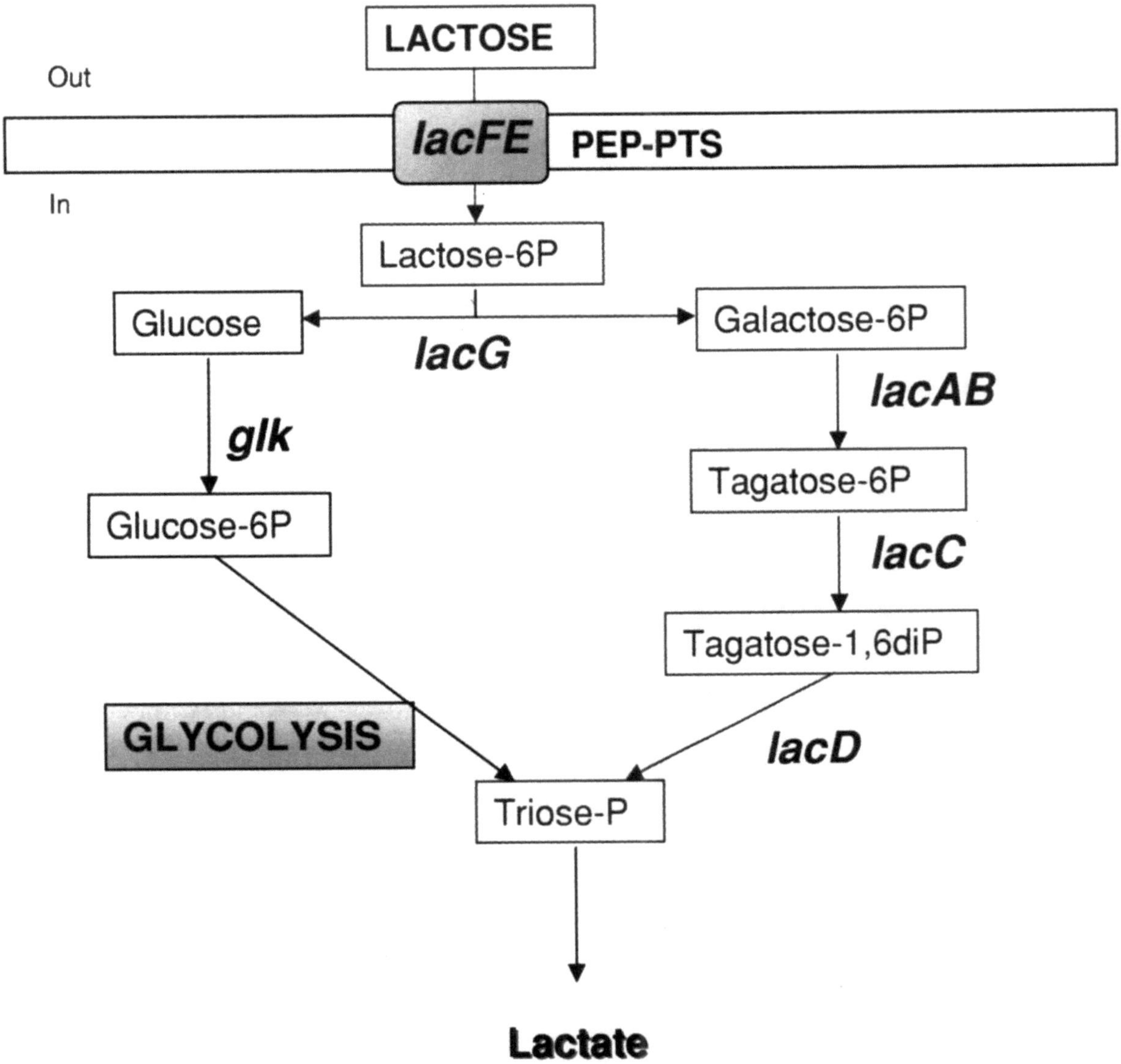

Figure 6. Pathway for lactose metabolism in *Lactococcus lactis* showing the tagatose-6-phosphate pathway encoded by the *lacABCD* genes (see text for more detail).

duction. Similar results were obtained when MPDH was overproduced in a strain with decreased phosphofructokinase (PFK) activity. Even higher production of mannitol by *Lactococcus lactis* can be expected if excretion of this polyol is facilitated, by introducing the mannitol-transporter present in *Leuconostoc mesenteroides*. Production of sorbitol can be induced in these bacteria in a similar way. By overexpression of the gene coding for sorbitol dehydrogenase, in combination with disruption of the MPDH and the LDH (Figure 5), considerable sorbitol production has been observed in *Lactobacillus plantarum* strains. Also in this case, even higher production of sorbitol is expected when the dephosphorylation and the transport (export) of sorbitol is enhanced.

Tagatose production

Tagatose is another carbohydrate that is considered as a potential sucrose replacement. It has almost equal sweetening power as sucrose – higher than similar components such as mannitol, sorbitol and erythritol – but much lower caloric value since it is poorly degraded by the human body (Zehner 1988). It has recently been launched on the food market as low-calorie sugar, as prebiotic, and as anti-plaque agent by MD-foods (www.tagatose.dk). Their is no biological source available for extraction of tagatose so the current production process is by enzymatic conversion using arabinose as a substrate (Kim et al. 2001). For many practical reasons, a fermentation process involving intact microbial cells, is almost always the preferred process over an enzymatic conversion. A biotransformation process, from galactitol to tagatose, has recently been described using acetic acid bacteria (Manzoni et al. 2001). However, economy-wise this does not seem to be applicable since galactitol, although much more abundant and cheaper than tagatose, is still a relatively expensive substrate. The use of lactic acid bacteria could, very well, be an attractive alternative. There is only one metabolic pathway that features tagatose and that is in the degradation of lactose as found in lactic acid bacteria and a few other micro-organisms. The so-called tagatose-6-P pathway (van Rooijen et al. 1991) is responsible for breakdown of the galactose moiety during lactose metabolism by bacteria such as *Lactococcus lactis*. After hydrolysis of lactose-P, through the action of phospho-β-galactosidase, the glucose is degraded through glycolysis. The galactose-P moiety is converted into tagatose-6-P catalysed by galactose-P isomerase, then phosphorylated forming tagatose-diphosphate catalysed by tagatose-phosphate kinase and finally hydrolysed, by tagatose-diphosphate aldolase, to form the glycolytic intermediates dihydroxyaceton-P and glyceraldehyde-P (Figure 6). The genes encoding for this tagatose pathway – *lacABCD* – in *Lactococcus lactis* can be found on the lactose operon (Figure 6). Using *L. lactis* as micro-organism with its relatively simple carbon metabolism, its broad arsenal of genetic tools, including an efficient gene integration system and its very efficient system (NICE) for overproduction of enzymes, this lactic acid bacterium was chosen as potential cell factory for the production of tagatose. To induce accumulation of tagatose-derivatives in lactococcal cells, the initial chosen strategy was to disrupt the *lacC* and/or *lacD* genes resulting in production of either tagatose-6-P or tagatose-1,6-diphosphate. Disruption of *lacD* was accomplished via a two step recombination process, involving integration of an erythromycin-resistance plasmid containing only the *lacC* and *lacF* genes via single crossing-over, followed by removal of *lacD* (or reversion to the wild-type) in a second, spontaneous, recombination event ('double cross-over'). The double crossing-over seemed to occur in almost the expected 50% of the cases, so the desired mutant was obtained relatively easy. Growth rates of the *lacD*-negative mutant on glucose were identical to the wild-type, but growth on lactose was severely compromised in the engineered strain. This was already a good indication that an unusual physiological response was occurring in the construct. HPLC-analysis focussing on phosphorylated metabolic intermediates, revealed the accumulation of a unique metabolite in the constructed strain, in comparison with the wild-type where basically the glucose- and fructose-phosphates were visible. Spiking the samples with specially for this purpose synthesised tagatose-phosphate and tagatose-biphosphate, gave strong indication that the unique metabolite was indeed tagatose-1,6-diphosphate. So, *L. lactis* was successfully transformed in a tagatose (phosphate) producing cell-factory. However, for the use in practical (food) situations, strategies will have to developed to dephosphorylate the tagatose-diphosphate and excrete the final product, tagatose.

Production of polysaccharides

Exopolysaccharides (EPS)

The exopolysaccharide (EPS) molecules produced by different lactic acid bacteria display a wide variety of chemical and biophysical properties, including variations in sugar composition, sugar linkage, polymer length and polymer branching. These polymers have an important role in the rheology and texture properties of fermented food products, and thus are of interest for food applications as *in situ* produced, natural bio-thickeners. Moreover, they could also be applied as natural, food-grade additives, replacing presently applied stabilisers and thickeners that are produced by non food-grade bacteria (Sutherland 1998; Becker et al. 1998). Interestingly, it has been suggested that some polysaccharides produced by lactic acid bacteria have prebiotic (Gibson & Roberfroid 1995), immunostimulatory (Hosono et al. 1997), antitumoral (Kitazawa et al. 1991) or cholesterol-lowering activity (Nakajima et al. 1992a). Over the last decade, the production of exopolysaccharides (EPS) by lactic acid bacteria has been extensively studied (for recent review see de Vuyst & Degeest 1999), generating important advances in our knowledge of EPS-genetics, biological diversity and distribution, biosynthetic pathways, metabolic models, and physics. This knowledge can now be applied to rationally design metabolic engineering studies to modify EPS production and composition.

The production of EPS by *L. lactis* is associated with strains that are isolated from highly viscous fermented milk products. The specific *eps* genes are encoded on large plasmids that can be conjugally transferred from one lactococcal strain to the next, thereby introducing the EPS-producing capacity in the recipient strain (Vedamuthu & Neville 1986; von Wright & Tynkkynen 1987; van Kranenburg et al. 1997). One of the *eps* gene clusters of *L. lactis* that is characterised in most detail, is that of strain NIZO B40 (for a review see Kleerebezem et al. 1999). This strain produces a polymer with a regular repeating unit, $\rightarrow$4)[α-L-Rha*p*-(1$\rightarrow$2)][α-D-Gal*p*-1-PO_4-3]-β-D-Gal*p*-(1$\rightarrow$4)-β-D-Glc*p*-(1$\rightarrow$4)-β-D-Glc*p*-(1$\rightarrow$, that is structurally identical to that produced by strain SBT 0495 (Nakajima 1992b; van Kranenburg et al. 1997). The NIZO B40 strain harbours a 42,180 bp EPS-plasmid pNZ4000 (van Kranenburg et al., 2000), containing the 12 kb NIZO B40 *eps* gene cluster with 14 co-ordinately expressed genes, *epsRXABCDEFGHIJKL* (Figure 7). Based on sequence comparisons putative functions could be assigned to several *eps* genes, predicting their involvement in biosynthesis of the repeating unit oligosaccharides by the sequential addition of sugars to a membrane-anchored lipid carrier, and subsequent export and polymerisation of these lipid-linked oligosaccharides. Functional analysis of the *epsDEFG* genes, encoding glycosyltransferases, generated a model for the biosynthesis of the B40-EPS repeating unit (van Kranenburg et al. 1999a).

Oligosaccharides

Current studies on *L. lactis* NIZO B40 *eps* genetics focus on the elucidation of the individual role of the genes that are predicted to be involved in export (*epsK*), polymerisation (*epsI*) and chain length determination (*epsA*, *epsB*) (van Kranenburg et al. 1997) (Figure 7). These studies might eventually allow directed manipulation of these functions, which should result in the production of polysaccharides that have shorter (oligosaccharides) or longer chain lengths compared to the native B40-polymer. Moreover, the strong increase in the available *eps* gene cluster sequences from various organisms, including several lactic acid bacteria, has generated a wealth in genetic information including a large number of glycosyltransferase encoding genes. The genetic information combined with biochemical information regarding donor- and acceptor-specificity of individual glycosyltransferase enzymes, could be exploited in a glycosyltransferase combinatorial approach aiming at construction of the biosynthetic machinery for the production of oligo- and/or polysaccharides in *L. lactis* that have a predetermined and novel structure. The oligosaccharide structure and composition will ultimately determine its functionality as a nutraceutical. For instance, fructose-containing oligosaccharides are claimed to, specifically, stimulate the growth of Bifidobacteria in the human intestine (Gibson & Wang 1994), while phosphorylated oligosaccharides display the best fiber-like properties in the protection of intestinal mucosa against oxidative damage (Kawaguchi et al. 1999). The latter approach requires that the export and polymerisation machinery do not display a unique specificity for its corresponding, native polysaccharide. Interestingly in this respect is the observation that heterologous expression of the *Streptococcus thermophilus* Sfi6 *eps*-gene cluster in *L. lactis* resulted in the production of a polysaccharide containing an altered repeating unit compared to the polymer produced in its original host (Stingele et al. 1999). These results show that neither the glycosyltransferases, nor the export and polymerisation machinery encoded by this *eps* gene clusters exhibit absolute specificity for their native substrate and product, suggesting that the approach above could be successful. Moreover, a first example of such an approach is the possibility to functionally exchange glycosyltransferases with identical substrate specificity originating from different genera in Gram-positive bacteria (van Kranenburg et al. 1999b). Current and future *eps* engineering strategies should clarify the possibilities of these approaches towards the production of 'designer'-polysaccharides.

Improving sugar conversion

Lactose/galactose removal

Lactose is present in amounts of 4–4.5% (w/v) in cow milk. In liquid fermented dairy products, such as yoghurt or buttermilk, usually less than half of this is fermented to lactic acid. This makes all these liquid dairy products unsuitable for the lactose-intolerant population and deprives them of all the nutritious components that are present in milk such as pro-

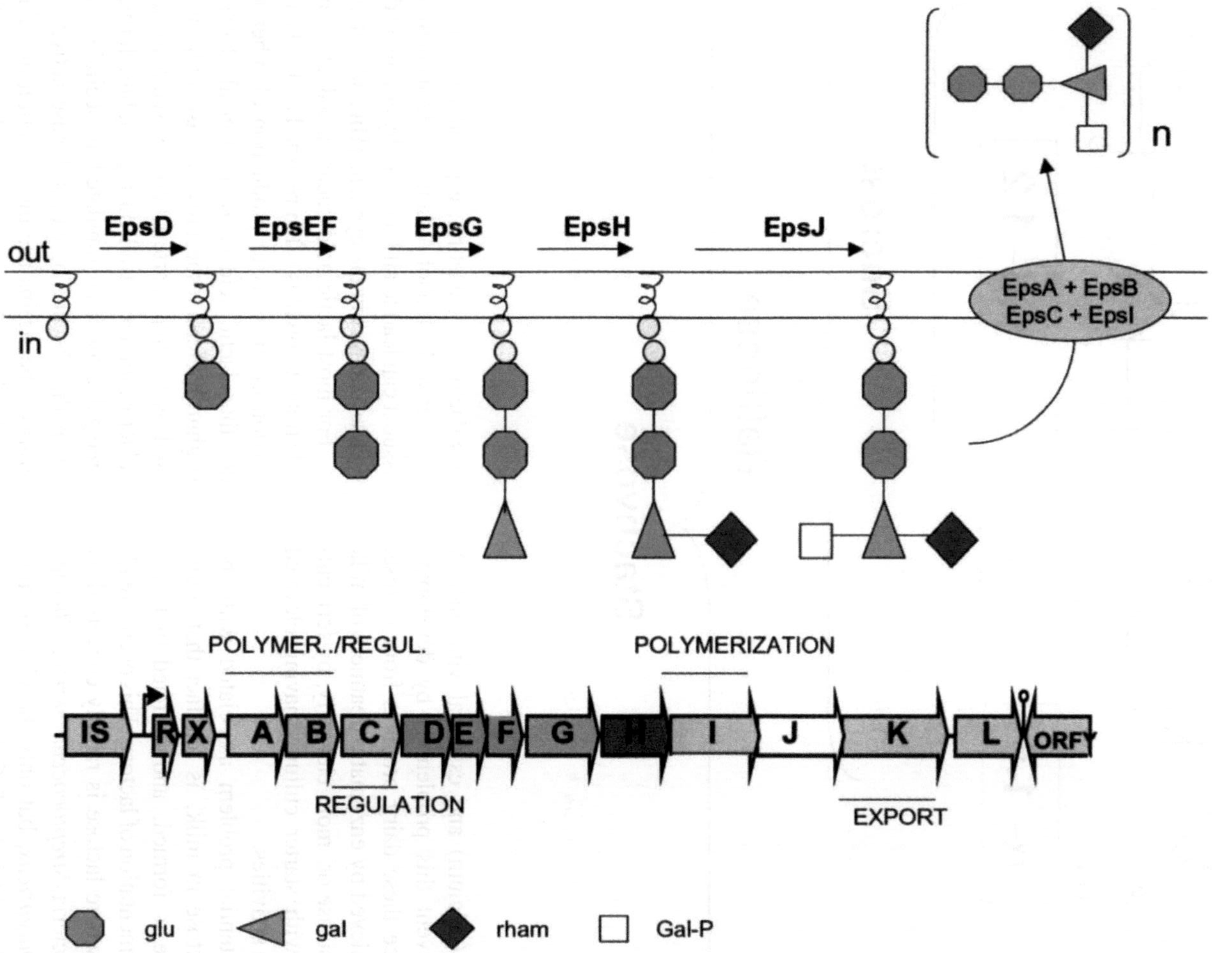

Figure 7. Schematic representation of the *L. lactis* NIZO B40 *eps* gene cluster and its flanking regions. The predicted individual roles of the *eps* genes in repeating unit biosynthesis (schematically indicated above the gene-cluster) and subsequent export, polymerisation and chain length determination are indicated; the role of *epsD*, *epsE*, *epsF* and *epsG* have been established, biochemically (see text for more detail).

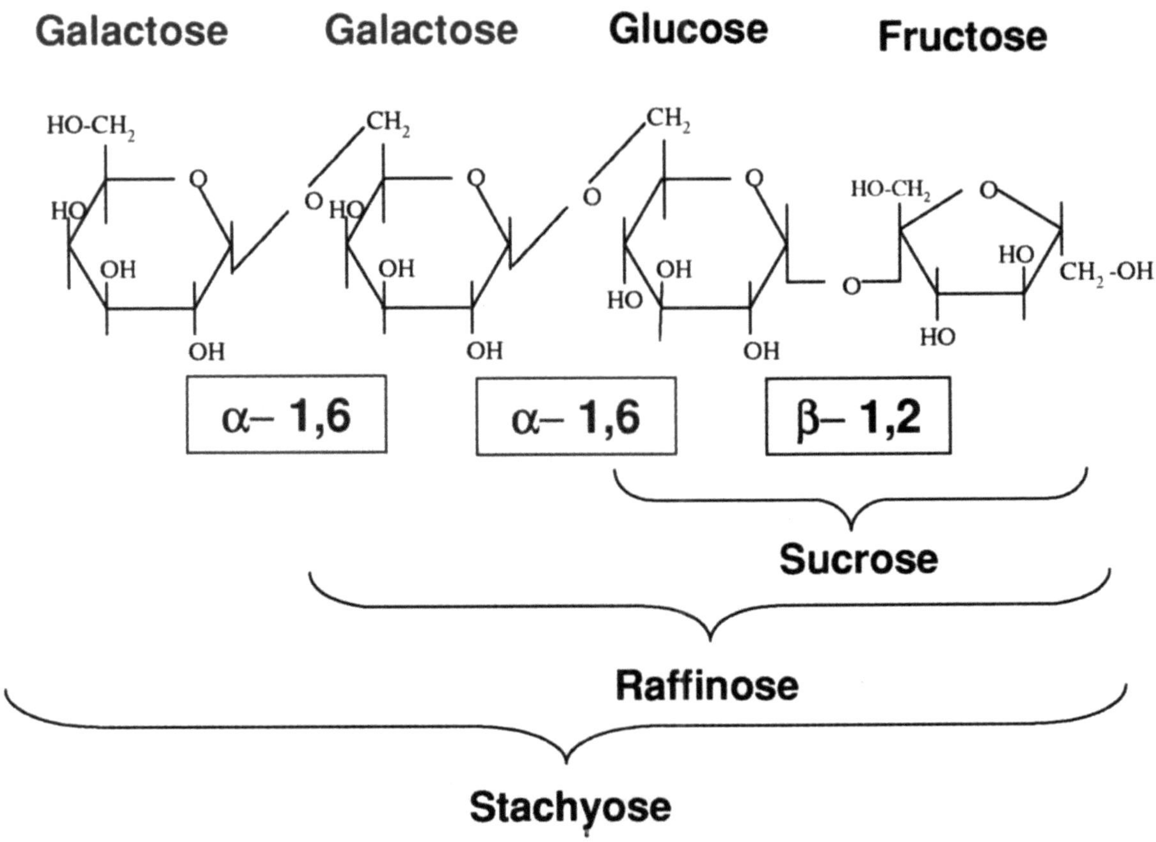

Figure 8. Structure of the (oligo)saccharides stachyose, raffinose and sucrose.

tein, minerals (calcium) and essential vitamins. A way to circumvent this problem is by developing processes to free these dairy products from lactose. This can be achieved by enzymatic treatment of milk with β-galactosidase or, more elegantly, by fermentation of milk with starter cultures having elevated β-galactosidase activities.

Another potential problem associated with the presence of lactose in milk, is the fact that, quite often, galactose is formed, and excreted in to the milk, during fermentation of lactose. This occurs readily in yoghurt where lactose is rapidly converted by both yoghurt bacteria, *Streptococcus thermophilus* and *Lactobacillus bulgaricus*, but only the glucose moiety appears to be metabolised. Galactose is excreted in almost stoichiometric amounts by the lactic acid bacteria (Veenhoff & Poolman 1999) resulting in accumulation of galactose upto concentrations of 1% and even higher. Besides being a poor sugar source for the human body, galactose presents another problem that too high consumption can lead to reduction of galactose to galactitol, especially when consumed together with alcohol. When galactitol accumulates in tissue it can lead to all kinds of physiological disorders such as cataract in the eye lens (Hirasuka et al. 1992).

For most lactic acid bacteria, galactose is a poor substrate. Growth is often retarded and galactose conversion often leads to production of other metabolites than lactic acid (Grossiord et al. 1998). Many yoghurt bacteria, especially the fast acidifying strains used as yoghurt starters, do not seem to utilise the galactose-moiety at all during yoghurt fermentation. Free galactose is accumulated intracellularly as a result of the absence of galactokinase activity in these strains. The strains use the excretion of galactose as actual driving force for the rapid uptake of lactose, in a one-to-one exchange reaction, allowing extremely fast growth of these bacteria on this sugar with doubling times that are even lower than 10 min. Molecular analysis of the galactose (and lactose) gene cluster in these strains of *Streptococcus thermophilus*, has shown that the gene for galactokinase is completely intact, but that one or more point mutations have taken place leading to a 'silent' phenotype (Vaughan et al. 2001). These mutations can, sometimes, spontaneously revert, especially when the bacteria are cultured for prolonged times on galactose. The use of these revertants in yoghurt production could be a way to produce galactose-free yoghurt, although it does lead to reduced growth rate, and more serious for practical purposes, reduced acidifcation rates.

Removal of raffinose

Not all oligosaccharides are beneficial in the human and animal diet. Soy- and pulse-derived food products contain high levels of α-galactosides such as stachyose and raffinose (Figure 8).

Humans and other monogastric animals do not produce intestinal α-galactosidase, the enzyme responsible for the degradation of these oligosaccharides. Thus, α-galactosides pass into the lower gut where they are fermented by gas-producing bacteria. Consequently, consumption of soy and pulses induces intestinal disturbance such as flatulence in all monogastric animals (Khane et al. 1994). Removal of the α-galactosides, through enzymatic treatment of the soy milk, has already been described successfully (Maity & Paul 1991), but does not seem to be an economically feasible process. The use of lactic acid bacteria, for prefermentation or as probiotic, could well present the perfect solution for removing undesirable sugars from soy.

The ability to hydrolyse α-galactosides has already been identified in a number of lactic acid bacteria (Ahrne & Molin 1991; Garro et al. 1996). By applying metabolic engineering strategies, lactic acid bacteria will be constructed with high α-galactosidase activities as starters for removal of α-galactosides during soy fermentation and as possible probiotics to deliver α-galactosidase activity in the gut for prevention of flatulence. The first step in this strategy has been the cloning of the relevant gene (*melA*) coding for α-galactosidase from *Lactobacillus plantarum* (Silvestroni et al. 2002). The gene encodes a 738-amino acid protein (theoretical MW=84 kDa) and is associated with genes involved in raffinose and galactose metabolism (*rafP* for a putative raffinose permease, *galM* for galactose epimerase and *lacLM* for β-galactosidase). The *melA* gene could complement an α-Gal deficient *Escherichia coli* strain strongly suggesting that the *melA* gene is functional in *Lb. plantarum*. Additional evidence for functionality is provided by regulation studies showing that *melA* is actually induced 4-fold in the presence of the α-galactoside melibiose as growth substrate in comparison to glucose as growth substrate. For construction of starter and probiotic bacteria with high α-galactosidase activity, the *melA* is being cloned in *Lactococcus lactis* in three different constructions resulting in: (1) expression of the enzyme in the cytoplasm for maximum protection of enzyme activity, (2) expression as a secreted enzyme for maximum exposure to the sugar substrate, or (3) expression on the surface but anchored to the surface of the cell. The first construct would provide a suitable starter for α-galactoside-removal during soy fermentation, the second construct would serve best as cell-factory for α-galactosidase production to be used as means for (bio)chemical removal of stachyose and raffinose from soy- and pulse-material, while the third construct will be applied as probiotic to supply high α-galactosidase activity in the gut for assistance in α-galactoside degradation.

Impact of genomics and related technologies

We have presented a number of examples where activity of lactic acid bacteria have lead or can lead to healthier fermented food products or to the production of food ingredients with a clear health benefit. This has been achieved through altered expression of relevant homologous or heterologous genes. With the rise of (functional) genomics research, we can expect a tremendous boost in this type of development. The current list of publicly available complete genome sequences already exceeds 60 microbes as indicated on the Comprehensive Microbial Resource homepage of the TIGR-Website (www.tigr.org) and is exponentially growing. Currently, only one LAB genome – *Lactococcus lactis* – has been published (Bolotin et al. 2001), but several others are expected shortly (see other reports LAB 7). The genome is the 'blueprint' of an organism and tells us directly what is the potential of the organism to produce certain metabolites and presents us with the relevant genes that could be targeted for overexpression or disruption in order to reach increased functionality. In addition, by comparative genomics it has become relatively easy to determine why certain bacteria cannot produce specific nutraceuticals and how production of these specific components can be introduced. For control or improvement of production levels, however, more is needed than only the genome sequence. Information is needed on how the flux through a specific pathway is controlled and how the activity of an individual pathway relates to other pathways. Kinetic, metabolic models have become available describing the dynamics of (pyruvate) metabolism in homofermentative lactic acid bacteria (Garrigues et al. 1997; Hoefnagel et al. 2002). However, these models seem, so far, only to be suitable for relatively simple metabolic pathways and are basically able to calculate, and simulate, the flux through the major metabolic pathways, but are much less accurate in describing pathways with much lower metabolic fluxes. The latter situation holds true, especially, for most biosynthetic pathways. The metabolic flux through these pathways is in most cases many orders of magnitude lower compared with the flux through catabolic pathways. Moreover, the enzymes involved generally have much lower activities and the metabolic intermediates and end-products are produced at much lower levels. In some cases this general problem can be solved by dissection of complicated (biosynthetic) pathways into several, simplified units and modelling each unit separately. However, in many cases this is just not possible. Even simple flux analysis can be an impossible task in metabolic networks involving loops and cycles. To analyse these networks, more advanced measurement techniques have been used successfully, such as the application of ^{13}C-NMR with metabolic substrates labelled at a specific position. Variation in the level of label-scrambling resulting from either enantiomer-specific or enantiomer-unspecific enzymes, will generate insight in control and distribution of different fluxes within a metabolic network (Rollin et al. 1995; Klapa et al. 1999; Park et al. 1999; Schmidt et al. 1999; de Graaf et al. 2000). However, it cannot provide insight in metabolic regulation, nor can it be used to predict fluxes under conditions that have not been tested. The optimistic vision that the knowledge of a (microbial) genome will automatically provide us with a comprehensive model of overall metabolism seems quite unrealistic for the reasons discussed above. Extensive experimentation on the physiology, enzymology and molecular biology of the organism is required before any practical modelling can be performed. The rapid development of genomics based high-throughput technologies, including transcriptomics, proteomics and especially metabolomics, might allow the effective analysis of the differential, global responses in micro-organisms upon the changing of environmental conditions or upon the engineering of specific enzyme activities.

With the genome sequence in hand of many industrial micro-organisms, traditional mutagenesis and natural selection of mutants has become a more directed and controlled activity and can now progress in parallel with the metabolic engineering approaches as described in this overview. It now seems possible to induce and select mutants that show f.i increased B-vitamin production, in a directed way without the use of recombinant DNA-techniques. The rise of high-throughput-screening facilities that, automatically and simultaneously, can sample and analyse thousands of samples is a powerful tool in this type of development. Many improved starter bacteria and cell factories for high nutraceutical production will be discovered by these natural selection strategies and have the huge advantage over engineered lactic acid bacteria that they can be implemented promptly in actual industrial fermentation. In addition, the molecular analysis of the selected mutants will provide very relevant information on how to effectively improve production of the desired nutraceutical. Many high-throughput molecular methods have been developed over the last few years to specifically pinpoint single nucleotide polymorphism (SNP) in these (natural) mutants, through direct Southern-blotting (Sokinenko 2001), through the use of whole-genome (Cutler et al. 2001) or pathway-specific DNA microarrays (Huber et al. 2002), or by specific mass-labelling techniques such as the Masscode System (Qiagen, Inc) that identify, by mass spectrometry, polymorphism directly in PCR-products (Kokoris et al. 2000).

Most examples of effective metabolic engineering in lactic acid bacteria discussed above deal with rerouting the major metabolic flux in these bacteria, i.e. the conversion of sugar into lactic acid. The results obtained include high production of metabolites that otherwise are not produced or produced in low amounts. For reasons discussed above, the engineering of more complex (biosynthetic) pathways and the analysis of the results of these engineering strategies is a much more complicated affair. Nevertheless, some promising results are already available for *Lactococcus lactis* in the case of folate and riboflavin biosynthesis. Future analyses of this and other complex pathways will provide us with valuable knowledge concerning the potential of LAB as cell factories for biologically active compounds like vitamins, low-calorie sugars or other nutraceuticals.

Acknowledgements

Part of the work described here, was supported by funding programs of the European Community (contracts: ERBFMBICT-950438, BIO4-CT98-0118, and QLK1-CT-2000-01376). We would like to thank Aurelio Silvestroni, Jose Luis Vera (both CERELA) and Cristelle Connes (INRA) for sharing unpublished results with us and Campina for financing the tagatose-work.

References

Ahrne S & Molin G (1991) Spontaneous mutations changing the raffinose metabolism of *Lactobacillus plantarum*. Antonie van Leeuwenhoek 60: 87–93.

Alm L (1980) Effect of fermentation on B-vitamin content of milk in Sweden. J. Dairy Sci. 65: 353–359.

Ames BN (1999) Micronutrient deficiencies cause DNA damage and cancer. Food Sc. Agric. Chem. 1: 1–15.

Bates CJ (1987) Human riboflavin requirements and metabolic consequences of deficiency in man and animals. World Rev. Nutr. Diet 50: 215–265.

Becker A, Katzen F, Pühler A & Ielpie L (1998) Xanthan gum biosynthesis and application: a biochemical/genetic perspective. Appl. Microbiol. Biotechnol. 50: 145–152.

Belko AZ, Obarzanek E, Kalkwarf HJ, Rotter MA, Bogusz S, Miller D, Haas JD & Roe DA (1983) Effects of exercise on riboflavin requirements of young women. Am. J. Clin. Nutr. 37: 509–517.

Bhandari SD & Gregory JF (1990) Inhibition by selected food components of human and porcine intestinal pteroylpolyglutamate hydrolase activity. Am. J. Clin. Nutr. 51: 87–94.

Bolotin A, Wincker P, Mauger S, Jaillon O, Malarme K, Weissenbach J, Ehrlich SD & Sorokin A (2001) The complete genome sequence of the lactic acid bacterium *Lactococcus lactis* ssp. *lactis* IL1403. Genome Res. 11: 731–753.

Boushey CJ, Beresford AA, Omenn GS & Moltulsky AG (1996) A quantitative assessment of plasma homocysteine as a risk factor for vascular disease. J. Amer. Med. Assoc. 274: 1049–1057.

Brattstrom L (1996) Vitamins as homocysteine-lowering agents. J. Nutr. 126 (4 Suppl): 1276S–1280S.

Covert MW, Schilling CH, Famili I, Edwards JS, Goryanin II, Selkov E & Palsson BO (2001) Metabolic modeling of microbial strains *in silico*. Trends Biochem. Sci. 26: 179–186.

Cutler DJ, Zwick ME, Carrasquillo MM, Yohn CT, Tobin KP, Kashuk C, Mathews DJ, Shah NJ, Eichler EE, Warrington JA & Chakravarti A (2001) High-throughput variation detection and genotyping using microarrays. Genome Res 11: 1913–1925

Debord B, Lefebvre C, Guyot-Hermann AM, Hubert J, Bouche R & Guyot JC (1987) Study of different forms of mannitol: Comparative behaviour under compression. Drug Develop. Ind. Pharm. 13: 1533–1546.

De Graaf AA, Mahle M, Mollney M, Weichert W, Stahmann P & Sahm M (2000) Determination of full 13C isotopomer distributions for metabolic flux analysis using heteronuclear spin echo difference NMR spectroscopy. J. Biotechnol. 77: 25–35.

de Ruyter PGGA, Kuipers OP & de Vos WM (1996) Controlled gene expression systems for *Lactococcus lactis* with the food-grade inducer nisin. Appl. Environ. Microbiol. 62: 3662–3667.

de Vuyst L & Degeest B (1999) Heteropolysaccharides from lactic acid bacteria. FEMS Microbiol. Rev. 21: 153–177.

de Vos WM (1996). Metabolic engineering of sugar catabolism in lactic acid bacteria. Antonie van Leeuwenhoek 70: 223–242.

Dwivedi BK (1978) Low Calorie and Special Dietary Foods. CRC Press, Inc, West Palm Beach.

Efiuvwevwere BJO, Gorris LG, Smid EJ & Kets EPW (1999) Mannitol-enhanced survival of *Lactococcus lactis* subjected to drying. Appl. Microbiol. Biotechnol. 51: 100–104

Ferain T, Schanck AN & Delcour J (1996) 13C nuclear magnetic resonance analysis of glucose and citrate end products in an *ldhL-ldhD* double-knockout strain of *Lactobacillus plantarum*. J. Bacteriol. 178: 7311–7315

Garrigues C, Loubiere P, Lindley ND & Cocaign-Bousquet M (1997) Control of the shift from homolactic acid to mixed acid fermentation in *Lactococcus lactis*: predominant role of $NADH/NAD^+$ ratio. J. Bacteriol. 179: 5282–5287.

Garro MS, Giori GS, de Valdez GF & Oliver G (1996) Characterization of a-galactosidase from *Lactobacillus fermentum*. J. Appl. Bacteriol. 75: 485–488.

Gibson GR & Roberfroid MB (1995) Dietary modulation of the human colonic microbiota: introducing the concept of prebiotics. J. Nutr. 124: 1401–1412.

Gibson GR & Wang X (1994) Bifidogenic properties of different types of fructo-oligosaccharides. Food Microbiol. 11(6): 491–498.

Griffin WC & Lynch MJ (1972) Polyhydric alcohols. In Furia TE (ed.) CRC Handbook of Food Additives, Vol 1 (2nd edn.), CRC-Press, Cleveland, OH (pp 431–455).

Grobben GJ, Peters SWPG, Wisselink HW, Weusthuis RA, Hoefnagel MHN, Hugenholtz J & Eggink G (2001) Spontaneous formation of a mannitol-producing variant of *Leuconostoc pseudomesenteroides* grown in the presence of fructose. Appl. Environ. Microbiol. 67: 2867–2870.

Grossiord B, Vaughan EE, Luesink E & de Vos WM (1998a) Genetics of galactose utilisation via the Leloir pathway in lactic acid bacteria. Lait 78: 77–84.

Hamm-Alvarez S, Sancar A & Rajagopalan KV (1989) Role of enzyme-bound 5,10-methenyltetrahydropteroylpolyglut amate in catalysis by *Escherichia coli* DNA photolyase. J. Biol. Chem. 264: 9649–9656.

Hassan RM & Thurnham DJ (1977) Effects of riboflavin deficiency on the metabolism of the red blood cell. Ind. J. Vit. Nutr. Res. 47: 349–355.

Hirasuka Y & Li G (1992) Alcohol and eye diseases: a review of epidemiologic studies. J. Stud. Alcohol 62: 397–402.

Hoefnagel MHN, Starrenburg MJC, Martens DE, Hugenholtz J, Kleerebezem M, van Swam II, Bongers R, Westerhoff HV & Snoep JL (2002) Metabolic engineering of lactic acid bacteria, the combined approach: kinetic modelling, metabolic control and experimental analysis. Microbiology, 148: 1003–1013.

Hols P, Kleerebezem M, Schank AN, Ferain T, Hugenholtz J, Delcour J & de Vos WM (1999a) Conversion of *Lactococcus lactis* from homolactic to homoalanine fermentation through metabolic engineering. Nat. Biotechnol. 17: 588–592.

Hosono A, Lee J, Ametani A, Natsume M, Hirayama M, Adachi T & Kaminogawa S (1997) Characterization of a water-soluble polysaccharide fraction with immunopotentiating activity from *Bifidobacterium adolescentis* M101-4. Biosci. Biotech. Biochem. 61: 312–316.

Huber M, Mundlein A, Dornstauder E, Schneeberger C, Tempfer CB, Mueller MW & Schmidt WM (2002) Accessing single nucleotide polymorhisms in genomic DNA by direct multiplex polymerase chain reaction amplification on nucleotide microarrays. Anal. Biochem. 303: 25–33.

Hugenholtz J, Hols P, Starrenburg MJC, de Vos, WM & Kleerebezem M (2000) Metabolic engineering of *Lactococcus lactis* leading to high diacetyl production. Appl. Environ. Microbiol. 66: 4112–4114.

Humbelin M, Griesser V, Keller T, Schurter W, Haiker M, Hohmann H-P, Ritz H, Richter G, Bacher A & van Loon APGM (1999) J. Ind. Microbiol. Biotechnol. 22: 1–7.

Kawaguchi M, Fujioka Y, Kishimoto M, Matsuma T & Ichikawa T (1999) Dietary fiber and phophorylated oligosaccharide protect small intestinal mucosa of rat from oxidative damages in vitro. J. Jap. Soc. Food Sci. Technol. 46: 487–490.

Khane SK, Jha K, Gandhi AP & Gupta MN (1994) Hydrolysis of flatulence-causing galactooligosaccharides by agarose-entrapped *Aspergillus oryzae* cells. Food Biochem. 51: 29–31.

Kim P, Yoon SH, Roh HJ & Choi JH (2001) High production of D-tagatose, a potential sugar substitute, using immobilized L-arabinose isomerase. Biotechnol. Prog. 17: 208–210.

Kitazawa H, Toba T, Itoh T, Kumano N, Adachi S & Yamaguchi T (1991) Antitumoral activity of slime-froming, encapsulated *Lactococus lactis* subsp. *cremoris* isolated from Scandinavian ropy sour milk, 'viili'. Animal Sci. Technol. 62: 277–283.

Klapa MI, Park SM, Sinskey AJ & Stephanopoulos G (1999) Metabolite and isotopomer balancing in the analysis of metabolic cycles: theory. Biotechnol. Bioeng. 62: 375–391.

Kleerebezem M, van Kranenburg R, Tuinier R, Boels IC, Zoon P, Looijesteijn E, Hugenholtz J & de Vos WM (1999) Exopolysaccharides produced by *Lactococcus lactis*: from genetic engineering to improved rheological properties? Antonie van Leeuwenhoek 76: 357–365.

Kleerebezem M, Beerthuyzen MM, Vaughan EE, de Vos WM & Kuipers OP (1997) Controlled gene expression systems for lactic acid bacteria: transferable nisin inducible expression cassettes for *Lactococcus, Leuconostoc* and *Lactobacillus* spp. Appl. Environ. Microbiol. 63: 4581–4584.

Kleerebezem M, Hols P & Hugenholtz P (2000) Lactic acid bacteria as a cell factory: rerouting of carbon metabolism in *Lactococcus lactis* by metabolic engineering. Enzyme Microb. Technol. 26: 840–848.

Kokoris M, Dix K, Moynahan K, Mathis J, Erwin B, Grass P, Hines B & Duesterhoeft A (2000) High-throughput SNP genotyping with the Masscode system. Mol. Diagn. 5: 329–340.

Kuipers OP, de Ruyter PGGA, Kleerebezem M & de Vos WM (1998) Quorum sensing-controlled gene expression in lactic acid bacteria. J. Biotech. 64: 15–21.

Lakshini AV (1998) Riboflavin metabolism – relevance to human nutrition. Ind. J. Med Res. 108: 182–190.

Levander F, Andersson U & Radstrom P (2001) Physiological role of beta-phosphoglucomutase in *Lactococus lactis*. Appl. Environ. Microbiol. 67: 4556–4553.

Levander F & Rådström P (2001) Requirement for phosphoglucomutase in exopolysaccharide biosynthesis in glucose- and lactose-utilizing *Streptococcus thermophilus*. Appl. Environ. Microbiol. 67: 2734–2738.

Looijesteijn PJ, Boels IC, Kleerebezem M & Hugenholtz J (1999) Regulation of exopolysaccharide production by *Lactococcus lactis* subsp. *cremoris* by the sugar source. Appl. Environ. Microbiol. 65: 5003–5008.

Lopez de Felipe F, Starrenburg MJC & Hugenholtz J (1997) The role of NADH-oxidation in acetoin and diacetyl production from glucose in *Lactococcus lactis*. FEMS Microbiol. Lett. 156: 15–19.

Lopez de Felipe F, Kleerebezem M, de Vos WM & Hugenholtz J (1998) Cofactor engineering: a novel approach to metabolic engineering in *Lactococcus lactis* by controlled overexpression of NADH oxidase. J. Bacteriol. 180: 3804–3808.

Maity TK & Paul SC (1991) Low-oligosaccharide soy milk: application of a-galactosidase for hydrolysing soy-oligosaccharide. Ind. Dairyman 43: 443–448.

Manzoni M, Rollini M & Bergomi S (2001) Biotransformations of D-galactitol to tagatose by acetic acid bacteria. Proc. Biochem. 36: 971–977.

McBurney MW & Whitmore GF (1974) Isolation and biochemical characterization of folate deficient mutants of Chinese hamster cells. Cell 2: 173–182.

Nakajima H, Suzuki Y, Kaizu H & Hirota T (1992a) Cholesterol lowering activity of ropy fermented milk. J. Food Sci. 57: 1327–1329.

Nakajima H, Hirota T, Toba T, Itoh T & Adachi S (1992b) Structure of the extracellular polysaccharide from slime-forming *Lactococcus lactis* subspecies *cremoris* SBT0495. Carbohydr. Res. 224: 245–253.

Neta T, Tabata K & Hirasawa M (2000) Low-cariogenicity of trehalose as a substrate. J. Dent 28: 571–576.

Neves AR, Ramos A, Shearman C, Gasson MJ, Almeida JS & Santos H (2000) Metabolic characterization of *Lactococcus lactis* deficient in lactate dehydrogenase using in vivo 13C-NMR. Eur. J. Biochem. 267: 3859–3868.

Park SM, Klapa MI, Sinskey AJ & Stephanopoulos G (1999) Metabolite and isotopomer balancing in the analysis of metabolic cycles: applications. Biotechnol. Bioeng. 62: 392–401.

Pavan S, Hols P, Delcour J, Geoffroy MC, Grangette C, Kleerebezem M & Mercenier A (2000) Adaptation of the nisin-controlled expression system in *Lactobacillus plantarum*: a tool to study *in vivo* biological effects. Appl. Environ. Microbiol. 66: 4427–4432.

Pszczola DE (1992) The nutraceutical initiative: a proposal for economic and regulatory reform. Food Biotechnol. 46(4): 77–79.

Ramos A, Boels IC, de Vos WM & Santos H (2001) Relationship between glycolysis and exopolysaccharides biosynthesis in *Lactococcus lactis*. Appl. Environ. Microbiol. 67: 33–41.

Rollin C, Morgant V, Guyonvarch A & Guerquin-Kern JL (1995) 13C-NMR studies of *Cyanobacterium melassecola* metabolic pathways. Eur. J. Biochem. 227: 488–493.

Rooijen van RJ, van Schalkwijk S & de Vos WM (1991) Molecular cloning, characterization and nucleotide sequence of the tagatose-6-phosphate pathway gene cluster of the lactose operon of *Lactococcus lactis*. J. Biol. Chem. 266: 7176–7181.

Rosenberg IH & Godwin HA (1971) Inhibition of intestinal gamma-glutamyl carboxypeptidase by yeast nucleic acid: an explanation of variability in utilization of dietary polyglutamyl folate. J. Clin. Investig. 50: 78a.

Ross NS & Klein MR (1990) Riboflavin deficiency in cultured rat hepatoma cells: a model for studying the hepatic effects of riboflavin deficiency. In Vitro Cell Dev. Biol. 26: 280–287.

Schmidt K, Nielsen J & Villadsen J (1999) Quantitative analysis of metabolic fluxes in *Escherichia coli*, using two-dimensional NMR spectroscopy and complete isotopomer models. J. Biotechnol. 71: 175–189.

Seyoum E & Selhub J (1998) Properties of food folates determined by stability and susceptibility to intestinal pteroylpolyglutamate hydrolase action. J. Nutr. 128: 1956–1960.

Shen B, Jensen RG & Bohnert HJ (1997) Mannitol protects against oxidation by hydroxyl radicals. Plant Phys. 115: 527–532.

Silvestroni A, Sesma F, Savoy G, Connes C & Piard J-C (2002) Structure, organization, and expression of the melA structural gene for α-galactosidase in *Lactobacillus plantarum* ATCC 8014. Submitted for publication.

Smid EJ, Starrenburg MJC, Mierau I, Sybesma W & Hugenholtz J (2001) Increase of folate levels in fermented foods. Innovations in Food Technology Feb/Mar, 13–15.

Smirnoff N & Cumbes QJ (1989) Hydroxyl radical scavenging activity of compatible solutes. Phytochemistry 28: 1057–1060.

Soetaert W, Buchholz K & Vandamme EJ (1995) Production of D-mannitol and D-lactic acid by fermentation with *Leuconostoc mesenteroides*. In: Agro Food Ind. High-Tech (pp 41–44).

Sokinenko EV (2001) Discovering the sweeping power of point mutations using a GIRAFF. Trends Microbiol. 9: 522–525.

Stingele F, Vincent SJF, Faber EJ, Newel JW, Kamerling JP & Neeser JR (1999) Introduction of the exopolysaccharide gene cluster from *Streptococcus thermophilus* Sfi6 into *Lactococcus lactis* MG1363: production and characterization of an altered polysaccharide. Mol. Microbiol. 32: 1287–1295.

Sutherland IW (1998) Novel and established applications of microbial polysaccharides. TIBTECH 16: 41–46.

Sybesma W, Hugenholtz J, Mierau I & Kleerebezem M (2001) Improved efficiency and reliability of RT-PCR using tag-extended RT primers and temperature gradient PCR. BioTechniques 31: 466–468.

Sybesma W, van den Born E, Starrenburg M, Mierau I, Kleerebezem M, de Vos WM & Hugenholtz J (2002) Increased production of bioavailable folate by engineering of *Lactococcus lactis*. Submitted for publication.

van Kranenburg R, Marugg JD, van Swam II, Willem NJ & de Vos WM (1997) Molecular characterization of the plasmid-encoded *eps* gene cluster essential for exopolysaccharide biosynthesis in *Lactococcus lactis*. Mol. Microbiol. 24: 387–397.

van Kranenburg R, van Swam II, Marugg JD, Kleerebezem M & de Vos WM (1999a). Exopolysaccharide biosynthesis in *Lactococcus lactis* NIZO B40: functional analysis of the glycosyltransferase genes involved in synthesis of the polysaccharide backbone. J. Bacteriol. 181: 338–340.

van Kranenburg R, Vos HR, van Swam II, Kleerebezem M & de Vos WM (1999b) Functional analysis of glycosyltransferase genes from *Lactococcus lactis* and other Gram-positive cocci: complementation, expression, and diversity. J. Bacteriol. 181: 6347–6353.

van Kranenburg R, Kleerebezem M & de Vos WM (2000) Nucleotide sequence analysis of the lactococcal EPS plasmid pNZ4000. Plasmid 43: 130–136.

Vaughan EE, van den Bogaard PT, Catzeddu P, Kuipers OP & de Vos WM (2001) J. Bacteriol. 183: 1184–1194.

Vedamuthu ER & Neville JM (1986) Involvement of a plasmid in production of ropiness (mucoidness) in milk cultures by *Streptococcus cremoris* MS. Appl. Environ. Microbiol. 51: 1385–1386.

Veenhoff LM & Poolman B (1999) Substrate recognition at the cytoplasmic and extracellular binding site of the lactose transport protein of *Streptococcus thermophilus*. J. Biol. Chem. 274: 33244–33250.

Von Wright A & Tynkkynen S (1987) Construction of *Streptococcus lactis* subsp. *lactis* strains with a single plasmid associated with mucoid phenotype. Appl. Environ. Microbiol. 53: 1385–1386.

Wald N (1991) Prevention of neural tube defects. Results of the Medical Research Council vitamin study. Lancet. 338: 131–137.

Weinstock GM, Smajs D, Hardham J & Norris SJ (2000) From microbial genome sequence to application. Res. Microbiol. 151: 158.

Wren BW (2000) Microbial genome analysis: insights into virulence, host adaptation and evolution. Nat. Rev. Genet. 1: 30–39.

Yao R, Schneider E, Ryan TJ & Galivan J (1996) Human gamma glutamyl hydrolase: cloning and characterization of the enzyme expressed in vitro. Proc. Natl. Acad. Sci. U.S.A. 93: 10134–10138.

Zehner LR (1988) D-tagatose as a low-calorie carbohydrate sweetener and bulking agent. European Patent Application.

Antonie van Leeuwenhoek **82:** 237–248, 2002.

Experimental determination of control of glycolysis in *Lactococcus lactis*

Brian J. Koebmann, Heidi W. Andersen, Christian Solem & Peter R. Jensen*
*Section of Molecular Microbiology, BioCentrum-DTU, Technical University of Denmark, Building 301, DK-2800 Kgs. Lyngby, Denmark (*Author for correspondence; Tel.: +45-45252510; Fax: +45-45932809; E-mail: prj@biocentrum.dtu.dk)*

Key words: ATPase, control, flux, glycolysis, Metabolic Control Analysis, metabolism

Abstract

The understanding of control of metabolic processes requires quantitative studies of the importance of the different enzymatic steps for the magnitude of metabolic fluxes and metabolite concentrations. An important element in such studies is the modulation of enzyme activities in small steps above and below the wild-type level. We review a genetic approach that is well suited for both Metabolic Optimization and Metabolic Control Analysis and studies on the importance of a number of glycolytic enzymes for metabolic fluxes in *Lactococcus lactis*. The glycolytic enzymes phosphofructokinase (PFK), glyceraldehyde-3-phosphate dehydrogenase (GAPDH), pyruvate kinase (PYK) and lactate dehydrogenase (LDH) are shown to have no significant control on the glycolytic flux in exponentially growing cells of *L. lactis* MG1363. Introduction of an uncoupled ATPase activity results in uncoupling of glycolysis from biomass production. With MG1363 growing in defined medium supplemented with glucose, the ATP demanding processes do not have a significant control on the glycolytic flux; it appears that glycolysis is running at maximal rate. It is likely that the flux control is distributed over many enzymes in *L. lactis*, but it cannot yet be excluded that one of the remaining glycolytic steps is a rate-limiting step for the glycolytic flux.

Introduction

The lactic acid bacterium *Lactococcus lactis* has been exploited for centuries in the production of food products, where the primary function is the homofermentative conversion of sugar through glycolysis to the major end product lactate. The resulting acidification preserves the fermented food against spoilage and contributes to the texture and organoleptic quality, and from an industrial point of view there is much interest to increase the overall flux through glycolysis and to control the production of other end products than lactate.

The living cell is a complex network of corresponding biochemical reactions, and a proper understanding of the interplay between these reactions is a pre-requisite for performing metabolic engineering. With respect to *L. lactis*, the metabolic pathways are known and a genome sequence is now available (Bolotin et al. 2001).

The free energy metabolism of *L. lactis* is relatively simple. During fermentation approximately 90% of the substrate is recovered as by-products, primarily lactate. This fermentation pattern shows that the role of glycolysis is almost exclusively to supply ATP for cell functions and growth (Figure 1a). Since oxidative phosphorylation does not normally occur in *L. lactis*, ATP is only generated by glycolysis. The $NADH/NAD^+$ redox balance closes by the conversion of pyruvate to the major end product lactate. This allows for an almost perfect splitting of the metabolism into a catabolic module generating ATP (ATP supply) and an anabolic module where building blocks are assembled into macromolecules and consuming ATP (ATP demand). The anabolic and catabolic modules are linked mainly by the ATP/ADP ratio (Figure 1b).

The regulation of glycolysis and the shift between different fermentation modes of *L. lactis* has been extensively studied (Crow & Pritchard 1977; Thompson 1978, 1987; Mason et al. 1981; Fordyce et al. 1982; Cocaign-Bousquet et al. 1996; Garrigues et al. 1997,

(a)

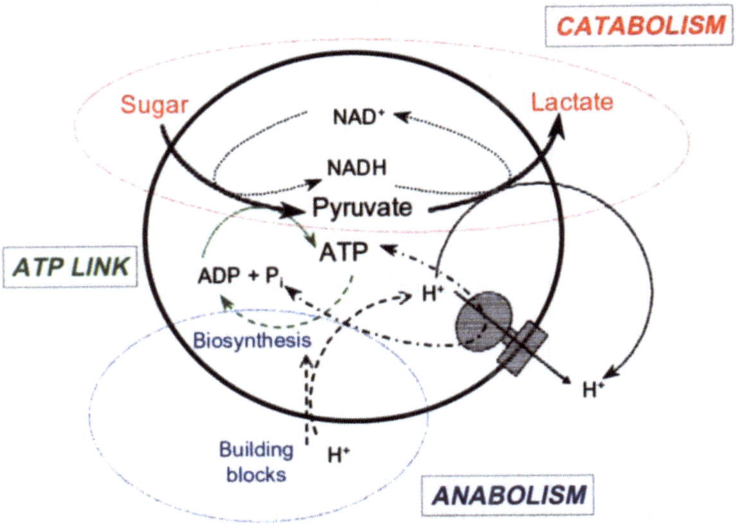

(b)

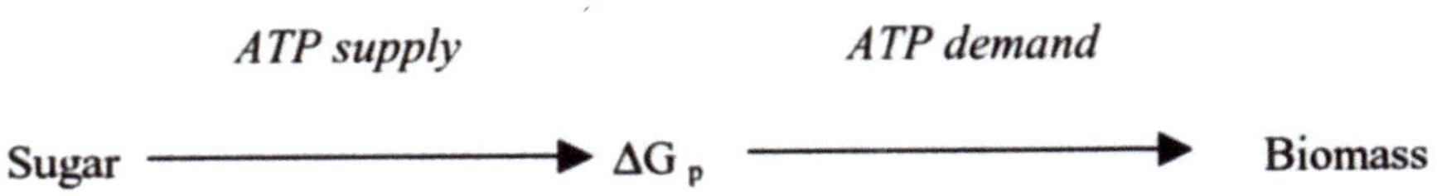

Figure 1. Free energy metabolism in *Lactococcus lactis*. (a) A simplified overview of the metabolism in *L. lactis*, where the synthesis and hydrolysis of ATP link catabolism and anabolism together. Primary energy is accomplished in the process of catabolism from sugar uptake to pyruvate and secondary energy is gained by excretion of charged compounds, symbolised as H^+. Dotted, full, dashed and half-dashed lines symbolize redox, energy generating, energy consuming, and synthesis/hydrolysis of ATP, respectively. (b) Simplified model of the energy metabolism in *L. lactis*, divided into a module of ATP supply (glycolysis) and ATP demand (anabolism, transport, maintenance). ΔG_p indicates the free-energy level.

2001a,b; Luesink et al. 1998; Jensen et al. 2001). However, the mapping of regulatory mechanisms does not necessarily lead to an understanding of which enzymes have *control* on the flux.

Metabolic Engineering has been applied in attempts to improve the fluxes and the performance of various microorganisms. The strategies have usually aimed at eliminating an undesired metabolic activity, introducing a new enzymatic activity or many fold overexpression of an enzyme thought to be the bottleneck of a metabolic pathway. Sometimes these strategies have indeed been successful in redirecting the flux and achieving an increased production of a desired product (e.g. Platteeuw et al. 1995; Hugenholtz et al. 2000; Kleerebezem et al. 2000). However, more often the results of such attempts have been disappointing (e.g. Schaaff et al. 1989; Snoep et al. 1995; Bakker et al. 1999; Smits et al. 2000), and the control of the flux through a metabolic pathway may well be distributed over many enzymatic steps.

Introduction to metabolic control analysis (MCA)

According to Metabolic Control Analysis (Kacser & Burns 1973; Heinrich & Rapoport 1974) flux control by an enzymatic step can be determined by changing the activity of the enzyme away from the normal enzyme activity and measure the effect on the metabolic flux. Preferably, one should achieve enzyme activities both above and below the wild-type level and fit a curve through the data points. If for instance 10% more enzyme results in 5% higher flux through a pathway, then the enzyme has a control coefficient of 0.5

-35 box **-10 box**

Oligoseq MCS-CATNNNNNAGTTTATTC**TTGACA**NNNNNNNNNNNNNNTGR**TATAAT**ANNWNAGTACTGTT-MCS

(b)

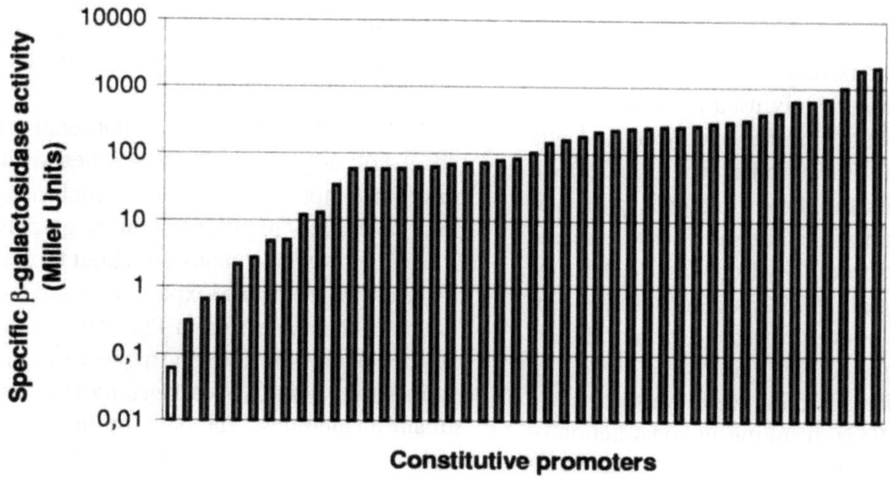

Figure 2. Construction of a library of synthetic constitutive promoters. (a) Part of oligosequence of the area from position −52 to +8 (relative to the putative transcription initiation site). The consensus sequences (−35 box and −10 box) are indicated in bold. MCS indicates multiple cloning sites. (b) β-galactosidase activities of CP promoters in *L. lactis*. The promoter activities were assayed from the expression of a reporter gene (*lacLM*) encoding β-galactosidase transcribed from the different synthetic promoter clones on the promoter cloning vector pAK80 (Israelsen et al. 1995). The specific β-galactosidase activity of pAK80 is indicated in the white column (from Jensen & Hammer 1998b).

with respect to that flux. Mathematically the flux control coefficient of an enzyme E with respect to a flux J is calculated from the equation:

$$C\frac{J}{E} = \frac{dJ}{J} / \frac{dE}{E}$$

The sum of flux control coefficients must always add up to 1, and the finding of low flux control by an enzyme then means that control must be elsewhere in the system. Even in branched pathways, where some enzymes may have negative control on a flux, the sum of flux control must still add up to 1 (Kacser 1983). Similarly, control of metabolite concentrations is defined as the relative change in a metabolite concentration resulting from a small change in the enzyme activity. The sum of control of all enzymes on a metabolite concentration must add up to 0.

Modulation of gene expression

In order to perform Metabolic Control Analysis (MCA) and Metabolic Optimization, it is an advantage to be able to *modulate* or *tune* the expression of genes. Genetic tools for inducible modulation of gene expression in *L. lactis* are available in term of the nisin-promoter (de Ruyter et al. 1996). However, a steady state of gene expression may be difficult to achieve with such systems, and the use of an inducible promoter only allows for changing the expression of a single gene or a set of genes. Metabolic optimization is likely to require the optimization of several genes individually (Jensen & Hammer 1998a).

In 1998 (Jensen & Hammer 1998b) a new approach for obtaining promoter libraries was reported, which is based on randomization of the DNA sequences (spacers) that separate the individual consensus sequences of promoters. The method takes advantage of the fact that actual sequences of bases in these spacer areas are less important for the strength of a particular promoter than the structure of the DNA. By randomizing many base pairs simultaneously in the vicinity of the consensus sequences, it is possible to change the DNA structure and the binding of transcription factors to the promoter sequences. The promoter libraries obtained by this approach contain promoters with virtually any activity and allows for accurate, simultaneous and individual modulation of many enzyme activities in a cell (Figure 2).

Replacement of the *las* promoter with synthetic promoters

The synthetic promoters described above (Jensen & Hammer 1998b) were recently used in an attempt to change the expression of the *las* operon in *L. lactis* subsp. *cremoris* MG1363 (Andersen et al. 2001a). This operon harbours the three genes *pfk*, *pyk*, and *ldh* coding for phosphofructokinase (PFK), pyruvate kinase (PYK) and lactate dehydrogenase (LDH), respectively (Llanos et al. 1993). Two constitutive promoters each flanked by the upstream region of the *las* operon and the truncated *pfk* region were cloned on an *E. coli* vector. The plasmids were transformed in MG1363 (Gasson 1983), resulting in construction of two mutant strains in which two synthetic constitutive promoters with different strengths had replaced the native *las* promoter (Andersen et al. 2001a). Surprisingly, the *las* mutants were found to have uncoordinated expression of *pfk*, *pyk* and *ldh* relative to the wild type. While the mutants had an approximately two-fold decrease in PFK activity, the activities of PYK and LDH remained closer to the wild-type level. A likely explanation for this phenomenon was that the functional half-life for the *pfk* part of the mRNA had been altered in the cloning process.

The lower PFK activities resulted in a reduction of the growth rate to 57 and 70% of wild-type rate and the glycolytic flux was reduced to 62 and 76% of wild-type levels in defined SA medium (Jensen & Hammer 1993) supplemented with glucose. The growth rate in complex M17 medium (Terzaghi & Sandine 1975) was even further reduced.

In spite of the significant effect of the low PFK activity on the glycolytic flux, the strains remained homolactic. This is interesting because the reduced catabolic activity was expected to result in energy limitation and a shift of the metabolism towards mixed acid fermentation (Thomas et al. 1979; Fordyce et al. 1984). An alternative explanation for the altered physiology could be increased pools of the hexose-phosphates, which were found to be increased two- to fourfold in the mutants. Elevated levels of sugar-phosphates have previously been reported to be toxic to the cells (Fraenkel 1968). Indeed, when the mutants were grown in a glucose-limited chemostat the growth rates could be gradually increased to 195% of the maximal growth rate observed in glucose-saturated batch cultures, which demonstrates that the negative effects caused by the low PFK activity can be relieved by restricting the sugar flux into the cells. Further support for this hypothesis was obtained when the mutants were grown on slow fermentable sugars with maltose as carbon/energy source (Andersen et al. 2001a). Here the growth rates of the PFK mutants were identical to the wild type strain MG1363.

The fact that an almost proportional reduction in the glycolytic flux is observed when reducing PFK makes it tempting to draw the conclusion that PFK must have a very high control on the glycolytic flux in *L. lactis*. However, conclusions about flux control cannot be drawn from these experiments and require modulation of the expression of PFK activity instead. The elevated pools of hexose phosphates indicate that PFK exerts strong control over the *concentration* of the up-stream metabolites. The data are in good agreement with a mathematic model in which the consequences of altering PFK activity is a poor homeostasis of the glycolytic intermediates (Thomas & Fell 1998).

Control by lactate dehydrogenase (LDH) activity on glycolysis in MG1363

The approach of modulating gene expression via synthetic promoters (Jensen & Hammer 1998b) has also been used to study the importance of LDH for metabolic fluxes in *L. lactis* MG1363. A full version of the *ldh* gene was cloned behind a set of constitutive promoters in a plasmid vector that allows for site-specific integration in a phage attachment site on the chromosome (Brøndsted & Hammer 1999). The plasmid constructions were then integrated in MG1363 and in an *ldh* disruption mutant, resulting in a series of mutant strains with modulated LDH activities ranging from 1 to 133% of the wild-type level (Andersen et al. 2001b). Changing the LDH activity from 59 to 133% of wild-type level had virtually no effect on the glycolytic flux and growth rate (Andersen et al. 2001b). Flux control coefficients were determined and it was found that LDH exerted virtually no control on the growth rate, glycolytic flux and lactate production at the wild type enzyme level (Figure 3). As expected, a shift towards the mixed acid products was observed in a strain deleted for LDH, but the enzyme was also found to have a high negative control on the flux to

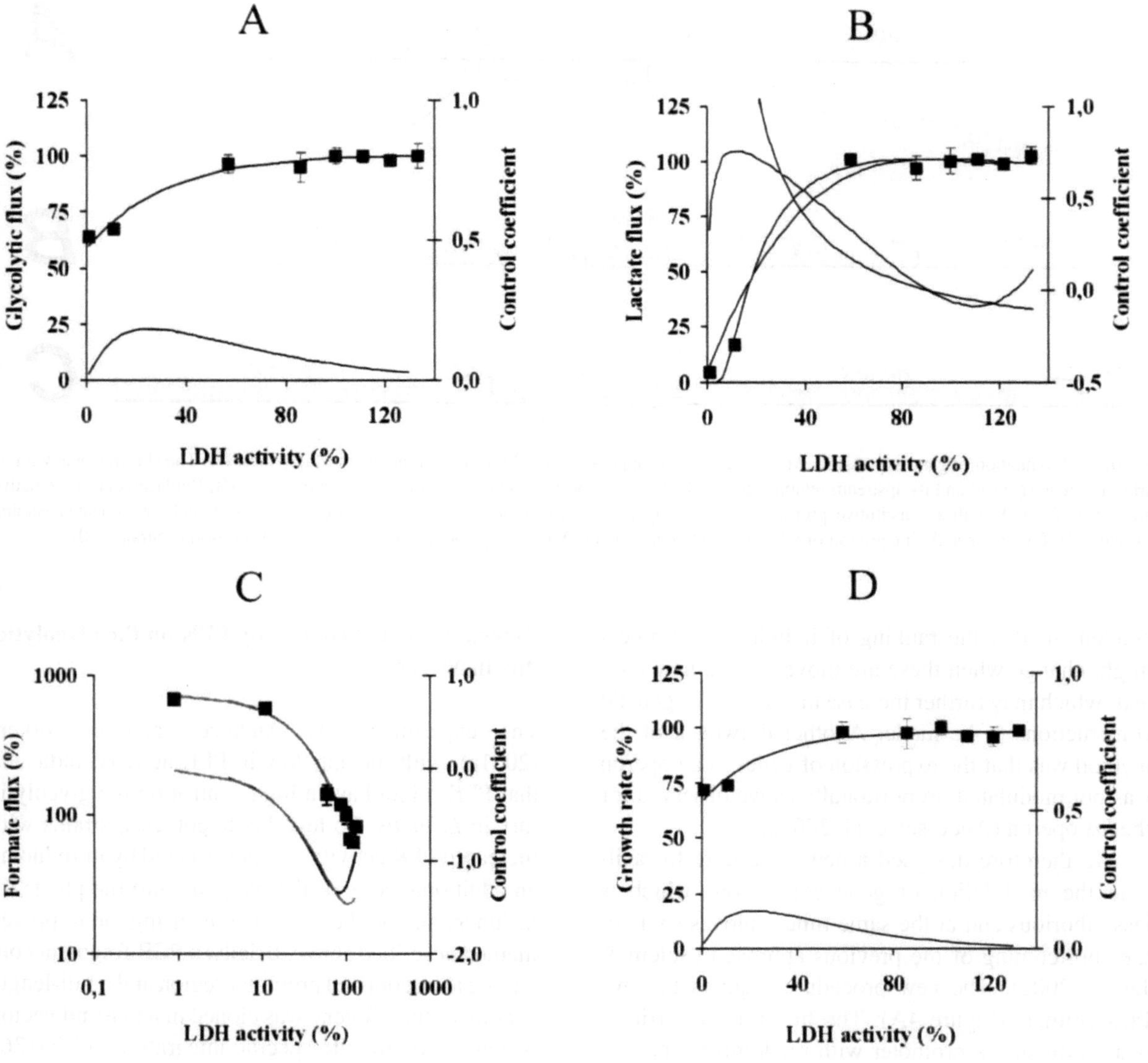

Figure 3. Control of lactate dehydrogenase on metabolic fluxes and growth rate. Glycolytic flux, lactate flux, formate flux, and growth rate are represented in A, B, C and D, respectively. The relative metabolic fluxes and the relative growth rate are shown as a function of the relative lactate dehydrogenase activity (squares). The experimental data points are fitted to curves shown by penetrating lines and lines without symbols show the calculated control coefficients (from Andersen et al. 2001b).

formate already at the wild-type level ($C = -1.3$; Figure 3c). This result is in good agreement with data obtained recently from a kinetic model, which showed that LDH has a high negative control on the flux to acetolactate in aerated *L. lactis* (Hoefnagel et al. 2002).

It has been suggested that shift from homolactic towards mixed acid fermentation is mainly regulated by the NADH/NAD^+ level (Garrigues et al. 1997; de Felipe et al. 1998). However, since the modulation of LDH close to the wild-type level did not affect the growth rate or the glycolytic flux, it is unlikely that changes in the NADH/NAD^+ratio played a role in the experiments with modulated LDH expression (Andersen et al. 2001b).

A new approach to ease gene modulation in *L. lactis*

Extensive cloning work is required to bring several individual promoters to express a chromosomal gene such as the *ldh* gene in the above experiments. Besides, the strength of promoters is often context de-

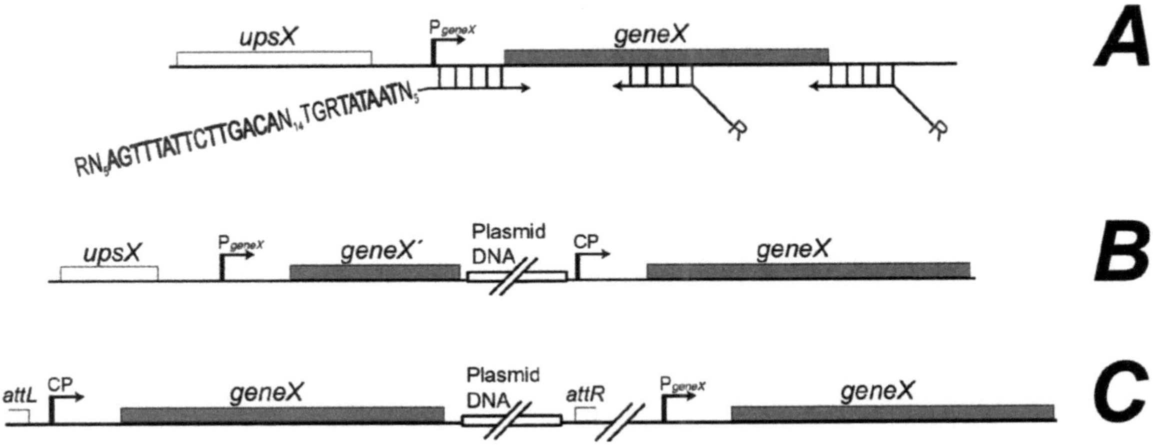

Figure 4. Modulation of gene expression. An overview of the approaches used for modulating enzyme activities. (A) The chromosome with an arbitrary gene (*geneX*) and its upstream region (*upsX*). The promoter primer and two reverse primers are shown. (B) Replacement of the native promoter of *geneX* with a constitutive promoter by homologous recombination between the chromosomal *geneX* and the truncated version (Strategy I) (C) Site-specific integration of a full-length version of *geneX* in a phage attachment site on the chromosome (Strategy II).

pendent so that the ranking of individual promoters might change when these are moved into a new context, which may further increase the number of genetic constructions to be made. Another drawback of the method was that the expression of genes in an operon was not modulated proportionally as we observed for the *las* operon (Andersen et al. 2001a).

We therefore designed a new procedure to facilitate the modulation of gene expression, which is less laborious and at the same time removes some of the shortcoming of the previous approach (Solem & Jensen 2002). The new procedure requires the two PCR-primers (Figure 4A): The first primer carries a restriction site, a promoter with randomized spacers, and a sequence with homology to the 5′-end of the mRNA of the target gene. The second primer also carries a suitable restriction site and a sequence with homology either within the gene in question or downstream of it. This allows for the generation of either a truncated or a full-length version of the gene behind synthetic promoters depending on whether or not an extra gene copy is to be introduced. The truncated version is used for replacing the chromosomal promoter with a synthetic promoter through homologous recombination (Figure 4B), while the full-length version can be used for introducing an extra gene copy on the chromosome, for instance by site-specific integration in a phage attachment site (Figure 4C). The method was applied to chromosomal genes in *Lactococcus lactis* and shown to generate libraries of clones with broad ranges of expression levels of target genes, see below.

Determination of control by PFK on the glycolytic flux in MG1363

The experiments by Andersen and co-workers (2001a), with mutants low in PFK activity, indicated that PFK might have a high control on the glycolytic flux in *L. lactis*. To test this hypothesis, strains with increased PFK activity was constructed by introducing an additional copy of the *pfk* gene into the phage attachment site on the chromosome using the improved method described above. Briefly, a PCR fragment containing a randomized promoter region and a full-length version of the *pfk* gene was cloned in a plasmid vector, which allows for site-specific integration in MG1363 (Brøndsted & Hammer 1999; Solem & Jensen 2002). The obtained plasmid library was then introduced to MG1363, which resulted in a series of strains with PFK activities from 1.4 to 11 times the wild-type activity (Figure 5). In this example the *pfk* gene was cloned upstream of a *gusA* gene encoding β-glucuronidase (Platteeuw et al. 1994), resulting in an operon structure with both genes transcribed from a common promoter. We found a linear correlation between the levels of expression of the *gusA* and *pfk* genes, which facilitates the screening for mutants with suitable enzyme activities (Figure 5a). The strains with increased PFK activity were then used to study the control of PFK on the glycolytic flux in cells of MG1363 growing exponentially in SA medium (Jensen & Hammer 1993) supplemented with glucose. We found that both the growth rate and glycolytic flux decreased slightly

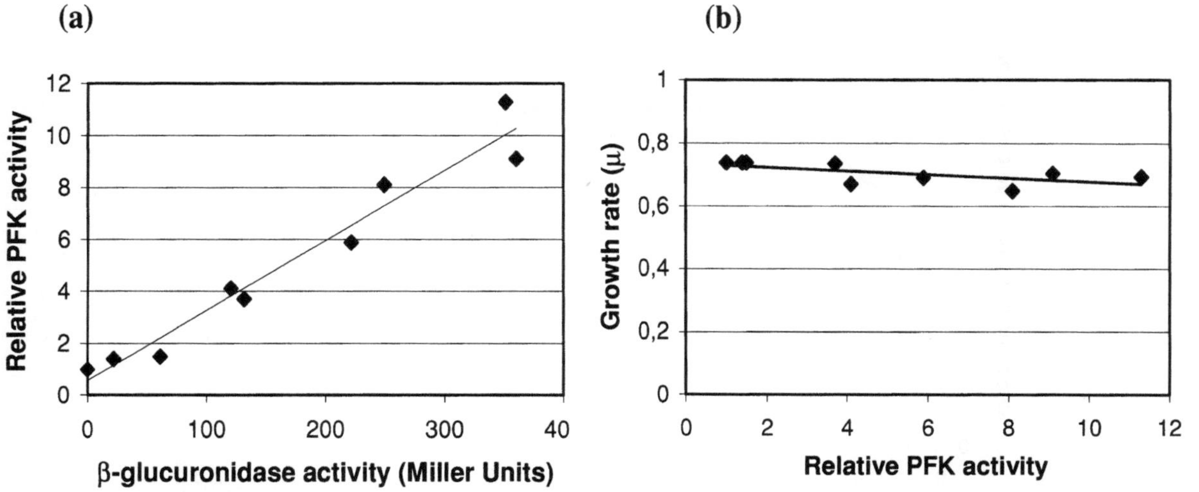

Figure 5. Modulation of phosphofructokinase activity and the effect on growth. (a) Relationship between the measured β-glucuronidase and PFK activities (from Solem & Jensen 2002). (b) Effect on growth rate by overexpression of PFK activity (from Koebmann et al., manuscript in preparation).

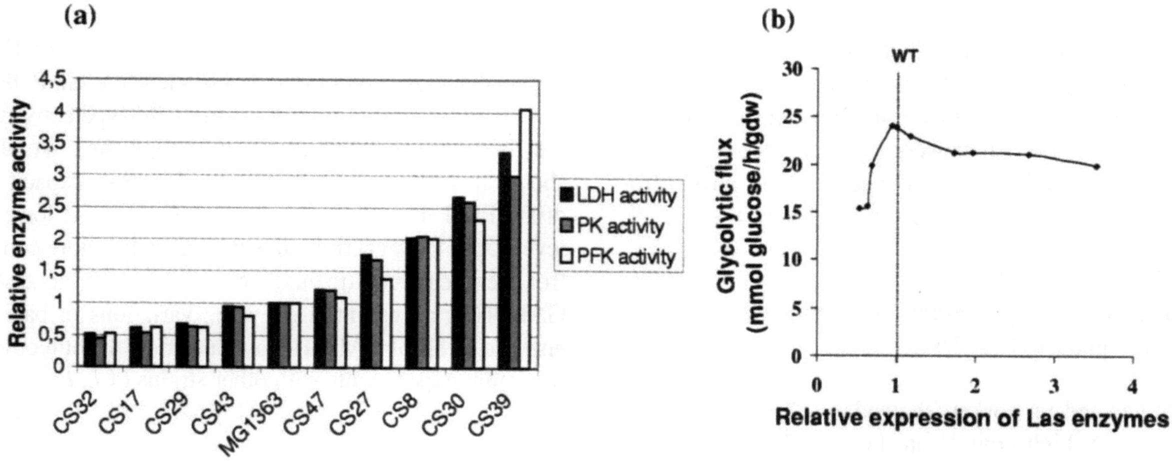

Figure 6. Modulation of the *las* operon. (a) Relative enzyme activity measured in strains with synthetic promoters in front of the *las* operon. (b) Effect of modulation of *las* enzymes on the glycolytic flux (from Koebmann et al., manuscript in preparation).

upon overexpression of *pfk*, and we conclude that PFK has no control on the glycolytic flux in MG1363 under these conditions (Figure 5b).

Determination of control by the three *las* enzymes

We also used the method for constructing a series of strains in which the expression of the *las*-operon, containing the genes *pfk*, *pyk* and *ldh*, was modulated. A truncated *pfk* fragment fused to a library of synthetic promoters was generated by PCR and cloned in an *E. coli* vector, which is unable to replicate in *L. lactis*. The resulting plasmid library was subsequently transformed into MG1363, and strains in which the plasmids had crossed into the chromosome through homologous recombination with the chromosomal *pfk* locus were obtained (Solem and Jensen 2002). Primer extension showed that the *las* mRNA in these mutants was initiated at the same position as in the wild-type and the modulation affected the activities of all three enzymes to the same extent (Figure 6a). The strains had enzyme activities ranging from 0.5 to 3.5 of the wild-type level and allowed us to determine how important the expression of the *las* enzymes is for glycolysis in *L. lactis* (Figure 6b). We found that the

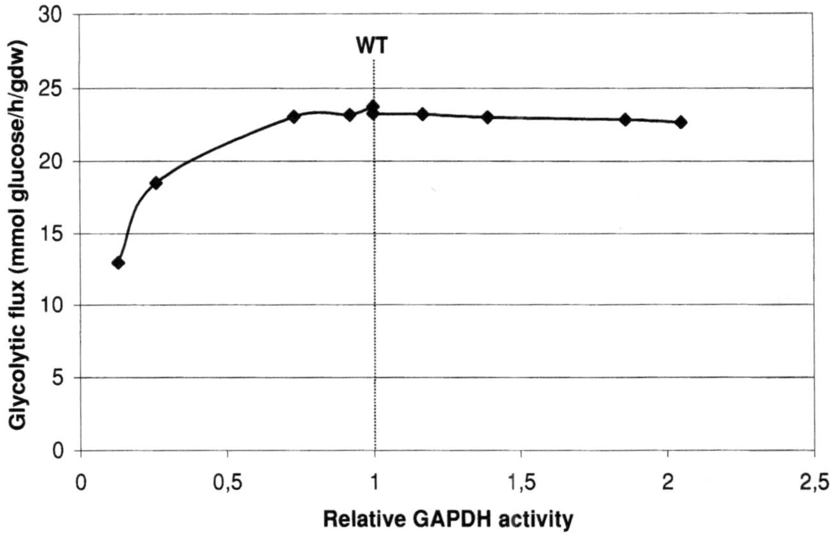

Figure 7. Effect of glyceraldehyde-3-phosphate dehydrogenase activity on the glycolytic flux. The glycoytic flux in steadily growing cells is shown as a function of the relative GAPDH activity (from Solem et al., manuscript in preparation).

highest flux was obtained with the strain, which had wild-type expression of the *las* enzymes, whereas both lower and higher activities resulted in a decreased flux (Koebmann et al., manuscript in preparation). Thus, the enzymes encoded by the *las* operon have zero control on the glycolytic flux in MG1363.

Modulation of glyceraldehyde-3-phosphate dehydrogenase (GAPDH)

Poolman and co-workers showed in 1987 that GAPDH had a very high control on the glycolytic flux in *L. lactis* subsp. *cremoris* Wg2 (Poolman et al. 1987). These experiments, which used the inhibitor titration method that allows the investigator to decrease enzyme activities, were performed with non-growing cells resuspended in buffer. In order to determine the control by GAPDH in growing cells we constructed a series of mutants of MG1363 with altered expression of GAPDH, by: (a) integrating a truncated version of the *gapB* locus on the chromosome as described for the *las* operon, and (b) by site-specific integration on the chromosome as described for PFK (Solem et al. manuscript in preparation). The resulting strains had GAPDH activities modulated from 13 to 210% of the activity in strain MG1363. Changes in GAPDH close to the wild-type level had no significant effect on the growth rate or on the glycolytic flux (Figure 7) when the cells where growing steadily in SA medium supplemented with glucose. We therefore conclude that GAPDH has virtually zero control on the glycolytic flux under these conditions. Even when the GAPDH activity was reduced to 25% of the wild-type activity the flux was still 80% of the wild-type flux, which shows that there is at least 3 fold excess capacity of GAPDH in wild-type cells of MG1363. At lower levels of GAPDH activity the enzyme gains more control over the glycolytic flux, which reflects the fact that GAPDH is an essential enzyme. Variations in basal enzyme levels of GAPDH may thus affect the outcome of similar experiments with other strains of *L. lactis.*

Control by the ATP consuming processes in *L. lactis*

According to Metabolic Control Analysis, the control of the flux through a pathway can also reside in processes outside the pathway itself, for instance in the processes that consume its products (Cornish-Bowden et al. 1995; Hofmeyr & Cornish-Bowden 2000). Indeed, when an additional ATP consuming process (F_1-ATPase) was expressed in steadily growing aerobic *Escherichia coli* cells, the glycolytic flux increased by 70% (Koebmann et al. 2002a) and it was calculated that the ATP demanding processes had almost full control over the glycolytic flux in these cells.

We recently used the same approach for studying the control exerted by the ATP consuming processes on glycolysis in *L. lactis*: The genes coding for the

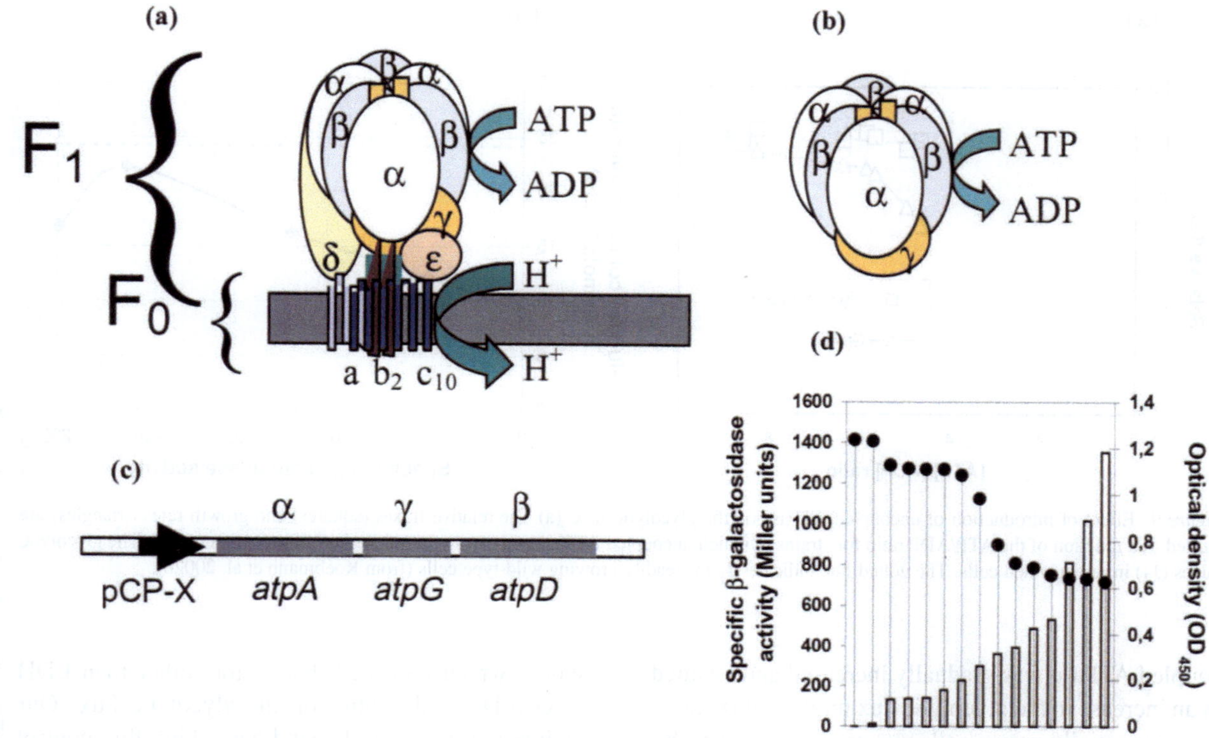

Figure 8. Introduction of uncoupled ATPase activity. (a) The functional (F_1F_0) H^+-ATPase consists of a membrane-bound domain, F_0, and a cytoplasmic domain, F_1. The arrows indicate the interconversion between the membrane proton gradient and the free energy potential in the form of ATP. (b) Uncoupled part of the F_1-domain possessing the catalytic site for ATP hydrolysis. (c) Construction of ATPase plasmids. The arrow indicates a constitutive promoter and the grey bars indicate the three genes *atpA*, *atpG* and *atpD*, coding for the α-, γ-, and β-subunits, respectively. (d) Library of *L. lactis* strains with different expression of uncoupled ATPase activity. The bars indicate the specific β-galactosidase activities. The lines indicate the corresponding biomass yields (from Koebmann et al. 2002b).

three F_1 subunits, α, β and γ from *L. lactis* were cloned into a library of synthetic constitutive promoters of a variety of strengths (Koebmann et al. 2002b) (Figure 8). The genes were cloned in such a way that they formed an operon together with the *lacLM* genes encoding β-galactosidase, and the specific β-galactosidase activity in these cells therefore reflected the expression of the F_1-ATPase genes (Figure 8c).

The uncoupled ATPase activity resulted in a decrease in the biomass yield of cells growing in SA medium supplemented with glucose (Figure 8d). This result shows that glycolysis is becoming gradually more and more uncoupled from biomass production as the expression of the ATPase genes increased. The increase in ATPase activity also resulted in a gradually lower intracellular ATP/ADP ratio, from 9 to 4, and this had a negative effect on the growth rate, which decreased to approximately 75% of the wild-type level (Figure 9A) (Koebmann et al. 2002b). The strains with uncoupled ATPase remained homolactic, except at very high expression level of ATPase where a slight tendency of a shift towards a mixed acid fermentation could be observed. Interestingly, the uncoupled ATPase activity had no significant effect on the glycolytic flux in MG1363 growing in defined medium supplemented with glucose (Figure 9a). The flux control by the ATP demanding processes was here close to zero throughout the range of ATP/ADP ratios.

The absence of control by ATP consuming processes in MG1363 indicated that under these conditions glycolysis is already running at maximal rate. If this is the case, then it might be possible to increase the glycolytic flux in cells, which have low glycolytic flux. To test this hypothesis, experiments were performed with non-growing cells (cells growing exponentially in SA medium with glucose, then harvested and resuspended in buffer). In such cells the glycolytic flux is lower than in growing cells, because ATP is not used for biomass production. The expression of the un-

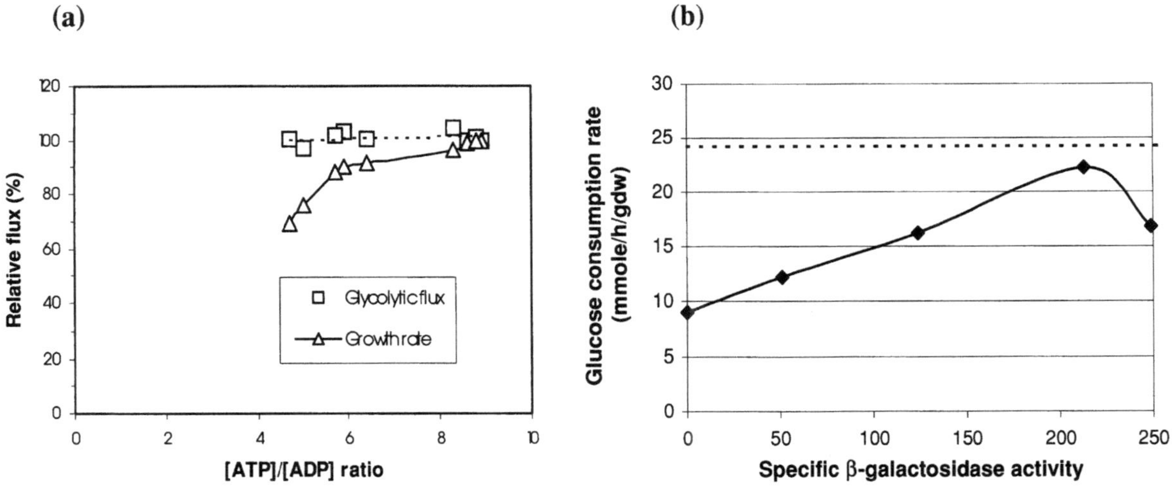

Figure 9. Effect of introduction of uncoupled ATPase on the glycolytic flux. (a) The relative fluxes (squares) and growth rates (triangles) are plotted as a function of the ATP/ADP ratio for strains in which uncoupled ATPase activity are introduced to different extents. (b) The glycolytic fluxes (J_g) in resuspended cells. The dotted line indicates J_g in steadily growing wild-type cells (from Koebmann et al. 2002b).

coupled ATPase was gradually increased and resulted in an increase in the glycolytic flux until the flux found in exponentially grown cells was reached (Figure 9b). Further increase in ATPase expression resulted in a decrease in the glycolytic flux, which indicates that the maximal capacity of glycolysis has been reached. The data show that in non-growing cells supplemented with excess glucose, the demand for ATP does have a high control on the glycolytic flux. This is in good agreement with a model for glycolysis in *L. lactis* that describes the sugar metabolism in *non-growing* cells (Neves et al. 1999) as being largely driven by ATP demand. It appears that in situations where there is an excess capacity of glycolysis the flux through glycolysis can be increased by increasing the ATP turnover in the cells.

In a recent publication it was found that the maximal glycolytic activity of re-suspended cells measured at 30 °C increased approximately 2.5-fold following a shift from 30 to 10 °C for 4 h (Wouters et al. 2000). The phosphoenolpyrvate-dependent sugar phosphotransferase system (PTS) subunit HPr and the catabolite control protein A (CcpA) appeared to be involved in the regulatory response of the stimulated glycolytic flux, and the production of these two proteins were induced two- to threefold upon the shift in temperature. Since a strain with disruption of PTS subunit enzyme I showed increased acidification similar to that in the wild-type strain the PTS uptake system was probably not involved in the response. Neither LDH nor GAPDH were induced upon cold shock, which indicated that factors other than LDH or GAPDH had control on the glycolytic flux. Our finding that the ATP demand has a high flux control in non-growing cells suggests that ATP consumption may be stimulated by the cold shock.

Conclusions and perspectives

Metabolic Engineering is likely to play an important role in the future development of lactic acid bacterial strains with improved properties. The development of a method to ease the modulation of gene expression allows for direct experimental determination of flux control in various microbial systems and should stimulate the interest in the more subtle approach of Metabolic Optimization when designing strains for specific purposes.

The fundamental question of what controls the glycolytic flux in *L. lactis* MG1363 cannot yet be answered unambiguously, except for those cases where the flux is much lower than the maximal capacity and where the flux appears to be controlled by the demand for ATP. When the glycolytic flux is already high, the glycolytic reactions, sugar uptake or lactate efflux is more likely to control the glycolytic flux. But virtually no flux control was found for the glycolytic enzymes PFK, PYK, GAPDH or LDH. We are currently continuing the measurements of individual flux control coefficients for the remaining glycolytic enzymes but a picture is emerging that the control may be distributed over many glycolytic steps, perhaps in combination with the ATP consuming processes.

The majority of the research presented in this review was performed with MG1363 as a model organism and with fixed growth conditions (SA medium, glucose, etc.). Metabolic control analysis requires that, for control coefficients to be comparable, the parameters of the system must be set to the same values. These parameters include both the external conditions for growth (concentrations of nutrients, pH, temperature, etc.) and the cellular make-up, i.e. the enzyme activities. Therefore, the conclusions and control coefficients can not be directly extended to include other LAB, other strains of *Lactococcus* or even other conditions for growth. But the results are likely to be sufficiently general to have some spin-off on future investigations of glycolysis in other setups.

Acknowledgements

The authors would like to thank Martin Bastian Pedersen for discussions and Chr. Hansen A/S, The Danish Academy of Technical Sciences (ATV), The Danish Research Agency, The Danish Dairy Research Foundation (Danish Dairy Board), and The Center of Advanced Food Studies (LMC) for financial support.

References

Andersen HW, Solem C, Hammer K & Jensen PR (2001a) Twofold reduction of phosphofructokinase activity in *Lactococcus lactis* results in strong decreases in growth rate and in glycolytic flux. J. Bacteriol. 183: 3458–3467.

Andersen HW, Pedersen MB, Hammer K & Jensen PR (2001b) Lactate dehydrogenase has no control on lactate production but has a strong negative control on formate production in *Lactococcus lactis*. Eur. J. Biochem. 268: 6379–6389.

Bakker BM, Michels PA, Opperdoes FR & Westerhoff HV (1999) What controls glycolysis in bloodstream form *Trypanosoma brucei*? J. Biol. Chem. 274: 14551–14559.

Bolotin A, Wincker P, Mauger S, Jaillon O, Malarme K, Weissenbach J, Ehrlich SD & Sorokin A (2001) The complete genome sequence of the lactic acid bacterium *Lactococcus lactis* ssp. *lactis* IL1403. Genome Res 11: 731–753.

Brøndsted L & Hammer K (1999) Use of the integration elements encoded by the temperate lactococcal bacteriophage TP901-1 to obtain chromosomal single-copy transcriptional fusions in *Lactococcus lactis*. Appl. Environ. Microbiol. 65: 752–758.

Cocaign-Bousquet M, Garrigues C, Loubiere P & Lindley ND (1996) Physiology of pyruvate metabolism in *Lactococcus lactis*. Antonie Van Leeuwenhoek 70: 253–267.

Cornish-Bowden A, Hofmeyr J-HS & Cardenas ML (1995) Strategies for manipulating metabolic fluxes in biotechnology. Bioorg. Chem. 23: 439–449.

Crow VL & Pritchard GG (1977) Fructose 1,6-diphosphate-activated L-lactate dehydrogenase from *Streptococcus lactis*: kinetic properties and factors affecting activation. J. Bacteriol. 131: 82–91.

de Felipe FL, Kleerebezem M, de Vos WM & Hugenholtz J (1998) Cofactor engineering: a novel approach to metabolic engineering in *Lactococcus lactis* by controlled expression of NADH oxidase. J. Bacteriol. 180: 3804–3808.

de Ruyter PG, Kuipers OP & de Vos WM (1996) Controlled gene expression systems for *Lactococcus lactis* with the food-grade inducer nisin. Appl. Environ. Microbiol. 62: 3662–3667.

Fordyce AM, Moore CH & Pritchard GG (1982) Phosphofructokinase from *Streptococcus lactis* Methods. Enzymol. 90: 77–82.

Fordyce AM, Crow VL & Thomas TD (1984) Regulation of product formation during glucose or lactose limitation in nongrowing cells of *Streptococcus lactis*. Appl. Environ. Microbiol. 48: 332–337.

Fraenkel DG (1968) The accumulation of glucose-6-phosphate from glucose and glucose-6-phosphate dehydrogenase. J. Biol. Chem. 243: 6451–6457.

Garrigues C, Loubiere P, Lindley ND & Cocaign-Bousquet M (1997) Control of the shift from homolactic acid to mixed-acid fermentation in *Lactococcus lactis*: Predominant role of the NADH/NAD^+ ratio. J. Bacteriol. 179: 5282–5287.

Garrigues C, Mercade M, Cocaign-Bousquet M, Lindley ND & Loubiere P (2001a) Regulation of pyruvate metabolism in *Lactococcus lactis* depends on the imbalance between catabolism and anabolism. Biotechnol. Bioeng. 74: 108–115.

Garrigues C, Goupil-Feuillerat N, Cocaign-Bousquet M, Renault P, Lindley ND & Loubiere P (2001b) Glucose metabolism and regulation of glycolysis in *Lactococcus lactis* strains with decreased lactate dehydrogenase activity. Metab. Eng. 3: 211–217.

Gasson MJ (1983) Plasmid complements of *Streptococcus lactis* NCDO 712 and other lactic streptococci after protoplast-induced curing. J. Bacteriol. 154: 1–9.

Heinrich R & Rapoport TA (1974) A linear steady-state treatment of enzymatic chains. General properties, control and effector strength. Eur. J. Biochem. 42: 89–95.

Hoefnagel MHN, Starrenburg MJC, Martens DE, Hugenholtz J, Kleerebezem M, Swam IIV, Bongers R, Westerhoff HV & Snoep JL (2002) Metabolic engineering of lactic acid bacteria, the combined approach: kinetic modelling, metabolic control and experimental analysis. Microbiology 148: 1003–1013.

Hofmeyr JS & Cornish-Bowden A (2000) Regulating the cellular economy of supply and demand. FEBS Lett. 476: 47–51.

Hugenholtz J, Kleerebezem M, Starrenburg M, Delcour J, de Vos W & Hols P (2000) *Lactococcus lactis* as a cell factory for high-level diacetyl production. Appl. Environ. Microbiol. 66: 4112–4114.

Israelsen H, Madsen SM, Vrang A, Hansen EB & Johansen E (1995) Cloning and partial characterization of regulated promoters from *Lactococcus lactis* Tn917-lacZ integrants with the new promoter probe vector, pAK80. Appl. Environ. Microbiol. 61: 2540–2547.

Jensen NB, Melchiorsen CR, Jokumsen KV & Villadsen J (2001) Metabolic behavior of *Lactococcus lactis* MG1363 in microaerobic continuous cultivation at a low dilution rate. Appl Environ Microbiol 67: 2677–2682.

Jensen PR & Hammer K (1993) Minimal requirements for exponential growth of *Lactococcus lactis*. Appl. Environ. Microbiol. 59: 4363–4366.

Jensen PR & Hammer K (1998a) Artificial promoters for metabolic optimization. Biotechnol. Bioeng. 58: 191–195.

Jensen PR & Hammer K (1998b) The sequence between the consensus sequences modulates the strength of prokaryotic promoters. Appl. Environ. Microbiol. 64: 82–87.

Kacser H & Burns JA (1973) The control of flux. Symp. Soc. Exp. Biol. 27: 65–104

Kacser H (1983) The control of enzyme systems *in vivo*: elasticity analysis of the steady state. Biochem. Soc. Trans. 11: 35–40.

Kleerebezem M, Hols P & Hugenholtz J (2000) Lactic acid bacteria as a cell factory: rerouting of carbon metabolism in *Lactococcus lactis* by metabolic engineering. Enz. Microbial. Tech. 26: 840–848.

Koebmann BJ, Westerhoff HV, Snoep JL, Nilsson D & Jensen PR (2002a) The glycolytic flux in *Escherichia coli* is controlled by the demand for ATP. J. Bacteriol. In press.

Koebmann BJ, Solem C, Pedersen MB, Nilsson D & Jensen PR (2002b) Expression of genes encoding F_1-ATPase results in uncoupling of glycolysis from biomass production in *Lactococcus lactis*. Appl. Environ. Microbiol., Manuscript submitted for publication.

Llanos RM, Harris CJ, Hillier AJ & Davidson BE (1993) Identification of a novel operon in *Lactococcus lactis* encoding three enzymes for lactic acid synthesis: phosphofructokinase, pyruvate kinase, and lactate dehydrogenase. J. Bacteriol. 175: 2541–2551.

Luesink EJ, van Herpen RE, Grossiord BP, Kuipers OP & de Vos WM (1998) Transcriptional activation of the glycolytic *las* operon and catabolite repression of the *gal* operon in *Lactococcus lactis* are mediated by the catabolite control protein CcpA. Mol Microbiol. 30: 789–798.

Mason PW, Carbone DP, Cushman RA & Waggoner AS (1981) The importance of inorganic phosphate in regulation of energy metabolism of *Streptococcus lactis*. J. Biol. Chem. 256: 1861–1866.

Neves AR, Ramos A, Nunes MC, Kleerebezem M, Hugenholtz J, de Vos WM, Almeida J & Santos H (1999) *In vivo* nuclear magnetic resonance studies of glycolytic kinetics in *Lactococcus lactis*. Biotechnol. Bioeng. 64: 200–212.

Platteeuw C, Simons G & de Vos WM (1994) Use of the *Escherichia coli* β-glucuronidase (*gusA*) gene as a reporter gene for analyzing promoters in lactic acid bacteria. Appl. Environ. Microbiol. 60: 587–593.

Platteeuw C, Hugenholtz J, Starrenburg M, vanAlen-Boerrigter I & de Vos WM (1995) Metabolic engineering of *Lactococcus lactis*: Influence of the overproduction of α-acetolactate synthase in strains deficient in lactate dehydrogenase as a function of culture conditions. Appl. Environ. Microbiol. 61: 3967–3971.

Poolman B, Bosman B, Kiers J & Konings WN (1987) Control of glycolysis by glyceraldehyde-3-phosphate dehydrogenase in *Streptococcus cremoris* and *Streptococcus lactis*. J. Bacteriol. 169: 5887–5890.

Schaaff I, Heinisch J & Zimmermann FK (1989) Overproduction of glycolytic enzymes in yeast. Yeast 5: 285–290.

Smits HP, Hauf J, Müller S, Hobley TJ, Zimmermann FK, Hahn-Hägerdahl B, Nielsen J & Olsson L (2000) Simultaneous overexpression of enzymes of the lower part of glycolysis can enhance the fermentative capacity of *Saccharomyces cerevisiae*. Yeast 16: 1325–1334.

Snoep JL, Yomano LP, Westerhoff HV & Ingram L (1995) Protein burden in *Zymomonas mobilis*: negative flux and growth control due to overproduction of glycolytic enzymes. Microbiology 141: 2329–2337.

Solem C & Jensen PR (2002) Modulation of gene expression made easy. Appl. Environ. Microbiol. 68 2397–2403.

Terzaghi BE & Sandine WE (1975) Improved medium for lactic streptococci and their bacteriophages. Appl. Microbiol. 29: 807–813.

Thomas TD, Ellwood DC & Longyear VM (1979) Change from homo- to heterolactic fermentation by *Streptococcus lactis* resulting from glucose limitation in anaerobic chemostat cultures. J. Bacteriol. 138: 465–476.

Thomas S & Fell DA (1998) A control analysis exploration of the role of ATP utilization in glycolytic-flux control and glycolytic-metabolite-concentraion regulation. Eur. J. Biochem. 258: 956–967.

Thompson, J (1978) In vivo regulation of glycolysis and characterization of sugar:phosphotransferase systems in *Streptococcus lactis*. J. Bacteriol. 136: 465–476.

Thompson, J (1987) Sugar transport in lactic acid bacteria. In: Reizer J. & Peterkofsky A. (Eds), Sugar Transport and Metabolism in Gram-Positive Bacteria. (pp 13–38). Ellis Horwood Limited, Chichester, UK.

Wouters JA, Kamphuis HH, Hugenholtz J, Kuipers OP, de Vos WM & Abee T (2000) Changes in glycolytic activity of *Lactococcus lactis* induced by low temperature. Appl. Environ. Microbiol. 66: 3686–3691.

Antonie van Leeuwenhoek **82:** 249–261, 2002.

Metabolism of lactic acid bacteria studied by nuclear magnetic resonance

Ana Ramos*, Ana Rute Neves & Helena Santos
*Instituto de Tecnologia Química e Biológica/Universidade Nova de Lisboa, Apt. 127, 2780-156 Oeiras, Portugal (*Author for correspondence; E-mail: ramos@itqb.unl.pt)*

Key words: LAB, *Lactococcus lactis*, glycolysis regulation, nuclear magnetic resonance

Abstract

The complexity of metabolic and regulatory networks presents a great scientific challenge to an integrated view of how individual components contribute to the overall function. Nuclear magnetic resonance (NMR) spectroscopy is undoubtedly a suitable technique for global investigations of microbial metabolism, since it allows a view into living cells without disturbing the cellular organisation. Therefore, metabolic processes can be monitored in real time under physiological conditions. In the present paper, examples of the application of NMR to study the metabolism of lactic acid bacteria will be given. These include the analysis of labelling patterns in end-products using ^{13}C as a tracer, thereby establishing metabolic pathways, the detection and quantification of intermediates in the pathway of exopolysaccharide biosynthesis, and *on line* monitoring of glycolytic kinetics to assess the effect of metabolic engineering strategies.

Introduction

NMR is a powerful analytical technique that allows the qualitative and quantitative characterisation of complex mixtures, the measurement of reaction rates in the steady state, the determination of the isotopic distribution within molecules, or the determination of the three-dimensional structure of molecules (van de Ven et al. 1991; Sheldon et al. 1996; Szyperski et al. 1999; Wand 2001). The tremendous improvement in the performance of spectrometers in the last three decades has made it possible for biochemists to benefit from the capabilities of NMR for carrying out measurements directly on living systems. The studies of Shulman and co-workers (den Hollander et al. 1986; Ugurbil et al. 1978; den Hollander et al. 1979) represent pioneering examples in this respect. An excellent introduction to *in vivo* NMR covering a wide range of methodological, instrumental, and physiological aspects, has been written by Gadian (1995). The basic concepts of the application of NMR to analyse microbial metabolism were recently reviewed (Grivet 2000).

Together with information about the kinetic properties of the enzymes in a given metabolic pathway, knowledge of the *in vivo* concentrations of intermediate metabolites is of fundamental importance for dynamic modelling and, consequently, in metabolic engineering. In this context, NMR spectroscopy has the advantage that a view into living cells is possible without disturbing the cellular organisation. Metabolic processes can be monitored in real time under physiological conditions and only the free metabolite pools are detected, since binding to macromolecules causes line broadening and hampers direct detection. Moreover, with the high number of nuclides observable, different aspects of cellular processes can be probed. Phosphorous 31 has been the most popular nucleus for studying biological systems *in vivo*; it allows monitoring the intracellular pH, the dynamics of intracellular phosphate pools, and the direct detection of important phosphorylated metabolites, such as ATP and ADP thus providing information about the energetic status of the cell (Navon et al. 1979). It enables also measurement of intracellular pH in intact cells, tissues, and organs. Carbon 13 has been widely used to establish metabolic routes and to follow the fate of individual carbon atoms through different pathways, thereby enabling the determination of carbon fluxes (London 1988).

An important drawback of NMR for *in vivo* applications is its intrinsic low sensitivity when observing

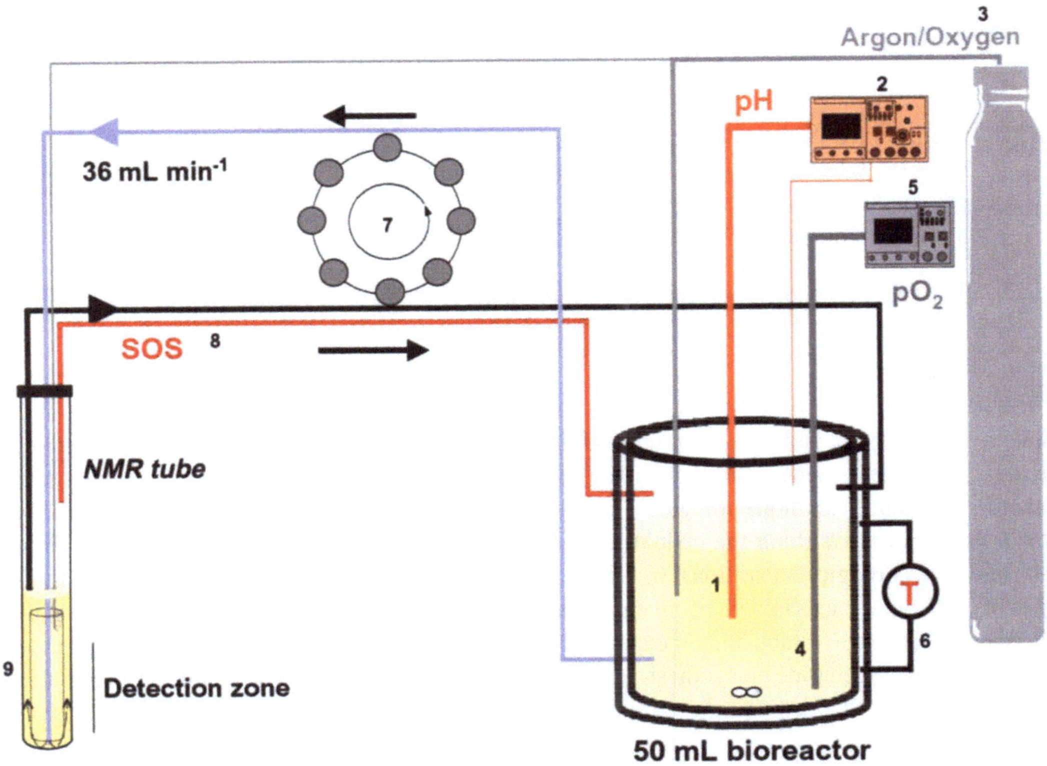

Figure 1. Schematic diagram of the experimental set-up used for *in vivo* NMR experiments. (1) pH electrode; (2) pH controller; (3) gas lines; (4) oxygen electrode; (5) oxygen controller; (6) water-bath; (7) circulation pump; (8) safeguard line; (9) 10-mm diameter NMR tube.

biologically relevant nuclei, which limits the detection to metabolites present in the millimolar range of concentration. This restriction imposes the use of dense cell suspensions (usually achieved with non-growing cells) in order to increase the total intracellular space available to detection.

Lactic acid bacteria (LAB) are ideal model systems for the study of the mechanisms involved in the regulation of sugar transport and metabolism. In particular, the economic value and the relative metabolic simplicity of *Lactococcus lactis*, a starter that metabolises sugars via the Embden-Meyerhof-Parnas pathway to pyruvate, generating energy mainly through substrate-level phosphorylation, makes it an attractive candidate for such studies. The large number of genetic tools available, including food-grade cloning vectors and selection markers (de Vos 1999) allowed significant breakthroughs in the construction of *L. lactis* strains with novel or improved properties (Hols et al. 1999; Hugenholtz et al. 2000). Moreover, the publication of the annotated genome sequence of *L. lactis* IL1403 (Bolotin et al. 2001) opened a wide range of possibilities for fundamental and applied research in this organism. However, efficient strain manipulation is not a straightforward process. The concept that a global understanding of metabolism is mandatory to achieve the desired outcome has been strengthened by several indications that in many cases single gene mutations do not affect the phenotype, or that the effect does not arise in a simple, obvious manner (Bailey 2001). Therefore, an integrated approach including contributions from functional genomics, proteomics and metabolomics will certainly provide new insight into the fine and coarse controls responsible for the modulation of activity of the sugar transport:glycolysis cycle and to extend our knowledge of metabolic regulation. This accomplishment will be essential for the rational design of *L. lactis* (and other LAB) optimised for the production of fine chemicals, flavours, exopolysaccharides or nutraceuticals. The purpose of this article is to summarise the contribution of NMR studies to our current understanding of carbon and energy metabolism in LAB.

Detection of intracellular components

Pools of phosphorylated metabolites determined by in vivo ^{31}P-NMR

In 1984, Thompson & Torchia reported the first NMR study of metabolic regulation in LAB. In this work, *in vivo* ^{31}P-NMR was used to measure intracellular phosphate (P_i) and the relative levels of the glycolytic intermediates fructose-1,6-bisphosphate (FBP) and 3-phosphoglycerate (3-PGA) in *L. lactis*, during the transition from starvation to glycolysis and in the reverse process. The results obtained *in vivo* with cell suspensions were complemented by ^{31}P-NMR and ^{14}C-fluorography measurements of those metabolites in cell extracts, and provided evidence for the antagonistic roles of FBP and P_i in modulating the activity of pyruvate kinase (PK). The authors encountered practical problems caused by peak broadening due to large pools of paramagnetic ions, such as manganese and also by the rapid glycolytic rates and concomitant acid production, which caused severe overlapping of relevant resonances, precluding their quantification. In the years thereafter, ^{31}P-NMR was used to quantify the metabolites accumulating after a lactose pulse in extracts of a *L. lactis* strain lacking glucokinase and mannose-phosphotransferase activities (Thompson et al. 1985), and to compare the levels of glycolytic intermediates during the metabolism of glucose or maltose in *L. lactis* (Lohmeier-Vogel et al. 1986). In the latter work, the shift to mixed acid fermentation observed during maltose metabolism was interpreted as being caused by high P_i concentrations and low FBP levels occurring during the utilisation of that sugar.

Maintaining controlled conditions inside the NMR tube over periods longer than a few minutes is troublesome, due to the high metabolic rates and to the utilisation of high cell densities. This problem triggered the development of various types of systems (hollow-fibers, air-lifts, or continuous flow/chemostats) to maintain the cells under specific conditions during the NMR experiment (Santos & Turner 1986; Weuster-Botz & de Graaf 1996; Shanks 2000). Our group developed a set-up for *in vivo* NMR studies of microbial metabolism (Figure 1) that consists of a reactor vessel containing the cell suspension coupled to NMR detection by a circulating system (Neves et al. 1999). Following the addition of a substrate, the time-course for its consumption, product formation, and the pools of intracellular metabolites can be monitored under controlled conditions of temperature, pH and gas atmosphere. The advantages of such a system become evident from the observation of the ^{31}P-NMR spectra obtained after cultivation of *L. lactis* in defined growth medium (Figure 2). In the spectrum of starved cells (A), the broad resonance in the phosphomonoester region is identified as 3-PGA; a sharp peak due to cytoplasmic inorganic phosphate (P_i) is also detectable, as well as the intense resonance of teichoic acid overlapping with the low-intensity signal of phosphoenolpyruvate (PEP). Once glucose is supplied, P_i and PEP disappear, and the dominant peak in the spectrum is assigned as FBP; the resonance of glucose-6-phosphate is also visible (B). After glucose depletion, cytoplasmic P_i increases while the FBP level decreases (C). The concentration of nucleosides triphosphate (NTP) rises from undetectable levels to a maximum of 8 mM and remains high even after glucose depletion (Neves et al. 1999). Measuring the chemical shift of the P_i resonance in the spectra and translating the corresponding pH from a titration curve for inorganic phosphate allows the determination of the intracellular pH. These experiments provide useful data on the intracellular pH and the evolution of P_i and NTP concentrations (Figure 2 – inset). However, there is considerable resonance overlapping, mainly in the region where resonances due to glycolytic intermediates appear. The presence of signals due to cell wall components, such as teichoic acid in Gram-positive organisms (Roberts et al. 1985), further limits the usefulness of phosphorous 31 to determine concentrations of individual metabolites *in vivo* (Galazzo et al. 1990). In fact, since ^{13}C has a larger chemical shift range, it is the nuclide of choice whenever ^{13}C-labelled substrates are available at a reasonable cost.

In vivo *^{13}C-NMR studies of glycolysis in* Lactococcus lactis

In spite of the wealth of information gathered from numerous studies, a comprehensive answer to the question 'how is the glycolytic flux controlled in *L. lactis*' has not been put forward. A satisfactory general response to that question is primarily hampered by the complexity of global metabolism, consisting of multiple interlocked pathways connected via common metabolites and cofactors through various degrees of genetic and metabolic regulation. Molecular biology has been a useful tool in dissecting complex systems into smaller parts that can then be analysed and understood; however, for obtaining a deeper perception of metabolic regulation, all the pieces must be

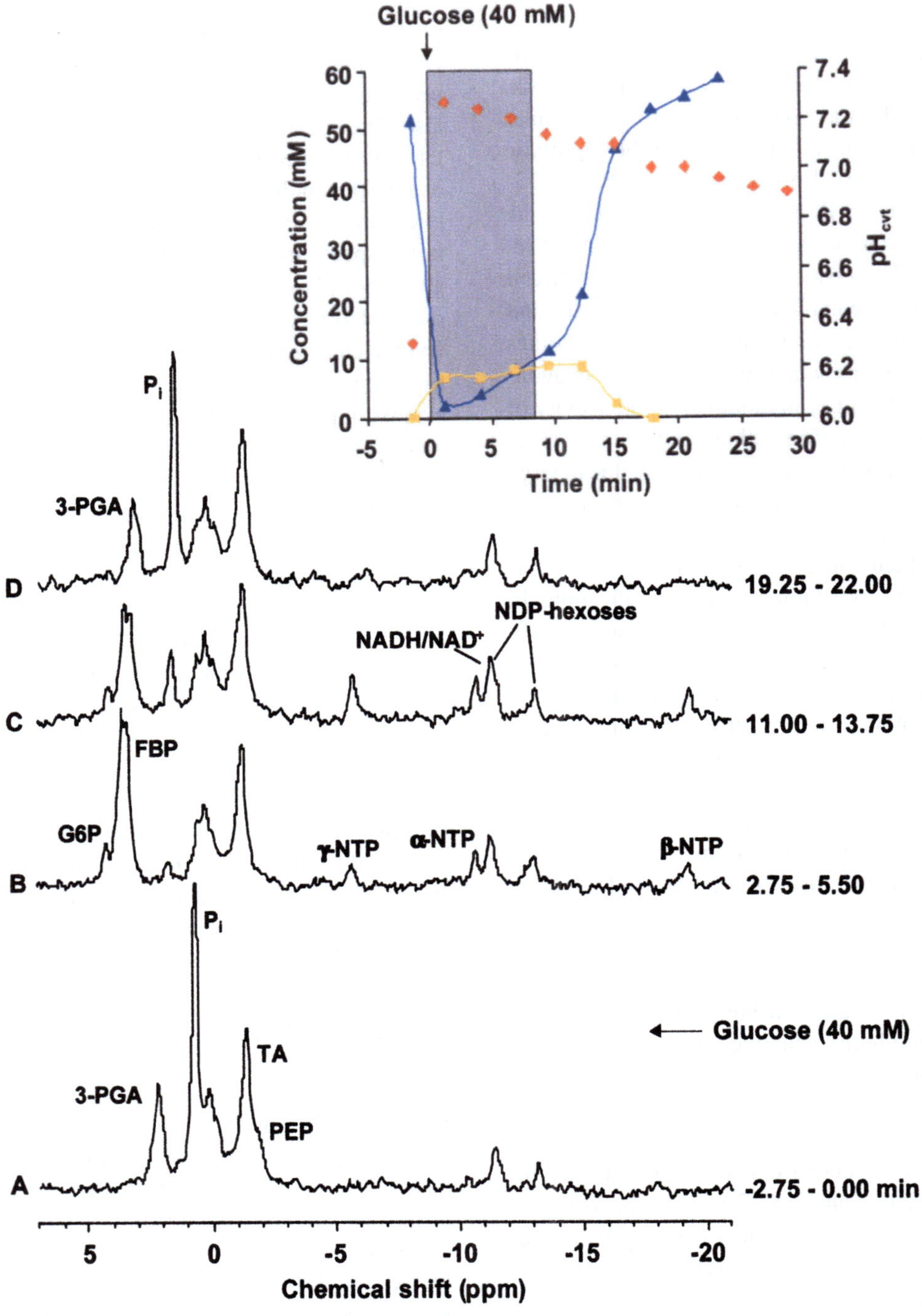

Figure 2. In vivo ^{31}P-NMR spectra acquired during the metabolism of glucose by strain MG1363 under anaerobic conditions at pH 6.5. Each spectrum represents 2.75 min of accumulation. Abbreviations: NTP, nucleosides triphosphate; TA, teichoic acid. Inset: Time course for the concentration of NTP (□), P_i (Δ) and intracellular pH (◊). The shaded area indicates glucose availability.

put together, and thus a comprehensive knowledge of the living cells as a whole is needed. In LAB, several lines of evidence gave rise to a simple model of glycolytic regulation that can be summarised as follows: during homolactic fermentation, the high levels of FBP, which activate lactate dehydrogenase (LDH) and PK, direct the flux towards the production of lactate. Accumulation of PEP and 3-PGA in starved cells is a consequence of PK inhibition by high P_i and low FBP concentrations, conditions associated with glucose depletion (Thompson 1987). The shift to mixed-acid fermentation (production of acetate, formate and ethanol) has been rationalised as resulting from a reduced FBP pool leading to LDH inactivation, in conjunction with modulation of the activity of pyruvate-formate lyase by glyceraldehyde-3-phosphate (Ga3P) and dihydroxyacetone-phosphate (Thomas et al. 1979). Nonetheless, the regulatory role of FBP was recently questioned, and it was proposed that inhibition or activation exerted by the ratio NADH/NAD^+on glyceraldehyde-3-phosphate dehydrogenase (Ga3PDH) or LDH is the main issue regulating glycolysis and the shift from homo- to heterolactic fermentation (Garrigues et al. 1997).

In our laboratory, NMR was used to obtain insight into the regulation of glucose metabolism in *L. lactis*. Since these studies were performed with resting cells, catabolism is decoupled from anabolism, and consequently the metabolic network is greatly simplified (Neves et al. 1999). Carbon 13 spectra acquired sequentially after the addition of [1-^{13}C]glucose to a cell suspension of MG1363 show, in addition to lactate, resonances due to intracellular FBP and 3-PGA (Figure 3). The FBP pool becomes labelled both on carbon 1 and carbon 6 (Figure 3 – inset), indicating the occurrence of scrambling of the label at the trioses phosphate pool and back flux through aldolase; the extent of the reverse flux is inversely correlated with the net rate of glucose consumption (Neves et al. 1999, 2000). In addition to FBP and 3-PGA, intracellular glucose-6-phosphate can also be detected and quantified in experiments where glucose labelled on carbon 6 is supplied. ^{13}C data provided a precise description of the dynamic behaviour of glucose, glycolytic intermediates and end-products with a time resolution of 30 s, and this feature is undoubtedly a major achievement of the experimental design used (Figure 4A). Interestingly, under an oxygen atmosphere, both the rate of FBP consumption and the magnitude of the 'PEP-potential' (3-PGA plus PEP) is considerably higher (Figure 4B), suggesting that the activity of NADH oxidases has to be taken into account in studies of metabolic regulation. In fact, the crucial role of the redox status is obvious in genetically modified *L. lactis* strains. For instance, a strain overproducing NADH oxidase, redirects the carbon flux towards the production of acetate and acetoin, since the redox balance is mostly fulfilled by NADH oxidase activity (Lopes de Felipe et al. 1998; Neves et al. submitted). A striking illustration of this metabolic adjustment is provided by a construct deficient in LDH, characterised by *in vivo* NMR. This mutant produces transiently high amounts of mannitol-1-phosphate and mannitol as an alternative route to maintain the redox balance in the absence of the main NAD^+-producing enzyme (Neves et al. 2000).

The alleged regulatory role of pyridine nucleotides in the central metabolism of *L. lactis* triggered a recent work in our group in which NAD^+ was enriched in carbon 13 by cultivation in defined medium containing its precursor, ^{13}C-nicotinic acid. This strategy was originally devised to monitor reduced and oxidised pyridine nucleotides in *Escherichia coli* and *Saccharomyces cerevisiae* (Unkefer et al. 1983; Unkefer & London 1984), and it has several advantages over the common quantification in cell extracts: (i) the problem of chemical instability of NADH is obviated; (ii) both the oxidised and reduced forms can be observed in the same sample; (iii) labelled glycolytic intermediates, NAD^+, and NADH can be detected simultaneously in a non-invasive manner.

Figure 5(A) shows the resonances of pyridine nucleotides before and after addition of glucose to a cell suspension of *L. lactis* MG1363 under anaerobic conditions. In starved cells, only the resonance of labelled NAD^+ is detected (concentration about 5 mM); after depletion of glucose, the resonance of NADH becomes visible, reaching a maximum of 3 mM. The graph (B) shows the kinetics of the pools of NAD^+, NADH and FBP.

The pattern of metabolite dynamics derived from ^{13}C and ^{31}NMR data acquired under equivalent conditions provided an integrated description of carbon and energy metabolism. Glucose, end-products, glycolytic intermediates, NAD^+ and NADH were monitored by carbon 13, whereas ^{31}P yielded NTP, Pi and the intracellular pH. As long as there is glucose available, PK is fully active due to low $P_i$and high FBP together with high phosphoenolpyruvate:phosphotransferase (PEP:PTS) activity, allowing for high-speed conversion of glucose to pyruvate. The NAD^+ pool remains constant, showing that

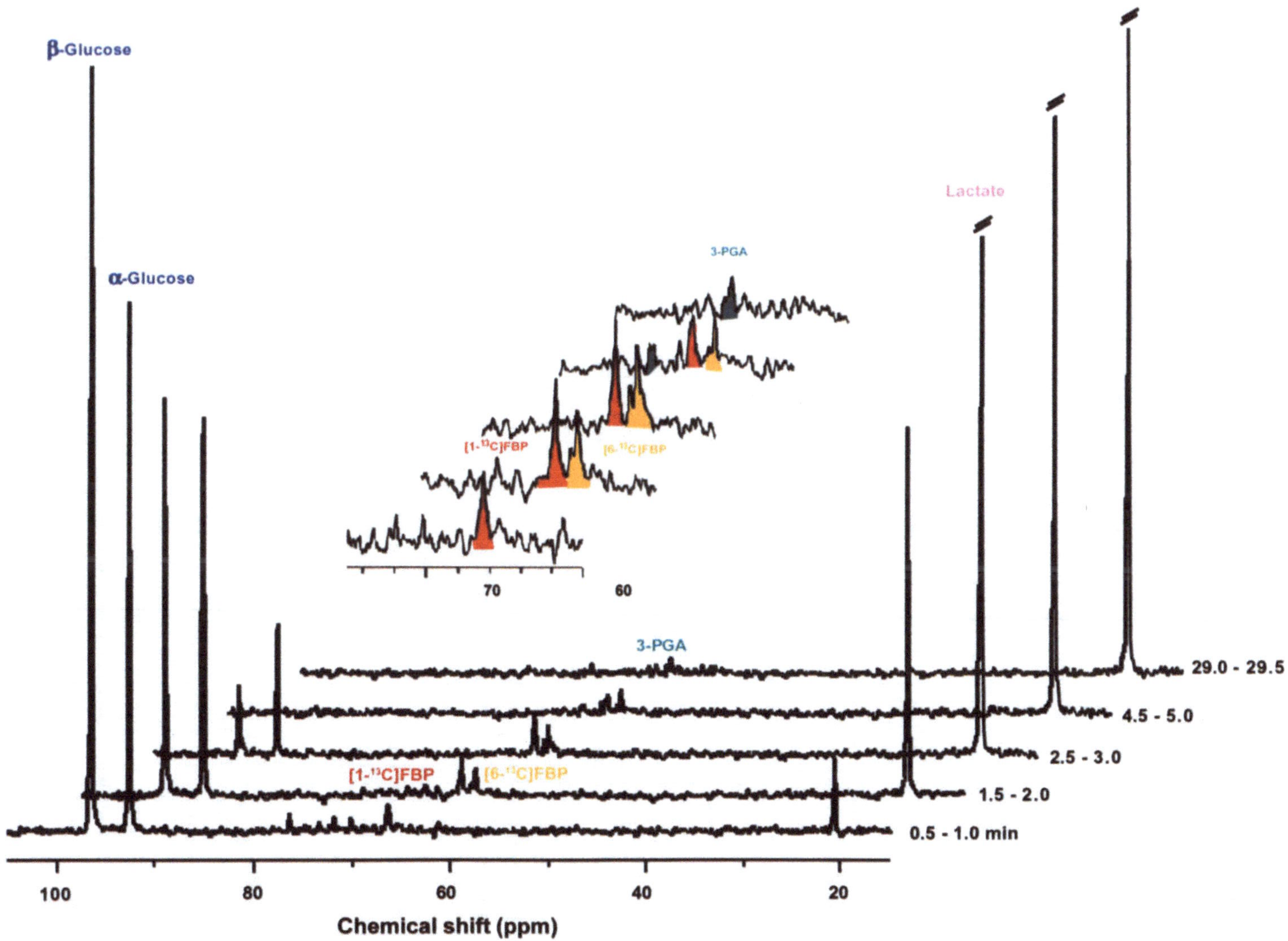

Figure 3. In vivo ^{13}C-NMR spectra obtained during the consumption of [1-^{13}C]glucose (20 mM) by strain MG1363 under anaerobic conditions at pH 6.5. Each spectrum represents 30 s of accumulation. Inset: horizontal magnification showing the region where both isotopomers of FBP and 3-PGA appear.

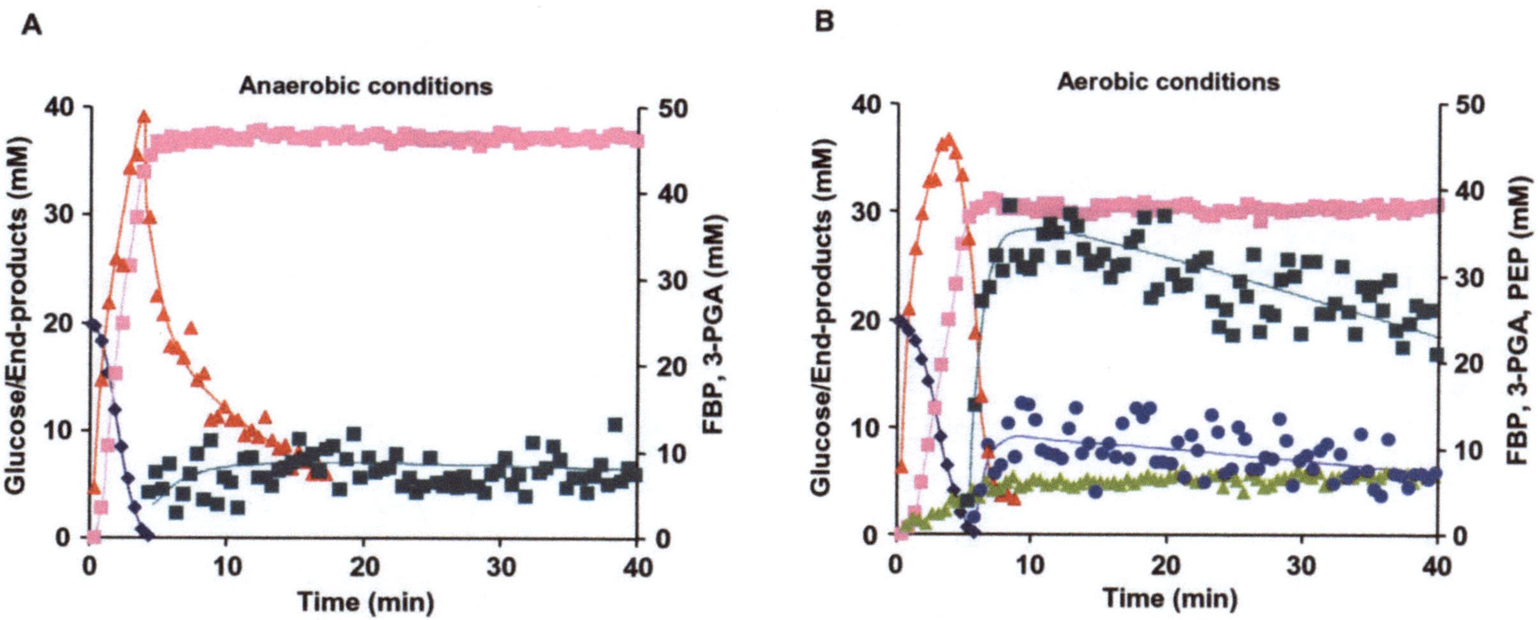

Figure 4. Kinetics of [1-^{13}C]glucose consumption/end-product formation and intracellular metabolites under anaerobic (A) or aerobic atmosphere (B) at pH 6.5. Metabolite concentrations were obtained from *in vivo* ^{13}C-NMR data. Symbols: (♦), glucose; (□), lactate; (Δ), acetate; (▲), total FBP; (■), 3-PGA; (●), PEP.

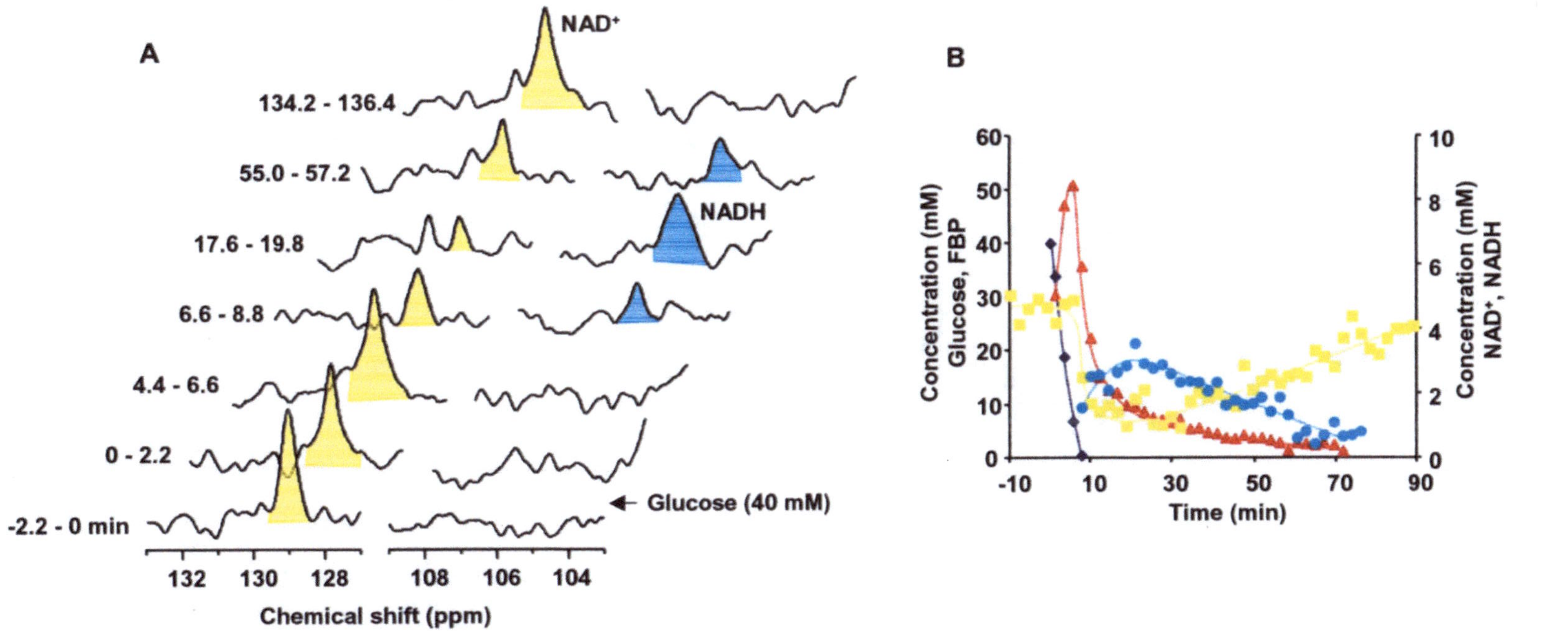

Figure 5. In vivo ^{13}C-spectra showing the evolution of NAD$^+$ and NADH resonances in strain MG1363 before and after addition of glucose under anaerobic conditions at pH 6.5 (A). Consumption of glucose and evolution of FBP, NAD$^+$ and NADH pools (B). Symbols: (♦), glucose; (Δ); FBP; (□), NAD$^+$; (●), NADH.

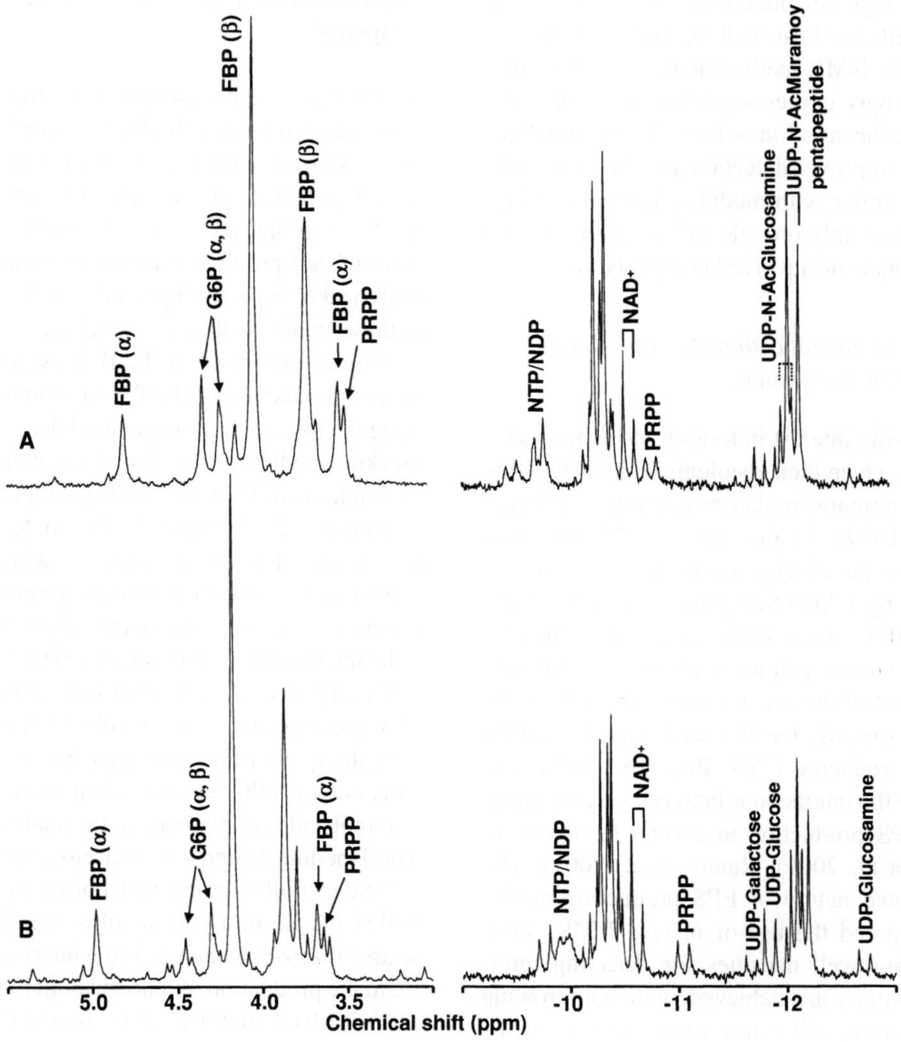

Figure 6. ^{31}P-NMR spectra of ethanol extracts obtained from cells of Eps$^-$ (A) and Eps$^+$ (B) *L. lactis*. Abbreviations: UDP-NAc-Glucosamine, UDP-*N*-Acetyl-Glucosamine; UDP-NAc-Muramoyl-pentapeptide, UDP-N-Acetyl-Muramoyl-pentapeptide; PRPP, 5-phosphorylribose-1-pyrophosphate; NTP, nucleosides triphosphate; NDP, nucleosides diphosphate; FBP, fructose-1,6-bisphosphate (α and β anomers); G6P, glucose-6-phosphate (α and β anomers). Resonances were assigned by spiking the extracts with the pure compounds with the exception of UDP-NAc-Muramoyl-pentapeptide, which was purified and identified by 2-dimensional NMR spectroscopy.

NADH is efficiently re-oxidised through LDH (Neves et al. submitted). Pyruvate kinase appears not to be engaged in the regulation at this stage, as evidenced by the observation that the flux to lactate does not increase in a mutant overproducing PK (Ramos et al. unpublished results). In contrast to anaerobiosis, under an oxygen atmosphere, NADH remains always below the detection limit (0.5 mm). Interestingly, in a strain defective in LDH, Ga3PDH was able to support a high flux even in the presence of a high NADH concentration, suggesting that the glycolytic flux is not primarily controlled at the level of Ga3PDH by NADH (Neves et al. submitted). After glucose depletion, FBP is consumed at a rate dependent on the capacity to regenerate NAD$^+$ (higher rate under aerobic conditions) (compare Figure 4A & 4B). The bigger 3-PGA/PEP pool under aerobiosis is explained by: (i) a stronger constriction of the flux through PK (caused by the combined action of a lower concentration of the activator, FBP, and a high pool of its strong inhibitor, P_i), (ii) the alleviated pressure to re-oxidise NADH downstream of pyruvate due to high NADH oxidase activity.

The unique opportunity provided by ^{13}C-NMR to measure *in vivo* levels of NAD$^+$ and NADH (simultaneously with glycolytic intermediates) represents a major breakthrough for understanding regulation of glycolysis. Further efforts will be focussed on the refinement of the NMR methodology in order to improve the sensitivity in their detection, since in some conditions the concentration of NADH was below the detection limit. Improved detection, together with data obtained from strains with modulated activity of key enzymes, will certainly provide a clear picture of the regulatory mechanisms involved in glycolysis.

Identification and quantification of intermediates in exopolysaccharide biosynthesis

There is a growing interest in food-grade exopolysaccharides (EPS), given their rheological characteristics and putative immunostimulatory properties (Kleerebezem et al. 1999). In this field, NMR has been mostly used for the elucidation of the structures of EPS produced by LAB (Nakajima et al. 1992; Germond et al. 2001). On the other hand, insight into the complex biosynthetic pathways involved in biosynthesis and export of these macromolecules represents a crucial step towards the biotechnological exploitation of EPS-producing LAB. Recently, NMR was used to analyse the interactions between glucose metabolism and EPS production in several strains of *L. lactis* (Boels et al. 2001); Ramos et al. 2001). The growth-associated nature of EPS production in this organism hampered the use of *in vivo* NMR, since the required high-cell densities for detecting intracellular metabolites are achieved with non-growing cultures. Therefore, cell extracts were used to measure by ^{31}P-NMR the intracellular pools of metabolic intermediates in the biosynthesis of EPS, cell wall or glycolysis (Figure 6). The concentrations of the EPS precursors, UDP-glucose and UDP-galactose were lower in the EPS-producing construct (A) when compared to the non-producer parent (B), whereas UDP-N-acetylmuramoyl-pentapeptide (an intermediate in peptidoglycan biosynthesis) was increased, indicating the existence of competition between EPS production and cell growth (Ramos et al. 2001). This work illustrates the power of NMR analysis for detecting new or unexpected components that otherwise would not be easily tracked. Moreover, it revealed the existence of multiple targets for enhancing EPS production; the flux to the synthesis of sugar nucleotides should be increased, as well as the pool of the lipid carrier (undecaprenol-phosphate, involved both in EPS export and in cell wall synthesis), so that a high productivity can be achieved without a severe effect on cell growth.

Analysis of labelling patterns on extracellular compounds

Tracer experiments using specifically ^{13}C labelled substrates combined to NMR detection have been used for the study of metabolic pathways in LAB; this approach provides details about the distribution of the label within the same molecule and allows quantification of end-products without resorting to elaborate analytical or separation procedures. Several examples of these studies will be given below.

Citrate catabolism in LAB is associated with enhanced production of the flavour compound diacetyl, since the increased pyruvate pool derived from citrate breakdown allows deviation of the carbon flux to the biosynthesis of the so-called C_4 compounds, diacetyl, acetoin and 2,3-butanediol. The route leading to the production of those metabolites, which remained a controversial subject until the early nineties (due to the chemical instability of α-acetolactate, a putative catabolic intermediate), was elucidated by ^{13}C-NMR. Two different pathways were proposed for the biosynthesis of C_4 compounds: one involving the condensation of hydroxyethyl-thiamine pyrophosphate with acetyl-CoA and the other encompassing the condensation of two molecules of pyruvate to originate α-acetolactate. The labelling patterns in α-acetolactate, acetoin and 2,3-butanediol after the metabolism of selectively labelled citrate or pyruvate plus acetyl-CoA demonstrated that α-acetolactate is the intermediate metabolite in the production of diacetyl both in *L. lactis* and in the wine bacterium *Oenococcus oeni* (Verhue & Tjan 1991; Ramos et al. 1994, 1995).

Whenever different substrates are converted to a common metabolic intermediate, the information supplied by analytical methods other than NMR is insufficient to characterize the co-metabolic processes. This is the case for citrate and glucose metabolism, where pyruvate is a central intermediate in the catabolic pathways of both sugar and citrate. By using glucose and citrate labelled on suitable positions, it was possible to determine precisely the origin of each carbon atom in the end-products. A strong interrelationship between citrate and glucose metabolism was observed. Carbon from citrate was channelled to lactate and 2,3-butanediol when glucose was present, whereas acetoin was the main end-product of the metabolism of citrate alone. These observations demonstrated that citrate metabolism provides alternative pathways of $NAD(P)^+$ regeneration (Ramos et al. 1994; Ramos & Santos 1996).

An identical approach was taken for elucidating the unforeseen utilisation of acetate in a *L. lactis* strain deficient in LDH; ^{13}C- and ^{1}H-NMR experiments using acetate and glucose labelled on appropriate positions demonstrated that acetate is exclusively converted to ethanol. This novel pathway provides an alternative route for NAD^+ regeneration in the absence of the primary NAD^+-producing enzyme (Hols et al. 1999). The end-products of citrate and glucose metabolism in a strain of *Lactobacillus plantarum* deficient in both D- and L-LDH were quantified by ^{13}C-NMR, showing that in the absence of LDHs the carbon flux is re-routed to the production of acetoin (Ferain et al. 1996).

NMR provides a straightforward way for detecting the unexpected; erythritol was identified as a product of the metabolism of glucose in *O. oeni* under anaerobic conditions (Veiga da Cunha et al. 1992, 1993). The use of glucose enriched with ^{13}C in different carbon atoms allowed the elucidation of the metabolic pathway of erythritol formation, which involves cleavage of fructose 6-phosphate by phosphoketolase, resulting in erythrose 4-phosphate and acetyl phosphate. The former compound is then dephosphorylated and reduced to erythritol with the concomitant oxidation of NADH.

Formation of methanethiol from methionine is believed to play a significant role in the development of cheddar cheese flavour. Two independent methionine catabolic pathways were presumed to exist in *L. lactis*, and carbon 13 NMR was used to establish that only the aminotransferase initiated pathway is involved in the formation of methanethiol (Gao et al. 1998). Recently, isotopic labelling coupled to NMR detection enabled the characterisation of the metabolism of glucose and fructose in species belonging to the genera *Lactobacillus* and *Leuconostoc* (Antunes et al. 2002).

Concluding remarks

Metabolites are the end-products of cellular regulatory processes, and their levels can be regarded as the ultimate response to genetic or environmental changes. Therefore, in parallel to the transcriptome and the proteome, the set of metabolites synthesized by a biological system (metabolome) is of utmost importance for acquiring an integrated description of how individual components contribute to the overall function.

NMR will certainly play an important role in global investigations of microbial metabolism. When compared to other analytical techniques, such as gas chromatography, high-performance liquid chromatography or mass spectrometry, NMR has the unique capability of providing structural information about the metabolites detected. NMR techniques are advancing rapidly, with increasing sensitivity and sophistication. In particular, the advent of sensitive and reliable HPLC–NMR systems will provide a powerful means of characterising complex mixtures (Wilson 2000).

In *L. lactis*, *in vivo* NMR provided an interpretation of glucose metabolism based on direct observations of relevant metabolic intermediates rather than on unreliable extrapolations from *in vitro* data, or speculative reasoning derived from the pattern of end-products. The metabolic networks in *L. lactis* are simplified by the fact that catabolism is uncoupled from anabolism to a significant extent. This makes it an adequate model-organism for *in vivo* NMR analysis, since the data obtained with resting cells are expected to be representative of the events that occur in growing cultures. Despite the useful results obtained with NMR to investigate cell metabolism, it is essential to increase the sensitivity and enable the detection of metabolites that exist at low intracellular concentrations. This accomplishment will be essential for the rational design of industrial microbial strains with improved properties and nutritional value.

Acknowledgments

Part of the work described here was funded by the European Community (contract BIOT-CT96-0498) and by Fundação para a Ciência e Tecnologia (FCT), Portugal (PRAXIS/BIO/39/96 and PRAXIS/P/BIA/11072/98).

References

Antunes A, Rainey FA, Nobre MF, Schumann P, Ramos A, Santos H & da Costa MS (2002) *Leuconostoc ficulneus* sp. nov., a new lactic acid bacterium isolated from ripened figs, and the reclassification of *Lactobacillus fructosus* as *Leuconostoc fructosus*. Int. J. Syst. Evol. Microbiol. 52: 647–655.

Bailey JE (2001) Reflections on the scope and the future of metabolic engineering and its connections to functional genomics and drug discovery. Metab. Eng. 3: 111–114.

Boels IC, Ramos A, Kleerebezem M & de Vos WM (2001) Functional analysis of the *Lactococcus lactis galU* and *galE* genes and their impact on sugar nucleotide and exopolysaccharide biosynthesis. Appl. Environ. Microbiol. 67: 3033–3040.

Bolotin A, Wincker P, Mauger S, Jaillon O, Malarme K, Weissenbach J, Ehrlich SD & Sorokin A (2001) The complete genome sequence of lactic acid bacterium *Lactococcus lactis* ssp. *lactis* IL1403. Genome Res. 11: 731–753.

den Hollander JA, Brown TR, Ugurbil K & Shulman RG (1979) ^{13}C nuclear magnetic resonance studies of anaerobic glycolysis in suspensions of yeast cells. Proc. Natl. Acad. Sci. USA 76: 6096–6100.

den Hollander JA, Ugurbil K, Brown TR, Bednar M, Redfield C & Shulman RG (1986) Studies of anaerobic and aerobic glycolysis in *Saccharomyces cerevisiae*. Biochemistry 25: 203–211.

Ferain T, Schanck AN & Delcour J (1996) ^{13}C nuclear magnetic resonance analysis of glucose and citrate end products in an *ldh*L-*ldh*D double knockout strain of *Lactobacillus plantarum*. J. Bacteriol. 178: 7311–7315.

Gadian DG (1995) NMR and Its Applications to Living Systems 2nd edn. Oxford Science Publications, Oxford.

Gallazo JL, Shanks JV & Bailey JE (1990) Comparison of intracellular sugar-phosphate levels from ^{31}P NMR spectroscopy of intact cells and cell-free extracts. Biotechnol. Bioeng. 35: 1164–1168.

Gao S, Mooberry ES & Steele JL (1998) Use of ^{13}C nuclear magnetic resonance and gas chromatography to examine methionine catabolism by lactococci. Appl. Environ. Microbiol. 64: 4670–4675.

Garrigues C, Loubiere P, Lindley ND & Cocaign-Bousquet M (1997) Control of the shift from homolactic acid to mixed-acid fermentation in *Lactococcus lactis*: predominant role of the NADH/NAD^{+} ratio. J. Bacteriol. 179: 5282–5287.

Germond JE, Delley M, D'Amico N & Vincent SJ (2001) Heterologous expression and characterization of the exopolysaccharide from *Streptococcus thermophilus* Sfi39. Eur. J. Biochem. 268: 149–156.

Grivet J-P (2000) NMR and Microorganisms. In: Barbotin J-N and Portais J-C (Eds) NMR in Microbiology: Theory and Applications (pp 27–46). Horizon Scientific Press, Wymondham.

Hols P, Kleerebezem M, Schanck AN, Ferain T, Hugenholtz J, Delcour J & de Vos WM (1999) Conversion of *Lactococcus lactis* from homolactic to homoalanine fermentation through metabolic engineering. Nat. Biotechnol. 17: 588–592.

Hols P, Ramos A, Hugenholtz J, Delcour J, de Vos WM, Santos H & Kleerebezem M (1999) Acetate utilization in *Lactococcus lactis* deficient in lactate dehydrogenase: a rescue pathway for maintaining redox balance. J. Bacteriol. 181: 5521–5526.

Hugenholtz J, Kleerebezem M, Starrenburg M, Delcour J, de Vos WM & Hols P (2000) *Lactococcus lactis* as a cell factory for high-level diacetyl production. Appl. Environ. Microbiol. 66: 4112–4114.

Kleerebezem M, van Kranenburg R, Tuinier R, Boels IC, Zoon P, Looijensteijn E, Hugenholtz J & de Vos WM (1999) Exopolysaccharides produced by *Lactococcus lactis*: from genetic engineering to improved rheological properties? Antonie van Leeuwenhoek 76: 357–365.

Lohmeier-Vogel EM, Hahn-Hägerdal B & Vogel HJ (1986) Phosphorous-31 NMR studies of maltose and glucose metabolism in *Streptococcus lactis*. Appl. Microbiol. Biotechnol. 25: 43–51.

London RE (1988) ^{13}C labeling studies of metabolic regulation. Progr. Nucl. Magn. Reson. Spectrosc. 20: 337–383.

Nakajima H, Hirota H, Toba T, Itoh T & Adachi S (1992) Structure of the extracellular polysaccharide from slime-forming *Lactococcus lactis* subsp. *cremoris* SBT 0495. Carbohydr. Res 224: 245–253.

Navon G, Shulman RG, Yamane T, Eccleshal TR, Lam KB, Baronofsky JJ & Marmur J (1979) ^{31}P NMR studies of wild-type and glycolytic pathway mutants of *Saccharomyces cerevisiae*. Biochemistry 18: 4487–4489.

Neves AR, Ramos A, Nunes MC, Kleerebezem M, Hugenholtz J, de Vos WM, Almeida JS & Santos H (1999) In vivo nuclear magnetic resonance studies of glycolytic kinetics in *Lactococcus lactis*. Biotechnol. Bioeng. 64: 200–212.

Neves AR, Ramos A, Shearman C, Gasson MJ, Almeida JS & Santos H (2000) Metabolic characterization of *Lactococcus lactis* deficient in lactate dehydrogenase using *in vivo* ^{13}C-NMR. Eur. J. Biochem. 267: 3859–3868.

Neves AR, Ventura R, Mansour N, Shearman C, Gasson MJ, Maycock C, Ramos A & Santos H (2002) Is the glycolytic flux in *Lactococcus lactis* primarily controlled by the redox charge? kinetics of NAD^{+} and NADH pools measured non-invasively by ^{13}C-NMR. Submitted.

Neves AR, Ramos A, Costa H, van Swam I, Hugenholtz J, Kleerebezem m, de Vos WM & Santos H (2002) Effect of different NADH oxidase levels on glucose metabolism of *Lactococcus lactis*: kinetics of intracellular metabolite pools by *in vivo* NMR. Submitted.

Ramos A & Santos H (1996) Citrate and sugar cofermentation in *Leuconostoc oenos*, a ^{13}C nuclear magnetic resonance study. Appl. Environ. Microbiol. 62: 2577–2585.

Ramos A, Boels IC, de Vos WM & Santos A (2001) Relationship between glycolysis and exopolysaccharide biosynthesis in *Lactococcus lactis*. Appl. Environ. Microbiol. 67: 33–41.

Ramos A, Jordan KN, Cogan TM & Santos H (1994) ^{13}C nuclear magnetic resonance studies of citrate and glucose cometabolism by *Lactococcus lactis*. Appl. Environ. Microbiol. 60: 1739–1748.

Ramos A, Lolkema JS, Konings WN & Santos H (1995) Enzyme basis for pH regulation of citrate and pyruvate metabolism by *Leuconostoc oenos*. Appl. Environ. Microbiol. 61: 1303–1310.

Ramos A, Neves AR, Lopez P & Santos H. Unpublished.

Roberts MF, Jacobson, GR, Scot PJ, Mimura CS & Stinson MW (1985) ^{31}P-NMR studies of the oral pathogen *Streptococcus mutans*: observation of lipoteichoic acid. Biochim. Biophys. Acta 845: 242–248.

Santos H & Turner DL (1986) Characterisation of the improved sensitivity obtained using a flow method for oxygenating and mixing cell suspensions in NMR. J. Magn. Reson. 68: 345–349.

Shanks JV (2000) *In situ* NMR systems. In: Barbotin, J-N & Portais J-C (Eds) NMR in Microbiology: Theory and Applications (pp 47–72). Horizon Scientific Press, Wymondham, UK.

Sheldon JG, Williams SP, Fulton AM & Brindle KM (1996) ^{31}P magnetization transfer study of the control of ATP turnover in *Saccharomyces cerevisiae*. Proc. Natl. Acad. Sci. USA 93: 6399–6404.

Szyperski T, Glaser RW, Hochuli M, Fiaux J, Sauer U, Bailey JE & Wüthrich K (1999) Bioreaction network topology and metabolic flux ratio analysis by biosynthetic fractional ^{13}C labeling and two-dimensional NMR spectroscopy. Metab. Eng. 1: 189–197.

Thomas TD, Elwood DC & Longyear VM (1979) Change from homo- to heterolactic fermentation by *Streptococcus lactis* resulting from fructose limitation in anaerobic chemostat cultures. J. Bacteriol. 138: 109–117.

Thompson J & Torchia DA (1984) Use of ^{31}P nuclear magnetic resonance spectroscopy and ^{14}C fluorography in studies of glycolysis and regulation of pyruvate kinase in *Streptococcus lactis*. J. Bacteriol. 158: 791–800.

Thompson J, Chassy BM & Egan W (1985) Lactose metabolism in *Streptococcus lactis*: studies with a mutant lacking glucokinase and mannose-phosphotransferase activities. J. Bacteriol. 162: 217–223.

Thompson J (1987) Sugar transport in lactic acid bacteria. In: Reizer J & Peterkofsky A (Eds) Sugar Transport and Metabolism in Gram-positive Bacteria (pp 13–38). Ellis Horwood Ltd., Chichester.

Ugurbil K, Brown TR, den Hollander JA, Glynn P & Shulman RG (1978) High-resolution ^{13}C nuclear magnetic resonance studies of glucose metabolism in *Escherichia coli*. Proc. Natl. Acad. Sci. USA 75: 3742–3746.

Unkefer CJ & London RE (1984) *In vivo* studies of pyridine nucleotide metabolism in *Escherichia coli* and *Saccharomyces cerevisiae* by carbon-13 NMR spectroscopy. J. Biol. Chem. 259: 2311–2320.

Unkefer CJ, Blazer RM & London RE (1983) *In vivo* determination of the pyridine nucleotide reduction charge by carbon-13 nuclear magnetic resonance spectroscopy. Science 222: 62–65.

van de Ven FJ, van den Hoven HW, Konings RN & Hilbers CW (1991) NMR studies of lantibiotics. The structure of nisin in aqueous solution. Eur. J. Biochem. 202: 1181–1185.

Veiga da Cunha M, Firme P, SanRomão MV & Santos H (1992) Application of ^{13}C nuclear magnetic resonance to elucidate the unexpected biosynthesis of erythritol by *Leuconostoc oenos*. Appl. Environ. Microbiol. 58: 2271–2279.

Veiga da Cunha M, Santos H & van Schaftingen E (1993) Pathway and regulation of erythritol formation in *Leuconostoc oenos*. J. Bacteriol. 175: 3941–3948.

Verhue WM & Tjan FB (1991) Study of the citrate metabolism of *Lactococcus lactis* subsp. *lactis* biovar *diacetylactis* by means of ^{13}C nuclear magnetic resonance. Appl. Environ. Microbiol. 57: 3371–3377.

Wand AJ (2001) Dynamic activation of protein function: a view emerging from NMR spectroscopy. Nat. Struct. Biol. 8: 926–931.

Weuster-Botz D & de Graaf AA (1996) Reaction engineering methods to study intracellular metabolite concentrations. In: Sahm H & Wandrey C (Eds) Advances in Biochemical Engineering Biotechnology, Vol. 54 (pp 75–108). Springer-Verlag, Berlin.

Wilson ID (2000) Multiple hyphenation of liquid chromatography with nuclear magnetic resonance spectroscopy, mass spectrometry and beyond. J. Chromatogr. A 892: 315–327

Antonie van Leeuwenhoek **82:** 263–269, 2002.

Respiration capacity and consequences in *Lactococcus lactis*

Philippe Gaudu, Karin Vido, Bénédicte Cesselin, Saulius Kulakauskas, Josselyne Tremblay, Lahcen Rezaïki, Gilles Lamberet, Sophie Sourice, Patrick Duwat & Alexandra Gruss*
*Génétique Appliquée – URLGA, Institut National de la Recherche Agronomique, Domaine de Vilvert, 78352 Jouy en Josas cedex, France (*Author for correspondence; Tel.: +33-1-34-65-21-68; Fax: +33-1-34-65-20-65; E-mail: gruss@biotec.jouy.inra.fr)*

Key words: cytochrome, heme, oxygen, respiration, superoxide dismutase, survival

Abstract

We recently reported that the well-studied fermenting bacterium *Lactococcus lactis* could grow via a respirative metabolism in the presence of oxygen when a heme source is present. Respiration induces profound changes in *L. lactis* metabolism, and improvement of oxygen tolerance and long-term survival. Compared to usual fermentation conditions, biomass is approximately doubled by the end of growth, acid production is reduced, and large amounts of normally minor end products accumulate. Lactococci grown via respiration survive markedly better after long-term storage than fermenting cells. We suggest that growth and survival of lactococci are optimal under respiration-permissive conditions, and not under fermentation conditions as previously supposed.

Our results reveal the uniqueness of the *L. lactis* respiration model. The well-studied 'aerobic' bacteria express multiple terminal cytochrome oxidases, which assure respiration all throughout growth; they also synthesize their own heme. In contrast, the *L. lactis cydAB* genes encode a single cytochrome oxidase (bd), and heme must be provided. Furthermore, *cydAB* genes mediate respiration only late in growth. Thus, lactococci exit the lag phase via fermentation even if heme is present, and start respiration in late exponential phase. Our results suggest that the spectacularly improved survival is in part due to reduced intracellular oxidation during respiration. We predict that lactococcal relatives like the Enterococci, and some Lactobacilli, which have reported respiration potential, will display improved survival under respiration-permissive conditions.

Introduction

Lactococcus lactis has long been used for its capacity to grow in milk and other foods for the production of fermented products. Its intensive uses in the food industry have channeled most studies of lactococci to the area of fermentation metabolism; studies on oxygen were focused on its effects on fermentation enzymes, and/or its toxicity due to radical formation.

We recently determined that different lactococcal strains undergo respiration and thrive when they are grown in the presence of oxygen and a heme source (Duwat et al. 1998, 1999, 2001); these observations are consistent with a generally overlooked finding in 1970 that gave evidence for respiration in *Lactococcus* (Sijpesteijn 1970). Note that lactococci were initially isolated from plants (Fenton 1987; Kelly et al. 1998). As such, and in retrospect, it would be surprising if these bacteria could not tolerate oxygen. Indeed, our results show that respiration is actually the *preferred* mode of *L. lactis* growth, as biomass is greatly improved, and survival is extended well beyond that of fermenting cultures (Figure 1A,B).

Here, we review the cumulative data that demonstrates respiration in *L. lactis*, and discuss the particularity of this system compared to other, well-studied respiring bacteria. Respiration has clear growth and survival advantages for *L. lactis*, and we predict that similar positive effects will be seen in other presumably respiring related species. Based on data from our laboratory, we propose that growth via respiration actually reduces intracellular oxidation, as compared to fermentative growth under static or aeration conditions. This may be a key factor promoting increased

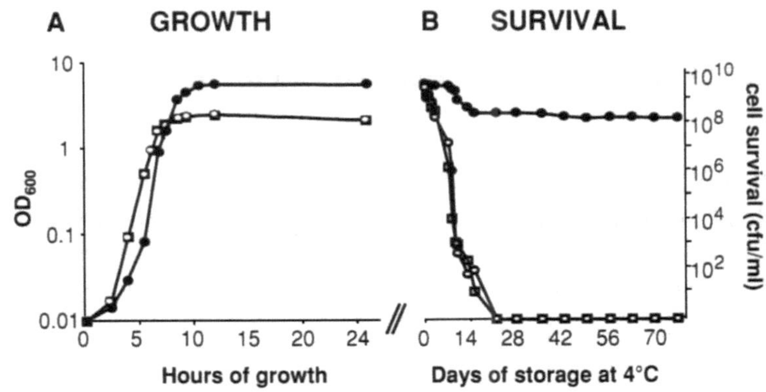

Figure 1. Improved growth and survival of *L. lactis* cultures grown in the presence of heme and oxygen. (A) *Growth:* Wild type *L. lactis* (MG1363) is grown (in M17 medium containing 1% glucose at 30 °C) under fermentation (open squares), aeration (open circles), or respiration (10 μm heme added) (closed circles) conditions. Growth is followed by OD_{600}. (B) *Survival*: Cells grown as in (A) for 24 h, and then stored at 4 °C. Samples were removed for plating at time '0' (corresponding to 24 h growth), and thereafter. Cell survival is expressed as colony forming units (cfu), symbols as above. The figure is adapted from previously published work (Duwat et al. 2001).

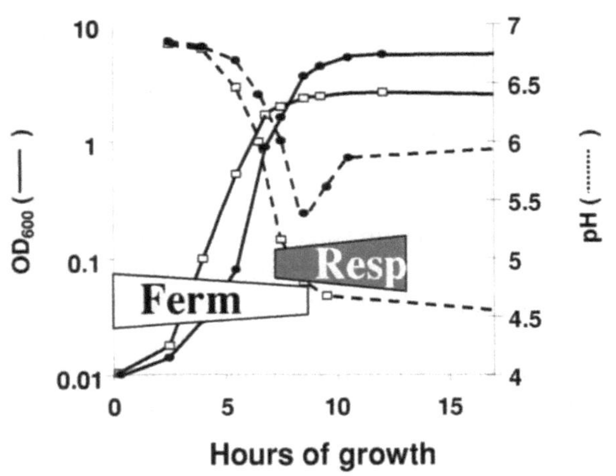

Figure 2. **Biphasic growth under respiration-permissive conditions.** Wild type *L. lactis* (MG1363) is grown under fermentation (open squares) or respiration-permissive (10 μm heme added) (closed circles) conditions as in Figure 1, legend. OD_{600} (solid lines), and pH (dashed lines) are followed during growth. Growth and acidification curves of aeration cultures (not shown) were comparable to those of fermentation cultures. Note that medium acidification during the first 8 h of growth is similar for all three cultures. A marked increase in pH of the respiration culture is observed after 9 h growth. The figure is modified from previously published work (Duwat et al. 2001).

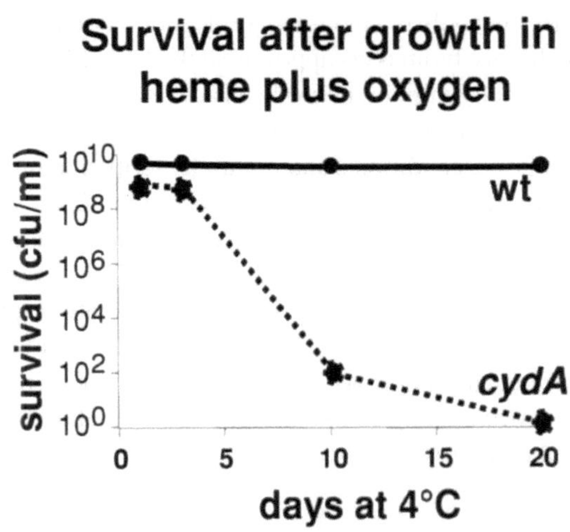

Figure 3. Improved growth and survival via respiration requires cytochrome bd oxidase. Cultures of *L. lactis* MG1363 (solid line) and the *cydA* mutant derivative (dotted line) were grown 24 h under respiration-permissive (10 μm heme added) conditions, as in Figure 1, legend, and then stored at 4 °C. Cell counts were determined starting at 24 h. The figure is modified from previously published work (Duwat et al. 2001).

survival of *L. lactis* (and possibly of other species that possess a similar gene complement) under respiration conditions.

Note that throughout this review, we distinguish in *L. lactis* between aeration or aerobic growth, and respiration or respirative growth (which requires exogenous heme). We refer to fermentation for static growth; however, *L. lactis* is also fermentative (i.e., converts glucose to lactic and other acids) under aeration conditions.

Proof of lactococcal respiration

Three key results confirmed that the presence of heme and oxygen promote *L. lactis* respiration (Duwat et al. 2001): (i) Fermentation products (mainly lactic acid) do not accumulate, and final culture pH is

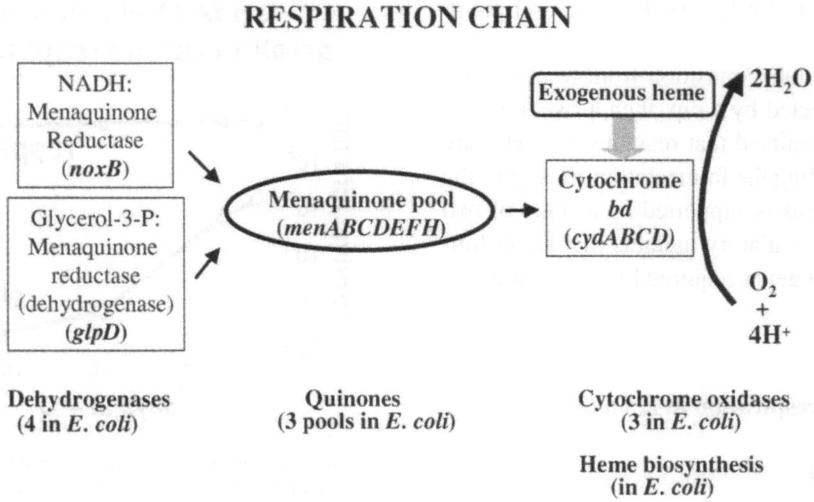

Figure 4. Non-flexible and non-redundant respiration machinery in *L. lactis*. A respiration chain requires dehydrogenases, quinones, and a terminal cytochrome oxidase. Putative components of the *L. lactis* respiration chain are presented, as derived from experimental results (Sijpesteijn 1970; Duwat et al. 1998, 2001) and deduced from available genome information (http://www.ncbi.nlm.nih.gov/BLAST/, (Bolotin et al. 2001)). A final F1Fo-ATPase, identified in *L. lactis* (Blank et al. 2001), is likely to pump protons in during respiration for energy production (not shown). The numbers of alternative components available in *E. coli* are presented below for comparison (Gennis & Stewart 1996). Note that heme is synthesized in nearly all characterized respiring bacteria, but not in *L. lactis*.

high (between 5.5 and 6 in M17 rich medium containing 1% glucose). Indeed, lactic acid is initially produced, and then decreases (it may be metabolized in the respiration process) during growth ((Duwat et al. 2001) (Kaneko et al. 1990); see Figure 2). (ii) The *cydAB* genes, encoding cytochrome bd oxidase, are required for improved growth and survival (Figure 3). Biomass is doubled for a wild type culture grown under respiration-permissive conditions; in contrast, no such effect is observed for a *cydA* mutant. The improved biomass is expected from the more energetically efficient respirative metabolism. (iii) Oxygen utilization in resting lactococci requires the presence of *cydA*. [Note that *L. lactis* encodes several NADH oxidase homologues (http://www.ncbi.nlm.nih.gov/BLAST/) (Bolotin et al. 2001), at least one of which consumes oxygen independently of respiration (Lopez de Felipe 1997)]. Putting together the genetic proof and recently available genomic data, we can reconstitute a complete respiratory chain, comprising dehydrogenases, potentially encoded by membrane proteins NoxB and/or GlpD, menaquinones, and a cytochrome oxidase (Figure 4).

Recently, the F_1F_0-ATPase of *L. lactis* was reported to be indispensable for *L. lactis* growth (Koebmann et al. 2000), as tested under fermentation conditions. The F_1F_0-ATPase was presumed to be needed for proton extrusion and to keep cytoplasmic pH up, particularly when external pH was low (Koebmann et al. 2000). It later turned out that the F_1F_0-ATPase is *not* necessary under respiration-permissive conditions (Blank et al. 2001). This would be expected, as proton extrusion occurs via respiration, and less acid is produced in respiration compared to fermentative growth. We speculate, however, that the absence of F_1F_0-ATPase reduces the capacity of *L. lactis* to produce ATP; at higher pH as in respiration, the pump is likely to pump protons in for energy production.

Respiration-permissive conditions lead to biphasic growth

Initial lactococcal growth out of lag phase occurs via fermentation, even when heme and oxygen are present (as examined in glucose-containing rich medium; (Figure 2). Lactic acid accumulates and the medium is acidified during the first several hours of aerated growth in the presence of heme, indicating that fermentation occurs. Transition to respiration occurs when the culture (grown in M17 plus 1% glucose) reaches approximately pH 5.3 (Duwat et al. 2001). At this point, the pH starts to increase. At this time in growth, a shift in the nature of metabolic products occurs, and genes required for respiration are induced ((Duwat et al. 2001); PG, KV, manuscripts in preparation).

In view of the sharp transition from fermentation to respiration, reflected by a dip, then a rise in pH of the cultures, we predicted that regulatory gene(s) are involved in controlling the fermentation-to-respiration shift. This hypothesis is supported by a study of two potential *L. lactis* regulatory mutants that affect timing of expression of genes required for respiration (our unpublished data).

Consequences of respiration in *L. lactis*

New genes induced

The profound changes in *L. lactis* growth, metabolic products, and survival, require the expression of an altered set of genes that take over during respiration. Proteome analyses indicate that growth via respiration results in increased expression of peptidases, amino acid biosynthesis genes, and redox factors (KV, PG, manuscript in preparation).

Increased long-term survival

A surprising phenotype of respiring *L. lactis* concerns the greatly improved long-term storage survival. Cell survival is improved up to 10^9-fold after six weeks of storage at 4 °C for respiration-grown, compared to fermentation-grown cultures (Figure 1B). [Note that good long-term survival of fermentative lactococci under certain conditions was recently reported (Kim et al. 2001). It was speculated that under the storage conditions used, the bacterial population was in constant flux between degradation and growth, and resulted in altered morphologies and genetic rearrangements.] We provide evidence below that a key factor of long term storage survival after respiration is the presence of cytochromes, which may protect cells from oxidative stress.

Respiratory growth leads to improved resistance to oxidative stress

In *E. coli*, respiration is a source of oxidative stress; oxygen radicals provoke DNA, protein, and membrane damage. The superoxide dismutases are necessary for survival of *Escherichia coli* respiring cells (Dukan & Nystrom 1999).

Our results suggest that growth of lactococci via respiration actually reduces oxidative stress, compared

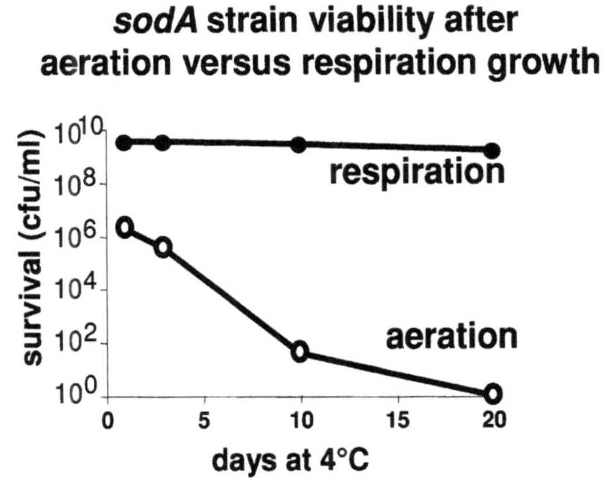

Figure 5. Respiration suppresses the need for oxygen radical elimination by SodA. The *L. lactis sodA* mutant strain (Sanders et al. 1995) was grown under aeration (open circles) or respiration-permissive (10 μm heme added) conditions (closed circles) for 24 h as described in Figure 1 legend, and then stored at 4 °C. Cell counts were determined starting at 24 h. The figure is modified from previously published work (Duwat et al. 2001).

to cells grown under aeration (no heme). The major proof of this comes from studies with a strain defective in the unique *L. lactis* superoxide dismutase, the manganese-dependent SodA (Sanders et al. 1995). SodA detoxifies the cytoplasm by removing oxygen radicals (Fridovich 1989). The *sodA* mutant is poorly viable under aeration conditions; after 24 h aerated growth, viability is only 10^6 cells per ml culture. In contrast, respiration-permissive growth conditions result in good yield ($\sim 10^9$ cells per ml) and long-term viability of the *sodA* strain (Figure 5) (Duwat et al. 2001). These results indicate that in *L. lactis*, compared to aeration conditions, respiration actually reduces intracellular oxidative stress.

Physiological cell state is altered by respiration

The green fluorescent protein (GFP) requires oxygen for activity (Hansen et al. 2001). We used a strain of *L. lactis* expressing the *gfp* gene (Gory et al. 2001) to compare GFP fluorescence when grown under fermentation, aeration, or respiration conditions. GFP expression levels were equivalent in the three growth conditions (as confirmed by Western blotting with GFP-specific antibodies). Late exponential phase cells were visualized by microscopy (Figure 6). Cells grown under fermentation conditions were not fluorescent, while cells grown under aeration conditions are strongly fluorescent, in keeping with the need for

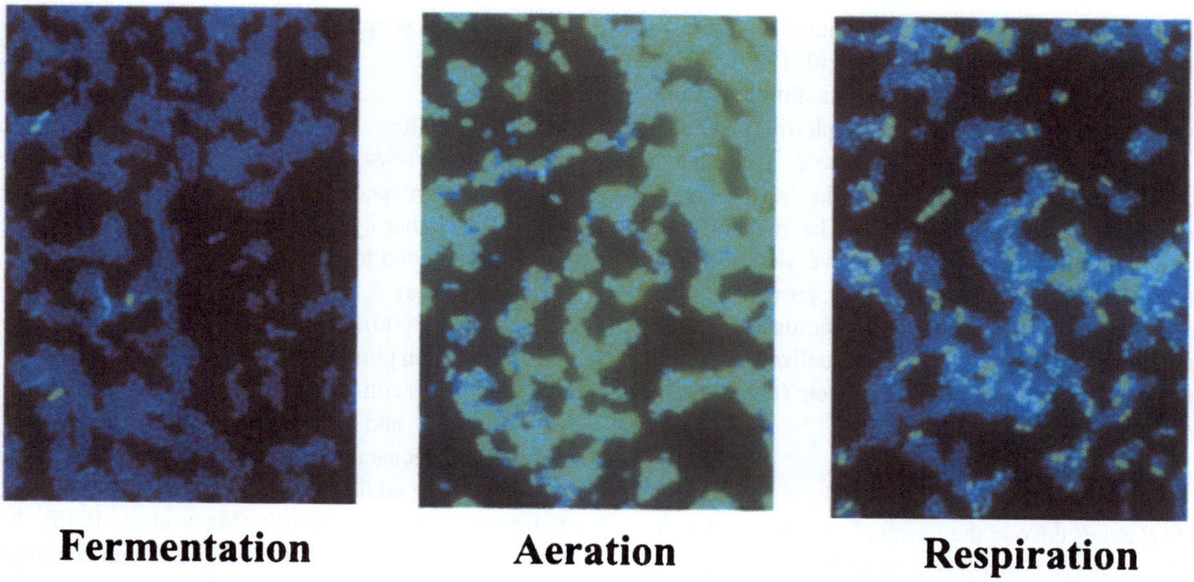

Figure 6. Physiology of *L. lactis* is markedly different under aeration and respiration conditions. *L. lactis* MG1363 expressing the *gfp* gene (Gory et al. 2001) was grown at 30 °C in rich medium under fermentation, aeration, or respiration-permissive (containing 10 μm heme) conditions in rich medium containing 1% glucose. Samples were taken in late exponential phase; this corresponds to an OD_{600} of 2 for fermentation and aeration cultures (pH ~4.9), and of 4 for the respiration culture (pH ~5.2). Cells were stained with DAPI and visualized by epifluorescence microscopy using one filter for GFP fluorescence (green cells), superimposed upon the image obtained by DAPI staining (blue cells).

oxygen for GFP fluorescence (Hansen et al. 2001). As lower pH (aerated and fermented cultures were at pH ~4.9 compared to pH ~5.2 for respiration cultures) reduces the GFP fluorescence signal (Hansen et al. 2001), the extent of oxidation in aerated cells may be underestimated. In contrast, cells cultivated under respiration conditions show very little fluorescence (KV, BC, SK, JT, PG, AG, manuscript in preparation). These results suggest that despite aeration of the culture, oxygen levels are maintained low in respiring cells, thereby reducing oxygen availability for the formation of toxic radicals.

We propose two explanations for improved oxygen tolerance by respiring cells. One is that a respiration component (e.g., cytochrome) scavenges and eliminates oxygen, thus thwarting reactivity with cell components (Goldman et al. 1996) (discussed below). The second is that the change in metabolism results in a more neutral intracellular pH [a low pH environment increases the production of toxic oxygen radicals; (Duwat et al. 1995)]. Growth and survival experiments comparing aeration *versus* respiration conditions in strongly buffered medium suggest that the presence of respiratory chain components is the *major* factor in reducing oxidative stress. Nevertheless, the higher pH of respiring cultures is likely to contribute to the reduction in oxidative stress.

A unique (and non-redundant) respiration model

Bacteria referred to as 'aerobic' are characterized as containing multiple and redundant respiratory components (see Poole & Cook 2000 for review), presumably to assure optimal growth under different environmental conditions. For example, aerobic bacteria *E. coli* and *Bacillus subtilis* encode three and four terminal cytochrome oxidases, respectively. As such, inactivation of a single cytochrome oxidase does not abolish respiration in those bacteria. In contrast, the *L. lactis* respiration pathway includes a minimal complement of genes; *cydAB* genes encode the sole cytochrome oxidase (bd), which is strictly required for respiration (Figure 3).

Aeration and respiration are separable states in *L. lactis*. This is because, in contrast to the characterized aerobic bacteria that synthesize their own heme for cytochrome activity, *L. lactis* requires exogenous heme addition (Figure 2). Part of the heme biosynthesis pathway is present (e.g., *L. lactis* can assimilate protoporphyrin IX and charge it with iron via ferrochelatase activity; (Duwat et al. 1999, 2001)). Respiration in *L. lactis* thus requires addition of heme or the immediate precursor, protoporphyrin IX (Duwat et al. 2001). Interestingly, wild type *L. lactis* has a *cydAB*-like phenotype when grown in the absence of heme (i.e., respiration-defective). The two different states, aeration and respiration, have very different consequences concerning physiology, growth, and survival. The striking difference in physiology of aerated versus respiring cultures can be visualized in strains expressing the green fluorescent protein (Figure 6).

CydAB *in respiration and survival*

CydAB genes encode the terminal cytochrome bd oxidase, an enzyme that has been well characterized in *E. coli* and *B. subtilis*. Its expression in these organisms is induced in late growth and under low oxygen availability (Georgiou et al. 1988). In *L. lactis*, *cydAB* is strictly required for respiration to occur, suggesting the absence of other cytochromes. Indeed, although cytochrome oxidases are well-conserved proteins, no other lactococcal homologues have been identified by genome searches.

We consider it likely that *cydAB* is directly involved not only in respiration but also in improved long-term survival of lactococci. In *E. coli* and other bacteria, *cydAB* genes are implicated in improving cell resistance to stress (Goldman et al. 1996; Siegele et al. 1996; Azarkina et al. 1999; Edwards et al. 2000). Bacterial mutants defective in cytochrome bd biosynthesis can be temperature and metal sensitive (Goldman et al. 1996; Edwards et al. 2000). Such mutants are also defective for exit from stationary phase (i.e., strain restart) (Siegele et al. 1996). These defects have been related to increased oxygen radical accumulation in the mutant strains (Goldman et al. 1996), although in one study superoxides were not increased in a cytochrome bd-deficient strain (Siegele et al. 1996). Difficulties in determining the role of *cydAB* in *E. coli* may be due to the functional genetic redundancy in this bacterium (Poole & Cook 2000).

We recently suggested that the uniqueness of genes encoding an important cell function was characteristic of bacteria having relatively small genomes (Poquet et al. 2000), and as such makes *L. lactis* a very interesting model for study. The non-redundancy of respiration genes is consistent with this hypothesis, and should simplify the dissection of this growth mode in lactococci.

Where else do 'surprising' respiration genes occur?

Biological studies and whole genome sequence searches (http://www.ncbi.nlm.nih.gov/BLAST/) suggest that certain species, like lactococci, encode *cydAB*, but not other cytochrome oxidase genes. These species correspond to: (i) *Enterococcus* (*faecalis* and *faecium*) (Ritchey & Seeley 1976; Whittenbury 1978; Winstedt et al. 2000), (ii) *Lactobacillus maltaromicus*; note that aeration plus heme lead to a 10-fold improvement in biomass compared to fermentation conditions (Meisel 1994), and (iii) *Bacteroides*. Interestingly, *Bacteroides* species are described as strict anaerobes; however, heme addition was previously reported to improve growth of *B. fragilis* (Macy et al. 1975). It remains to be seen whether *Bacteroides* can actually undergo respiration, and whether in consequence, its oxygen tolerance is improved, as might be suggested from our studies.

Why should *L. lactis* undergo respiration?

A minimal respiratory chain is conserved in lactococci. If one considers *Lactococcus* as a dairy organism, there is no reason to maintain these genes, as neither oxygen, nor a heme source, is freely available in that environment. However, milk cannot be a primary niche for lactococci, as it is present in an enclosed, sterile environment (Hirsch 1952), nor is it a significant source of heme. We propose that lactococci might benefit from a respiratory metabolism when they grow among plants and animals. Both these niches can contain oxygen, and may provide sources of iron-charged porphyrin.

Acknowledgements

We thank our colleagues Stig Lykke Iversen, Eric Johansen, Dan Nilsson (Chr Hansen, Denmark), Claus Maxel Henriksen, and Egon Bech Hansen, for discussions and warm support. Patrick Duwat (deceased January 5, 2000) initiated the work linking respiration to good growth and long term survival. We thank the 'Génétique Appliquée' team for support throughout this work. We acknowledge the valuable comments of S.L. Iversen, and the anonymous referee. Part of the work presented here was supported by a research grant from Chr Hansen Company, Denmark.

References

Azarkina N, Siletsky S, Borisov V, von Wachenfeldt C, Hederstedt L & Konstantinov AA (1999) A cytochrome bb'-type quinol oxidase in *Bacillus subtilis* strain 168. J. Biol. Chem. 274: 32810–32817.

Blank LM, Koebmann BJ, Michelsen O, Nielsen LK & Jensen PR (2001) Hemin reconstitutes proton extrusion in an H(+)-ATPase-negative mutant of *Lactococcus lactis*. J. Bacteriol. 183: 6707–6709.

Bolotin A, Wincker P, Mauger S, Jaillon O, Malarme K, Weissenbach J, Ehrlich SD & Sorokin A (2001) The complete genome sequence of the lactic acid bacterium *Lactococcus lactis* ssp. *lactis* IL1403. Genome Res. 11: 731–753.

Dukan S & Nystrom T (1999) Oxidative stress defense and deterioration of growth-arrested *Escherichia coli* cells. J. Biol. Chem. 274: 26027–26032.

Duwat P, Ehrlich SD & Gruss A (1995) The recA gene of *Lactococcus lactis*: characterization and involvement in oxidative and thermal stress. Mol. Microbiol. 17: 1121–1131.

Duwat P, Gruss A, LeLoir Y, Gaudu P (1999) Bactéries lactiques transformées pour leur conférer un métabolisme respiratoire, et levains comprenant lesdites bactéries. French patent application FR2798670. http://ep.espacenet.com/

Duwat P, Sourice S, Cesselin B, Lamberet G, Vido K, Gaudu P, Le Loir Y, Violet F, Loubiere P & Gruss A (2001) Respiration capacity of the fermenting bacterium *Lactococcus lactis* and its positive effects on growth and survival. J. Bacteriol. 183: 4509–4516.

Duwat P, Sourice S & Gruss A (1998) Process for preparing starter cultures of lactic acid bacteria. French patent application FR9809463. http://ep.espacenet.com/

Edwards SE, Loder CS, Wu G, Corker H, Bainbridge BW, Hill S & Poole RK (2000) Mutation of cytochrome bd quinol oxidase results in reduced stationary phase survival, iron deprivation, metal toxicity and oxidative stress in Azotobacter vinelandii. FEMS Microbiol. Lett. 185: 71–77.

Fenton MP (1987) An investigation into the sources of lactic acid bacteria in grass silage. J. Appl. Bacteriol. 62: 181–188.

Fridovich I (1989) Superoxide dismutases. An adaptation to a paramagnetic gas. J. Biol. Chem. 264: 7761–7764.

Gennis RB & Stewart V (1996) Respiration. In: Neidhardt FC (Ed) *Escherichia coli* and Salmonella: Cellular and Molecular Biology Vol. I (pp 217–261). ASM Press, Washington, DC.

Georgiou CD, Dueweke TJ & Gennis RB (1988) Regulation of expression of the cytochrome d terminal oxidase in *Escherichia coli* is transcriptional. J. Bacteriol. 170: 961–966.

Goldman BS, Gabbert KK & Kranz RG (1996) The temperature-sensitive growth and survival phenotypes of *Escherichia coli* cydDC and cydAB strains are due to deficiencies in cytochrome bd and are corrected by exogenous catalase and reducing agents. J. Bacteriol. 178: 6348–6351.

Gory L, Montel MC & Zagorec M (2001) Use of green fluorescent protein to monitor *Lactobacillus sakei* in fermented meat products. FEMS Microbiol. Lett. 194: 127–133.

Hansen MC, Palmer RJ Jr, Udsen C, White DC & Molin S (2001) Assessment of GFP fluorescence in cells of *Streptococcus gordonii* under conditions of low pH and low oxygen concentration. Microbiology 147: 1383–1391.

Hirsch A (1952) The evolution of the lactic acid bacteria. J. Dairy Res. 19: 290–293

Kaneko T, Tagahashi M & Suzuki H (1990) Acetoin fermentation by citrate-positive *Lactococcus lactis* subsp. *lactis* 3022 grown aerobically in the presence of hemin or Cu2+. Appl. Environ. Microbiol. 56: 2644–2649.

Kelly WJ, Davey GP & Ward LJ (1998) Characterization of lactococci isolated from minimally processed fresh fruit and vegetables. Int. J. Food Microbiol. 45: 85–92.

Kim WS, Park JH, Ren J, Su P & Dunn NW (2001) Survival response and rearrangement of plasmid DNA of *Lactococcus lactis* during long-term starvation. Appl. Environ. Microbiol. 67: 4594–4602.

Koebmann BJ, Nilsson D, Kuipers OP & Jensen PR (2000) The membrane-bound H(+)-ATPase complex is essential for growth of *Lactococcus lactis*. J. Bacteriol. 182: 4738–4743.

Lopez de Felipe F, Starrenburg M J C & Hugenholtz J (1997) The role of NADH oxidation in acetoin and diacetyl production from glucose in *Lactococcus lactis* subsp. *lactis* MG1363. FEMS Microbiol. Lett. 156: 15–19

Macy J, Probst I & Gottschalk G (1975) Evidence for cytochrome involvement in fumarate reduction and adenosine 5′-triphosphate synthesis by *Bacteroides fragilis* grown in the presence of hemin. J. Bacteriol. 123: 436–442.

Meisel J, Wolf, G & Hammes WP (1994) Heme-dependent cytochrome formation in *Lactobacillus maltaromicus*. System. Appl. Microbiol. 17: 20–23

Poole RK & Cook GM (2000) Redundancy of aerobic respiratory chains in bacteria? Routes, reasons and regulation. Adv. Microb. Physiol. 43: 165–224.

Poquet I, Saint V, Seznec E, Simoes N, Bolotin A & Gruss A (2000) HtrA is the unique surface housekeeping protease in *Lactococcus lactis* and is required for natural protein processing. Mol. Microbiol. 35: 1042–1051.

Ritchey TW & Seeley HWJ (1976) Distribution of cytochrome-like respiration in Streptococci. J. Gen. Microbiol. 93: 195–203.

Sanders JW, Leenhouts KJ, Haandrikman AJ, Venema G & Kok J (1995) Stress response in *Lactococcus lactis*: cloning, expression analysis, and mutation of the lactococcal superoxide dismutase gene. J. Bacteriol. 177: 5254–5260.

Siegele DA, Imlay KR & Imlay JA (1996) The stationary-phase-exit defect of cydC (surB) mutants is due to the lack of a functional terminal cytochrome oxidase. J. Bacteriol. 178: 6091–6096.

Sijpesteijn AK (1970) Induction of cytochrome formation and stimulation of oxidative dissimilation by hemin in *Streptococcus lactis* and *Leuconostoc mesenteroides*. Antonie Van Leeuwenhoek 36: 335–348

Whittenbury R (1978) Biochemical characteristics of Streptococcus species. Soc. Appl. Bacteriol. Symp. Ser. 7: 51–69

Winstedt L, Frankenberg L, Hederstedt L & von Wachenfeldt C (2000) *Enterococcus faecalis* V583 contains a cytochrome bd-type respiratory oxidase. J. Bacteriol. 182: 3863–3866.

Antonie van Leeuwenhoek **82:** 271–278, 2002.

Glutamate dehydrogenase activity: a major criterion for the selection of flavour-producing lactic acid bacteria strains

Catherine Tanous, Agnieszka Kieronczyk, Sandra Helinck, Emilie Chambellon & Mireille Yvon*
*INRA, Unité de Biochimie et Structure des Protéines, 78352 Jouy-en-Josas (*Author for correspondence; E-mail: yvon@jouy.inra.fr)*

Key words: amino acid catabolism, lactic acid bacteria, *Lactococcus lactis*, lactobacilli, glutamate dehydrogenase, flavour formation

Abstract

Lactic acid bacteria (LAB) have the enzyme potential to transform amino acids into aroma compounds that contribute greatly to cheese flavour. Generally, amino acid conversion by LAB is limited by their low production of α-ketoglutarate since this α-ketoacid is essential for the first step of the conversion. Indeed, we have demonstrated that adding exogenous α-ketoglutarate to cheese curd, as well as using a genetically modified *L. lactis* strain capable of producing α-ketoglutarate from glutamate, greatly increased the conversion of amino acid to potent aroma compounds in cheese. Here we report the presence of glutamate dehydrogenase (GDH) activity required for the conversion of glutamate to α-ketoglutarate in several 'natural' LAB strains, commonly used in cheese manufacturing. Moreover, we show that the ability of LAB to produce aroma compounds from amino acids is closely related to their GDH activity. Therefore, GDH activity appears to be a major criterion for the selection of flavour-producing LAB strains, which could be used as a starter or as an adjunct to intensify flavour formation in some cheeses.

Introduction

Amino acid catabolism is a major process for flavour formation in cheese, especially in cheeses containing only lactic acid bacteria (LAB). Indeed, in these cheeses ripened without surface flora, flavour is largely due to aldehydes, alcohols and acids derived from aromatic amino acids and branched-chain amino acids and to sulphur compounds derived from methionine (Christensen & Reineccius 1995). The compounds derived from aromatic amino acids have a rather floral or fruity note, those derived from branched-chain amino acids have generally malty or animal note and the compounds derived from methionine have a characteristic sulphur aroma reminiscent of cabbage, cauliflower or garlic (Molimard & Spinnler 1996; Kubickova & Grosch 1997; Milo & Reineccius 1997; Friedrich & Acree 1998).

Over the past few years, several research groups have focused on amino acid catabolism in LAB, with the aim of elucidating the reactions involved in aroma formation and identifying the enzymes and factors playing a key role. The discovery of a relationship between some enzyme activities and high flavouring potential or the specific production of a desired aroma, could enable the selection of LAB strains in order to accelerate, intensify or diversify flavour formation in cheese. This would be of particular interest for cheeses without surface flora, in which flavour formation requires a very long ripening time. Indeed, until today the selection of flavouring strains used as starters or as adjuncts (non-starter lactic acid bacteria, NSLAB) has been empirical and not always successful (Crow et al. 2001).

In this paper, we do not intend to review the literature on the flavour formation from amino acid catabolism by LAB since an overview on the subject has been recently published (Yvon & Rijnen 2001), but we focus on a key step of the amino acid conversion to aroma compounds by LAB: the transamination step, and the major role of glutamate dehydrogenase activity for this reaction.

Reactions involved in amino acid conversion to aroma compounds by LAB

Amino acid transamination is catalysed by aminotransferases which transfer the amino group of an amino acid to an α-ketoacid. The α-ketoacid acceptor is thus transformed into the corresponding amino acid while the amino acid is deaminated to the corresponding α-ketoacid. This reaction is a key step in the formation of aroma compounds from amino acids by LAB. Indeed, catabolism of aromatic amino acids (ArAAs), branched-chain amino acids (BcAAs) and methionine (Met) is essentially initiated by a transamination reaction since amino acid degradation in LAB occurs only in the presence of an α-ketoacid. This was demonstrated in lactococci (Gao et al. 1997, 1998; Yvon et al. 1997), in mesophilic lactobacilli such as *Lactobacillus paracasei*, *Lactobacillus casei*, *Lactobacillus plantarum* and *Lactobacillus rhamnosus* (Gummalla & Broadbent 1996; Tammam et al. 2000; Kieronczyk et al. 2001), and also in thermophilic LAB such as *Streptococcus thermophilus*, *Lactobacillus helveticus*, *Lb. delbruckii bulgaricus* and *Lb. delbruckii lactis* (Gummalla & Broadbent 1999).

Aminotransferase activities towards aromatic amino acids, branched-chain amino acids and methionine have been detected in most LAB (Amarita et al. 2001; Williams et al. 2001) and two aminotransferases, AraT and BcaT, were isolated from *L. lactis* (Yvon et al. 1997, 2000). Inactivation of both AraT and BcaT in *L. lactis* TIL 46 totally prevented the degradation of aromatic amino acids, branched-chain amino acids and methionine, confirming the key role of the transamination reaction in amino acid conversion (unpublished results). Aminotransferases of LAB mainly use α-ketoglutarate as the α-ketoacid acceptor (Williams et al. 2001; Yvon et al. 1997, 2000). However, some aminotransferases of NSLAB may also use other α-ketoacids such as pyruvate, glyoxylate or oxaloacetate (Amarita et al. 2001). But to our knowledge, glyoxylate is not produced by LAB, and pyruvate and oxaloacetate are rapidly used by other enzymes than aminotransferases.

The α-ketoacids resulting from amino acid transamination are then degraded in one or two additional steps into the different aroma compounds, which are aldehydes, alcohols, carboxylic acids and sulphur compounds (Figure 1). In *L. lactis*, the major final products of enzymatic conversion are hydroxyacids and carboxylic acids (Gao et al. 1998; Yvon et al. 1997). Generally, aldehyde and alcohol production by *L. lactis* is low but some particular strains produce a large amount of aldehydes mainly from branched-chain amino acids (Tucker & Morgan 1967; Morgan 1976; Weerkamp et al. 1996; Ayad et al. 1999). These particular strains were often named 'maltigenes' due to the malty aroma of these aldehydes (MacLeod & Morgan 1958; Turker & Morgan 1967). *L. lactis* and *Lb. plantarum* also generate methylthioacetaldehyde and benzaldehyde from the α-ketoacids derived from methionine and phenylalanine respectively (Nierop-Groot & de Bont 1998; Rijnen et al. 2000; Yvon et al. 2001). These aldehydes result from chemical oxidation of the α-ketoacids, which is catalysed by bivalent cations (Nierop-Groot & de Bont 1999). The aldehydes and acids derived from methionine are quite unstable and can be spontaneously degraded into methanethiol and dimethyldisulfide (DMDS), which were found to be major degradation products of methionine (Weimer et al. 1999). Before focusing on the transamination step we have to mention that methanethiol can also be produced by LAB from methionine directly via a demethiolation catalysed by cystathionine-lyases (Alting et al. 1995; Dias & Weimer 1998; Bruinenberg et al. 1997).

α-Ketoglutarate is a limiting factor for amino acid conversion to aroma compounds by LAB

Although lactococci have high aminotransferase activity, only very low and slow amino acid degradation occurs in semi-hard cheese made with lactococci. Since amino acid transamination also requires the presence of an α-ketoacid, we assumed that the amount of available α-keto acid could be the limiting factor for amino acid degradation. We therefore tried to add exogenous α-ketoglutarate to the cheese curd or alternatively use a strain capable of producing α-ketoglutarate to increase amino acid degradation during cheese ripening.

Adding α-ketoglutarate to St Paulin cheese (1–4 g per kg cheese) made with various starters greatly increased the conversion of ArAAs, BcAAs and methionine to potent aroma compounds which enhanced the intensity of cheese aroma (Rijnen et al. 1999; Yvon et al. 1998). These results were recently confirmed by Banks et al. who added α-ketoglutarate to Cheddar cheese (5–6 g per kg cheese) (Banks et al. 2001). Moreover, this study has shown the effect of α-ketoglutarate addition on the aroma characteristics of cheese. Addition of α-ketoglutarate significantly increased aroma intensity, creamy aroma and fruity

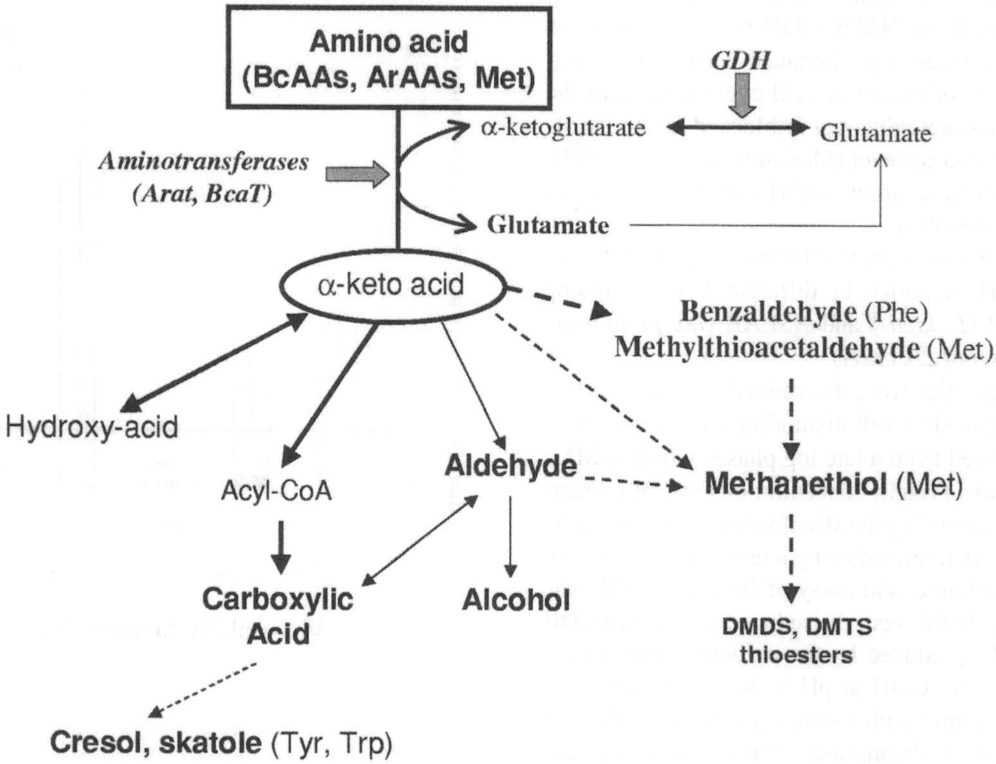

Figure 1. Main amino acid catabolism pathway found in LAB and some chemical reactions (dotted arrows) occurring in cheese. Aroma compounds appear in bold characters.

aroma. The results clearly indicate that the aroma intensity of 6- (or 12)-week-old Cheddar containing α-ketoglutarate was equivalent to the aroma intensity of a control cheddar (without α-ketoglutarate) ripened for 6 months. However, the organoleptic qualities of these cheeses remain to be assessed since so far only sniffing tests were performed.

As an alternative to adding exogenous α-ketoglutarate to the cheese, we suggested using, as a starter, a strain capable of producing α-ketoglutarate from glutamate that is always present in large quantities in cheese. To test this hypothesis, we expressed a gene from *Peptostreptococcus asaccharolyticus,* encoding a catabolic glutamate dehydrogenase, in *L. lactis* (Rijnen et al. 2000). This enzyme catalyses the deamination of glutamate to α-ketoglutarate. The GDH-producing lactococcal strain (TIL323) degraded amino acids without added α-ketoglutarate in a cheese model to the same extent that the wild-type strain degraded amino acids with added α-ketoglutarate. Interestingly, the GDH-producing strain produced a higher proportion of carboxylic acids, which are major aroma compounds. These results suggest that cheese aroma is greatly enhanced by using a strain with GDH activity. However, in that study, organoleptic analysis of model cheeses was not performed since the strain used was a genetically modified organism (GMO).

Considering the high ability of the modified *L. lactis* strain overexpressing GDH to degrade amino acids into aroma compounds, we have searched for 'natural' (wild type) LAB with GDH activity.

Glutamate dehydrogenase (GDH) activity in LAB

GDHs are ubiquitous enzymes that catalyse the reversible oxidative deamination of glutamate to α-ketoglutarate and ammonium, utilising NAD, NADP or both as a cofactor. Generally, microbial enzymes are specific for one of the two cofactors. It is believed that NADP-GDH has a role in glutamate biosynthesis while NAD-GDH is involved in glutamate catabolism (Smith et al. 1975). NADP-GDH activity has only been detected in a few LAB strains including *Lb. fermentum* IFO 3071, *Lb. plantarum* ATCC 8014, *Lb. casei* ATCC 7469 and *Streptococcus lactis* IFO 12546

(Misono et al. 1985; Morishita & Yajima 1995). However none of these NADP-GDH positive strains are capable of synthesising glutamate because the oxidative pathway of the citric acid cycle leading to the synthesis of α-ketoglutarate is blocked at the isocitrate dehydrogenase level (Morishita & Yajima 1995). Therefore, in these strains, GDH can only be used for glutamate catabolism.

Here, we have searched for both NAD-GDH and NADP-GDH activities in different LAB including starter LAB (*L. lactis*) and NSLAB (*Lb. plantarum*, *Lb. paracasei* or *Lb. casei*).

GDH activities were determined in fresh cell-free extracts obtained by cell disruption. Cells were previously harvested from a late log phase culture in M17, MRS or milk. Protein concentrations in cell extracts were measured using the Bradford method, and GDH activity was determined using a test based on the colorimetric glutamic acid assay of Boehringer (Misono et al. 1985). In this test, the reduced cofactor (NADH or NADPH), produced by the oxidative deamination of glutamate by GDH at pH 9, is measured by colorimetric reaction with iodonitrotetrazolium (INT) in the presence of diaphorase. The reaction produces formazan which is measured by absorbance at 492 nm.

GDH activities were detected in several LAB but the presence was highly species- and strain-dependent. Occurrence of GDH activities in the tested LAB is shown in Table 1. For each species, strains were classified as positive (GDH+) or negative (GDH-), assuming that an activity inferior to 0.5 units (nmol NAD(P)H formed/min mg of protein) is not significant. The strains with an activity of 0.5 ± 0.1 units were classified as ±.

NADP-dependent activity was found in most *Lb. plantarum* strains and in half of *Lb. paracasei* and *L. lactis* strains, but a few strains did not exhibit any activity. Only one strain of *L. lactis,* isolated from peas, exhibited both significant NAD- and NADP-dependent activities, but a comparison showed that its NAD-GDH activity was 300-fold lower than the activity of the modified *L. lactis* TIL323 overproducing heterologous GDH (Figure 2b). *L. lactis* IL1403 which lacks a GDH gene (Bolotin et al. 2001) did not exhibit any GDH activity. The detected dehydrogenase activities were specific for glutamate since no NAD or NADP reduction occurred when using phenylalanine or leucine instead of glutamate as substrates.

From the first screening, we selected two or three strains of each species with different levels of GDH

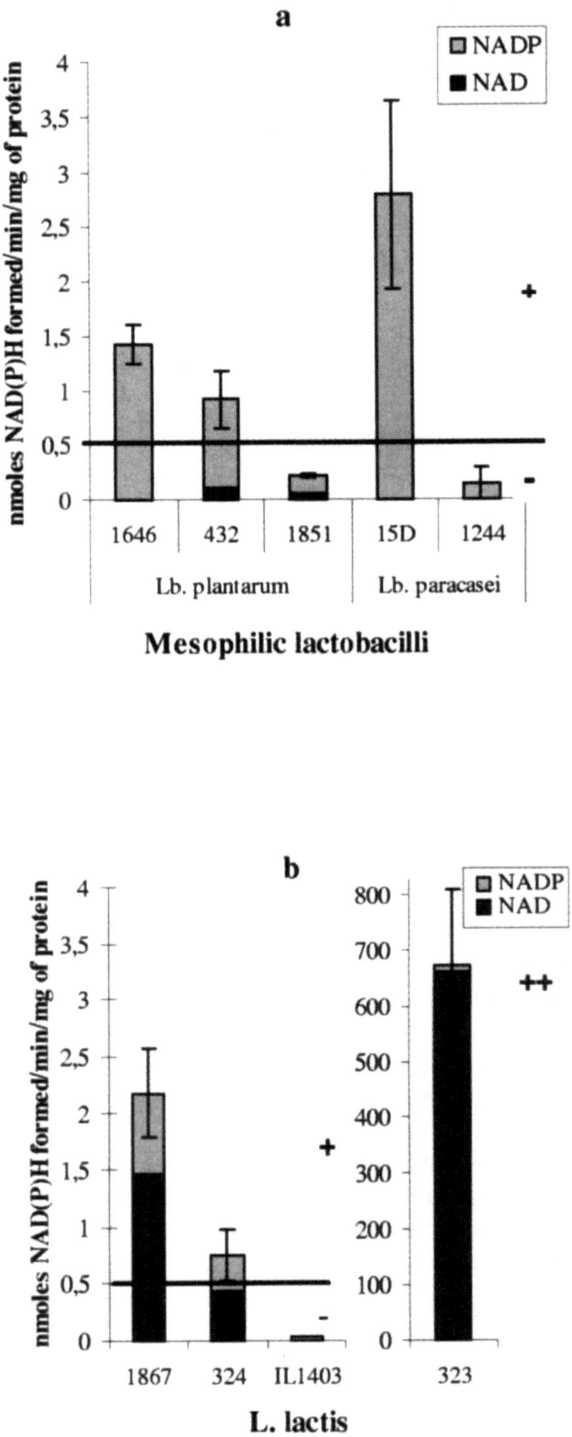

Figure 2. NAD and NADP-specific GDH activities of mesophilic lactobacilli grown in MRS (a) and of *L. lactis* grown in M17 (b). For each species the strains were classified according to the level of total GDH activity as follows: highly positive (++) (only modified strain TIL 323), positive (+) or negative (–), the line indicates the threshold used to distinguish between positive and negative strains. GDH activity was determined as described in the text. Results are the means of two determinations and error bars indicate both values.

Table 1. Occurrence of NAD and NADP-specific glutamate dehydrogenase activities in lactic acid bacteria. The strains analysed in more detail (Figures 2 and 3) appear in bold characters

Lb plantarum strains (from cheese)	GDH activity		*Lb paracasei* strains (from cheese	GDH activity		*L. lactis* strains	GDH activity	
	NADP	NAD		NADP	NAD		NADP	NAD
CNRZ 184	+	–	INF15D	+	–	NCDO 2118 (peas)	–	–
CNRZ 424	+	–	**CNRZ 1244**	–	–	NCDO 2633 (cow)	+	–
CNRZ 432	+	–	CNRZ 1219	–	–	NCDO 2146 (cow)	+	–
CNRZ 738	+	–	CNRZ 1244	–	–	NCDO 2091(radish)	–	–
CNRZ 764	+	–	CNRZ 383	–	–	**NCDO 1867** (peas)	+	+
CNRZ 1229	+	–	CNRZ 1830	+	–	NCDO 2054 (cheese)	+	–
CNRZ 1362	+	–	CNRZ 2756 [a]	+	–	**IL 1403** (laboratory)	–	–
CNRZ 1628	+	–	ATCC 316	+	–	**Til 324**[b]	–	+/–
CNRZ 1646	+	–	INF76	–	–	**Til 323**$_{Gdb++}$ [b]	+	++
CNRZ 1851	–	–	INF478	–	–			
CNRZ 1887	+	–						
CNRZ 1890	+	–						
CNRZ 1892	+	–						
ATCC 8014	+	–						
Total of GDH	**13/14**	**0/14**		**4/10**	**0/10**		**5/8**	**3/8**

[a] *Lb casei* strain. [b] Strains described in Rijnen et al. (2000). The origin of *L. lactis* strains is indicated in parenthesis.

activity (Figure 2) to compare their ability to degrade amino acids into aroma compounds.

Relationship between GDH activity and amino acid conversion to aroma compounds by LAB

Ability of the selected LAB strains to degrade amino acids was studied as previously described (Rijnen et al. 2000), by incubating resting cells in a reaction medium at pH 5.5, the pH of cheese. Cells were the same as those used for GDH determination. They were introduced into the reaction medium (phosphate buffer, pyridoxal phosphate, leucine or phenylalanine and glutamate) containing radiolabelled leucine or phenylalanine as the tracer. Aliquots of the reaction mixtures were analysed at zero time and after incubation for 40 h at 37 °C by reverse-phase and ion-exclusion HPLC with both UV detection and radioactivity detection.

The GDH-negative strains did not degrade both amino acids significantly (degradation lower than 2%). In contrast, all the GDH-positive strains were capable of significantly degrading leucine and phenylalanine in the reaction medium containing glutamate (Figure 3). We also observed amino acid degradation even when glutamate was not added to the medium, probably because most of the cells contained glutamate that had accumulated during growth and that was used by the GDH. But for most of the strains, amino acid degradation was greater when glutamate was added to the reaction medium, indicating that GDH is really involved in degradation.

Taking both lactococci and lactobacilli into account, the percentage of amino acid degradation is not totally proportional to NAD or NADP-GDH activities. But taking only mesophilic lactobacilli strains into account, amino acid degradation appears to be closely related to their GDH activities (Figure 4). Thus it appears that it is possible to connect GDH activity (determined at optimal pH (pH 9) with amino acid degradation within a species.

In future work, the ability of natural GDH positive strains to degrade amino acids and to produce aroma compounds in cheese should be verified. We can already note that the GDH-positive *Lactobacillus paracasei* ssp. *paracasei* INF 15D was previously associated with good quality Norvegia cheese (Skeie et al. 2001). If a positive effect of GDH activity on aroma formation is confirmed in cheese, GDH activity could be used as a criterion for the selection of flavour-producing LAB strains.

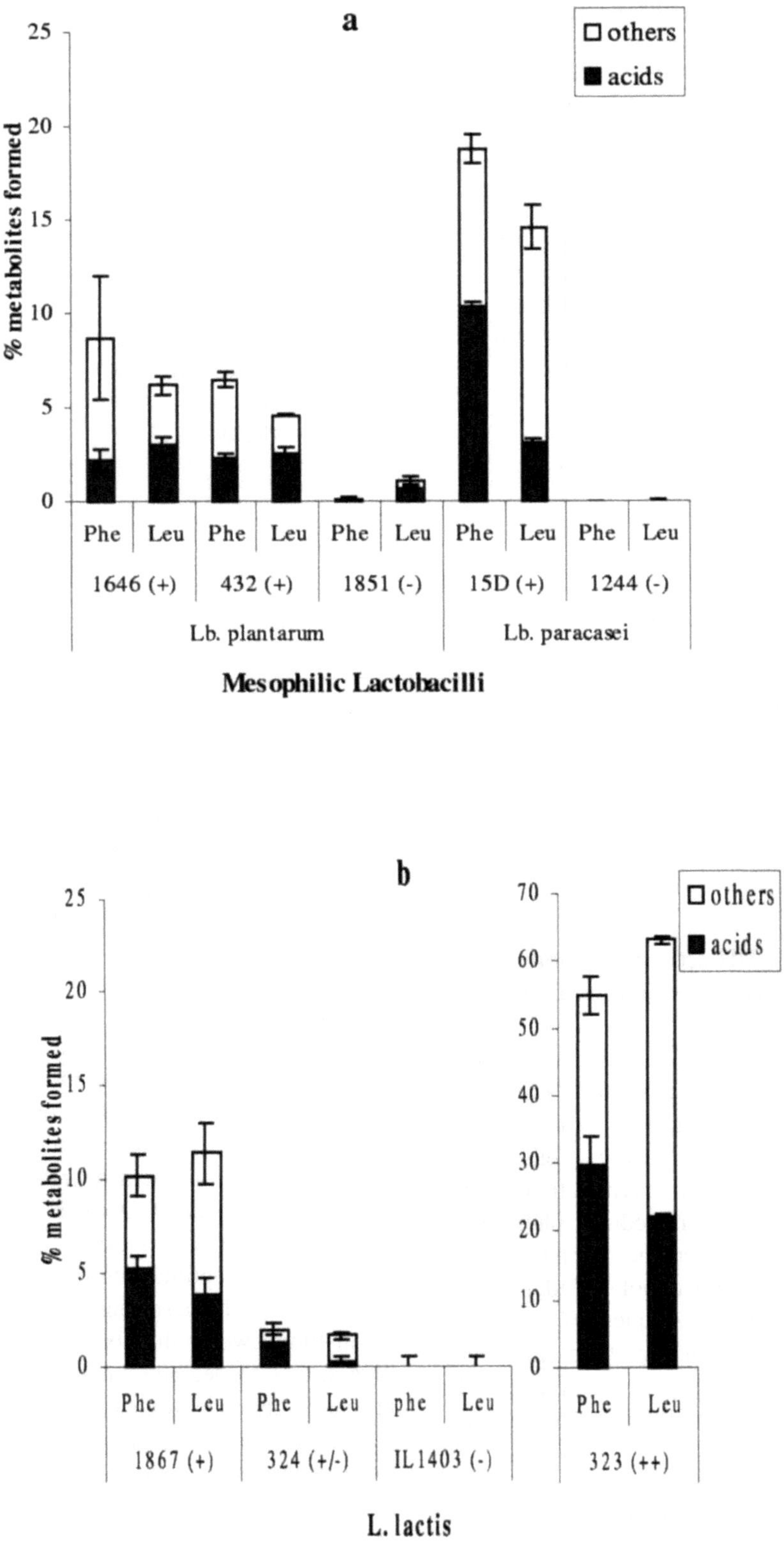

Figure 3. Degradation of phenylalanine (Phe) and leucine (Leu) by resting cells of mesophilic lactobacilli grown in MRS (a) and *L. lactis* grown in M17 (b). Carboxylic acids formed (phenylacetate from Phe and isovalerate from Leu) appear in black and other metabolites, such as keto acids, hydroxy acids and alcohols in some cases, appear in white. (++), (+) (+/–) and (–) indicate the level of GDH activity of the strains according to the classification of Figure 2. The results are the means of two determinations and error bars indicate the two values.

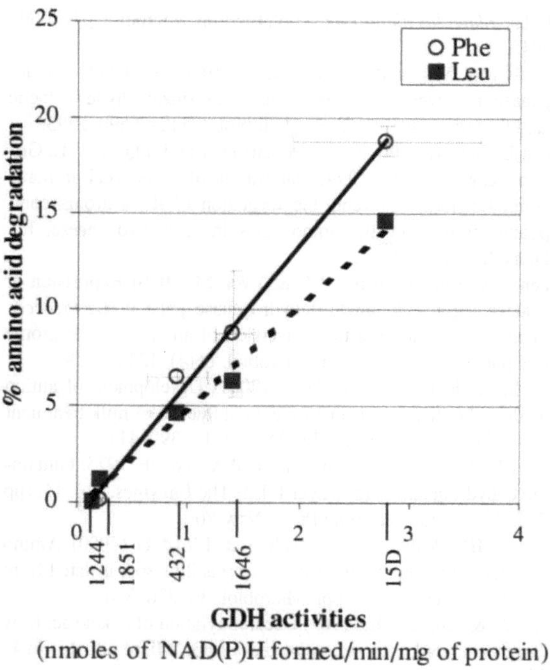

Figure 4. Relationships between percentages of phenylalanine (Phe) and leucine (Leu) degradation and GDH activities of mesophilic lactobacilli. Strain numbers are shown on the horizontal axis. The results are the means of two determinations and error bars indicate both values.

Concluding remarks

Over the past few years, we have demonstrated that it is possible to enhance aroma formation in cheese by intensifying the first step of amino acid catabolism by LAB, i.e. amino acid transamination. Two possibilities for stimulating amino acid transamination were demonstrated. The first one consists of the exogenous addition of α-ketoglutarate before cheese maturation, and the second one consists of the use of a strain capable of producing α-ketoglutarate from the glutamate present in cheese.

Only a few natural LAB strains possess the GDH activity required for glutamate conversion to α-ketoglutarate. However we found several GDH-positive strains especially within the *Lactobacillus plantarum* and the *Lactobacillus paracasei* species. These NSLAB species are sometimes used in Cheddar and Gouda manufacturing as flavour-producing adjuncts (lactobacilli). The relationship between flavouring properties of the strains used and their GDH activity has never been reported but our results suggest that this relationship is highly likely. We detected also low GDH activity in some *L. lactis* strains, especially in one strain isolated from peas, suggesting that a lactococcal *gdh* gene exists, at least in a few strains, even if it was not present in the genome of *L. lactis* IL1403.

Moreover, the association of a GDH-positive NSLAB strain with the starter *L. lactis* in cheese should be beneficial in aroma formation. Indeed, we observed *in vitro* that *L. lactis* NCDO 763 transformed the α-ketoacids produced from amino acids by the GDH-positive *Lb. paracasei* strains, into carboxylic acids which are potent aroma compounds in cheese (Kieronczyk, submitted.).

To conclude, selecting LAB strains with high GDH activity as a starter or as an adjunct appears to be an interesting way to intensify aroma formation in some cheeses.

Acknowledgements

The research described was supported in part by the EC FAIR contract CT97-3173 and TMR grant ERB 4001 GT95421 from the Commission of European Communities. The authors would like to thank A.-M. Wall (INRA Translation Unit, Jouy-en-Josas, France) for revising the English version of the manuscript and V. Monnet for critically reading the manuscript.

References

Alting AC, Engels WJM, Schalkwijk van S & Exterkate FA (1995) Purification and characterization of cystathionine β-lyase from *Lactococcus lactis* subsp. *cremoris* B78 and its posible role in flavor development in cheese. Appl. Environ. Microbiol. 61(11): 4037–4042.

Amarita F, Requena T, Taborda G, Amigo L & Pelaez C (2001) *Lactobacillus casei* and *Lactobacillus plantarum* initiate catabolism of methionine by transamination. J. Appl. Microbiol. 90: 971–978.

Ayad E, Verheul A, de Jong C, Wouters JTM & Smit G (1999) Flavour forming abilities and amino acid requirements of *Lactococcus lactis* strains isolated from artisanal and non-dairy origin. Int. Dairy J. 9: 725–735.

Banks JM, Yvon M, Gripon JC, Fuente MAdL, Brechany EY, Williams AG & Muir DD (2001) Enhancement of amino acid catabolism in cheddar cheese using α-ketoglutarate: amino acid degradation in relation to volatile compounds and other aroma character. Int. Dairy J. 11(4-7): 235–243.

Bolotin A, Wincker P, Mauger S, Jaillon O, Malarme K, Weissenbach J, Ehrlich SD & Sorokin A (2001) The complete genome sequence of the lactic acid bacterium *Lactococcus lactis* ssp.*lactis* IL1403. Genome Res. 11(5): 731–753

Bruinenberg PG, Roo Gd & Limsowtin GKV (1997) Purification and characterization of cystathionine γ-lyase from *Lactococus lactis* subsp. *cremoris* SK11: possible role in flavor compound formation during cheese maturation. Appl. Environ. Microbiol. 63: 561–566.

Christensen KR & Reineccius GA (1995) Aroma extract dilution analysis of aged Cheddar cheese. J. Food Sci. 60: 218–220.

Crow V, Curry B & Hayes M (2001) The ecology of non-starter lactic acid bacteria (NSLAB) and their use as adjuncts in New Zealand Cheddar. Int. Dairy J. 11: 275–283.

Dias B & Weimer B (1998) Conversion of methionine to thiols by Lactococci, Lactobacilli, and Brevibacteria. Appl. Environ. Microbiol. 64: 3320–3326.

Friedrich JE & Acree TE (1998) Gas chromatography olfactometry (GC/O) of dairy products. Int. Dairy J. 8: 235–241.

Gao S, Oh DH, Broadbent JR, Johnson ME, Weimer BC & Steele JL (1997) Aromatic amino acid catabolism by lactococci. Lait. 77: 371–381.

Gao S, Mooberry ES & Steele JL (1998) Use of 13C nuclear magnetic resonance and gas chromatography to examine methionine catabolism by Lactococci. Appl. Environ. Microbiol. 64: 4670–4675.

Gummalla S & Broadbent JR (1996) Indole production by *Lactobacillus* spp. in cheese: a possible role for tryptophanase. J. Dairy Sci.: 101.

Gummalla S & Broadbent JR (1999) Tryptophan catabolism by *Lactobacillus casei* and *Lactobacillus helveticus* cheese flavor adjuncts. J. Dairy Sci. 82(10): 2070–2077.

Kieronczyk A, Skeie S, Olsen K & Langsrud T (2001) Metabolism of amino acids by resting cells of non-starter lactobacilli in relation to flavour development in cheese. Int. Dairy J. 11: 217–224.

Kubickova J & Grosch W (1997) Evaluation of potent odorants of Camembert cheese by dilution and concentration techniques. Int. Dairy J. 7: 65–70.

MacLeod P & Morgan ME (1958) Differences in the ability of lactic streptococci to form aldehydes from certain amino acids. J. Dairy Sci. 41: 908–913.

Milo C & Reineccius GA (1997) Identification and quantification of potent odorants in regular-fat and low-fat mild Cheddar cheese. J. Agric. Food Chem. 45: 3590–3594.

Misono H, Goto N & Nagazaki S (1985) Purification, crystallisation and properties of NADP+-specific glutamate dehydrogenase from *Lactobacillus fermentum*. Agric. Biol. Chem. 49: 117–123.

Molimard P & Spinnler HE (1996) Review: Compounds involved in the flavor of surface mold-ripened cheeses: origins and properties. J. Dairy Sci. 79: 169–184.

Morgan ME (1976) The chemistry of some microbially induced flavor defects in milk and dairy foods. Biotechnol. Bioeng. 18: 953–965.

Morishita T & Yajima M (1995) Incomplete operation of biosynthetic and bioenergetic functions of the citric acid cycle in multiple auxotrophic lactobacilli. Bios. Biotech. Biochem. 59(2): 251–255.

Nierop-Groot MN & de Bont JAM (1998) Conversion of phenylalanine to benzaldehyde initiated by an aminotransferase in *Lactobacillus plantarum*. Appl. Environ. Microbiol. 64: 3009–3013.

Nierop-Groot MN & de Bont JAM (1999) Involvement of manganese in conversion of phenylalanine to benzaldehyde by lactic acid bacteria. Appl. Environ. Microbiol. 65(12): 5590–5593.

Rijnen L, Delacroix-Buchet A, Demaizières D, Le Quéré J-L, Gripon J-C & Yvon M (1999) Inactivation of lactococcal aromatic aminotransferase prevents the formation of floral aroma compounds from aromatic amino acids in semi-hard cheese. Int. Dairy J. 9: 877–885.

Rijnen L, Courtin P, Gripon J-C & Yvon M (2000) Expression of a heterologous glutamate dehydrogenase gene in *Lactococcus lactis* highly improves the conversion of amino acids to aroma compounds. Appl. Environ. Microbiol. 66(4): 1354–1359.

Skeie S, Lindberg C & Narvhus J (2001) Development of amino acids and organic acids in Norvegia, influence of milk treatment and adjunct *Lactobacillus*. Int. Dairy J. 11: 399–411.

Smith EL, Austen BM, Blumenthal KM & Nyc JF. 1975. Glutamate dehydrogenases. In: Boyer P Ed) The Enzymes. Vol. 11. (pp 293–367). 3 ed. Academic Press, New York.

Tammam JD, Williams AG, Noble J & Lloyd D (2000) Amino acid fermentation in non-starter Lactobacillus spp. isolated from Cheddar cheese. Lett. Appl. Microbiol. 30: 370–374.

Tucker JS & Morgan ME (1967) Decarboxylation of α-keto acids by *Streptococcus lactis* var. *maltigenes*. Appl. Microbiol. 15: 694–700.

Weerkamp AH, Klijn N, Neeter R & Smit G (1996) Properties of mesophilic lactic acid bacteria from raw milk and naturally fermented raw milk products. Neth. Milk Dairy J. 50: 319–332.

Weimer B, Seefeldt K & Dias B. 1999. Sulfur metabolism in bacteria associated with cheese. Antonie van Leeuwenhoek 76: 247–261.

Williams AG, Noble J & Banks JM (2001) Catabolism of amino acids by lactic acid bacteria isolated from Cheddar cheese. Int. Dairy J. 11(4–7): 203–215.

Yvon M & Rijnen L (2001) Cheese flavour formation by amino acid catabolism. Int. Dairy J. 11(4–7): 185–201.

Yvon M, Thirouin S, Rijnen L, Fromentier D & Gripon JC (1997) An aminotransferase from *Lactococcus lactis* initiates conversion of amino acids to cheese flavor compounds. Appl. Environ. Microbiol. 63: 414–419.

Yvon M, Berthelot S & Gripon JC (1998) Adding α-ketoglutarate to semi-hard cheese curd highly enhances the conversion of amino acids to aroma compounds. Int. Dairy J. 8: 889–898.

Yvon M, Chambellon E, Sorokine A & Roudot-Algaron F (2000) Characterization and role of the branched-chain aminotransferase (BcaT) isolated from *Lactococcus lactis* subsp. *cremoris* NCDO 763. Appl. Environ. Microbiol. 66(2): 571–577.

Yvon M, Bonnarme P, Chambellon E, Semon E & Spinnler HE 2001 Transamination reaction initiates the methionine conversion to methylthioacetaldehyde by *Lactococcus lactis*. NIZO dairy conference on food microbes – From knowledge to application. Ede, The Netherlands: 36.

Antonie van Leeuwenhoek **82:** 279–289, 2002.

Probiotics: an overview of beneficial effects

Arthur C. Ouwehand[1,*], Seppo Salminen[1] & Erika Isolauri[2]
[1]*Department of Biochemistry and Food Chemistry, University of Turku, FIN-20014 Turku, Finland;* [2]*Department of Paediatrics, Turku University Central Hospital, FIN-20500 Turku, Finland (*Author for correspondence; Tel.: +358-2-333-6894; Fax: +358-2-333-6860; E-mail: arthur.ouwehand@utu.fi)*

Key words: health effects, lactic acid bacteria, probiotics

Abstract

Food products fermented by lactic acid bacteria have long been used for their proposed health promoting properties. In recent years, selected probiotic strains have been thoroughly investigated for specific health effects. Properties like relief of lactose intolerance symptoms and shortening of rotavirus diarrhoea are now widely accepted for selected probiotics. Some areas, such as the treatment and prevention of atopy hold great promise. However, many proposed health effects still need additional investigation. In particular the potential benefits for the healthy consumer, the main market for probiotic products, requires more attention. Also, the potential use of probiotics outside the gastrointestinal tract deserves to be explored further. Results from well conducted clinical studies will expand and increase the acceptance of probiotics for the treatment and prevention of selected diseases.

Abbreviations: IBD – Inflammatory bowel disease; CD – Crohn's disease; UC – Ulcerative colitis

Introduction

The development of probiotics during the past decade has signalled an important advance in the food industry transferring to towards the development of functional foods. The term probiotic, popularised by R. Fuller in 1989, was defined recently by an Expert Committee as 'Living micro-organisms which upon ingestion in certain numbers exert health benefits beyond inherent general nutrition' (Guarner & Schaafsma 1998). Such a definition does not require changes in intestinal microflora or so-called 'colonisation' or temporary colonisation of the human gastrointestinal tract as the probiotic organism can exert its effects locally or during transient passage through the gastrointestinal system. This definition, however, still sets the requirements that the micro-organisms must be alive, not pasteurised or otherwise inactivated. Although specific numbers are not mentioned in the definition, generally it is thought that at least 10^9 colony forming units per day need to be ingested. Health benefits must be scientifically established by clinical studies in humans performed by several independent research groups and published in peer-reviewed journals.

The definition may be changing little by little as especially Japanese scientists have shown that also inactivated probiotic micro-organisms or their cell structures may have beneficial effects on human health. This has led to new definitions of probiotics and may change the way we look at probiotics in the future (Lee et al. 1999; Salminen et al. 1999).

Micro-organisms used as probiotics

Microbes from many different genera are being used as probiotics (Table 1). The most commonly used strains are members of the heterogeneous group of lactic acid bacteria; lactobacilli, enterococci and bifidobacteria. In particular lactobacilli are generally used as probiotics. This may have historical reasons since Metchnikoff proposed that the lactobacilli present in yoghurt would have a health promoting effect. Also, the most common means of administration is still a fermented dairy product. However, other microbes and even yeasts have been developed as potential probiotics during recent years (Table 1).

The choice what microbe to use as a probiotic is determined by many different factors (Table 2).

Table 1. Microbes used as probiotics and their documented health benefits in human clinical trials

Genus	Species	Example strains	Health benefit	Reference
Lactobacillus	*acidophilus*	La5	Reduced antibiotic associated diarrhoea	Black et al. 1991
	casei	Shirota	Shortening of rotavirus diarrhoea	Sugita & Togawa 1994
			Reduced recurrence of superficial bladder cancer	Aso et al. 1995
			Immune modulation	Nagao et al. 2000
	crispatus			
	fermentum	KLD		
	johnsonii	La1	Improved oral vaccination	Link-Amster et al. 1994
			Reduced colonisation by *Helicobacter pylori*	Felley et al. 2001
	paracasei	F19		
	plantarum	299v	Relief of irritable bowel syndrome	Niedzielin et al. 2001
			Reduction of LDL-cholesterol	Bukowska et al. 1998
	reuteri	SD2112	Shortening of rotavirus diarrhoea	Shornikova et al. 1997
	rhamnosus	GG	Shortening of rotavirus diarrhoea	Guandalini et al. 2000
			Immune modulation	Kaila et al. 1992
			Relief of inflammatory bowel disease	Gupta et al. 2000
			Treatment and prevention of allergy	Kalliomäki et al. 2001b; Majamaa & Isolauri 1997
	salivarius	UCC118	Reduced symptoms of inflammatory bowel disease	Mattila-Sandholm et al. 1999
Bifidobacterium	*breve*		Reduced symptoms of irritable bowel disease	Brigidi et al. 2001
	longum	BB536		
	lactis	Bb12	Treatment of allergy	Isolauri et al. 2001
			Shortening of rotavirus diarrhoea	Saavedra et al. 1994
			Reduced incidence of travellers diarrhoea	Black et al. 1989
			Improved oral vaccination	Link-Amster et al. 1994
Propionibacterium	*freudenreichii*	JS		
Bacillus	*subtilis*			
	cereus	toyoi		
Escherichia	*coli*	Nissle 1917	Fewer relapses of inflammatory bowel disease	Malchow 1997
Enterococcus	*faecium*	SF68		
Saccharomyces	*cerevisiae*	*boulardii*	Fewer relapses of inflammatory bowel disease	Guslandi et al. 2000

Table 2. Main properties for probiotic bacteria

Property	Benefit
Resistance to pancreatic enzymes, acid and bile	Survival of passage through the intestinal tract
Adhesion to the intestinal mucosa	Immune modulation
	Pathogen exclusion
	Enhanced healing of damaged mucosa
	Prolonged transient colonisation (?)
Human origin	Species specific interactions with the host
Documented health effects	Proposed health effects are 'true'
Safe	No health risk to consumer
Good technological properties	Strain stability
	Production at large scale
	Oxygen tolerance

In order to survive passage through the gastrointestinal tract, resistance to low pH, bile and pancreatic enzymes are important. Acid and bile tolerance can be easily monitored and they are considered intrinsic properties of lactic acid bacteria. Thus, in fermented milks acid stability is already required during the fermentation. Adhesion to the intestinal mucosa is considered important for immune modulation (the intestine is the largest immune organ of the body), pathogen exclusion, enhanced healing of damaged mucosa and prolonged transient colonisation. To obtain reasonable assurance on adherence, the use of at least two different test systems is required to describe both mucus and epithelial adhesion which represent the early and late stages of adherence to the mucosa. Human origin is thought to be important for host specific interactions by the probiotic, although e.g. *S. cerevisiae* (*boulardii*) is not of human origin. The microbes administered should obviously be safe. This is, however, often not specifically assessed. Lactobacilli and bifidobacteria are simply considered safe based on their taxonomic position. Although this may seem improper, it is difficult to assess the safety of generally non-pathogenic species. In practice, the first human feeding trial will also be the first safety trial, although this is often not recognised as such. Finally, potential probiotics need to have good technological properties so that they can be cultured on large scale, have an acceptable shelf life and, in case of application in fermented products, contribute to a good taste.

Lactobacilli have often good resistance to the *in vivo* stresses, as described in the next section, and several strains have good technological properties. This may, in addition to the historical reasons, explain their frequent use as probiotics. Bifidobacteria are also commonly used, though less then lactobacilli. They are sensitive to oxygen and have more strict growth requirements. This makes them technologically more difficult to use. The other probiotic species are, with the exception of propionibacteria and enterococci, usually not used in fermented products but as dietary supplements, in capsules, powders, etc.

Gut mucosal barrier: first line in host defence

The gastrointestinal tract is a complex microenvironment where the cells of the largest lymphoid organ of the human body interface with a myriad of endogenous and exogenous stimuli. The intestinal mucosa provides protective host defence to the constant presence in the gut lumen of antigens from food and the normal microflora.

Protection against potentially harmful agents is ensured by a number of factors including saliva, gastric acid, peristalsis, mucus, intestinal proteolysis, intestinal flora, and epithelial cell membranes with the intercellular junctional complexes (Sanderson & Walker 1993). Together with the well-functioning immunological defence, these processes provide antigen exclusion in the gut. However, there are specialised antigen transport mechanisms in the villous epithelium (Ducroc et al. 1983). Antigens are absorbed across the epithelial layer by transcytosis, and here the main degradative pathway entails lysosomal processing of the antigen. This second line of defence, immune elimination, is directed towards removal of antigens that have penetrated the mucosa. A minor pathway allows the transport of unprocessed antigens (Isolauri 1999; Heyman & Desjeux 2000). The immunological regulation takes place in several compartments: aggregations of lymphoid cells in follicles and the Peyerþs patches, distributed within the mucosa and in the intestinal epithelium, as well as in secretory sites (Brandtzaeg 1995). The intraepithelial T-lymphocytes have mainly a suppressor/cytotoxic phenotype, while the lamina propria cells show the helper/inducer phenotype. Peyerþs patches, crucial in determining the subsequent immune responses to the antigen, are covered by the M-cells. In general, antigen transport across this epithelium is characterised by rapid uptake and reduced degradation (Ducroc et al. 1983).

The lamina propria is also endowed with lymphocytes belonging to the B-cell lineage. IgA antibody production is abundant at mucosal surfaces, where secretory IgA is present in dimeric or polymeric form. Secretory IgA is relatively resistant to intra-luminal proteolysis and does not activate complement or inflammatory responses. There are differences between the upper and lower parts of the human gut-associated immune system in the isotype distribution of immunoglobulin-producing cells (Brandtzaeg 1995; Salminen et al. 1998). IgA1 immunocytes predominate in the small intestine while IgA2-producing cells are most frequent in the colon, the latter being more resistant to bacterial proteases. The secretory IgA antibodies in the gut are part of the common mucosal immune system including respiratory tract and lacrimal, salivary and mammary glands. Consequently, an immune response initiated in the gut-associated lymphoid tissue can affect immune responses at other mucosal surfaces.

The intestine's mucosal surface provides a defence barrier against antigens encountered by the enteric route. As a result of the barrier function, systemic hyporesponsiveness to antigens such as food proteins, oral tolerance, is a hallmark of the intestinal immune system. In this system also a balance is generated and maintained between the host and the normal resident microflora. In addition to antigen degradation and thereby participating in tolerance induction, intestinal colonisation acts as an important endogenous stimulus for the maturation of the gut-associated lymphoid tissue (Helgeland et al. 1996). So far, the human gut microflora is still an unexplored organ of host defence and its impact in health and disease may be stronger than currently known. As stated by MacDonald (2001): "It is likely that the normal flora also produces immunoregulatory molecules and it is not entirely unfeasible that the disease-free state of the gut in normal individuals is caused by the flora and not by sophisticated immunoregulatory circuits".

Health effects of probiotics

Probiotic therapy and modulation of the intestinal microflora

The original idea with probiotics has always been to change the composition of the normal intestinal microflora from a potentially harmful composition towards a microflora that would be beneficial for the host. In general this would mean a reduction in the number of, e.g. coliforms and clostridia and an increase in lactobacilli and/or bifidobacteria. Probiotics that survive gastrointestinal transit are likely to cause an increase in faecal levels of that particular genus, especially when initial levels were low. Due to competition for adhesion sites and nutrients, and possibly the production of antimicrobial substances, levels of certain less desirable genera can decrease. A concomitant increase in faecal levels of genera other than the probiotic consumed has also been observed for certain probiotics. E.g. consumption of *L. rhamnosus* GG has been observed to be associated with an increase in faecal bifidobacteria (Benno et al. 1996) and consumption of *L. salivarius* UCC118 caused an increase in faecal *Enterococcus* levels (Mattila-Sandholm et al. 1999).

It is obvious that avoiding colonisation by pathogens and reducing the risk for over growth of potential pathogenic bacteria is beneficial to the host. However, in some cases too much emphasis is placed on this change in microflora composition without considering the actual health benefit. A mere change in intestinal microflora composition is not a sufficient biomarker for a potential health benefit of a given probiotic strain. Moreover, for some health effects, like immune modulation, it may not be necessary to obtain a measurable modification of the intestinal microflora composition.

Immune modulation by probiotics

The demonstration that in the absence of the intestinal microflora antigen transport is increased indicates that the gut microflora is an important constituent in the intestines defence barrier. In affecting the development of gut-associated lymphoid tissue at an early age the gut microflora directs the regulation of systemic and local immune responsiveness, including hyporesponsiveness to antigens derived from micro-organisms and food. Experimental animals lacking interleukin-10 or transforming growth factor-β generate a mucosal inflammatory response to the resident gut microflora (Groux et al. 1999). The role of the intestinal microflora in oral tolerance induction has been investigated in germ-free mice (Sudo et al. 1997). In contrast to control mice, germ-free animals were seen to maintain their tendency to a systemic immune response, for example production of IgE antibodies, upon oral antigen administration. Abrogation of oral tolerance was due to the absence of intestinal flora. The aberrant IgE response could be corrected by reconstitution of the microflora at the neonatal stage, but not at a later age. In human infants, colonisation has been associated with the maturation of humoral immune mechanisms, particularly of circulating IgA- and IgM-secreting cells (Grönlund et al. 2000), reflecting the dependency of the regulation of the mucosal immune response on the normal gut microflora.

In several gut-related inflammatory conditions the healthy host-microbe interaction is disturbed and inflammation is accompanied by imbalance in the intestinal microflora in such a way that an immune response may be induced by resident bacteria (Isolauri 1999). Normalisation of the properties of unbalanced indigenous microflora by specific strains of the healthy gut microflora constitutes the rationale in probiotic therapy. The success of probiotic therapy manifests itself in normalisation of the increased intestinal permeability and altered gut microecology, improvement of the intestine's immunological barrier functions and alleviation of the intestinal inflammatory response. The targets for probiotic therapy are identified as clinical conditions involving impaired mucosal barrier function, particularly infectious and inflammatory diseases (Isolauri 2001).

Probiotics and allergic disease

The prevalence of atopic diseases has been progressively increasing in Western societies. The hygiene hypothesis conceives the rapid increase in atopy to be related to reduced exposure to microbes at an early age and subsequent lower number of infections in early life (Strachan 1989). This is related to smaller family size, vaccinations, consumption of almost sterile food and over hygienic practices in Western societies, which may cause the infants immune system to develop an inflammatory response. The earliest and most massive source of such exposure is associated with the establishment of the gut microflora. Indeed, differences in the neonatal gut microecology were recently documented as being associated with the development of atopic diseases (Kalliomäki et al. 2001a).

The T helper (TH) 2 responder phenotype is associated with enhanced production of IgE antibodies against ubiquitous environmental antigens, eosinophilia, and consequently constitutes a hallmark of atopic diseases. Specific strains of the gut microflora have been shown to contribute to the generation of counter-regulatory TH1- and TH3-type immune responses (Isolauri et al. 2001). In addition, these contribute to the processing of food antigens in the gut and reduce their immunogenicity *in vitro* and *in vivo*, together with a potential to dampen inflammatory responses to these antigens (Sütas et al. 1996; Majamaa et al. 1997; Isolauri et al. 2000; Pessi et al. 2000a).

The regulatory role of probiotics in allergic disease was first emphasised in a demonstration of a suppressive effect on lymphocyte proliferation and interleukin-4 generation *in vitro* (Sütas et al. 1996). Subsequently, the immunoinflammatory responses to dietary antigens in allergic individuals were shown to be alleviated by probiotics, this being partly attributable to enhanced production of anti-inflammatory cytokines, e.g. interleukin-10 (Pessi et al. 2000b) and transforming growth factor-β (Haller et al. 2000), and partly due to control of allergic inflammation in the gut (Majamaa & Isolauri 1997). The mucosal dysfunction caused by inflammation, characterised by the altered rate, route and mode of antigen presentation, is stabilised by probiotics (Isolauri 2001). So far, clinical effects have been seen as a significant improvement in the course of atopic eczema in infants given probiotic-supplemented elimination diets (Majamaa & Isolauri 1997; Isolauri et al. 2000). The preventive potential of probiotics in atopic disease has recently been demonstrated in a double-blind, placebo-controlled study (Kalliomäki et al. 2001b). Probiotics administered pre- and postnatally for 6 months to children at high risk of atopic diseases succeeded in reducing the prevalence of atopic eczema to half as compared with that in infants receiving placebo.

Probiotics in diseases of the gut

Probiotics have traditionally been used to treat disease related to the gastrointestinal tract, although other diseases have also been suggested to be relieved by the use of probiotics.

Lactose intolerance

Lactose intolerance, or more correctly lactose maldigestion, is caused by a reduced production of β-galactosidase. This is a normal condition in all adult mammals, with the exception of people from north-west European decent, and should therefore not be considered a disease as such. In these subjects, consumption of lactose leads to an increased osmotic load in the small intestine with subsequent secretion of fluids which leads to loose stools (Launiala 1968). The origin of the abdominal pain that is associated with the consumption of lactose by lactose maldigesting subjects is not well understood though it does not appear to relate to the production of gasses from the fermentation of lactose by the intestinal microflora (Lasser et al. 1975). Fermented milk products have been observed to be tolerated well by lactose maldigesters as compared to milk. This can be explained by the presence of β-galactosidase in the bacteria fermenting the milk. Upon ingestion, the bacteria are lysed by bile in the small intestine, the enzyme is released and degrades lactose. In addition to this, the more viscous properties of fermented milks, compared to plain milk, gives them a longer gastro-caecal transit time, thus further aiding digestion of lactose (Vesa et al. 2000). This beneficial effect is usually more associated with products fermented with *L. delbrueckii* subsp. *bulgaricus* and *S. thermophilus*. To what extend probiotics contribute to relief of lactose intolerance symptoms is uncertain, some probiotics like, e.g. *L. rhamnosus* GG are not able to ferment lactose.

Acute gastro-enteritis

Acute gastro-enteritis may have bacterial or viral origin. Rotavirus is one of the most common causes of acute childhood diarrhoea in industrial countries (Claeson & Merson 1990). Rotavirus invades and replicate in the differentiated absorptive columnar cells of the small intestinal epithelium. This results in partial disruption of the intestinal mucosa with loss of microvilli, a decrease in the villus/crypt ratio and an increased intestinal permeability (Salim et al. 1990). Several studies have shown that selected probiotics, such as *L. rhamnosus* GG, *L. reuteri*, *L. casei* Shirota and *B. lactis* Bb12, can shorten the duration of rotavirus diarrhoea by approximately 1 day (Kaila et al. 1992; Saavedra et al. 1994; Sugita & Togawa 1994; Shornikova et al. 1997). Several mechanisms maybe behind this favourable out come. The production of rotavirus specific IgA has been observed to be enhanced in response to treatment with certain probiotics (Kaila et al. 1992), the permeability of the intestinal mucosa has been observed to be reduced (Isolauri et al. 1993) and the composition of the intestinal microflora normalised (Salminen et al. 1996).

Antibiotic associate diarrhoea (AAD) is mainly due to an overgrowth of *Clostridium dificille*. In particular, *Saccharomyces cerevisiae* (*boulardii*) has been observed to reduce the risk for AAD (Surawicz et al. 1989). The incidence of AAD was less than half or a third in the *S. cerevisiae* (*boulardii*) group compared to the control group. Also other probiotics like *Lactobacillus rhamnosus* GG, *L. acidophilus* and *Enterococcus faecium* SF68 have been observed to prevent or treat AAD (Gismondo et al. 1999).

Inflammatory bowel disease

Inflammatory bowel disease (IBD) is clinically characterised by two overlapping phenotypes, Crohn's disease (CD) and ulcerative colitis (UC), which predominantly affect the colon (UC and CD) and/or the distal small intestine (CD). The aetiology of the disease is not completely understood, but a genetic predisposition and the normal intestinal microflora are thought to play an important role. Modifying the composition and activity of the normal microflora may thus improve the disease. Indeed selected probiotics have been observed to reduce the number of relapses and prolong the period of remission. Interestingly, not only lactic acid bacteria, *L. salivarius* UCC118 and *L. rhamnosus* GG, but also *S. cerevisiae* (*boulardii*) and a strain of *E. coli* (Nissle) have been observed to be effective in alleviating the symptoms of IBD (Mattila-Sandholm et al. 1999; Gupta et al. 2000; Guslandi et al. 2000; Hamilton-Miller 2001).

Colorectal cancer

The aetiology of colorectal cancer is diverse and diet has clearly been indicated to be involved (Greenwald et al. 2001). Diets, especially high in meat and fat or low in fibre, have been observed to cause changes in the composition of the intestinal microflora, with increasing levels of *Bacteroides* and*Clostridium* and decreased levels of *Bifidobacterium* (Benno et al. 1991). This change in microflora composition is associated with an increase in faecal enzyme activity, β-glucuronidase, azoreductase, urease, nitroreductase and glycocholic acid reductase. These enzymes convert procarcinogens into carcinogens and may thus contribute to an increased risk for colorectal cancer. The consumption of selected lactobacilli have been observed to reduce this faecal enzyme activity. Whether this also reduces the actual risk for colorectal cancer remains to be proven. However, most, but not all, epidemiological studies suggest that regular consumption of fermented dairy products are related to lower risk for certain types of cancer (Hirayama & Rafter 2000). Some positive effect of probiotic lactic acid bacteria on the risk for colorectal cancer can therefore be anticipated although definite proof remains to be presented.

Constipation

Constipation is a major digestive complaint among the elderly, in particular the institutionalised. Although also, otherwise healthy, adults and hospitalised subjects may experience constipation. Constipated subjects have been observed to have a modified faecal microflora with reduced levels of bifidobacteria, *Bacteroides* and, in particular, reduced levels of clostridia (Shimoyma et al. 1984). Probiotics have been suggested to relieve constipation (Goldin 1998; Lee et al. 1999). However, review of the literature does not substantiate this claim. This may relate to the causes of constipation; physical inactivity, low-fibre diets, insufficient liquid intake and some drugs. The altered microflora composition is more likely to be a consequence than the cause of constipation, correcting the microflora composition may therefore not be of help.

Benefits for healthy subjects

Determining the potential health effects of probiotics for healthy subjects is difficult although this is of major importance since probiotics are mainly marketed for healthy subjects. The health effects of probiotics on healthy subjects are likely to be limited to risk reduction. As mentioned above, consumption of fermented dairy products maybe related to a reduced risk for colorectal cancer. However, that evidence is rather circumstantial. More direct evidence suggests that, at least in children, long term consumption of probiotics in non-fermented milk may reduce the risk for infections, absence from day care due to illness and the use of antibiotics (Hatakka et al. 2001). This study indeed indicates that probiotics can also be of benefit to the healthy consumer. Probiotics are often marketed as 'boosting the immune system'. For healthy individuals this may not be the case, since the immune system is likely to be working optimally (Spanhaak et al. 1998). However, in combination with oral vaccination, improved antibody titres have been observed with probiotics (Link-Amster et al. 1994).

Developing future probiotic strains

Quality of probiotic strains

Probiotics are special ingredients that are used in both foods and pharmaceutical or special dietary applications. They have been selected to express strain specific properties which are important for their proposed health effects. Such characteristics should be retained through food and pharmaceutical processes and storage to be of benefit to the consumer.

The most important factor is to retain the strain characteristics and the purity of the preparation. It has been reported that especially dried probiotic preparations may have contaminants. This sets the requirements for hygienic preparation of the products and careful identification of the strains used. All commercial strains should be placed in an international type culture collection for future comparison of the properties and identity. Some probiotic preparations may also mislabel the strains they contain, using old or non-existing nomenclature.

Unlike pharmaceuticals or food chemicals such as additives, the quality criteria for probiotics are largely undefined. This is a key factor for health effects as long term transfer of probiotic lactic acid bacteria or bifidobacteria in food processing along with the storage may result in changes in their characteristics and health properties. To control these properties, criteria for assessing such changes should be included in functional food regulations. The criteria currently used for selecting new probiotics have been suggested as the optimal quality control measures to be used in industrial practise (Tuomola et al. 2001).

In recent studies the necessity of testing the stability of strain characteristics was established for model bacteria and common probiotics used in foods (Tuomola et al. 2000, 2001). Adherence properties can be considered as the main selection criterion for current probiotics and adherence is important for both local colonisation and immune modulation through contact with the gut associated lymphoid tissue. Adherence varies greatly among the current probiotic strains and adherence characteristics can vary in two different *in vitro* models. Processing and gastric secretions also influence adhesion and they should be taken into consideration (Tuomola et al. 2000; Ouwehand et al. 2001).

Early reports have documented that adherence properties depend on culture conditions, the number of transfers in industrial scale fermentation and use of cryoprotectants in freeze-drying (Elo et al. 1991). Transfer of cultures in processing over a period of 3 years decreased adhesion and changing the culture medium could also result in diminished adhesion properties (Elo et al. 1991; Tuomola et al. 2001). We have also shown that specific probiotics isolated from a similar product from different countries show very different adhesion properties. Viability is crucial for lactic acid bacteria used as starter cultures or probiotics.

Viability may be important for health effects as currently most of the clinical evidence has been reported for viable strains and relatively few effects have been documented for non-viable strains (Ouwehand & Salminen 1998). There are recent reports on the viability of probiotic formulations in Britain and the United States and they demonstrate a lack of quality control in this respect. In the US, of 30 supplements tested 11 contained no viable bacteria and in Britain only six out of 13 formulations were satisfactory in terms of viability (Temmerman et al. 2001; Hamilton-Miller 2001). Viability can relatively easily be assessed by the culture method or by flow cytometry (Virta et al. 1998; Bunthof et al. 2001). It is important to guarantee the viability of probiotics in the final product especially when viability has been documented as one of prerequisites for immune effects (Gill & Rutherfurd 2001).

Future probiotics

Probiotics for specific target groups

Current probiotics have mainly been selected based on the common criteria as outlined in Table 2. To refine the selection criteria, understanding of the mechanisms of probiotic action is necessary. This will make it possible to select future strains with more specific characteristics, to suit the needs of specific age and patient groups. This need is clearly indicated by the difference in mucosal adhesion of probiotic bifidobacteria to mucus from different age groups (Ouwehand et al. 1999) and the influence of disease on mucosal adhesion of selected probiotics (Ouwehand et al. 2002). The use of specially selected probiotics for particular subject groups may provide more specific health effects.

Non-viable probiotics

Most definitions of probiotic bacteria stress the importance of the viability of the microbes. However, very little research has been done on non-viable probiotics. Non-viable probiotics would have several advantages over viable ones: longer shelf life, improved safety and no need for refrigerated storage or transport. Review of the literature suggests that non-viable probiotics may have positive health effects as well (Ouwehand & Salminen 1998). This has been shown for shortening of rotavirus diarrhoea (Kaila et al. 1995) and alleviation of lactose intolerance (Vesa et al. 2000). Although viable probiotics appear to have more health effects than non-viable ones, the latter are not always without health effects. This also implicates that heat-inactivated products can not be used as controls without verifying their lack of activity.

Alternative applications

Probiotics are mainly used to influence the composition or activity of the intestinal microflora. However, in principle any part of the body which harbours a normal microflora can be a potential target for specific probiotics.

The oral cavity has a microflora that equals the intestinal microflora in complexity. Here too, some of the members of the normal microflora have a detrimental effect on the host, causing, e.g. dental caries or periodontal disease. Probiotics could have potential applications in the oral cavity. Yoghurt have been observed to reduce the colonisation by mutans streptococci, which are responsible for dental caries (Petti et al. 2001). While a specific probiotic *Lactobacillus* strain has been detected in saliva samples (Meurman et al. 1994). Although there is a considerable potential for probiotic use in the oral cavity, very little work has been done in this area.

The normal microflora of the urogenital tract is less complex than the microflora of the intestine and the oral cavity. However, more than 50 species are thought to colonise the urogenital tract and in health, H_2O_2-producing lactobacilli predominate (Redondo-Lopez et al. 1990). Disturbances in the *Lactobacillus* flora are thought to be related to the risk for urinary tract infections. Some work has therefore been done on the use of probiotics for urogenital tract infections. Selected *Lactobacillus* strains have been observed to reduce the recurrence of urinary tract infections (Reid et al. 1992) and reduce the risk for vaginitis (Hilton et al. 1992; Reid et al. 2001). Much work has also been done on the mechanisms of probiotic lactobacilli on urinary tract infections; production of hydrogen peroxide and of biosurfactants appear to be important factors contributing to the efficacy of the probiotic strains for use in the urogenital tract (Reid 2001). The probiotic *L. casei* Shirota has been observed to reduce the recurrence of superficial bladder cancer (Aso et al. 1995). These findings indicate that use of probiotics for the urogenital tract is a promising future area.

The skin has a normal microflora which is different depending on the site of the body. The most common genera found in the microflora of the skin are propionibacteria, *Staphylococcus, Micrococcus, Corynebacterium* and the yeast *Malassezia*. Several species within these genera can be opportunistic pathogens. However, the potential use of probiotics for the skin has been considered little to non (Barefoot & Ratnam 1998).

Also, the nasopharynx has a normal microflora, *Streptococcus pneumoniae* being frequently one of its normal members. Even lactobacilli have been isolated from the upper respiratory tract. Their potential use as probiotics in there has only recently been considered and may have interesting applications (Cangemi de Gutierrez et al. 2001).

Thus, there are many potential applications for probiotics which have received little attention but which may provide significant health effects.

Conclusion

The specific health effects of selected probiotic strains are becoming increasingly accepted thanks to an expanding volume of documentation from double-blind, placebo-controlled, clinical studies. In particular, relief of lactose intolerance symptoms, by yoghurt cultures, shortening of rotavirus diarrhoea and treatment of allergies are now well established. Also, the mechanisms behind these health effects are being elucidated through *in vitro* and animal studies, this can be expected to lead to more carefully formulated selection criteria for probiotics. However, many proposed beneficial health effects of probiotics still need further investigation, in particular the potential benefits for healthy consumers. For this, it is important to use well defined strains, since each strain has to be judged on its own merits, and that appropriate biomarkers are used for the evaluation of the effects. In addition to this, well selected target groups are needed. Such studies may indicate additional areas for probiotic use and further consolidate the acceptance of probiotics.

References

Aso Y, Akaza H, Kotake T, Tsukamoto T, Imai K, Naito S & BLP Study Group (1995) Preventive effect of a *Lactobacillus casei* preparation on the recurrence of superficial bladder cancer in a double-blind trial. Eur. J. Urol. 27: 104–109.

Barefoot SF & Ratnam P (1998) Composition for treating acne. US patent WO98/10743.

Benno Y, Mitsuoka T & Kanazawa K (1991) Human faecal flora in health and colon cancer. Acta Chirurgica Scand. 521: 15–23.

Benno Y, He F, Hosoda M, Hashimoto H, Kojima T, Yamazaki K, Iino H, Mykkänen H & Salminen S (1996) Effects of *Lactobacillus* GG yogurt on human intestinal microecology in Japanese subjects. Nutrition Today 31: 9S–11S.

Black FT, Andersen PL, Örsskov J, Örskov F, Gaarslev K & Laulund S (1989) Prophylactic efficacy of lactobacilli on travelers diarrhea. Travel Med. 7: 333–335.

Black F, Einarsson K, Lidbeck A, Orrhage K & Nord CE (1991) Effect of lactic acid producing bacteria on the human intestinal microflora during ampicillin treatment. Scand. J. Infect. Dis. 2: 247–254.

Brandtzaeg P (1995) Molecular and cellular aspects of the secretory immunoglobulin system. APMIS 103: 1–19.

Brigidi P, Vitali B, Swennen E, Bazzocchi G & Matteuzzi D (2001) Effects of probiotic administration upon the composition and enzymatic activity of human fecal microbiotia in patients with irritable bowel syndrome or functional diarrhea. Res. Microbiol. 152: 735–741.

Bunthof CJ, Bloemen K, Breeuwer P, Rombouts FM & Abee T (2001) Flow cytometric assessment of viability of lactic acid bacteria. Appl. Environ. Microbiol. 67: 2326–2335.

Cangemi de Gutierrez R, Santos V & Nader-Macías, ME (2001) Protective effect of intranasally inocculated *Lactobacillus fermentum* against *Streptococcus pneumoniae* challenge on the mouse respiratory tract. FEMS Immunol. Med. Microbiol. 31: 187–195.

Claeson M & Merson MH (1990) Global progress in the control of diarrheal disease. Pediatr. Infect. Dis. J. 9: 345–355.

Ducroc R, Heyman M, Beaufrere B, Morgat JL & Desjeux JF (1983) Horseradish peroxidase transport across rabbit jejunum and Peyer's patches in vitro. Am. J. Physiol. 245: G54–G58.

Elo S, Saxelin M & Salminen S (1991) Attachment of *Lactobacillus casei* strain GG to human colon carcinoma cell line Caco-2: comparison with other dairy strains. Lett. Appl. Microbiol. 13: 154–156.

Felley CP, Corthesy-Theulaz I, Rivero JL, Sipponen P, Kaufmann M, Bauerfeind P, Wiesel PH, Brassart D, Pfeifer A, Blum AL & Michetti P (2001) Favourable effect of an acidified milk (LC-1) on *Helicobacter pylori* gastritis in man. Eur. J. Gastroenterol. Hepatol. 13: 25–29.

Fuller R (1989) Probiotics in man and animals. J. Appl. Bacteriol. 66: 365–378.

Gill HS & Rutherfurd KJ (2001) Probiotic supplementation to enhance natural immunity in the elderly: effects of a newly characterized immunomodulatory strain *Lactobacillus rhamnosus* HN001 (DR20^{TM}) on leucocyte phagocytosis. Nutr. Res. 21: 183–189.

Gismondo MR, Drago L & Lombardi A (1999) Review of probiotics available to modify gastrointestinal flora. Int. J. Antimicrobial Agents 12: 287–292.

Goldin BR (1998) Health benefits of probiotics. Br. J. Nutr. 80: S203–S207.

Greenwald P, Clifford CK & Milner JA (2001) Diet and cancer prevention. Eur. J. Cancer 37: 948–965.

Grönlund MM, Arvilommi H, Kero P, Lehtonen OP & Isolauri E (2000) Importance of intestinal colonisation in the maturation of humoral immunity in early infancy: a prospective follow up study of healthy infants aged 0–6 months. Arch. Dis. Childhood 83: F186–F192.

Groux H & Powrie F (1999) Regulatory T cells and inflammatory bowel disease. Immunol. Today 20: 442–446.

Guandalini S, Pensabene L, Zikri MA, Dias JA, Casali LG, Hoekstra H, Kolacek S, Massar K, Micetic-Turk D, Papadopoulou A, de Sousa JS, Sandhu B, Szajewska H & Weizman Z (2000) *Lactobacillus* GG administered in oral rehydration solution to children with acute diarrhea: a multicenter European trial. J. Ped. Gastroenterol. Nutr. 30: 54–60.

Guarner F & Schaafsma GJ (1998) Probiotics. Int. J. Food Microbiol. 39: 237–238.

Gupta P, Andrew H, Kirschner BS & Guandalini S (2000) Is *Lactobacillus* GG helpful in children with Crohn's disease? Results of a preliminary, open-label study. J. Ped. Gastroenterol. Nutr. 31: 453–457.

Guslandi M, Mezzi G, Sorghi M & Testoni PA (2000) *Saccharomyces boulardii* in maintenance treatment of Crohn's disease. Digestive Dis. Sci. 45: 1462–1464.

Haller D, Bode C, Hammes WP, Pfeifer AMA, Schiffrin EJ & Blum S (2000) Non-pathogenic bacteria elicit differential cytokine response by intestinal epithelial cell/Leukocyte co-cultures. Gut 47: 79–87.

Hamilton-Miller JMT (2001) A review of clinical trials of probiotics in the management of inflammatory bowel diesease. Infect. Dis. Rev. 3: 83–87.

Hamilton-Miller JMT & Shah S (2002) Deficiencies in microbiological quality and labellinf of probiotic supplements. Int. J. Food Microbiol. 72: 175–176.

Hatakka K, Savilahti E, Pönkä A, Meurman JH, Poussa T, Näse L, Saxelin M & Korpela R (2001) Effect of long term consumption of probiotic milk on infections in children attending day care centres: double blind randomised trial. Br. Med. J. 322: 1327–1329.

Helgeland L, Vaage JT, Rolstad B, Midtvedt T & Brandtzaeg P (1996) Microbial colonization influences composition and T-cell receptor V beta repertoire of intraepithelial lymphocytes in rat intestine. Immunology 89: 494–501.

Heyman M & Desjeux JF (2000) Cytokine-induced alteration of the epithelial barrier to food antigens in disease. Ann. NY Acad. Sci. 915: 304–311.

Hilton E, Isenberg HD, Alperstein P, France K & Borenstein MT (1992) Ingestion of yoghurt containing *Lactobacillus acidophilus* as prophylaxis for candidal vaginitis. Ann. Int. Med. 116: 353–357.

Hirayama K & Rafter J (2000) The role of probiotic bacteria in cancer prevention. Microbes Infect. 2: 681–686.

Isolauri E (1999) Probiotics and gut inflammation. Curr. Opin. Gastroenterol. 15: 534–537.

Isolauri E (2001) Probiotics in human disease. Am. J. Clin. Nutr. 73: S1142–S1146.

Isolauri E, Kaila M, Arvola T, Majamaa H, Rantala I, Virtanen E & Arvilommi H (1993) Diet during rotavirus enteritis affects jejunal permeability to macromolecules in suckling rats. Ped. Res. 33: 548–553.

Isolauri E, Arvola T, Sütas Y , Moilanen E & Salminen S (2000) Probiotics in the management of atopic eczema. Clin. Exp. Allergy 30: 1605–1610.

Isolauri E, Sütas Y, Kankaanpää P, Arvilommi H & Salminen S (2001) Probiotics: effects on immunity. Am. J. Clin. Nutr. 73: S444–S445.

Kaila M, Isolauri E, Soppi E, Virtanen E, Laine S & Arvilommi H (1992) Enhancement of the circulating antibody secreting cell response in human diarrhea by a human *Lactobacillus* strain. Ped. Res. 32: 141–144.

Kaila M, Isolauri E, Saxelin M, Arvilommi H & Veskari T (1995) Viable versus inactivated *Lactobacillus* strain GG in acute rotavirus diarrhoea. Arch. Dis. Childhood 72: 51–53.

Kalliomäki M, Kirjavainen P, Eerola E, Kero P, Salminen S & Isolauri E (2001a) Distinct patterns of neonatal gut microflora in infants in whom atopy was and was not developing. J. Allergy Clin. Immunol. 107: 129–134.

Kalliomäki M, Salminen S, Arvilommi H, Kero P, Koskinen P & Isolauri E (2001b) Probiotics in primary prevention of atopic disease: a randomised placebo-controlled trial. Lancet 357: 1076–1079.

Lasser RB, Bond JH & Levitt MD (1975) The role of intestinal gas in functional abdominal pain. New Engl. J. Med. 293: 524–526.

Launiala K (1968) The effect of unabsorbed sucrose and mannitol on the small intestinal flow rate and mean transit time. Scand. J. Gastroenterol. 3: 665–671.

Lee Y-K, Nomoto K, Salminen S & Gorbach SL (1999) Handbook of Probiotics. John Wiley & Sons, Inc., New York.

Link-Amster H, Rochat F, Saudan KY, Mignot O, Aeschlimann JM (1994) Modulation of a specific humoral immune response and changes in intestinal flora mediated through fermented mik intake. FEMS Immunol. Med. Microbiol. 10: 55–64.

MacDonald T (2001) The reaction of the immune system to pathogens but not food antigens and commensal bacteria. Seminars Immunol. 13: 159–161.

Majamaa H & Isolauri E (1997) Probiotics: a novel approach in the management of food allergy. J. Allergy Clin. Immunol. 99: 179–186.

Malchow HA (1997) Crohn's disease and *Echerichia coli*. J. Clin. Gastroenterol. 25: 653–658.

Mattila-Sandholm T, Blum S, Collins JK, Crittenden R, de Vos W, Dunne C, Fondén R, Grenov G, Isolauri E, Kiely B, Marteau P, Morelli L, Ouwehand A, Reniero R, Saarela M, Salminen S, Saxelin M, Schiffrin E, Shanahan F, Vaughan E & von Wright A (1999) Probiotics: towards demonstrating efficacy. Trends Food Sci. Technol. 10: 393–399.

Meurman JH, Antila H & Salminen S (1994) Recovery of *Lactobacillus* strain GG (ATCC 53103) from saliva of healthy volunteers after consumption of yoghurt prepared with the bacterium. Microbial Ecol. Health Dis. 7: 295–298.

Nagao F, Nakayama M, Muto T & Okumura K (2000) Effects of a fermented milk drink containing *Lactobacillus casei* strain Shirota on the immune system in healthy human subjects. Biosci. Biotechnol. Biochem. 64: 2706–2708.

Niedzielin K, Kordecki H & Birkenfeld B (2001) A controlled, double-blind, randomized study on the efficacy of *Lactobacillus plantarum* 299v in patients with irritable bowel syndrome. Eur. J. Gastroenterol. Hepatol. 13: 1143–1147.

Ouwehand AC & Salminen SJ (1998) The health effects of cultured milk products with viable and non-viabel bacteria. Int. Dairy J. 8: 749–758.

Ouwehand AC, Isolauri E, Kirjavainen PV & Salminen SJ (1999) Adhesion of four *Bifidobacterium* strains to human intestinal mucus from subjects in different age groups. FEMS Microbiol. Lett. 172: 61–64.

Ouwehand AC, Tölkkö S & Salminen S (2001) The effect of digestive enzymes on the adhesion of probiotic bacteria *in vitro*. J. Food Sci. 66: 856–859.

Ouwehand AC, Salminen S, Tölkkö S, Roberts PJ, Ovaska J & Salminen E (2002) Disease dependent adhesion of lactic acid bacteria to colonic tissue *in vitro*. Microecol. Ther. In press.

Pessi T, Isolauri E, Sütas Y, Kankaanranta H, Moilanen E & Hurme M (2000a) Suppression of T cell activation by *Lactobacillus rhamnosus* GG-degraded bovine casein. Immunopharmacology 1: 211–218.

Pessi T, Sütas Y, Hurme M & Isolauri E (2000b) Interleukin-10 generation in atopic children following oral *Lactobacillus rhamnosus* GG. Clin. Exp. Allergy 30: 1804–1808.

Petti S, Tarsitani G & Simonetti D'Arca A (2001) A randomized clinical trial of the effect of yoghurt on the human salivary microflora. Arch. Oral Biol. 46: 705–712.

Redondo-Lopez V, Cook RL & Sobel JD (1990) Emerging role of lactobacilli in the control and maintenance of the vaginal bacterial microflora. Rev. Infect. Dis. 12: 856–872.

Reid G, Bruce AW & Taylor M (1992) Influence of three day antimicrobial therapy and *Lactobacillus* vaginal suppositories on recurrence of urinary tract infections. Clin. Therapy 14: 11–16.

Reid G (2001) Probiotic agents to protect the urogenital tract against infection. Am. J. Clin. Nutr. 73: 437S–443S.

Reid G, Bruce AW, Fraser N, Heinemann C, Owen J & Henning B (2001) Oral probiotics can resolve urogenital infections. FEMS Immunol. Med. Microbiol. 30: 49–52.

Saavedra JM, Bauman NA, Oung I, Perman JA & Yolken RH (1994) Feeding of *Bifidobacterium bifidum* and *Streptococcus thermophilus* to infants in hospital for prevention of diarrhoea and shedding of rotavirus. Lancet 344: 1046–1049.

Salim AF, Phillips AD & Farthing MJ (1990) Pathogenesis of gut virus infection. Baillières Clin. Gastroenterol. 4: 593–607.

Salminen S, Isolauri E & Salminen E (1996) Clinical uses of probiotics for stabilizing the gut mucosal barrier: successful strains and future challenges. Antonie Van Leeuwenhoek 70: 347–358.

Salminen S, Bouley C, Boutron-Ruault M-C, Cummings JH, Franck A, Gibson GR, Isolauri E, Moreau M-C, Roberfroid M & Rowland I (1998) Functional food science and gastrointestinal physiology and function. Br. J. Nutr. 80: S147–S171.

Salminen S, Ouwehand A, Benno Y & Lee YK (1999) Probiotics: how should they be defined? Trends Food Sci. Technol. 10: 107–110.

Sanderson IR & Walker WA (1993) Uptake and transport of macromolecules by the intestine: possible role in clinical disorders (an update). Gastroenterology 104: 622–639.

Shimoyama T, Hori S, Tamura K, Yamamura M, Tanaka M & Yamazaki K (1984) Microflora of patients with stool abnormality. Bifidobacteria and Microflora 3: 35–42.

Shornikova A-V, Casas I, Mykkänen H, Salo E & Vesikari T (1997) Bacteriotherapy with *Lactobacillus reuteri* in rotavirus gastroenteritis. Ped. Infect. Dis. J. 16: 1103–1107.

Spanhaak S, Havenaar R & Schaafsma G (1998) The effect of consumption of milk fermented by *Lactobacillus casei* strain Shirota on the intestinal microflora and immune parameters in humans. Eur. J. Clin. Nutr. 52: 899–907.

Strachan DP (1989) Hay fever, hygiene, and household size. Br. Med. J. 299: 1259–1260.

Sudo N, Sawamura S, Tanaka K, Aiba Y, Kubo C & Koga Y (1997) The requirement of intestinal bacterial flora for the development of an IgE production system fully susceptible to oral tolerance induction. J. Immunol. 159: 1739–1745.

Sugita T & Togawa M (1994) Efficacy of *Lactobacillus* preparation Biolactis powder in children with rotavirus enteritis. Jpn. J. Pediatr. 47: 2755–2762.

Surawicz CM, Elmer GW, Spleeman P, McFarland LV, Chinn J & van Belle G (1989) Prevention of antibiotic associated diarrhoea by *Saccharomyces boulardii*: a prospective study. Gastroenterology 96: 981–988.

Sütas Y, Hurme M & Isolauri E (1996) Downregulation of antiCD3 antibody-induced IL-4 production by bovine caseins hydrolysed with *Lactobacillus* GG-derived enzymes. Scand. J. Immunol. 43: 687–689.

Temmerman R, Huys G, Pot B & Swings J (2001) Identification and antibiotic resistance of isolates from probiotic products. Abstracts of 101st ASM General Meeting, C-289.

Tuomola EM, Ouwehand AC & Salminen SJ (2000) Chemical, physical and enzymatic pre-treatments of probiotic lactobacilli alter their adhesion to human intestinal mucus glycoproteins. Int. J. Food Microbiol. 60: 75–81.

Tuomola E, Crittenden R, Playne M, Isolauri E & Salminen S (2001) Quality assurence criteria for probiotic bacteria. Am. J. Clin. Nutr. 73: S393–S398.

Vesa T, Marteau P & Korpela R (2000) Lactose intolerance. J. Am. Coll. Nutr. 19: 165S–175S.

Virta M, Lineri S, Kankaanpää P, Karp M, Peltonen K, Nuutila J & Lilius E-M (1998) Determination of complement-mediated killing of bacteria by viability staining and bioluminescence. Appl. Environ. Microbiol. 64: 515–519.

Antonie van Leeuwenhoek **82:** 291–302, 2002.

Product development strategies for foods in the era of molecular biotechnology

Jeffrey K. Kondo[1,*] & Eric Johansen[2]
[1]*Chr. Hansen, Inc., 9015 W. Maple St., Milwaukee, WI 53214, USA;* [2]*Chr. Hansen A/S, Bøge Allé 10-12, Hørsholm, Denmark (*Author for correspondence; Tel.: +1-414-607-5936; Fax: +1-414-607-5709; E-mail: jkondo@chr-hansen-us.com)*

Key words: bioengineering, product development, crop production, commercialization

Abstract

Breakthroughs in science and technology are accelerating development of new products that are impacting our regulatory systems. Genetically modified or bioengineering plant varieties have entered the food supply on a global basis, especially in the U.S. The U.S. Food and Drug Administration regulates on the premise of 'substantial equivalence' and has developed premarket notification procedures and voluntary labeling guidelines. Japan's Ministry of Health, Labor and Welfare regulates biotechnology products and has imposed biotechnology labeling regulations. However, the EU continues to be in a regulatory gridlock between member states and has proposed strict traceability and labeling guidelines. These requirements are currently restricting imports of bioengineered foods and are creating an international debate. In contrast to bioengineered plant varieties, to our knowledge, there are no strains of lactic acid bacterial starter cultures on the market that contain rDNA. The majority of strains have been improved via selection and mutagenesis. However, conjugation and electroporation have been used to transfer native lactococcal phage resistance plasmids to industrial strains. In addition, plasmids have been introduced to allow for selection of certain characteristics and then been eliminated by curing. The potential benefits of bioengineered foods are far reaching and are one of the most important opportunities of this century. However, bioengineered foods remain an emotional debate that is affecting world trade.

Introduction

Advances and innovation in science are revolutionizing our 21st century world. Our future will be driven by a multi-disciplinary scientific approach to knowledge, which is information intensive, increasingly complex and rapidly changing. New perspectives and approaches are occurring because of breakthroughs in information technology, nanotechnology, mathematics, physics and genomics.

As we move into the future, this rapidly changing technology will create new and conflicting value propositions that will affect our social and commercial patterns. A good example is the field of genomics and elucidation of the DNA sequence of the human genome, which is often hailed as the greatest scientific achievement of the last century. As a result, we have entered the 'omics' era of genomics, proteomics, metabolomics, etc. The use of microarrays is revolutionizing the way we study genetics and microbial physiology but one of the challenges will be to validate the techniques and results from these studies.

Although there may be far reaching benefits, a delicate balance between science and society needs to be achieved through science-based discussions and policies. However, from the regulatory standpoint, we are just beginning to understand and address the complex issues surrounding these new technologies.

In foods, molecular biotechnology and genomics are currently driving innovation, where invention has found application in commercial markets. It is estimated that the global area for planting genetically modified (GM) crops in 2001 was about 130 million acres (52.6 million hectares) in 13 countries. This is a 19% increase from the year 2000. Four countries, the U.S. (68%, 88.2 million acres), Argentina (22%,

29.1 million acres), Canada (6%, 7.9 million acres), and China (3%, 3.7 million acres) accounted for 99% of the global GM crop area (ISAAA 2002). However, negative views of modern biotechnology exist and emotion rather than science-based analysis is driving opposition on the use of recombinant DNA technology (rDNA) and other methods of modern biotechnology. Classical genetics has been widely used for centuries to improve plants and animals in our food supply, and improvements may now be looked at as sustaining innovations (Christensen 1997). With the use of rDNA technology and other tools of modern molecular biotechnology, improvements in our food supply have created new and conflicting social value propositions characteristic of disruptive innovation (Christensen 1997).

In this paper, we will discuss the limited information on product development strategies for the development of bioengineered lactic acid bacteria in foods and use current trends in bioengineered plant varieties to discuss the influence of regulatory laws and consumer opinion. We will not attempt to do an in depth regulatory review but will draw upon relevant regulations from the development of new plant varieties as there is little or no published information on the use of bioengineered lactic acid bacteria in foods. We will discuss regulatory policy from markets that influence world policy, the European Union (EU), Japan and U.S.A., since they are at different stages from a regulatory standpoint. Also, we will focus on the U.S. regulatory process since it is most conducive to the introduction of bioengineered foods. Our discussion will not be focused on why we should not use rDNA technology or the emotional issues surrounding use of bioengineered foods but on product development strategies to introduce new and safe products into the marketplace, i.e. we will take a proactive look at use of the technology based on the current perception and regulatory constraints.

As a preface to this discussion, we need to define the various terminology that has been developed and the background behind the terminology. The EU views and regulates the safety of genetically modified organisms (GMO's) or genetically modified microorganisms (GMM's) with a component that takes into consideration the processes and techniques (Commission of the European Community 2000). Therefore, any genetic modification that does not occur naturally, i.e. *in vitro* genetic manipulation, would be considered a GMO or GMM even though the food is entirely safe. From a scientific basis, genetic modification occurs in classical techniques such as natural selection, and we have in a sense been consuming genetically modified foods for centuries, which can make these terms ambiguous and confusing. However, as mentioned, these terms take into consideration techniques and thus GMO's distinguish between classical vs. insertion of DNA. The U.S. Food and Drug Administration (FDA) is using the term 'bioengineered foods' (FDA 2001c) to distinguish foods derived through the use of recombinant DNA and other modern molecular methods from foods developed through classical or traditional genetic methods. In either case, traditional vs. bioengineered, genetic modification has occurred. As far as regulatory rules are concerned, the method of deriving the new plant variety does not matter in the safety evaluation, but instead the safety of the final product is given the highest consideration.

U.S. Regulatory platform

The Food and Drug Administration (FDA) regulates foods under the 1938 Federal Food, Drug and Cosmetics Act (FFDCA) with signficant amendments to the Act (FDA 1992a): the Pesticide Amendment (1954), the Food Additives Amendment (1958), and the Color Additive Amendment (1960). FDA is responsible for ensuring the safety and proper labeling of food products for human consumption, except for meat and poultry products, which are regulated by the United States Department of Agriculture (USDA). Also, FDA regulates the safety and labeling of animal feed. Within USDA, the Animal and Plant Health Inspection Service (APHIS) governs field testing of agricultural biotechnology crops. The Environmental Protection Agency (EPA) regulates pesticides and has jurisdiction over new insect protected biotechnology crops.

The FFDCA specifically prohibits the adulteration of any food item, i.e. if they contain certain poinsonous or deleterious substances. The FFDCA defines a food additive as any substance not 'generally recognized as safe' (GRAS) by qualified experts for its intended use. New food additives must go through a FDA petition and approval process, in which the company has the responsibility to establish the safety of the new substance. In contrast, if a food manufacturer believes that the food substance is GRAS, they may market the ingredient without seeking FDA approval, with the risk that FDA will disagree and take legal action to remove the product from the marketplace

(FDA 1997b). Key points from an industry viewpoint is that if a substance was not on the original 1958 Food Additive Amendment GRAS list, it could still lawfully market a substance without a food additive petition if it was GRAS by common knowledge for a particular use. The common knowledge element of the GRAS exemption standard relies on the premise that data and information are available and that there is a consensus among qualified experts about the safety of the substance in its intended use. Data from publications in peer-reviewed journals, e.g. toxicology studies, is often used for developing an expert consensus. In addition, documentation of the opinion of an 'expert' panel or a recommendation of an authoritative body, such as the National Academy of Sciences, is usually performed by companies to establish expert consensus. In the past, companies have been able to market GRAS substances without FDA notification, although many companies consulted with FDA and obtained an FDA opinion on whether their determination was valid. Opinion letters from FDA were issued in these cases (FDA 1997).

In 1997, FDA published a proposed rule for Substances Generally Recognized as Safe (FDA 1997b). In this proposal, FDA clarified the criteria for any company to determining if a product is GRAS using a notification procedure. Under this process, FDA evaluates whether the notice provides sufficient basis for a GRAS determination and whether any information raises issues that might lead the agency to question if the use of the substance is GRAS. FDA would respond to the developer within 90 days. By providing guidelines and streamlining the GRAS process, it is thought that companies would be more open to inform FDA of their internal company GRAS determinations and to allow for an increased awareness by FDA on ingredients that are entering the food supply (FDA 1997b).

To address bioengineered plant varieties, FDA published its "Statement of Policy: Foods Derived from New Plant Varieties" in 1992 (FDA 1992b). In the 1992 policy, interpretation of the FFDCA was clarified and provided guidance to companies with respect to human foods and animal feeds derived from new plant varieties. This policy applied to foods derived from all new plant varieties, including those that were developed using rDNA technology. FDA defined foods derived from rDNA technology as 'bioengineered foods'. It also recommended that developers consult with FDA about bioengineered foods under development. However, it is important to note that FDA was specific in its scope stating that "This notice does not address foods and food ingredients regulated by FDA that have been derived from algae, microorganisms and other non-plant organisms, including: (1) Foods produced by fermentation, where microorganisms are essential components of the food (e.g., yogurt and single cell protein); (2) food ingredients produced by fermentation, such as many enzymes, flavors, amino acids, sweeteners, thickeners, antioxidants, preservatives, colors and other substances; (3) substances produced by new plant varieties whose purpose is to color food, and (4) foods derived from animals that are subject to FDA's authority, including seafood". FDA is considering whether to address these issues in future *Federal Register* notices. Therefore, use of genetically engineered lactic acid bacteria (LAB) starter cultures technically are not regulated by this document but would most likely be considered on a case by case basis.

The FDA 1992 policy on new plant varieties defines certain safety criteria, such as transfer of an allergen or increased levels of a natural toxicant which allow additional review by FDA. The policy includes flow diagrams or decision trees on the issues of safety and nutrition. Assessments are based on risk-based considerations that include: (1) toxicants known to be characteristic of the host and donor, (2) food allergens that are transferred, (3) concentration or bioavailability changes in key nutrients, (4) the safety and nutritional value of newly introduced proteins, (5) the identity, nutritional value, and composition of carbohydrates, oils and fats that may be modified (FDA 1992b).

FDA also conducted three public meetings in 1999 to discuss issues with consumers, industry, etc. and based on discussions developed proposed rules for Premarket Notification (FDA 2001c) and guidelines for voluntary labelling (FDA 2001a). The Premarket Notification proposal requires developers to submit a scientific and regulatory assessment of the bioengineered food 120 days before the bioengineered food is marketed. It was recommended that developers continue to practice of consulting with FDA before submitting the required premarket biotechnology notice (PBN) as discussed in FDA's 1997 guidance procedures (FDA 1997a). In the consultation process, FDA does not perform a comprehensive scientific review but considers whether there are any unresolved issues, e.g. significant increases in plant toxicants or new allergens. Consultation requests are sent to the Center for Food Safety and Applied Nutrition (CFSAN), which in turn coordinates with the Center for Veterinary Medicine on consultations for FDA on foods for humans, animal feed, or both. After consultations, FDA recommends that the PBN be sent to CFSAN, which will coordinate with CVS, for the submission. CFSAN would then evaluate and respond to the developer by letter. FDA would also require notification if the bioengineered food has been modified to contain a pesticidal substance. Although the Environmental Protection Agency regulates all pesticides, new plant varieties that contain a pesticidal substance will be evaluated by both agencies. To date, FDA has issued 52 letters on submissions (FDA 2001b).

Foods derived from rDNA technology are subject to existing requirements of the FFDCA governing food additives and substances that are GRAS. This means that if a rDNA food is used to produce a GRAS substance or an approved food additive, the substance would continue to be regulated as it has been regulated historically. Transfer of native plasmids among lactococci for improved phage resistance was probably handled by expert review panels within a company's GRAS review process. If a company wanted to market a new strain developed through rDNA, consultation with CFSAN would be the first step in the notification process.

Issues relevant to safety evaluation and labelling

Safety evaluation of FDA's 1992 policy focuses on the characteristics of the food product which has led to the idea of 'substantial equivalence' that compares a new plant variety to its traditional counterpart. A determination of substantial equivalence takes into consideration the intentional and unintentional effects of genetic modification and evaluates the phenotypic and compositional changes, e.g. the quantitative assessment of the concentration of the constituents in the modified food compared to the wide range of variation typified by traditional counterparts. FDA identified characteristics of bioengineered foods that would initiate further review by the agency to establish safety. These include substances that are completely new to the food supply, allergens that are present in an unexpected circumstance, changes in the concentrations of important dietary nutrients, and increased concentrations of antinutritional factors and toxins that are characteristic to the food. Using substantial equivalence to assist in the evaluation of bioengineered foods has support by the Food and Agriculture Organization of the United Nations (FAO) and the World Health Organization (WHO) who has considered this concept (FAO/WHO 2000), and in a May 2000 report by the Organization for Economic Cooperation and Development (OECD 2000).

The FDA policy also evaluates the safety of the introduced DNA and gene product to determine if there is any potential risk. This includes the evaluation of the DNA sequence of the genetic material that is used, the proteins involved, the regulatory sequences, and the presence of additional coding sequences in the DNA. The source of the genetic material is also an important variable in the analysis, especially if the DNA originates from a pathogenic, toxin-producing, or allergen source. FDA also considered that DNA occurs in all foods and that it is rapidly digested in the gastrointestinal tract and that individuals consume large quantitities of DNA when eating conventional foods (IFBC 1990; Miraglia et al. 1998). Using this analysis, FDA allowed the use of kanamycin resistance genes in the development of Calgene's rDNA-derived FLAVR SAVR™ tomatoes. FDA concluded that the DNA for kanamycin resistance was not different from other DNA in its digestibility and thus does not pose a substantial food safety risk. This was later supported by FAO/WHO (2000) who also concluded that use of an antibiotic resistance marker does not pose a health risk to humans. However, antibiotic resistance genes with specific medical uses or limited alternative therapies should not be used in widely disseminated bioengineered foods. FDA also takes into consideration the appropriateness of the antibiotic resistance marker and considers the origin and type of antibiotic resistance genes that might be used.

The 1992 policy also addresses labeling of bioengineering foods from new plant varieties and does not establish special labeling because labeling requirements that apply for foods in general, also apply for bioengineered foods. Therefore, if a bioengineered food is significantly different from its traditional counterpart, the new food must be given a name that describes the difference. If an issue exist for constituents of the food, a statement must be made on the label to describe the issue. If a bioengineered food has a significantly different nutritional property or includes an allergen that consumers would not expect, these must be labeled.

The FDA also is providing assistance to manufacturers who wish to voluntarily label foods that are made with or without bioengineered ingredients and published guidelines (FDA 2001a). Interestingly, consumer focus group data shows that consumers do not have a full understanding of the terms 'GMO' and 'GM' and actually prefer label statements with words that mean bioengineering (Levy & Derby 2000). As mentioned previously, terminology such as 'not genetically modified' and 'GMO free', that include the word 'modified' are not technically accurate because 'Genetic modification' means the alteration of the genotype using any technique, whether it be traditional or bioengineered. Consumers do not have a good understanding that there are many ways to perform genetic modification and that essentially all food crops have been genetically modified. Although it is accurate to say that a bioengineered food was 'genetically modified', it would be inaccurate to state that a food that had been developed through traditional plant breeding was 'not genetically modified' without providing a clear meaning so that the consumer can understand that the statement applies to bioengineering (FDA 2001a).

EU Regulatory platform

The EU uses a precise definition of 'genetically modified organism' and has very strict regulations, regardless of the properties of the resulting strains (Directive 2001/18/EC). In other parts of the world, it is the properties of the strains, not how it was constructed, which determines the regulations which are applicable. The net result is that a strain that might be considered a GMO in Europe can be GRAS in the USA.

In the EU, GMOs and GMMs are defined as "organisms (and micro-organisms) in which the genetic material (DNA) has been altered in a way that does not occur naturally by mating and/or natural recombination" (Commission of the European Communities 2000). The EU directives also includes a list of techniques that constitute genetic modification and further state that the techniques of genetic modification include "Recombinant nucleic acid techniques involving the formation of new combinations of genetic material by the insertion of nucleic acid molecules produced by whatever means outside an organism" and "Techniques involving the direct introduction into a micro-organism of heritable material prepared outside the micro-organism". Natural processes such as conjugation, transduction and transformation are not considered to result in genetic modification, on the condition that they do not involve the use of recombinant nucleic acid molecules. Thus, in the EU it is the method used to produce a strain that determines the type of strain.

Regulatory laws are undergoing change with the proposal for regulations "concerning the traceability and labeling of genetically modified organisms and traceability of food and feed products produced from genetically modified organisms and amending Directive 2001/18/EC" (Commission of the European Communities 2001). When adopted, the new regulations will repeal three of the four laws currently regulating GMO's in the EU. Until then, Directive 2001/18 governs the approval for environmental release and commercialization of GMO's; the Novel Foods Regulation 258/97 governs food safety assessments and labeling for most genetically modified foods; Commission Regulation 49/2000, which sets a 1% threshold for accidental contamination with GMO crops; and Commission regulation 50/2000 that outlines specific labeling requirements for food and food ingredients containing additives and/or flavorings that have been genetically modified or have been produced from GMO's. The EU regulations also use the idea of 'substantial equivalence' that allows for use of a simplified procedure for foods derived from GMO's but not longer contains GMO's. In these cases, applications go through a notification process very similar to the U.S. policy.

Since 1998, there has essentially been an ad hoc moratorium on new product approvals due to a blocking minority of member states, which has effectively limited entrance or use of GMO products. This political gridlock has received a great deal of attention from the U.S. Under Secretary of State, Alan Larson states "We have pressed the EU over the last 3 years to restart the review and approval process, and we have cautioned Europe on the potential consequences for world trade and the development of this technology posed by its recent proposals on traceability... this is a critical moment ripe for U.S.–EU cooperation" (Larson 2002). Segregation of GM vs. non-GM crops and testing to determine conformance to the 1% rule have brought up global debate. Quantitative and accurate methods need to be further validated as this point often enters into the debate.

Currently, there are two genetically modified food plants that were authorized for use in foods under Directive 90/220/EEC, a soya variety and a maize variety (Belgian Biosafety Server 2001). The EU Commission has been notified in 11 petitions as being substantially equivalent and there are ten applications pending authorization (Belgian Biosafety Server 2001). However, the moratorium will likely not be lifted until the Commission's traceability and labeling proposals take effect.

Regulations in Japan

In Japan, the Ministry of Health, Labor and Welfare (MHLW) reviews and approves biotechnology products. The MHLW's Inspection and Safety Division ensures that applications are properly prepared and the application is submitted to the Expert Panel, Biotechnology Subcommittee, Food Sanitation Committee. The Expert panel makes the recommendation to the Biotechnology Subcommittee, which makes the recommendation to the Food Sanitation Committee. The Food Sanitation Committee makes the decision whether to recommend or not recommend the proposal to the MHLW Minister. If approved, then it is published in the Japanese Governments 'Gazette'. As of August 2001, the MHLW has approved 37 petitions (USDA 2001). On April 1, 2001 new legislation took effect, which makes it illegal to import bioengineered foods that are not yet approved in Japan and provides labeling standards. Thus, MHLW samples and tests imported food at ports of entry and local health authorities test processed foods at the retail level. Since the new law, only one product has been found that was unapproved. It was a potato snack that contained New Leaf plus potatoes, a variety that has been approved in the U.S. and Canada, but not Japan (USDA 2001).

The Ministry of Agriculture, Forestry, and Fisheries (MAFF) and the MHLW regulates biotechnology labeling for foods under the Food Sanitation Law and the Japan Agricultural Standards Law. Legislation for labeling biotech foods took effect as a result of the April 1, 2001 legislation, which requires labeling of bioengineered foods if the DNA or protein can be detected in the final food. For both ministries, labeling requirements are the same. Labels are required if the content of the bioengineered food exceeds 5%. To be labeled non-GM, certification must be provided based on production of the food from production through processing. Special label requirements apply to 24 foods made from corn or soybeans. If the 5% threshold is exceeded, the label must state 'GM Ingredients Used' or 'GM Ingredient Not Segregated'. The MHLW and MAFF both randomly test the 24 foods currently under the labeling program. Recently, it was announced that 59 food items were tested and in one case, the 5% threshold was exceeded. The manufacturer was notified and had to correct the labeling to show the product was made with 'GM Ingredients (USDA 2001).

Consumer attitudes on bioengineered foods

Consumer attitudes vary widely as culture and social values are impacted by bioengineered foods. Many surveys have been performed with sometimes different results as the method of asking questions can influence answers. Trust in the government also influences consumer attitudes toward modern biotechnology (IFT 2000). In one of the recent International Food Information Council (IFIC) surveys (IFIC 2001), 65% of the consumers support or do not oppose the FDA policy on the labeling of foods produced through biotechnology, which shows a general trust in the FDA. In other questions, 61% of Americans believe that biotechnology will benefit them or their families in the next 5 years; 65% would be likely to purchase a variety of produce that have been modified through biotechnology to be protected from insect damage and require fewer pesticide applications; 52% were likely to purchase produce if it had been modfied to taste better or fresher. Of consumers knowing that foods produced by biotechnology are in the supermarket, they anticipated benefits of the following: 39% improved health and nutrition, 33% improved taste and quality, 21% reduced chemical and pesticide use on plants, 9% reduced cost of food and 9% improved crops and crop yields. When asked "What, if anything are you concerned about when it comes to food safety", only 2% cited altered or engineered foods.

In another U.S. survey recently conducted by the Food Policy Instutute of Rutgers University (Rutgers News 2001), it was found that only 40% of Americans could correctly answer seven out of ten questions on a ten item true/false quiz, showing that consumers are not very familiar with modern biotechnology. However 58% of Americans approve of using genetic modification techniques to produce new plants but only 28% approve of producing new animals. High approval rates for plant products that have clear benefits to society were obtained: 85% would approve of using genetic modification to create more nutritious grains that could feed people in poor countries and 80% approve of a genetically modified rice with enhanced vitamin A to prevent blindness. Nearly 75% also say they would approve of the use of genetic modification to create less expensive or better tasting produce. Food Policy Institute Associate Director Hallman states "people are willing to consider the characteristics of the products of biotechnology rather than deciding that all biotechnology is good or bad". In contrast to the IFIC survey, 60% believe that the government does not have the tools to properly regulate GM foods and 90% believe that food created through biotechnology should have special labeling.

Attitudes and knowledge in the European Union have been surveyed at regular intervals, most recently in 2000 (Eurobarometer 52.1, 2000). A differentiation was made between genetic engineering and biotechnology with genetic engineering receiving less positive feedback than biotechnology. The expectation that these fields will have a positive impact on the quality of life in Europe are low, but positive. A considerable decrease since previous survey in 1996 was noted, however, with the positive responses for biotechnology decreasing from 50 to 45% and for genetic engineering decreasing 43 to 37%. The opinion on genetically modified food varies considerably across the EU, being neutral or slightly positive in some countries and fairly negative in others. The overall European average is somewhat negative. Public opinion in the European Union is against genetically modified food primarily because the perceived risks are too great. The majority of those interviewed would pay more for non-GMO food and over 80% do not feel adequately informed about biotechnology or genetic engineering. This latter opinion is borne out by the fact that to the statement "Ordinary tomatoes do not contain genes, while genetically modified tomatoes do", 35% responded True, 35% responded False, 30% responded Do Not Know.

Overall, consumers world-wide are not that informed about biotechnology and surveys are useful to look at trends. World-wide, most consumers, when asked, would like special labeling for bioengineered foods so that they can make choices. People are still deciding about modern biotechnology, but lack of knowledge provides opportunities for opponents of GM foods to influence public opinion. However, as many of the dramatic claims from opponents of bioengineered foods have not been true, their credibility on these issues seems to be decreasing as people are becoming more informed on these issues. Clearly, education and a science-based approach on the benefits of the technology will have a positive influence on the perception of bioengineered foods.

Product strategies and molecular biotechnology

The contradictions of the agricultural market place in the debate over bioengineered foods is a dilemna for industry. Companies must assess the risk of potential ramifications from consumer opinion and lobby groups that contend that these foods are potentially dangerous vs. FDA opinion and data, which has determined that bioengineered foods are safe and are substantially equivalent to foods that are not bioengineered. Adding to the debate, are environmental release issues with plants that get more attention than recombinant enzymes that are purified and produced in closed systems. Many of the activist environmental groups are more focused on the environmental impact rather than on safety and health. The dilemna becomes greater with large multinational companies that provide products over a wide range. For example, Frito-Lay announced in January 2000 that it would stop using bioengineered corn and Gerber indicated it will not used bioengineered ingredients in their baby foods (Barboza 2000). However, PepsiCo, Frito-Lay's parent company uses corn syrup derived brom bioengineered crops and Novartis, the parent company of Gerber, is one of the leading producers of bioengineered seeds (Barboza 2000).

Although most food companies in the U.S. support the technology and believe in the FDA policies, there is a wide range of product strategies. Some companies are open in using bioengineered plant varieties, some are using but avoiding comment, others are contradictory in their marketing strategies announcing that they are not using bioengineered foods for some products, while producing other products that utilize bioengineered foods, and other companies are dropping bioengineered foods completely for the time being. The dilemna also occurs because it is virtually impossible to segregate raw materials especially bioengineered corn and soybeans that were planted on over 60 million acres in the U.S. (Barboza 2000). As many companies source raw materials on the open market, it is very difficult to ensure that all raw materials do not contain bioengineered sources as most commodity suppliers do not segregate bioengineered grains. In addition, testing methods are still being validated for accuracy and sensitivity.

Global companies also have internal debates on the use of rDNA technology between their regional affiliates. Countries in the EU, having to deal with a different set of rules and sentiments that discourage use of the technology, often have different opinions on the use of bioengineered foods than their U.S. counterparts that are promoting bioengineered foods because FDA has determined that the use of bioengineered foods is safe. Innovation tends to be extended for internal company policies to be analyzed thoroughly and implemented. This may be true for many of the European based companies that provide lactic acid bacteria to the food industry as there are essentially no bioengineered strains on the market. Within this decision making process, there are economic considerations that factor in when comparing LAB's and bioengineered plant varieties. Overall, the economic benefits in plant agriculture are much greater than LAB usage in foods, so there is more impetus to bring bioengineered plant varieties products onto the market. It would follow that for LAB in foods, decisions are often made conservatively with a debate between public relations people and the scientific people. In the end, the economic risk of adverse consumer perception has a key influence in the decision making process.

To our knowledge, there are no lactic acid bacterial starter culture strains in the market place that contain rDNA to improve their characteristics, although the tools of molecular biology have been used either for generating information necessary to develop the new strain or as a selection process. The majority of strains that have been developed have used classical techniques of strain selection, screening, and mutagenesis. In some cases conjugation of plasmids, e.g. for phage resistant dairy starter cultures, has been used globally since the pioneering work of Sanders et al. (1986). Electroporation of native plasmids has been used in the U.S. to develop phage resistant lactococcal strains and no DNA has been introduced into strains that are not native to the species. In other cases such as development of MC010 discussed below, plasmids have been introduced for selection purposes but eliminated in the final strain.

Over the last 25 years, molecular biology has played the key role in our understanding of LAB used in foods. We are now entering the era of genomics which will serve as an accelerator to our knowledge base, capabilities, and our ability to improve LAB strains used in the industry. Since the sequence of *Lactococcus lactis* in 1999 (Bolotin et al. 1999), many of genome sequences of food LAB's are being elucidated as it is becoming easier and more inexpensive to sequence a bacterial genome. It is now possible to sequence a bacterial genome in one day as did the U.S. Department of Energy Joint Genome Institute (Anonymous 2000). LAB genome sequences that are cited by the National Center for Biotechnology Information (NCBI 2002) include *Lactococcus lactis* subsp. *lactis, Enterococcus faecium, Lactobacillus acidophilus/gasseri, Lactobacillus delbruekii* subsp. *bulgaricus* and *Streptococcus thermophilus.* Over the next year, additional genome sequences for *Oenococcus oeni, Leuconostoc mesenteroides, Lactobacillus gasseri, Lactobacillus brevis, Lactobacillus delbrueckii* ssp. *bulgaricus, Lactobacillus helveticus, Bifidobacterium longum, Bifidobacterium animalis, Brevibacterium linens, Lactobacillus casei, Pediococcus pentosaceus*, and *Lactobacillus plantarum* will be available (Dr James Steele, University of Wisconsin, Madison and Dr Jeffery Broadbent, Utah State University, personnal communication). Many of the studies will be the result of the Lactic Acid Bacteria Genomics Consortium in conjunction with the U.S. Department of Energy, Joint Genome Institute (Klaenhammer 2002). This effort will serve to look at the diversity of LAB's in milk, wine, meat, plants and the gastrointestinal tract and their functions in food fermentation and probiotics. The influx of information will truly usher in the 'omics' age of genomics, functional genomics, proteomics, metabolomics, etc. for LAB's.

As the technology surges ahead, it is disappointing that so few applications of this technology has found its way to the market place. As discussed, industry is taking a careful and safe approach to commercialization even though regulatory laws provide many opportunities, especially in the U.S. Again, this will be the dilemna that we all face whether it be commercialization or seeking research funding for studies and applications in LAB's. As this is the Seventh Symposium, many applications have been discussed and we will not attempt to describe all of the opportunities that abound. However, the future derived from these studies will have applications in food safety, human and animal health, food fermentation improvement and increased quality and nutrition of foods. Discussed below are some examples on development of strains that would be appropriate to enter the market place with minimal questions on the safety of the strain. The type of strain improvements discussed are likely to be the first examples of LAB's introduced into the market place.

The concept of food-grade GMO's in the EU

As scientific information and genetic capabilities using rDNA technology on LAB's is increasing exponentially, applications of the technology is still taking small steps. Therefore, we have elaborated a definition of food-grade GMO which describes what sort of genetic modifications of lactic acid bacteria we believe will be acceptable in Europe and which can readily attain GRAS status in the USA (Johansen 1999). In its strict sense, food-grade GMO's, has essentially no meaning in the U.S., as the regulatory laws do not recognize GMO as a term that relates to use of rDNA vs. traditional methods. However, this terminology frequently finds its way into discussions and it would be relevant in the EU, so we should be clear on the exact constraints of the definition. This definition states that a food-grade GMO may only contain DNA from the same genus and possibly small stretches of synthetic DNA. Thus, transgenic strains are specifically excluded by this strict definition while acceptable strains include those in which the copy number of a gene has been increased by inclusion on a food-grade cloning vector (Sørensen et al. 2000). From a scientific viewpoint, a broader definition that allows the inclusion of DNA from other genera of food microorganisms should be included provided these organisms have a safe history of use in foods. An example would be a *Lactococcus lactis* strain modified so that it expresses an amino-peptidase from *Lactobacillus helveticus* (Joutsjoki et al. 2002). A number of genetic alterations resulting in strains satisfying the strict definition of food-grade GMO have been developed and some of these are discussed below.

Strains developed with chromosomal alterations

Genes can be precisely deleted *in vitro* and the genetic alteration introduced into the chromosome by homologous recombination (Johansen 1999). Using this technique, a series of strains each lacking one or more aminopeptidase gene was constructed (Mierau et al. 1996). Deletion endpoints can be precisely defined and using specific PCR primers, it is possible to delete a gene from its start codon (ATG) to its stop codon (often TAA) eliminating any concern regarding the properties of a truncated protein that may be produced. Similarly, a single base pair or a few base pairs in a gene can be changed. Sørensen et al. (2000) constructed food-grade GMO's of a number of industrial *Lactococcus* strains with amber mutations in their *pyrF* gene. In these strains, a serine codon (TCT) is changed to an amber stop codon (TAG). Addition of a number of base pairs is also possible to do in a food-grade manner. Janzen et al. (2002) inserted four base pairs into the middle of the *pip* gene of *L. lactis* resulting in a +1 frameshift mutation. This strain has the expected resistance to prolate-headed bacteriophage.

The food-grade GMO's described above could also have been constructed by traditional genetic techniques. For the ones in which there is not a direct selection available, this would have required mutagenesis followed by extensive screening. In the case of strains with multiple deletions, this would have required multiple rounds of mutagenesis. The major problem with mutagenesis is that an effective mutagenesis results in multiple mutations in each strain. Most of these mutations will be unknown and their effects on the behavior of the resulting strain unpredictable. Thus, the use of more precise recombinant DNA techniques is preferable.

In the case of *pip* mutants, these can be obtained by direct selection in the presence of prolate-headed bacteriophage. When this is done, the majority of the mutants obtained result from the spontaneous insertion of IS-elements into the *pip* gene (Janzen et al. 2002). Since IS-elements contribute to the plasticity of bacterial genomes, these strains are expected to be less stable genetically than the food-grade GMO described above.

Strains with additional genes

In order to increase the expression of a desired gene, it is often sufficient to increase the number of copies of the relevant gene in the strain. This is typically done by placing the gene of interest on a plasmid vector. Introduction of a plasmid into a strain is normally done by electroporation, a relatively inefficient process. Consequently, plasmids must contain selectable markers. Traditional cloning vectors usually contain antibiotic-resistance genes as selectable markers but these are clearly unacceptable for strains intended for use in food. Food-grade selection systems have been developed and have been used to construct strains overexpressing various genes. To date, none of these strains have been commercialized as starter cultures.

The first food-grade selection system developed was based on nisin resistance (von Wright et al. 1990). However, since many industrial strains of *L. lactis* are intrinsically resistant to nisin, this system has not been further developed.

Later systems were based on deletion of a gene from a strain and complementing the phenotype by introducing the same gene on a plasmid (e.g. *lacF*, MacCormick et al. 1995). These systems work quite well but have so far only been developed in laboratory strains.

One system for use in industrial strains has been described (Sørensen et al. 2000). It is based on the use of a nonsense suppressor tRNA on a plasmid suppressing a nonsense mutation in the *pyrF* gene on the chromosome. The disadvantage of this system, as well as of the complementation systems described in the previous section, is that specific strains must be developed to act as hosts for the food-grade vectors. The $pyrF_{\mathrm{amber}}$ mutants described in the previous section were developed for this purpose. Using this system or its predecessor (Dickely et al. 1995), strains overexpressing various aminopeptidases have been constructed and tested in small-scale cheese productions (Guldfeldt et al. 2001). Commercialization of these strains has not yet taken place.

Strains obtained using gene technology, which do not contain any recombinant DNA

Careful use of gene technology allows the production of strains, which do not contain any recombinant DNA, and as such these strains should not be considered to be GMO's.

One well-known example is the diacetyl-overproducing strain MC010 (Curic et al. 1999) that contains a nonsense mutation in the *aldB* gene encoding α-acetolactate decarboxylase. In *L. lactis*, the activity of α-acetolactate decarboxylase is allosterically activated by leucine. Thus, in the presence of leucine there is a high enzyme activity and the cells have a very low level of α-acetolactate (ALA). In fact, the level of ALA is insufficient to support the biosynthesis of leucine, valine and isoleucine. Selection of strains able to grow in the presence of leucine results in *aldB* mutants. This selection only works if the strain is able to synthesize branched chain amino acids from ALA. Since most dairy *Lactococcus* strains have lost this biosynthetic pathway, the selection can only be performed if the relevant genes are reintroduced, for example on a recombinant plasmid. This approach was followed by Curic et al. (1999). A plasmid containing the relevant biosynthetic genes was introduced into a wild-type *diacetylactis* strain, leucine-resistant mutants were isolated and characterized to find the desired *aldB* mutant and the recombinant plasmid was eliminated. The resulting strain, MC010, contains no recombinant DNA and an expert panel has concluded that the strain is safe. Clarification of the status of MC010 in Europe is pending, while MC010 is currently on the market in the U.S.

Co-electroporation of a plasmid with an antibiotic resistance marker with a native plasmid with a desirable property (in this case bacteriophage resistance; Emond et al. 2001) allows the introduction of both plasmids into a strain. Transformants containing both plasmids are obtained by a combination of selection for the antibiotic resistance and screening for the desired bacteriophage resistance. Subsequently, curing eliminated the plasmid containing the antibiotic-resistance gene. The high stability of the phage-resistance plasmid ensures its maintenance for many generations without selection. The resulting strains do not contain any recombinant DNA and an expert panel consensus concluded that the strain is safe. The status in the EU is expected to be the same as MC010.

Conclusion

Science and technology has accelerated our knowledge, capabilities, and product development. Bioengineered plant varieties now account for over 130 million acres globally and have put our global regulatory system under tremendous pressure. As more bioengineered foods enter the approval process, countries are scrambling to develop policies for approval, traceability and labeling.

As public and government opinions vary, use of rDNA continues to be an emotional issue. In the U.S., FDA has developed a science-based policy that allows for premarket notification of bioengineered foods and for voluntary labeling. The system appears to be working well and it is estimated that 60–70% of the food on U.S. store shelves contain ingredients derived from bioengineered foods (IFIC 2001a). Japan has implemented approval and labeling laws that are now in effect. However, the EU continues with an ad hoc moratorium on the approval of bioengineered foods until traceability and labeling proposals are approved in 2003 or 2004. Because of the differences in policies, there is global trade disharmony. From the U.S. perspective, it wants to ensure that markets are open to bioengineered foods in a non-discriminatory manner. It is thought that forced labeling would give these products an undesirable connotation that the consumer does not really understand. In addition, costs for bioengineered foods would increase significantly due to traceability and segregation issues.

LAB starter culture strains that contain rDNA have not reached the market as companies have taken a more conservative approach. In the future, FDA policy is more compliant towards this new technology and the success of new strains will be likely be determined and tested in the U.S. market place. A variety of molecular methods and techniques can be used to develop strains that are completely safe and that will provide benefits to consumers and processors. Genomics and functional genomics will play a large role in the future on the development of LAB strains for the food industry. The information explosion will no doubt drive innovation with LAB's in the future but it is still uncertain on the timing for introduction of these microorganisms in the global market place.

References

Anonymous (2000) JGI sequences 'Supergerm' genome in 1 day. http://jgi.doe.gov/News/news_5_9_00.htm

Barboza D (2000) Modified foods put companies in a quandary. http://www.biotech-info.net/quandry.html

Belgian Biosafety Server (2001) Placing genetically modified foods (GM foods) on the EU Market. Regulation (EC) 258/97: placing GM foods on the market. http://www.biosafety.be/NF/GMfood/Gmfood_ma rket.html

Bolotin A, Mauger S, Malarme K, Dusko Ehrlich S & Sorokin A (1999) Low-redundancy sequencing of the entire *Lactococcus lactis* IL1403 genome. Antonie van Leeuwenhoek 76: 27–76.

Christensen CM (1997) The Innovators Dilemma. Harvard Business School Press, Boston, MA, USA.

Commission of the European Communities (2000) Facts on GMOs and GMMs. Memo/00/43, Brussels 13 July 2000. http://europa.en.int/comm/dgs/health_consumer/library/press/press63_en.pdf

Commission of the European Communities (2001) A proposal for a regulation of the European parliament and the council, concerning traceability and labeling of genetically modified organisms and traceability of food and feed products produced from genetically modified organisms and amending Directive 2001/18/EC. Brussels, 25.7.2001, COM (2001) 182 final, 2001/0180 (COD). http://europa.eu.int

Curic M, Stuer-Lauridsen B, Renault P & Nilsson D (1999) A general method for selection of α-acetolactate decarboxylase-deficient *Lactococcus lactis* mutants to improve diacetyl formation. Appl. Environ. Microbiol. 65: 1202–1206.

Dickely F, Nilsson D, Hansen EB & Johansen E (1995) Isolation of *Lactococcus lactis* nonsense suppressors and construction of a food-grade cloning vector. Molec. Microbiol. 15: 839–847.

Directive 2001/18/EC of the European Parliament and of the Council of 12 March 2001 on the deliberate release into the environment of genetically modified organisms and repealing Council Directive 90/220/EEC. Official Journal L 106, 17/04/2001 pp 0001–0039.

Emond É, Lavall E, Drolet G, Moineau S & LaPointe G (2001) Molecular characterization of a theta replication plasmid and its use for development of a two-component food-grade cloning system for Lactococcus lactis. Appl. Environ. Microbiol. 67: 1700–1709.

Eurobarometer 52.1 (2000) The Europeans and Biotechnology. http://europa.eu.int/comm/research/pdf/euro barometer-en.pdf

FAO/WHO (2000) Safety aspects of genetically modified foods of plant origin. Report of a Joint FAO/WHO expert consultation of foods derived from Biotechnology. Food and Agriculture Organization of the United Nations and World Health Organization. WHO, Geneva, Switzerland.

FDA (1992a) Food Additives. FDA/IFIC Brochure. http://www.cfsan.fda.gov~lrd/foodaddi.html

FDA (1992b) Statement of policy: foods derived from new plant varieties; notice. Food and Drug Administration. Fed. Reg. 57: 22984–23005.

FDA (1997a) Guidance on consultation procedures, foods derived from new plant varieties. U.S. Food and Drug Administration, Center for Food Safety and Applied Nutrition. http://www.cfsan.fda.gov October, 1997.

FDA (1997b) Substances generally recognized as safe. Food and Drug Administration Fed. Reg. 62: 18937–18964.

FDA (2001a) Guidanace for Industry: voluntary labeling indicating whether foods have or have not been developed using bioengineering. Draft guidance. U.S. Food and Drug Administration, Center for Food Safety and Applied Nutrition. http://vm.cfsan.fda.gov/~dms/biolabgu.html

FDA (2001b) List of completed consultations on bioengineered foods. Food and Drug Administration, Center for Food Safety and Applied Nutrition, Office of Food Additive Safety, July 2001. http://www.cfsan.fda.gov/~lrd/biocon.html

FDA (2001c) Premarket notice concerning bioengineered foods. U.S. Food and Drug Administration Fed. Reg. 66: 4706–4738.

Guldfeldt LU, Sørensen KI, Strøman P, Behrndt H, Williams D & Johansen E (2001) Effect of starter cultures with a genetically modified peptidolytic or lytic system on Cheddar cheese ripening. Int. Dairy J. 11: 373–382.

IFBC (1990) Biotechnologies and food: assuring the safety of food produced by genetic modification. International Food Biotechnology Council. Regulatory Toxicol Pharmacol. 12: Part 2.

IFIC (2001a) Agricultural biotechnology: myths & facts. International Food Information Council, Food Safety & Nutritional Information, IFIC Food Insight, October, 2001. http://ific.org

IFIC (2001b) Most Americans can articulate expected benefits of food biotechnology. International Food Information Council, Food Safety & Nutrition Information, IFIC Background, November, 2001. http://ific.org.

IFT (2000) IFT expert report on biotechnology and foods. Labeling of rDNA biotechnology-derived foods. Food Technol. 54: 63-74.

ISAAA (2002) Global GM crop area continues to grow and exceeds 50 million hectares for first time in 2001. International Service for the Acquisition of Agri-biotech Applications. January 10, 2002. http://www.isaaa.org/press%20release/Global%20Area_Jan 2002.htm

Janzen T, Stuer-Lauridsen B, Rasmussen P, Schnabl J & Riisberg L (2002) Inactivation of the *pip* gene is the dominating mutation type for mutants from *Lactocococus lactis* resistant towards c2-

like phages. Poster presentation at the Seventh Symposium on Lactic Acid Bacteria, the Netherlands.

Johansen E (1999) Genetic engineering (b) Modification of bacteria, In: Robinson R, Batt C & Patel P (Eds) Encyclopedia of Food Microbiology, Academic Press, London (pp 917–921).

Joutsjoki V, Louoma S, Tamminen M, Kilpi M, Johansen E & Palva A (2002) Recombinant *Lactococcus* starters as a potential source of additional peptidolytic activity in cheese ripening. J. Appl. Microbiol. In press.

Klaenhammer TR (2002) Lactobacillus gasseri genome project. http://jgi.doe.gov/JGI_microbial/html/lactobacillus_gas_content .html

Larson A (2002) Larson on trade liberalization, biotechnology and growth. The United States Mission to the European Union, Brussels, Belgium, January 8, 2002. http://www.useu.be/Categories/Biotech/Jan0802BiotechTrade LarsonEUe.html

Levy AS & Derby BM (2000) Report on Consumer Focus Groups on Biotechnology, Consumer Studies Team, Center for Food Safety and Nutrition, Food and Drug Administration, Washington, DC.

MacCormick C, Griffin H & Gasson M (1995) Construction of a food-grade host/vector system for *Lactococcus lactis* based on the lactose operon. FEMS Microbiol. Lett. 127: 105–109.

Mierau I, Kunji E, Leenhouts K, Hellendoorn M, Haandrikman A, Poolman B, Konings W, Venema G & Kok J (1996) Multiple-peptidase mutants of *Lactococcus lactis* are severely impaired in their ability to grow in milk. J. Bacteriol. 178: 2794–2803.

Miraglia M, Onori R, Brera C & Cava E (1998) Safety assessment of genetically modified food products: An evaluation of developed approaches and methodologies. Microchem. J. 59: 154–159.

NCBI (2002) Microbial genomes. Completed and ongoing projects. http://www.ncbi.nlm.nih.gov/PMGifs/Genomes/bact.html

OECD (2000) Report of the task force for the safety of novel foods and feeds. Organization for Economic Cooperation and Development, Paris. 86/ADDA, May 17

Rutgers News (2001) To the point: Rutgers' Food Policy Institute report shows that Americans confused and undecided about biotechnology. http://njaes.rutgers.edu/www/news/pressreleases/2001/1115-biotechnology.html

Sanders ME, Leonhard PJ, Sing WE & Klaenhammer TR (1986) Conjugal strategy for construction of fast acid-producing, bacteriophage-resistant lactic streptococci for use I dairy fermentations. Appl. Environ. Microbiol. 52: 1001–1007.

Sørensen K, Larsen R, Kibenich A, Junge M & Johansen E (2000) A food-grade cloning system for industrial strains of *Lactococcus lactis*. Appl. Environ. Microbiol. 66: 1253–1258.

USDA (2001) Japan Biotechnology. Update of Japan's Biotechnology Safety Approval and Labeling Policies 2001. Foreign Agricultural Service Global Agriculture Information Network Report. GAIN Report #JA1080, August 29, 2001.

von Wright A, Wessels S, Tynkkynen S & Saarela M (1990) Isolation of a replication region of a large lactococcal plasmid and use in cloning of a nisin resistance determinant. Appl. Environ. Microbiol. 56: 2029–2035.

Antonie van Leeuwenhoek **82:** 303–321, 2002.

Bacteriophage-resistance systems in dairy starter strains: molecular analysis to application

Aidan Coffey[1] & R. Paul Ross[2]
[1]*Department of Biological Sciences, Cork Institute of Technology, Rossa Avenue, Cork, Ireland (E-mail: acoffey@cit.ie);* [2]*Dairy Products Research Centre, Teagasc (Irish Agricultural and Food Development Authority), Moorepark, Fermoy, Co. Cork, Ireland (E-mail: pross@moorepark.teagasc.ie)*

Key words: bacteriophage, starter culture, cheese, phage-resistance, restriction-modification, abortive infection

Abstract

Starter inhibition by bacteriophage infection in dairy fermentations can limit the usage of specific bacterial strains used in the manufacture of Cheddar, Mozzarella and other cheeses and can result in substantial economic losses. A variety of practical measures to alleviate the problem of phage infection have been adopted over the years but has invariably resulted in a very limited number of strains which can withstand intensive usage in industry. The application of genetic techniques to improve the phage-resistance of starter cultures for dairy fermentations has been intensively studied for the last 20 years to a point where this approach now has significant potential to alleviate the problem. This paper highlights the recent findings and developments that have been described in the literature that will have an impact on improvement of the phage-resistance of starter cultures.

Introduction

Bacteriophage inhibition during dairy fermentations is an ever-present threat for industry. Thus, phage resistance in dairy starter strains is a property, which has attracted significant scientific interest world-wide, particularly since the early 1980s when the earlier investigations into the genetic basis for this property in lactic acid bacteria were recognized (Limsowtin et al. 1978; Boussemaer et al. 1980; Sanders and Klaenhammer, 1981). This scientific interest has of course been complimented by the implementation of a variety of practical measures to minimize the phage problem in industry as discussed later. Research progress in phage-resistance has been well documented both by this series of symposia (Klaenhammer, 1987; Hill, 1993; Daly et al. 1996; Forde & Fitzgerald, 1999; Moineau, 1999) and a variety of other comprehensive reviews in recent years (Klaenhammer 1984, 1991; Sanders, 1988; Sing & Klaenhammer 1990b; Coffey et al. 1991; Klaenhammer & Fitzgerald 1994; Garvey et al. 1995a; Dinsmore & Klaenhammer 1995; Nyengaard et al. 1995; Allison & Klaenhammer 1998; Josephsen & Neve 1998; Boucher & Moineau 2001; Coffey et al. 2001). The highlights of this research, which has been mainly focused on *Lactococcus lactis* and more recently *Streptococcus thermophilus*, include the identification, molecular characterization, and in some cases exploitation of genes encoding 21 phage abortive infection systems (Abi) and 29 restriction modification (R/M) systems, and the partial characterization of a variety of other phage resistance systems which include adsorption inhibition (Ads) mechanisms, injection blocking systems and additional less well-characterized Abi and R/M systems (Figure 1). (Note: the majority of the R/M systems from both genera are also described in the New England Biolabs web site REBASE.NEB.COM). These phage-resistance systems are predominantly linked to plasmid DNA, a property, which can be exploited sometimes to horizontally transfer them between starter strains. Indeed bacterial conjugation has been used to transfer a number of the plasmid-encoded systems to starter strains, which can be used in commercial dairy fermentations (Sanders et al. 1986; Sing & Klaenhammer 1986; Jarvis 1988, 1989; Kelly et al. 1990; Powell et al. 1990; Harrington & Hill 1991; Gireesh et al. 1992; Ward et al. 1992; Coakley et al. 1997; Pillidge et al. 2000;

Trotter et al. 2001; Mills et al. 2002), a task which is made easier by the availability of an efficient selectable marker on the plasmid. In parallel with this, impressive advances have been made on phage characterization and this has greatly benefited the development of new phage-resistance systems. Nevertheless, despite the intensive research efforts, which has given detailed scientific insights about the operation of resistance mechanisms and methods which phage may use to circumvent them, it is fair to say that the full potential of this research has not yet been realized and thus far, industry has not benefited to the extent that it might. This is due to a variety of reasons including the lack of progress in the development of legislation regarding genetically manipulated starter strains. It is noteworthy that many food-grade cloning systems are continuously being developed to fit within self-cloning guidelines. Another reason is that much of the research work has been done using a small subset of lactococcal strains and not the wide range of industrial strains. The purpose of this paper is to detail the main tools that could be used in effective defense strategies in the future and to suggest that molecular dissection of phage-resistance systems permits them to be exploited to a greater degree.

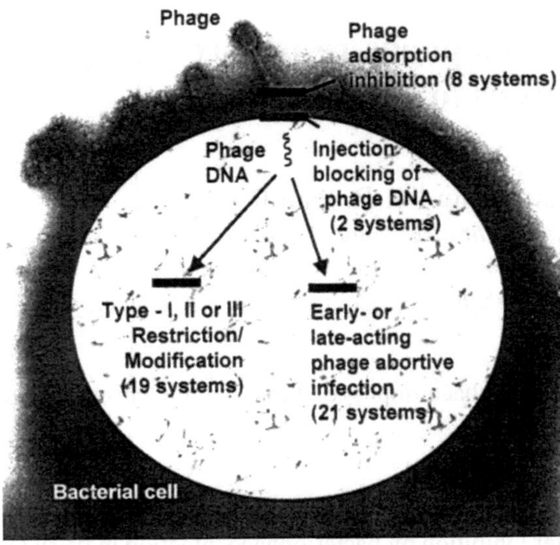

Figure 1. Traditional phage-resistance systems identified in lactococci and current number of different mechanisms reported (adsorption inhibition, injection blocking, abortive infection and restriction modification). Combinations of various systems can be found in (or can be introduced to) many industrial starter strains.

Defining the phage problem

Nowadays, it is true to say that there is relatively little actual product loss due to phage infection since failed vats are a rarity in modern cheese plants. However, to our knowledge, phage are still highly responsible for variability in cheese quality, especially where bulk starter systems are employed and this is a significant problem for manufacturers who require consistent product quality. This situation is generally associated with slow vats where acidity development is sub-optimal, resulting in a cheese of higher pH with associated defects in flavor, texture and even safety. Another big problem is that phage limit the variety of starter cultures which can be used for manufacture and also limit the extent to which specific cultures can be used in a given production schedule. Given that even small to medium sized cheese plants process thousands of liters of milk to cheese per day, obviously only the most robust strains can withstand the stress of continuous use. The adoption of newly isolated strains and cultures associated with improved flavor (such as autolytic strains) as acid producers is often impractical since many of them succumb to phage readily upon introduction to a cheese plant. Given the relationship between highly phage-sensitive strains and good flavored cheese, it is interesting to speculate at the relationship between phage-induced lysis and flavor (Feirtag & McKay 1987; O'Sullivan et al. 2000a). Indeed, research has demonstrated that phage infection can be associated with improved flavor (Crow et al. 1995). While it is difficult to quantify the precise economic losses mediated by phage due to confidentiality within dairy companies, it is true to say that researchers involved in starter culture developments in many countries, and who have close links with the industry, continue to cite phage disturbances during fermentation as a key problem. The importance of phage-resistance of starters should not be underestimated since many of the sophisticated developments in starter strains (including metabolic engineering, cell factories, probiotic adjuncts, pathogen biocontrol cultures and molecular improvement of cheese flavor-related determinants, etc.) will require associated phage defence strategies to ensure they withstand the rigors of commercial fermentations. The biology of phages of various dairy lactic acid bacteria is comprehensively discussed elsewhere in this symposium and also in a recent review by Brüssow (2001) and thus, will not be covered in this paper.

What phage types are involved?

Phage disturbances in industry are most noticeable in fermentations concerning lactococci (e.g., Cheddar production) and *S. thermophilus* (e.g., Mozzarella and yogurt production) where these bacteria are the principal acid producers. Briefly, the lactococcal phages have been divided into 12 species (Jarvis 1989; Jarvis et al. 1991) and the majority identified world-wide in the context of dairy fermentations, belong to three of these, namely the 936 species (the type phage being P008), the P335 species (the type phage being P335) and c2 species (the type phage being c6A, Moineau, 1999). These species belong to the *Siphoviridae* family (Ackermann & DuBow 1987) with the 936/P008 and P335 species belonging to morphotype B1 and c2/c6A species belonging to morphotype B2 (Jarvis et al. 1991). Generally, lactococcal phage genomes consist of double-stranded DNA, which varies in size from 19 to 21 kb for prolate phages, 28 to 38 kb for small isometric- and 150 to 155 kb for large isometric-headed phage. There is a high level of diversity between lactococcal phage in terms of host sensitivity, morphology and DNA homology. Generally, the phage population is a dynamic state which should be carefully monitored to ensure that appropriate phage defense measures are employed (Boucher & Moineau 2000). An epidemiological study by Moineau et al. (1990) found that the c2-type phages were the most frequently found in Canadian dairy plants. In a similar study in US plants, Moineau et al. (1996) found that 80% of phages classified belonged to the 936 group, with the remaining belonging to the other two groups. In an additional study by Bissonette et al. (2000), 936 phages were also found to dominate in cheese factories using *L. lactis* ssp. *cremoris* mixed strain starter cultures. Dominance of phages of the 936 species was also noted by Jarvis et al. (1991) with respect to isolates from New Zealand, US and Ireland. Both the c2 and the 936 groups are exclusively lytic and have presented a broad host spectrum (Moineau et al. 1992; Bissonette et al. 2000). In recent years P335 phages, which include both temperate and lytic members, have been isolated from cheese factories with increasing frequency (Moineau et al. 1993; Josephsen et al. 1994; Durmaz & Klaenhammer 2000; Madsen et al. 2001). This increase in incidence of P335 lytic phages coupled with their homology to temperate phages indicates that prophages or remnants of prophages in lactococcal genomes play a role in the development of new lytic phages. Recently, Labrie & Moineau (2000) developed a multiplex PCR method for the detection and identification of lactococcal phages of the three dominant species above which had a detection limit as high as 10^3 to 10^5 PFU/ml. The method, which was adapted to detect the presence of any of the three species in whey samples or in phage lysates is based on three sets of primers, one for each species. These were designed based on conserved regions of their genomes and have considerable potential for rapid classification of industrial isolates. The authors of the work observed a high genomic diversity among structural proteins of the P335-like phages suggesting that the classification of lactococcal phages within this group should be revised.

S. thermophilus phage which have emerged in the dairy fermentation industry are also members of the *Siphoviridae* family and have an isometric-head (45–60 nm) and a long non-contractile tail (180–270 nm) (Neve et al. 1989; Brüssow et al. 1998). Typically, *S. thermophilus* phages have a narrow host range, they all share some degree of DNA homology and can be divided into two groups based on the different numbers of major structural proteins and the mode of DNA packaging. All phages, which had two major structural proteins, were shown to have cohesive ends (*cos* type). Phages with three major capsid proteins were shown to have no cos sites and packaged their DNA via the headful mechanism (*pac*-type). Phages from the different groups were shown to share a lesser degree of DNA homology with each other than those from the same group (Le Marrec et al. 1997).

In conclusion, phage identification, classification and characterization is extremely important for implementation of meaningful control of the phage problem. Such studies have also lead to the development of novel phage-resistance mechanisms as discussed elsewhere in this paper.

Traditional approaches to minimize phage problems

(a) Practical strategies

Even though phages are generally ubiquitous, phage infections of starter cultures can be limited by adopting appropriate working practices in the dairy factory. Over the years, many elaborate procedures have been implemented to help control phage proliferation (Huggins & Sandine 1979; Klaenhammer 1984; Sandine 1989). Development and implementation of good manufacturing practices (GMP) to limit the effect of phage in cheese plants have met with significant success and can still be regarded as crucial in controlling the phage problem. Improved plant design, strict asepsis in culture propagation, Direct vat inoculation (DVI), superior sanitation procedures, the use of closed cheese vats, and the careful handling and disposal of whey have also limited but not eliminated phage proliferation. In addition, the almost universal changeover to defined strain cultures provided cheesemakers with a significant degree of control (Daly et al. 1996). These starter cultures usually contain two or three well-characterized strains whose cheesemaking attributes and phage sensitivity patterns have been carefully and scientifically defined (Thunell et al. 1981). This knowledge facilitates the withdrawal of sensitive strains on emergence of a homologous phage and substitution with a phage-unrelated strain of similar fermentative properties. The use of phage inhibitory media (PIM) was advocated as a solution to the phage problem in the late 1970s and was widely adopted by the U.S. Dairy industry (Bester & Lombard 1975; Gulstrom et al. 1979). In addition to growth-promoting ingredients such as milk-solids, yeast extract, protein hydolysates, etc., these media contain components such as phosphates and citrates to bind calcium which is considered necessary for phage proliferation (reviewed by Whitehead et al. 1993). Richardson and co-workers at Utah State University introduced low-cost formulations (based on whey) and also the concept of continuous neutralization of the bulk tank culture to provide higher numbers of more active cells (Richardson et al. 1977). In the early 1990s, Sandine and co-workers at Oregon State University developed 'Economy 6', arguably the most sophisticated phage inhibitory medium, that boasts, in addition to its phage inhibitory action, an internal buffering system which ensures the pH of the bulk culture never drops below 5.0 (Khosravi et al. 1991).

Although bulk propagation of thermophilic starters in some cases is still performed in reconstituted skim milk, much of the media used is whey-based and is designed specifically to maintain a balanced rod:coccus ratio prior to inoculation, while also protecting cultures from phage (Whitehead et al. 1991; Sandine, 1995). Lactobacilli typically tolerate lower pH values and higher temperatures than *S. thermophilus* and, thus, it has been suggested that the rod:coccus ratio in thermophilic bulk cultures can be balanced by controlling pH and temperature or alternatively by propagating the lactobacilli and *S. thermophilus* separately (Whitehead et al. 1991; Oberg & Broadbent 1993). Streptococci can be inhibited by the inorganic phosphate concentrations generally encountered in mesophilic bulk starter media used for lactococci, which can also be exploited for control of population dynamics in situations where streptococci are combined with lactococci for fermentation rescue (Stokes et al. 2001). Despite the implementation of GMP and the extensive use of PIM, exclusion of phage from cheese fermentations remains difficult to achieve in practice. Consequently, the exploitation of genetically improved starters with regard to phage-resistance in commercial production would undoubtedly enhance strain performance during intensive use.

'Fire-fighting' approach

The experience at Moorepark indicates that when an important component strain(s) associated with a specific cheese has 'phaged out' or failed in industry, the immediate result is that a constraint arises on the amount or quality of that cheese which can be made, given that specific flavor attributes are often associated with particular strains. Frequently, a phage-unrelated strain of similar fermentative or flavor properties may not be available with the result that the producer will not be able to meet the market demand for the cheese in question. A rapid response by the group or company providing the starter for the industry is necessary. A typical effective response solution, although nowadays less frequently adopted, is to generate bacteriophage insensitive mutants (BIMs) of the starter strain in question (Coffey et al. 1998). BIMs are spontaneous phage-resistant mutant strains that survive extended exposure to high levels of lytic phage in factory whey (see review by Daly et al. 1996). Thus, their resistance is designed to be effective against the phage present in the vat at the time of isolation. Acquisition of phage insensitivity in BIMs is most often effected by non-specific point mutations in genes encoding cell receptor sites. At Moorepark, the approach has been useful down through the years for both lactococci (Cheddar cheese) and thermophilic streptococci (Mozzarella cheese). Although this technique has the convenience of simplicity and speed, drawbacks may include narrow phage-specificity, reversion to the phage-sensitive phenotype, and alterations to cellular growth (slow acid production) and flavor formation. Moreover, it has proven impossible to generate BIMs of some starter strains by the approach described. However, the analysis of BIMs has been important in generating information regarding the nature of phage receptors, with most reports indicating that carbohydrate and/or protein moieties play a key role in phage adsorption (Forde & Fitzgerald 1999). At present, in the context of starter strain improvement, BIM formation is considered to be of relatively limited value in minimizing the phage problem in dairy fermentations.

Natural host-encoded phage-resistance systems

Individual members of the four well-known phage-resistance systems have been discussed in depth on several occasions in the recent past. This section therefore, will only mention these systems briefly and will draw attention to the more recent findings.

(a) Phage adsorption and injection: new developments

Over the years, a number of instances of host-mediated inhibition of phage adsorption have been described (Figure 1) and these have been comprehensively reviewed recently (Forde & Fitzgerald 1999). In addition, to date, two host-mediated systems that inhibit phage DNA injection have been reported, one plasmid encoded, namely pNP40 (Garvey et al. 1996), and one chromosomally encoded system in the lactococcal strain PLM-18 (Akçelik 1999). The former has by now been transferred to a variety of strains used in industrial fermentations (Harrington & Hill 1991; Pillidge et al. 2000; Trotter et al. 2001).

While the genetic processes directing phage adsorption blocking are not well understood, the biochemistry of the phenomenon has received some attention (Dinsmore & Klaenhammer 1995; Forde & Fitzgerald 1999; Quiberoni et al. 2000) and has been attributed to various substances including galactosyl-containing lipoteichoic acid, galactose/rhamnose or galactose/glucuronic acid polymer, and cell wall proteins. The physical mechanism of adsorption blocking is typically due to phage receptor masking or competition. Prevention of phage DNA penetration has also been accomplished resulting from studies on the cell surface. This was achieved by Garbutt et al. (1997) by the mutation of the determinant for phage infection protein (*pip*) to render the host strain insensitive to phage infection without growth rate being affected. More recently, the latter group also reported the cloning of host DNA which seemed to suppress a phage resistance system and thus restored phage sensitivity to phage sk1 resistant BIM strain (Kraus & Geller 2001).

Another advance in the area of phage adsorption has been the recent identification of the genetic determinant encoding phage anti-receptor, i.e., the phage structure involved in host recognition. This was identified by Duplessis & Moineau (2001) in the *S. thermophilus* phage DT1 on the basis of resemblance with the phage lambda anti-receptor gene and also on the basis of similarity in organisation with the adhesin of the T-even coliphages. The gene was detected in seven different *S. thermophilus* phages and the region within the encoded protein, which was responsible for host recognition was identified. Gene replacement studies using the anti-receptor determinant from different *S. thermophilus* phages provided biological evidence for its role in host receptor specificity. Aside from its scientific value, the latter study has considerable importance for *S. thermophilus* starter strain rotation in industry in order to minimize problems with phage infection.

Recently, McGrath et al. (2002) reported a new phage-resistance system manifested as a DNA injection blocking and which represented a novel superinfection exclusion mechanism. This was linked to a lactococcal prophage gene designated sie_{2009}, which is situated between the genes encoding the repressor and integrase, on the lysogeny module of phage Tuc2009. The Sie_{2009} protein is associated with the cell membrane and its expression left phage adsorption, transfection and plasmid transformation unaffected, but it interfered with plasmid transduction, as well as phage replication. This report was the first description of a phage-encoded superinfection exclusion/injection blocking mechanism in the genus *Lactococcus*.

Studies on phage-adsorption inhibition and phage-injection blocking have considerable importance, given that these mechanisms are the first two lines of defense, which the bacterial strain may use to prevent phage DNA entry. Consequently, a variety of lactococcal starter strains have been improved by the incorporation of either of these phage-resistance mechanisms.

(b) Overview of abortive infection and restriction/modification

Twenty-one abortive infection systems have been identified in *Lactococcus* by researchers worldwide over the last 20 years, the first being AbiA discovered by Klaenhammers group in the US (Klaenhammer & Sanozky 1985; Hill et al. 1990b). These are listed in Table 1. The phage types against which they exhibit activity are included although it is apparent from the literature that in most cases, representatives of all three major groups of phages may not have been available when the data was reported. Elucidation of the mode of action of Abi has been an important focus of research interest and much of the information has been reviewed in depth previously (see bibliography listed in Table 1). Briefly, Abi A, F, K and R interfere with phage DNA replication; Abi B and G interfere with RNA transcription; AbiC reduces synthesis of major capsid protein of the phage; AbiD1 causes interference of a phage ORF necessary for phage development; AbiQ causes accumulation of the replicative form of phage DNA, preventing maturation; AbiS functions by sequestering factors necessary for phage development; AbiU appears to delay phage transcription. This mechanistic information, while it is not available for every Abi system, indicates that the mechanisms act on different targets and at different stages in phage development and thus may be combined to act in cooperation for controlling phage.

Restriction modification systems, which in the case of *Lactococcus* and *S. thermophilus* collectively fall into classification Types I, II and III, are scientifically better understood than Abi (see review by Forde & Fitzgerald 1999; Coffey et al. 2001). Those systems capable of mediating a reduction in EOP of four log cycles or greater are more desirable when applied to controlling phage in industry (Moineau 1999).

Generally, Abi and R/M have received the most detailed research attention than the phage-resistance mechanisms which act earlier in the lytic cycle. There has been, and undoubtedly continues to be considerable interest in exploiting these mechanisms either individually or together, naturally or through artificial recombination, for strain improvement as discussed below.

Exploitation of plasmid encoded anti-phage systems

Natural phage defence mechanisms are widespread particularly in the lactococci and the mechanisms can co-exist and complement each other in industrial hosts, helping prevent the infection and proliferation of phage. The linkage of plasmid DNA to the majority of these defence mechanisms, coupled with the observation that many of these plasmids can be conjugally transferred, has led to strain construction strategies for obtaining phage-insensitive starters. Conjugation has been employed successfully to construct transconjugant starters that performed reliably in industrial environments. The first plasmid to be exploited extensively was pTR2030, which encodes both R/M and Abi (see Table 1 and 2). A number of transconjugants of this plasmid were generated and subsequently survived prolonged exposure to phage in milk fermentations under commercial cheese-making conditions (Steenson & Klaenhammer 1985; Sanders et al. 1986; Sing & Klaenhammer 1986). Since that time, the approach has been widely adopted world-wide with considerable success (Jarvis 1988, 1989; Kelly et al. 1990; Powell et al. 1990; Harrington & Hill 1991; Gireesh et al. 1992; Ward et al. 1992; Coakley et al. 1997, Pillidge et al. 2000; Trotter et al. 2001; Mills et al. 2002), although details on the level of commercial usage of transconjugant strains can be difficult to obtain.

There are limitations to using bacterial conjugation for the transfer of phage-resistance plasmids to industrial starter strains. One in particular has been the absence of dominant selectable markers on the majority of these plasmids requiring that resistance to phage is used to select for transconjugants (O'Sullivan et al. 1998). The problem with this is that phage-insensitive mutants will usually outnumber genuine phage-resistant transconjugants. Additionally, introduction of some phage-resistant plasmids (e.g., those encoding R/M) may not give complete resistance to phages used in the selection. This makes it even less likely that transconjugants would be successfully identified on the basis of a phage selection.

The genetic linkage of anti-phage genes with bacteriocin production and/or immunity/resistance determinants on conjugative plasmids in lactococci has proven particularly useful in strain construction over the last 10 years. Early reports include the exploitation of nisin resistance associated with the plasmid pNP40 encoding three anti-phage genes (McKay & Baldwin 1984; Garvey et al. 1995b, 1996, 1997) to assist in the selection transconjugants of the lactic butter starter strain *L. lactis* ssp. *lactis* biovar *diacetylactis* 425A (Harrington & Hill 1991). Interestingly, cadmium resistance determinants were later identified on this plasmid permitting a cleaner selection than that permitted by nisin resistance (Trotter et al. 2001). Dricin resistance was used by Ward et al. (1992) to facilitate selection for the 131-kb conjugative plasmid, pMU1311. This plasmid originated in *L. lactis* subsp.

lactis DRC1 and carries genes responsible for R/M and Abi. In the same laboratory, Gireesh et al. (1992) mated nisin-producing strains of *L. lactis* subsp. *lactis* with derivatives of *L. lactis* subsp. *lactis* LM0230. Phage resistant and nisin-producing transconjugants were generated where selection was performed on plates containing nisin. Interestingly, in this case, sucrose-utilizing ability and reduced phage sensitivity were co-transferred with the nisin-producing characteristic. Analysis of donor, recipient and transconjugant genomic DNA indicated that 68 kb of DNA had been transferred from the chromosome of the donor into the chromosome of the transconjugant. More recently, at Moorepark, immunity to lacticin 3147 was used to select for transconjugants of the phage-resistance plasmid pMRC01. Over 30 transconjugants of different lactococcal starter strains have been conveniently made to date using this plasmid combined with bacteriocin-assisted selection. As part of this work, Coakley et al. (1997) showed that not all recipient lactococcal strains are amenable to plasmid uptake (due to plasmid incompatibility most likely) and also, that not all of those which do incorporate the plasmid show improved resistance against available phages. To date, other markers used successfully for selection include immunity to lacticin 481 (Pillidge et al. 2000; Mills et al. 2002), lactose metabolism (Sanders et al. 1986; Jarvis, 1988, Coffey et al. 1989), and cadmium resistance (Deng et al. 1997; Trotter et al. 2001). The availability of these conjugative systems containing selective markers, in our experience, provides a more effective response when phage disturbances of starter strains limit manufacturing potential in industry.

Conjugation has many advantages as a delivery strategy for anti-phage determinants; the transconjugant strains have food-grade status; using different plasmids/mechanisms a number of phage-insensitive variants can be made of the same strain; and rotation of these strains can be used to reduce/eliminate phage problems. However, the fact that many phage-resistance systems are not conjugative, the potential for plasmid incompatibility, and the paucity of dominant selectable markers for many anti-phage systems continues to present challenges in adopting this approach for food-grade strain improvement. Plasmid donation approaches for the delivery of non-self-transmissible phage-resistance to industrial strains can also be useful as demonstrated by Trotter et al. (2001) who delivered two type-I R/M systems to the industrial starter DPC4268.

Despite these successes, the extended industrial use of phage-resistant transconjugant strains has led to the emergence of new phages which can overcome both the innate and acquired defense mechanisms. This can be counteracted by the stacking of multiple phage-resistance systems in a single starter strain (Klaenhammer 1987, 1991; Josephsen & Klaenhammer, 1990; Kelly et al. 1990; Sing & Klaenhammer 1991, 1993; Durmaz & Klaenhammer 1995; O'Sullivan et al. 1998) since this leads to a significantly reduced possibility of emergence of phage mutants which are tolerant to the phage-resistance systems as a result of mutation or recombination. Despite the large number of phage-resistance systems documented, patent restriction do limit their availability for general use and thus many research groups have and continue to accumulate their own varied selection of phage-resistance systems for strain improvement. Where conjugation is concerned, there are obvious constraints to the amount of phage-resistance systems (including paucity of selectable markers) which can be stacked in a food grade manner. In addition, the introduction of bacteriocin-producing transconjugants in starter mixtures may kill companion strains required for starter performance. This points to the need for implementation of self-cloning recombinant methodology to achieve the full potential of introducing and stacking of phage-resistance systems in industrial strains.

Considerable progress in the development of food-grade vector systems has been made in recent years, and the vectors which are currently available include those based on dominant selectable markers such as bacteriocin resistance (Hughes & McKay 1992) or heavy metal resistance (Liu et al. 1996). In addition, complementation has been exploited as a mechanism to select and maintain plasmids (Dickely et al. 1995; Platteeuw et al. 1996) this approach has the disadvantage of requiring recipient strains to be mutated to facilitate selection. Another recently developed two-plasmid system involves a food-grade theta-replicating lactococcal recombinant plasmid lacking a selectable marker but carrying the gene of interest, and a rep-deficient plasmid encoding erythromycin resistance which depends on introduction of the former plasmid for maintenance. The latter plasmid is readily cured from the recipient after introduction of the gene of interest (Emond et al. 2001). All these approaches require that recipient strains must be capable of being electroporated. Looking ahead, the exploitation of cloned *oriT* determinants (origin of transfer) to facilitate the delivery of non-self-conjugative DNA as described by Hickey et al. (2001) will be important for the delivery of multiple recombinant phage-resistance systems to industrial starter strains where electroporation may not be feasible.

In conclusion, natural plasmid-encoded systems have been exploited successfully in starter strain improvement for over 15 years around the world. Their usage has become increasingly convenient with the improvement of conjugation protocols and selection procedures. The successes experienced to date in using these natural plasmid-encoded systems has undoubtedly played an important role in stimulating further research in the area of phage-resistance.

Traditional versus innovative strain rotational strategies

Some cheese plants operate strain rotation by manufacturing different cheese varieties with their associated starters on different days and this helps to prevent excessive buildup of phage against individual starter strains. Thus, in principal, the more varieties of cheese being produced by an individual plant with a concomitant diversity of starter strains being employed, the better this system operates. Nevertheless, if this approach is carried through, and if manufacturing schedules for individual cheese types increase, an important disadvantage is that a heterologous population of phages is likely to emerge in these cheese plants.

Since phage-resistance plasmids were deliberately introduced to industrial cheese starters during the 1980s, phages have proven remarkably flexible in their capacity to evolve and overcome the anti-phage systems in use. This situation shows that application of phage-resistance systems has not overcome the need for strain rotation and replacement in some form. However, many cheese manufacturers point out that traditional strain rotation approaches can lead to variation in quality and flavor for a specific fermented product. Novel rotational strategies, whereby phage-resistance mechanisms are rotated within an isogenic starter strain, devised by Klaenhammer's group in the early 1990s, can now eliminate the disadvantages of product variation and in-plant phage diversification (Sing & Klaenhammer 1993; Durmaz & Klaenhammer 1995). The rationale is that protection against new phage is possible because the emerging phage capable of overcoming one mechanism will have to confront a different resistance mechanism in the next cycle of rotation. Introduction and rotation of diverse anti-phage systems based on the flexibility of plasmid transfer in an isogenic background are now utilized in commercial practice. This is referred to as Phage Defense Rotation Strategy (PDRS) and is particularly useful at protecting new, specialized strains that are being used to meet specific product and process requirements during continuous and extended use.

Moineau et al. (1996) proposed an alternative rotational strategy whereby strains were rotated based on their phage-species sensitivity rather than on phage isolates. From a study on phages affecting lactococcal cultured buttermilk fermentations throughout the US, these authors concluded that many phage-sensitive strains were infected by phages from the same species rather than by phages from distinct species. Strains sensitive to phages from two species were isolated but not as frequently supporting the view that new phages develop from existing phages within a species either by point mutation or recombination (Bouchard & Moineau 2000; Boucher & Moineau 2000; Durmaz & Klaenhammer 2000). Thus, a rotation strategy based on phage species sensitivity is likely to be beneficial (Moineau et al. 1996; Boucher & Moineau 2001). Recently, some starter culture suppliers have extended their activities to isolate problematic phages and identify their species through molecular techniques (Moineau 1999).

In recent years, also, it has become clear that the rotation of type-I R/M specificity (*hsdS*) determinants can also play a role in modern rotational strategies (Schouler et al. 1998, O'Sullivan et al. 2000b). Type I R/M systems combine the functions of site-specific methylation and restriction activity in one large complex multimeric protein. Type I restriction and modification (methylase) enzymes consist of three types of subunit, encoded by the genes *hsdR, hsdM* and *hsdS*. The HsdS subunit determines the specificity of the endonuclease or methylase on the target DNA. When the HsdM subunits combine with the HsdS subunit (in the ratio of 2:1), it forms a modification enzyme that methylates two adenine residues in the recognition sequence. The inclusion of HsdR subunits in this complex ($R_2M_2S_1$) produces a restriction enzyme that binds to the DNA sequence determined by the HsdS subunit. A single subunit that concomitantly confers sequence specificity to both restriction and modification facilitates the acquisition of new specificities (Murray et al. 1993). Type I R/M systems are the only examples known to date of proteins that recognize specific DNA sequences and that can change their sequence specificities by well-understood natural processes (Bickle & Kruger 1993). The first reports of type I R/M systems in *Lactococcus* were by Schouler et al. (1998a,b; see Table 2). Furthermore, this group also demonstrated that the interaction of HsdS subunits between different R/M systems resulted in enhanced phage restriction. Schouler et al. (1998a) described the acquisition of new R/M specificities by transferring naturally occurring plasmids bearing only genetic determinants for the HsdS subunit to a lactococcal strain with a resident, but different *hsd* locus. This 'combinational variation' of type I R/M systems represents a general strategy in which lactococci can acquire R/M systems with novel specificities, enabling them to restrict a wider range of phage and to cope with the evolution of phage variants.

At Moorepark, O'Sullivan et al. (2000b) demonstrated that homologous recombination between the *hsdS* determinants of plasmids pAH33 and pAH82 facilitated the formation of the natural co-integrate plasmid pAH90 and the evolution of two novel HsdS subunits with altered R/M specificity. Evidence was also provided that this phenomenon may be common in lactococci and that homologous recombination between type I *hsdS* genes can permit new restriction specificities to rapidly evolve in lactococci. This phenomenon where, a natural genetic rearrangement generated the ability in the R/M system to adapt and target newly evolving phage, thereby protecting the starter culture (Figure 2) is rather like a primitive immune system acquiring immunity against new phage. Recently, Trotter et al. (2001) exploited this system for strain rotation by constructing two food-grade derivatives of the Cheddar cheese starter *L. lactis* DPC4268 containing either pAH90 or pAH82, which resulted in up to a 4 log reduction in EOP. These isogenic derivatives of the industrial starter only differed with respect to *hsdS* genes. It demonstrated that if the recipient starter strain harbors an effective type I R/M system, different *hsdS* genes can be introduced to modulate or change the restriction specificity against phage. This approach was based on the identification of different HsdS subunit determinants associated with the R/M plasmids pAH90 and pAH82, which also encode the selectable marker cadmium resistance (O'Sullivan et al. 2000b, 2001).

In the recent past, a number of remote *hsdS* genes have been identified on small plasmids in lactococci and streptococci with the ability to alter the specificity of indigenous type I R/M systems (O'Sullivan et al. 1999, 2000b; Seegers et al. 2000; Madsen et al. 2000; Boucher et al. 2001; Solow & Somkuti; 2001). Considering the apparently ubiquitous nature of type I specificity subunits in starter strains, these elements may serve as a reservoir for novel phage-resistance, which may provide an evolutionary reason for maintenance of these small, hitherto cryptic plasmids in lactococci.

In conclusion, the rotation of individual phage-resistance mechanisms within individual starter strains represents an important advance in the area of starter strain improvement. Significantly this advance has not necessarily involved the use of artificial recombinant technology. As a consequence of this development, improved starter strains have been developed which can permit prolonged use of some starter cultures in industry.

Recombinant intelligent approaches

Recombinant DNA techniques in tandem with the increased knowledge of phage-host interactions have facilitated the design of novel tactics for combating phage proliferation. The identification of genes responsible for many phage-resistance systems and the elucidation of the mode of action are important for the design of efficient recombinant anti-phage systems. Engineered or 'intelligent' systems have been successfully developed that exploit the phage genome as a means of inhibiting the intracellular development of the parent phage or related phage types in a manner similar to abortive infection. They are based on the principle that the phage life cycle can be interrupted if, in the host, its gene products are present in a dysfunctional form, present in excess, or expressed at the wrong stage of development (Sanford & Johnson 1985). Such a system, designated PER (phage encoded resistance), was devised by Hill et al. (1990a) and its efficacy resulted from the placing of a phage origin of replication (*ori*) on a multi-copy plasmid within a host cell. Infecting phage begin the process of phage DNA replication at the *ori* site. However, in a PER+ host, the phage replication factors are titrated by the multi-copy *ori* present on the plasmid, leading to induction of plasmid rather than phage replication. As the plasmid copy increases, so does the effective resistance, resulting in a dramatic decrease in phage DNA replication. Thus, PER is an extremely energy efficient anti-phage system where uptake and retention of a single copy of the PER locus is all that is required of the host cell. O'Sullivan et al. (1993) who

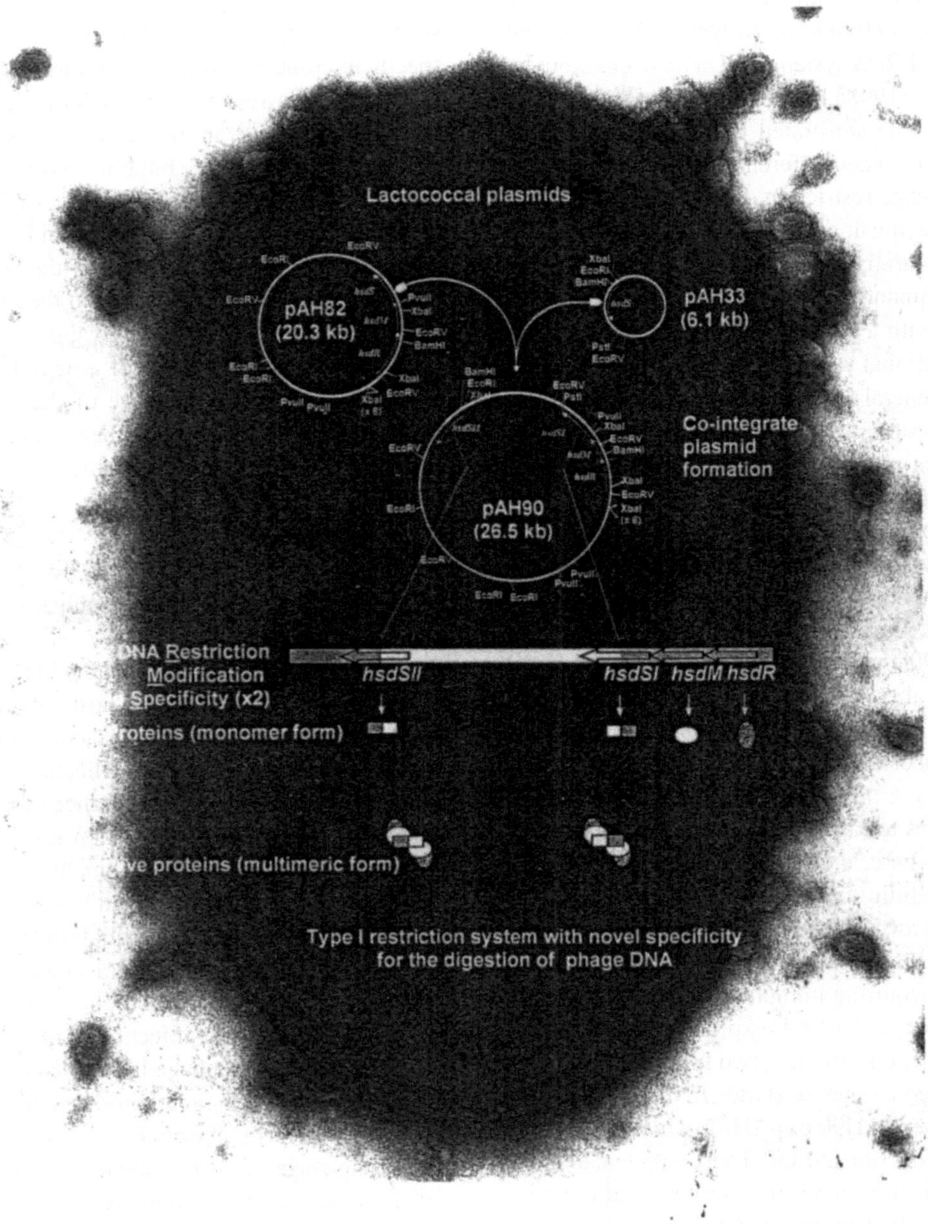

Figure 2. Homologous recombination between conserved domains of Type-I *hsdS* determinants gives rise to alternative anti-phage restriction activity. Plasmids are food-grade and have been delivered to an industrial *L. lactis* strain permitting rotation of *hsdS* genes in an isogenic background (O'Sullivan et al. 2000b; Trotter et al. 2001).

showed that increasing copy number of the plasmid vector augmented the efficiency of PER subsequently applied the approach to other phages. More recently, McGrath et al. (1999, 2001) presented similar data and also showed that cloning multiple copies of the phage replication determinants resulted in increased effectiveness of the PER phenotype.

An alternative recombinant technique for engineering phage-resistance is based on anti-sense RNA technology. This strategy utilizes transcription of the non-coding strand, usually of conserved phage genes, which bind to sense RNA resulting in unstable, non-transcribable RNA hybrid structures that are rapidly degraded. Knowledge of the organization and sequence of the target phage genome is obviously a

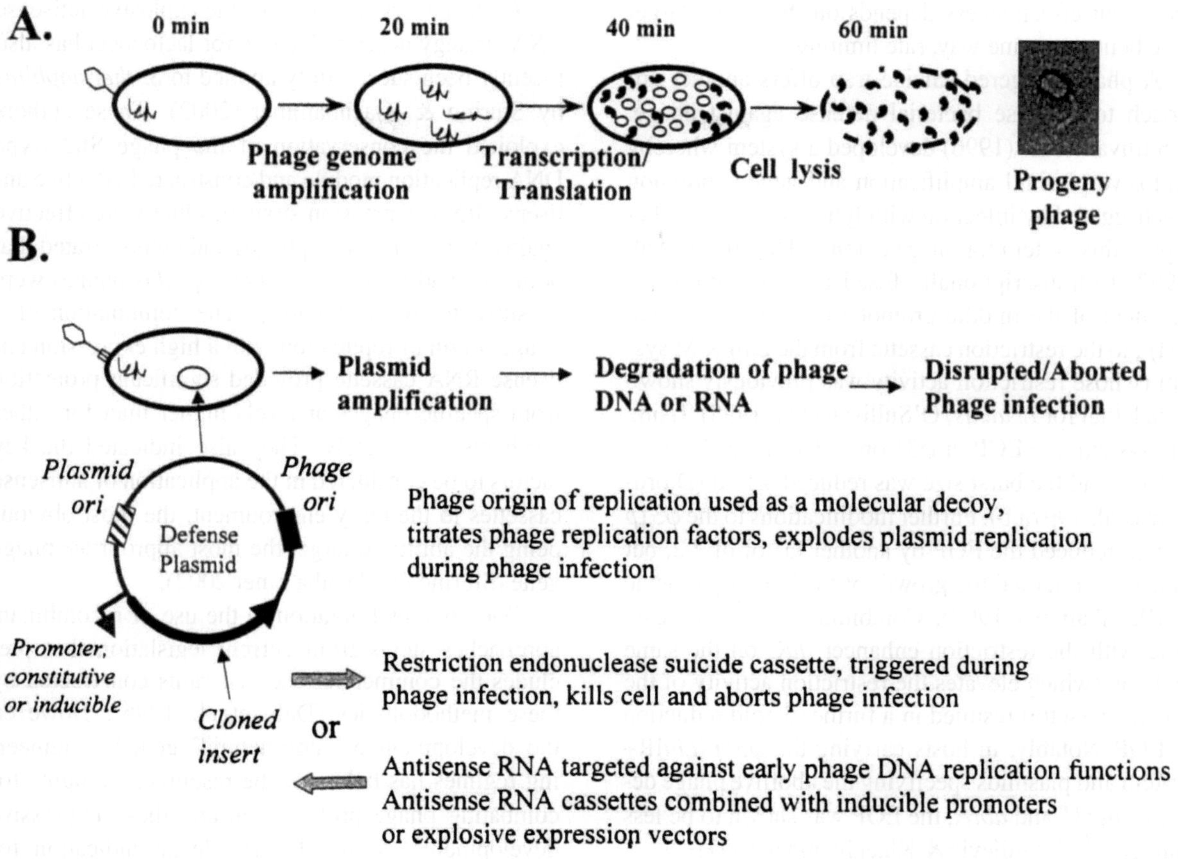

Figure 3. Inducible phage defense systems for lactic acid bacteria. (A) Normal phage development cycle. (B) Disruption of intracellular development by explosive molecular decoys, inducible suicide systems, or antisense RNA targeted to early DNA replication functions. Adapted from O'Sullivan et al. (1993, 1996), Djordjevic et al. (1997a,b) and Sturino & Klaenhammer (2002) (figure provided by J Sturino and TR Klaenhammer).

pre-requisite for the design and development of antisense regimes. Its exploitation for inhibition of phage was initially described by Coleman et al. (1985) for *E. coli*. The approach was subsequently evaluated for lactococcal phage by a group in the US in the late 1980s and early 1990s (Kim & Batt 1991; Chung et al. 1991, 1992; Kim et al. 1992). More recently, Polzin et al. (1996) developed antisense RNA against a number of ORFs of phage c2 but no significant increase in phage-resistance was observed, indicating that effective implementation of this approach was somewhat more complex than was first assumed. Similarly, Walker & Klaenhammer (1998) demonstrated that anti-sense RNA, directed against the activator of the middle genes of $\phi31$ (P335 type), was ineffective against $\phi31$ propagation. However, these authors recently reported a refinement of this system that was successfully applied to inhibit $\phi31$ replication (Walker & Klaenhammer 2000). By subcloning the antisense cassette to pTRK360, a low copy number vector containing the $\phi31$ origin of replication, it was observed that the levels of antisense RNA transcripts later in the lytic cycle was substantially increased, since *ori*31 allows for explosive amplification of the low-copy number vector upon phage infection (Figure 3). In addition, the presence of *ori*31 also lowered the burst size of $\phi31$ 4-fold, resulting in fewer sense (target) mRNAs being expressed from the phage genome. Growth curves performed in the presence of varying levels of $\phi31$ showed that this system offered significant protection to *L. lactis*, even at multiplicities of infection of 0.01 and 0.1 (Walker & Klaenhammer 2000). Recently also, McGrath et al. (2001) constructed antisense systems which targeted the replication module of four P335 type phages and demonstrated efficient inhibition of replication of these phages. These studies indicate that the antisense RNA approaches have significant potential for phage-resistance in lactococci, but effectiveness depends on the phage target gene being, in some way, rate limiting.

A phage-triggered suicide trap offers another approach to increase bacterial defense against phage. O'Sullivan et al. (1996) developed a system whereby explosive plasmid amplification and gene expression was triggered by infection with lytic phage. To further exploit this system for phage defense, Djordjevic et al. (1997a,b) transcriptionally fused a 239-bp promoter fragment of the middle promoter of ϕ31, designated ϕ31*p*, to the restriction cassette from the *Lla*I R/M system (whose restriction activity was previously shown to be lethal for *L. lactis*; O'Sullivan et al. 1995). Using this system, the EOP of ϕ31 on *L. lactis* was lowered to 10^{-5} and the burst size was reduced 4-fold (Djordjevic et al. 1997a,b). Further modifications to the ϕ31*p* system reduced the EOP by another factor of 2.2, but somewhat retarded the growth of the host (Djordjevic & Klaenhammer 1997). Combining the suicide cassette with the restriction enhancer *lla*C on the same replicon (which elevates the restriction activity of the suicide cassette) resulted in a further 5-fold reduction in EOP. Notably, in hosts carrying the ϕ31*p*-*Lla*IR+ system and plasmids specifying the abortive phage defenses *per*31 and *abi*A, the EOP was shown to be less than 10^{-10} (Djordjevic & Klaenhammer 1997).

Progress in the area of phage-resistance within the genus *Lactococcus* down through the years has been very helpful for the development of anti-phage strategies in thermophilic streptococci. Interestingly, despite the many reports of native R/M systems in *S. thermophilus* strains (Table 2) to date, none of these are reported to have been used to enhance the level of phage-resistance of industrial (e.g., yogurt or Mozzarella) starter strains, although notably, the introduction the lactococcal R/M system *Lla*DCHI to the genus by Moineau et al. (1995) did confer broad-range phage-resistance in various strains of *S. thermophilus*. Another advance in the area was made at Nestlé, Switzerland, by Lucchini et al. (2000) who used a thermolabile insertional vector for mutagenesis in *S. thermophilus* resulting in a phage-resistance phenotype, which was linked to inactivation of a transmembrane protein involved in the phage DNA injection process. This is reminiscent of the study of Garbutt et al. (1997), who demonstrated inactivation of the lactococcal *pip* protein necessary for infection by c2-type phages. At Nestlé also, the PER system described above for lactococci was successfully applied to *S. thermophilus* by Foley et al. (1998) and this approach was also adopted by Stanley et al. (2000) at University College Cork. In addition, the explosive antisense RNA strategy described above for lactococci has also recently been successfully applied to *S. thermophilus* by Sturino & Klaenhammer (2002). These authors exploited the conservation of the phage Sfi21-type DNA replication module and constructed effective antisense RNA expression strategies that were effective against *S. thermophilus* phages and demonstrated that both *cos*- and *pac*- type *S. thermophilus* phages were sensitive to this technology. The combination of a phage origin of replication with a high expression antisense RNA cassette provided significant protection from specific phages at levels higher than for either mechanism separately. They also indicated the key factors to be considered in the application of antisense cassettes to the dairy environment, the most obvious being the ability to target the most appropriate phage gene (Sturino & Klaenhammer 2002).

One obvious limitation to the use of recombinant approaches stems from current legislation that precludes the commercial use of strains constructed by these methodologies (Daly et al. 1996). However, the development of 'non-natural' genetic engineering regimes has bolstered the resources available for combating phage proliferation and these impressive developments, at worst, provide an indication for the potential of modern biotechnological practices for the generation of highly effective anti-phage systems (see paper on LAB-GMO in food by J. Kondo, this proceedings).

In conclusion, a variety of sophisticated highly effective recombinant anti-phage mechanisms have been devised. These have resulted from the exploitation of both phage genome research and the molecular manipulation of host-encoded phage-resistance genes. As a result of their recombinant nature, these systems are the most versatile, but will require sophisticated food-grade cloning vectors and consumer acceptance to permit their usage in industrial strains.

Phage counter defence

Knowledge is currently being accumulated regarding the mechanisms of evolution that lead to the genetic diversity of lactococcal phages in the dairy factory environment and this is covered elsewhere in this symposium proceedings. Nonetheless, to date, the response of a phage to a particular phage-resistance mechanism cannot be easily predicted. In general, this response depends on the ability of the phage to rep-

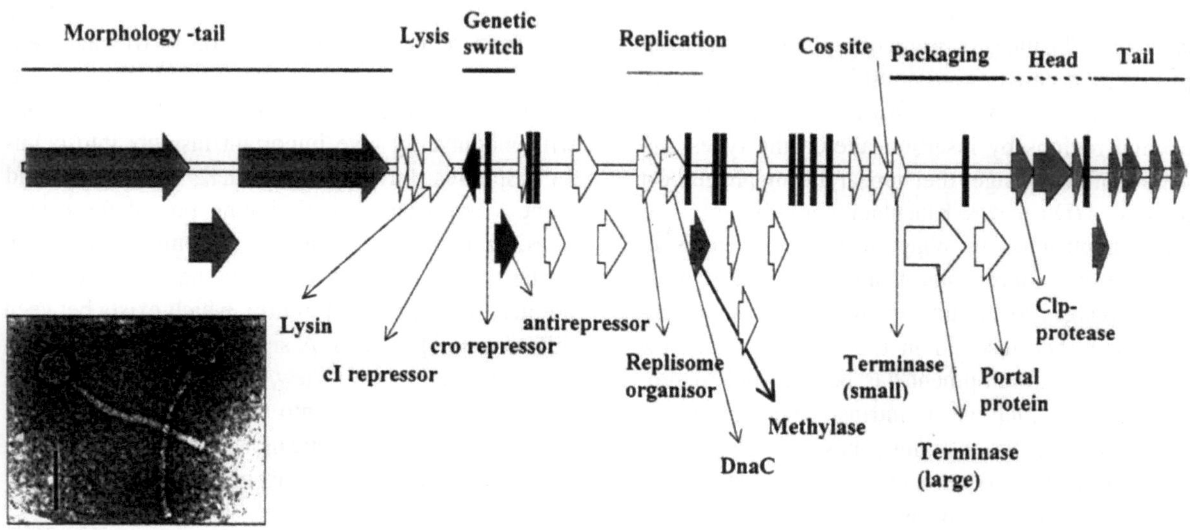

Figure 4. Genome (36596 bp) of a P335-type phage 4268, which is lytic for the cheese starter *L. lactis* DPC4268. The phage first emerged in industry 1994 and exhibited high virulence against the DPC4268 which is a 'workhorse' strain for up to 80% of Cheddar cheese production in some Irish cheese plants. The strain possesses chromosomal and plasmid-encoded R/M. A homologue of the chromosomal methylase is also present on the phage. The phage genome also contains a lytic switch with incomplete lysogenic module which suggests that it has incorporated DNA from a prophage during its evolution. Inset: electron micrograph of phage 4268. Bar=100 nm.

licate and to mutate to variants that can overcome the particular phage-resistant mechanism. The mechanisms by which this is achieved vary. They include non-specific point mutations in the phage genome, uptake of DNA from co-infecting phages or from functional or defective prophages, or various other regions of host DNA, all of which can assist in phage survival (see Figure 4). The latter phenomenon was first described by Hill et al. (1991b) and concerned the uptake by a phage genome of a methylase gene from a host R/M system. Phage uptake of host DNA was also later reported by Moineau et al. (1994) in the case of the lytic phage ϕul36 (P335 type). This phage was inhibited by *abi*C encoded by pTRK99 but a variant of the phage, ϕul37, evolved that was resistant to this mechanism. It was demonstrated that ϕul37 had incorporated a large fragment from the chromosome of pTR2030 transconjugants. This variant was also shown to be insensitive to *per*50, while disruption of the specific chromosome sequences in the host (*L. lactis* ssp. *lactis* NCK203) prevented the evolution of phage mutants corresponding to ϕul37 during subsequent infections (Moineau et al. 1994). More recently, Bouchard & Moineau (2000) described two new DNA exchanges involving ϕul36 and its *L. lactis* host. They occurred by homologous recombination with phage-related sequences present in the host chromosome *L. lactis* SMQ-88 (a strain containing the abortive infection mechanism AbiK). Both mutants (designated ϕul36.1 and ϕul36.2) generated by these recombination events were insensitive to AbiK and one had a reduced burst size as well as a new origin of replication. These authors proposed that DNA exchange with prophages or remnants of prophages occurs frequently within the P335 species and they supported this theory with DNA–DNA comparisons between P335-like phages. Given that phage-related sequences seem to be widespread in *L. lactis* (Bolotin et al. 2001), this mode of DNA replacement may be of great significance for the evolution of lactococcal phage variants. Moineau et al. (1993) also suggested that resistance of phages to host endonucleolytic activity may be due to the paucity or loss of restriction sites during the course of phage evolution. Resistances to *abi*A, *abi*C, *abi*D1, *abi*F, *per*50, and *per*31 have been shown to involve specific mutational responses and/or gross genetic rearrangements (O'Sullivan et al. 1993; Dinsmore & Klaenhammer 1994, 1997; Moineau et al. 1994; Bidnenko et al. 1995; Rince et al. 1999). A recent study by Durmaz & Klaenhammer (2000) characterized four recombinant phage variants that arose from infection of *L. lactis* NCK203 (Abi+) with ϕ31 (P335-type). Their results demonstrated that these phages had acquired DNA cassettes of various sizes from two different loci of the chromosome in order to overcome Abi defense, and that this may to have resulted from homologous recombination between ϕ31 and defective or non-inducible prophages from the chromosome of the host strain. In one of the recombinant phages (ϕ31.1), the ϕ31 origin of replication had been replaced by a new origin permitting it to circumvent the phage defense system. Disruption of these chromosomal regions by insertion altered the types and diversity of new phages that appear during phage-host interactions (Durmaz & Klaenhammer 2000).

The application of phage-resistance systems in dairy starter cultures places an evolutionary pressure on phages to circumvent the resistance mechanism. This has resulted in the emergence of new phages whose development has been linked to prophage DNA, which is an intrinsic element of the chromosomes of starter strains. These new phages will continue to pose an economic threat to the cheese industry. Given the intensive way that starter strains are grown in the presence of industrial phages, the dairy fermentation environment is an ideal niche to monitor phage evolution from a generic point of view. It can be expected that these studies are likely to result in novel precise methods to counteract adaptation to phage-resistance mechanisms.

Conclusions and future outlook

It is possible to argue that phage are as much a problem now as they ever have been, given that they interfere with acidification and consequently product quality, they force the use of strain rotation and also make it impossible to make cheese with very desirable strains such as autolytic bacteria. We are in a very opportune time at present, because of the depth of understanding that has resulted from phage-related research, particularly over the last 10 years. One could argue that the level of transfer of information to the industrial sector has been disappointing, but it is clearly evident that the complexity of phage–host interaction dictates that there is not a simple solution to the problem. We are now entering the post genomic era, a time where the genomes of almost all the species of lactic acid bacteria used in food production will have been sequenced. Indeed, since the landmark publications from INRA (Bolotin et al. 1999, 2001) detailing the entire genome of *Lactococcus lactis*, a number of other genomes have been are in the process of being sequenced (see papers on LAB genomes by T.R. Klaenhammer et al., this proceedings). This almost exponential increase in genetic information will provide an indispensable resource when it comes to understanding phage/host interplay, especially given the number of phage genomes that have also been sequenced (see paper on (pro)phage genomics by F. Desire, this proceedings). In this respect, microarrays covering entire starter bacterial genomes will undoubtedly give important insights into cellular responses during phage attack, which may lead to new approaches for designing host defense systems. In addition, comparative genomics of different starter strains may help to explain the tremendous difference in phage susceptibility, which exists between different starter strains. A significant challenge, from a commercial point of view, is the translation of this wealth of information into the rational design of a new generation of genetically modified starters with programmed defenses against phage attack.

Acknowledgements

We extend our gratitude to Gerald Fitzgerald for his ongoing cooperation over the years. We are also grateful to Denis Twomey for helpful discussions, to various colleagues overseas who helpfully responded to E-mail messages during the writing of this manuscript, to Todd Klaenhammer and Joe Sturino for providing Figure 3, to Elke Arendt for electron microscopy of phage adsorption (Figures 1 and 2), and to Susan Mills and Maeve Trotter for their critical review of the manuscript.

References

Ackermann H-W & DuBow MS (1987) Bacteriophage taxonomy. In: Viruses of Prokaryotes, Vol. 1 (pp 1–11). CRC Press, Boca Raton, FL.

Akçelik M (1999) A phage DNA injection-blocking type resistance mechanism encoded by chromosomal DNA in *Lactococcus lactis* subsp. *lactis* PLM-18. Milchwissenschaft 53: 619–622.

Allison GE & Klaenhammer TR (1998) phage-resistance mechanisms in lactic acid bacteria. Int. Dairy J. 8: 207–226.

Anba J, Bidnenko E, Hillier A, Ehrlich SD & Chopin M-C (1995) Characterisation of the lactococcal *abiD1* gene coding for phage abortive infection. J. Bacteriol. 177: 3818–3823.

Batt CA, Erlandson K & Bsat N (1995) Design and implementation of a strategy to reduce bacteriophage infection of dairy starter cultures. Int. Dairy J. 5: 949–962.

Benbadis L, Garel JR & Hartley DL (1991) Purification, properties and sequence specificity of *Ssl*I, a new typeII restriction endonuclease from *Streptococcus salivarius* ssp. *thermophilus*. Appl. Environ. Microbiol. 57: 3677–3678.

Bester BH & Lombard SH (1975) Protection of starter cultures against bacteriophages by propagation in a phage-resistant medium. South African J. Dairy Technol. 7: 235–240.

Bickle TA & Kruger DH (1993) Biology of DNA restriction. Microbiol. Rev. 57: 434–450.

Bidnenko E, Ehrlich SD & Chopin M-C (1995) Phage operon involved in sensitivity to the *Lactococcus lactis* abortive infection mechanism AbiD1. J. Bacteriol. 177: 3824–3829.

Bissonnette F, Labrie S, Deveau H, Lamoureux M & Moineau S (2000) Characterization of mesophilic mixed starter cultures used for the manufacture of aged Cheddar cheese. J. Dairy Sci. 83: 620–627.

Bolotin A, Mauger S, Malarme K, Ehrlich SD & Sorokin A (1999) Low-redundancy sequencing of the entire *Lactococcus lactis* IL1403 genome. Antonie Van Leeuwenhoek 76: 27–76.

Bolotin A, Wincker P, Mauger S, Jaillon O, Malarme K, Weissenbach J, Ehrlich SD & Sorokin A. (2001) The complete genome sequence of the lactic acid bacterium *Lactococcus lactis* ssp. *lactis* IL1403. Genome Res. 11: 731–753.

Bouchard JD & Moineau S (2000) Homologous recombination between a lactococcal bacteriophage and the chromosome of its host strain. Virology 270: 65–75.

Bouchard JD, Dion E & Moineau S. (2000) Characterization of a novel phage-resistance mechanism in *Lactococcus lactis*. 95th Annual Meeting of the American Dairy Science Association. July 24–28. Baltimore, MD. J. Dairy Sci. 83 (Suppl. 1): 132.

Boucher I & Moineau S (2001) Phages of *Lactococcus lactis*: an ecological and economical equilibrium. Recent Res. Dev. Virol. 3: 243–256.

Boucher I, Emond E, Dion E, Montpetit D & Moineau S (2000) Microbiological and molecular impacts of AbiK on the lytic cycle of *Lactococcus lactis* phages of the 936 and P335 species. Microbiol. 146: 445–453.

Boucher I, Ēmond Ē, Parrot M & Moineau S (2001) DNA sequence analysis of three *Lactococcus lactis* plasmids encoding phage-resistance mechanisms. J. Dairy Sci. 84: 1610–1620.

Boussemaer JP, Shrauwen PP, Sourrouille JL & Guy P (1980) Multiple modification/restriction systems in lactic streptococci and their significance in defining a phage typing system. J. Dairy Res. 47: 401–409.

Brüssow H (2001) Phages of dairy bacteria. Annu. Rev. Microbiol. 55: 283–303.

Brüssow H, Bruttin A, Dèsiere F, Lucchini S & Foley S, (1998) Molecular ecology and evolution of *Streptococcus thermophilus* bacteriophages – a review. Virus Genes 16: 95–109.

Burrus V, Bontemps C, Decaris B & Guedon G (2001) Characterization of a novel type II restriction-modification system, Sth368I, encoded by the integrative element ICESt1 of *Streptococcus thermophilus* CNRZ368. Appl. Environ. Microbiol. 67: 1522–1528.

Butler D & Fitzgerald GF (2001) Transcriptional analysis and regulation of expression of the *ScrF*I restriction-modification system of *Lactococcus lactis* subsp. *cremoris* UC503. J. Bacteriol. 183: 4668–4673.

Chung DK, Kim JH & Batt CA (1991) Cloning and nucleotide sequence of the major capsid protein from *Lactococcus lactis* ssp. *cremoris* bacteriophage F4-1. Gene 101: 121–125.

Chung DK, Chung SK & Batt CA (1992) Antisense RNA directed against the major capsid protein from *Lactococcus lactis* ssp. *cremoris* bacteriophage F4-1 confers partial resistance to the host. Appl. Microbiol. Technol. 37: 79–83.

Cluzel P -J, Chopin A, Ehrlich SD & Chopin M-C (1991) Phage abortive infection mechanism from *Lactococcus lactis* ssp. *lactis*, expression of which is mediated by an iso-ISS*1* element. Appl. Environ. Microbiol. 57: 3547–3551.

Coakley M, Fitzgerald GF & Ross RP (1997) Application and evaluation of phage-resistance- and bacteriocin-encoding plasmid pMRC01 for the improvement of dairy starter cultures. Appl. Environ. Microbiol. 63: 1434–1440.

Coffey AG, Fitzgerald GF & Daly C (1989) Identification and characterisation of a plasmid encoding abortive infection from *Lactococcus lactis* ssp. *lactis* UC811. Neth. Milk. Dairy J. 43: 229–244.

Coffey AG, Fitzgerald GF & Daly C (1991a) Cloning and characterisation of the determinant for abortive infection from the lactococcal plasmid pCI829. J. Gen. Microbiol. 143: 1355–1362.

Coffey AG, Costello V, Daly C & Fitzgerald GF (1991b) Plasmid encoded bacteriophage insensitivity in members of the genus *Lactococcus* with special reference to pCI829. *In:* Dunny GM, Cleary PP & McKay LL (Eds.) Genetics and Molecular Biology of Streptococci, Lactococci and Enterococci (p 131–135). American Society for Microbiology, Washington, DC.

Coffey A, Stokes D, Fitzgerald GF & Ross RP (2001) Traditional and molecular approaches to improving bacteriophage-resistance of Cheddar and Mozzarella starters. Ir. J. Agric. Fd. Res. 40: (2) (in press).

Coleman J, Hirashima A, Inokuchi Y, Green PJ & Inouye M (1985) A novel immune system against bacteriophage infection using complementary RNA (micRNA). Nature 315: 601–603.

Crow V, Martley FG, Coolbear T & Roundhill S (1995) The influence of phage-assisted lysis of *Lactococcus lactis* subsp. *lactis* ML8 on Cheddar cheese ripening. Int. Dairy J. 5: 451–472.

Dai G, Su P, Allison GE, Geller BL, Zhu P, Kim WS & Dunn NW (2001) Molecular characterization of a new abortive infection system (AbiU) from *Lactococcus lactis* LL51-1. Appl. Environ. Microbiol. 67: 5225–5232.

Daly C, Fitzgerald GF & Davis R (1996) Biotechnology of lactic acid bacteria with special reference to bacteriophage-resistance. Antonie van Leeuwenhoek 70: 99–110.

Davis R, van der Lelie D, Mercenier A, Daly C & Fitzgerald GF (1993) *Scr*FI restriction-modification system of *Lactococcus lactis* subsp. *cremoris* UC503: cloning and characterization of two ScrFI methylase genes. Appl. Environ. Microbiol. 59: 777–785.

Deng Y-M, Harvey ML, Liu CQ & Dunn NW (1997) A novel plasmid-encoded phage abortive infection system for *Lactococcus lactis* biovar *diacetylactis*. FEMS Microbiol. Lett. 146: 149–154.

Deng YM, Liu CQ & Dunn NW (1999) Genetic organisation and functional analysis of a novel abortive infection system, AbiL, from *Lactococcus lactis*. J. Biotechnol. 67: 135–149.

Deng YM, Liu CQ & Dunn NW (2000) LldI, a plasmid-encoded type I restriction and modification system in *Lactococcus lactis*. DNA Seq. 11: 239–245.

Dickely F, Nilsson D, Hansen EB & Johansen E (1995) Isolation of *Lactococcus lactis* nonsense suppressors and construction of a food-grade cloning vector. Mol. Microbiol. 15: 839–847.

Dinsmore PK & Klaenhammer TR (1994) Phenotypic consequences of altering the copy number of *abiA*, a gene responsible for aborting bacteriophage infections in *Lactococcus lactis*. Appl. Environ. Microbiol. 60: 1129–1136.

Dinsmore PK & Klaenhammer TR (1995) Bacteriophage-resistance in *Lactococcus*. Mol. Biotechnol. 4: 297–314.

Dinsmore PK & Klaenhammer TR (1997) Molecular characterisation of a genomic region in a *Lactococcus* bacteriophage that is involved in its sensitivity to the phage defence mechanism AbiA. J. Bacteriol. 179: 2949–2957.

Dinsmore PK, O' Sullivan DJ & Klaenhammer TR (1998) A leucine repeat motif in AbiA is required for resistance of *Lactococcus lactis* to phages representing three species. Gene 212: 5–11.

Djordjevic GM, O' Sullivan DJ, Walker SA, Conkling MA & Klaenhammer TR (1997a) Bacteriophage-triggered defence systems: phage adaptation and design improvements. Appl. Environ. Microbiol. 63: 4370–4376.

Djordjevic GM, O'Sullivan DJ, Walker SA, Conkling MA & Klaenhammer TR (1997b) Triggered-suicide system designed as a defense against bacteriophages. J. Bacteriol. 179: 6741–6748.

Duplessis M & Moineau S (2001) Identification of a genetic determinant responsible for host specificity in *Streptococcus thermophilus* bacteriophages. Mol. Microbiol. 41: 325–336.

Durmaz E & Klaenhammer TR (1995) A starter culture rotation strategy incorporating paired restriction-modification and abortive infection bacteriophage defences in a single *Lactococcus lactis* strain. Appl. Environ. Microbiol. 61: 1266–1273.

Durmaz E & Klaenhammer TR (2000) Genetic analysis of chromosomal regions of *Lactococcus lactis* acquired by recombinant lytic phages. Appl. Environ. Microbiol. 66: 895–903.

Durmaz E, Higgins DL & Klaenhammer TR (1992) Molecular characterisation of a second abortive phage resistance gene present in *Lactococcus lactis* ssp. *lactis* ME2. J. Bacteriol. 174: 7463–7469.

Ēmond Ē, Holler BJ, Boucher PA, Vandenbergh P, Vedamuthu ER, Kondo JK & Moineau S (1997) Phenotypic and genetic characterisation of the bacteriophage abortive infection mechanism AbiK from *Lactococcus lactis*. Appl. Environ. Microbiol. 63: 1274–1283.

Ēmond Ē, Dion E, Walker SA, Vedamuthu ER, Kondo JK & Moineau S (1998) AbiQ, an abortive infection mechanism from *Lactococcus lactis*. Appl. Environ. Microbiol. 64: 4748–4756.

Ēmond Ē, Lavall EER, Drolet G, Moineau S & LaPointe G (2001) Molecular characterization of a theta replication plasmid and its use for development of a two-component food-grade cloning system for *Lactococcus lactis*. Appl. Environ. Microbiol. 67: 1700–1709.

Feirtag JM & McKay LL (1987) Thermoinducible lysis of temperature-sensitive *Streptococcus cremoris* strains. J. Dairy Sci. 70: 1779–1784.

Fitzgerald GF, Daly C, Brown LR & Gingeras TR (1982) *Scr*FI: a new sequence-specific endonuclease from *Streptococcus cremoris*. Nucleic Acid Res. 10: 8171–8179.

Fitzgerald GF, Twomey DP, Daly C & Coffey AG (1995) Bacteriophage-resistance in *Lactococcus*: molecular characterization of the ScrFI restriction/modification system from *Lactococcus lactis* subsp. *cremoris* UC503. Dev Biol Stand. 85: 581–590.

Foley S, Lucchini S, Zwahlen MC & Brussow H (1998) A short noncoding viral DNA element showing characteristics of a replication origin confers bacteriophage-resistance to *Streptococcus thermophilus*. Virology 250: 377–387.

Forde A & Fitzgerald GF (1999) Bacteriophage defence systems in lactic acid bacteria. Antonie van Leeuwenhoek 76: 89–113.

Forde A, Fitzgerald GF & Daly C (1999) Identification of four phage-resistance plasmids from *Lactococcus lactis* ssp. *cremoris* HO2. Appl. Environ. Microbiol. 65: 1540–1547.

Garbutt KC, Kraus J & Geller BL (1997) Bacteriophage-resistance in *Lactococcus lactis* engineered by replacement of a gene for a bacteriophage receptor. J. Dairy Sci. 80: 1512–1519.

Garvey P, van Sinderen D, Twomey DP, Hill C & Fitzgerald GF (1995a) Molecular genetics of bacteriophage and natural phage defence systems in the genus *Lactococcus*. Int. Dairy J. 5: 905-947.

Garvey P, Fitzgerald GF & Hill C (1995b) Cloning and DNA sequence analysis of two abortive infection phage resistance determinants from the lactococcal plasmid pNP40. Appl. Environ. Microbiol. 61: 4321–4328.

Garvey PA, Hill C & Fitzgerald GF (1996) The lactococcal plasmid pNP40 encodes a third bacteriophage resistance mechanism, one which affects phage DNA penetration. Appl. Environ. Microbiol. 62: 676–679.

Garvey P, Rince A, Hill C & Fitzgerald GF (1997) Identification of a RecA homolog ($RecA_{LP}$) on the conjugative lactococcal phage-resistance plasmid pNP40: evidence of a role for chromosomally encoded $RecA_L$ in abortive infection. Appl. Environ. Microbiol. 63: 1244–1251.

Gautier M & Chopin MC (1987) Plasmid-determined systems for restriction and modification activity and abortive infection in *Streptococcus cremoris*. Appl. Environ. Microbiol. 53: 923–927.

Gireesh T, Davidson BE & Hillier AJ (1992) Conjugal transfer in *Lactococcus lactis* of a 68-kb pair chromosomal fragment containing the structural gene for the peptide bacteriocin nisin. Appl. Environ. Microbiol. 58: 1670–1676.

Guimont C, Henry P & Linden G (1993) Restriction-modification in *Streptococcus thermophilus*: isolation and characterisation of a type II restriction endonuclease Sth455I. Appl. Microbiol. Biotechnol. 39: 216–220.

Gulstrom TJ, Pearce L E, Sandine W E & Elliker PR (1979) Evaluation of commercial phage inhibitory media. J. Dairy Sci. 62: 208–221.

Harrington A & Hill C (1991) Construction of a bacteriophage-resistance derivative of *Lactococcus lactis* ssp. *lactis* 425A by using the conjugal plasmid pNP40. Appl. Environ. Microbiol. 57: 3405–3409.

Harrington A & Hill C (1992) Plasmid involvement in the formation of a spontaneous bacteriophage insensitive mutant *Lactococcus lactis*. FEMS Microbiol. Lett. 96: 132–142.

Hickey RM, Twomey DP, Ross RP & Hill C (2001) Exploitation of plasmid pMRC01 to direct transfer of mobilizable plasmids into commercial lactococcal starter strains. Appl. Environ. Microbiol. 67: 2853–2858.

Hill C (1993) Bacteriophage and bacteriophage-resistance in lactic acid bacteria. FEMS Microbiol. Lett. 12: 87–108.

Hill C, Pierce K & Klaenhammer TR (1989) The conjugative plasmid pTR2030 encodes two bacteriophage defence mechanisms in lactococci, restriction-modification (R^+/M^+) and abortive infection (Hsp^+). Appl. Environ. Microbiol. 55: 2416–2419.

Hill C, Miller LA & Klaenhammer TR (1990a) Cloning, expression, and sequence determination of a bacteriophage fragment encoding bacteriophage-resistance in *Lactococcus lactis*. J. Bacteriol. 172: 6419–6426.

Hill C, Miller LA & Klaenhammer TR (1990b) Nucleotide sequence and distribution of the pTR2030 resistance determinant (*hsp*) which aborts bacteriophage infection in lactococci. Appl. Environ. Microbiol. 56: 2255–2258.

Hill C, Miller LA & Klaenhammer TR (1991) *In vivo* genetic exchange of a functional domain from a type II A methylase between lactococcal plasmid pTR2030 and a virulent bacteriophage. J. Bacteriol. 173: 4363–4370.

Huggins AR & Sandine WE (1979) Selection and charaterisation of phage insensitive lactic streptococci. J. Dairy Sci. 62: 70–71.

Hughes BF & McKay LL (1992) deriving phage insensitive lactococci using a food-grade vector encoding phage and nisin resistance. J. Dairy Sci. 75: 914–923.

Jarvis AW (1988) Conjugal transfer in lactic streptococci of plasmid-encoded insensitivity to prolate- and small isometric-headed bacteriophages. Appl. Environ. Microbiol. 54: 777–784.

Jarvis AW (1989) Bacteriophages of lactic acid bacteria. J. Dairy Sci. 72: 3406–3428.

Jarvis AW (1992) Analysis of phage-resistance mechanisms encoded by lactococcal plasmid pAJ2074. Can. J. Microbiol. 39: 252–258.

Jarvis AW, Fitzgerald GF, Mata M, Mercenier A, Neve H, Powell IB, Ronda C, Saxelin M & Teuber M (1991) Species and types of phages of lactococcal bacteriophages. Intervirology 32: 2–9.

Josephsen J & Klaenhammer TR (1990) Stacking of three different restriction and modification systems in *Lactococcus lactis* by co-transformation. Plasmid 23: 71–75.

Josephsen J & Neve H (1998) Bacteriophages and lactic acid bacteria. In: Salminen S & van Wright A (Eds.) Lactic Acid Bacteria: Microbiology and Functional Aspects, 2nd Edn. (pp 385–436). Marcel Dekker, New York.

Josephsen J, Andersen N, Behrndt E, Brandsborg E, Christiansen G, Hansen MB, Hansen S, Nielsen EW & Vogensen (1994) An ecological study of of lytic bacteriophages of lactococcal bacteriophages of *Lactococcus lactis* subsp. *cremoris* isolated in a cheese plant over a 5-year-period. Int. Dairy J. 4: 123–140.

Josephsen J, Jorgen-Jensen B & Nyengaard NR (1998) Determination of the recognition sequence of the type II restriction endonuclease, *Lla*CI, from *Lactococcus lactis* W15. FEMS Microbiol. Lett. 163: 25–29.

Kelly W, Dobson J, Jorck-Ramberg D, Fitzgerald GF & Daly C (1990) Introduction of bacteriophage-resistance plasmids into commercial *Lactococcus* starter cultures. FEMS Microbiol. Rev. 87: Abst. C20.

Khosravi L, Sandine WE & Ayres JW (1991) Evaluation of a newly-formulated bacteriophage inhibitory medium for cultivation of thermophilic lactic acid bacteria. Cultered Dairy Products 29: 4–9.

Kim SG & Batt CA (1991) Antisense mRNA-mediated bacteriophage-resistance in *Lactococcus lactis* ssp. *lactis*. Appl. Environ. Microbiol. 57: 1109–1113.

Kim SG, Bor Y-C & Batt CA (1992) Bacteriophage-resistance in *Lactococcus lactis* ssp. *lactis* using antisense ribonucleic acid. J. Dairy Sci. 75: 1761–1767.

Klaenhammer TR (1984) Interactions of bacteriophage with lactic streptococci. Adv. Appl. Microbiol. 30: 1–29.

Klaenhammer TR (1987) Plasmid-directed mechanisms for bacteriophage defence in lactic streptococci. FEMS Microbiol. Rev. 46: 313–325.

Klaenhammer TR (1991) Development of bacteriophage resistant strains of lactic acid bacteria. Biochem. Soc. Trans. 19: 675–681.

Klaenhammer TR & Fitzgerald GF (1994) Bacteriophage and bacteriophage-resistance. In: Gasson MJ & de Vos WM (Eds.) Genetics and Biotechnology of Lactic Acid Bacteria (pp 106–168). Blackie Academic and Professional, Glasgow.

Klaenhammer TR & Sanozky RB (1985) Conjugal transfer from *Streptococcus lactis* ME2 of plasmids encoding phage-resistance, nisin resistance and lactose-fermenting ability: evidence for a high-frequency conjugative plasmid responsible for abortive infection of virulent bacteriophage. J. Gen. Microbiol. 131: 1531–1541.

Kraus J & Geller BL (2001) Cloning of genomic DNA of *Lactococcus lactis* that restores phage sensitivity to an unusual bacteriophage sk1-resistant mutant. Appl. Environ. Microbiol. 67: 791–798.

Labrie S & Moineau S (2000) Multiplex PCR for detection and identification of lactococcal bacteriophages. Appl. Environ. Microbiol. 66: 987–994.

Le Marrec C, van Sinderen D, Walsh L, Stanley E, Viegels E, Moineau S, Heinze P, Fitzgerald GF & Fayard B (1997) Two groups of bacteriophages infecting *Streptococcus thermophilus* can be distinguished on the basis of mode of packaging and genetic determinants for major structural proteins. Appl. Environ. Microbiol. 63: 3246–3253.

Lepeuple A-S, Vassal L, Cesselin B, Delacroix-Buchet A, Gripon J-C & Chapot-Chartier M-P (1998) Involvement of a prophage in the lysis of *Lactococcus lactis* subsp. *cremoris* AM2 during cheese ripening. Int. Dairy J. 43: 301–311.

Limsowtin GKY, Heap H & Lawrence RC (1978). Heterogeneity among strains of lactic streptococci N.Z. J. Dairy Sci. 13: 1–8.

Liu C-Q, Leelawatcharamas V, Harvey ML & Dunn NW (1996) Cloning vectors for lactococci based on a plasmid encoding resistance to cadmium. Curr. Microbiol. 33: 35–39.

Lucchini S, Sidoti J & Brussow H. (2000) Broad-range bacteriophage-resistance in *Streptococcus thermophilus* by insertional mutagenesis. Virology 275: 267–277.

Madsen A & Josephsen J (1998a) Characterisation of *Lla*CI, a new restriction-modification system from *Lactococcus lactis* ssp. *cremoris* W15. Biol. Chem. 379: 443–449.

Madsen A & Josephsen J (1998b) Cloning and characterisation of the lactococcal plasmid-encoded type II restriction-modification system, *Lla*DII. Appl. Environ. Microbiol. 64: 2424–2431.

Madsen A & Josephsen J (2001) The *Lla*GI restriction and modification system of *Lactococcus lactis* W10 consists of only one single polypeptide. FEMS Microbiol. Lett. 200: 91–96.

Madsen A, Westphal C & Josephsen J (2000) Characterization of a novel plasmid-encoded HsdS subunit, S.LlaW12I, from *Lactococcus lactis* W12. Plasmid 44: 196–200.

Madsen SM, Mills D, Djordjevic G, Israelsen H & Klaenhammer TR (2001) Analysis of the genetic switch and replication region of a P335-type bacteriophage with an obligate lytic lifestyle on *Lactococcus lactis*. Appl. Environ. Microbiol. 67: 1128–1139.

Mayo B, Hardisson C & Brana AF (1991) Nucleolytic activities in *Lactococcus lactis* ssp. *lactis* NCDO497. FEMS Microbiol. Lett. 79: 195–198.

McGrath S, Seegers JF, Fitzgerald GF & van Sinderen D (1999) Molecular characterization of a phage-encoded resistance system in *Lactococcus lactis*. Appl. Environ. Microbiol. 65: 1891–1899.

McGrath S, Fitzgerald GF & van Sinderen D (2001) Improvement and optimization of two engineered phage resistance mechanisms in *Lactococcus lactis*. Appl. Environ. Microbiol. 67: 608–616.

McGrath S, Fitzgerald GF & van Sinderen D (2002) Identification and characterization of phage-resistance genes in temperate lactococcal bacteriophages. Mol. Microbiol. 43: 509–520.

McLandsborough LA, Kolaetis KM, Requena T & McKay LL (1995) Cloning and characterisation of the abortive infection genetic determinant *abiD* isolated from pBF61 of *Lactococcus lactis* ssp. *lactis* KR5. Appl. Environ. Microbiol. 61: 2023–2026.

Mills S, Coffey A, O'Sullivan L, Stokes D, Hill C, Fitzgerald GF & Ross RP (2002) Use of lacticin 481 to facilitate delivery of the bacteriophage-resistance plasmid, pCBG104 to cheese starters. J. Appl. Microbiol. 92: 238–246.

Moineau S (1999) Applications of phage-resistance in lactic acid bacteria. Antonie van Leeuwenhoek 76: 377–382.

Moineau S, Fortier J, Ackermann H-W & Pandian S (1992) Characterization of lactococcal bacteriophages from Québec cheese plants. Can. J. Microbiol. 38: 875–882.

Moineau S, Pandian S & Klaenhammer TR (1993a) Restriction-modification systems and restriction endonucleases are more effective on lactococcal bacteriophages that have emerged recently in the dairy industry. Appl. Environ. Microbiol. 59: 197–202.

Moineau S, Pandian, S & Klaenhammer TR (1993b) Evolution of a lytic bacteriophage via DNA acquisition from the *Lactococcus lactis* chromosome. Appl. Environ. Microbiol. 60: 1832–1841.

Moineau S, Walker SA, Vedamuthu ER & Vandenbergh PA (1995a) Cloning and sequencing of LlaDCHI [corrected] restriction/modification genes from *Lactococcus lactis* and relatedness of this system to the *Streptococcus pneumoniae* DpnII system. Appl. Environ. Microbiol. 61: 2193–2202.

Moineau S, Walker SA, Holler BJ, Vedamuthu ER & Vandenbergh PA (1995b) Expression of a *Lactococcus lactis* phage-resistance mechanism by *Streptococcus thermophilus*. Appl. Environ. Microbiol. 61: 2461–2466.

Moineau S, Borkaev M, Holler BJ, Walker SA, Kondo JK, Vedamuthu ER & Vandenbergh PA (1996) Isolation and characterization of lactococcal bacteriophages from cultured buttermilk plants in the United States. J. Dairy Sci. 79: 2104–2111.

Murray NE, Daniel AS, Cowan GM & Sharp PM (1993) Conservation of motifs within the unusually variable polypeptide sequences of type I restriction and modification enzymes. Mol. Microbiol. 9: 199–143.

Neve H, Krusch U & Teuber M (1989) Classification of virulent bacteriophages of *Streptococcus salivarius* subsp. *thermophilus* isolated from yoghurt and Swiss-type cheese. Appl. Microbiol. Biotechnol. 30: 624–629.

Nyengaard N, Vogensen FK & Josephsen J (1993) *Lla*AI and*Lla*BI, two type II restriction endonucleases from *Lactococcus lactis* ssp. *cremoris* W9 and W56 recognising, respectively, 5′-/GATC-3′ and 5′-C/TRYAG-3′. Gene 136: 371–372.

Nyengaard N, Vogensen FK & Josephsen J (1995) Restriction-modification systems in *Lactococcus lactis*. Gene 157: 13–18.

Nyengaard NR, Falkenberg-Klok J & Josephsen J (1996) Cloning and analysis of the restriction-modification system *Lla*BI, a bacteriophage-resistance system from *Lactococcus lactis* ssp. *cremoris* W56. Appl. Environ. Microbiol. 62: 3494–3498.

O' Connor L, Coffey A, Daly C & Fitzgerald GF (1996) AbiG, a genotypically novel abortive infection mechanism encoded by plasmid pCI750 of *Lactococcus lactis* ssp. *cremoris* UC653. Appl. Environ. Microbiol. 62: 3075–3082.

O' Connor L, Tangney M & Fitzgerald GF (1999) Expression, regulation, and mode of action of the AbiG abortive infection system of *Lactococcus lactis* subsp. *cremoris* UC653. Appl. Environ. Microbiol. 65: 330–335.

O' Sullivan D, Coffey A, Fitzgerald GF, Hill C & Ross RP (1998) Design of a phage-insensitive lactococcal dairy starter via sequential transfer of naturally occurring conjugative plasmids. Appl. Environ. Microbiol. 64: 4618–4622.

O' Sullivan D, Ross RP, Fitzgerald GF & Coffey A (2000a) Investigation of the relationship between lysogeny and lysis of *Lactococcus lactis* in cheese using prophage-targeted PCR. Appl. Environ. Microbiol. 66: 2192–2198.

O' Sullivan D, Twomey DP, Coffey A, Hill C, Fitzgerald GF & Ross RP (2000b) Novel type I restriction specificities through domain shuffling of HsdS subunits in *Lactococcus lactis*. Mol. Microbiol. 36: 866–875.

O' Sullivan D, Ross RP, Twomey DP, Fitzgerald GF, Hill C & Coffey A (2001) Naturally occurring lactococcal plasmid pAH90 links bacteriophage-resistance and mobility functions to a food-grade selectable marker. Appl. Environ. Microbiol. 67: 929–937.

O' Sullivan DJ, Hill C & Klaenhammer TR (1993) Effect of increasing the copy number of bacteriophage origins of replication, *in trans*, on incoming-phage proliferation. Appl. Environ. Microbiol. 59: 2449–2456.

O' Sullivan DJ, Zagula K & Klaenhammer TR (1995) In vivo restriction by *Lla*I is encoded by three genes, arranged in an operon with *llaIM*, on the conjugative *Lactococcus* plasmid pTR2030. J. Bacteriol. 177: 134–143.

O' Sullivan DJ, Walker SA, West SG & Klaenhammer TR (1996) Development of an expression system using a lytic phage to trigger explosive plasmid amplification and gene expression. Biotechnology 14: 82–87.

O' Sullivan DJ & Klaenhammer TR (1998) Control of expression of *Lla*I restriction in *Lactococcus lactis*. Mol. Microbiol. 27: 1009–1020.

O' Sullivan T, van Sinderen D & Fitzgerald G (1999) Structural and unctional analysis of pCI65st, a 6.5 kb plasmid from *Streptococcus thermophilus* NDI-6. Microbiology 145: 127–134.

Oberg CJ & Broadbent JR (1993) Thermophilic starter cultures: another set of problems. J. Dairy Sci. 76: 2392–2406.

Parreira R, Ehrlich SD & Chopin M-C (1996) Dramatic decay of phage transcripts in lactococcal cells carrying the abortive infection determinant AbiB. Mol. Microbiol. 19: 221–230.

Pillidge CJ, Collins LJ, Ward, LJH, Cantillon BM, Shaw BD, Timmins MJ, Heap HA & Polzin KM (2000) Efficacy of four conjugal lactococcal phage-resistance plasmids against phage in commercial *Lactococcus lactis* subsp. *cremoris* cheese starter strains. Int. Dy J. 10: 617–625.

Platteeuw C, van Alen-Boerrigter I, van Schalkwijk S & de Vos WM (1996) Food-grade cloning and expression system for *Lactococcus lactis*. Appl. Environ. Microbiol. 62: 1008–1013.

Poch MT, Somkuti GA & Solaiman DKY (1997) *Sth*132I, a novel class IIS restriction endonuclease of *Streptococcus thermophilus* ST132. Gene 195: 201–206.

Polzin KM, Collins LJ, Lubbers MW & Jarvis AW (1996) Effect of various mRNAs on bacteriophage c2 replication. In: Proceedings of 5th Symposium on Lactic Acid Bacteria, Genetics, Metabolism and Applications, Veldhoven, The Netherlands. Abs F2.

Prevots F & Ritzenthaler P (1998) Complete sequence of the new lactococcal abortive phage-resistance gene *abiO*. J. Dairy Sci. 81: 1483–1485.

Prevots F, Daloyau M, Bonin O, Dumont X & Tolou S (1996) Cloning and sequencing of the novel abortive infection gene *abiH* of *Lactococcus lactis* ssp. *lactis* biovar. *diacetylactis* S94. FEMS Microbiol. Lett. 142: 295–299.

Prevots F, Tolou S, Delpech B, Kaghad M & Daloyau M (1998) Nucleotide sequence and analysis of the new chromosomal abortive infection gene *abiN* of *Lactococcus lactis* ssp. *cremoris* S114. FEMS Microbiol. Lett. 159: 331–336.

Quiberoni A, Stiefel JI & Reinheimer JA (2000) Characterization of phage receptors in *Streptococcus thermophilus* using purified cell walls obtained by a simple protocol. J. Appl. Microbiol. 89: 1059–1065.

Richardson GH, Cheng CT & Young R (1977) Lactic bulk culture system utilising a whey-based bacteriophage inhibitory medium and pH control. J. Dairy Sci. 60: 378–386.

Rince A, Tangney M & Fitzgerald GF (2000) Identification of a DNA region from lactococcal phage sk1 protecting phage 712 from the abortive infection mechanism AbiF. FEMS Microbiol. Lett. 182: 185–191.

Sanders ME (1988) phage-resistance in lactic acid bacteria. Biochimie 70: 411–422.

Sanders ME & Klaenhammer TR (1981) Evidence for plasmid linkage of restriction and modification in *Streptococcus cremoris* KH. Appl. Environ. Microbiol. 42: 944–950.

Sanders ME, Leonard PJ, Sing WD & Klaenhammer TR (1986) Conjugal strategy for the construction of fast-acid producing, bacteriophage-resistant lactic streptococci for use in dairy fermentations. Appl. Environ. Microbiol. 52: 1101–1107.

Sandine WE (1989) Use of bacteriophage-resistant mutants of lactococcal starters in cheese-making. Neth. Milk Dairy J. 43: 211–219.

Sandine WE (1995) Commercial production of dairy starter cultures. In: Cogan TM, Accolas JP (Eds.) Dairy Starter Cultures, 3rd Edn. (pp 191–206) VCH Publishers, New York.

Sanford JC & Johnston SA (1985) The concept of parasite-derived resistance- deriving resistance genes from the parasites own genome. J. Theor. Biol. 113: 395–405.

Schouler C, Gautier S, Ehrlich SD & Chopin M-C (1998a) Combinational variation of restriction-modification specificities in *Lactococcus lactis*. Mol. Microbiol. 28: 169–178.

Schouler C, Clier F, Lerayer AD Ehrlich SD & Chopin M-C (1998b) A type IC restriction-modification system in *Lactococcus lactis*. J. Bacteriol. 180: 407–411.

Seegers JF, van Sinderen D & Fitzgerald GF (2000) Molecular characterization of the lactococcal plasmid pCIS3: natural stacking of specificity subunits of a type I restriction/modification system in a single lactococcal strain. Microbiology 146: 435–443.

Sing WD & Klaenhammer TR (1986) Conjugal transfer of bacteriophage-resistance determinants on pTR2030 into *Streptococcus cremoris* strains. Appl. Environ. Microbiol. 51: 1264–1271.

Sing WD & Klaenhammer TR (1990b) Plasmid encoded abortive infection in lactococci: a review. J. Dairy Sci. 73: 2239–2251.

Sing WD & Klaenhammer TR (1991) Characterization of restriction-modification plasmids from *Lactococcus lactis* ssp. *cremoris* and their effects when combined with pTR2030. J. Dairy Sci. 74: 1133–1144.

Sing WD & Klaenhammer TR (1993) A strategy for rotation of different bacteriophage defences in a lactococcal single-strain starter culture system. Appl. Environ. Microbiol. 59: 365–372.

Solaiman DKY & Somkuti GA (1990) Isolation and characterisation of a type II restriction endonuclease from *Streptococcus thermophilus* ST117. FEMS Microbiol. Lett. 80: 261–266.

Solaiman DKY & Somkuti GA (1991) A type II restriction endonuclease from *Streptococcus thermophilus*. FEMS Microbiol. Lett. 67: 261–266.

Solow BT & Somkuti GA (2001) Molecular properties of *Streptococcus thermophilus* plasmid pER35 encoding a restriction modification system. Curr. Microbiol. 42: 122–128.

Stanley E, Walsh L, van der Zwet A, Fitzgerald GF & van Sinderen D (2000) Identification of four loci isolated from two *Streptococcus thermophilus* phage genomes responsible for mediating bacteriophage resistance. FEMS Microbiol. Lett. 182: 271–277.

Stokes D, Ross RP, Fitzgerald GF & Coffey A (2001). Application of *Streptococcus thermophilus* DPC1842 as an adjunct to counteract phage in a predominantly lactococcal Cheddar cheese starter: use in bulk starter culture systems. Le Lait 81: 327–334.

Sturino JM & Klaenhammer TR (2002) Expression of antisense RNA against *Streptococcus thermophilus* bacteriophages. Appl. Environ. Microbiol. 68: 588–596.

Su P, Harvey M, Im HJ & Dunn NW (1997) Isolation, cloning and characterisation of the *abiI* gene from *Lactococcus lactis* ssp. *lactis* M138 encoding abortive phage infection. J. Biotechnol. 54: 95–104.

Su P, Im H, Hsieh H, Kang'A S & Dunn NW (1999) *Lla*FI, a type II restriction and modification system in *Lactococcus lactis*. Appl. Environ. Microbiol. 65: 686–693.

Thunell RK, Sandine WE & Bodyfelt FW (1981) Phage insensitive multiple strain starter approach to Cheddar cheese making. J. Dairy Sci. 64: 2270–2277.

Trotter M, Mills S, Ross RP, Fitzgerald GF & Coffey A (2001) The use of cadmium resistance on the phage-resistance plasmid pNP40 facilitates selection for its horizontal transfer to industrial dairy starter lactococci. Lett. Appl. Microbiol. 33: 409–414.

Twomey DP, Gabillet N, Daly C & Fitzgerald GF (1997) Molecular characterisation of the restriction endonuclease gene (*scrFIR*) associated with the *Scr*FI restriction-modification system from *Lactococcus lactis* ssp. *cremoris* UC503. Microbiology 143: 2277–2286.

Twomey DP, McKay LL & O'Sullivan DJ (1998) Molecular characterisation of the *Lactococcus lactis Lla*KR2I restriction-modification system and effect of an IS*982* element positioned between the restriction and modification genes. J. Bacteriol. 180: 5844–5854.

Twomey DP, De Urraza PJ, McKay LL & O'Sullivan DJ (2000) Characterization of AbiR, a novel multicomponent abortive infection mechanism encoded by plasmid pKR223 of *Lactococcus lactis* subsp. *lactis* KR2. Appl. Environ. Microbiol. 66: 2647–2651.

Walker SA & Klaenhammer TR (1998) Molecular characterization of a phage-inducible middle promoter and its transcriptional activator from the lactococcal bacteriophage phi31. J. Bacteriol. 180: 921–931.

Walker SA & Klaenhammer TR (2000) An explosive antisense RNA strategy for inhibition of a lactococcal bacteriophage. Appl. Environ. Microbiol. 66: 310–319.

Ward AC, Davidson BE, Hillier AJ & Powell IB (1992) Conjugally transferable phage-resistance activities from *Lactococcus lactis* DRC1. J. Dairy Sci. 75: 683–691.

Whitehead WE, Ayres JW & Sandine, WE (1993) A review of starter media for cheese making. J. Dairy Sci. 76: 2344–2353.

Antonie van Leeuwenhoek **82:** 323–331, 2002.

In situ delivery of cytokines by genetically engineered *Lactococcus lactis*

Lothar Steidler
VIB – Ghent University, Department of Molecular Biology, KL Ledeganckstraat 35, B-9000 Gent, Belgium (E-mail: Lothar.Steidler@DMB.RUG.AC.BE)

Key words: cytokine, delivery, IBD, interleukin-10, *Lactococcus lactis*

Abstract

The development of novel approaches that allow for accurate targeting of therapeutics to the bowel mucosa is a priority in the research on inflammatory bowel disease. We have engineered *Lactococcus lactis* to secrete soluble, fully active, correctly processed cytokines. We have used these live, recombinant strains for the *in situ* delivery of mouse interleukin (mIL)-2, -6 and -10 at airway mucosa or mucosa of the colon. Strains that secrete mIL-2 or mIL-6 and produce TTFC intracellular show a higher level of anti-TTFC induction in mice following intranasal inoculation. We showed that mIL-10 producing *L. lactis* can prevent and cure enterocolitis in mice. The daily ingestion of this strain leads to the prevention of colitis in IL-10 –/– 129 Sv/Ev mice. The repeated addition of DSS to the drinking water of Balb/c mice leads to the induction of chronic colitis with a typical mean histological score of five points. Subsequent daily treatment with 10^8 IL-10 producing *L. lactis* reduced the inflammation to a score of approximately 1 in 40% of the treated mice, which is a status equal to that of healthy control mice. Most other animals from the treated group only showed minor patchy remnants of the inflammation. Killing of the IL-10 producing bacteria by UV irradiation immediately prior to inoculation abrogates this therapeutic effect. Therefore it can be attributed to the active *in vivo* delivery of IL-10. We have further documented this by demonstrating *in situ de novo* synthesis of IL-10 in the colon of IL-10 –/– mice.

Abbreviations: α-GalCer – α-galactosylceramide; CD – Crohn's disease; CMV – cytomegalovirus; COX – cyclooxygenase; DNB – dinitrobenzene; DSS – dextran sulphate sodium; IBD – inflammatory bowel disease; IEC – intestinal epithelial cells; iIEL – intraepithelial lymphocytes; IL – interleukin; LPS – lipopolysacharide; MCP-1 – monocyte chemoattractant protein-1; NF-κB – nuclear factor-κB; NK T-cells – natural killer T- cells; SCID – severe combined immune deficient; TCR – T-cell receptor; TGF-β – transforming growth factor-β; Th1 – T-helper1-cell; Th2 – T-helper2-cell; Th3 – T-helper3-cell; TNBS – 2,4,6-trinitrobenzene sulphonic acid; TNF – tumour necrosis factor; Tr1 – T-regulatory1-cell; TTFC – tetanus toxin fragment C; UC – ulcerative colitis

Introduction

Inflammatory bowel disease (IBD) is a group of chronic intestinal inflammatory diseases, which consists of ulcerative colitis (UC), an inflammation of the large intestine, and Crohn's disease (CD), which can affect any part of the gastrointestinal tract. IBD affects one to two in every 1000 individuals in western countries and represents a genuine problem in public health because of the absence of etiological treatment. The clinical image is characterized by recurrent inflammation, often resulting in a chronic, unpredictable course. The symptoms are extremely unpleasant and impact all aspects of quality of life. They include diarrhea, abdominal pain, rectal bleeding, fever, nausea, weight loss, lethargy and loss of appetite. If left untreated, malnutrition, dehydration and anemia follow, which, in extreme cases, can even lead to death. IBD is often found as a complication in spondyloarthropathy (De Keyser et al. 1998), a syndrome of locomotive defects.

The cause of IBD remains unknown. The pathogenesis of CD and UC probably involves interaction between genetic and environmental factors, such as bacterial agents (recently reviewed; Sartor 1997). The main theory, supported by findings in experimental models (Rath et al. 1996; Sellon et al. 1998), is that abnormal immune responses, driven by the intestinal microflora, occur in IBD. Most experimental models for IBD cannot be established in germ-free animals and IL-10 -/- mice show the appearance of mucosal adherent colonic bacteria causative for the development and maintenance of the inflammation (Madsen et al. 2000).

Long term anti-inflammatory medication is often the fate of patients suffering from IBD. Although many patients are managed successfully with conventional medical therapy, such as anti-inflammatory corticosteroid treatment, some stay refractory to treatment, most will have recurrent activity of disease and two-thirds will require surgery. Administered orally or by injection, only a fraction of the active components of most conventional drugs reaches the intended target site, the inflamed intestinal lining. This is not only an inefficient way to deliver drugs, but, more important, means that patients are often subject to a spectrum of unpleasant side effects that result from the high levels of the drugs in other, otherwise healthy tissues and organs of the body. For these reasons researchers have sought to develop methods which allow for more localized delivery of therapeutics.

Novel approaches for the localised delivery of therapeutics

The use of enemas for the localised delivery of conventional chemical therapy has gained interest in recent years (Lang & Peppercorn 1999). This technique involves colonic instillation through a rectally introduced catheter. The pressure applied allows reaching any part of the digestive tract as required. This method, however, is very unpleasant for the patient, especially when the colon is actively inflamed, and is unsuited for delivery in a day by day routine.

Cytokines are small soluble proteins that, together with numerous growth factors and chemokines, provide the means by which cells of the immune system communicate with each other and with most other tissues in the body. As such, these molecules are able to regulate many aspects of the immune response in which numerous cells and tissues may be involved at any one time. It is therefore not surprising that cytokines form the basis of many novel anti-inflammatory therapies. A fascinating aspect is that they can be synthesized at distal sites by recombinant vectors.

Kitani et al. (2000) reported a DNA borne delivery system for regulatory cytokines. They constructed a plasmid that carries the transforming growth factor (TGF)-β gene under control of the cytomegalovirus (CMV) promoter, a promoter widely used in DNA vaccination. A single intranasal dose of the plasmid can prevent the onset or abrogate the establishement of 2,4,6-trinitrobenzene sulphonic acid (TNBS) colitis. The mechanism of action may imply the induction of regulatory – TGF-β secreting – Th3 immune cells. Homing of these cells to the intestine through the routes of the common mucosal immune system may then allow the observed effect in distal mucosa. Most interestingly, neither prevention nor curing was observed in this model when TGF-β protein was administerd through intraperitoneally injection. A drawback of this method may be the high amount of plasmid DNA required.

Exogenous IL-4 can be introduced by gene transfer using a recombinant human type 5 adenovirus (Ad5) vector (Hogaboam et al. 1997). Two injections of Ad5IL-4 into rats caused high expression of IL-4, and inhibited tissue damage in TNBS-induced colitis. A single injection however has no effect. Gene therapy using an adenovirus-IL-10 construct was successful in preventing but not in reversing experimental DNB colitis in rats (Barbara et al. 2000). This method was effective only when applied before or during the acute phase of induction. Moreover, when adenoviral vectors are administerd through the systemic (intravenous or intraperitoneal) route this often leads to high expression of the recombinant gene in the liver and the spleen, which is a highly undesirable lack of organ specificity. Wirtz et al. demonstrated that rectal delivery of an adenoviral vector could lead to localised expression of the marker β-galactosidase (Wirtz et al. 1999). Human trials using local liposome-mediated gene transfer of two antiinflammatory cytokines, IL-4 and IL-10, in patients with severe inflammatory bowel disease of the rectum are planned, the rationale for the simultaneous use of IL-4 and IL-10 being to shift the immune environment in favour of Th2 immune responses (Rogy et al. 2000).

We have investigated the feasibility of delivering cytokines in a localised manner by the action of recombinant *L. lactis*, a non-pathogenic and non-colonising bacterium used mainly in dairy industries. As viable cells are regular constituents of the human diet, their harmless nature is generally accepted. They have accordingly been attributed the GRAS status by the US FDA. For this reason, the organism has been considered a good candidate to serve as a live vaccine delivery vehicle for mucosal immunisation. We have demonstrated the potential of *L. lactis* cells to secrete

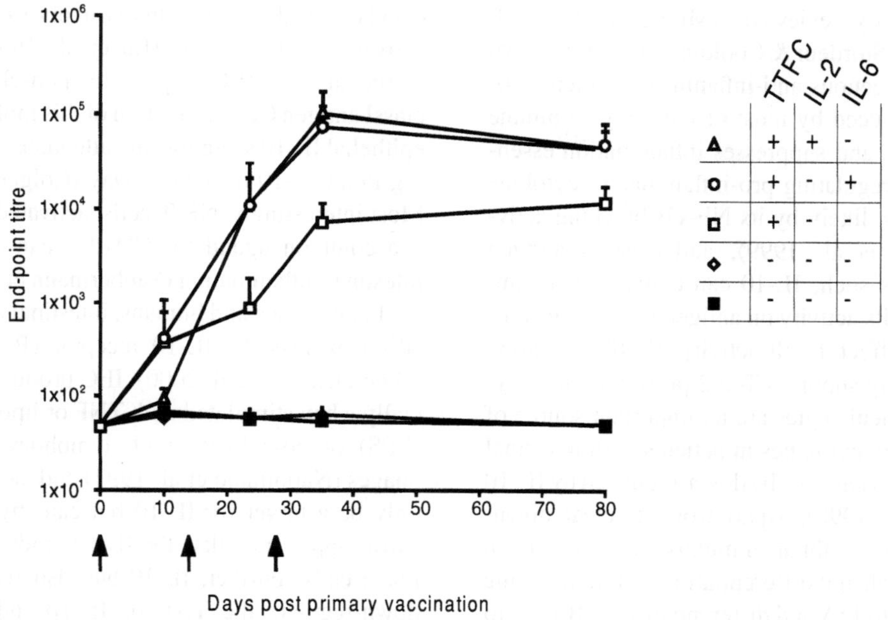

Figure 1. Anti-TTFC Serum IgG levels following immunisation with strains expressing the proteins indicated.

fully biologically active cytokines, such as murine interleukins (mIL) -2, -6 and -10, human interleukins -2, -6 and -10 and murine soluble type 1 and type 2 tumour necrosis factor (TNF) receptors (Steidler et al. 1995, 1996; unpublished data). In the course of the development of these recombinant, cytokine producing *L. lactis* strains we were struck by the apparent ease with which this organism can secrete the abovementioned factors without any apparent inhibition of its growth. Therefore we have taken up the project of developing genetically engineered *L. lactis* strains for the delivery of immune modulating cytokines at mucosal surfaces. Live recombinant microorganisms have also been used for the delivery of single-chain antibodies at the vaginal mucosa to cure rats from experimental *Candida albicans* infections (Beninati et al. 2000).

Delivery of IL-2 or IL-6 at the respiratory mucosa stimulates immunisation

It has been shown that intra-nasal and oral immunisation with recombinant *L. lactis* constitutively expressing tetanus toxin fragment C (TTFC) elicits protective high level immune responses against a challenge with the complete toxin (Robinson et al. 1997). In general, both IL-2 and IL-6 act as potent stimulators in the onset and maintenance of immune reactions. To investigate whether the mucosal and systemic responses to TTFC can be enhanced or modulated by these cytokines we have constructed strains of *L. lactis* that produce TTFC intracellularly and secrete functional murine IL-2 or IL-6. This was done by engineering recombinant operons, which drive both the expression of TTFC and of the desired cytokine. Groups of six female Balb/c mice were immunised with these recombinant strains as well as with the relevant control strains. Immunisations were performed for three times, with intermediate 2-week spacing. Anti-TTFC serum antibody responses were significantly higher, up to 15-fold, in mice immunised intra-nasally with those strains *of L. lactis* secreting IL-2 or IL-6 (Figure 1). Also, the concentration of serum IgA reactive with TTFC was considerably higher after immunisation with the IL-6 secreting strain (Steidler et al. 1998). This reflects the biological properties of IL-6 as a B-cell growth and IgA secretion-stimulating factor.

Interleukin-10, a key factor in the down regulation of immunity at the intestinal mucosa

As it persistently faces up to the discrimination between 'harmless' antigen – food or enteric bacteria – and pathogenic challenge, the intestine is a site of immune activation that is critically controlled for the non-inflammatory state. Any misinterpretation may lead to either invasion of enteric pathogens or hyperreaction to the normal luminal content.

IL-10 (recently reviewed: (Moore et al. 1993; Schreiber 1997; Stordeur & Cooldmay 1998) is known as a major endogenous anti-inflammatory intercessor and can be produced by most of the body's immune cells. It controls and suppresses inflammation essentially by down-regulating pro-inflammatory cytokine production, most likely by its NF-κB blocking activity (Schottelius et al. 1999), and class II antigen presentation. As such, IL-10 can counteract inflammation through its activity on antigen presenting cells, which in turn affect T-cell activity. IL-10 can, however, also directly suppress T-cell proliferation. Polymorphonuclear leukocytes are an important source of pro-inflammatory cytokines in patients with intestinal inflammation and can also be down-regulated by IL-10 (Nikolaus et al. 1998). Apart from its clear down-regulatory role on cellular immunity, IL-10 is known as a B-cell growth and differentiation factor, involving class switching to IgA and differentiation of B-cells to antibody producing plasma cells. This is an especially interesting feature in view of the observation that IgA antibodies that react with the commensal microflora provide a primitive, T-cell independent mechanism of tolerance (Macpherson et al. 2000).

Mice that cannot make IL-10 spontaneously develop IBD (Kuhn et al. 1993), suggesting an important role for this natural anti-inflammatory in protection against intestinal inflammation. This was further confirmed by the finding that, following the induction of dextran sulphate sodium (DSS) colitis and the increase in TNF and IL-1β there is a marked increase of IL-10. Neutralisation of IL-10 showed that its increased synthesis was an essential component for the down-regulation of inflammation (Tomoyose et al. 1998). Further, abundant model studies have shown the efficacy of IL-10 for the curing of enterocolitis. This has lead to the knowledge that IL-10 has a very distinct role in the surveillance of the intestinal mucosa, especially for maintenance of tolerance and non-T-cell-mediated immunity. This is well illustrated by the finding that lamina propria T-cells secrete significantly higher amounts of IL-10 than periferal blood T-cells upon stimulation (Braunstein et al. 1997).

The intestinal epithelial lining is a principal barrier against the uncontrolled influx of luminal antigen in the mucosa. It is therefore not surprising that intrusion of the epithelium is a key aspect in the development of IBD. IL-10 is involved in the maintenance of epithelial barrier integrity as IL-10 gene-deficient mice show an increase in ileal and colonic permeability prior to the development of mucosal inflammation (Madsen et al. 1999). This is likely to be caused by activated immune cells (Lu et al. 1998). Intestinal epithelial cells (IEC) express the non-classical MHC-classI antigen CD1d. Ligation of this molecule induces epithelial IL-10 secretion and attenuates IFN-γ signaling in an IL-10-dependent way (Colgan et al. 1999). Most interestingly, NK T-cells, activated by α-GalCer – a common ligand for CD1d – can down-regulate intestinal inflammation (Saubermann et al. 2000).

In both mice and humans, intestinal epithelial cells (IEC) express the IL-10 receptor (Bourreille et al. 1999; Denning et al. 2000). IEC produce IL-10, especially when stimulated with TNF or lipopolysacharide (LPS) or co-cultured with lymphocytes or macrophages (Napolitano et al. 1997) and can therefore not only be a target for IL-10 but can, by providing an extra supply, modulate the IL-10 production from immune cells. Further, IL-10 but also IL-4, and IL-13 down-regulate the TNF or IL-1β induced production of the monocyte chemotactant protein-1 (MCP-1) from human IEC (Kucharzik et al. 1998).

The intestinal epithelium is intercalated with T-cells, the so-called intestinal intraepithelial lymphocytes (iIEL) which can be stimulated *in vitro* to express IL-10 (Lundqvist et al. 1996). The iIEL are mainly CD8+ and these cells serve to lyse stressed IEC and to provide anti-inflammatory signals. A considerable part of iIEL are $\gamma\delta$ T-cells. These regulatory $\gamma\delta$ T-cells can secrete interleukin-10. Outside the gut, $\gamma\delta$ T-cells mediate IL-10 dependant prevention of death in acute infection models. This attributes a role to $\gamma\delta$ T-cells in the control of Th1 responses (Hsieh et al. 1996). Low doses of oral antigen induce antigen-specific regulatory T-cells in the gut, which act by releasing inhibitory cytokines such as TGF-β, interleukin-4, and interleukin-10 at the target organ (Weiner 1997). As low dose oral tolerance is defective in TCR δ –/– mice and in animals treated with anti-TCR $\gamma\delta$, $\gamma\delta$ T-cells present in the gut play a significant immunoregulatory role in IL-10-mediated, low-dose oral tolerance induction (Fujihashi et al. 1999).

IL-10 does not solely display inhibitory effects on T-cells but also up-regulates the IL-2 receptor and associated IL-2-dependent proliferation (Cohen et al. 1994) and exerts positive influences on clonal expansion of subsets of preactivated T-cells (Pawelec et al. 1995). Groux et al. (1997) showed that chronic activation of both human and murine CD4+ T-cells in the presence of IL-10 gives rise to so-called Tr1 cells, CD4+ T-cell clones with low proliferative capacity, producing high levels of IL-10, low levels of IL-2 and no IL-4. Co-transfer of these cells with CD4+CD45Rbhi T-cells to severe combined immune deficient (SCID, containing neither B- nor T-cells) mice prevents the onset of colitis. IL-10 can therefore be called a growth factor of Tr1 cells, an IL-10 producing cell type capable of preventing intestinal inflammation (Asseman & Powrie 1998). A similar, TGF-β- and IL-10- dependent feature resides in CD45Rblo T-cells, cells capable of preventing the onset of colitis in the SCID CD4+CD45Rbhi transfer colitis model (Asseman et al. 1999). The down-regulatory effects of Tr1 cells appear not restricted to Th1 but also affect Th2 driven immunity (Cottrez et al. 2000). Recently Tr1 cells have been isolated from humans (Stephens et al. 2001).

Infringement of tolerance towards commensal microflora may be the driving force behind IBD. The absence of tolerance to the indigenous microflora also appears in TNBS induced colitis. Tolerance can however be restored upon treatment with IL-10. This treatment leaves immune reactivity towards heterologous bacterial antigen unaffected (Duchmann et al. 1996). It may be interesting here to bear in mind that IEC may act as antigen-presenting cells (Bland & Warren 1986; Brandeis et al. 1994), that IL-10 blocked the IFN-γ-induced expression of class II MHC molecules on IEC (Denning et al. 2000) and, when combined with TGF-β, reduced HLA-DR expression (Zimmerman et al. 1998). Staphylococcal enterotoxin B can break self-tolerance at the intestinal epithelium. IL-10 can neutralize this by preventing the activation of T-cells that contribute to epithelial cell damage (Ito et al. 2000). In humans, the activation and proliferation of CD8+ T-cells from the epithelium and lamina propria by the lectin phytohaemagglutinin can be blocked by IL-10. The IL-2-induced cytotoxicity reported in this context may help to maintain host defence (Ebert 2000).

Prostaglandins are a group of pro-inflammatory mediators of particular interest in intestinal inflammation. IL-10 and IL-4 inhibit cyclooxygenase (COX)-2, the inducible part of the COX pathway for prostaglandin synthesis (Niiro et al. 1995). This is reflected in the early up regulation of COX-2 in IL-10 –/– mice (Shattuck-Brandt et al. 2000).

A number of cytokines and certain groups of immune cells have been identified as key propagators of IBD. It is however imperative to note that neither the immune cells infiltrating the mucosa nor the pro-inflammatory cytokines produced by them during the course of inflammation cause tissue disturbance as observed in IBD. By digesting the extracellular matrix, matrix metalloproteinases are the eventual principal mediators of tissue architecture disturbance in the inflamed intestine (MacDonald et al. 1999). IL-10 down-regulates the expression of this class of Zn-dependent endopeptidases (Pender et al. 1998).

Topical IL-10 enema treatment of patients with ulcerative colitis unveiled that IL-10 is effective in down-regulating pro-inflammatory cytokine synthesis from IBD monocytes/lymphocytes both *in vitro* and *in vivo* (Schreiber et al. 1995). In this study, equal concentrations of IL-10 appear detectable in both normal and IBD intestinal lamina propria biopsy homogenates. This seems surprising in view of other results (Gasche et al. 2000) but Autschbach et al. 1998) also argue against a general deficiency in IL-10 production in IBD. They in fact observed an increase in IL-10 producing cells and rather suggest that in IBD the production of IL-10 is dislocated and insufficient to down-regulate pro-inflammatory cytokines in the lamina propria compartment.

Delivery of IL-10 cures and prevents enterocolitis

With regard to its manifest role in establishing and maintaining tolerance, as demonstrated by the overwhelming evidence presented above, IL-10 is a good candidate as therapy in the treatment of IBD and has in fact been demonstrated effective in clinical trials (van Deventer et al. 1997). When, however, administered by injection, the high levels of IL-10 that are distributed throughout the body can lead to a number of side effects and so IL-10 is no better in this respect than conventional synthetic immunosuppressive therapies.

We have investigated the feasibility of delivering this anti-inflammatory protein in a localised manner by the action of recombinant *L. lactis* (Steidler et al. 2000). We constructed a recombinant *L. lactis* that secretes mIL-10 (Schotte et al. 2000) by joining coding sequences for the usp45 secretion leader (van Asseldonk et al. 1990) and for mature mIL-10 (Thompson-Snipes et al. 1991). This construct was placed down-stream from the lactococcal P1 promotor (Waterfield et al. 1995), as present in the pTREX1 expression vector (Wells & Selafield 1996). Growth of this recombinant strain in adequately buffered culture medium leads to the accumulation of mIL-10 up to a level of 1 mg ml^{-1} culture. The recombinant mIL-10 was correctly processed and displayed full biological activity.

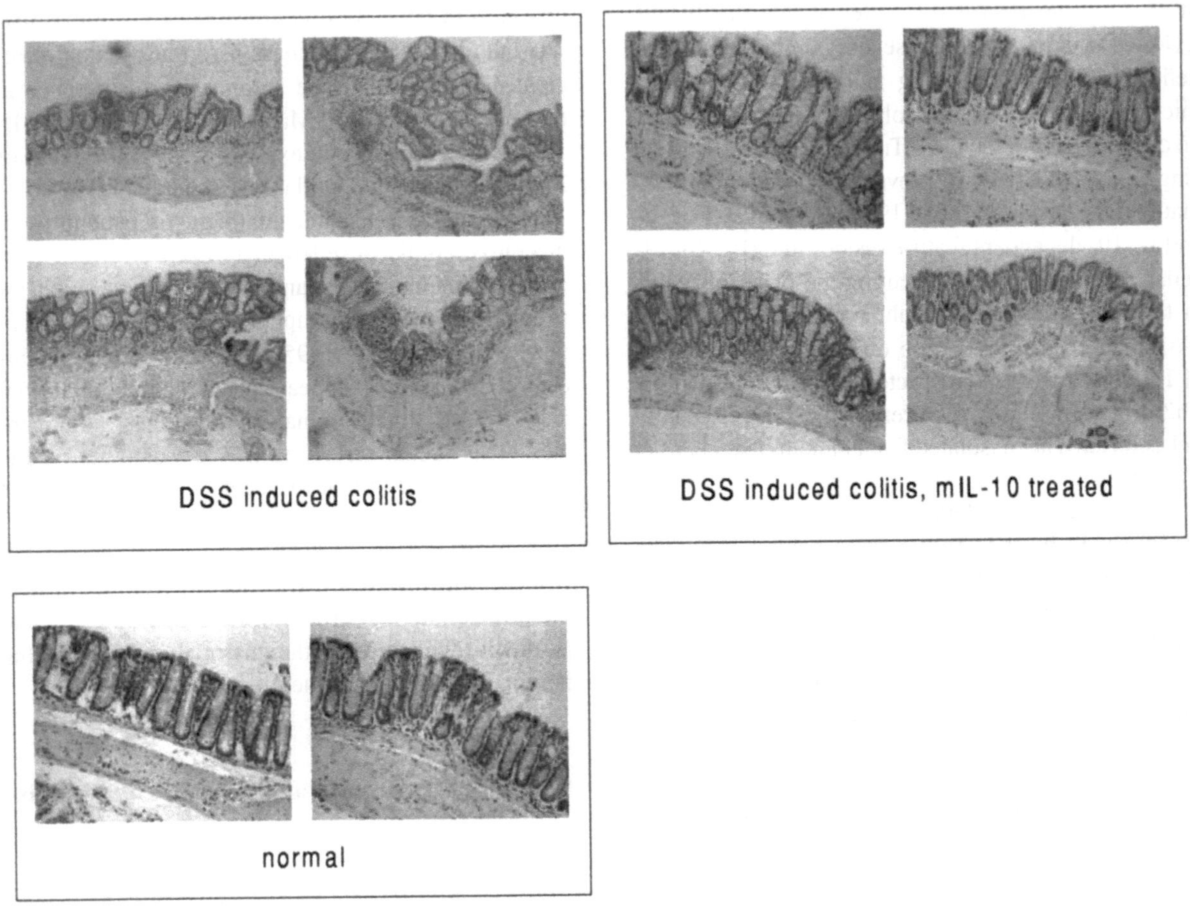

Figure 2. Haematoxylin and eosin-stained histological slides from colon tissues. The pictures show representative images of untreated mice (normal), mice in which colitis has been induced by the addition of DSS to the drinking water and mice which have undergone the same induction followed by a treatment with a mIL-10 producing *L. lactis* strain.

Oral inoculation of this strain leads to 50% curing of DSS-induced colitis and completely prevents the onset of colitis in IL-10 –/– mice (Steidler et al. 2000). The repeated addition of DSS to the drinking water of Balb/c mice leads to the induction of chronic colitis (Okayasu et al. 1990). The histology of the colon is characterised by a gradual decrease of the number of goblet cells in the epithelial layer, disturbance of crypt architecture, infiltration of lymphocytes in the lamina propria and in the submucosa, thickening of the mucosa, appearance of crypt abscesses, swollen lymph follicles and ulceration (Figure 2). Occasionally we have observed the formation of polyps, adenoma and squamous metaplasia. Inflammation in the colon can be rated by notching up the appearance of the above-mentioned phenomena in blinded histological slides. Typically, a mean score of 5 was recorded on a scale ranging between 0 and 8. The inflammation is further associated with the upregulation of several pro-inflammatory cytokines such as IFNγ, IL-12 and TNF and can be cured by the systemic administration of neutralising antibodies towards these proteins. Recombinant IL-10 is very acid sensitive and is therefore rapidly degraded when given orally. However, when given systemically, 5 consecutive doses of 5 μg daily will substantially decrease the inflammation. The daily ingestion, for 2 weeks, of mIL-10 producing *L. lactis* cells results in a histological score of approximately 1 in 40% of the treated mice, which is a status equal to that of healthy control mice. Most other animals only showed minor patchy remnants of the inflammation (Figure 2). Killing of the IL-10 producing bacteria by UV radiation resulted in an immediate block of the mIL-10 accumulation in the culture, stagnation of the increase in turbidity and 6 log decrease in viable cell count. Such killing of the bacteria led to the abrogation of the therapeutic effects on enterocolitis reported here. Therefore these can be attributed strictly to the active *in vivo* delivery of IL-10. The degree of the observed healing is comparable to systemic treatment with prominent anti-inflammatory drugs such as dexamethasone, neutralizing antibodies to IL-12 or mIL-10 but the amount of IL-10 required to achieve this effect is 10 000 fold higher than when delivered through *in situ* synthesis by *L. lactis*. This shows that a significantly improved delivery had been achieved. Reduction of the required amount of therapeutic is a major objective in the search for methods that bear less risk to side effects.

Daily treatment with IL-10 producing *L. lactis* leads to the prevention of colitis, normally associated with the IL-10 –/– genotype in 129 Sv/Ev mice. When left untreated, 129 Sv/Ev IL10 –/– mice developed colitis from week 3 on, resulting in a mean histological score of 6 at week 7. Littermates treated from week 3 on, stalled at a histological score of approximately 1.5 points.

Discussion and conclusions

Genetically modified *L. lactis* that secrete cytokines can be used to modulate immune responses. We demonstrate that *L. lactis* can deliver both IL-2 and IL-6 at the respiratory mucosa in quantities, which substantially enhance immune responses to a co-expressed antigen. Furthermore we have shown the possibility to deliver IL-10 at the mucosa of the colon following intragastric inoculation of suitably engineered *L. lactis*. The observed curing of established chronic enterocolitis is at least as effective as prominent anti-inflammatory therapies. We were able to reduce the amount of therapeutic required by four orders of magnitude. The delivery of cytokines through this technology is strictly dependent on the viability of the bacteria, from which we conclude that engineered *L. lactis* actively synthesize and secrete cytokines *in situ*. We have confirmed this by measurements of *de novo* IL-10 synthesis in the colon.

Conventional anti-inflammatory and immunosuppressive therapies used for the treatment of inflammatory bowel diseases are, in general, effectual. Some patients however, particularly in the case of Crohn's disease, remain refractory to such treatments. Often, the side effects of the anti-inflammatories and immunosuppressives used impact quality of life to the same degree as the symptoms of the disease. Thus, the development of improved or novel therapeutics for the treatment of inflammatory bowel disease is a priority. As we learn more about the pathophysiology of Crohn's disease and ulcerative colitis, the central role of cytokines therein becomes increasingly apparent. Therapeutic intervention by targeting these soluble inflammatory factors to redress the imbalance of inflammatory and regulatory cytokines and to restore immune homeostasis in the gut wall is therefore an obvious choice. In a number of animal models and in some human studies, blocking the bioactivity of the inflammatory factors by administering neutralising anti-cytokine antibodies, anti-cytokine receptor antibodies or soluble forms of the cytokine receptor has proved to be beneficial. So too has the administration of recombinant regulatory cytokines. Cytokines are powerful, multi-potent, pleiotropic agents, acting to regulate host immunity as components of a complex communicatory network. Interference with individual component of this process may, while having acute beneficial clinical effects, result in deleterious effects from its down-stream actions on other immune processes. It is of paramount importance to take these factors into consideration when evaluating efficacy and safety data from longer-term animal studies and clinical trials.

Cytokine, chemokine and growth factor-based therapies may provide alternative or complementary option for the treatment of inflammatory bowel disease. Targeting these therapeutics by *in situ* synthesis through recombinant vectors may be a means of treating such conditions safely, effectively and without the complications and side effects of conventional therapies, overall, improving the quality of life for those affected with inflammatory bowel disease.

References

Asseman C, Mauze S, Leach MW, Coffman RL & Powrie F (1999) An essential role for interleukin 10 in the function of regulatory T cells that inhibit intestinal inflammation. J. Exp. Med. 190: 995–1004.

Asseman C & Powrie F (1998) Interleukin 10 is a growth factor for a population of regulatory T cells. Gut 42: 157–158.

Autschbach F, Braunstein J, Helmke B, Zuna I, Schurmann G, Niemir ZI, Wallich R, Otto HF & Meuer SC. (1998) In situ expression of interleukin-10 in noninflamed human gut and in inflammatory bowel disease. Am. J. Pathol. 153: 121–130.

Barbara G, Xing Z, Hogaboam CM, Gauldie J & Collins SM (2000) Interleukin 10 gene transfer prevents experimental colitis in rats. Gut 46: 344–349.

Beninati C, Oggioni MR, Boccanera M, Spinosa MR, Maggi T, Conti S, Magliani W, De Bernardis F, Teti G, Cassone A, Pozzi G & Polonelli L (2000) Therapy of mucosal candidiasis by expression of an anti-idiotype in human commensal bacteria [In Process Citation]. Nat. Biotechnol. 18: 1060–1064.

Bland PW & Warren LG (1986) Antigen presentation by epithelial cells of the rat small intestine. II. Selective induction of suppressor T cells. Immunology 58: 9–14.

Bourreille A, Segain JP, Raingeard de la Bletiere D, Siavoshian S, Vallette G, Galmiche JP & Blottiere HM (1999) Lack of interleukin 10 regulation of antigen presentation-associated molecules expressed on colonic epithelial cells. Eur. J. Clin. Invest. 29: 48–55.

Brandeis JM, Sayegh MH, Gallon L, Blumberg RS & Carpenter CB (1994) Rat intestinal epithelial cells present major histocompatibility complex allopeptides to primed T cells. Gastroenterology 107: 1537–1542.

Braunstein J, Qiao L, Autschbach F, Schurmann G & Meuer S (1997) T cells of the human intestinal lamina propria are high producers of interleukin-10 [see comments]. Gut 41: 215–220.

Cohen SB, Katsikis PD, Feldmann M & Londei M (1994) IL-10 enhances expression of the IL-2 receptor alpha chain on T cells. Immunology 83: 329–332.

Colgan SP, Hershberg RM, Furuta GT & Blumberg RS (1999) Ligation of intestinal epithelial CD1d induces bioactive IL-10: critical role of the cytoplasmic tail in autocrine signaling. Proc. Natl. Acad. Sci. U.S.A. 96: 13938–13943.

Cottrez F, Hurst SD, Coffman RL & Groux H (2000) T regulatory cells 1 inhibit a Th2-specific response in vivo. J. Immunol. 165: 4848–4853.

De Keyser F, Elewaut D, De Vos M, De Vlam K, Cuvelier C, Mielants H & Veys EM (1998) Bowel inflammation and the spondyloarthropathies. Rheum. Dis. Clin. North Am. 24: 785–813, ix-x.

Denning TL, Campbell NA, Song F, Garofalo RP, Klimpel GR, Reyes VE & Ernst PB (2000) Expression of IL-10 receptors on epithelial cells from the murine small and large intestine. Int. Immunol. 12: 133–139.

Duchmann R, Schmitt E, Knolle P, Meyer zum Buschenfelde KH & Neurath M (1996) Tolerance towards resident intestinal flora in mice is abrogated in experimental colitis and restored by treatment with interleukin-10 or antibodies to interleukin-12. Eur. J. Immunol. 26: 934–938.

Ebert EC (2000) IL-10 enhances IL-2-induced proliferation and cytotoxicity by human intestinal lymphocytes. Clin. Exp. Immunol. 119: 426–432.

Fujihashi K, Dohi T, Kweon MN, McGhee JR, Koga T, Cooper MD, Tonegawa S & Kiyono H (1999) gammadelta T cells regulate mucosally induced tolerance in a dose- dependent fashion. Int. Immunol. 11: 1907–1916.

Gasche C, Bakos S, Dejaco C, Tillinger W, Zakeri S & Reinisch W (2000) IL-10 secretion and sensitivity in normal human intestine and inflammatory bowel disease [In Process Citation]. J. Clin. Immunol. 20: 362–370.

19. Groux H, O'Garra A, Bigler M, Rouleau M, Antonenko S, de Vries JE & Roncarolo MG (1997) A CD4+ T-cell subset inhibits antigen-specific T-cell responses and prevents colitis. Nature 389: 737–742.

Hogaboam CM, Vallance BA, Kumar A, Addison CL, Graham FL, Gauldie J. & Collins SM (1997) Therapeutic effects of interleukin-4 gene transfer in experimental inflammatory bowel disease. J. Clin. Invest. 100: 2766–2776.

Hsieh B, Schrenzel MD, Mulvania T, Lepper HD, DiMolfetto-Landon L & Ferrick DA (1996) In vivo cytokine production in murine listeriosis. Evidence for immunoregulation by gamma delta+ T cells. J. Immunol. 156: 232–237.

Ito K, Takaishi H, Jin Y, Song F, Denning TL & Ernst PB (2000) Staphylococcal enterotoxin B stimulates expansion of autoreactive T cells that induce apoptosis in intestinal epithelial cells: regulation of autoreactive responses by IL-10. J. Immunol. 164: 2994–3001.

Kitani A, Fuss IJ, Nakamura K, Schwartz OM, Usui T & Strober W. (2000) Treatment of experimental (trinitrobenzene sulfonic acid) colitis by intranasal administration of transforming growth factor (TGF)-beta1 plasmid: TGF-beta1-mediated suppression of T helper cell type 1 response occurs by interleukin (IL)-10 induction and IL-12 receptor beta2 chain down-regulation. J. Exp. Med. 192: 41–52.

Kucharzik T, Lugering N, Pauels HG, Domschke W & Stoll R (1998) IL-4, IL-10 and IL-13 down-regulate monocyte-chemoattracting protein-1 (MCP-1) production in activated intestinal epithelial cells. Clin. Exp. Immunol. 111: 152–157.

Kuhn R, Lohler J, Rennick D, Rajewsky K & Muller W (1993) Interleukin-10-deficient mice develop chronic enterocolitis [see comments]. Cell 75: 263–274.

Lang KA & Peppercorn MA (1999) Promising new agents for the treatment of inflammatory bowel disorders. Drugs R D 1: 237–244.

Lu J, Philpott DJ, Saunders PR, Perdue MH, Yang PC & McKay DM (1998) Epithelial ion transport and barrier abnormalities evoked by superantigen-activated immune cells are inhibited by interleukin-10 but not interleukin-4. J. Pharmacol. Exp. Ther. 287: 128–136.

Lundqvist C, Melgar S, Yeung Mm, Hammarstrom S & Hammarstrom ML. (1996) Intraepithelial lymphocytes in human gut have lytic potential and a cytokine profile that suggest T helper 1 and cytotoxic functions. J. Immunol. 157: 1926–1934.

MacDonald TT, Bajaj-Elliott M & Pender SL (1999) T cells orchestrate intestinal mucosal shape and integrity. Immunol. Today 20: 505–510.

Macpherson AJ, Gatto D, Sainsbury E, Harriman GR, Hengartner H & Zinkernagel RM (2000) A primitive T cell-independent mechanism of intestinal mucosal IgA responses to commensal bacteria. Science 288: 2222–2226.

Madsen KL, Malfair D, Gray D, Doyle JS, Jewell LD & Fedorak RN (1999) Interleukin-10 gene-deficient mice develop a primary intestinal permeability defect in response to enteric microflora. Inflamm. Bowel Dis. 5: 262–270.

Madsen KL, Doyle JS, Tavernini Mm, Jewell LD, Rennie RP & Fedorak RN. (2000) Antibiotic therapy attenuates colitis in interleukin 10 gene-deficient mice. Gastroenterology 118: 1094–1105.

Moore KW, O'Garra A, de Waal Malefyt R, Vieira P & Mosmann TR (1993) Interleukin-10. Annu. Rev. Immunol. 11: 165–190.

Napolitano LM, Buzdon Mm, Shi HJ & Bass BL (1997) Intestinal epithelial cell regulation of macrophage and lymphocyte interleukin 10 expression. Arch. Surg. 132: 1271–1276.

Niiro H, Otsuka T, Tanabe T, Hara S, Kuga S, Nemoto Y, Tanaka Y, Nakashima H, Kitajima S, Abe M, et al. (1995) Inhibition by interleukin-10 of inducible cyclooxygenase expression in lipopolysaccharide-stimulated monocytes: its underlying mechanism in comparison with interleukin-4. Blood 85: 3736–3745.

Nikolaus S, Bauditz J, Gionchetti P, Witt C, Lochs H & Schreiber S (1998) Increased secretion of pro-inflammatory cytokines by circulating polymorphonuclear neutrophils and regulation by interleukin 10 during intestinal inflammation. Gut 42: 470–476.

Okayasu I, Hatakeyama S, Yamada M, Ohkusa T, Inagaki Y & Nakaya R. (1990) A novel method in the induction of reliable experimental acute and chronic ulcerative colitis in mice. Gastroenterology 98: 694–702.

Pawelec G, Pohla H, Schlotz E, Rehbein A, Schaudt K, Hambrecht A & Friccius H (1995) Interleukin 10 is a human T cell growth factor in vitro. Cytokine 7: 355–363.

Pender SL, Breese EJ, Gunther U, Howie D, Wathen NC, Schuppan D & MacDonald TT (1998) Suppression of T cell-mediated injury in human gut by interleukin 10: role of matrix metalloproteinases. Gastroenterology 115: 573–583.

Rath HC, Herfarth HH, Ikeda JS, Grenther WB, Hamm TE Jr, Balish E, Taurog JD, Hammer RE, Wilson KH & Sartor RB (1996) Normal luminal bacteria, especially Bacteroides species, mediate chronic colitis, gastritis, and arthritis in HLA-B27/human beta2 microglobulin transgenic rats. J. Clin. Invest. 98: 945–953.

Robinson K, Chamberlain LM, Schofield KM, Wells JM & Le Page RW. (1997) Oral vaccination of mice against tetanus with recombinant *Lactococcus lactis*. Nat. Biotechnol. 15: 653–657.

Rogy MA, Beinhauer BG, Reinisch W, Huang L & Pokieser P (2000) Transfer of interleukin-4 and interleukin-10 in patients with severe inflammatory bowel disease of the rectum. Hum. Gene. Ther. 11: 1731–1741.

Sartor RB (1997) Pathogenesis and immune mechanisms of chronic inflammatory bowel diseases. Am. J. Gastroenterol. 92: 5S–11S.

Saubermann LJ, Beck P, De Jong YP, Pitman RS, Ryan MS, Kim HS, Exley M, Snapper S, Balk SP, Hagen SJ, Kanauchi O, Motoki K, Sakai T, Terhorst C, Koezuka Y, Podolsky DK & Blumberg RS (2000) Activation of natural killer T cells by alpha-galactosylceramide in the presence of CD1d provides protection against colitis in mice. Gastroenterology 119: 119–128.

Schotte L, Steidler L, Vandekerckhove J & Remaut E (2000) Secretion of biologically active murine interleukin-10 by Lactococcus lactis. Enzyme Microb. Technol. 27: 761–765.

Schottelius AJ, Mayo MW, Sartor RB & Baldwin AS Jr (1999) Interleukin-10 signaling blocks inhibitor of kappaB kinase activity and nuclear factor kappaB DNA binding. J. Biol. Chem. 274: 31868–31874.

Schreiber S (1997) Interleukin-10 in the intestine [comment]. Gut 41: 274–275.

Schreiber S, Heinig T, Thiele HG & Raedler A (1995) Immunoregulatory role of interleukin 10 in patients with inflammatory bowel disease. Gastroenterology 108: 1434–1444.

Sellon RK, Tonkonogy S, Schultz M, Dieleman LA, Grenther W, Balish E, Rennick DM & Sartor RB (1998) Resident enteric bacteria are necessary for development of spontaneous colitis and immune system activation in interleukin-10-deficient mice. Infect. Immun. 66: 5224–5231.

Shattuck-Brandt RL, Varilek GW, Radhika A, Yang F, Washington MK & DuBois RN (2000) Cyclooxygenase 2 expression is increased in the stroma of colon carcinomas from IL-10 (–/–) mice [see comments]. Gastroenterology 118: 337–345.

Steidler L, Fiers W & Remaut E (1996). In: Metabolism and Application of Lactic Acid Bacteria. Ray B & Bozoglu F (Eds) (pp 63–79). (Springer, Berlin, Heidelberg New York.

Steidler L, Hans W, Schotte L, Neirynck S, Obermeier F, Falk W, Fiers W & Remaut E (2000) Treatment of murine colitis by Lactococcus lactis secreting interleukin-10 [see comments]. Science 289: 1352–1355.

Steidler L, Robinson K, Chamberlain L, Schofield KM, Remaut E, Le Page RW & Wells JM (1998) Mucosal delivery of murine interleukin-2 (IL-2) and IL-6 by recombinant strains of Lactococcus lactis coexpressing antigen and cytokine. Infect. Immun. 66: 3183–3189.

Steidler L, Wells JM, Raeymaekers A, Vandekerckhove J, Fiers W & Remaut E. (1995) Secretion of biologically active murine interleukin-2 by Lactococcus lactis subsp. lactis. Appl. Environ. Microbiol. 61: 1627–1629.

Stephens LA, Mottet C, Mason D & Powrie F (2001) Human CD4(+)CD25(+) thymocytes and peripheral T cells have immune suppressive activity in vitro. Eur. J. Immunol. 31: 1247–1254.

Stordeur P & Goldman M (1998) Interleukin-10 as a regulatory cytokine induced by cellular stress: molecular aspects. Int. Rev. Immunol. 16: 501–522.

Thompson-Snipes L, Dhar V, Bond MW, Mosmann TR, Moore KW & Rennick DM (1991) Interleukin 10: a novel stimulatory factor for mast cells and their progenitors. J. Exp. Med. 173: 507–510.

Tomoyose M, Mitsuyama K, Ishida H, Toyonaga A & Tanikawa K (1998) Role of interleukin-10 in a murine model of dextran sulfate sodium-induced colitis. Scand. J. Gastroenterol. 33: 435–440.

van Asseldonk M, Rutten G, Oteman M, Siezen RJ, de Vos WM & Simons G (1990) Cloning of usp45, a gene encoding a secreted protein from *Lactococcus lactis* subsp. lactis Mg1363. Gene 95: 155–160.

van Deventer SJ, Elson CO & Fedorak RN (1997) Multiple doses of intravenous interleukin 10 in steroid-refractory Crohn's disease. Crohn's Disease Study Group. Gastroenterology 113: 383–389.

Waterfield NR, Le Page RW, Wilson PW & Wells JM (1995) The isolation of lactococcal promoters and their use in investigating bacterial luciferase synthesis in *Lactococcus lactis*. Gene 165: 9–15.

Weiner HL (1997) Oral tolerance for the treatment of autoimmune diseases. Annu. Rev. Med. 48: 341–351.

Wells JM & Schofield KM (1996) Cloning and expression vectors for Lactococci. In: Nato ASI Series H 98: 37–62.

Wirtz S, Galle PR & Neurath MF (1999) Efficient gene delivery to the inflamed colon by local administration of recombinant adenoviruses with normal or modified fibre structure. Gut 44: 800–807.

Zimmerman MJ, Radford-Smith GR & Jewell DP (1998) The effect of interleukin-10 and transforming growth factor beta-1 on HLA-DR expression in colonic epithelial cells. Mediators Inflamm. 7: 7–11.

Antonie van Leeuwenhoek **82:** 333–340, 2002.

Anti-hypertensive activity of fermented dairy products containing biogenic peptides

Dr. T. Takano
Frontiers Laboratory, Calpis Co. Ltd, R&D Center, 11-10, 5-chome Fuchinobe, Sagamihara-shi, Kanagawa 229-0006, Japan (E-mail: toshiaki.takano@calpis.co.jp)

Key words: biogenics, ACE-angiotensin-I converting enzyme, anti-hypertension, lactic acid bacteria

Abstract

A considerable amount of research has been conducted to study the health benefits of fermented milks. One possible mechanism for the beneficial effects is the involvement of biogenic compounds, which are produced by the action of microorganisms and make the products directly beneficial without the need for live bacteria. Research on the health effects of a pasteurized sour milk, which is fermented by a starter culture containing *Lactobacillus helveticus* as the predominant microorganism, indicated clear evidence for the involvement of biogenic compounds. In this paper, beneficial effects of sour milk are reviewed with special attention to the effect on anti-hypertension and angiotensin-I converting enzyme inhibitory peptides identified in sour milk.

Introduction

Fermented dairy products are made from the milk of domesticated mammals through the action of microorganisms, mainly lactic acid bacteria. Fermented milk has been consumed throughout the world for over thousands of years as valuable preserved foods and their intermediate products. Fermented milk has also been recognized to be beneficial for human health.

Since the beginning of 20th century, industrial products of fermented milk became popular in many countries. Current interest in the health benefits of fermented milk is believed to derive from the ideas of Metchnikoff based on his observations of lactic acid bacteria in the human digestive tract in the early 20th century. His hypothesis proposed that people who regularly consume fermented milk live longer, as the ingested lactic acid bacteria (LAB) colonize the intestine and inhibit putrefaction caused by harmful bacteria, thus slowing the aging process. One group of modern products utilizes microorganisms isolated from traditional products, for example *Lactobacillus delbrueckii* subsp. *bulgaricus* and *Streptococcus thermophilus*, *L. helveticus*, etc. However, studies following the work of Metchnikoff proved that the LAB from yogurt do not colonize the intestinal tract. Based on the idea that LAB originating from the intestinal tract may colonize, many experiments have been undertaken, focusing initially on *L. acidophilus* or bifidobacteria as advocated from the 1920s.

Since the 1960s, scientific studies on the health benefits of both types of products, using LAB from traditional products and from intestines, have been conducted by many groups in the world (reviewed by Rasic & Kurmann 1978). An early research objective was to determine if consumption of fermented milk relative to unfermented milk was nutritionally beneficial. Improvement of nutritional composition and growth efficiency of experimantal animals, as well as alleviation of lactose intolerance was reported. More recently, scientific studies have examined the physiologic mechanisms for the putative health benefits, including inhibition of infections, inhibition of tumors, regulation of immune function, reduction of serum cholesterol, and reduction of hypertension.

In theory, microorganisms can exert beneficial effects through two mechanisms: direct effects of the live microbial cells or indirect effects via metabolites of these cells (Mitsuoka 1976). The former are called 'probiotics' (Fuller 1989). Much research has been done on bacterial strains with the capacity to improve the intestinal microflora, immune function, etc. For the latter, the term 'biogenics' has been proposed (Mitsuoka 2000). 'Biogenics' are defined as food components derived through microbial activity which provide health benefits without involving intestinal microflora. The most important biogenics in fermented milk may be peptides which are not present prior to fermentation.

This paper correlates the anti-hypertensive activity of a specific fermented milk product called 'Sour milk' with biogenic peptides isolated from it.

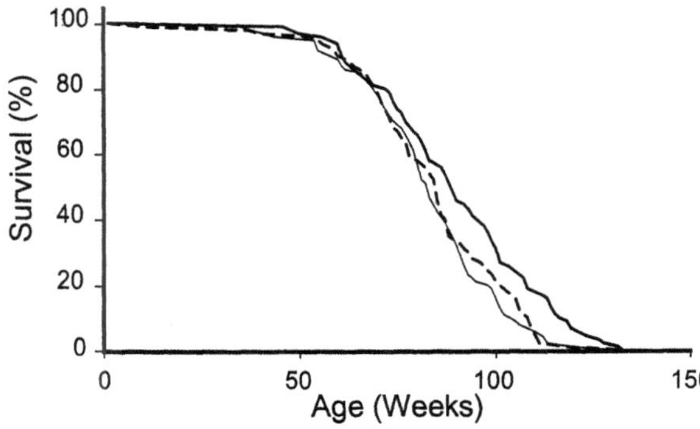

Figure 1. Changes in the survival ratio of mice administered with Sour milk (—), whole milk (- - -) or control diet (—) for their whole life (Arai et al. 1980).

Milk proteins as sources of biogenics

The main chemical change in milk during fermentation by lactic acid bacteria is hydrolysis of some milk proteins as well as formation of lactic acid from lactose. As many lactobacilli are unable to synthesize some amino acids, they hydrolyze milk proteins and utilize some of the resultant amino acids and peptides for growth. As a result, the residual level of peptides and free amino acids increases in the fermented milk. Milk proteins serve as nutrients for animals and humans, and also have potential physiological effects. Peptides have been discovered that exert opioid agonistic and antagonistic activity, that inhibit angiotensin-I converting enzyme (ACE), that have immunomodulatory effects, antimicrobial activity, antithrombotic activity, and mineral binding properties (reviewed by Meisel 1998). Peptides having ACE-inhibitory activity have been the subject of recent research and industrial application.

ACE is important in the regulation of blood pressure. In the renin angiotensin system, ACE acts on angiotensin I to hydrolyze His-Leu from its C-terminus to produce a potent vasopressor, angiotensin II (Skeggs et al. 1956). In addition, in the kinin-kallilrein system, ACE inactivates bradykinin, which exhibits vasodilation activity (Erdos 1975). By these actions, ACE elevates blood pressure. Some inhibitors of ACE show antihypertensive effects *in vivo* (Ondetti et al. 1977) and are used as pharmaceuticals. Extensive efforts have been made to find such substances in foods, i.e., proteolytic digests of wheat zein, soybean, tuna, bonito, sardine and milk proteins (reviewed by Yamamoto 1997). The milk proteins include α_{S1}-casein, ß-casein, κ-casein as well as the whey proteins α-lactalbumin, ß-lactoglobulin and serum albumin (reviewed by Takano 1998; FitzGerald & Meisel 2000). For many peptides, only *in vitro* ACE-inhibitory activity has been reported; some show no *in vivo* effect. The effects of consuming val-pro-pro and ile-pro-pro in experimental animals and in humans has been reported. These studies, including research on the health effects of sour milk, from which these tripeptides were discovered, are summarized below.

Anti-hypertensive activity of Calpis^{TM} sour milk

Prolongation of life span in mice

A series of studies were performed on pasteurized Calpis^{TM} sour milk (Sour milk) which is an intermediate product in the production of the pasteurized soft drink Calpis^{TM}, that has been consumed in Japan as a commercial product for over 80 years. Sour milk is made by inoculating skim milk with a starter culture containing *L. helveticus* and *S. cerevisiae* as the predominant microorganisms, and fermenting at 37 °C for 24 h. A portion of Sour milk is used as a starter culture.

In the initial study, dietary Sour milk was administered to mice and correlated with life-span (Arai et al. 1980). Female ICR mice were divided into three groups, and fed diets containing: (1) 14% Sour milk, (2) 1.6% whole milk powder, or (3) a control diet, from 4 weeks of age until death. Death of the mice was monitored and the cause of death was analyzed by necropsy. After 70 weeks of age, the survival ratio in the Sour milk group was higher than that of the other groups (Figure 1). The average life span of the control group, whole milk group and the Sour milk group was 84.9, 84.4 and 91.8 weeks, respectively. The life span of the Sour milk group was significantly longer than that of the other groups. However, there was little difference in the maximum life span among the three groups. The prolongation of the average life span was considered to be a result of the preventive effect of

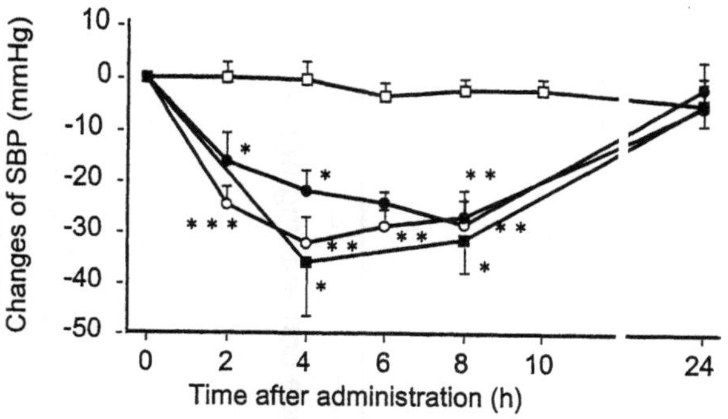

Figure 2. Changes of blood pressure after single oral administration of Sour milk (■), Val-Pro-Pro (○), Ile-Pro-Pro (●) or PBS-casein (□) on spontaneously hypertensive rats (Nakamura et al. 1995b).

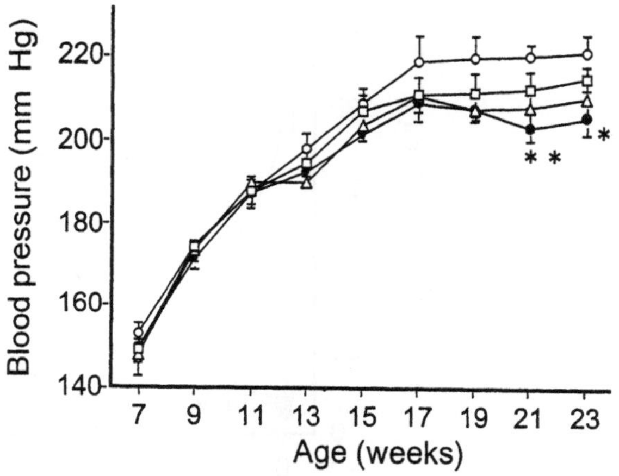

Figure 3. Changes of blood pressure during feeding the diet containing 0.25% (w/w) Sour milk powder (□), diet containing 1.2.5% (w/w) sour milk powder (Δ), diet containing 2.50% (w/w) sour milk powder (●) or standard laboratory diet (○) in spontaneously hypertensive rats (Nakamura et al. 1996).

the Sour milk on death caused by certain diseases. The major causes of death were found to be renal atrophy, tumors and pneumonia. However, there was little difference in the ratio of the death causes among the groups. These findings suggest that the Sour milk inhibited the incidence of these diseases.

Animal models using transplantable tumors and chemical tumorigenesis suggested that Sour milk has anti-tumor activity (Takano et al. 1985). In a further study, the progression of renal lesions induced by a high meat diet in APA hamsters was suppressed by feeding Sour milk (Wada et al. 1984), suggesting that the Sour milk might affect the circulatory system, including blood pressure regulation.

Decrease of blood pressure in SHR rats

In a second study, the *in vivo* anti-hypertensive effect of Sour milk was examined in spontaneously hypertensive rats (SHR). The effect of a single oral administration (5 ml of Sour milk per kg of body weight by gastric intubation) in 22 to 26-week-old rats is shown in Figure 2 (Nakamura et al. 1995b). The mean systolic blood pressure (SBP) before administration of the Sour milk was 213.7 ± 3.4 mm Hg, and decreased significantly 4 to 8 h after administration. The SBP returned to the initial level 24 h after administration. In contrast, the SBP of rats given 5 ml of unfermented milk per kg body weight did not change significantly.

The change in SBP during long-term feeding of Sour milk was also tested in SHR rats (Nakamura et al. 1996; Figure 3). Lyophilized Sour milk powder was mixed with the standard commercial diet (F-2, Funabashi Farms, Chiba, Japan) to contain 0.25, 1.25, 2.50% Sour milk powder. Feeding of the Sour milk diets began at 7 weeks of age. The SBP of all rats gradually increased during the course of the experiment. The SBP of the control rats fed a standard commercial diet reached about 220 mm Hg at 17 weeks of age. After 13 weeks of age, the SBP of rats in the Sour milk groups tended to be lower than in the control group, in a dose-dependent manner. That is, the greater the amount of Sour milk in the diet, the lower the SBP. In the 2.50% Sour milk group, the SBP was significantly lower than that of the control group after 21 weeks of age.

These results suggest that Sour milk has both a short-term antihypertensive effect by a single oral administration and a preventive effect during long-term feeding of the Sour milk.

Anti-hypertensive peptides found in Sour milk

To determine if Sour milk contained ACE inhibitory activity, the change in activity during fermentation was investigated (Nakamura et al. 1995a; Figure 4). During fermentation of milk with the Sour milk starter, cell density, pH, and peptide content changed significantly during the log phase. The ACE inhibitory activity increased during the late log phase to stationary phase, suggesting that the ACE inhibitory substances were produced by fermentation.

Two kinds of ACE inhibitory substances were purified from the Sour milk by using a four-step reversed phase HPLC procedure and by gel filtration columns (Nakamura et al. 1995a). Val-pro-pro and ile-pro-pro were identified by amino acid analysis and by [^{1}H] and [^{13}C] nuclear magnetic resonance analyses. The concentration of val-pro-pro and ile-pro-pro providing

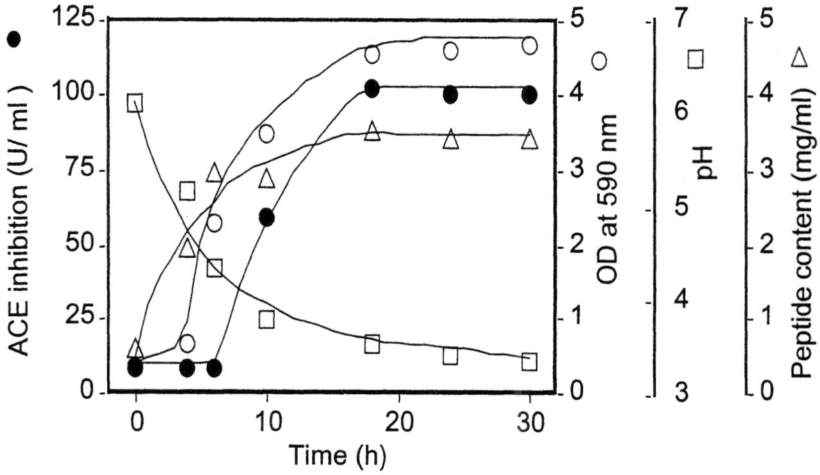

Figure 4. Change in inhibitory activity of angiotensin I-converting enzyme (●), Cell density at 590 nm (○), pH (□) and peptide content (Δ) during fermentation of milk by the Sour milk starter (Nakamura et al.).

50% inhibition of ACE was found to be 9 and 5 μmol L^{-1}, respectively. Most inhibitory activity in the Sour milk was attributed to these two tripeptides.

The first ACE inhibitors described were found in snake venom and had pro-pro sequences at their C-termini (Ferreira et al. 1970; Kato et al. 1971). Several investigators have reported that the C-terminal amino acids of inhibitory peptides are important in binding to the active site of ACE, and that aromatic (trp, tyr, and phe) and imino acids (pro) are favored (Cheung et al. 1980). The sequences of the two Sour milk tripeptides val-pro-pro and ile-pro-pro are in consistent with these observations.

To establish that val-pro-pro and ile-pro-pro indeed have anti-hypertensive activity, the two peptides were chemically synthesized and tested for its blood pressure activity using SHR rats and normotensive Wistar Kyoto (WKY) rats (Nakamura et al. 1995b). Rats were given the Sour milk or the synthesized peptides dissolved in 0.2 mmol L^{-1} phosphate-buffered saline (PBS) (pH 7.2) containing 0.5% sodium caseinate. Figure 2 shows the changes of SBP after single oral administration of val-pro-pro (0.6 mg kg^{-1} of body weight) or ile-pro-pro (0.3 mg kg^{-1} of body weight) to SHR rats. These doses were predicted to correspond to the ACE inhibitory activity observed following administration of 5 ml of Sour milk per kg of body weight. Although, the SBP of the control rats given PBS-casein did not change significantly, the SBP decreased significantly several hours after administration of the tripeptides, as observed following administration of the Sour milk. In the long-term study, development of hypertension was also attenuated in SHR rats given val-pro-pro and ile-pro-pro dissolved in water (Sipola et al. 2001).

To determine dose–response relationships, the changes of SBP were measured 6 h after a single oral administration of several doses of val-pro-pro or ile-pro-pro (Nakamura et al. 1995b). Both peptides showed dose-dependent activity up to 5 mg per kg of body weight. These data suggest that both val-pro-pro and ile-pro-pro play essential roles in the antihypertensive activity of Sour milk.

In contrast, a single oral administration of either Sour milk or tripeptides did not change the SBP of normotensive WKY rats, even though the doses of the Sour milk and tripeptides administered were far higher than the doses found to be effective in SHR rats. Thus, the antihypertensive effect of the Sour milk and tripeptides was specific to hypertension.

Production of antihypertensive peptides by L. helveticus

To determine which lactic acid bacteria were responsible for antihypertensive activity, milks fermented separately by sixteen strains of seven species of lactic acid bacteria commonly used in fermented dairy products were tested (Yamamoto et al. 1994). SHR rats were given 5 ml of fermented milk per kg of body weight by gastric intubation. The milks fermented by several strains of *L. helveticus* showed antihypertensive activity. However, milks fermented by other species of lactic acid bacteria, *L. delbrueckii* subsp. *bulgari-*

cus, L. casei, L. acidophilus, L. delbrueckii subsp. lactis, Streptococci thermophilus, L. lactis subsp. lactis and L. lactis subsp. cremoris did not show significant antihypertensive activity.

The milks fermented by *L. helveticus* which exhibited antihypertensive activity were found to contain the val-pro-pro and ile-pro-pro peptides (unpublished observation). The amino acid sequence of val-pro-pro is found in the primary structure of bovine ß-casein (84–86) and ile-pro-pro found in ß-casein (74–76) and κ-casein (108–110). *L. helveticus* fermented milk which showed the antihypertensive effect, had higher ACE inhibitory activity and proteinase activity (Yamamoto et al. 1994). Milk fermented by *L. helveticus* CP791, a proteinase-defective variant derived from *L. helveticus* CP790, did not exhibit antihypertensive activity (Yamamoto et al. 1994a). These observations suggest that val-pro-pro and ile-pro-pro are produced from milk casein by proteolytic activity of *L. helveticus*.

The first step in casein hydrolysis by *L. helveticus* is cleavage by a cell-wall associated proteinase. However, val-pro-pro and ile-pro-pro were not found in the hydrolysate of α_{S1}- and β-casein by a purified cell wall-associated proteinase from *L. helveticus* CP790 (Yamamoto et al. 1993). Further hydrolysis may be required to liberate these tripeptides. Interestingly, casein hydrolyzed by the same cell wall-associated proteinase exhibited ACE inhibitory activity and decreased the blood pressure of SHR rats, while that hydrolyzed by trypsin did not (Yamamoto et al. 1994a). An anti-hypertensive heptapeptide purified from casein hydrolyzed by the cell wall-asociated proteinase had weak ACE inhibitory activity. However, after digestion of the heptapeptide by pancreatin, one C-terminal residue was removed yielding a hexapeptide with high ACE activity and potent antihypertensive activity (Maeno et al. 1996). This suggests possible activation of the peptide in the digestive tract. Also, milk fermented by a low-acid mutant strain of *L. helveticus* CP790 with impaired activity of a membrane bound ATPase did not contain val-pro-pro or ile-pro-pro, but exhibited anti-hypertensive activity in SHR rats. In this case, the active anti-hypertensive substance was a dipeptide with low ACE inhibitory activity (Yamamoto et al. 1999). These studies further document the existence of biogenic peptides in fermented milk.

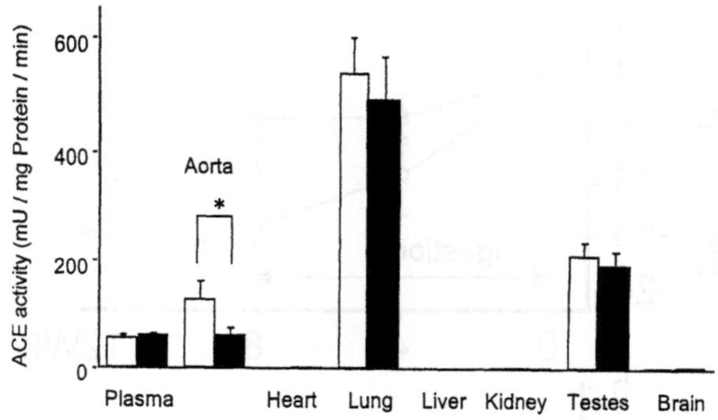

Figure 5. The activity of angiotensin I-converting enzyme in various tissues from spontaneously hypertensive rats administered with saline (□) or Sour milk (■) (Nakamura et al. 1996).

In vivo evidence for mode of action

It has been suggested that inhibition of ACE in the vascular wall, kidney, brain and lung is important for the antihypertensive activity of Captopril, an ACE inhibitory medicine (Unger et al. 1981; Velletri & Bean 1982; Unger et al. 1985). To further investigate the antihypertensive activity of Sour milk, ACE activity in various tissues was measured after long-term feeding of Sour milk to SHR rats (Nakamura et al. 1996; Figure 5). Both SBP and ACE activity of the aorta in SHR rats fed with the Sour milk diet were significantly lower than those in rats fed a control diet. Similarly, the specific activity of ACE in an aorta extract was significantly lower in rats 6 h after a single oral administration of Sour milk than in those given saline (Masuda et al. 1996). In this study, the activity in aorta and lung was significantly higher in SHR rats than that observed in WKY rats. These results suggest that inhibition of ACE in aorta may play an important role in the antihypertensive effect of Sour milk.

Furthermore, val-pro-pro and ile-pro-pro were detected by HPLC analysis in the heat-treated, solubilized fraction from the aorta of SHR rats after administration of the Sour milk (Masuda et al. 1996). However, the tripeptides were not detected in rats given saline, or in normotensive WKY rats given Sour milk or saline. Small peptides, such as di- or tripeptides, are more easily absorbed in the intestine than are amino acids or larger oligopeptides (Craft et al. 1968; Adibi 1971; Hara et al. 1984). In addition, peptides containing pro are generally resistant to degradation by digestive enzymes (Kim et al. 1972; Adibi & Kim 1981). Although several peptidases that are specific for pro have been reported, tripeptides that have a C-terminal pro-pro sequence are resistant to de-

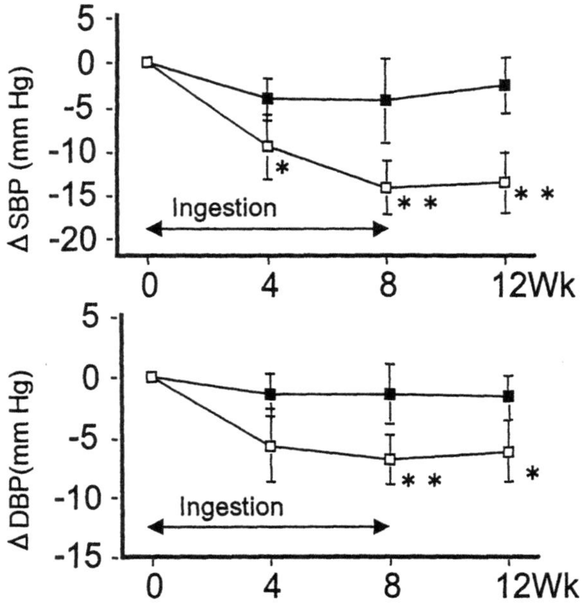

Figure 6. Changes in systolic blood pressure (SBP) and diastolic blood pressure (DBP) of hypertensive subjects by treatment for 8 weeks with Sour milk (■) and artificially acidified milk as placebo (□) (Hata et al. 1996).

gradation by peptidases (Kim et al. 1976; Yoshimoto et al.: 1978; Yoshioka et al. 1987; Mock et al. 1990). Val-pro-pro has been shown to be transported across a monolayer of Caco-2 cells, presumably by paracellular diffusion (Satake et al. 2002). This suggests that val-pro-pro and ile-pro-pro in Sour milk may be absorbed directly in SHR rats without being further hydrolyzed by digestive enzymes, and may reach the abdominal aorta, and exhibit antihypertensive effects through inhibition of ACE.

Clinical efficacy

A placebo-controlled study was conducted to examine the effect of Sour milk on blood pressure in 30 hypertensive outpatients (Hata et al. 1996).

Most of the subjects were receiving concurrent treatment with antihypertensive medicines, but the doses and medicines used were not changed during the trial. Subjects were randomly assigned to two groups. One group ingested 95 ml of Sour milk daily for 8 weeks, and the other group ingested the same amount of artificially-acidified milk as a placebo for 8 weeks. Val-pro-pro and ile-pro-pro were present in the Sour milk drink at 1.5 and 1.1 mg, respectively, but were not detectable in the placebo drink. Blood pressure was measured every 4 weeks at the hospital. Figure 6 shows the changes in SBP, diastolic blood pressure (DBP), and pulse rate during the experiment. At 4 and 8 weeks, SBP decreased significantly relative to the initial SBP in the Sour milk group. Similarly, the DBP decreased significantly relative to the initial DBP. In contrast, no significant changes in blood pressure were observed in the placebo group. In both groups, no marked change was observed in other indexes, including pulse rate, body weight, and blood serum variables. Although undesirable side effects of ACE inhibitors (cough and adverse effects on serum lipid metabolism) have been reported (Ames 1983; Sesoko & Kaneko 1985; Nakamura 1987), no adverse side effects were observed in any subjects in this study. In a similar study, these results were confirmed in untreated hypertensive subjects (Itakura et al. 2001). In healthy volunteers with normal blood pressure, neither Sour milk nor a placebo displayed significant effects on blood pressure (Itakura et al. 2001).

Sour milk effectively decreased hypertension in 8 wk ingestion. The dose of ACE inhibitory peptides in sour milk is relatively low compared to the recommended dose of ACE inhibitory pharmaceuticals, for example 25–150 mg per day in the case of captopril. The efficacy at low dose and the lack of adverse side effects seem desirable but raise questions about the mechanism of action of the Sour milk peptides. However, in studies on pharmaceuticals, it is recognized that *in vitro* ACE inhibition does not clearly correlate with *in vivo* anti-hypertensive activity (Pong et al. 1985). Involvement of additional mechanisms of action are postulated for their anti-hypertensive activity (Borgaert 1986). Further pharmacological studies are necessary to understand more clearly the detailed mechanism(s) of action of val-pro-pro and ile-pro-pro and the Sour milk.

Concluding remarks

One possible mechanism for the beneficial effects of fermented milk is involvement of biogenics, which makes the products directly beneficial without need for live bacteria. Research on the health effects on Calpis™ Sour milk indicate clear evidence for biogenics. Mice that consumed a life-long diet of pasteurized Sour milk lived longer than those who received control diets. Analysis of the cause of death in the treated animals suggested beneficial effects on circulatory diseases. The Sour milk inhibited the hypertension of SHR rats. Two kinds of tripeptides, val-pro-pro and ile-pro-pro, which inhibit angiotensin-I converting enzyme, were identified in the Sour milk. The chemically synthesized peptides showed anti-hypertensive activity after oral administration to SHR rats. The peptides were recovered from the aorta of SHR rats after administration of the Sour milk. In clinical trials, the Sour milk decreased the blood pressure of hypertensive subjects, but had no significant effects in volunteers with normal blood pressure. Based on this evidence, a functional food product containing Sour milk as a significant component was approved as 'Food for Specified Use' in Japan with an appropriate health claim on the label. Further studies on biogenic peptides will provide the necessary scientific evidence for new health benefits and will likely lead to development of new functional food products.

References

Adibi SA (1971) Intestinal transport of dipeptides in man: relative importance of hydrolysis and intact absorption. J. Clin. Invest. 50: 2266–2275.

Adibi SA & Kim YS (1981) Peptide absorption and hydrolysis. In: Johnson LR (Ed) Physiology of the Gastrointestinal Tract (pp 1097–1122). Raven Press, New York, USA.

Ames RP (1983) Negative effects of diuretic drugs on metabolic risk factors for coronary heart disease. Am. J. Cardiol. 51: 632–638.

Arai K, Murota I, Hayakawa K, Kataoka M & Mitsuoka T (1980) Effects of administration of pasteurized fermented milk to mice on the life-span and intestinal flora. J. Jap. Soc. Nutr. Food Sci. 33: 219–223.

Cheung HS, Wang FL, Ondetti MA, Sabo EF & Cushman DW (1980) Binding of peptide substrates and inhibitors of angiotensin-converting enzyme. J. Biol. Chem. 255: 401–407.

Craft IL, Geddes D, Hyde CW, Wise IJ & Matthews DM (1968) Absorption and malabsorption of glycine and glycine peptides in man. Gut 9: 425–437.

Erdos EG (1975) Angiotensin I converting enzyme. Circul. Res. 36: 247–255.

Ferreira SH, Bartelt DC & Greene LJ (1970) Isolation of bradykinin-potentiating peptides from *Bothrops jararaca* venom. Biochemistry 9: 2583–2592.

FitzGerald RJ & Meisel H (2000) Milk protein-derived peptide inhibitors of angiotensin-I-converting enzyme. Br. J. Nutr. 84 (Suppl 1: S33–S37.

Fuller R (1989) Probiotics in man and animals. J. Appl. Bacteriol. 66: 365–378.

Hara H, Funabiki R, Iwata M & Yamazaki K (1984) Portal absorption of small peptides in rats under unrestrained condition. J. Nutr. 114: 1122–1129.

Hata Y, Yamamoto M, Ohni M, Nakajima K, Nakamura Y & Takano T (1996) A placebo-controlled study of the effect of sour milk on blood pressure in hypertensive subjects. Am. J. Clin. Nutr. 64: 767–771.

Itakura H, Ikemoto S, Terada S & Kondo K (2001) The effect of sour milk on blood pressure in untreated hypertensive and normotensive subjects. J. Jap. Soc. Clin. Nutr. 23: 26–31.

Kato H & Suzuki T (1971) Bradykinin-potentiating peptides from the venom of *Agkistrodon haly blomhoffi*. Isolation of five bradykinin potentiators and the amino acid sequence of two of them, potentiators B and C. Biochemistry 10: 972–980.

Kim YS, Bertwhistle W & Kim YW (1972) Peptide hydrolyses in the brush border and soluble fractions of small intestinal mucosa of rat and man. J. Clin. Invest. 51: 1419–1430.

Kim YS, Brophy EJ & Nicholson JA (1976) Rat intestinal brush border membrane peptidases. 2. Enzymatic properties, immunochemistry and interactions with lectins of two different forms of the enzymes. J. Biol. Chem. 251: 3206–3212.

Maeno M, Yamamoto N & Takano T (1996) Identification of antihypertensive peptides from casein hydrolysate produced by a proteinase from *Lactobacillus helveticus* CP790. J. Dairy Sci. 73: 1316–1321.

Masuda O, Nakamura Y & Takano T (1996) Antihypertensive peptides are present in aorta after oral administration of sour milk containing these peptides to spontaneously hypertensive rats. J. Nutr. 126: 3063–3068.

Meisel H (1998) Overview on milk protein-derived peptides. Int. Dairy J. 8: 363–373.

Mitsuoka T (1976) Taxonomy of lactic acid bacteria and medical effect of sour milk Jap. J. Dairy Food Sci. 25: A170–176.

Mitsuoka T (2000) Significance of dietary modification of intestinal flora and intestinal environment. Biosci. Microflora 19: 15–25.

Mock WL, Green PC & Boyer KD (1990) Specificity and pH dependence for acylproline cleavage by prolidase. J. Biol. Chem. 265: 19600–19605.

Nakamura H (1987) Effects of antihypertensive drugs on plasma lipid. Am. J. Cardiol. 60: 24E–28E.

Nakamura Y, Masuda O & Takano T (1996) Decrease of tissue angiotensin I-converting enzyme activity upon feeding sour milk in spontaneously hypertensive rats. Biosci. Biotech. Biochem. 60: 488–489.

Nakamura Y, Yamamoto N, Sakai K, Okubo A, Yamazaki S & Takano T (1995a) Purification and characterization of angiotensin I-converting enzyme inhibitors from a sour milk. J. Dairy Sci. 78: 777–783.

Nakamura Y, Yamamoto N, Sakai K & Takano T (1995b) Antihypertensive effect of sour milk and peptides isolated from it that are inhibitors to angiotensin I-converting enzyme. J. Dairy Sci. 78: 1253–1257.

Ondetti MA, Rubin B & Cushman DW (1977) Design of specific inhibitors of angiotensin converting enzyme: a new class of orally active antihypertensive agents. Science 196: 441–444.

Pong SF, Brooks RR, Huang CT & Moore AF (1985) Correlation between anti-hypertensive activity in rats and plasma angiotensin I-converting enzyme (ACE) inhibition in mice following oral administration of ACE inhibitors. Pharmacol. Res. Comm. 17: 331–344.

Rasic JL & Kurmann JA (1978) The nutritional-physiological value of yoghurt.: In Yoghurt (pp 99–137), Tech Dairy Publ. House, Copenhagen, Denmark.

Satake M, Enjoh M, Nakamura Y, Takano T, Kawamura Y, Arai S & Shimizu M (2002) Transepithelial transport of the bioactive tripeptide, Val-Pro-Pro, in human intestinal Caco-2 cell monolayers. Biosci. Biotechnol. Biochem. 66: 378–384.

Sipola M, Finckenberg P, Santisteban J, Korpela R, Vapaatalo H & Nurminen ML (2001) Long-term intake of milk peptides attenuates development of hypertension in spontaneously hypertensive rats. J. Physiol. Pharmacol. 52: 745–754.

Skeggs LT, Kahn JE & Shumway NP (1956) The preparation and function of the angiotensin-converting enzyme. J Exper. Med. 103: 295–299.

Seseko S & Kaneko Y (1985) Cough associated with the use of captopril. Arch. Int. Med. 145: 1524.

Takano T (1998) Milk derived peptides and hypertension reduction. Int. Dairy J. 8: 375–381.

Takano T, Arai K, Murota I, Hayakawa K, Mizutani T & Mitsuoka T (1985) Effects of feeding sour milk on longevity and tumorigenesis in mice and rats. Bifidobacteria Microflora 4: 31–37.

Unger T, Ganten D, Lang RE & Scholkens BA (1985) Persistent tissue converting enzyme inhibition following chronic treatment with Hoe498 and MK421 in spontaneously hypertensive rats. J. Cardiovascul. Pharmacol. 7: 36–41.

Unger T, Kaufmann-Buhler I, Scholkens B & Ganten D (1981) Brain converting enzyme inhibition: a possible mechanism for the antihypertensive action of Captopril in spontaneously hypertensive rats. Euro. J. Pharmacol. 70: 467–478.

Velletri P & Bean BL (1982) The effect of Captopril on rat aortic angiotensin-converting enzyme. J. Cardiovascul. Pharmacol. 4: 315–325.

Wada K, Arai K, Takano T, Murota I, Hayakawa K, Mizutani T & Mitsuoka T. (1984) Effect of sour milk on the renal disorder and intestinal flora. In: Mitsuoka T (Ed) Intestinal Flora and Geriatric Diseases (pp 175–197). Gakkai-Shuppan Center, Tokyo, Japan.

Yamamoto N (1997) Antihypertensive peptides derived from food proteins. Biopolymer 43: 129–134.

Yamamoto N, Akino A & Takano T (1993) Purification and specificity of a cell-wall-associated proteinase from *Lactobacillus helveticus* CP790. J. Biochem. 114: 740–745.

Yamamoto N, Akino A & Takano T (1994) Antihypertensive effects of different kinds of fermented milk in spontaneously hypertension rats. Biosci. Biotech. Biochem. 58: 776–778.

Yamamoto N, Akino A & Takano T (1994a) Antihypertensive effects of peptides derived from casein by an extracellular proteinase from *Lactobacillus helveticus* CP790. J. Dairy Sci. 77: 917–922.

Yamamoto N, Maeno M & Takano T (1999) Purification and characterization of an antihypertensive peptide form a yogurt-like product fermented by Lactobacillus helveticus CPN4. J. Dairy Sci. 82: 1388–1393.

Yoshimoto T, Fischl M, Orlowski RC & Walter R (1978) Post-proline cleaving enzyme and post-proline dipeptidyl aminopeptidase comparison of two peptidases with high specificity foe proline residues. J. Biol. Chem. 253: 3708–3716.

Yoshioka M, Erickson RH, Woodley JF, Gulli R, Guan D & Kim YS (1987) Role of rat intestinal brush-border membrane angiotensin-converting enzyme in dietary protein digestion. Am. J. Physiol. 253: G781–G786.

Antonie van Leeuwenhoek **82:** 341–352, 2002.

The Intestinal LABs

Elaine E. Vaughan[1,2,*], Maaike C. de Vries[1,2], Erwin G. Zoetendal[1,2], Kaouther Ben-Amor[1], Antoon D.L. Akkermans[1] & Willem M. de Vos[1,2]
[1]*Laboratory of Microbiology, Wageningen University, Hesselink van Suchtelenweg 4, 6703 CT Wageningen, The Netherlands;* [2]*Wageningen Centre for Food Sciences, P.O. Box 557, 6700 AN Wageningen, The Netherlands (*Author for correspondence; E-mail: elaine.vaughan@algemeen.micr.wau.nl)*

Key words: bifidobacteria, intestine, lactic acid bacteria, microbiota, molecular methods, 16S rRNA

Abstract

The complete gastrointestinal (GI) tract of humans is colonised soon after birth by a myriad of microbial species with a characteristic distribution depending on the location. GI-tract ecology has been experiencing a revival due to the development of molecular techniques, especially those based on 16S RNA (zRNA) genes. A richer ecosystem than previously imagined of novel species is being discovered that is significantly influenced by our host genotype. Special attention has been focused on the bifidobacteria and the lactic acid bacterial (LAB) populations, both those that are naturally present within this complex ecosystem and those that are ingested as probiotics in functional foods. Overall this interest stems from a increasing awareness of interplay between microflora, diet and the health of the host, and is further stimulated by an increasing incidence of gastrointestinal illnesses and atopy. Substantial documentation of benefits to host health has especially distinguished the LAB for multidisciplinary research aimed to determine the molecular mechanisms involved. Recent advances in molecular technologies, including high-throughput genomics-based approaches, can significantly advance our understanding of the microbe–diet–host interactions and offer valuable information for design and application of health-targeted microbes.

Introduction

The microbiota are present in vast numbers in the human gastrointestinal (GI) tract where they are closely associated with each other and the intestinal epithelial cells of the host. Colonisation of the germfree infant commences rapidly after birth, initially strongly influenced by the manner of birth and environment, and subsequently by the diet, genetic background, and environment of the individual. Lactic acid bacteria (LAB) and bifidobacterial populations have been detected as part of this complex ecosystem in the majority of humans. These unique autochthonous or indigenous populations may be supplemented with LAB from dairy or probiotic products that are generally allochthonous, due to their presence in the GI-tract for a limited time. The overall activity of the commensal bacteria in the GI-tract includes protection against pathogens, development of the immune system, and positive effects on colonic health and host nutrition (Wilson 1995; Falk et al. 1998; Cebra 1999). Furthermore, a range of documented or potential benefits have been ascribed to the ingestion of probiotic LAB including modulation of the GI-tract, antagonism against pathogenic microbes and maintaining the intestinal mucosal barrier; there are many recent excellent reports on this subject with different degrees of scepticism (Salminen et al. 1998; Dunne et al. 1999; Bengmark 2000; de Roos & Katan 2000; Simmering & Blaut 2001). In the future turbo-probiotic LAB (Shanahan 2001), such as *Lactococcus lactis* genetically engineered to secrete interleukin-10 for treatment of inflammatory bowel disease (IBD) may also become transient members of the intestinal community (Steidler et al. 2000). The largely anaerobic GI-tract community, comprising many undescribed members has presented a challenge to investigations of the host–microbe interactions, which is exacerbated by difficulties to obtain samples in many regions of the GI-tract. The emergence of culture independent methods for studying diversity and functionality of complex ecosystems has created an exciting era for GI-tract research (Tannock 1999; Vaughan et al. 1999, 2000). Increased acknowledgement of the link between health and food consumption and rise in intestinal disorders has enhanced collaboration of the medical profession in this research area. In Europe, the significance of the food and health link has been recognised by the formation of the Food, GI-tract Functionality and Human Health cluster (PROEUHEALTH) subsidised by the European Community that disseminates information through science, industry and consumer platforms (Mattila-Sandholm et al. 2002a).

In the present review, recent results from a variety of culture independent techniques and molecular tools, some undergoing high-throughput development, that are having a major impact on investigating the composition, dynamics and activity of the microbiota including the intestinal LAB will be presented. Furthermore, the use of complete genome sequences for LAB, DNA microarrays and 'omics' technologies that generate insight relevant to the intestinal LAB will also be addressed.

Members of the intestinal LAB

A brief overview of the taxonomy of the LAB is provided because of its relevance to many of the molecular techniques discussed below. Comparative sequence analysis of the 16S ribosomal RNA (rRNA) gene has become a very important tool in classifying microorganisms as it reflects the natural evolutionary relationships or phylogeny. The 16S rRNA is universally distributed and a reasonable size for sequence analysis. The highly conserved and variable domains make excellent primer and probe targets. The Gram-positive bacteria form two main lines of descent, the *Clostridium* phylum with a DNA base composition of generally less than 50 mol% guanine plus cytosine (G + C), and the *Actinomycetes* phylum with greater than 50 mol% G + C. The typical LAB, such as the *Lactobacillus, Lactococcus* and *Leuconostoc* genera, belong to the low G + C phylum. The low G + C LAB constitute three phylogenetic clusters, the *Lactobacillus delbrueckii* group, the *Lactobacillus casei-Pediococcus* group, and the *Leuconostoc* group (Collins et al. 1991). The lack of a clear monophyletic origin and the strong relationship to other LAB has resulted in quite some taxonomic difficulties (Figure 1) (Schleifer & Ludwig 1995). The *Bifidobacterium* genus is phylogenetically distinct from the LAB as it belongs to the *Actinomycetes* phylum. However, it shares phenotypic features typical of LAB, such as acid production, and for both traditional as well as practical purposes are regarded as part of the LAB. Bifidobacteria sequences form a coherent phylogenetic unit and their 16S rRNA sequences share over 93% similarity (Miyake et al. 1998).

Overall the taxonomy of the LAB based on the 16S rRNA has presented some challenges to microbiologists studying and identifying the intestinal tract LAB but various interesting molecular approaches have been developed some of which are described below. The most common LAB so far detected in human intestinal samples are the bifidobacteria, lactobacilli and the leuconostoc, and it is seldom that LAB are not detected in an individual. Members of the *Lactobacillus* are found not only in the GI-tract of humans and animals but also in diverse plant material such as silage and grains. So far, bifidobacteria have only been found in the intestinal ecosystem of humans and other warm-blooded animals, and honey bees (Biavati et al. 1991). Of course the LABs that stir the most interest for the consumer are the lactobacilli and bifidobacteria because of their documented health promoting potential. There are several excellent recent reviews on culture based knowledge of the GI-tract microflora including LAB (Mackie et al. 1999; Kleessen et al. 2000; Reuter 2001). The emphasis here will be on results generated by culture independent or novel molecular methods.

Succession and dynamics

Comparison between microscopic and plate counts, group-specific dot-blot hybridisation and fluorescent *in situ* hybridisation (FISH) has indicated that a significant fraction of the microbiota can escape cultivation (Langendijk et al. 1995; Suau et al. 1999; Sghir et al. 2000). Fluorescent *in situ* hybridisation (FISH) is a technique using a specific fluorescent oligonucleotide probe binding to rRNA in whole cells to detect, identify and enumerate bacteria in complex ecosystems. Using this technique levels of bifidobacteria between 10^9 and 10^{10} per gram dry weight of adult faeces could be detected using the Bif164-probe (Langendijk et al. 1995; Franks et al. 1998; Harmsen et al. 1999b). Using the Lab158 *Lactobacillus*-enterococci targeted probe, numbers of 10^7 to 10^8per gram dry weight were estimated (Harmsen et al. 1999a,b). Because of the previous underestimation of the total microbiota counts when culturing, the bifidobacterial and lactobacilli numbers in adult faeces formerly es-

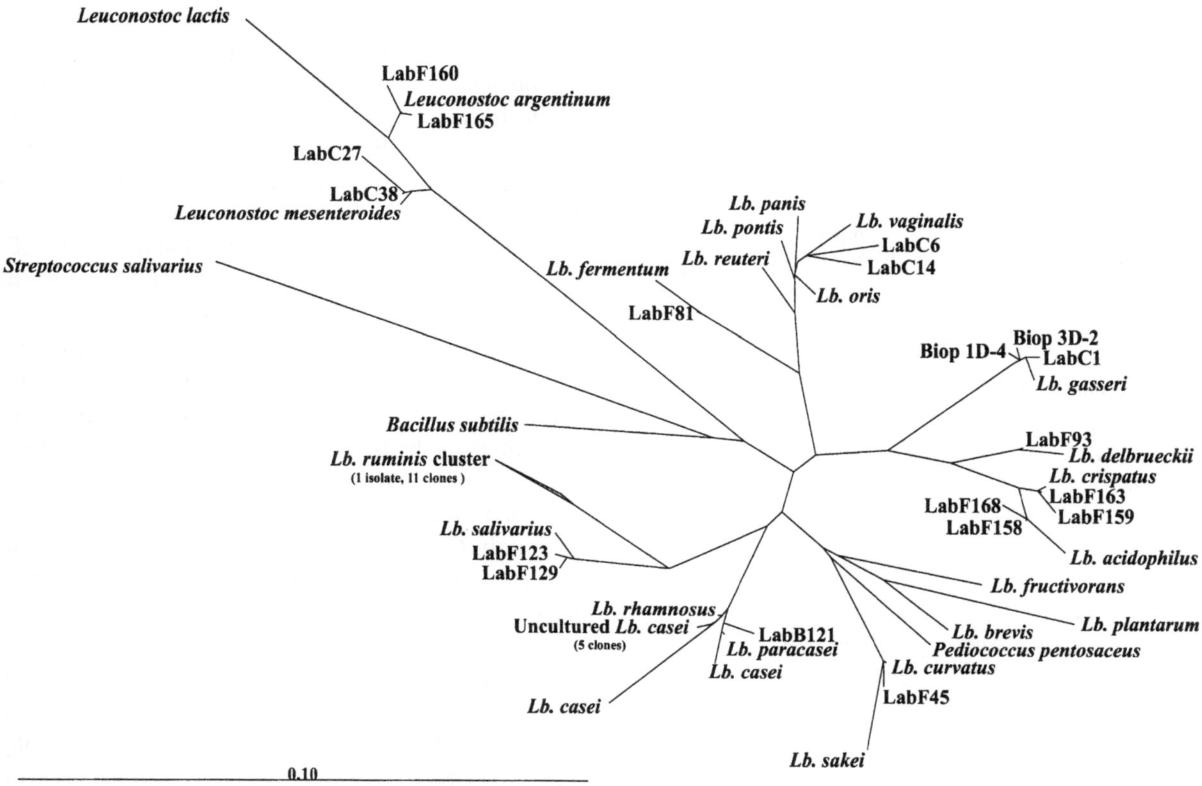

Figure 1. Neighbour-joining tree of partial 16S rDNA sequences (*E. coli* positions 108–648) showing the phylogenetic relationship between cultured and uncultured lactic acid bacteria. *Lactobacillus* is indicated by *Lb*. Sequences encoded with Lab'X' were derived from cloned 16S rDNA PCR products obtained from various GI-tract samples (Heilig et al. 2002). Sequences with the code Biop 'X' were cloned 16S rDNA PCR products from biopsy samples of the descending colon (Zoetendal et al. 2002). The other sequences originate from cultured bacteria. *Bacillus subtilis* was used as an outgroup to root the tree. The scalebar represents the calculated distance between the sequences.

timated to be approximately 10 and 2% (Moore & Holdeman 1974; Mitsuoka 1992), respectively, are now estimated to be 3 and <1% using the FISH or quantitative rRNA dot blot hybridisation techniques (Sghir et al. 2000). The values for lactobacilli in baby faeces are similar to adults, but the bifidobacteria can range from 60 to 90% of the total microbiota in breast fed infants, and an average of 50% in formula-fed infants due to the influence of diet (Harmsen et al. 2000).

The study of the microbial diversity and community behaviour over time of the GI-tract has been stimulated by the advent of fingerprinting techniques, such as DGGE or TGGE (denaturing or temperature gradient gel electrophoresis), and terminal-restriction fragment length polymorphism (T-RFLP) (Muyzer & Smalla 1998; Kitts 2001; Zoetendal et al. 2001a). TGGE/DGGE of 16S rRNA PCR products are especially suited to study diversity in samples with largely unknown microbial content, the GI-tract being a classic case, without cultivation. Briefly, in D/TGGE the 16S rRNA genes are amplified using a primer pair, one of which has a G+C 'clamp' attached to the 5′ end that prevents the two DNA strands from completely dissociating even under strong denaturing conditions. During electrophoresis through a polyacrylamide gel containing denaturants, the migration of the double stranded DNA is prevented once a domain in the PCR product reaches its melting conditions. Following staining of the DNA, a banding pattern emerges that represents the diversity of the different rRNA gene sequences present in the sample. The intensity of a band is a semi-quantitative measure for the relative abundance of this sequence in the population. Bands can be excised and sequenced, and the identity determined by comparison to the databases.

Application of DGGE for GI-tract samples indicated that the total microbiota of each adult individual had a unique pattern reflecting their differences in composition and partly dependent on the host genotype (Zoetendal et al. 1998, 2001b). By comparing samples over time, the predominant microbiota was

Table 1. LAB species detected in the human GI-tract by molecular methods

Low G+C Gram-positive LAB	Bifidobacteria
[a]*L. acidophilus*	[b]*B. adolescentis*
L. crispatus	*B. angulatum*
L. brevis	*B. bifidus*
L. casei	*B. breve*
L. delbrueckii group	*B. catenulatum*
L. fermentum	*B. infantis*
L. gasseri	*B. longum*
L. paracasei	*B. pseudocatenulatum*
L. plantarum group	
L. rhamnosus	
L. ruminis	
L. sakei	
L. salivarius	
L. vaginalis	
L. curvatus	
L. fructivorans	
Pediococcus pentosaceous	
Leuconostoc argentinum	
Leuconostoc mesenteroides	

[a]*L, Lactobacillus.*
[b]*B, Bifidobacterium.*

shown to be remarkably stable and very complex in adults, less complex and more unstable in children, and was unstable and developing in infants (Zoetendal et al. 1998; Vaughan et al. 2000; Favier et al. 2002). The detection limit of PCR-DGGE was estimated at 10^5 cells per ml bacteria, which was influenced by the DNA isolation protocols using a dilution series of pure cultures and flow cytometry for counting (Zoetendal et al. 2001c). The sensitivity of the DGGE technique for LAB has been vastly improved by combining specific primers for bifidobacteria and lactobacilli groups (Satokari et al. 2001a; Walter et al. 2001; Heilig et al. 2002). Analysis of the composition of LAB by both culturing (McCartney et al. 1996; Kimura et al. 1997) and the culture independent PCR-DGGE methods indicated a unique strain collection for each individual. The bifidobacteria were generally stable over time and comprised often greater than 4 strains in adults. In infants, the number of bifidobacterial strains is usually lower, generally 1 to 3 (Matsuki et al. 1998; Satokari et al. 2002). The lactobacilli-leuconostoc group showed more variation especially in the relative proportions of strains and usually comprised less than 4 strains.

A list of typical LAB species found in the human GI-tract is presented in Table 1. *Bifidobacterium adolescentis* and the *Bifidobacterium catenulatum* group appear to be most common in adults, while *Bifidobacterium infantis* and *Bifidobacterium breve* are often found in infant faeces. Interestingly, allergic infants were reported to be colonised mainly by *B. adolescentis* species with a lower mucous binding capacity than bifidobacteria from healthy infants (He et al. 2001a). Another interesting point is that recent DGGE and sequencing of 16S rRNA indicated that the shift in species composition of bifidobacteria does not appear to occur in response to weaning as has often been assumed (Satokari et al. 2002). Similar bifidobacterial species were obtained by culturing compared to molecular methods. Culturing studies indicated that *Lactobacillus reuteri* and *Lactobacillus gasseri* were dominant in adult faeces, and *Lactobacillus ruminis* (formerly *Catenabacterium catenaforme*) was also a regular component. Culture independent monitoring so far pointed to *Lactobacillus ruminis* as the dominant type in the adult GI-tract, and *Lactobacillus gasseri* were common in colonic biopsies (Figure 1) (Tannock et al. 2000; Heilig et al. 2002; Zoetendal et al. 2002), but geographic differences may play a role here. *Lactobacillus acidophilus (sensu lato)* appears to be a dominant type in faeces of breast or formula-fed babies, especially as determined by molecular studies (Morelli et al. 1998; Satokari et al. 2002).

PCR-DGGE has been used to monitor faecal bifidobacterial and lactobacilli populations, including the presence of the LAB probiotic itself, in a prebiotic and probiotic clinical trials, and may also be used to monitor the presence of LAB during manufacture of a food product (Satokari et al. 2001b; Randazzo et al. 2002). In addition to tracking of the probiotic *Lactobacillus paracasei* F19 strain, DGGE profiles supported the presence of this strain as autochthonous within the intestinal community of a proportion of individuals (Crittenden et al. 2002; Heilig et al. 2002).

To further facilitate analysis of the composition and dynamics of microorganisms various high-throughput approaches are being developed. FISH using an array of 16S rRNA oligonucleotide probes combined with microscopy-based enumeration of intestinal bacteria has been automated and is analysed by computer software (Jansen et al. 1999). FISH in combination with flow cytometry (FCM) can be applied to identify and enumerate fluorescently labelled microbes hybridised with 16S rRNA-targeted probes in faecal samples as illustrated in Figure 2, as well as in biopsy samples (Zoetendal et al. 2001a). The FCM technique supports high speed and resolution,

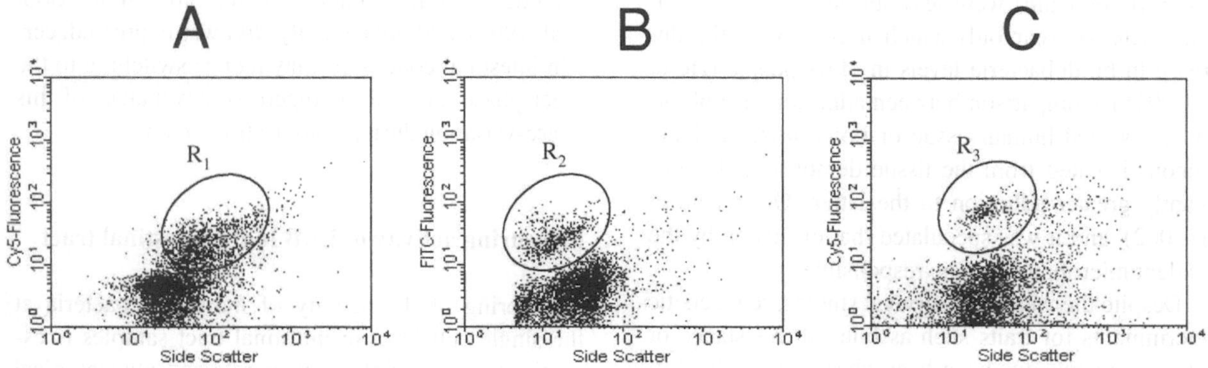

Figure 2. Flow cytometric analysis of an adult faecal sample hybridised with different 16S rRNA-targeted oligonucleotide probes. The faecal sample was hybridised with (A) Cy5-labelled Bac303 probe specific for the *Bacteroides* group, (B) FITC-labelled Erec482 probe that targets the *Clostridium coccoides* group, and (C) Cy5-labelled Bif164 probe specific for *bifidobacteria*. The positively hybridised cells with the Bac303, Erec482 and Bif164 probes, represented in region R_1, R_2 and R_3, respectively, are specifically enumerated by the flow cytometer.

and therefore can be automated more readily than microscopic quantification of stained microorganisms. Furthermore, FCM and activated cell sorting (FACS) offers the possibility to separate bacterial cells for subsequent analysis and physically enrich defined population with low frequency in a microbial ecosystem (Winson & Davey 2000). DNA arrays also have great potential for the analysis of microbial diversity by rapidly analysing RNA abundance or DNA homology of genes (reviewed by Ye et al. 2001).

Survival and colonisation of intestinal and probiotic LAB

The autochthonous microbiota persist and colonise various niches within the intestinal tract despite the harsh and changing environment. For ingested probiotic LAB, they must endure a rigorous passage through regions with different adversities or factors that can affect their survival (reviewed by Bezkorovainy 2001). Various studies have quantified the survival of LAB or probiotic strains during passage through the GI-tract using experiments involving intestinal intubation and culturing. *Lactobacillus plantarum* showed a high survival rate (25±29%) among several LAB strains being tested for their pharmacokinetics in the GI-tract (Vesa et al. 2000). In the upper intestinal tract, acidity in the stomach, and bile salts and pancreatic secretions in the small intestine are key antagonistic agents. Consequently, bile salts and acid resistance have become selection criteria for probiotic and vaccine delivery LAB strains. In addition the resident LAB either have the capacity to attach to the intestinal tissue or grow faster than the intestinal transit time in order to effectively colonise. Thus, adherence to intestinal cells is also considered a very valuable criterion for a new probiotic, and this ability is regarded as a prerequisite to exert certain beneficial effects, such as immunomodulation of the host or enteropathogenic exclusion (Schriffin et al. 1995; Mack et al. 1999). However, it is plausible that direct contact alone may be sufficient to provide probiotic effects as several studies indicate that ingested strains do not easily colonise the GI-tract (Vesa et al. 2000). Due to the limited availability of human intestinal tissue, *in vitro* propagated human intestinal cell lines, especially Caco-2, HT-29 and HT-29 MTX (mucous producing) are extensively used to study adhesion (Blum et al. 1999). Although derived from colonic carcinomas, these cell lines exhibit traits of differentiated absorptive enterocytes, and while a useful *in vitro* model also to study microbe-host interactions, their predictive value for *in vivo* adhesion remains questionable. In the intestine, the host tissue is covered with a protective layer of mucous that is composed of large glycoproteins called mucins of which different types are isolated from different tissues (Neutra & Forstner 1987). The mucous produced by the HT-29 MTX cells most closely resembles that of gastric mucins (Lesuffleur et al. 1993).

Besides the tissue culture cells, an intestinal mucous model using mucins isolated from faeces or ileostomy effluent and intestinal tissue have been used to a lesser extent (Ouwehand et al. 2001). The latter has the advantage that effects on host factors such as individuality, age and health may be studied. For example, *Bifidobacterium* species isolated from seniors (>70 year old) were less adhesive than those of adults (30–40 year old) which may explain the decrease in bifidobacteria levels in older people (He et al. 2001b). Comparison between adhesion of probiotics to resected human tissue or immobilised colonic mucous isolated from the tissue demonstrated significantly greater adhesion to the latter (Ouwehand et al. 2002), and it was speculated that exclusion by the resident microbiota may be responsible.

Despite the wealth of survival studies, few genetic determinants for traits such as bile salt resistance or adhesion to mucous have been characterised for human LAB isolates. Bile salt hydrolyase (*bsh*) genes, that may play a role in increased resistance to the toxic levels of bile salts in the gastrointestinal environment have been characterised for *Lactobacillus plantarum*, *Lactobacillus johnsonii* and *Bifidobacterium longum* (Christiaens et al. 1992; Ëlkins et al. 1998; Tanaka et al. 2000). The *bsh* negative mutants constructed in some cases may be used in animal model experiments to better ascertain their *in vivo* function. Potential cell surface adhesion determinants have been characterised for a *Lactobacillus reuteri* pig isolate, products of the *aggH* and *mub* genes responsible for an autoaggregating phenotype and adhesion to mucous, respectively (Roos et al. 1999; Roos & Jonsson 2002). Lipoteichoic acid was shown to be involved in adhesion *of Lactobacillus johnsonii* to the Caco-2 cell line (Granato et al. 1999). Other potential adhesin determinants include a collagen-binding protein (CnBP) of the human *L. reuteri* strain, and MAPP of *Lactobacillus fermentum*, that have strong similarities to the binding protein component in ATP-binding cassette (ABC) transporters suggesting that these proteins may have a double role (Patent 1996; Roos et al. 1996). It is anticipated that the completion of genome sequences of LAB should substantially advance our knowledge of GI-tract colonisation and survival factors.

The active mucosal immune system is another potential obstacle for intestinal LAB. Interestingly, *Bacteroides fragilis*, a well-known commensal microbe in the human colon was recently demonstrated to be able to modulate its surface antigenicity by displaying an array of different surface polysaccharide combinations (Krinos et al. 2001). The phase variation presumably allows it to circumvent the host immune response and to maintain its human colon niche. There are hints that similar mechanisms may occur in LAB intestinal commensals. Two S-layer encoding (*slp)* genes have been characterised for *Lactobacillus acidophilus,* and alternate expression of the two *slp* genes has been demonstrated under laboratory conditions (Boot et al. 1996). Although mainly *slpA* was expressed, certain intestinal conditions may induce switching to the other phase and offer competitive advantages of this phase-variation during colonisation of a host.

Measuring activity of LAB in the intestinal tract

Monitoring of the activity of individual bacteria at the single cell level in intestinal tract samples is essential to gain insight in their role and physiological state. Many studies confirm that allochthonous probiotic bacteria can pass through the complete GI-tract and be retrieved in reasonable viable numbers in faecal samples, and technologies to improve this survival are being developed (Tuomola et al. 2001; Mattila-Sandholm et al. 2002b). Growth and activity of the autochthonous LAB is supported by several studies including a recent comparison of the caecal (initial part of the right colon) and the faecal microbiota that revealed substantial differences both quantitatively and qualitatively including the LAB (Marteau et al. 2001). Bifidobacterial numbers, as determined by plate counts, were consistently lower in the caecal samples of 8 healthy subjects compared to faeces, 6.7 *versus* 8.9 (log 10 per ml), indicating their growth within or downstream of the right colon. On the other hand, using hybridisation with rRNA probes, the % RNA of bifidobacteria and the *Lactobacillus-Enterococcus* groups had decreased from 5.8 to 3.3 and 22.8 to 6.6, respectively, indicating their overall lower numbers and activity in faeces relative to the high density of the dominant microbiota (Marteau et al. 2001). Still information on the activity or state of viability of an indigenous or probiotic LAB at a particular location in the GI-tract is not clear. Several approaches to investigate the activity of microorganisms in a complex ecosystem based on various principles are possible. Examples of techniques adapted for LAB or where their applicability has been demonstrated for intestinal samples are described below.

The combination of different probes and flow cytometry has been applied to assess the viability of various LAB including lactobacilli and bifidobacteria during stress conditions (Bunthof et al. 2001; Ben-Amor et al. 2002). Carboxyfluorescein diacetate, bis-(1,3-dibutylbarbutiric acid) trimethine oxonol, propidium iodide, and TOTO-1 were used to monitor esterase activity, membrane potential, membrane integrity and/or membrane permeability of different

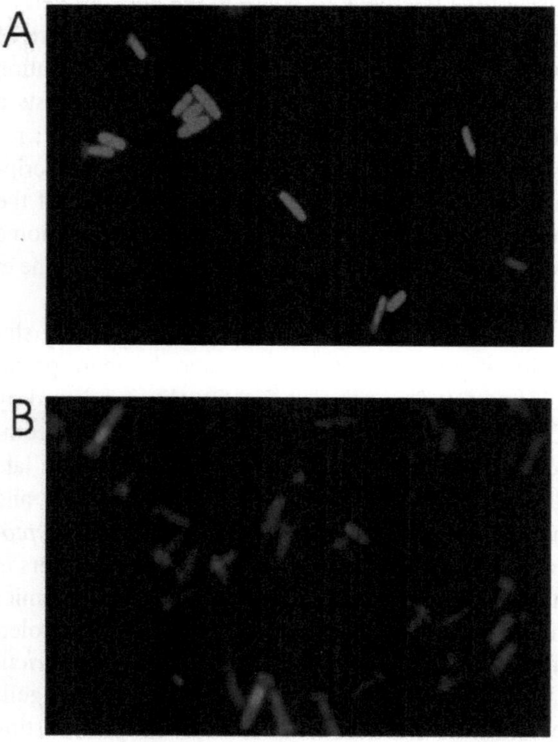

Figure 3. The variation in fluorescent intensity due to rRNA expression at early and late stationary growth stages, A and B, respectively, in *Lactobacillus plantarum* was performed by hybridisation with the universal bacterial oligonucleotide probe EUB338 labelled with FITC, and visualised by the fluorescent microscope. The fluorescence intensity is measured by image analysis software.

LAB. These physiological parameters served as good indicators to monitor viable, injured and dead cells exposed to heat, acid and bile salt stress.

Quantitative *in situ* hybridisation with fluorescent oligonucleotide probes (FISH) was used to estimate the growth rates of single *E. coli* cells in faeces, mucous and caecal contents from the intestinal tract of mouse models (Rang et al. 1999). Essentially, growth was measured by correlating the rRNA cellular content, as determined by fluorescent intensity, with the cell growth rate. Fluorescent intensity can be determined by a combination of microscopy and image analysis or flow cytometry. An example of the variation in fluorescent intensity due to rRNA expression in a pure culture of *Lactobacillus plantarum* as measured at different growth phases by FISH and image analysis is presented in Figure 3. Although there are many reports of quantitative *in situ* reverse-transcriptase-PCR of messenger RNAs (mRNA) using fluorescence for detection either by microscopy or flow cytometry, these techniques have not yet been applied to a complex ecosystem.

Several elegant recombinant DNA techniques using reporter genes including the Green Fluorescent Protein (GFP) and luciferase have been used to follow LAB such as *Lactococcus lactis* and *Lactobacillus plantarum* in the GI-tract of mice or rats (Corthier et al. 1998; Drouault et al. 1999; Geoffroy et al. 2000). Several other techniques involving flow chambers and microscopy have excellent potential to be applied to the GI-tract ecosystem. Biofilms cultured in flow chambers, in combination with scanning confocal laser microscopy allowed direct monitoring of promoter activity with GFP genes, and may be used for diverse applications such as monitoring rates of rRNA synthesis and visualisation of pheromone-mediated cell–cell communication between bacteria (Sternberg et al. 1999; Steidle et al. 2001).

In addition to the above single cell activity methods, gene expression may also be measured using RT-PCR of specific mRNA transcripts from total RNA preparations from intestinal samples. The expression of several specific *Helicobacter pylori* genes as well as 16S rRNA were monitored by real time RT-PCR in stomach biopsies during the course of a 6-month infection in mice. mRNA of the same genes could be detected in biopsies of the gastric mucosa from 15 infected persons (Rokbi et al. 2001). While presumably in the latter study the stomach environment was not so complex, mRNA of the adenosine-5′-phosphosulfate reductase gene of sulphate reducing bacteria was detected in many regions of mice GI-tract also by RT-PCR (DePlancke et al. 2000). A method to isolate RT-PCR grade total RNA from human faeces has been developed and validated by spiking fecal samples with *Lactobacillus acidophilus* and subsequent detection of the S-layer*(slpA)* transcript (N. Fitzsimons et al. unpublished results). The latter has potential for assessing specific functions of an ingested probiotic strain and may be useful for *in situ* monitoring of phase variation by the *slp* genes in *Lactobacillus acidophilus*.

Genomics based approaches and prospects

We are now at the onset of a genomics revolution that offers great potential to address more rapidly the interaction of the microbiota with the host (Hooper & Gordon 2001). The human GI-tract probably contains more than a 1000 microbial species with an average genome size of 5 megabases. Hence, it can be calculated that the size of the collective microbial genomes, also known as the microbiome (De Vos 2001; Hooper & Gordon 2001) is approaching that of the human genome (5 gigabases). There is an urgent need for genome sequences of commensal microbes but this is still a formidable task considering the numbers and the difficulties in culturing. Fortunately because of the significance of probiotic bacteria, genome sequencing of select LAB strains is ongoing or completed (reviewed by De Vos 2001; Klaenhammer et al. this issue). Some of these species are intestinal isolates including *Lactobacillus plantarum*, *Lactobacillus acidophilus*, *Bifidobacterium breve* and *Bifidobacterium longum*. A significant supplement to sequencing genomes of commensals is the construction of metagenomic libraries, consisting of large cloned DNA fragments in absence of culturing (Handelsman et al. 1998). This could represent an approach for cloning the microbiome and develop into a valuable strategy for accessing the genetic and functional diversity of the substantially uncultured GI-tract microorganisms.

Intestinal epithelial cells are constantly exposed to the resident microbiota of the GI-tract but still little is known about the interactions between the intestinal epithelial cells and the microbiota and their high concentrations of lumen products. Many remarkable studies are appearing that reveal signalling between the intestinal cells, microbiota and microbial products, and insight into the mechanisms and accompanying molecules and receptors involved (Haller et al. 2000; Cario et al. 2002; Vidal et al. 2002). For example, the probiotic strain *Lactobacillus plantarum* 299v and its cell-free supernatant induced intestinal mucin gene expression in the cell line HT-29, that in turn diminished the binding of enteric pathogens to the mucosal epithelial cells (Mack et al. 1999). Studies on host microbiota crosstalk may be powerfully enhanced by the development of high-throughput DNA arrays, and this can also significantly speed up research on effects of intestinal LAB and probiotics on the host. This approach was elegantly demonstrated in a landmark study of intestinal transcriptional responses of germfree mice with the commensal *Bacteroides thetaiotaomicron* using DNA arrays, and supported by laser capture microdissection (Hooper et al. 2001). Overall, DNA array technology will allow determination and parallel analyses of a large number of biomarkers or genes that are indicative of microbe–host interaction. The accumulation of genomic sequences for the intestinal and probiotic LAB will also allow analysis of the microbial response to the host, as well as microbe–microbe crosstalk. The ubiquitous nature of two-component systems involved in communication between Gram-positive intestinal bacteria suggests a relevant *in vivo* function (Nakayama et al. 2002; reviewed by Swift et al. 2000). Mutation and transcriptional profiling studies are aiding the analysis of the functionality of these systems and revealing genome wide effects of signal peptides (reviewed by Sturme et al. 2002).

The short half life of mRNA and contrasting stability of proteins suggests they may be more promising targets for analysing functionality (reviewed by Graves & Haystead 2002). The potential of proteomics for the LAB has been shown already in the late 1990s during study of the expression of cold- and heat-shock proteins in *Lactococcus lactis* and *Streptococcus thermophilus* (Kilstrup et al. 1997; Wouters et al. 1999a,b). Since then several reports of proteomics in lactobacilli have appeared concerning thermotolerance in *Lactobacillus delbrueckii* subsp. *bulgaricus* (Lim et al. 2000; Rechinger et al. 2000; De Angelis et al. 2001; Gouesbet et al. 2002). Many interesting applications of proteomics have been described that may be adapted to determine factors involved in the interaction of LAB commensals and probiotics with their host. Proteomics techniques are proving to be valuable for identification of proteins present in particular subcellular locations and even extracellularly. Studies investigating host–pathogen interactions may be especially heuristic such as the extracellular protein expression patterns described for *Streptococcus pyogenes* (Chaussee et al. 2001). Analysis of the outer surface proteins of *Streptococcus agalactiae* revealed that several enzymes with a presumed cytoplasmic location appeared instead on the surface (Hughes et al. 2002), indicating alternative or multiple functions for some proteins. Metabolomics (Phelps et al. 2002) or measurement of metabolites synthesised by the GI-tract ecosystem is another powerful analytic tool but a challenge will be to determine whether the metabolites are host or microbe in origin.

The relationship between the autochthonous microbiota and the host can change from a commensal if not symbiotic association toward pathogenicity under circumstances, in some instances which we do not yet comprehend. There is an increasing prevalence of intestinal disorders such as inflammatory bowel disease as well as allergies for which the microbiota are suspected or have been demonstrated to be involved in the aetiology. While we know that administration of therapeutic or prophylactic agents alters the microbial composition or balance, again there are many unanswered questions regarding the longterm effects. Moreover, the mechanisms of documented probiotic benefits for the most part require serious clarification. The identification of biomarker genes for microbe host gut interactions is necessary to aid in diagnoses of microbiota imbalances. Concerted multidisciplinary research performed in intestinal laboratories around the world is providing a greater understanding of the human microbiota in health and disease. The intestinal and probiotic LAB, along with health designed LAB, constitute a major part of this research that can have generous rewards concerning the quality of life for consumers and patients.

Acknowledgements

This work has been carried out with financialy support of the European Community specific RTD programme 'Quality of Life and Management of Living Resources' research projects Microbe Diagnostics (QLK1-2000-00108), Progid (QKL1-2000-00536) and EU Microfunction (QKL1-2001-00135).

References

Ben Amor K, Breeuwer P, Verbaarschot P, Rombouts FM, Akkermans ADL, de Vos WM & Abee T (2002) Multiparametric flow cytometry and cell sorting for the assessment of the viable, injured and dead cells of bifidobacteria during bile salt stress. Submitted.

Bengmark S (2000) Bacteria for optimal health. Nutr. 16: 611–615.

Bezkorovainy A (2001) Probiotics: determinants of survival and growth in the gut. Am. J. Clin. Nutr. 73S: 399S–405S.

Biavati B, Sorbati B & Scardovi V (1991) The genus *Bifidobacterium*. In: Balows A, Truper HG, Dworkin M, Harder W & Schleifer KH (Eds) The Prokaryotes, 2nd edn. (pp 816–833). Springer-Verlag, New York.

Blum S, Reniero R, Schiffrin EJ, Crittenden R, Mattila-Sandholm T, Ouwehand AC, Salminen S, von Wright A, Saarela M, Saxelin M, Collins K & Morelli L (1999) Adhesion studies for probiotics: need for validation and refinement. Trends Food Sci. Technol. 10: 405–410.

Boot HJ & Pouwels PH (1996) Expression, secretion and antigenic variation of bacterial S-layer proteins. Mol. Microbiol. 21: 1117–1123.

Bunthof CJ, Bloemen K, Breeuwer P, Rombouts FM & Abee T (2001) Flow cytometric assessment of viability of lactic acid bacteria. Appl. Environ. Microbiol. 67: 2326–2335.

Cario E, Brown D, McKee M, Lynch-Devaney K, Gerken G & Podolsky DK (2002) Commensal-associated molecular patterns induce selective toll-like receptor-trafficking from apical membrane to cytoplasmic compartments in polarized intestinal epithelium. Am. J. Path. 160: 165–173.

Cebra JJ (1999) Influences of microbiota on intestinal immune system development. Am. J. Clin. Nutr. 69: 1046S–1051S.

Chaussee MS, Watson RO, Smoot JC & Musser JM (2001) Identification of Rgg-regulated exoproteins of *Streptococcus pyogenes*. Infect. Immun. 69: 822–831.

Christiaens H, Leer RJ, Pouwels PH & Verstraete W (1992) Cloning and expression of a conjugated bile acid hydrolase gene from *Lactobacillus plantarum* by using a direct plate assay. Appl. Environ. Microbiol. 58: 3792–3798.

Collins MD, Rodrigues U, Aguirre M, Farrow JAE, Martinez-Murcia A, Philips BA, Williams AM & Wallbanks S (1991) Phylogenetic analysis of the genus *Lactobacillus* and related lactic acid bacteria as determined by reverse transcriptase sequencing of 16S rRNA. FEMS Microbiol. Lett. 77: 5–12.

Corthier G, Delorme C, Ehrlich SD & Renault P (1998) Use of luciferase genes as biosensors to study bacterial physiology in the digestive tract. Appl. Environ. Microbiol. 64: 2721–2722.

Crittenden R, Saarela M, Mättö J, Ouwehand AC, Salminen S, Pelto L, Vaughan EE, de Vos WM, von Wright A, Fondén R & Mattila-Sandholm T (2002) *Lactobacillus paracasei* F19: survival, ecology and safety in the human intestinal tract. Microbial Ecol. Health Dis. S3: 22–26.

De Angelis M, Bini L, Pallini V, Cocconcelli PS & Gobbetti M (2001) The acid-stress response in *Lactobacillus sanfranciscensis* CB1. Microbiology 147: 1863–1873.

DePlancke B, Hristova KR, Oakley HA, McCracken VJ, Aminov R, Mackie RI & Gaskins HR (2000) Molecular ecological analysis of the succession and diversity of sulfate-reducing bacteria in the mouse gastrointestinal tract. Appl. Environ. Microbiol. 66: 2166–2174.

De Roos NM & Katan MB (2000) Effects of probiotic bacteria on diarrhea, lipid metabolism, and carcinogenesis: a review of papers published between 1988 and 1998. Am. J. Clin. Nutr. 71: 405–411.

De Vos (2001) Advances in genomics for microbial food fermentations and safety. Curr. Opin. Biotechnol. 12: 493–498.

Drouault S, Corthier G, Ehrlich SD, Renault P (1999) Survival, physiology, and lysis of *Lactococcus lactis* in the digestive tract. Appl. Environ. Microbiol. 65: 4881–4886.

Dunne C, Murphy L, Flynn S, O'Mahony L, O'Halloran S, Feeney M, Morrissey D, Thornton G, Fitzgerald G, Daly C, Kiely B, Quigley EM, O'Sullivan GC, Shanahan F & Collins JK (1999) Probiotics: from myth to reality. Demonstration of functionality in animal models of disease and in human clinical trials. Antonie Van Leeuwenhoek 76: 279–292.

Ëlkins CA & Savage DC (1998) Identification of genes encoding conjugated bile salt hydrolase and transport in *Lactobacillus johnsonii* 100–100. J. Bacteriol. 180: 4344–4349.

Falk PG, Hooper LV, Midtvedt T & Gordon JI (1998) Creating and maintaining the gastrointestinal ecosystem: what we know and need to know from gnotobiology. Microbiol. Mol. Biol. Rev. 62: 1157–1170.

Favier C, Vaughan EE, De Vos WM & Akkermans ADL (2002) Molecular Monitoring of succession of bacterial communities in human neonates. Appl. Envir. Microbiol. 68: 219–226.

Franks AH, Harmsen HJM, Raangs GC, Jansen GJ, Schut F & Welling GW (1998) Variations of bacterial populations in human feces measured by fluorescent in situ hybridization with group-specific 16S rRNA-targeted oligonucleotide probes. Appl. Envir. Microbiol. 64: 3336–3345.

Geoffroy M-C, Guyard C, Quatannens B, Pavan S, Lange M & Mercenier A (2000) Use of green fluorescent protein to tag lactic acid bacterium strains under development as live vaccine vectors. Appl. Environ. Microbiol. 66: 383–391.

Gouesbet G, Jan G & Boyaval P (2002) Two-dimensional electrophoresis study of *Lactobacillus delbrueckii* subsp. *bulgaricus* thermotolerance. Appl. Environ. Microbiol. 68: 1055–1063.

Granato D, Perotti F, Masserey I, Rouvet M, Golliard M, Servin A & Brassart D (1999) Cell surface-associated lipoteichoic acid acts as an adhesion factor for attachment of *Lactobacillus johnsonii* La1 to human enterocyte-like Caco-2 cells. Appl. Environ. Microbiol. 65: 1071–1077.

Graves PR & Haystead TA (2002) Molecular biologist's guide to proteomics. Microbiol. Mol. Biol. Rev. 66: 39–63.

Haller D, Bode C, Hammes WP, Pfeifer AM, Schiffrin EJ & Blum S (2000) Non-pathogenic bacteria elicit a differential cytokine response by intestinal epithelial cell/leucocyte co-cultures. Gut 2000. 47: 79–87.

Handelsman J, Rondon MR, Brady SF, Clardy J & Goodman RM (1998) Molecular biological access to the chemistry of unknown soil microbes: a new frontier for natural products. Chem. Biol. 5: R245–249.

Harmsen HJ, Wildeboer-Veloo AC, Raangs GC, Wagendorp AA, Klijn N, Bindels JG & Welling GW (2000) Analysis of intestinal flora development in breast-fed and formula-fed infants by using molecular identification and detection methods. J. Pediatr. Gastroenterol. Nutr. 30: 61–67.

Harmsen HJM, Elfferich P, Schut P & Welling GW (1999a) A 16S rRNA-targeted probe for detection of lactobacilli and enterococci in faecal samples by fluorescent in situ hybridisation. Microbial Ecology in Health and Disease 11: 3–12.

Harmsen HJM, Gibson GR, Elfferich P, Raangs GC, Wildeboer-Veloo ACM, Argaiz A, Roberfroid MB & Welling GW (1999b) Comparison of viable cell counts and fluorescence in situ hybridisation using specific rRNA-based probes for the quantification of human fecal bacteria. FEMS Microbiol. Lett. 183: 125–129.

He F, Ouwehand AC, Isolauri E, Hashimoto H, Benno Y & Salminen S (2001a) Comparison of mucosal adhesion and species identification of bifidobacteria isolated from healthy and allergic infants. FEMS Immunol. Med. Microbiol. 30: 43–47.

He F, Ouwehand A, Isolauri E, Hosoda M, Benno Y & Seppo S (2001b) Differences in composition and mucosal adhesion of bifidobacteria isolated from healthy adults and healthy seniors. Curr. Microbiol. 43: 351–354.

Heilig GHJ, Zoetendal EG, Vaughan EE, Marteau P, Akkermans ADL & de Vos WM (2002) Molecular diversity of *Lactobacillus* spp. and other lactic acid bacteria in the human intestine as determined by specific amplification of 16S ribosomal DNA Appl. Envir. Microbiol. 68: 114–123.

Hooper LV & Gordon JI (2001) Commensal host-bacterial relationships in the gut. Science 292: 1115–1118.

Hooper LV, Wong MH, Thelin A, Hansson L, Falk PG & Gordon JI (2001) Molecular analysis of commensal host–microbial relationships in the intestine. Science 291: 881–884.

Hughes MJ, Moore JC, Lane JD, Wilson R, Pribul PK, Younes ZN, Dobson RJ, Everest P, Reason AJ, Redfern JM, Greer FM, Paxton T, Panico M, Morris HR, Feldman RG & Santangelo JD (2002) Identification of major outer surface proteins of *Streptococcus agalactiae*. Infect. Immun. 70: 1254–1259.

Jansen GJ, Wildeboer-Veloo AC, Tonk RH, Franks AH & Welling GW (1999) Development and validation of an automated, microscopy-based method for enumeration of groups of intestinal bacteria. J. Microbiol. Methods 37: 215–221.

Kilstrup M, Jacobsen S, Hammer K & Vogensen FK (1997) Induction of heat shock proteins DnaK, GroEL, and GroES by salt stress in *Lactococcus lactis*. Appl. Environ. Microbiol. 63: 1826–1837.

Kimura K, McCartney AL, McConnell MA & Tannock GW (1997) Analysis of fecal populations of bifidobacteria and lactobacilli and investigations of the immunological responses of their human hosts to the predominant strains. Appl. Environ. Microbiol. 63: 3394–3398.

Kitts LC (2001) Terminal restriction fragment patterns: a tool for comparing microbial communities and assessing community dynamics. Curr. Issues Intest. Microbiol. 2: 17–25.

Klaenhammer TRK, Altermann E, Arigoni F, Bolotin A, Breidt F, Broadbent J, Cano R, Chaillou S, Deutscher J, Gasson M, van de Guchte M, Guzzo J, Hawkins T, Hols P, Hutkins R, Kleerebezem M, Kok J, Kuipers O, Lubbers M, Maguin E, McKay L, Mills D, Nauta A, Overbeek R, Pel H, Pridmore D, Saier M, van Sinderen D, Sorokin A, Steele J, O'Sullivan D, de Vos W, Weimer B, Zagorec M & Siezen R (2002) Discovering lactic acid bacteria by genomics Antonie van Leeuwenhoek. This issue.

Kleessen B, Bezirtzoglu & Matto J (2000) Culture based knowledge on biodiversity, development and stability of human gastrointestinal microflora. Microb. Ecol. Health Dis. Suppl. 2: 53–63.

Krinos CM, Coyne MJ, Weinacht KG, Tzianabos AO, Kasper DL & Comstock LE (2001) Extensive surface diversity of a commensal microorganism by multiple DNA inversions. Nature 414: 555–558.

Langendijk PS, Schut F, Jansen GJ, Raangs GC, Kamphuis G, Wilkinson MHF & Welling GW (1995) Quantitative fluorescence *in situ* hybridisation of *Bifidobacterium* spp. with genus-specific 16S rRNA targetted probe and its application in fecal samples. Appl. Environ. Microbiol. 61: 3069–3075.

Lesuffleur T, Porchet N, Aubert JP, Swallow D, Gum JR, Kim YS, Real FX & Zweibaum A (1993) Differential expression of the human mucin genes MUC1 to MUC5 in relation to growth and differentiation of different mucus-secreting HT-29 cell subpopulations. J. Cell. Sci. 106: 771–83.

Lim EM, Ehrlich SD & Maguin E (2000) Identification of stress-inducible proteins in *Lactobacillus delbrueckii* subsp. *bulgaricus*. Electrophoresis 21: 2557–2561.

Mack DR, Michail S, Wei S, McDougall L & Hollingsworth MA (1999) Probiotics inhibit enteropathogenic *E. coli* adherence in vitro by inducing intestinal mucin gene expression. Amer. J. Physiol. 276: G941–G950.

Mackie RI, Sghir A & Gaskins HR (1999) Developmental microbial ecology of the neonatal gastrointestinal tract. Am. J. Clin. Nutr. 1999 69: 1035S–1045S.

Marteau P, Pochart P, Dore J, Bera-Maillet C, Bernalier A & Corthier G (2001) Comparative study of bacterial groups within the human cecal and fecal microbiota. Appl. Environ. Microbiol. 67: 4939–4942.

Matsuki T, Watanabe K, Tanaka R & Oyaizu H (1998) Rapid identification of human intestinal bifidobacteria by 16S rRNA-targeted species- and group-specific primers. FEMS Microbiol. Lett. 167: 113–121.

Mattila-Sandholm T, Blaut M, Daly C, De Vuyst L, Dore J, Gibson G, Goossens H, Knorr D, Lucas J, Lähteenmaki L, Mercenier A, Saarela M, Shahanan F & de Vos WM (2002a) Food, GI-tract functionality and Human health cluster, PROEUHEALTH. Microbial Ecol. Health Dis. In press.

Mattila-Sandholm T, Myllärinen, Crittenden R, Mogensen G, Fondén R & Saarela M, (2002b) Technological challenges for future probiotic foods. Int. Dairy J. 12: 173–182.

McCartney AL, Wenzhi W & Tannock GW (1996) Molecular analysis of the composition of the bifidobacterial and lactobacillus microflora of humans. Appl. Environ. Microbiol. 62: 4608–4613.

Mitsuoka T (1992) The human gastrointestinal tract. In: Wood BJB (Ed) The Lactic Acid Bacteria. Vol.1, The Lactic Acid Bacteria in Health and Disease (pp 69–114). Elsevier Applied Science, London, UK.

Miyake T, Watanabe K, Watanabe T & Oyaizu H (1998) Phylogenetic analysis of the genus *Bifidobacterium* and related genera based on 16S rDNA sequences. Microbiol. Immunol. 42: 661–667.

Moore WE & Holdeman LV (1974) Human fecal flora: the normal flora of 20 Japanese-Hawaiians. Appl. Microbiol. 27: 961–979.

Morelli L, Cesena C, de Haen C & Gozzini L (1998) Taxonomic *Lactobacillus* composition of feces from human newborns during the first few days. Microb. Ecol. 35: 205–212.

Muyzer G & Smalla K (1998) Application of denaturing gradient gel electrophoresis (DGGE) and temperature gradient gel electrophoresis (TGGE) in microbial ecology. Antonie van Leeuwenhoek 73: 127–141.

Nakayama J, Akkermans ADL & de Vos WM (2002) Genomic survey of two-component regulatory systems putatively involved in peptide pheromone mediated quorum sensing of low G+C gram-positive bacteria. Submitted.

Neutra RN & Forstner JF (1987) Gastrointestinal mucus: synthesis, secretion and function. In: Johnson LR (Ed) Physiology of the Gastrointestinal Tract, 2nd edn. Raven Press, New York.

Ouwehand AC, Salminen S, Tölkkö S, Roberts P, Ovaska J & Salminen E (2002) Resected human colonic tissue: new model for characterizing adhesion of lactic acid bacteria. Clin. Diagn. Lab. Immunol. 9: 184–186.

Ouwehand AC, Tuomola EM, Lee YK & Salminen E (2001) Microbial interactions to intestinal mucosal models. Methods Enzymol. 337: 200–212.

Patent, TNO, International Patent (1996) PCT/NL96/00409.

Phelps TJ, Palumbo AV & Beliaev AS (2002) Metabolomics and microarrays for improved understanding of phenotypic characteristics controlled by both genomics and environmental constraints. Curr. Opin. Biotechnol. 13: 20–24

Randazzo CL, Torriani S, Akkermans ADL, de Vos WM & Vaughan EE (2002) Diversity, dynamics and activity of bacterial communities during production of an artisanal sicilian cheese as evaluated by 16S rRNA analysis. Appl Environ. Microbiol. 68: 1882–1892.

Rang CU, Licht TR, Midtvedt T, Conway PL, Chao L, Krogfelt KA, Cohen PS & Molin S (1999) Estimation of growth rates of *Escherichia coli* BJ4 in streptomycintreated and previously germfree mice by in situ rRNA hybridisation. Clin. Diag. Lab. Immunol. 6: 434–436.

Rechinger KB, Siegumfeldt H, Svendsen I & Jakobsen M (2000) 'Early' protein synthesis of *Lactobacillus delbrueckii* ssp. *bulgaricus* in milk revealed by [35S] methionine labeling and two-dimensional gel electrophoresis. Electrophoresis 21: 2660–2669.

Reuter G (2001) The *Lactobacillus* and *Bifidobacterium* microflora of the human intestine: composition and succession. Curr. Issues Intest. Microbiol 2: 43–53.

Rokbi B, Seguin D, Guy B, Mazarin V, Vidor E, Mion F, Cadoz M & Quentin-Millet M-J (2001) Assessment of *Helicobacter pylori* gene expression within mouse and human gastric mucosae by real time reverse transcriptase PCR. Infect. Immun. 69: 4759–4766.

Roos S & Jonsson H (2002) A high-molecular-mass cell-surface protein from *Lactobacillus reuteri* 1063 adheres to mucus components. Microbiology 148: 433–442.

Roos S, Lindgren S & Jonsson H (1999) Autoaggregation of *Lactobacillus reuteri* is mediated by a putative DEAD-box helicase. Mol Microbiol. 32: 427–436.

Roos S, Aleljung P, Robert N, Lee B, Wadstrom T, Lindberg M & Jonsson H (1996) A collagen binding protein from *Lactobacillus reuteri* is part of an ABC transporter system? FEMS Microbiol. Lett. 144: 33–38.

Salminen S, Deighton MA, Benno Y & Gorbach SL (1998) Lactic acid bacteria in health and disease. In: Salminen S & von Wright A (Eds) Lactic Acid Bacteria. Microbiology and Functional Aspects, 2nd edn. (pp 211–253). Marcel Dekker, Inc., New York.

Satokari RM, Vaughan EE, Akkermans AD, Saarela M & de Vos WM (2001a) Bifidobacterial diversity in human feces detected by genus-specific PCR and denaturing gradient gel electrophoresis. Appl. Environ. Microbiol. 67: 504–513.

Satokari RM, Vaughan EE, Akkermans AD, Saarela M & De Vos WM (2001b) Polymerase chain reaction and denaturing gradient gel electrophoresis monitoring of fecal *Bifidobacterium* populations in a prebiotic and probiotic feeding trial. Syst. Appl. Microbiol. 24: 227–231.

Satokari RM, Vaughan EE,Favier CF,Doré J, Edwards C& de Vos WM(2002) Diversity of *Bifidobacterium* and *Lactobacillus* spp. in breast-fed and formula-fed infants as assessed by 16S rDNA sequence differences. Microbiol. Ecol. Health Dis. In press.

Schiffrin EJ, Rochat F, Link-Amster H, Aeschlimann JM & Donnet-Hughes A (1995) Immunomodulation of human blood cells following the ingestion of lactic acid bacteria. J. Dairy Sci. 78: 491–497.

Schleifer KH & Ludwig W (1995) Phylogenetic relationships of lactic acid bacteria. In: Wood BJB & WH Holzapfel (Eds) The Genera of Lactic Acid Bacteria: The Lactic Acid Bacteria, Vol. 2 (pp 7–17). Chapman and Hall, Glasgow.

Sghir A, Gramet G, Suau A, Rochet V, Pochart P & Dore J (2000) Quantification of bacterial groups within human fecal flora by oligonucleotide probe hybridization. Appl. Environ. Microbiol. 66: 2263–2266.

Shanahan F (2001) Turbo probiotics for IBD. Gastroenterology 120: 1297–1298.

Simmering R & Blaut M (2001) Pro- and prebiotics – the tasty guardian angels? Appl. Microbiol. Biotechnol. 55: 19–28.

Steidle A, Sigl K, Schuhegger R, Ihring A, Schmid M, Gantner S, Stoffels M, Riedel K, Givskov M, Hartmann A, Langebartels C & Eberl L (2001) Visualization of N-acylhomoserine lactone-mediated cell-cell communication between bacteria colonizing the tomato rhizosphere. Appl. Environ. Microbiol. 67: 5761–5770.

Steidler L, Hans W, Schotte L, Neirynck S, Obermeier F, Falk W, Fiers W & Remaut E (2000) Treatment of murine colitis by *Lactococcus lactis* secreting interleukin-10. Science 289: 1352–1355.

Sternberg C, Christensen BB, Johansen T, Toftgaard Nielsen A, Andersen JB, Givskov M & Molin S (1999) Distribution of bacterial growth activity in flow-chamber biofilms. Appl. Environ, Microbiol. 65: 4108–4117.

Sturme MHJ, Kleerebezem M, Nakayama J, Akkermans ADL, Vaughan EE & de Vos WM (2002) Cell to cell communication by autoinducing peptides in gram-positive bacteria. Antonie Van Leeuwenhoek. In press.

Suau A, Bonnet R, Sutren M, Godon JJ, Gibson GR, Collins MD & Dore J (1999) Direct analysis of genes encoding 16S rRNA from complex communities reveals many novel molecular species within the human gut. Appl. Environ. Microbiol. 65: 4799–4807.

Swift S, Vaughan EE & de Vos WM (2000) Quorum sensing within the gut ecosystem. Microbial Ecol. Health Dis. 2S: 88–92.

Tanaka H, Hashiba H, Kok J & Mierau I (2000) Bile salt hydrolase of *Bifidobacterium longum* – biochemical and genetic characterization. Appl. Environ. Microbiol. 66: 2502–2512.

Tannock GW (1999) Analysis of the intestinal microflora: a renaissance. Antonie Van Leeuwenhoek 76: 265–278.

Tannock GW, Munro K, Harmsen HJM, Welling GW, Smart J & Gopal PK (2000) Analysis of the faecal microflora of human subjects consuming a probiotic product containing *Lactobacillus rhamnosus* DR20. Appl. Environ. Microbiol. 66: 2578–2588.

Tuomola E, Crittenden R, Playne M, Isolauri E & Salminen S (2001) Quality assurance criteria for probiotic bacteria. Am. J. Clin. Nutr. 73(2S): 393S–398S.

Vaughan EE, Mollet B & de Vos WM (1999) Functionality of probiotics and intestinal lactobacilli: light in the intestinal tract tunnel. Curr. Opin. Biotech. 10: 505–510.

Vaughan EE, Schut F, Heilig GHJ, Zoetendal EG, de Vos WM & Akkermans ADL (2000) A molecular view of the intestinal ecosystem. Curr. Issues Intest. Microbiol. 1: 1–12.

Vesa T, Pochart P & Marteau P (2000) Pharmacokinetics of *Lactobacillus plantarum* NCIMB 8826, *Lactobacillus fermentum* KLD, and *Lactococcus lactis* MG1363 in the human gastrointestinal tract. Aliment Pharmacol Ther. 14: 823–828.

Vidal K, Donnet-Hughes A & Granato D (2002) Lipoteichoic acids from *Lactobacillus johnsonii* strain La1 and *Lactobacillus acidophilus* strain La10 antagonize the responsiveness of human intestinal epithelial HT29 cells to lipopolysaccharide and gram-negative bacteria. Infect. Immun. 70: 2057–2064.

Walter J, Hertel C, Tannock GW, Lis CM, Munro K & Hammes WP (2001) Detection of *Lactobacillus, Pediococcus, Leuconostoc,* and *Weissella* species in human feces by using group-specific PCR primers and denaturing gradient gel electrophoresis. Appl. Environ. Microbiol. 67: 2578–2585.

Wilson (1995) The gastrointestinal microflora. In Yamada T (Ed) Textbook of Gastroenterology (pp 607–615). JP Lippincott, Philadelphia.

Winson MK & Davey HM (2000) Flow cytometric analysis of microorganisms. Methods 21: 231–240.

Wouters JA, Jeynov B, Rombouts FM, de Vos WM, Kuipers OP & Abee T (1999a) Analysis of the role of 7 kDa cold-shock proteins of *Lactococcus lactis* MG1363 in cryoprotection. Microbiology 145: 3185–3194.

Wouters JA, Rombouts FM, de Vos WM, Kuipers OP & Abee T (1999b) Cold shock proteins and low-temperature response of *Streptococcus thermophilus* CNRZ302. Appl. Environ. Microbiol. 65: 4436–4442.

Ye RW, Wang T, Bedzyk L & Croker KM (2001) Applications of DNA microarrays in microbial systems. J. Microbiol. Methods 47: 257–272.

Zoetendal EG, Akkermans ADL & de Vos WM (1998) Temperature gradient gel electrophoresis analysis of 16S rRNA from human faecal samples reveals stable and host specific communities of active bacteria. Appl. Environ. Microbiol. 64: 3854–3859.

Zoetendal EG, Akkermans ADL & de Vos WM (2001a) Molecular characterisation of bacterial communities in the human gastrointestinal tract. PhD thesis, Wageningen University, The Netherlands.

Zoetendal EG, Akkermans ADL, Akkermans-van Vliet WM, de Visser JAGM & de Vos WM (2001b) The host genotype affects the bacterial community in the human gastrointestinal tract. Microbial Ecol. Health Dis. 13: 129–134.

Zoetendal EG, Ben-Amor K, Akkermans ADL, Abee T & de Vos WM (2001c) DNA isolation protocols affect the detection limit of PCR approaches of bacteria in samples from the human gastrointestinal tract. System. Appl. Microbiol. 24: 405–410.

Zoetendal EG, von Wright A, Vilpponen-Salmela T, Ben-Amor K, Akkermans ADL & de Vos WM (2002) Mucosa-associated bacteria in the human gastrointestinal tract are uniformly distributed along the colon and differ from the community recovered from feces. Appl. Environ. Microbiol. 68: 3401–3407.

Antonie van Leeuwenhoek **82:** 353–360, 2002.

Lactic acid bacteria in a changing legislative environment

Jean Feord
TNO Bibra International Ltd., Woodmansterne Road, Carshalton, Surrey, SM5 4DS UK

Key words: dietary supplement, feed, food, genetically modified organism, lactic acid bacteria, microorganism

Abstract

The benefits of using lactic acid bacteria in the food chain, both through direct consumption and production of ingredients, are increasingly recognised by the food industry and consumers alike. The regulatory environment surrounding these products is diverse, covering foods and food ingredients, processing aids, feed additives and dietary supplements. On a global basis, there are different approaches taken by the various regulatory authorities. While in Europe, the national legislation is gradually being harmonised, predominantly through the Novel Foods Regulation, there is still a wide disparity between the stringency of regulation of microbial products fed to animals and the comparatively relaxed approach to 'non-novel' microbial products intended for human consumption. In the United States, the onus is on self-regulation of the manufacturer, with the Generally Recognised As Safe (GRAS) and Dietary Supplement Health Education Act (DSHEA) notification schemes encouraging industry to be more open about the ingredients they market. In Japan, the Foods for Special Health Use system continues to gain recognition as more products are approved, and is a potential model for other countries in regulating functional foods. Despite the different approaches to regulating these products, safety of microorganisms such as lactic acid bacteria in the food chain is paramount in all countries. This paper discusses the regulatory requirements of microbial products, predominantly lactic acid bacteria within the global markets, focusing mainly on the developments in Europe.

Introduction

Lactic Acid Bacteria play a prominent role in many aspects of food development and health. They have a history of use in dairy, meat and fermentation products, where their properties have been harnessed to manufacture products such cheese, yoghurts, fermented milk products, beverages, sausage and other meat products. More recently, with the advent of so-called 'probiotics', lactic acid bacteria have received serious attention from those interested in the health benefits that have been purported to be derived from them.

While consumers are readily adopting these developing concepts of nutrition, the legislative environment is slowly adapting to encompass all potential applications of these products. Applications are abundant within the food chain, and with a worldwide view, their categorisations are diverse: food ingredients, processing aids, food supplements, novel foods, functional foods and feed additives. It is the aim of this paper to examine the regulatory frameworks within which these products are categorised in the main global markets, including Europe, USA and Japan.

Safety

Europe

The European regulatory environment for lactic acid bacteria is probably one of the most diverse, due mainly to the fact that legislation in individual Member States is only gradually being harmonised, the first example of this in the food area being the Novel Foods Regulation (258/97/EC) in 1997. Alongside this regulation, which is only applicable to newly discovered strains, or genetically modified lactic acid bacteria, national legislation continues to be enforced concerning existing strains. The term 'national legislation' is used loosely here, as in many countries microorganisms used in foods are not regulated as such – microorganisms are categorised as either ingredients or as processing aids.

Ingredients and processing aids

Within the EU, there has never been a harmonised authorisation process for placing microorganisms into foods, or using them as processing aids in the manufacture of a variety of products. Where a microorganism is considered to be a food ingredient, it is required to be labelled on the final food product. In countries that classify microorganisms as processing aids, there is no requirement to indicate their use on the label of the final food. Prior to the introduction of the Novel Foods Regulation, the manufacture of new microorganism strains was required to be notified to the authorities in some countries, for example in Denmark. The notification was based on the strain characterisation, and assessment of the safe use of the strain in food. While this process has more or less been encompassed into 258/97/EC, applicable only to microorganisms not previously consumed in Europe, it highlights the seeming lack of regulation of microorganisms in general food use prior to 1997, in many European countries.

Novel foods

Regulation (EC) No 258/97 of the European Parliament and of the Council of 27 January 1997 concerning novel foods and novel food ingredients encompasses microorganisms under two potential categories:

(a) foods and food ingredients containing or consisting of genetically modified organisms within the meaning of 90/220/EEC (Council Directive 90/220/EEC is repealed by Directive 2001/18/EC).

(b) foods and food ingredients consisting of or isolated from microorganisms, fungi or algae.

The definition of a novel food or food ingredient being given as:

> *"foods and foods ingredients which have not hitherto been used for human consumption to a significant degree within the Community"*

This means that newly identified strains of lactic acid bacteria which do not have a history of use as ingredients or processing aids in the EU should be treated as novel foods. However, given the complex nature of microorganism classification schemes, and changes in classification over time, it is unlikely that completely new, wild type lactic acid bacteria strains would be identified, and need to be approved.

It is instead the approval of lactic acid bacteria which have been genetically modified which is anticipated to be the most likely reason that the microorganism industry will come within the scope of the Novel Foods Regulation. At the time of submission of this paper, no genetically modified microorganism (GMM) has been approved for use as a food ingredient or processing aid. While this may be due more to current consumer feeling towards genetically modified organisms (GMOs) in general than to specific safety issues of new products, safety of new microorganisms, both GMO and non-GMO is key to the approval as a novel food.

The primary factors to be assessed for a novel food are addressed in Commission Recommendation of 29 July 1997 concerning the scientific aspects and the presentation of information necessary to support applications for the placing on the market of novel foods and novel food ingredients and the preparation of initial assessment reports under Regulation (EC) No 258/97 of the European Parliament and of the Council (97/618/EC):

- Effect of the genetic modification on the properties of the host organism
- Genetic stability of the GMO as a novel food source
- Specificity of expression of novel genetic material
- Transfer of genetic material from the GMO
- Ability of the GMM to survive and colonise in the human gut
- Anticipated intake /extent of use of the novel food
- Information from previous exposure to the novel food or its source
- Nutritional information on the novel food
- Microbiological information on the novel food
- Toxicological information (toxicity studies *in vitro* and *in vivo* including mutagenicity, 90-day feeding study, metabolism, toxicokinetics, chronic toxicity/carcinogenicity, reproduction and teratogenicity, neurotoxicity, immunotoxicity, allergenicity)

Many lactic acid bacteria are used not only as direct ingredients, but also as production vehicles for other ingredients. For example, products such as polysaccharides, protein compounds or fat products could be produced from microbial fermentation, which may also be treated as novel foods, especially if the microorganism has been genetically modified, or is in itself novel by definition. In cases where microorganisms are used as a vehicle for production, the safety of the microorganism is required to be evaluated (as described below) alongside that of the final product.

Some ingredients produced this way could potentially be shown to be substantially equivalent to similar ingredients produced by different methods, especially if the producing microorganism has been modified to receive genetic material from a 'non-novel' organism. However, the product would still be classed as novel due to the novel production process. The final product would be subject to the minimum safety package (subchronic toxicity and mutagenicity), developing into a more extensive package if safety concerns are discovered during the initial studies.

The regulation recognises that if a microorganism has no history of safe use in Europe, it is unlikely to be substantially equivalent to an existing strain, therefore safety must be assessed (although for self-cloned microorganisms substantial equivalence may be applicable). The key criteria for the safety assessment of microorganisms include containment, potential to colonise the mammalian gut, potential for toxicity and pathogenicity in mammals and potential of transfer of genetic material from genetically modified micoorganisms.

In the case of genetically modified microorganisms both the segregational properties, the safety of the gene product and its effect on the novel food and the properties of the microorganism, and the horizontal properties, the potential for gene transfer to occur in the gut, are required to be examined.

For genetically modified microorganisms, the evaluation procedure will be simplified if the parent (recipient) microorganism is compliant with one of the two following criteria:

- A microorganism with a tradition of food fermentation in the Community
  - o Non-pathogenic
  - o Biologically advantageous human intestinal commensal

or

- A microorganism traditionally used as a production organism for foods, food additives or technical aids.

In such cases, the safety of the parent microorganism can be assumed based on its history of safe use in food, and therefore the safety evaluation focuses on the resultant genetically modified microorganism. In other cases, where the parent microorganism does not have a history of safe use, both the parent microorganism and that of the resultant genetically modified microorganism will be subject to evaluation for being novel.

The other factors which are particularly applicable to genetically modified microorganisms are related to the stability of the novel microorganism, both under the conditions of use, and if the microorganism is intended to colonise the human gut. This is interrelated with the ability of the microorganism to transfer genetic material to other microorganisms within the human host, and safety concerns related to this occurrence. This point is especially pertinent to lactic acid bacteria as some strains are known to possess antibiotic resistance marker genes, which may be transferred due to their position on plasmids and transposons.

This particular trait of lactic acid bacteria has received much attention in the field of animal nutrition, where lactic acid bacteria are receiving increasing interest as feed additives, due to their unique properties in offering potential alternatives to antibiotic growth promoters, which are being phased out in the EU.

Animal feeds

Lactic acid bacteria fall within the scope of Council Directive 70/524/EEC concerning additives in feedingstuffs. Any microorganism intended for use as a feed additive is required to be assessed according to Council Directive 87/153/EEC fixing guidelines for the assessment of additives in animal nutrition. From a safety perspective, the assessment of microorganisms is somewhat different from other additives (in food and feed), due to the nature of the products. At present, no genetically modified microorganisms have been authorised as feed additives, but the diversity of naturally occurring microorganisms that may potentially be utilised in animal feeds is wide. As a consequence, awareness of potential safety concerns related to certain microorganisms has increased, and there is now much focus on the safety of microorganisms prior to authorisation.

To date, there have been 12 feed additives authorised containing lactobacillus species (Salminen & Wright 1998) (*L. farciminis, L. rhamnosus, L. casei, L. plantarum, Enterococcus faecium, Pediococcus acidilactici, Streptococcus infantarius*). Compare this to the 45 Identification Notes that were submitted for products containing lactobacillus species, under Council Directive 93/113/EEC, permitting temporary authorisation of products already on the market (published in the Official Journal of the European Communities, 96/C 263.03), prior to the dossier authorisation process, according to guidelines given in Commission Directive 94/40/EEC. The species covered by the 45 submissions included *L. plantarum, L.*

farciminis, L. bulgaricus, L. rhamnosus, L. acidophilus, L. brevis, L. reuteri, L. casei, L. delbrueckii spp. *bulgaricus, L. salivarius, L. fermentum, L. amylovorus, Pediococcus acidilactici, Streptococcus thermophilus, Streptococcus salivarius, Streptococcus infantarius, Enterococcus faecium.*

While it is not possible to know the reason for the failure to obtain authorisation for the other 33 products, it is interesting to note the species of interest in the field of animal nutrition. Many of these are commonly used in the food industry, and Lactobacillus acidophilus is a very common product in health food shops across Europe. Whether the failure to authorise these 33 products reflects on the safety of these microorganisms, or simply the inability of the companies concerned to generate the required data, the moves towards stricter regulation of such products in animal feedingstuffs may signal similar moves in foodstuffs in the future.

In order to gain an authorisation for a microorganism as a feed additive, the following safety issues must be addressed:

- Genetic stability
- Toxins and virulence factors
- Antibiotic production and antibiotic resistance
- Tolerance in target species
- Effect of the microflora in the digestive tract
- Genotoxicity
- Oral toxicity
- Worker safety
- Environmental risk assessment

In addition to this, the efficacy of the microorganism must be demonstrated in the target species, thus preventing entrance to the market of products which have no measurable effect. The strains registered must also be deposited at an internationally recognised culture collection, thereby adding a further level of control on strains in the feed chain.

In the recently updated guidelines for microorganism feed additives, it is now a requirement for microorganisms to be subject to a 90-day toxicity study. This is probably due to raised awareness of potential safety concerns related to certain microorganism species. As already mentioned, the aspect of safety which is receiving a great deal of attention in the EU at present is that of the naturally occurring presence of antibiotic resistance markers. The Scientific Committee for Animal Nutrition (SCAN) recently addressed this issue in its paper "Criteria for assessing the safety of microorganisms resistant to antibiotics of human clinical and veterinary importance". The presence of antibiotic resistance markers does not automatically exclude the microorganism from authorisation – the risk assessment is based on the genetic basis of the antibiotic resistance properties of the microorganism (acquired or intrinsic), and the type of antibiotic for which resistance is expressed. A scheme of the assessment is given in Figure 1.

When considering the lengths to which the European Commission have gone to in order to ensure safe use of microbial species in animal feeds, which has resulted in the exclusion of many products from the market, it begs the question as to why this has not been conducted in the field of human nutrition. Aside from the novel microorganisms, microbial preparations have a long history of use in food production, and more recently, in development of probiotics. Probiotics are used as food ingredients (increasingly popular in dairy products such as yoghurts and drinks) and food supplements, which fall within a grey area between food and medicines in many European countries. The European Commission has recognised the need to harmonise the control of food supplements, and has drafted a proposed directive on the approximation of laws of the Member States on food supplements (COM/2001/0159 final – COD 2000/0080*/). The work has started in assigning safe intake levels of vitamins and minerals. It has not been indicated when, or if, probiotics might be examined under this proposed legislation.

The increased general interest in probiotics in foods and supplements has inevitably greatly increased human consumption of live microorganisms, without any structured approach to assessing the safety of the products at intended levels of consumption. The recent SCAN paper cites precisely these concerns, and urges the European Commission to adopt a consistent approach to all microorganisms entering the food chain.

USA

Food

Safe microorganisms are traditionally used as ingredients in foods such as dairy products and beverages. Most lactic acid bacteria used in food have a history of safe use, for which the manufacturer is permitted to make a self-affirmation of GRAS status (Generally Regarded As Safe), although there is no 'approved list' of self-affirmed GRAS ingredients. The process of submitting a GRAS petition (in order to obtain

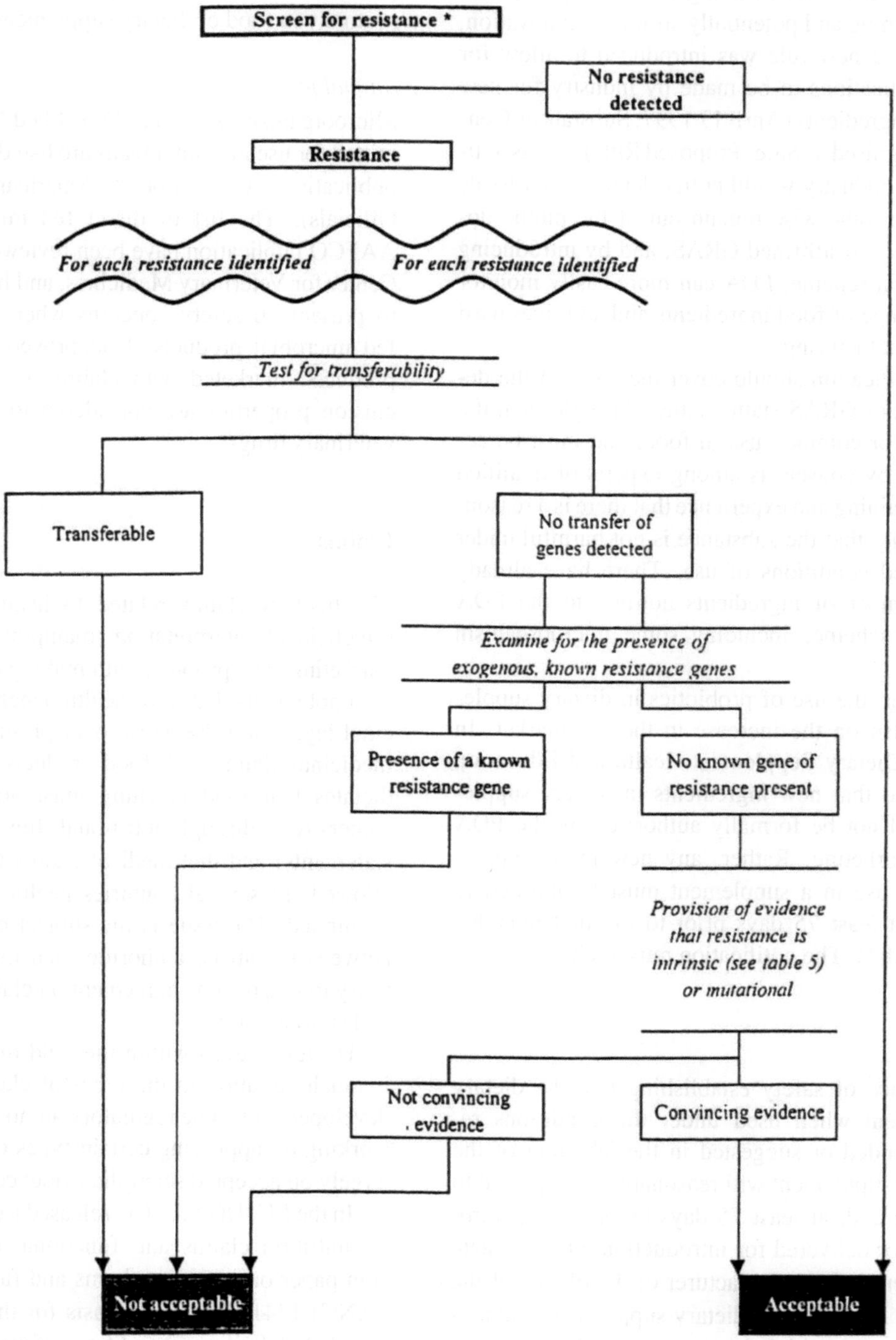

*: For organisms intrinsically resistant to antibiotics, test of which is not required (see table 1), check the presence of exogenous and known genes, expression of which may be masked by intrinsic resistance

Figure 1. From the opinion of the Scientific Committee on Animal Nutrition on the criteria for assessing the safety of microorganisms resistant to antibiotics of human clinical and veterinary importance (adopted 3 July 2001).

authorisation by the FDA) for an ingredient was recognised by the FDA as being particularly onerous, time-consuming and potentially stifling to innovation, so in 1997, a new rule was introduced to allow for GRAS notifications to be made by industry for new foods and ingredients (April 17 1997, Substances Generally Recognised as Safe; Proposed Rule). It was anticipated that industry would notify details about foods which might otherwise remain out of the public domain, under self-affirmed GRAS, and by introducing a notification scheme, FDA can more easily monitor the wide range of food ingredients and additives used in food manufacturing.

The notification should cover the basis of the determination of GRAS status, either through scientific procedures or common use in food, and must be accompanied by consensus among experts of qualified scientific training and experience that there is a reasonable certainty that the substance is not harmful under the intended conditions of use. There have already been a number of ingredients notified to the FDA under this scheme, including some microorganism strains.

However, the use of probiotics in dietary supplements is also on the increase in the US market. In 1994, the Dietary Supplement Health and Education Act decreed that new ingredients in dietary supplements need not be formally authorised by the FDA prior to marketing. Rather, any new ingredient intended for use in a supplement must be notified to the FDA at least 75 days prior to the first intended marketing date. The notification must include:

> "evidence of safety establishing that the dietary ingredient when used under the conditions recommended or suggested in the labelling of the dietary supplement will reasonably be expected to be safe and, at least 75 days before being introduced or delivered for introduction into interstate commerce, the manufacturer or distributor of the dietary ingredient or dietary supplement provides the Secretary with information, including any citation to published articles, which is the basis on which the manufacturer or distributor has concluded that a dietary supplement containing such dietary ingredient will reasonably be expected to be safe". (US Food and Drug Administration, Dietary Supplement Health and Education Act of 1994, Section 8)

This is applicable for new strains of microorganisms which have not previously been included routinely in food or dietary supplements.

Animal feeds

Microorganisms known as Direct Fed Microbials, permitted for use in animal feeds are listed in the AAFCO publication (Association of American Feed Control Officials). The list of direct fed microbials in the AAFCO publication have been reviewed by the FDA, Center for Veterinary Medicines, and have been found to present no safety concerns when used in direct-fed microbial products. Unapproved microorganism products, marketed with claims of competitive exclusion properties are considered to be unapproved veterinary drugs.

Claims

The field of claims related to health benefits is a minefield of interpretation, manipulation and clever marketing. The problem with making claims on foods or supplements that have health benefits is that medicinal legislation the world over prohibits associating medicinal claims with food products. Food law also dictates that food labelling must not mislead consumers (e.g. through nutritional, functional or health statements) and that medical claims for food are not allowed. In several countries health claims are also prohibited. The issue is the subject of strong debate between scientists, authorities and industry, in order to try to reach some agreement on claim acceptability and substantiation.

However, even within the field of 'claims', there is much variation in the types of claims that can be developed, but some regulatory authorities are already working on approving certain types of claims, based largely on accepted scientific evidence.

In the EU, DG SANCO released a discussion paper on nutrition claims and functional claims (Discussion paper on nutrition claims and functional claims, SANC) 1341/2001). The basis for this paper was to establish definitions, conditions of making claims and authorisation of claims, for nutritional labelling and for functions of nutrients on certain normal bodily functions. While the paper was greeted as a recognition that such claims need to be controlled in order to protect the consumer, it was also commented by the majority of interested parties that the fact that the paper specifically excluded health claims was a serious shortfall in attempts to introduce legislation to control the use of health claims. DG SANCO has responded to these requests by pledging to address the issue of health claims as a high priority in this area.

Prior to this paper from the Commission, some Member States had already developed their own bodies to oversee the area of health claims on a voluntary basis. Sweden, The Netherlands, Belgium and the UK all have successful schemes in place which define Codes of Practice used by industry when generating evidence to substantiate health claims. The bodies have no legal standing, and can only serve to assess claims on the basis that they are substantiated by scientific evidence, and are not medicinal in content.

As an example of the requirements for substantiating health claims, the UK Joint Health Claims Initiative, addresses 'Innovative Health Claims'. This type of claim would appear to be the most appropriate for probiotics. While there are differences in the way that claims are presented in different countries, the fundamental requirement for the types of data needed to substantiate claims is the same. Using the UK Code of Practice as a guide, substantiation is critical when making health claims, and particularly for innovative health claims, it is strongly recommended that pre-market advice be sought. The health claim should be based on a review of all available scientific data, including published literature. Experimental data should include:

- Human studies, including epidemiological and clinical studies and chemical extrapolation from food usage in disease status.
- Animal studies.
- Biochemical and cellular studies.
- Any other relevant source.

Human studies are always required as support for a health claim.

In the US, the FDA has published a list of authorised health claims for certain foodstuffs. Many of these claims are related to substantial scientific evidence of dietary factors related to health, such as the role of calcium in osteoporosis. For new claims, however, the format of the food is important when deciding the type of claim that can be made. The FDA recognises three types of health claims that can be made for foods and dietary supplements:

1. NLEA (Nutrition Labelling and Education Act) authorised health claims. This is based on extensive review of the scientific literature, and can be applied to both foods and dietary supplements.
2. Health claims based on authoritative statements (from a scientific body of the US Government or the National Academy of Sciences), as defined in the 1997 Food and Drug Administration Modernization Act (FDAMA). This type of claim is only permissible for foods, excluding dietary supplements.
3. Qualified health claims, based on the weight of scientific evidence, as a result of the US Court of Appeals for the DC Circuit 1999 decision in the case of Pearson v. Shalala. These types of claims do not carry the same validity standard as the claims authorised under NLEA, and are only applicable to dietary supplements.

In Japan, certain foods have been authorised under FOSHU (Foods for Specific Health Use). Since the introduction of FOSHU, 259 products have been authorised (by December 2001), and the Japanese authorities have listed foods and food ingredients which may be included in FOSHU foods. Included in this list are over 13 probiotic species, including lactic acid bacteria.

How does all this relate to lactic acid bacteria, and potential health claims? It appears that the concept of certain probiotics having a general benefit to digestive health is well known, and advertising for these products suggests as much. However, as there are many strains of probiotics available, data needs to be generated on the specific health benefit related to specific strains/final products. As the number of products containing probiotics, both as food ingredients and dietary supplements, increases, concerns have been raised related to the potential for all probiotics to become 'generic' in the perception of the public. For this reason, the need for harmonised legislation in the area of claims is paramount, but in the absence of such legislation, the recommendations of the voluntary codes are even more important. However, until the authorities resolve the larger problem of distinguishing between a health claim and a medicinal claim, it is likely that certain countries will continue to work along voluntary lines, while other countries forbid all types of health claims.

Conclusion

In conclusion, it is evident that regulatory processes are gradually catching up with the fast development of new products appearing in our food, and is particularly pertinent to microorganisms in general, and especially lactic-acid bacteria which have a general history of acceptance in food use. It remains to be seen whether the more stringent regulatory process that has been applied to microorganisms for feed use will be applied to the food area, as recommended by the EU Scientific Committee for Animal Nutrition, and whether the use of microorganisms in food supplements in the EU will come under greater scrutiny in the future. However, the European food industry should be aware of the developments in regulatory status of lactic acid bacteria in other fields of nutrition, and other countries.

References

Amended proposal for a Directive of the European Parliament and of the Council on the approximation of the laws of the member states relating to food supplements/* COM/2001/0159 final – COD 2000/0080*/.

April 17 1997, Substances Generally Recognized as Safe; Proposed Rule. U.S. Food and Drug Administration.

Association of American Feed Control Officials, Official Publication 2002.

Commission Directive 94/40/EC of 23 July 1994 amending Council directive 87/153/EEC fixing guidelines for the assessment of additives in animal nutrition.

Commission Recommendation of 29 July 1997 concerning the scientific aspects and the presentation of information necessary to support applications for the placing on the market of novel foods and novel food ingredients and the preparation of initial assessment reports under Regulation (EC) No 258/97 of the European Parliament and of the Council (97/618/EC).

Commission Regulation (EC) No 2200/2001 of 17 October 2001 concerning provisional authorisations of additives in feedingstuffs.

Council Directive 70/524/EEC of 23 November 1970 concerning additives in feedingstuffs.

Council Directive 87/153/EEC of 16 February 1987 fixing guidelines for the assessment of additives in animal nutrition.

Council Directive 90/220/EEC of 23 April 1990 on the deliberate release into the environment of genetically modified organisms.

Council Directive 93/113/EC of 14 December 1993 concerning the use and marketing of enzymes, micro-organisms and their preparations in animal nutrition.

Dietary Supplement Health and Education Act of 1994. U.S. Food and Drug Administration.

Directive 2001/18/EC of the European Parliament and of the Council of 12 March 2001 on the deliberate release into the environment of genetically modified organisms and repealing Council Directive 90/220/EEC.

Discussion Paper on nutrition claims and functional claims (SANCO 1341/2001), prepared by Directorate General Health and Consumer Protection.

Joint Health Claims Initiative, Code of Practice on Health Claims on Food, UK.

List of Enzymes, Microorganisms and their preparations in animal nutrition permitted for use in individual member states (96/C 263/03).

Nutrition Labelling and Education Act (NLEA) of 1990, U.S. Food and Drug Administration.

Opinion of the Scientific Committee on Animal Nutrition on the criteria for assessing the safety of mircoorganisms resistant to antibiotics of human clinical and veterinary importance (adopted 3 July 2001).

Regulation (EC) No 258/97 of the European Parliament and of the Council of 27 January 1997 concerning novel foods and novel food ingredients.

Salminen S & von Wright A (1998) Current probiotics – safety assured? Microbial Ecol. Health Dis. 10: 68–77.

Strategy for Implementation of Pearson Court Decision December 1 1999, Food Labeling: Health Claims and Label Statements for Dietary Supplements.

The Food and Drug Administration Modernization Act (FDAMA) of 1997. U.S. Food and Drug Administration.

Antonie van Leeuwenhoek **82:** 361–366, 2002.

Genetically modified *Streptococcus mutans* for the prevention of dental caries

Jeffrey D. Hillman
University of Florida College of Dentistry, Box 100424, 1600 SW Archer Road, Gainesville, FL 32610, USA (Tel.: +1-352-846-0792; Fax: +1-352-392-7357; E-mail: jhillman@dental.ufl.edu)

Key words: bacterial interference, dental caries, effector, replacement therapy

Abstract

There are many examples of positive and negative interactions between different species of bacteria inhabiting the same ecosystem. This observation provides the basis for a novel approach to preventing microbial diseases called replacement therapy. In this approach, a harmless effector strain is permanently implanted in the host's microflora. Once established, the presence of the effector strain prevents the colonization or outgrowth of a particular pathogen. In the case of dental caries, replacement therapy has involved construction of an effector strain called BCS3-L1, which was derived from a clinical *Streptococcus mutans* isolate. Recombinant DNA technology was used to delete the gene encoding lactate dehydrogenase in BCS3-L1 making it entirely deficient in lactic acid production. This effector strain was also designed to produce elevated amounts of a novel peptide antibiotic called mutacin 1140 that gives it a strong selective advantage over most other strains of *S. mutans*. In laboratory and rodent model studies, BCS3-L1 was found to be genetically stable and to produce no apparent deleterious side effects during prolonged colonization. BCS3-L1 was significantly less cariogenic than wild-type *S. mutans* in gnotobiotic rats, and it did not contribute at all to the cariogenic potential of the indigenous flora of conventional Sprague-Dawley rats. And, its strong colonization properties indicated that a single application of the BCS3-L1 effector strain to human subjects should result in its permanent implantation and displacement over time of indigenous, disease-causing *S. mutans* strains. Thus, BCS3-L1 replacement therapy for the prevention of dental caries is an example of biofilm engineering that offers the potential for a highly efficient, cost effective augmentation of conventional prevention strategies. It is hoped that the eventual success of replacement therapy for the prevention of dental caries will stimulate the use of this approach in the prevention of other bacterial diseases.

Introduction

As reviewed by Florey (1946), the use of beneficial bacteria to fight harmful bacteria was first attempted over a century ago when Cantani employed a harmless organism referred to as 'Bacto. Termo' to treat tuberculosis. Since then, there have been dozens of reports describing both positive and negative bacterial interactions in which the presence of a particular indigenous microorganism promotes or deters the presence of a pathogen. The reason for the abiding interest in this area of microbiology is the prospect of preventing an infection by an approach traditionally called 'replacement therapy', or, more recently, 'probiotics'. In this approach, a naturally occurring or laboratory-derived effector strain is used to intentionally colonize the niche in susceptible host tissues that is normally colonized by the pathogen. By being better adapted than the pathogen, a well-designed effector strain will prevent colonization or outgrowth of the pathogen by blocking attachment sites, competing for essential nutrients, or other mechanisms. In this fashion, the host is protected for as long as the effector strain persists as a member of the indigenous flora, which, ideally, is for the lifetime of the host.

Application of replacement therapy to various diseases

As a rule, instances of naturally occurring bacterial interactions are detected during *in vitro* cultivation, and their occurrence *in vivo* is inferred from correlations provided by cultivation studies. For example, the predominant microorganisms in the pharynx of healthy neonates are one or more species of alpha-hemolytic streptococci, and the absence of these species was shown to correlate with a significantly increased risk of infections including sepsis, meningitis, pneumonia, and cystitis (reviewed by Sprunt & Leidy 1988). Strain 215 is a naturally occurring alpha-hemolytic Streptococcus that was chosen to serve as an effector strain based largely on its *in vitro* ability to inhibit a variety of common pathogens that initially colonize the pharynx. When given to a group of infants lacking an indigenous alpha-hemolytic Streptococcus in their throat flora, strain 215 usually became the predominant microorganism in pharyngeal cultures, and the numbers of potential pathogens declined to low or undetectable levels. These infants suffered a significantly lower incidence of infections than uninoculated controls (Sprunt et al. 1980). Also, in older individuals, natural or antibiotic-induced low levels of alpha-hemolytic streptococci in the pharynx have been shown to correlate with increased susceptibility to *Streptococcus pyogenes* infections (Sanders 1969; Crowe & Sanders 1973; Sanders et al. 1976; Fujimori et al. 1997). This suggests the potential for replacement therapy in the prevention of streptococcal pharyngitis in susceptible subjects.

Replacement therapy using strain 502A, a naturally occurring *Staphylococcus aureus* strain of low virulence, was very successful in curtailing various diseases caused by this species (Shinefield et al. 1971; Perl & Golub 1998). After gaining relatively widespread acceptance in the 1960s and 1970s, 502A has been little used partly because of lack of need and partly because of infections that have been reported following its use (Drutz et al. 1966; Blair & Tull 1969; Houck et al. 1972). Although the benefits of using strain 502A far outweigh the hazards (Light et al. 1967; Houck et al. 1972), this work emphasized the potential difficulty of using naturally occurring effector strains that may have residual pathogenic potential.

Application of replacement therapy to dental caries

A large body of literature implicates *Streptococcus mutans* as being the principle etiologic agent of dental caries (reviewed by Anderson 1992). *S. mutans* strain BCS3-L1 is a genetically modified effector strain designed for use in replacement therapy of dental caries (Hillman et al. 2000). Construction of BCS3-L1 took into account the following logical prerequisites:

1. It must have a significantly reduced pathogenic potential;
2. It must persistently and preemptively colonize the *S. mutans* niche, thereby preventing colonization by wild-type (disease-causing) strains whenever the host comes into contact with them;
3. Ideally, it should be able to aggressively displace indigenous strains of *S. mutans*, thereby allowing even previously infected subjects to be treated with replacement therapy;
4. It must be generally safe and not predispose the host to other disease conditions.

Pathogenicity

Several different approaches have been examined to satisfy the first of these prerequisites for effector strain construction. In accord with the acidogenic theory of dental caries, lactic acid production by *S. mutans* has long been considered to be the main pathogenic mechanism for production of caries lesions. Consequently, lactate dehydrogenase (LDH) deficiency was chosen as the approach for reducing acidogenicity in construction of BCS3-L1. Earlier work (Johnson et al. 1980) with a closely related *S. rattus* strain had provided convincing evidence for the effectiveness of this approach. LDH-deficient mutants were virtually non-cariogenic in gnotobiotic rats and did not contribute significantly to the cariogenic potential of the indigenous flora in conventional pathogen-free rats. Attempts to transfer these findings directly to *S. mutans* proved to be difficult. LDH-deficient mutants of various strains of *S. mutans* were not found using the same screening methods used to isolate mutants of *S. rattus.* Cloning the structural gene encoding the *S. mutans* LDH (Hillman et al. 1990) provided the basis for solving this puzzle. Standard insertional mutagenesis methods failed to yield LDH-deficient clones (Duncan et al. 1991; Hillman et al. 1994), suggesting that LDH-deficiency was a lethal mutation in most *S. mutans* strains. This hypothesis was definitively proven by creation of a temperature sensitive LDH mutant (Chen et al. 1994). This isolate grew well at 30 °C but did not grow at 42 °C under a variety of cultivation conditions. Chemostat studies indicated that some aspect of glucose metabolism was toxic during growth under the non-permissive condition (Hillman et al. 1996). The toxic effect could be overcome by limiting the amount of environmental glucose. This and other data accorded with studies of *S. mutans* central intermediary metabolism indicating that this organism has enzymatic activities, including pyruvate formate-lyase (Abbe et al. 1982; Takahashi et al. 1982) and pyruvate dehydrogenase (Carlsson et al. 1985; Hillman et al. 1987a), for pyruvate dissimilation. However, at high sugar concentrations, the levels of activity of these enzymes are apparently insufficient to compensate for the absence of LDH. It was found (Hillman et al. 1996) that a supplemental alcohol dehydrogenase (ADH) activity, when expressed in the temperature sensitive LDH mutant, could complement LDH deficiency.

With this background of information, BCS3-L1 construction started with the *ldh* gene cloned into an appropriate suicide vector for *S. mutans.* Essentially the entire gene except for transcription and translation signal sequences was deleted and replaced with the *Zymomonas mobilis* open reading frame for alcohol dehydrogenase (ADH) II. Transformation of the recombinant molecule into the *S. mutans* starting strain, JH1140, and allelic exchange resulted in the isogenic mutant, BCS3-L1. This effector strain had no measurable LDH activity and ca. 10-fold elevated levels of ADH activity relative to its parent. Fermentation end-product analysis showed that BCS3-L1 made no detectable lactic acid. As predicted from earlier work with *S. rattus*, much of the metabolized carbon was converted to the neutral end-products, ethanol and acetoin. Under various cultivation conditions, including growth on a variety of sugars and polyols, such as sucrose, fructose, lactose, mannitol, and sorbitol, BCS3-L1 yielded final pH values that were 0.4 to 1.2 pH units higher than those of its parent, JH1140. The reduced acidogenic potential of BCS3-L1 was reflected in its dramatically decreased cariogenic potential as shown in several animal models (Hillman et al. 2000). The results of these studies provided strong evidence that an LDH-deficient *S. mutans* strain such as BCS3-L1 has significantly reduced pathogenic potential, and thus satisfies the first prerequisite for use as an effector strain in replacement therapy for dental caries.

Colonization

Transmission of mutans streptococci within the human population has been extensively studied. Most studies support the idea that this organism is usually transmitted from mother (primary caretaker) to child within a several year period following the onset of tooth eruption (Davey & Rogers 1984; Berkowitz & Jones 1985; Caufield & Walker 1989; Alaluusua 1991; Genco & Loos 1991). Numerous studies (Krasse et al. 1967; Jordan et al. 1972; Ruangsri & Orstavik 1977; Svanberg & Krasse 1981) have documented the difficulty of persistently introducing laboratory strains of mutans streptococci into the mouths of humans, particularly if they already harbored an indigenous strain of this organism.

From the standpoint of replacement therapy of caries, these results suggest that implantation of an effector strain would best be accomplished in children immediately after the onset of tooth eruption and before their acquisition of a wild-type strain. In order to prevent supercolonization by wild-type strains when the host comes in contact with them, an effector strain should have some significant selective advantage to colonization. This would also enable subjects who have already been infected with wild-type *S. mutans* to be treated by replacement therapy. The ability of an effector strain to preemptively colonize the human oral cavity and aggressively displace indigenous wild-type strains was initially thought to be a complex phenomenon dependent on a large number of phenotypic properties. However, it was discovered that a single phenotypic property could provide the necessary selective advantage. A naturally occurring strain of *S. mutans* was isolated from a human subject that produces a bacteriocin called mutacin 1140 that is capable of killing virtually all other strains of mutans streptococci against which it was tested (Hillman et al. 1984). Mutants were isolated that produced no detectable mutacin 1140 or that produced approximately three-fold elevated amounts. The mutants were used in a rat model to correlate mutacin production to preemptive colonization and aggressive displacement.

A correlation was also made between mutacin 1140 production and the ability of *S. mutans* to persistently colonize the oral cavities of human subjects and aggressively displace indigenous mutans streptococci (Hillman et al. 1985, 1987b). Three years following a single, 3 min infection regimen involving brushing and flossing of a concentrated cell suspension onto and between the teeth, all of the subjects remained colonized by the mutant strain producing 3-fold elevated amounts of mutacin 1140 (Hillman et al. 1989). No other strains of mutans streptococci were observed in saliva and plaque samples of these colonized volunteers. The same results were found recently, 14 years after colonization, for at least two of three subjects who are still available for testing. These results indicate that this strain of *S. mutans* succeeded in satisfying the second and third prerequisites for use as an effector strain in replacement therapy: It persistently and preemptively colonized the *S. mutans* niche in the human oral cavity and it aggressively displaced indigenous strains of this organism. Consequently, *S. mutans* strain JH1140, which has a spontaneous mutation resulting in ca. 3-fold elevated production of mutacin 1140, served as the starting strain for construction of BCS3-L1 described above. Introduction of the *ldh* mutant allele into JH1140 had no measurable effects on phenotypic properties known to be important in the natural history of infection by *S. mutans.* This includes intracellular polysaccharide storage and aciduricity. Interestingly, production of extracellular polysaccharide (glucan), which serves as an important holdfast during colonization, was somewhat increased during *in vitro* cultivation in the presence of sucrose, probably because of the pH dependence of the glucosyl transferase activities. It is not known if this effect is also manifested during *in vivo* growth.

Safety and stability

To serve as an effector strain in replacement therapy of dental caries, BCS3-L1 must be safe in several important regards. First, it must be genetically stable. In the case of BCS3-L1, reacquisition of an acidogenic phenotype by spontaneous reversion is extremely unlikely because construction of BCS3-L1 involved deletion of essentially the entire *ldh* open reading frame. Although horizontal transmission of an *ldh* gene is possible, repeated attempts have failed to demonstrate transduction of *S. mutans* by bacteriophage (A. Delisle, personal communication). In addition, transformability is known to be very strain dependent (Perry & Kuramitsu 1981; Westergran & Emilson 1983), and although it has been shown (Li et al. 2001) that BCS3-L1 has a low transformation frequency due to a natural mutation in its gene for competence stimulating peptide, allelic exchange has been used to delete the *comE* gene to provide added assurance that exogenous DNA will not transform this strain.

Mutacin 1140 has been shown (Hillman et al. 1998) to be a member of a small class of antibiotics called lantibiotics. It is a small (2263 Da), very stable peptide containing modified amino acids, lanthionine, methyllanthione, didehydroalanine and didehydrobutyrine, characteristic of lantibiotics. Sufficient mutacin 1140 has not been purified to directly test its toxicity. However, the prototype lantibiotic, nisin, is known to have extremely low toxicity (Hurst 1981), and has been developed and used for decades as a food preservative that is generally recognized as safe. It was shown (Hillman et al. 2000) that after colonization for 6 months, the mean weights of conventional Sprague-Dawley rats colonized with BCS3-L1 did not differ significantly from animals colonized with JH1140 or *S. mutans*-free control animals. Histopathological examination revealed no treatment-related lesions in any of the major organs examined.

Mutacin production by BCS3-L1 and the change in fermentation products resulting from LDH deficiency could conceivably upset plaque ecology and lead to the bloom of another microorganism with pathogenic potential. Recent studies (reviewed by Costerton 1995) have provided an appreciation for the complicated structural architecture of biofilms, presumably including dental plaque. Following specific initial attachment to a surface, the growth of cells leads to the formation of a thick layer of differentiated mushroom- and pillar-like structures consisting of cells embedded in their extracellular polysaccharide matrix. Between these cellular structures are water-filled spaces that serve as channels for the introduction of nutrients and the elimination of waste products. In biofilms consisting of two or more species of bacteria, each cellular structure may be pure or mixtures of cells depending on the pressures imposed by positive and negative bacterial interactions. These interactions may also extend over a finite area to affect the general composition of plaque in a particular habitat. Clearly, however, there is a physical limit to this sphere of influence. The mutacin 1140 up-producing strain of *S. mutans* eliminated mutacin-sensitive indigenous strains of *S. mutans* but had no effect on indigenous *S. oralis* strains that were equally sensitive to mutacin killing *in vitro* (Hillman et al. 1987b). These results indicate that *S. mutans* has a physically distinct habitat that is separated from the *S. oralis* habitat by a distance sufficient for dilution to reduce the concentration of mutacin below its minimal inhibitory concentration. A similar explanation could account for the failure to observe qualitative or quantitative changes in the plaque of rats following long-term infection with an LDH deficient mutant, even though the mutant's metabolic end-products are certain to be different from those of the wild-type strain (Stashenko & Hillman 1989).

A final aspect of replacement therapy safety is the requirement for controlled spread of the effector strain within the population. Mutacin 1140 up-production clearly provides a selective advantage to BCS3-L1 colonization, but the minimal infectious dose has not been determined for this strain or any *S. mutans* strain in humans. As described above, horizontal transmission of natural strains appears to be a rare event, but mutacin up-production may promote its occurrence. Wives and children of the two subjects infected with the mutacin up-producing *S. mutans* strain were not colonized when tested 14 years after the initial infection regimen (J.D. Hillman, unpublished result). Clearly, additional studies with larger populations will have to be performed to properly measure the potential for horizontal transmission. It is expected that, like wild-type strains of *S. mutans*, vertical transmission of BCS3-L1 from mother to child will occur at a high frequency.

Concluding remarks

In general, replacement therapy using a carefully constructed effector strain provides a number of advantages over conventional prevention strategies. In the case of dental caries, a single colonization regimen that leads to persistent colonization by the effector strain should provide lifelong protection. In the event that the effector strain does not persist indefinitely in some subjects, reapplication could be performed as the need arises without added concern for safety or effectiveness.

One of the greatest advantages of replacement therapy is that there is no need for patient compliance. Conscientious use of conventional prevention methods (brushing, flossing, topical fluoride, controlled sugar consumption, etc.) is sufficient in most cases to maintain the *S. mutans* level below its minimal pathogenic dose. The fact that dental caries remains as one of the most common infectious diseases afflicting humans is a clear indication that a truly effective prevention strategy must avoid the need to rely on patient compliance.

The current resurgence of various infectious diseases indicates that traditional and antibiotic-based therapies alone will not suffice. The continued study of bacterial interactions as they occur *in vivo* will inevitably lead to the identification of naturally occurring effector strains for the replacement therapy of various infections. If ultimately successful, the use of genetic engineering to tailor an effector strain for replacement therapy of dental caries would encourage similar efforts to prevent other infectious diseases.

Acknowledgments

I would like to thank J.L. Snoep, J. Novak, K. Stashenko, S. Andrews, and M. Duncan for their many years of help in the design and execution of these studies. I am also very grateful for the financial support provided by the United States National Institutes of Health, Procter & Gamble Co. and OraGen, Inc.

References

Abbe K, Takahashi S & Yamada T (1982) Involvement of oxygen-sensitive pyruvate formate-lyase in mixed-acid fermentation by *Streptococcus mutans* under strictly anaerobic conditions. J. Bacteriol 152: 175–182.

Alaluusua S (1991) Transmission of mutans streptococci. Proc. Finnish Dent. Soc. 87: 443–447.

Anderson MH (1992) Changing paradigms in caries management. Curr. Opin. Dent. March: 2157–2162.

Berkowitz RJ & Jones P (1985) Mouth-to-mouth transmission of the bacterium *Streptococcus mutans* between mother and child. Arch. Oral Biol. 30: 377–379.

Blair EB & Tull AH (1969) Multiple infections among newborns resulting from colonization with *Staphylococcus aureus* 502A. Amer. J. Clin. Path. 52: 42–49.

Carlsson J, Kujala U & Edlund M-BK (1985) Pyruvate dehydrogenase activity in *Streptococcus mutans*. Infect. Immun. 49: 674–678.

Caufield PW & Walker TM (1989) Genetic diversity within *Streptococcus mutans* evident from chromosomal DNA restriction fragment polymorphisms J. Clin. Microbiol. 27: 274–278.

Chen A, Hillman JD & Duncan M (1994) L(+)-lactate dehydrogenase deficiency is lethal in *Streptococcus mutans*. J. Bacteriol. 176: 1542–1545.

Costerton JW (1995) Overview of microbial biofilms. J. Indust. Microbiol. 15: 137–140.

Crowe CC & Sanders Jr ES (1973) Bacterial interference II. Role of the normal throat flora in prevention of colonization by group A *streptococcus*. J. Infect. Dis. 128: 527–532.

Davey AL & Rogers AH (1984) Multiple types of the bacterium *Streptococcus mutans* in the human mouth and their intra-family transmission. Arch. Oral Biol. 29: 453–460.

Drutz DJ, van Way MH, Schaffner W & Koenig MG (1966) Bacterial interference in the therapy of recurrent staphylococcal infections: multiple abscesses due to implantation of the 502A strain of staphylococcus. New Eng. J. Med. 275: 1161–1165.

Duncan MJ & Hillman JD (1991) DNA sequence and *in vitro* mutagenesis of the gene encoding the fructose-1,6-diphosphate-dependent L-(+)-lactate dehydrogenase of *Streptococcus mutans*. Infect. Immun. 59: 3930–3934.

Florey HW (1946) The use of micro-organisms for therapeutic purposes. Yale J. Biol. Med. 19: 101–117.

Fujimori I, Kikushima K, Hisamatsu K, Nozawa I, Goto R & Murakami Y (1997) Interaction between oral alpha-streptococci and group A streptococci in patients with tonsillitis. Ann. Oto. Rhino. Laryng. 106: 571–574.

Genco RJ & Loos BG (1991) The use of genomic DNA fingerprinting in studies of the epidemiology of bacteria in periodontitis. J. Clin. Periodont. 18: 396–405.

Hillman JD, Johnson KP & Yaphe BI (1984) Isolation of a *Streptococcus mutans* strain producing a novel bacterioc. Infect. Immun. 44: 141–144.

Hillman JD, Yaphe BI & Johnson KP (1985) Colonization of the human oral cavity by a strain of *Streptococcus mutans.* J. Dent. Res. 64: 1272–1274.

Hillman JD, Andrews SW & Dzuback AL (1987a) Acetoin production by wild-type strains and a lactate dehydrogenase-deficient mutant of *Streptococcus mutans.* Infect. Immun. 55: 1399–1402.

Hillman JD, Dzuback AL & Andrews SW (1987b) Colonization of the human oral cavity by a *Streptococcus mutans* mutant producing increased bacteriocin. J. Dent. Res. 66: 1092–1094.

Hillman JD, Andrews SW, Painter S & Stashenko P (1989) Adaptive changes in a strain of *Streptococcus mutans* during colonization of the human oral cavity. Microb. Ecol. Hlth. Dis. 2: 231–239.

Hillman JD, Duncan MJ & Stashenko KP (1990) Cloning and expression of the gene encoding the fructose-1,6-diphosphate-dependent L-(+)-lactate dehydrogenase of *Streptococcus mutans.* Infect. Immun. 58: 1290–1295.

Hillman JD, Chen A, Duncan M & Lee S-W (1994) Evidence that L(+)-lactate dehydrogenase deficiency is lethal in *Streptococcus mutans.* Infect. Immun. 62: 60–64.

Hillman JD, Chen A & Snoep JL (1996) Genetic and physiological analysis of the lethal effect of L-(+)-lactate dehydrogenase deficiency in *Streptococcus mutans*: complementation by alcohol dehydrogenase from *Zymomonas mobilis.* Infect. Immun. 64: 4319–4323.

Hillman JD, Novak J, Sagura E, Gutierrez JA, Brooks TA, Crowley PJ, Azziz A, Leung K-P, Cvitkovitch D & Bleiweis AS (1998) Genetic and biochemical analysis of mutacin 1140, a lantibiotic from *Streptococcus mutans.* Infect. Immun. 66: 2743–2749.

Hillman JD, Brooks T, Michalek SM, Harmon CC & Snoep JL (2000) Construction and characterization of an effector strain of *Streptococcus mutans* for replacement therapy of dental caries. J. Dent. Res. 68: 543–549.

Houck PW, Nelson JD & Kay JL (1972) Fatal septicemia due to *Staphylococcus aureus* 502A. Amer. J. Dis. Child. 123: 45–48.

Hurst A (1981) Nisin. In: Perlman and Laskin (Eds) Advances in Applied Microbiology Vol. 27 (pp 85–123). Academic Press, London.

Johnson CP, Gross SM & Hillman JD (1980) Cariogenic potential *in vitro* in man and *in vivo* in the rat of lactate dehydrogenase mutants of *Streptococcus mutans.* Arch. Oral Biol. 25: 707–713.

Jordan HV, Englander HR, theEngler WO & Kulczyk S (1972) Observations on implantation and transmission of *Streptococcus mutans* in humans. J. Dent. Res. 51: 515–518.

Krasse B, Edwardsson S, Svensson I & Trell L (1967) Implantation of caries-inducing streptococci in the human oral cavity. Arch. Oral Biol. 12: 231–236.

Li Y-H, Lau PCY, Lee JH, Ellen R & Cvitkovitch DG (2001) Natural genetic transformation of *Streptococcus mutans* growing in biofilms. J. Bacteriol. 183: 897–908.

Light IJ, Walton L, Sutherland JM, Shinefield HR & Brackvogel V (1967) Use of bacterial interference to control a staphylococcal nursery outbreak: deliberate colonization of all infants with the 502A strain of *Staphylococcus aureus.* Amer. J. Dis. Child. 113: 291–300.

Perl TM & Golub JE (1998) New approaches to reduce *Staphylococcus aureus* nosocomial infection rates: treating *S. aureus* nasal carriage. Ann. Pharmacother. 32: S7–16.

Perry D & Kuramitsu HK (1981) Genetic transformation of *Streptococcus mutans.* Infect. Immun. 32: 1295–1297.

Ruangsri P & Orstavik D (1977) Effect of the acquired pellicle and of dental plaque on the implantation of *Streptococcus mutans* on tooth surfaces in man. Caries Res. 11: 204–210.

Sanders E (1969) Bacterial interference I. Its occurrence among the respiratory tract flora and characterization of inhibition of group A streptococci by viridans streptococci. J. Infect. Dis. 120: 698–707.

Sanders CC, Sanders WE & Harrowe DJ (1976) Bacterial interference: effects of oral antibiotics on the normal throat flora and its ability to interfere with group A streptococci. Infect. Immun. 13: 808–812.

Shinefield HR, Ribble JC & Boris M (1971) Bacterial interference between strains of *Staphylococcus aureus,* 1960 to 1970. Amer. J. Dis. Child. 121: 148–152.

Sprunt K, Leidy G & Redman W (1980) Abnormal colonization of neonates in an ICU: conversion to normal colonization by pharyngeal implantation of alpha hemolytic streptococcus strain 215. Ped. Res. 14: 308–313.

Sprunt K & Leidy G (1988) The use of bacterial interference to prevent infection. Can. J. Microbiol. 34: 332–338.

Stashenko KP & Hillman JD (1989) Microflora of plaque in rats following infection with an LDH-deficient mutant of *Streptococcus mutans.* Caries Res. 23: 375–377.

Svanberg M & Krasse B (1981) Oral implantation of saliva-treated *Streptococcus mutans* in man. Arch. Oral Biol. 26: 197–201.

Takahashi S, Abbe K & Yamada T (1982) Purification of pyruvate formate-lyase from *Streptococcus mutans* and its regulatory properties. J. Bacteriol. 149: 1034–1040.

Westergren G & Emilson CG (1983) Prevalence of transformable *Streptococcus mutans* in human dental plaque. Infect. Immun. 41: 1386–1388.

Antonie van Leeuwenhoek **82:** 367–374, 2002.

Exploiting exopolysaccharides from lactic acid bacteria

Laure Jolly*, Sébastien J. F. Vincent, Philippe Duboc & Jean-Richard Neeser
*Nestlé Research Center, Nestec Ldt, Vers-chez-les-Blanc, PO Box 44, CH-1000 Lausanne 26, Switzerland (*Author for correspondence; Tel: +41-21-785-81-37; Fax: +41-21-785-85-49; E-mail: laure.jolly@rdls.nestle.com)*

Key words: exopolysaccharides, food industry, lactic acid bacteria, milk fermentation, texture; yoghurt

Abstract

Microbial exopolysaccharides (EPSs) synthesized by lactic acid bacteria (LAB) play a major role in the manufacturing of fermented dairy products. EPS production is characterized by a large variety in terms of quantity, chemical composition, molecular size, charge, type of sidechains and rigidity of the molecules. Monosaccharide unit's composition, linkages, charge and size determine the EPS' intrinsic properties and their interactions with other milk constituents. EPSs contribute to texture, mouthfeel, taste perception and stability of the final product. Furthermore, it was reported that EPS from food grade organisms, particularly LAB, have potential as food additives and as functional food ingredients with both health and economic benefits. A better understanding of structure-function relationships of EPS in a dairy food matrix and of EPS biosynthesis remain two major challenges for further applications of EPS and the engineering of functional polysaccharides.

Abbreviations: CPS – capsular polysaccharide; EPS – exopolysaccharide; LAB – lactic acid bacteria; LPS – lipopolysaccharide; LTA – lipotechoïc acid

Introduction

Many bacteria are known to produce cell-surface polysaccharides involved in a wide variety of biological functions including prevention of desiccation, protection from environmental stresses, adherence to surfaces, pathogenesis and symbiosis (Whitfield & Valvano 1993; Roberts 1996). Cell-surface polysaccharides comprise *O*-antigens lipopolysaccharides (LPSs), lipoteichoic acids (LTAs), capsular polysaccharides (CPSs) and exopolysaccharides (EPSs). LPS, LTA and CPS are linked to the surface, while EPSs are generally loosely attached or excreted into the environment.

Several LABs are known to produce EPSs (Sutherland 1998; De Vuyst et al. 2001). The precise function of these EPSs remains unclear and complex, although it is considered to be related to cell adhesion and cell protection in different environment, e.g., biofilm formation (Cerning & Marshall 1999). EPSs also play a major industrial role in the production of fermented dairy products, in particular for the production of yoghurt, drinking yoghurt, cheese, fermented cream and milk-based desserts. Their presence significantly contributes to texture, mouthfeel, taste perception and stability of the final products (Crescenzi 1995; Bouzar et al. 1997; Cerning & Marshall 1999; Christiansen et al. 1999)

EPSs produced by LAB display a great variety of structures (reviewed by De Vuyst & Degeest 1999; Ricciardi & Clementi 2000; De Vuyst et al. 2001). They can be divided into two groups: homopolysaccharides (EPSs composed of one monosaccharide moiety) synthesized by conversion of extracellular substrates by secreted enzymes, e.g., dextransucrase, levansucrase. Homopolysaccharides have been reviewed elsewhere (Monsan et al. 2001) and will not be discussed in this review. The other type of EPSs (composed of different sugar moieties, e.g. glucose, galactose, rhamnose, mannose, *N*-acetyl glucucosamine, *N*-acetyl galactosamine, glucuronic acid) is produced in the cytoplasm by using glycosyltransferases and then polymerized on the outer-cell before secretion (reviewed by Ricciardi & Clementi 2000; De Vuyst et al. 2001). This review will focus on the applications of these heteropolysaccharides as food ingredients. The genetic determinants of several EPSs have been identified (Jolly & Stingele 2001; Jolly et al. 2002). The goal in the currently increasing understanding of EPS biosynthesis is to link genes, protein functions and EPS chemical structure to a given physico-chemical behavior. This will allow one to select and customize EPS-producing bacteria for several different industrial applications. An important aspect is that the properties of an EPS in a fermented milk product may differ considerably from its properties in purified form. Nevertheless, determining the structure, composition and rheological properties of a purified EPS is a necessary first step towards the investigation of the molecular interactions with other constituents from the medium.

Challenges for the dairy industry

A very large variety of fermented dairy products are available to consumers. They vary as a function of the type of milk used, i.e. cow, goat, buffalo or defatted milk, the bacterial starter culture, the additives and the process. Yoghurts are produced by fermentation of milk with EPS-producing strains *Streptococcus thermophilus* and *Lactobacillus delbrueckii* subsp. *bulgaricus* (Berkman et al. 1990; Tamime & Robinson 1999). Nordic ropy milk is the generic name for fermented milks with slime-producing mesophilic *cocci* (Macura & Townsley 1984). Kefir is produced by a mixture of bacteria including several LAB, acetic acid bacteria and lactose-fermenting yeasts, while Finnish viili contain amongst others *Geotrichum candidum* moulds. EPS-producing LABs are also important for cheese production, several strains of EPS-producing *Lactobacillus helveticus* being used in mozzarella production where EPS were shown to contribute to water retention in the product (Kosikowski 1982; Kojic et al. 1992; McMahon et al. 1993; Perry et al. 1997; Low et al. 1998).

The microstructure of yoghurts consists of a matrix of aggregated casein particles and it changes according to the type of starter bacteria. Ropy bacteria surround themselves by an EPS layer capable of binding the protein matrix and of modifying its structure (Kalab et al. 1983; Teggatz & Morris 1990; van Marle & Zoon 1995). Use of these ropy bacteria leads to a decrease in firmness and cohesiveness of the gel structure. EPSs have also been shown to act as a stabilizer that retains water and hence decreases susceptibility to syneresis.

In the yoghurt business, there is an increasing consumer demand for stirred yoghurt products with a smooth and creamy texture. Texture and stability

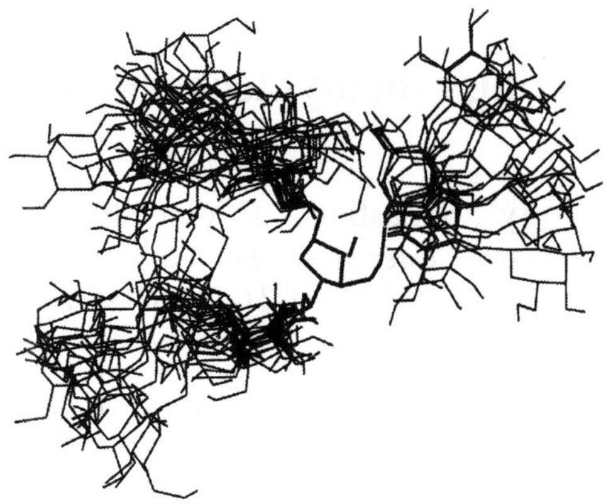

Figure 1. Snapshots taken from a 5 ns molecular dynamic of the repeating unit of the EPS from *Lactobacillus helveticus* Lh59 performed in water. The primary structure is the following: →3)-α-D-Galp-(1→3)- α-D-Glcp-(1→3)-β-D-Glcp -(1→5)-[β-D-Galp-(1→4)-β-D -Glcp-(1→3)-]-β-D-Galf-(1→. The pictures were aligned on the branching β-D-Galf monosaccharide. Courtesy of Dr A. van Kuik, University of Utrecht, The Netherlands.

are affected by the culture conditions (Rohm & Kovac 1994; Sebastiani & Zelger 1998) as well as the manufacturing procedure. Well-known technological approaches to improve the texture of products include the use of an increased content of milk solids such as fat, proteins (Rohm & Schmid 1993) or sugars (sucrose, fructose), addition of stabilizers such as pectin, starch, alginate, gelatin, when permitted by national legislation, and heat treatment. However, these approaches have shown their limitations when other strong consumer demands are taken into account, namely low (or reduced) fat, low sugar, low cost and a minimal amount of food additives. An answer to this challenge could be to take advantage of the EPSs naturally produced by starter culture LAB.

Structure–function relationships

A large variety of different heteropolysaccharides are produced by LAB at levels of 10–1000 mg/L with high molar masses, typically above 10^6 g/mol. Several structures were determined as recently reviewed (Ricciardi & Clementi 2000; De Vuyst et al. 2001). It has been observed that the structure of a heteropolysaccharide is strain specific. The EPSs often differ by monosaccharide composition, linkages between units, presence of repeated sidechains and substitutions. Moreover, length and composition of branching strongly affects the rheological properties since

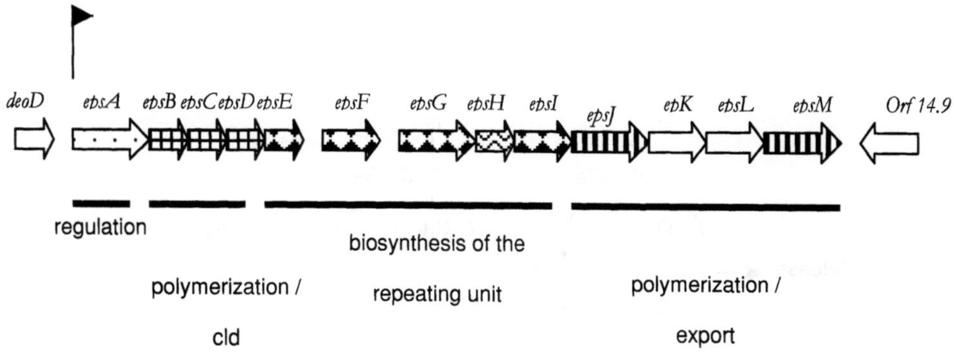

Figure 2. Eps gene cluster organization of *Streptococcus thermophilus* Sfi6 (Stingele 1996). → Regulation; → polymerization and chain length determination (cld); → biosynthesis of the repeating unit (glycosyltransferases); → polymerization and export.

it affects the compactness of the EPS (Vincent et al. 2001).

EPSs have an important thickening power and influence the rheology and texture of fermented products at low concentrations, favorably compares with other thickeners (Duboc & Mollet 2001). Since the polysaccharides of different LAB greatly vary in composition, charge, spatial arrangement, rigidity, and ability to interact with proteins, it is not surprising that no clear correlation between observed EPS concentrations and apparent viscosities of the products could be established. However, some trends have been traced. To obtain a high viscosity, the molar mass should be high and the chain should be relatively stiff. Faber et al. (1998) have shown that milks fermented by two different strains secreting EPSs in the same amount, with the same repeating unit but with different molar masses were more viscous with the EPS that had the highest molecular mass. Some general features on the relation between polysaccharide chemistry and chain stiffness can also be outlined. $\beta(1,4)$ linkages result in stiffer chains than $\alpha(1,4)$ or $\beta(1,3)$, as well as the much more flexible $\alpha(1,6)$ and $\beta(1,6)$ linkage (Tuinier et al. 1999). Sidechains can also affect the chain stiffness and hence the thickening properties (Tuinier et al. 2001). Van den Berg et al. (1995) showed that neutral EPS contributed to the viscosity but not to the elasticity. On the other hand, negatively charged polysaccharides contributed to the elasticity, but not to the viscosity.

More research is needed to fully understand the physical functionality of EPSs in different matrices. In recent work the available information is transformed into structure–function relationships (Jay et al. 1998; Vincent et al. 2001; Duboc et al. 2002). Beyond these structure-based relationships, molecular modeling is appearing as a growingly important tool for the development of atomic-scale models (Pérez et al. 1996; Faber et al. 2002) presenting a link to macroscopic properties such as rheology and the capacity for interactions with other molecules present in the medium (Figure 1).

The next logical step will be the establishment of correlations between atomic structural properties and physico-chemical properties of an EPS in a complete and complex food matrix rather than in aqueous solution. Different parameters to be studied are the influence of pH, proteins, interactions between several EPSs and, for example, the influence of heat treatment. Molecular modeling is likely to play a major role in these investigations.

Genetic engineering of EPSs

In order to improve properties of EPSs, genetic engineering could result in the production of designed polysaccharides targeting predefined macroscopic properties. The biosynthesis of heteropolysaccharides involves several steps directly connected to the cell's carbon metabolism. Significant progresses have been made in the genetic analysis of EPS biosynthesis of various LABs. The genes required for the biosynthesis, have been identified in different clusters (Cieslewicz et al. 2001; Jolly & Stingele 2001; Jolly et al. 2002). Genetic analysis of EPS-producing strains has revealed a common operon structure, strongly suggesting that a common mechanism might be involved in their biosynthesis (Jolly & Stingele 2001). The gene sequence in all the Gram-positive cell-surface polysaccharide-biosynthesis clusters identified to date is as follows: regulation, chain length determination, biosynthesis of the repeating unit, polymerization and export (Figure 2). The sequential transfer of sugar

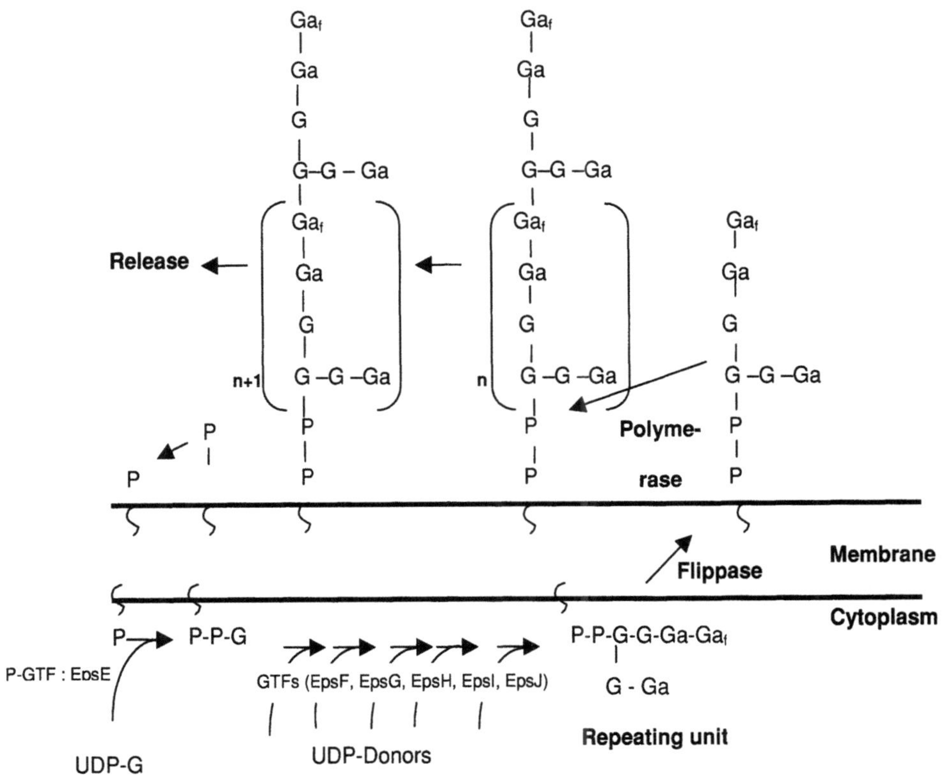

Figure 3. Model of the biosynthesis of the EPS of *Lactobacillus helveticus* NCC2745 (Jolly 2001, Jolly 2002).

residues from activated sugar donors onto a lipophilic carrier is directly followed by assembly of the oligosaccharide-repeating unit by specific glycosyltransferases. Subsequently, the completed repeating unit is exported, polymerized and released (in the case of secreted polysaccharides) (Figure 3).

Based on the available genetic information, genetic modification of *eps* genes should lead to EPSs with a different repeating unit or with a different chain length. Based on the available genetic information, heterologous production of an EPS has been successful by transfer of the complete *eps* gene cluster alone of a LAB strain into a non EPS-producing heterologous host, provided that the heterologous host possessed all necessary genetic information for precursor synthesis (Germond et al. 2001). Another experiment involving gene cluster transfer showed that a different polysaccharide could be produced from a cluster of genes, illustrating a certain broad selectivity in the export and polymerization system and some limitations for the supply of the precursors (Stingele et al. 1999). Gene shuffling with glycosyltransferases is a very promising tool to produce different linkages or use different sugars. Heterologous expression of glycosyltransferases could also allow to add other sugars at strategic positions to generate EPSs with new properties.

The glycosyltransferases are key enzymes for the biosynthesis of the EPS repeating unit, since they catalyze the transfer of sugar moieties from activated donor molecules to specific acceptor molecules, thereby forming glycosidic linkages. Reaction mechanisms of these enzymes still remain unclear. However, a few crystal structures have been published in the last few years and an increasing number of studies are dedicated to these enzymes in order to better understand their catalytic mechanism and hence to modify their properties by mutagenesis (Unligil & Rini 2000; Breton et al. 2001; Persson et al. 2001).

Improvement of EPS yields

EPSs from food grade microorganisms like LAB have potential as food ingredients or as molecule produced *in situ*, with both health functionality and economic benefits. For heteropolysaccharides, efficient production and recovery costs require knowledge of biosynthesis and adoption of appropriate bioprocess technology.

The production of EPS is influenced by bacterial growth phase, medium composition (carbon source and nitrogen source), pH and temperature (De Vuyst & Degeest 1999; Petry et al. 2000). Since the biosynthesis of EPS is linked to the primary carbohydrate metabolism of the producing cells (Escalante et al. 1998; De Vuyst & Degeest 1999; Boels et al. 2001; Levander et al. 2002), EPS production is expected to take place during active sugar consumption, since it requires large numbers of activated nucleotide sugars, energy for building the repeating units, for polymerization, for transmembrane translocation and for polymerization. EPS production was indeed shown to be growth-associated. Several factors influence growth and EPS production, in particular medium composition (De Vuyst & Degeest 1999; Duboc & Mollet 2001): the presence of casein hydrolysate (Cerning et al. 1986), co-cultures (Cerning et al. 1990) and the amount of glucose (Cerning et al. 1994). Furthermore, different EPSs have been produced by one strain in different media (Grobben et al. 1995; van Geel-Schutten et al. 1999).

The optimal pH for EPS production was found to be close to the optimal pH for growth in continuous cultures of *Lactobacillus delbrueckii* subsp. *bulgaricus* (about pH 6.5). Concerning temperature, however, sub-optimal conditions for the bacterial growth were often reported to be favorable for EPS synthesis (Cerning et al. 1992; Gancel & Novel 1994; Mozzi et al. 1995; Moreira et al. 2000), although for *Lactobacillus delbrueckii* subsp. *bulgaricus* the optimal temperature for EPS production was close to the optimal temperature for growth (Mozzi et al. 1994; Grobben et al. 1995).

Heterologous EPS overexpression was demonstrated in *Lactococcus lactis* (van Kranenburg et al. 1999; Stingele et al. 1999; Germond et al. 2000). The *epsD* gene, encoding a priming glucosyltransferase, and part of the plasmid-located operon was placed under control of the nisin-controlled expression (NICE) system. After induction with nisin A, EPS production of a transformed *Lactococcus lactis* strain was higher than that of a wild-type (van Kranenburg et al. 1999). Although these results are promising for increased heteropolysaccharides production, most LAB strains are not suitable for profitable production of heteropolysaccharides and their application as potential thickener or food-grade ingredients require significant yield improvements.

However, these food-grade organisms show a clear potential for *in situ* EPS production for creating fermented dairy products with improved rheological properties.

Improvement of EPS functionality

Consumers' requests point towards natural dairy products with a smooth and creamy texture, low in fat and sugars. These could be satisfied by a judicious use of LAB producing EPS *in situ*. The success of the application of an EPS is determined by its ability to bind water, interact with proteins and increase milk serum phase viscosity. EPS may act as texturizers and stabilizers, and consequently prevent the use of food additives (Hess et al. 1997). The benefits of EPS are detectable at extremely low concentrations: the visual appearance (gloss) of a product can be appealing, syneresis prevented, a creamy and firm texture generated and an overall pleasant mouthfeel obtained with less than 1% EPS. However, since production of one kind of EPS may not satisfy all texture specifications, the production of several EPSs, by one or several starter cultures, may be required. As such, it will be possible to precisely tailor the texture of an end product and to match significantly varying consumer preferences, for example between different countries or between different lines of products. EPSs produced by LAB have no intrinsic taste. But since a fermented milk product is becoming more viscous, its residence time in the mouth upon consumption and time of contact with the palate and taste receptors are increased and therefore the taste of the product can change significantly upon texturization by an EPS.

Structure–function relationships linking EPS primary structure to macroscopic properties carry an important potential for the controlled use of EPS into products tailored to specific needs in the food business. Using molecular modeling, it is possible to translate this atomic scale information into three-dimensional models, which will further enable a predictive approach. *In silico* prediction of macroscopic properties will make the bed for truly predictive design of properties such as rheology and targeted intermolecular interaction.

Certain EPSs produced by LAB are also claimed to have beneficial physiological effects. It is speculated that the increased viscosity of EPS containing foods may increase the residence time of ingested fermented milk in the gastrointestinal tract and, coupled to a low degradability of EPSs (Ruijssenaars et al. 2000), might therefore be beneficial to a transient colonization by probiotic bacteria (German et al. 1999). A further example of a suggested health benefit of some EPSs is the generation of short-chain fatty acids (SCFAs) upon degradation in the gut by the colonic microflora. SCFAs provide energy to epithelial cells and some have been claimed to play a role in the prevention of colon cancer (Harris & Ferguson 1993; Cummings & Englyst 1995). *In vivo* studies by oral administration of EPSs will be crucial to demonstrate clearly the different health beneficial properties that have been mentioned in the literature so far (e.g., antitumor, antiulcer, immunomodulating or cholesterol-lowering activities).

It is now admitted that bacteria must attach to biological surfaces in order to cause infections. In some cases, binding occurs via carbohydrate interactions or can also be prevented by adding carbohydrates (Choudhury et al. 1999; Reid et al. 2001). Bacterial adhesion is today a major field of investigation. Modification of EPSs into biologically active carbohydrates can be envisaged. Recently, Vincent et al. (2001) determined the structure of a polysaccharide produced by *Streptococcus macedonicus* Sc136 that contains the trisaccharide sequence β-D-Glc*p*NAc-(1$\rightarrow$3)-β-D-Gal*p*-(1$\rightarrow$4)-β-D-Glc*p*. This corresponds to an internal EPS backbone of lacto-*N*-tetraose and lacto-*N*-neotetraose. In fact, the same trioses have been identified in the structure of several human milk oligosaccharides that are important for infant nutrition. Deleting and inserting new genes coding for glycosyltransferases can be one way to engineer functional EPSs, based on an existing template such as the EPS produced by *S. macedonicus* Sc136, similarly to the approach based on LPS in *Escherichia coli* that led to *in vivo* neutralization of the Shiga toxin in mice (Paton et al. 2000). The use of EPS-producing probiotics to display *in situ* health beneficial polysaccharides is a very promising approach for a new generation of functional food.

Low amounts of EPS produced by the LAB reduce their potential use as food-grade ingredients. However, EPSs offer new opportunities to targeted applications for improved consumer health benefits going in the direction of consumer-specific nutrition.

References

Berkman T, Bozoglu TF & Özilgen M (1990) Mixed culture growth kinetics of *Streptococcus thermophilus* and *Lactobacillus bulgaricus*. Enzyme Microbial Technol. 12: 138–140.

Boels IC, van Kranenburg R, Hugenholtz J, Kleerebezem M & de Vos WM (2001) Sugar catabolism and its impact on the biosynthesis and engineering of exopolysaccharide production in lactic acid bacteria. Int. Dairy J. 11: 723–732.

Bouzar F, Cerning J & Desmazeaud M (1997) Exopolysaccharide production and texture-promoting abilities of mixed-strain starter cultures in yogurt production. J. Dairy Sci. 80: 2310–2317.

Breton C, Mucha J & Jeanneau C (2001) Structural and functional features of glycosyltransferases. Biochimie 83: 713–718.

Cerning J, Bouillanne C, Desmazeaud MJ & Landon M (1986) Isolation and characterization of exocellular polysaccharide produced by *Lactobacillus bulgaricus*. Biotechnol. Lett. 8: 625–628.

Cerning J, Bouillanne C, Landon M & Desmazeaud MJ (1990) Comparison of exocellular polysaccharide production by thermophilic acid bacteria. Science des Aliments 10: 443–451.

Cerning J, Bouillanne C, Landon M & Desmazeaud MJ (1992) Isolation and characterization of exopolysaccharides from slime-forming mesophilic lactic acid bacteria. J. Dairy Sci. 75: 692–699.

Cerning J & Marshall VME (1999) Exopolysaccharides produced by the dairy lactic acid bacteria. Recent Results Develop. Microbiol. 3: 195–209.

Cerning J, Renard CMGC, Thibault JF, Bouillanne C, Landon M, Desmazeaud MJ & Topisirovic L (1994) Carbon source requirements for exopolysaccharide production by *Lactobacillus casei* CG11 and partial structure analysis of the polymer. Appl. Environ. Microbiol. 60: 3914–3919.

Christiansen PS, Madeira AIMR & Edelstein D (1999) The use of ropy milk as stabilizer in the manufacture of ice cream. Milchwissenschaft 54: 138–140.

Choudhury D, Thompson A, Stojanoff V, Langermann S, Pinkner J, Hultgren SJ & Knight SD (1999) X-ray structure of the FimC-FimH chaperone-adhesin complex from uropathogenic *Escherichia coli*. Science 285: 1061–1066.

Cieslewicz MJ, Kasper DL, Wang Y & Wessels MR (2001) Functional analysis in type Ia group B *Streptococcus* of a cluster of genes involved in extracellular polysaccharide production by diverse species of streptococci. J. Biol. Chem. 276: 139–146.

Crescenzi V (1995) Microbial polysaccharides of applied interest: on going research activities in Europe. Biotechnol. Progr. 11: 251–259.

Cummings JH & Englyst HN (1995) Gastrointestinal effects of food carbohydrate. Am. J Clin. Nutr. 61: 938–945.

De Vuyst L, De Vin F, Vaningelgem F & and Degeest B. (2001) Recent developments in the biosynthesis and applications of heteropolysaccharides from lactic acid bacteria. Int. Dairy J. 11: 687–707.

De Vuyst L & Degeest B (1999) Heteropolysaccharides from lactic acid bacteria. FEMS Microbiol. Rev. 23: 153–177.

Duboc P, Fischer M & Vincent SJF (2002) Characterization of gluconacetan, a new texturizing carbohydrate polymer: a basis for a structure-function relationship in polysaccharides. Submitted for publication.

Duboc P & Mollet B (2001) Applications of exopolysaccharides in the dairy industry. Int. Dairy J. 11: 759–768.

Escalante A, Wacher-Rodarte C, Garcia-Garibay M & Farrés A (1998) Enzymes involved in carbohydrate metabolism and their role on exopolysaccharide production in *Streptococcus thermophilus*. J. Appl. Microbiol. 84: 108–114.

Faber EJ, van Kuik JA, Kamerling JP & Vliegenthart JF (2002) Modeling of the structure in aqueous solution of the exopolysaccharide produced by *Lactobacillus helveticus* 766. Biopolymers 63: 66–76.

Faber EJ, Zoon P, Kamerling JP & Vliegenthart JF (1998) The exopolysaccharides produced by *Streptococcus thermophilus* Rs and Sts have the same repeating unit but differ in viscosity of their milk cultures. Carbohydr. Res. 310: 269–276.

Gancel F & Novel G (1994) Exopolysaccharide production by *Streptococcus salivarius* ssp. *thermophilus* cultures. Conditions of production. J. Dairy Sci. 77: 685–688.

German B, Schiffrin EJ, Reniero R, Mollet B, Pfeifer A & Neeser JR (1999) The development of functional foods: lessons from the gut. Trends Biotechnol. 17: 492–499.

Germond JE, Delley M, D'Amico N & Vincent SJ (2001) Heterologous expression and characterization of the exopolysaccharide from *Streptococcus thermophilus* Sfi39. Eur. J. Biochem. 268: 5149–5156.

Grobben GJ, Sikkema J, Smith MR & de Bont JAM (1995) Production of extracellular polysaccharides by *Lactobacillus delbrueckii* ssp. *bulgaricus* NCFB 2772 grown in a chemically defined medium. J. Appl. Bacteriol. 79: 103–107.

Harris PJ & Ferguson LR (1993) Dietary fibre: its composition and role in protection against colorectal cancer. Mutation Res. 290: 97–110.

Hess SJ, Roberts RF & Ziegler GR (1997) Rheological properties of nonfat yogurt stabilized using *Lactobacillus delbrueckii* ssp. *bulgaricus* producing exopolysaccharide or using commercial stabilizer systems. J. Dairy Sci. 80: 252–263.

Jay AJ, Colquhoun IJ, Ridout MJ, Brownsey GJ, Morris VJ, Fialho AM, Leito JH & Sa-Correira I (1998) Analysis of structure and function of gellans with different substitution patterns. Carbohydr. Polym. 35: 179–188.

Jolly L, Newell J, Porcelli I, Vincent SJF & Stingele F (2002) *Lactobacillus helveticus* glycosyltransferases: from genes to carbohydrate synthesis. Glycobiol. In press.

Jolly L & Stingele F (2001) Molecular organization and functionality of exopolysaccharide gene clusters in lactic acid bacteria. Int. Dairy J. 11: 733–745.

Kalab M, Allan-Wojtas P & Phipps-Todd BE (1983) Development of microstructure in set-style nonfat yoghurt . A review. Food Microstructure 2: 51–66.

Kojic M, Vujcic M, Banina A, Cocconcelli P, Cerning J & Topisirovic L (1992) Analysis of exopolysaccharide production by *Lactobacillus casei* CG11, isolated from cheese. Appl. Environ. Microbiol. 58: 4086–4088.

Kosikowski FV (1982) Cheese and Fermented Milk Foods. 2nd edn.

Levander F, Svensson M & Radstrom P (2002) Enhanced exopolysaccharide production by metabolic engineering of *Streptococcus thermophilus*. Appl. Environ. Microbiol. 68: 784–790.

Looijesteijn PJ, Boels IC, Kleerebezem M & Hugenholtz J (1999) Regulation of exopolysaccharide production by *Lactococcus lactis* subsp. *cremoris* by the sugar source. Appl. Environ. Microbiol. 65: 5003–5008.

Low D, Ahlgren JA, Horne D, McMahon DJ, Oberg CJ & Broadbent JR (1998) Role of *Streptococcus thermophilus* MR-1C capsular exopolysaccharide in cheese moisture retention. Appl. Environ. Microbiol. 64: 2147–2151.

Macura D & Townsley PM (1984) Scandinavian ropy milk: identification and characterization of endogenous ropy lactic streptococci and their extracellular excretion. J. Dairy Sci. 67: 735–744.

McMahon DJ, Oberg CJ & McManus W (1993) Functionality of mozzarella cheese. Austr. J. Dairy Technol. 48: 99–104.

Monsan P, Bozonnet S, Albenne C, Joucla G, Willemot RM & Remaud-Siméon M (2001) Homopolysaccharides from lactic acid bacteria. Int. Dairy J. 11: 675–685.

Moreira LM, Becker JD, Puhler A & Becker A (2000) The *Sinorhizobium meliloti* ExpE1 protein secreted by a type I secretion system involving ExpD1 and ExpD2 is required for biosynthesis or secretion of the exopolysaccharide galactoglucan. Microbiology 146: 2237–2248.

Mozzi F, Olivier G, Savyo de Giori GS & Font de Valdez GF (1995) Influence of temperature on the production of exopolysaccharides by thermophilic lactic acid bacteria. Milchwissenschaft 50: 80–82.

Mozzi F, Savyo de Giori GS, Olivier G & Font de Valdez GF (1994) Effect of culture pH on the growth characteristics and polysaccharide production by *Lactobacillus casei*. Milchwissenschaft 49: 667–670.

Paton AW, Morona R & Paton JC (2000) A new biological agent for treatment of Shiga toxigenic Escherichia coli infections and dysentery in humans. Nat. Med. 6: 265–270.

Perry DB, McMahon DJ & Oberg CJ (1997) Effect of exopolysaccharide producing cultures on moisture retention in low-fat mozzarella cheese. J. Dairy Sci. 80: 799–805.

Persson K, Ly HD, Dieckelmann M, Wakarchuk WW, Withers SG & Strynadka NC (2001) Crystal structure of the retaining galactosyltransferase LgtC from *Neisseria meningitidis* in complex with donor and acceptor sugar analogs. Nat. Struct. Biol. 8: 166–175.

Petry S, Furlan S, Crepeau MJ, Cerning J & Desmazeaud m (2000) Factors affecting exocellular polysaccharide production by *Lactobacillus delbrueckii* subsp. *bulgaricus* grown in a chemically defined medium. Appl. Environ. Microbiol. 66: 3427–3431.

Pérez S, Kouwijtzer M, Mazeau K & Engelsen SB (1996) Modeling Polysaccharides: Present Status and Challenges. J. Model. Graph. 307–321.

Reid G, Howard J & Gan BS (2001) Can bacterial interference prevent infection? Trends Microbiol. 9: 424–428.

Ricciardi A & Clementi F (2000) Exopolysaccharides from lactic acid bacteria: structure, production and technological applications. Ital. J. Food. Sci. 23–45.

Roberts IS (1996) The biochemistry and genetics of capsular polysaccharide production in bacteria. Annu. Rev. Microbiol. 50: 285–315.

Rohm H & Kovac A (1994) Effects of starter cultures on linear viscoelastic and physical properties of yogurt gels. J. Texture Studies 25: 311–329.

Rohm H & Schmid W (1993) Influence of dry matter fortification on flow properties of yogurt. 1. Evaluation of flow curves. Milchwissenschaft 48: 556–560.

Ruijssenaars HJ, Stingele F & Hartmans S (2000) Biodegradability of food-associated extracellular polysaccharides. Curr. Microbiol. 40: 194–199.

Sebastiani H & Zelger G (1998) Texture formation by thermophilic lactic acid bacteria. Milchwissenschaft 53: 15–20.

Stingele F, Neeser JR & Mollet B (1996) Identification and characterization of the *eps* (Exopolysaccharide) gene cluster from *Streptococcus thermophilus* Sfi6. J. Bacteriol. 178: 1680–1690.

Stingele F, Vincent SJ, Faber EJ, Newell JW, Kamerling JP & Neeser JR (1999) Introduction of the exopolysaccharide gene cluster from *Streptococcus thermophilus* Sfi6 into *Lactococcus lactis* MG1363: production and characterization of an altered polysaccharide. Mol. Microbiol. 32: 1287–1295.

Sutherland IW (1998) Novel and established applications of microbial polysaccharides 181. Trends Biotechnol. 16: 41–46.

Tamime AY & Robinson RK (1999) Yoghurt Science and Technology.

Teggatz JA & Morris HA (1990) Changes in the rheology and microstructure of ropy yogurt during shearing. Food Structure 9: 133–138.

Tuinier R, ten Grotenhuis E, Holt C, Timmins PA & de Kruif CG (1999) Depletion interaction of casein micelles and an exocellular polysaccharide. Physical Review 60: 848–856.

Tuinier R, van Casteren WH, Looijesteijn PJ, Schools HA, Voragen AG & Zoon P (2001) Effects of structural modifications on some physical characteristics of exopolysaccharides from *Lactococcus lactis*. Biopolymers 59: 160–166.

Unligil UM & Rini JM (2000) Glycosyltransferase structure and mechanism. Curr. Opin. Struct. Biol. 10: 510–517.

Van den Berg DJC, Robijn GW, Janssen AC, Giuseppin MLF, Vreeker R, Kamerling JP, Vliegenthart JFG, Ledeboer AM & Verrips CT (1995) Production of a novel extracellular polysaccharide by *Lactobacillus sake* 0-1 and characterization of the polysaccharide. Appl. Environ. Microbiol. 61: 2840–2844.

Van Geel-Schutten GH, Faber EJ, Smit E, Bonting K, Smith MR, Ten Brink B, Kamerling JP, Vliegenthart JF & Dijkhuizen L (1999) Biochemical and structural characterization of the glucan and fructan exopolysaccharides synthesized by the *Lactobacillus reuteri* wild-type strain and by mutant strains. Appl. Environ. Microbiol. 65: 3008–3014.

van Kranenburg R, Boels IC, Kleerebezem M & de Vos WM (1999) Genetics and engineering of microbial exopolysaccharides for food: approaches for the production of existing and novel polysaccharides. Curr. Opin. Biotechnol. 10: 498–504.

van Marle ME & Zoon P (1995) Permeability and rheological properties of microbially and chemically acidified skim-milk gels. Netherlands Milk Dairy J. 49: 47–65.

Vincent SJ, Faber EJ, Neeser JR, Stingele F & Kamerling JP (2001) Structure and properties of the exopolysaccharide produced by *Streptococcus macedonicus* Sc136. Glycobiology 11: 131–139.

Whitfield C & Valvano MA (1993) Biosynthesis and expression of cell-surface polysaccharides in gram- negative bacteria. Adv. Microb. Physiol. 35: 135–246.

Pay-Per-View
is here!
$21.50
Every article
Every journal
Everyday
Vist
www.kluweronline.com
Then go to Journal Contents and click on the article you wish to view. After a quick registration process, you can purchase the paper of your choice.
kluwer
the language of science

MIX
Papier aus verantwortungsvollen Quellen
Paper from responsible sources
FSC® C105338

If you have any concerns about our products,
you can contact us on
ProductSafety@springernature.com

In case Publisher is established outside the EU,
the EU authorized representative is:
Springer Nature Customer Service Center GmbH
Europaplatz 3, 69115 Heidelberg, Germany

Printed by Libri Plureos GmbH
in Hamburg, Germany